The ASQ Certified Reliability Engineer Handbook

The ASQ Certified Reliability Engineer Handbook

Fourth Edition

Karen Hulting and Mary McShane-Vaughn, Editors

ASQExcellence
Milwaukee, Wisconsin

Published by ASQExcellence, Milwaukee, WI

Produced and distributed by Quality Press, ASQ, Milwaukee, WI

Publisher's Cataloging-in-Publication data

Names: McShane-Vaughn, Mary A., 1963-, editor. | Hulting, Karen J., 1961–, editor.
Title: The ASQ certified reliability engineer handbook, fourth edition / Karen Hulting and Mary McShane-Vaughn, Editors
Description: Includes bibliographical references and index. | Milwaukee, WI: ASQExcellence, 2025.
Identifiers: LCCN: 2025941992 | ISBN: 9781636941783 (hardcover) | 9781636941790 (pdf) | 9781636941806 (epub)
Subjects: LCSH Production management—Quality control—Handbooks, manuals, etc. | Reliability (Engineering)—Handbooks, manuals, etc. | BISAC BUSINESS & ECONOMICS / Industries / Manufacturing | BUSINESS & ECONOMICS / Quality Control | BUSINESS & ECONOMICS / Total Quality Management | MATHEMATICS / Optimization | TECHNOLOGY & ENGINEERING / Industrial Engineering | TECHNOLOGY & ENGINEERING / Manufacturing
Classification: LCC TS156 .A77 2025 | DDC 658.4013—dc23

ASQ and ASQExcellence advance individual, organizational, and community excellence worldwide through learning, quality improvement, and knowledge exchange.

Bookstores, wholesalers, schools, libraries, businesses, and organizations: Quality Press and ASQExcellence books are available at quantity discounts for bulk purchases for business, trade, or educational uses. For more information, please contact Quality Press at 800-248-1946 or books@asq.org.

To place orders or browse the selection of all Quality Press titles, visit our website at: http://www.asq.org/quality-press.

30 29 28 27 26 25 GP 7 6 5 4 3 2 1

Limit of Liability/Disclaimer of Warranty

For all those who depend on the reliability of medical devices.—KJH

For Jim, the most reliable engineer I know.—MMV

Table of Contents

List of Tables and Figures

Part I

Part II

Part III

Part IV

Part V

Preface to the Fourth Edition

We are honored to present the fourth edition of *The ASQ Certified Reliability Engineer Handbook*. The handbook has been updated to reflect the Reliability Engineering Body of Knowledge released in 2024.

We want to thank the previous editor, Mark Allen Durivage, for his work on the third edition. We appreciate the breadth and depth of the source material we had for the editing effort.

CHANGES TO THE BODY OF KNOWLEDGE

There have been some significant changes to the 2024 Certified Reliability Engineer Body of Knowledge (BoK). These changes stem from the results of the ASQ Job Survey completed by certified reliability engineers in late 2023. Respondents were asked how important each tool in the BoK was in their work and how often they used it. In addition, they were asked if any tools or topics should be added to the BoK. The Certified Reliability Engineer Body of Knowledge is thus a living document that reflects the current state of the practice of reliability engineering in a wide range of industries. The updated version of the BoK can be found at the end of the book.

The new BoK has added new subtopics, including project management in reliability engineering, the quality triangle, risk assessment, and software/firmware reliability. In addition, almost every carry-over subtopic has added tools and techniques.

There were also changes in the Bloom's cognition levels for several topics, as shown in Table 0.1.

EDITORIAL GOALS

As editors, our goals for this edition were to produce a handbook that:

- Adheres to the updated Body of Knowledge
- Reflects the latest technology and applications of reliability engineering
- Presents thorough explanations of concepts in a straightforward way
- Offers more examples illustrating key concepts

Table 0.1 Changes in Bloom's cognitive levels for the BoK.

2018 BoK code	Topic	2018 Bloom's level	2024 Bloom's level
II.A.1	Risk management techniques	Analyze	↑Evaluate
III.A.4	Probability functions	Apply	↑Analyze
III.A.6	Statistical process control (SPC) and process capability studies (C_p, C_{pk}, P_p, and P_{pk})	Understand	↑Apply
IV.B.5	Software testing	Evaluate	↓Understand
IV.C.4	Reliability prediction methods	Apply	↓Understand
V.A.7	Design for reliability	Evaluate	↓Analyze

- Provides an excellent resource for the certification exam, both for study prep and on exam day
- Serves as a valuable addition to any reliability engineer's professional library

Please let us know if we have achieved these goals by dropping a note to books@asq.org.

WHAT'S NEW?

There are some changes to the look and feel of the handbook. We reordered some features, refining and reordering the appendices to better follow the order of the text. The appendices also feature new standard normal, Student's *t*, *F*, and normal tolerance interval factor tables that include more values.

To help candidates prepare for the exam, we numbered key equations for easier reference and focused the examples on hand calculations as much as possible. We also present computer output using Minitab®, JMP®, and DesignEase® software. Here is a summary of the content changes in each chapter.

Chapter 1 Reliability Fundamentals
Additions include:

- Reliability strategy, including its mission, vision, objective, and requirements;
- A new project management subtopic along with an example; and
- Defect tracking and reliability growth monitoring using key performance indicators, the Duane growth and Crow-AMSAA models, and overall equipment effectiveness (OEE)

Chapter 2 Reliability Foundations
Additions include:

- The concepts of durability, mean time between critical failure (MTBCF), mean time between maintenance (MTBM), and mean downtime (MDT)
- Operational, achieved, and inherent availability
- Environmental, social, and governance (ESG) policies
- Root cause analysis tools such as fishbone diagrams, 5 Whys, and 8D
- A new subtopic on the quality triangle
- Principles of the Six Sigma DMAIC project cycle and lean methodologies

Chapter 3 Risk Identification
Additions include:

- Environmental, liability, and security concerns
- *p*-diagrams and use cases
- A new subtopic on risk assessment
- Environmental, strategic, cybersecurity, and analytical risks

Chapter 4 Risk Analysis
Action priority has been added, and the existing failure mode and effects analysis (FMEA) section has been expanded.

Chapter 5 Risk Mitigation
Additions include:

- ISO 31000 and ISO 55000 standards
- The reliability goals of as low as reasonably possible (ALARP), as low as reasonably achievable (ALARA), and as low as possible (ALAP)

Chapter 6 Probability and Statistics for Reliability, Basic Concepts
Additions include:

- Kaplan–Meier analysis
- Probability trees
- Chi-square and Student's *t* distributions
- The capability concepts of P_p and P_{pk}
- Confidence and tolerance intervals using the Weibull, normal, and log-normal distributions

Chapter 7 Data Management
Additions include:

- Big data, the internet of things (IoT), and artificial intelligence (AI)
- Normalizing datasets and data transformations

- Scales of measurement
- Nevada chart analysis and survival analysis
- The Cox proportional hazards model
- Data precision and integrity
- Bad actor analysis
- Pareto charts, scatter plots, and box-and-whisker plots
- Failure analysis methods to include scanning acoustic microscopy (SAM), infrared inspection, and radiography analysis as well as material testing and measurement techniques like tensile testing; shear testing; viscosity measurement; vibration testing; visual, ultrasonic, and electrical analysis; cycle testing; nondestructive testing (NDT); and predictive technologies

Chapter 8 Reliability Planning
Additions include:

- Customer profiles
- "Zero failure" tests
- Multiple stress factors (e.g., highly accelerated life testing (HALT))
- Customer requirements and user needs
- Validating test capability

Chapter 9 Reliability Testing
Additions include:

- A new software/firmware reliability subtopic
- Software validation and verification
- Qualitative versus quantitative reliability testing
- Built-in testing
- Regression testing to verify code integrity

Chapter 10 Reliability Modeling
Additions include:

- K out of N and shared load models
- Physics of failure topics of excessive deformation, creep, and delamination
- The Coffin–Manson model for temperature cyclic stress
- Markov analysis
- Neural networks and machine learning for reliability prediction
- Rapid prototyping technologies, phase diagrams, and digital twins

Chapter 11 Reliability Design Techniques
Additions include:

- Stress–strength analysis for lognormal and Weibull distributions
- Analysis of variance (ANOVA), blocking, replication, and randomization for designed experiments
- Taguchi and combined array techniques for robust design
- Finite element analysis (FEA)

Chapter 12 Parts and Systems Development
Additions include:

- Safe operating area (SOA) criteria
- Software selection for product and evaluation equipment functionality
- Reliability-centered maintenance (RCM)

Chapter 13 Maintainability
Material refreshed and expanded with new examples.

THE HANDBOOK TEAM

We were lucky enough to assemble a diverse team who applied their years of experience in reliability engineering, leadership, software development, statistics, and training to help us craft this new edition of the handbook.

Mr. David Auda is the past chair of the ASQ Reliability & Risk Division (RRD) and presently serves as the vice chair for RRD. He is the conference chair for the 2025 ASQ Reliability, Maintenance & Managing Risk Conference (RMMR). David is also executive producer of the ASQRRD Global webinar series. He is an ASQ fellow and holds CRE and CMQ/OE certifications.

Dr. Mark Bailey works in statistical software development at JMP Statistical Discovery LLC. He pursued research and development of medical diagnostics at Eastman Kodak and Abbott Laboratories and helped Abbott Diagnostics adopt Six Sigma in R&D activity. He also taught training courses in data visualization and analysis, statistical programming and modeling, design of experiments, and predictive modeling for more than 20 years at SAS Institute.

Ms. Kaitlyn Bellucci is a senior reliability engineer at Medtronic. She earned a BS in chemical engineering from the University of Connecticut. She has over 10 years of reliability engineering experience in reliability testing and the reliability discipline. At Medtronic, she has developed a variety of training courses for her peers on reliability engineering and its importance in the medical device field to further educate all levels of learning.

Dr. Daniel C. Conrad is director of design quality, reliability, and testing at Hussmann Corporation, a Panasonic company. He is the recipient of the ASQ 2024 Shainin Medal for Innovation in Quality & Reliability. He is past chair of the Reliability, Maintenance, and Risk Management Conference (RMMR 2024). Prior to joining Hussmann, he was the global design for reliability leader with Whirlpool Corporation. He earned his BS from the University of Kentucky in mechanical engineering and his MS and PhD from Purdue University in engineering mechanics. He holds multiple US and foreign patents, and he has published numerous papers in journals and conferences on new technology and processes for appliance applications.

Mr. Tim Gaens is a project manager at Jabil for Regulated Industries, Automotive and Transportation segments. Previously, he was a lead product validation engineer performing and managing reliability projects toward the highest quality levels and standards. During this time, he was also the Quality Lab manager of the ISO 17025 accredited lab in Hasselt. Tim is past chair and still an active volunteer of the ASQ Reliability and Risk Division and is currently the director of membership of the Project Management Institute (PMI) in Belgium.

Ms. Michelle L. Grasso has more than 30 years of reliability engineering experience. She earned a BS in industrial technology from Central Connecticut State University and is an ASQ Certified Reliability Engineer. Michelle worked for Medtronic and Pitney Bowes in the reliability discipline, where she developed many training programs. She currently owns her own consulting company and is employed by Cooper Surgical.

Ms. Mindy Hotchkiss is an independent consultant operating as Enquery Research LLC. She was technical specialist and enterprise-wide subject matter expert in statistics at Aerojet Rocketdyne, now an L3Harris Technologies company. She has over 25 years of experience as a statistical consultant between Pratt & Whitney and Aerojet Rocketdyne, supporting engineering, operations, and technology development across the enterprise, including hypersonics and additive manufacturing. Mindy has BS degrees in mathematics and statistics and an MBA from the University of Florida, and a masters of statistics from North Carolina State University. She is an ASQ-Certified Reliability Engineer, Quality Engineer, and Six Sigma Black Belt and is a past chair of the ASQ Statistics Division, serving in many roles since 2008, including co-chairing the ASQ/ASA Fall Technical Conference (FTC). She is a frequently invited speaker at statistics and reliability conferences. Mindy is currently serving on the Governing Board of the Ellis R. Ott Scholarship for Applied Statistics and Quality Management and also represents ASQ RRD on the Board of Directors for the Reliability and Maintainability Symposium (RAMS).

Dr. Fred Hulting is a private consultant, recently retired from a 30-year career in industry. At General Mills, he led a large international and interdisciplinary Knowledge Services team that provided statistics, computing, and data science expertise to the R&D and Quality organizations. His publications and invited presentations include new methods for system reliability and life data analysis. He is a fellow of the American Statistical Association.

Ms. Michelle Jirak works as a design assurance quality engineer in the medical device industry and is a senior member of ASQ. She is an ASQ Certified Quality Engineer, Certified Reliability Engineer (CRE), Certified Quality Auditor (CQA), and Certified Medical Device Auditor (CMDA). Michelle holds an MBA from the Keller Graduate School of Management and a BS in chemical engineering from Iowa State University, where she graduated summa cum laude.

Dr. Thomas Jones teaches industrial engineering at Florida Polytechnic University. He brings a focus on digital technologies into the classroom with the goal of preparing graduates to deploy proven methods effectively in a modern business landscape. He holds a bachelor of science, master of science, and PhD, all in industrial engineering, with specializations in quality systems and digital transformation. Thomas is a senior member of ASQ and holds CRE, CQE, CSSBB, CMQ/OE, and CQA certifications. Dr. Jones has worked as a quality engineer and industrial engineer in the aerospace and defense and automotive industries, leading change and improvement. He has extensive experience in new product development and introduction, operational excellence including Lean Six Sigma, and experience in stabilizing processes. His research has centered on digital transformation and the implementation of Industry 4.0, with an emphasis on developing systems that are user friendly and easy to work with, and through this research, he coauthored the book *Total Digital Transformation* set to release in 2025.

Dr. Peng Liu is a statistician and software engineer at JMP Statistical Discovery LLC. He holds a PhD in statistics and a bachelor's degree in computer science. Since 2007, he has developed most of the JMP reliability analysis software products and brought JMP from its early stages to a leading vendor in the reliability analysis software market. Besides reliability, he also specializes in time series analysis, nonlinear statistical modeling, Bayesian inference and computing, machine learning methods and system development, and graphical user interface design and implementation on multiple platforms. He is a member of the American Statistical Association (ASA) and IEEE.

Dr. Mohammad Pourgol is the director for design for safety and reliability at Schneider Electric. He is also an adjunct associate professor of mechanical engineering at the University of Maryland. Previously, Dr. Pourgol held the position of associate professor of reliability engineering at Sahand University of Technology (SUT). Dr. Pourgol's 20+ year career encompasses industrial applications, research, and teaching in safety applications and reliability engineering for Teradyne Semiconductor, Johnson Controls, Sahand University of Technology, FM Global, Daikin Comfort, UMD, and the Massachusetts Institute of Technology (MIT). Dr. Pourgol is an IEEE Senior Member, an ASQ Fellow, and an ASME Fellow. He served as the ASME Safety Engineering and Risk/Reliability Analysis Division (SERAD) chair from 2017 to 2022. Dr. Pourgol earned his PhD in reliability engineering from the University of Maryland (UMD) and is a registered Professional Engineer (PE) in the State of Massachusetts. He is an ASQ certified Reliability Engineer, Six Sigma Black Belt, and Manager of Quality/Organizational Excellence. Dr. Pourgol has authored over 160 papers and is currently the associate editor for the ASME-ASCE *Journal of Risk and Uncertainty in Engineering Systems.*

As editors of the handbook, any errors that may appear in the text are all ours. Please submit suggestions, additions, corrections, or deletions for the handbook to books@asq.org.

—*The Editors*

Acknowledgments

We want to thank Michael Byrnes, executive director of certification operations at ASQE, for offering us the opportunity to edit this handbook. The editing exercise stretched our project management skills, writing ability, and knowledge of the field.

Most importantly, we were able to surround ourselves with a team of professionals who had both depth and breadth of experience in reliability engineering. For seven months, the team mapped out changes we needed to make based on the new BoK and then went about editing, rewriting, commenting, cross-checking, re-reading, and finalizing each chapter. Each of us was motivated by our love of the subject as well as a profound sense of responsibility to the ASQ reliability engineering community. We are extremely proud of the outcome.

Thanks to the stellar team of subject-matter experts who volunteered their time, intellect, and energy to make this handbook possible:

Mr. David Auda, ASQ Fellow, CRE, CMQ/OE

Dr. Mark Bailey

Ms. Kaitlyn Bellucci

Dr. Daniel Conrad, CRE

Mr. Tim Gaens

Ms. Michelle Grasso, CRE

Ms. Mindy Hotchkiss, CRE, CQE, CSSBB

Dr. Fred Hulting, ASA Fellow

Ms. Michelle Jirak, CRE, CQE, CQA, CMDA

Dr. Tom Jones, CRE, CQE, CSSBB, CMQ/OE, CQA

Dr. Peng Liu

Dr. Mohammad Pourgol, PE, ASQ Fellow, CRE, CSSBB, CMQ/OE

Team member contributions are shown in Table 0.2.

Thank you to Paul Burte, immediate past chair of the reliability engineering exam committee, and Mike Sage, psychometrician at ASQE, for fielding questions we had about the changes in the Body of Knowledge.

Table 0.2 Team member contributions.

Chapter	Contributor
1 Leadership Foundations	Kaitlyn Bellucci Michelle Grasso
2 Reliability Foundations	Tom Jones
3 Risk Identification	David Auda
4 Risk Analysis	David Auda
5 Risk Mitigation	Michelle Jirak
6 Probability and Statistics for Reliability, Basic Concepts	Mindy Hotchkiss
7 Data Management	Tom Jones Fred Hulting
8 Reliability Planning	Mark Bailey Mohammad Pourgol
9 Reliability Testing	Mark Bailey Mohammad Pourgol
10 Reliability Modeling	Peng Liu Fred Hulting Kaitlyn Bellucci Michelle Grasso
11 Reliability Design Techniques	Daniel Conrad Mindy Hotchkiss Michelle Jirak
12 Parts and Systems Development	Tim Gaens
13 Maintainability	Michelle Jirak

In addition, we would like to thank the Quality Press management staff, Lillian McAnally and Erica Barse, for their cheerful support during the entire project and for providing references when they were needed. An extra shout-out to Fred Hulting, who was both qualified and nearby, so couldn't escape being drawn into the review process. Fred checked examples, tables and figures throughout the book and identified valuable opportunities for improvement. Finally, we would like to thank the staff at Westchester for steering this handbook from draft to final form. Their efforts have produced a final product worthy of the ASQ Quality Press family of publications.

—The Editors

CONTRIBUTORS' ACKNOWLEDGMENTS

The American Society for Quality and Quality Press would like to thank the Quality Press Peer Review Committee for its invaluable volunteer participation and contributions to this work. Without our volunteers' subject matter expertise, time, and passion for creating content, none of our efforts would be possible.

QUALITY PRESS PEER REVIEW COMMITTEE MEMBERS

Scott A. Laman, Chair

Peter Pylipow, Vice Chair

Melvin Alexander

Lance Coleman

Ahmad Elshennawy

Marc Hamilton

Gary Jing

Trevor Jordan

Jane Keathley

Betsy Macht

Mary McShane-Vaughn

Jayet Moon

G.S. Sureshchandar

Tiea Theurer

Part I

Reliability Fundamentals

Chapter 1
Leadership Foundations

Reliability is defined as the probability that an item will perform its intended function without failure under stated conditions for a specified period. A statement of reliability has four key components:

1. *Probability.* Probability is a number between 0 and 1 that expresses the likelihood of an outcome. In this case, the outcome is that the item performs its intended function over time. For example, a pump's reliability goal could be 0.9995. When the goal is achieved, this means that at least 99.95% of the pumps will function for a specified period under the stated conditions. The reliability goal could be stated as: The pump is 99.95% reliable for the X intended function, under Y conditions, for a specified period of time t.
2. *Intended function.* This should be defined for every part, subassembly, and product. The statement of the intended function should explicitly state or imply a failure definition. For example, a pump's intended function requires it to have a minimum flow rate of 20 gallons per minute. If the pump does not meet that minimum flow rate, then that would be considered a failure.
3. *Stated conditions.* These include environmental conditions, maintenance conditions, usage conditions, and storage and moving conditions, among others. Any condition that could affect the performance of the product should be identified. Examples would be the pump's ability to function with different temperature fluids flowing through the pump (hot/cold) or under the warehouse conditions that would be part of its normal use.
4. *Specified period of time.* The time period can be measured by a number of button presses, actuations, cycles, hours, days, or another metric. An example of this could be a pump that must be able to function for 10,000 hours. Other units of measurement may be more appropriate to use depending on the product. The specified period of time for an automotive tire might be stated in terms of miles; for a laundry appliance, in terms of wash cycles; or for a phone battery, in terms of number of recharge cycles.

These four components of the definition of reliability are illustrated in Example 1.1.

EXAMPLE 1.1

A statement of reliability for a pump is as follows. At least 95% of pumps must be able to move water without failure for 10,000 hours while being run at temperatures ranging from 10° F to 105° F. We observe:

The probability is 0.95.

The intended function is "move water."

The stated condition is "temperatures ranging from 10°F to 105°F."

The specified period of time is "10,000 hours."

BENEFITS OF RELIABILITY ENGINEERING

Describe the value that reliability has on achieving company goals and objectives, and how reliability engineering techniques and methods improve programs, processes, products, systems, and services. (Understand)

Body of Knowledge I.A.1

Reliability engineering is key to designing a safe and reliable product. The following items are some influences that have increased the importance of the study of reliability engineering:

- Customers expect products not only to meet the specified parameters upon delivery but to function throughout what they perceive as a reasonable lifetime. A product that is not reliable for the expected lifetime of the device can lead to poor customer satisfaction and a negative effect on the brand.
- Reliability values are used in marketing and warranty material.
- Competitive pressures require increased emphasis on reliability.
- It is more common that contracts specify reliability requirements.
- As products become more complex, the reliability requirements of the components increase.
- An unreliable product often has safety and health hazards.

Reliability engineering addresses each of these influences by helping designers determine and increase the useful lifetime of products, processes, and services. Reliability can be improved by implementing "design for reliability" practices early in the product design process. By understanding the reliability needs at the beginning of a new product development cycle, design decisions to achieve reliability goals can be made earlier and at a lower cost. For example, when designing a medical product, the reliability of the product and all its components must be very high. In that case, it will be cost-effective to invest in reliability modeling and testing to achieve these reliability goals. However, if the product under development is a reusable water bottle, then the reliability needs might be much lower, and the investment in reliability engineering for the product would be less.

INTERRELATIONSHIP OF SAFETY, QUALITY, AND RELIABILITY

> Describe the relationship of and distinguish between reliability and quality. Describe the importance of safety in reliability engineering and how reliability impacts safety. (Understand)
>
> **Body of Knowledge I.A.2**

In most organizations, the quality assurance function is responsible for continually improving the organization's ability to produce products and services that meet or exceed the customer requirements. All quality engineers should have a working knowledge of reliability engineering. What, then, is the distinction between these two fields?

Reliability can be thought of as "how quality changes over time." The difference between quality and reliability is that quality shows how well the item initially performs its intended function, while reliability shows how well the item maintains its original level of performance over time through various conditions. In general, the higher the level of quality, the greater the probability that the product will be safe and reliable. Reliability engineering seeks answers to questions such as these:

- Are components failing prematurely?
- Is burn-in needed, and if so, what length of time is sufficient?
- Is this failure rate acceptable?
- What changes in design, manufacturing, installation, operation, or maintenance would improve reliability?

Another way to describe the difference between quality and reliability is that quality is how the product performs at a snapshot in time, while reliability is the quality of the product over its lifetime. For example, the quality department inspects products using sampling plans during manufacturing to ensure the product is meeting its requirements, either for process monitoring or product acceptance testing. The reliability department may take random samples and test them over time as they would be used in the field to ensure that they will continue to meet the requirements for their lifetime.

Quality and reliability engineers provide different inputs into the design process. Quality engineers suggest changes that permit the item to be produced within tolerance at a reasonable cost. Reliability engineers make recommendations that permit the item to function correctly for its intended lifetime.

The preceding comparisons show that although the roles of quality and reliability are different, they are interrelated. For example, in the product design phase, both quality and reliability functions have the goal of proposing cost-effective ways to satisfy and exceed customer expectations. This often requires that the two functions work together to produce a design that both works correctly initially and performs for an acceptable period. When processes are designed and operated, the quality and reliability engineers work together to determine the process parameters that impact the performance and longevity of the product so that those parameters can be appropriately controlled. A similar interrelationship holds as specifications are developed for packaging, shipment, installation, operation, and maintenance.

Reliability will be impacted by product design and by the processes used in the product's manufacturing process. Therefore, the designers of products and processes must understand and use reliability data as design decisions are made. The earlier that reliability data are considered in the design process, the more efficient and effective their impact will be.

When conducting risk assessments, all failure modes should be investigated for possible safety risks. When a reliable product is designed, quality engineering techniques should ensure that the manufacturing processes do, in fact, produce a reliable product.

Safety considerations pervade all aspects of both the quality engineering and reliability engineering fields. When a process or product change is proposed, the proposal should be accompanied by a thorough study of the impact the change will have on safety. Some questions to investigate include the following:

- Could this change make the production process less safe? How will this be mitigated?

  Example: Workers accustomed to doing things the old way may be more at risk with the proposed changes.

- Could this change make the use of the product less safe? How will this be mitigated?

  Example: The new dishwasher latch, if not engaged properly, allows steam to escape into the electronic timer, causing a fire hazard.

- With the failure of another component now more likely, are there new safety risks?

  Example: Proposals for increasing the useful life of a component should be accompanied by a study on the effect the increased life will have on other components.

- As the product reaches its wear-out phase, could this change introduce safety risks? How will this be mitigated?

  Example: The new lighting system contains chemical compounds that are toxic when improperly disposed.

RELIABILITY ENGINEER LEADERSHIP RESPONSIBILITIES

> Describe how to be a reliability champion by influencing program decisions and facilitating cross-functional communication. Understand the fundamentals of reliability strategy, including its mission, vision, objectives, and requirements. (Understand)
>
> **Body of Knowledge I.A.3**

"If your actions inspire others to dream more, learn more, do more, and become more, you are a leader." – John Quincy Adams

It is the reliability leader's primary job to influence and educate the new product development team on reliability engineering discipline. This will allow the team to make effective decisions on safety and business impacts, which can include development timelines, costs, and acceptable risks associated with the product.

A reliability champion should:

- Instill a vision of "giving the most accurate answer with the data available."
- Influence the behavior of individuals or groups using cross-functional communication to achieve reliability goals.
- Understand the business consequences of a reliability question or solution.
- Explain in clear terms the reliability and safety risks involved in a management decision so that managers can make an informed choice.
- Remind team members of their ethical responsibilities in documentation and verbal or written communication.
- Support the decision of the reliability position to management.

A large part of influencing and educating the team on reliability is to develop a reliability strategy for the new product. This may include a reliability mission and vision statement for the project unless there is an overarching company reliability mission and vision.

Developing the reliability strategy includes drafting reliability objectives and requirements for the product that align with the mission and vision of the organization. Once created, the design team will work toward fulfilling the objectives and requirements. Reliability leaders communicate the decision-making process to management and educate their peers to ensure informed decisions are made.

RELIABILITY ENGINEER ROLE AND RESPONSIBILITIES IN THE PRODUCT LIFECYCLE

> Describe how the reliability engineer influences the product lifecycle, including their role in anticipating the impact of reliability on risk and costs in the design review process and ensuring performance over time. (Understand)
>
> **Body of Knowledge I.A.4**

Producing a product that is safe must be a top priority for every organization. The responsibility of the reliability engineer in meeting this priority includes the following actions:

1. Understand the design of the new product and its intended function.
2. Gather information on the use conditions to test the product design appropriately. If the use conditions are not understood correctly, then the team may miss possible failures or create ones that will not be seen in the field.
3. Collect and analyze data regarding failures and failure rates.
4. Make reliability predictions during the development process to aid in the decision-making process.
5. Present data and analyses in an understandable format.
6. Make sure that the key decision makers understand the analyses.

These actions are illustrated in Example 1.2.

> **EXAMPLE 1.2**
>
> During in-house testing, a failure mode is found. It is the reliability engineer's (RE) responsibility to understand the failure mode, what induced the failure mode, and what is needed to investigate the failure mode. The RE then gets the respective cross-functional disciplines together to understand the impact of the failure mode compared to the design failure mode effect analysis (DFMEA) and the potential failure rate in the field. The cross-functional disciplines must also include, at a minimum, the quality and design engineers. Once the risks are understood, a proposed plan of action is presented to the product development team so they can understand and agree upon the impact on the program if the proposed actions are taken. If the risk of the failure mode is very low, including the risk to the business, the team may choose to tolerate the failure mode as opposed to implementing a change to mitigate the risk.

In fulfilling these responsibilities, the following questions are among the additional items that must be considered:

- Could the failure of the product cause some chain of events with safety or liability implications? This analysis should be done with the safety organization since they are responsible for defining risk (human injury or death).

  Example: The product is installed as part of a system in which the failure of the product was not contemplated in system design.

- What aspects of the product could possibly cause safety or liability hazards even though the product has not failed?

  Example: During normal maintenance, the product must be partially disassembled, which may expose energized electrical conductors.

- What misuse of the product might cause safety or liability issues?

  Examples: During product development, the product was tested for shipping and storage. It passed the test. If the user then does not follow the storage instructions and stores the product in a different manner than was tested, damage can occur to the product. If the product is stacked too high, it could cause the entire stack to fall over and damage the product. If the product is exposed to temperatures that differ from the tested conditions, the seals could fail. If the pH of the solvent used in the product is below what was expected for the design, the product could develop hazardous leaks or degrade over time. When used on a windy day, a product functions correctly but endangers downwind organisms. In any of these situations, the reliability engineer must work with the safety and quality organizations to perform a safety hazard analysis.

- Can the final disposition of the product present safety or liability issues?

  Example: The product, when crushed for recycling, releases gases that produce a reaction in some people.

- Can the malfunction of other parts of the system cause safety or liability issues for the product?

  Example: When exposed to fluid pressures outside its operating range, the product will act unpredictably.

- What is the impact of government regulation, current or contemplated, on safety and liability issues?

  Example: Several states are contemplating legislation that will declare some types of metallurgical content of a component hazardous. In the medical device field, the review of government regulations can vary by country and should be considered when designing a product or making a design change.

- Does the product design compromise the reliability of components?

  Examples: An electronic component has an acceptable reliability based on a minimum level of air circulation, but its enclosure is not properly ventilated. Or an electrical component, a sensor, has acceptable reliability based on its intended use but fails when unexpected fluids come in contact with it.

PROJECT MANAGEMENT IN RELIABILITY ENGINEERING

> Apply key practices in requirements management, use Gantt charts, and understand the critical path method (CPM). (Apply)
>
> **Body of Knowledge I.A.5**

Without effective project management in new product development, the topic of reliability engineering can become an afterthought. The project management team might overlook reliability engineering until they realize they may be required to have statistical evidence to show that the product can reliably perform its intended function for the expected life of the product. In this case, this realization comes late in the project cycle and impacts the product's time to market. To prevent this situation, reliability engineering should be involved early in the project planning phase. The project management team can use Gantt charts and the critical path method (CPM) to make timeline decisions.

Reliability Engineering Gantt Chart Considerations

A *Gantt chart* is a project management tool that provides a high-level overview of the tasks needed to complete a project and the status of these tasks. The chart displays planned and completed tasks in relation to time. As such, the Gantt chart provides an excellent visualization of time-based progress. Each sub-team in the project should use the same project management methods for their respective teams; this includes the reliability engineering sub-team.

Developing the reliability engineering Gantt chart is a team effort. At a minimum, the team should include the reliability engineer and the test engineers. The team will brainstorm possible approaches to meet test goals. If various reliability tests are needed, then those tests often become a part of the critical path due to the extended nature of the testing. By implementing this project management tool, the critical path may be identified and shortened. Some items to consider when creating the Gantt chart are as follows:

- Discovery testing
  - Is discovery testing needed for the program? Testing such as the Highly Accelerated Life Test (HALT) can aid in discovering latent defects, primarily in electronics.
- Test methods
  - Can accelerated life testing (ALT) be implemented to get test data more quickly and then correlate the results to the reliability goals?

  - Do the needs of the program call for implementing reliability growth?
  - Is a full system test required to satisfy the program needs or can subsystem analysis and predictions suffice? For example, the Food and Drug Administration (FDA) may require that a full system test be implemented on a medical device to statistically validate the reliability requirements. In contrast, if a highly complex product, say an airplane, is being developed, a full system reliability test most likely would not be feasible. Instead, reliability will be confirmed using subsystem and component testing.
  - To what environments will the product be exposed? For example, a reusable surgical product that will be used for a patient in a hospital setting must be sterile. This will cause the product to go through the central processing area in the hospital. It can take more than 4 hours for the product to go through the central processing area. If the product must be reliable for 100 use cases, this means that the central processing aspect alone can take 400 hours per unit on test. This may have a significant impact on the timeline if it is not understood early in the product development planning process.
- Resources
  - What resources are needed for analysis?
    - What skill level is needed? How many resources are needed?
  - What resources are needed for test implementation?
    - What skill level is needed? How many testers are needed?
    - Are test fixtures needed? If so, how many? What is the timeline to receive and install them? What is the overall cost of these fixtures?
- Test duration
  - What is the cycle time for the fixtures?
  - What types of environments will the product need to be exposed to?
  - How many units are needed for the test?
  - Will periodic functional testing be needed? How long will this testing take? How often does it need to be implemented?
- Analysis methods
  - Are reliability data available for analysis for each component or subsystem? Or do products need to be tested to understand the reliability? If data do not exist, testing can impact the overall schedule.

Connecting all this information into the planning process is important to accurately predict the timeline and understand critical paths.

Example 1.3 illustrates the process of reliability engineering project planning.

EXAMPLE 1.3

A reliability team is working on a project to test a reusable surgical stapler and has identified the following activities and dependencies.

A. Design, procure, and approve test fixtures.

The first step in developing a test strategy is to understand if test fixtures are needed. For this example, we will assume that they are needed.

B. Develop test protocols and procedures for HALT testing; implement HALT testing.

Test protocols can be developed while fixtures are being procured. Since HALT testing is a discovery test, it can start during the protocol development, as feedback from the test will be put into the protocols. HALT testing is usually implemented on a few subsystems, so not all of the fixtures are needed in-house.

C. Develop test protocols and procedures for the first phase of reliability growth testing.

Test protocols can be developed while fixtures are being procured.

D. Develop test protocols and procedures for final system acceptance testing.

E. Implement reliability growth testing (phases):

Reliability growth testing can be implemented on subsystems as well as fully assembled systems. In this project, the reliability growth testing will occur on fully assembled systems. Dependencies include activities B and C: HALT testing and fixtures need to be finalized and in-house, and the test protocol and procedures need to be approved prior to the start of the test.

Develop test protocols and procedures for final system acceptance testing: dependency activity D. Test protocols need to be developed before the phase 2 reliability growth testing can begin.

F. Continue reliability growth testing phases 2+.

G. Implement final system acceptance test:

Dependencies: Test protocols and reliability growth testing must be complete before final acceptance testing. This is the final acceptance test, and therefore all other testing needs to be completed and accepted.

H. Write final report.

Data are analyzed throughout testing. Reports are drafted throughout testing with the final report released and approved after test completion.

Figure 1.1 shows the project Gantt chart. Activities (also known as objectives or tasks) are listed in the first column, followed by the department responsible for the activity. The grid shows the duration of each activity and the projected start and end dates. Arrows connecting activity durations denote dependencies. We see that the project duration is 240 days, assuming a five-day work week.

Continued

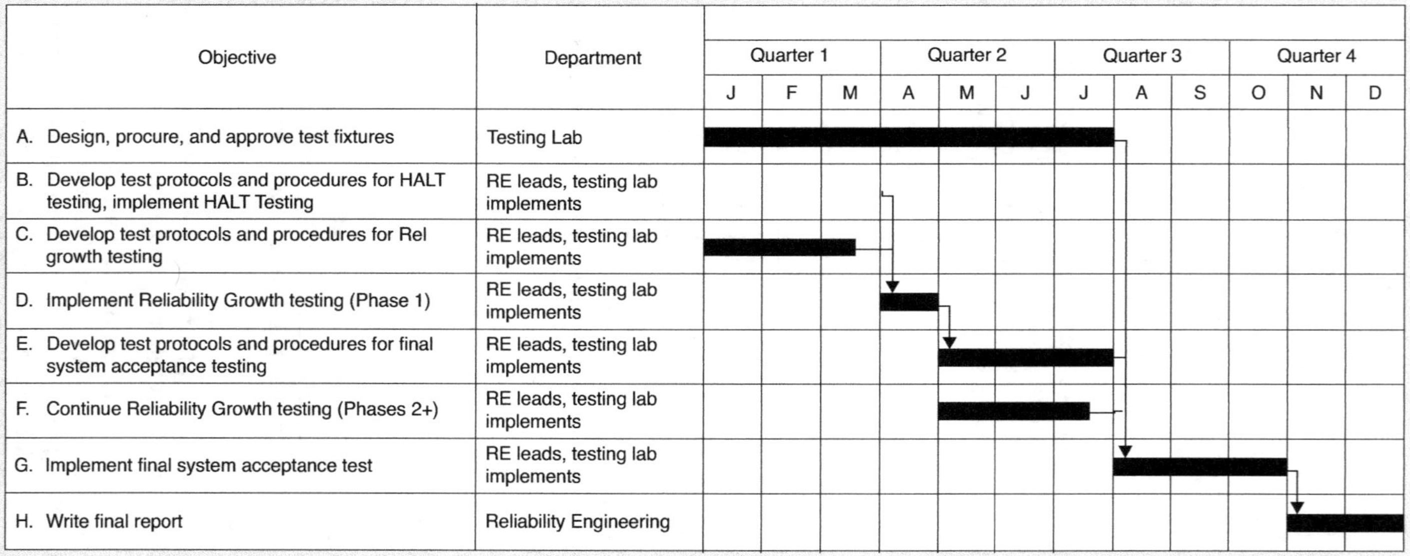

Figure 1.1 Gantt chart for surgical stapler project.

The Critical Path Method

The critical path is the longest sequence of activities in a project. If an activity on the critical path is delayed, the entire project will be delayed. The *activity network diagram* is a tool used to illustrate a sequence of events or activities (nodes) and the interconnectivity of such nodes. It is used for scheduling and especially for determining the critical path through nodes. The basic procedure for developing the activity network diagram follows:

Note: The following procedure is adapted from Tague (2023).

1. List all the necessary tasks in the project or process. If the diagram is being created manually, one convenient method is to write each task on the top half of a card or sticky note. Across the middle of the card, draw a horizontal arrow pointing right. Otherwise, project management software can be used.
2. Determine the correct sequence of the tasks. Do this by asking three questions for each task:
    - Which tasks must happen before this one can begin?
    - Which tasks can be done at the same time as this one?
    - Which tasks must happen after this task is complete?
3. Diagram the network of tasks. Arrange the tasks in sequence. Time should flow from left to right, and concurrent tasks should be vertically aligned.
4. When the network is correct, label all tasks in sequence with event numbers in the circles. It can be useful to label all tasks in sequence using letters.
5. Determine task times—the best estimate of the time that each task should require. Use one measuring unit (hours, days, or weeks) throughout for consistency. Write the time on each task's arrow.
6. Determine the critical path—the longest path from the beginning to the end of the project. Mark the critical path with a heavy line or color. Calculate the length of the critical path: the sum of all the task times on the path.
7. Calculate the earliest times each task can start and finish, based on how long preceding tasks take. These are called earliest start (ES) and earliest finish (EF). Start with the first task, where ES = 0, and work forward. Draw a square divided into four quadrants, as in Figure 1.2. Write the ES in the top left box and the EF in the top right.

    For each task:

    - Earliest start (ES) = The largest EF of the tasks leading into this one
    - Earliest finish (EF) = ES + Task time for this task
8. Calculate the latest times each task can start and finish without upsetting the project schedule, based on how long later tasks will take. These are called latest start (LS) and latest finish (LF). Start from the last task, where the latest finish is the project deadline, and work backward. Write the LS in the lower left box and the LF in the lower right box.

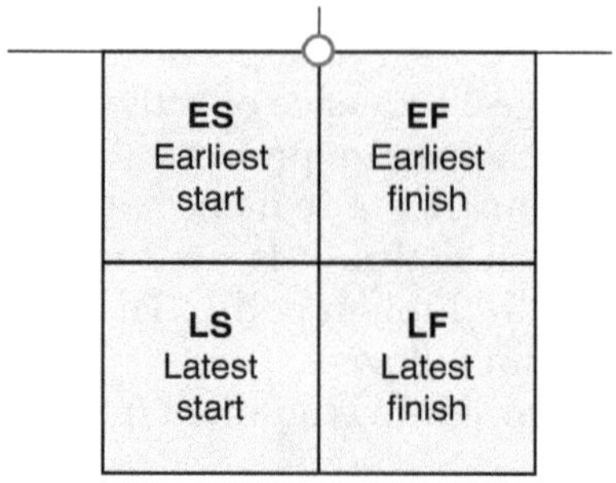

Figure 1.2 Arrow diagram time box.

- Latest finish (LF) = The smallest LS of all tasks immediately following this one
- Latest start (LS) = LF – Task time for this task

9. Calculate slack times for each task and for the entire project.
   - Slack time for an activity is defined as:
     - Slack time = Latest start time – Earliest start time
   - Total slack is the time a job could be postponed without delaying the project schedule.
     - Total slack = LS – ES = LF – EF
   - Free slack is the time a task could be postponed without affecting the early start of any job following it.
     - Free slack = The earliest ES of all tasks immediately following this one – EF

   Figure 1.3 shows a schematic way to remember which numbers to subtract.

Project management software is available to identify and calculate the critical path for projects with multiple paths. The longest path in the network is the time required to complete the project. If activities on the critical path are delayed, the entire project will be delayed.

One way to complete a project in less time is to decrease the time for at least one of the activities. This is usually accomplished by putting more resources into one or more activities on the critical path. This is sometimes referred to as *crashing the project*.

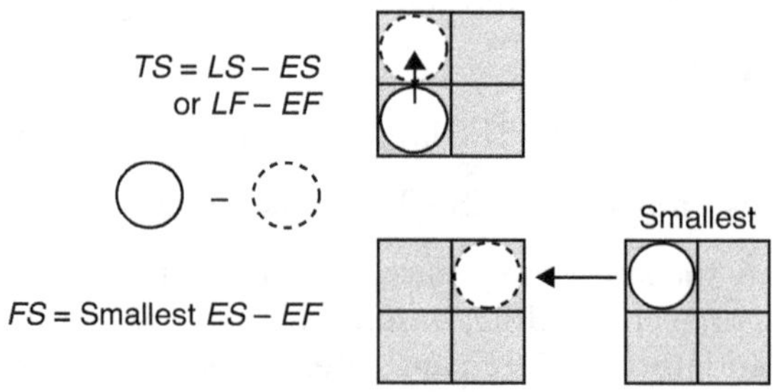

Figure 1.3 Guide for slack calculations.

EXAMPLE 1.4

Recall the reliability project described in Example 1.3. The required activities (tasks) were:

A. Design, procure, and approve test fixtures.

B. Develop test protocols and procedures for HALT testing; implement HALT testing.

C. Develop test protocols and procedures for reliability growth testing.

D. Implement reliability growth testing phase 1.

E. Develop test protocols and procedures for final system acceptance testing.

F. Continue reliability growth testing phase 2+.

G. Implement final system acceptance test.

H. Write the final report.

The order of these tasks and their dependencies is shown in Table 1.1.

From this table, we can see, for example, that activities A, B, and C can begin simultaneously and run concurrently. Activity D occurs only after activities B and C are complete. Activities E and F can only begin after activity D is complete, and activity G occurs only after activities A, E, and F are complete. Finally, activity H can begin after activity G. The time in working days for each activity is also listed.

From this table, we can sketch out the order of the tasks, as shown in Figure 1.4. In addition, the entries in the time box for each activity have been calculated to show early start, late start, early finish, and late finish times.

Table 1.1 Activity precedence table.

Activity	Predecessor	Time (working days assuming 5 days a week)
A		120
B		60
C		50
D	B, C	20
E	D	60
F	D	50
G	A, E, F	60
H	G	40

The program evaluation and review technique (PERT) and CPM have essentially merged in current software packages. The *critical path* is the path from start to finish that requires the most time. In Figure 1.4, there are five paths:

Continued

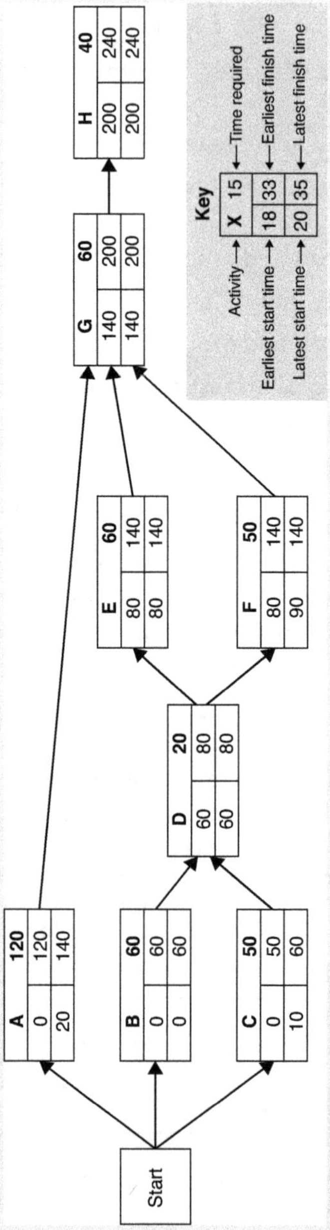

Figure 1.4 Activity network diagram for the surgical stapler project.

- Path AGH requires 120 + 60 + 40 = 220 days
- Path BDEGH requires 60 + 20 + 60 + 60 + 40 = 240 days
- Path BDFGH requires 60 + 20 + 50 + 60 + 40 = 230 days
- Path CDEGH requires 50 + 20 + 60 + 60 + 40 = 230 days
- Path CDFGH requires 50 + 20 + 50 + 60 + 40 = 220 days

Therefore, BDEGH is the critical path. If any of the tasks on this path are delayed, the entire project will be delayed.

FUNCTION OF RELIABILITY IN ENGINEERING

> Describe how reliability techniques can be used to apply best practices in engineering (e.g., measuring reliability early), how industry standards can impact reliability, and how reliability can inform the decision analysis process. (Analyze)
>
> **Body of Knowledge I.A.6**

Reliability engineering is usually undertaken to determine and improve the useful lifetime of products. Data are collected on the failure rates of the components and products, including those produced by suppliers. Competitors' products may also be subjected to reliability testing and analysis for comparison and benchmarking.

Every facet of an organization, including product design, manufacturing, marketing and advertising, legal, purchasing, quality assurance, packaging, field service, and logistics, can benefit from a knowledge of reliability engineering. An understanding of the lifecycles of the products and equipment these departments use and handle can improve the effectiveness and efficiency of their function.

For example, reliability analysis can improve product design. Reliability predictions provide guidance as components are selected. Derating techniques aid in increasing a product's useful lifetime. Reliability improvements can be achieved through component redundancy.

In the manufacturing area, processes can benefit from reliability tools in the following ways:

- The impact of process parameters on product failure rates can be studied.
- Alternative processes can be compared for their effect on reliability.
- Reliability data for process equipment can determine preventive maintenance schedules and spare parts inventories.
- Using parallel process streams to improve process reliability can be evaluated.
- Safety can be enhanced through the understanding of equipment failure rates.
- Vendors can be evaluated more effectively.
- End-of-line inspection or run-in processes can be implemented in manufacturing if needed due to the identification of potential failure modes.

Marketing and advertising can be enhanced as the warranty and other documents that inform customer expectations are prepared. Warranties that are not supported by reliability data can incur extra costs for the organization and damage customer perceptions and brand loyalty.

It is increasingly important to detect and prevent or mitigate liability issues. Effective warnings and alarms should be incorporated into the design when hazards cannot be eliminated. Products whose failure can introduce safety and health hazards need to be analyzed for reliability so that procedures or design improvements can be put in place to reduce the probability that those hazards will occur. Failure rates typically escalate in the final phase of a product's life. If these failures are found to occur after the product's useful life, then procedures or design improvements need to be implemented so that the product cannot be used beyond its useful lifetime. Components with useful lifetimes that are shorter than that of the product's lifetime should be replaced on a schedule that can be determined through reliability engineering techniques.

ETHICS IN RELIABILITY ENGINEERING

> Identify appropriate ethical behaviors for a reliability engineer in various situations. (Evaluate)
>
> **Body of Knowledge I.A.7**

The ASQ Code of Ethics, shown in Figure 1.5, provides guidelines on how reliability engineers and quality professionals should conduct themselves.

ASQ CODE OF ETHICS

Fundamental Principles

ASQ requires its representatives to be honest and transparent. Avoid conflicts of interest and plagiarism. Do not harm others. Treat them with respect, dignity, and fairness. Be professional and socially responsible. Advance the role and perception of the quality professional.

Expectations of a Quality Professional

A. Act with Integrity and Honesty

1. Strive to uphold and advance the integrity, honor, and dignity of the quality profession.
2. Be truthful and transparent in all professional interactions and activities.
3. Execute professional responsibilities and make decisions in an objective, factual, and fully informed manner.
4. Accurately represent and do not mislead others regarding professional qualifications, including education, titles, affiliations, and certifications.
5. Offer services, provide advice, and undertake assignments only in your areas of competence, expertise, and training.

B. Demonstrate Responsibility, Respect, and Fairness

1. Hold paramount the safety, health, and welfare of individuals, the public, and the environment.
2. Avoid conduct that unjustly harms or threatens the reputation of the Society, its members, or the quality profession.
3. Do not intentionally cause harm to others through words or deeds. Treat others fairly, courteously, with dignity, and without prejudice or discrimination.

4. Act and conduct business in a professional and socially responsible manner.
5. Allow diversity in the opinions and personal lives of others.

C. Safeguard Proprietary Information and Avoid Conflicts of Interest
1. Ensure the protection and integrity of confidential information.
2. Do not use confidential information for personal gain.
3. Fully disclose and avoid any real or perceived conflicts of interest that could reasonably impair objectivity or independence in the service of clients, customers, employers, or the Society.
4. Give credit where it is due.
5. Do not plagiarize. Do not use the intellectual property of others without permission. Document the permission as it is obtained.

Figure 1.5 ASQ Code of Ethics.

The following are some examples that illustrate the ASQ Code of Ethics.

A.2. Be truthful and transparent in all professional interactions and activities.

Engineers are frequently called on to apply their expertise to issues not directly related to their employer. These opportunities vary from service on a committee in a professional organization to advising on public works projects. When it is necessary to issue a statement in this capacity, the Code of Ethics requires a disclaimer separating one's views from those of the employer. Conversely, when the engineer is asked to speak for the employer, the statement should make that fact clear as well.

The reliability engineer is required to present both the good and bad news scenarios when making recommendations. This will help equip the decision maker with informed options and their potential outcomes. If hypothesis tests were used to reach conclusions, the significance level should be disclosed. For sampling reports, the confidence level and margin of error should be included.

A.4. Accurately represent and do not mislead others regarding professional qualification, including education, titles, affiliations, and certifications.

This phrase requires that all who subscribe to this code of ethics never exaggerate their qualifications, for example, on a resume, in a research proposal, or in the introduction to a professional presentation.

B1. Hold paramount the safety, health, and welfare of individuals, the public, and the environment.

This phrase indicates that the reliability engineer's responsibilities are not limited to crunching numbers and producing good analyses but also includes the promotion of product reliability and safety. This includes discussing any issues that occur and presenting the risks that could occur to determine acceptability. Safety is extremely important and goes hand-in-hand with reliability.

C.1. Ensure the protection and integrity of confidential information.

This clause says that even in the absence of a confidentiality agreement, the individual is honor bound to act as if one is in place. As a practical matter, it may be advisable to have a signed statement from the former employer or client releasing the information.

C.3. Fully disclose and avoid any real or perceived conflicts of interest that could reasonably impair objectivity or independence in the service of clients, customers, employers, or the Society.

Professionals of all types make value judgments as part of their responsibilities. This section of the code of ethics requires a conscious search to identify any connections that might bias conclusions. In some situations, especially public service, any connection that could be perceived as a conflict of interest should be divulged.

C.4 Give credit where credit is due.

This clause requires action by the person preparing or presenting a report. Rather than leaving the report uncredited, which might imply that the credit is due to the presenter, the "take care" phrase requires an acknowledgment of those involved. If a team is due credit, the team members should usually be named.

The entire ASQ Code of Ethics should be studied and used as a basis for action by all in the reliability engineering field.

SUPPLIER RELIABILITY ASSESSMENTS

> Explain how supplier reliability impacts the overall reliability program. Describe key reliability concepts that should be included in supplier reliability assessments. Continuously assess supplier reliability to ensure they meet established reliability standards. (Analyze)
>
> **Body of Knowledge I.A.8**

In an ideal world, every supplier would have an excellent reliability engineering program with regular, dependable reports delivered to customers. As this is not always the case, it is essential to have the customer choose between three scenarios when contracting with suppliers:

1. The customer assumes all responsibility for the reliability engineering function and requires supplier compliance with all specifications. With this arrangement, the supplier must report any proposed changes in the process or product so that the potential impact on reliability can be studied. This option is more common in situations where the customer has full design responsibilities and outsources relatively minor components.
2. The supplier assumes responsibility for reliability engineering and reports its analysis and decision-making process to the customer for agreement. The customer is ultimately responsible to its customers, but the supplier may share financial responsibility for warranty claims and so on. This option is more common when the supplier has design control of the supplied component.

3. Some sort of shared responsibility for the reliability engineering analysis and interpretation exists, perhaps involving a third party. Third-party involvement is more common when the supply chain is geographically long or complex.

With any of these options, the arrangement must be clearly spelled out in the contractual agreement between the parties. The customer will want to conduct assessments customized to that agreement. For example:

- For suppliers with a long-term relationship based on mutual trust and understanding, the reliability functions conducted by the supplier can be verified at the time of a quality audit. At least one auditing team member should be familiar with reliability engineering functions. The supplier's collection and analysis of lifecycle cost data should be studied, and the mechanism for feedback of this information to the manufacturing and design functions should be confirmed.
- For suppliers without a strong favorable history with the components involved, the customer should consider performing actual testing and evaluation of the products. This could vary from a full-fledged reliability program to less elaborate programs, depending on the situation.

The full reliability program would begin with establishing goals and translating goals into product and process design requirements and continue through validating production output. Less elaborate programs could comprise monitoring the reliability engineering function at the supplier's location, training supplier personnel, and/or testing random samples from production to ensure that reliability requirements are being met.

The performance of the suppliers should be monitored much like the performance monitoring used for the overall process described in "Performance Monitoring." The suppliers need to be continuously assessed in a manner that is acceptable by both parties to ensure that the reliability of the supplied items meets requirements over the lifetime of the product.

PERFORMANCE MONITORING

> Describe the importance of performance monitoring in maintaining compliance with product reliability and safety standards. Identify key points in the product lifecycle where both process and product reliability data are collected and evaluated. Integrate essential reliability elements like defect tracking and reliability growth monitoring (e.g., key performance indicators [KPIs], overall equipment effectiveness [OEE], Duane growth model, and Crow-AMSAA Model) to ensure continuous adherence to these requirements. (Understand)
>
> **Body of Knowledge I.A.9**

Performance monitoring involves periodically measuring a project's progress toward explicit short- and long-term reliability, maintainability, and safety (RMS) objectives and giving feedback on the results to decision makers who can use the information in various ways to improve performance.

Uses of Performance Indicators

Performance indicators can be used in strategic planning, performance accounting, forecasting and early warning efforts, measuring program results, program marketing and public relations, benchmarking, and quality management.

Strategic Planning

For any program, incorporating performance measurement forces greater consideration of the critical assumptions of the program and the paths it is following. Performance indicators provide feedback that can help clarify the RMS objectives and logic in the program.

Performance Accounting

Performance indicators can help inform resource allocation decisions, direct RMS resources to the most successful activities, and thereby promote the most efficient use of those resources.

Forecasting and Early Warning During Program Implementation

Measuring progress against indicators points toward future performance, providing feedback that can be used for planning, identifying areas needing RMS improvement, and suggesting what can be done.

Measuring Program Results

Good performance indicators measure what a program has achieved relative to its RMS objectives, not just what it has completed, thus promoting RMS accountability.

Program Marketing and Public Relations

Performance indicators can demonstrate program RMS results to satisfy an external audience or customer. RMS performance data can communicate the value of a program or project.

Benchmarking

Performance indicators can generate RMS data against which to measure other projects or programs. They also provide a way to identify good RMS applications, thus learning from experience how to improve the RMS performance of other projects or programs.

Quality Management

Performance indicators can measure customer satisfaction relative to customer requirements. Performance can be monitored through administering satisfaction surveys and evaluating complaints or product reviews.

How Are RMS Performance Indicators Developed?

Performance indicators must be based on the unique objectives of individual projects and the customer's requirements. Identifying the RMS objectives—indeed all project objectives—can flow from the customer's requirements via a quality function deployment (QFD).

As the project's RMS objectives are developed, the best mix of outputs to achieve these objectives and components is derived. Tracking parameters for these outputs (e.g., mean time between failures, reject rates, number of defectives, aborts, material defects per item, etc.) can be set up for regular, timely reporting.

For example, for a typical product development cycle, the reliability and safety tools and methods are used, modified, and developed for the product.

Key performance indicators (KPIs) are the key metrics determined by the program team and monitored throughout the product's lifecycle. The following are a few of the performance indicators that can be used to measure the program's progress toward reliability goals.

- Mean time between failures (MTBF)
- Time to failure over time
- Return rate percentage over time
- Warranty costs
- Open and closed FRACAS issues (see next section)
- Reliability growth
- Overall equipment efficiency (OEE)

Performance Monitoring Phases and Expectations

The project team should agree upon when to implement performance monitoring. A failure reporting and corrective action (FRACAS) system, described in Chapter 11, can provide data throughout the phases of a product's lifecycle. Trends should be analyzed and actions taken as necessary. Throughout the new product development cycle, it is helpful to have visibility to the trends in open and closed reliability issues. Figure 1.6 gives an example of monitoring a FRACAS system graphically.

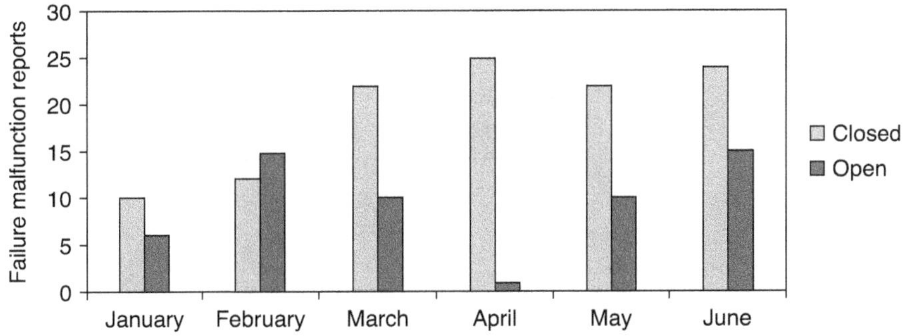

Figure 1.6 FRACAS issue status by month.

Performance Monitoring During New Product Development

New product development teams establish performance goals, including reliability goals, early in the development process. These goals are usually based on the performance of comparable products in the field. Therefore, field monitoring is important. As the product is designed and improved, the goals are continually reviewed, and in-house tests or analysis results are compared to them. A team that includes, at a minimum, the reliability engineer and/or quality engineer, design engineering, and manufacturing engineering reviews each failure as it occurs. The team analyzes the failures for root cause and proposes corrective actions. From this analysis, reliability predictions can be established. As testing progresses, actual results are reviewed against the predicted results and actions are taken as necessary.

One analysis method used during new product development is *reliability growth modeling*. Reliability growth is defined as the positive improvement in a reliability parameter over time due to implementation of corrective actions to system design, operation or maintenance procedures, or the associated manufacturing process. Reliability growth is an iterative design process, sometimes referred to as "test, analyze, fix," or TAF, which includes (MIL HDBK 189-C, 2011):

1. Failure mode discovery
2. Feedback of problems identified
3. Failure mode root cause analysis and proposed corrective action
4. Approval and implementation of proposed corrective action
5. Fabrication of hardware

A reliability growth program finds reliability problems by testing, implementing corrective actions, and monitoring the increase of the product's reliability throughout the test phases. The term "growth" is used since it is assumed that the reliability of the product will increase over time as design changes and fixes are implemented.

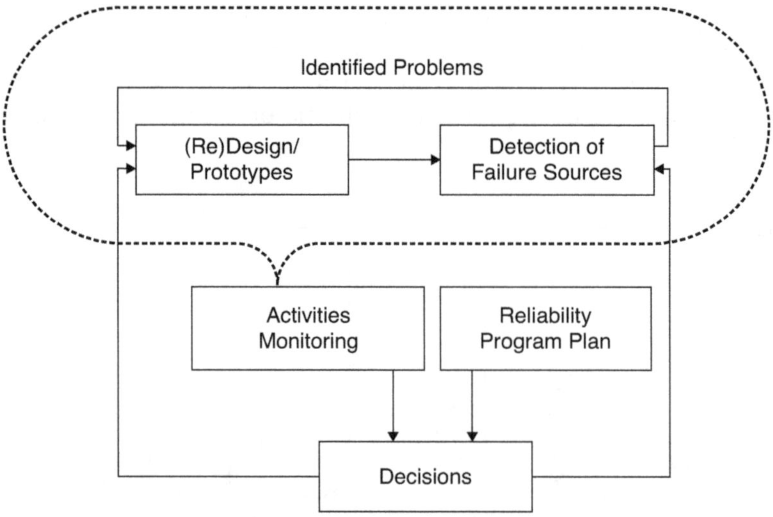

Figure 1.7 Reliability growth management model (monitoring approach).
Source: MIL HDBK 189-C (2011). Public domain.

The reliability growth management model shown in Figure 1.7 emphasizes the iterative nature of finding and fixing problems in the design phase.

Benefits of the reliability growth management include the following (MIL HDBK 189-C, 2011):

- Finding unforeseen deficiencies
- Designing-in improvement through surfaced problems
- Reducing the risks associated with final demonstration testing
- Increasing the probability of meeting reliability objectives

Reliability growth monitoring tracks the improvement of product reliability across the phases of product development. In the early phases, prototypes may be made with limited tooling and will often have design, manufacturing, and/or quality deficiencies. As a result, early product reliability may be well below the reliability goal. Reliability growth modeling then concentrates on design improvements. To identify and correct the early deficiencies, the prototypes are subjected to a rigorous testing program. During testing, problem areas are identified, and appropriate corrective actions (or redesigns) are implemented.

Two reliability growth models will be presented here, the Duane model and the Crow-AMSSA model. Reliability growth models can be used for planning, tracking, and projecting reliability parameters. These models are further discussed in Chapter 8.

Figure 1.8 shows nine possible growth patterns for two test phases in three rows and columns. Row 1 shows phase 1 as having all fixes delayed until the end of the testing phase. Row 2 shows having some fixes inserted during test, with some

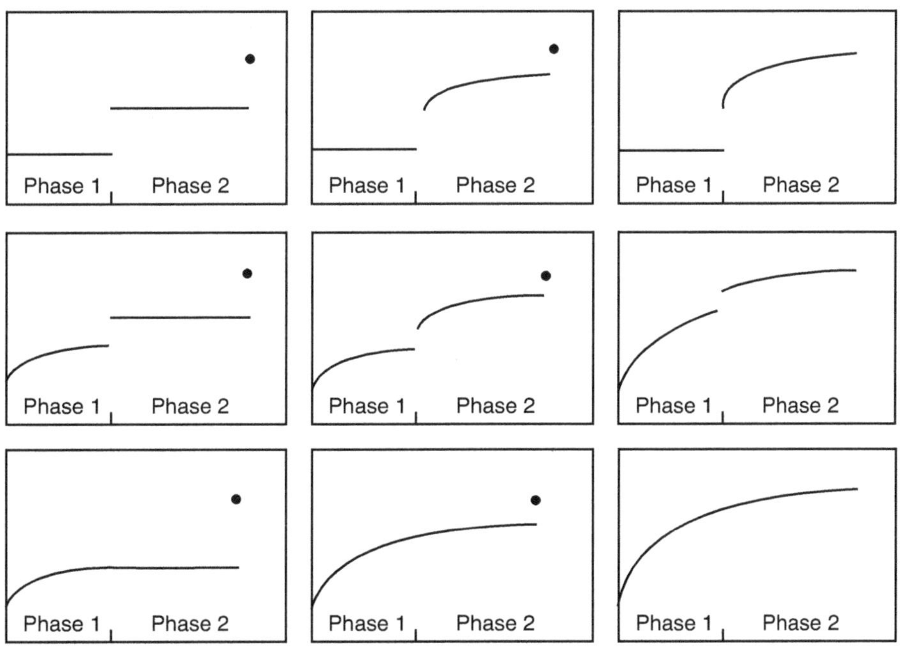

Figure 1.8 The nine possible general growth patterns for two test phases.
Source: MIL-HDBK-189C (2011). Public domain.

being delayed. Row 3 shows phase 1 as having all fixes inserted during test with none delayed.

Column 1 of Figure 1.8 shows phase 2 as having all the fixes delayed until the end of the testing phase. Column 2 shows phase 2 as having some fixes inserted during test and some delayed. Finally, Column 3 shows phase 2 as having all fixes inserted during test, with none delayed. The graphs in the top left (all fixes delayed) and bottom right (all fixes inserted during test) represent the two extremes in possible growth test patterns (MIL HDBK 186-C, 2011).

Duane (1964) observed that if fixes to improve reliability are implemented during the design phase, the graph of the cumulative failure rate versus cumulative test time is linear when plotted on log-log paper. Since the plot is a straight line, we can write out the simple linear regression model:

$$\log_{10}\left(\frac{n(T)}{T}\right) = -\alpha \log_{10}(T) + b$$

where T is the cumulative test time, $n(T)$ is the cumulative number of failures to time T, and $n(T)/T$ is the cumulative failure rate.

When the slope parameter α is zero, this implies that there is no improvement in reliability. When the slope is positive, the failures occur farther and farther apart as test time increases. An upper limit to the slope is 1, meaning that the number of failures is independent from the length of the test (Breneman et al., 2022). The log-log plot and the regression model comprise what is known as the *Duane postulate*.

An example of a Duane growth model curve is shown in Figure 1.9 (Breneman et al., 2022). Here we see that the slope parameter, representing the reliability growth coefficient, is 0.34.

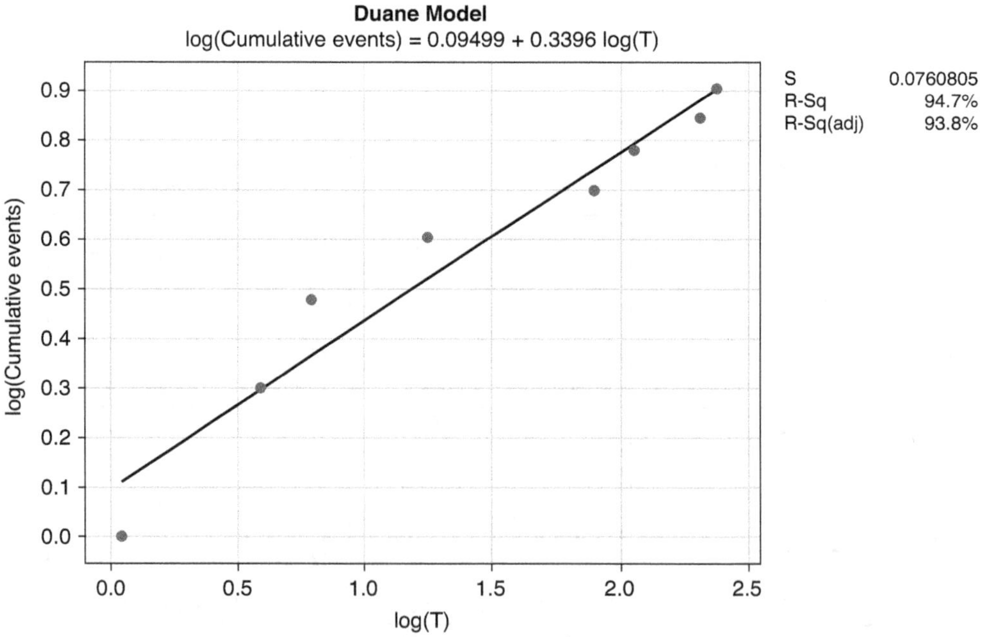

Figure 1.9 Duane model, cumulative events versus cumulative time.
Source: Breneman et al. (2022).

The Duane model can be applied using different forms for the dependent variable, depending on reporting needs. Choices for the dependent variable include the following:

- Cumulative events
- Cumulative rate
- Instantaneous rate
- Cumulative MTBF
- Instantaneous MTBF.

Figure 1.10 shows an example of a Duane model plot using cumulative MTBF. In Chapter 8, the Duane model is discussed in the reliability planning context, using cumulative MTBF as the dependent variable.

The major drawback of the Duane model is that it is deterministic, meaning that the parameters of the growth model cannot change over time. Crow (1975) improved upon Duane's power law model by incorporating a stochastic element. He determined that the reliability growth pattern could be modeled as a non-homogeneous Poisson process (NHPP) with a Weibull intensity function. This modeling technique allows for statistical procedures to be used in the application of reliability growth. In 1981, Crow extended this model to produce what is now known as the Crow-AMSAA model. As an extension to the Duane model, in the Crow-AMSAA model, the failure intensity of the underlying NHPP is linear when plotted on log-log scale.

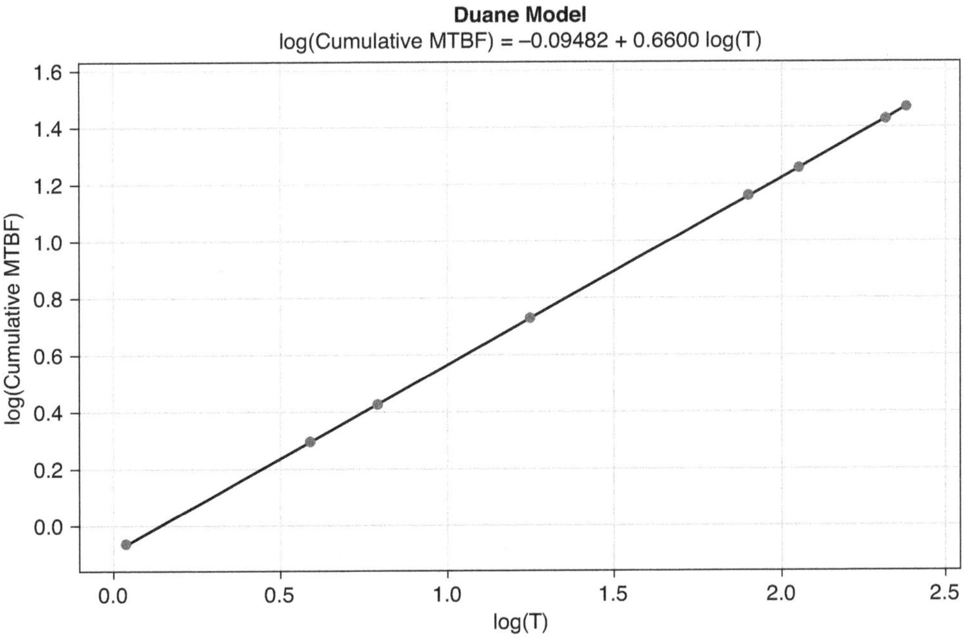

Figure 1.10 Duane reliability growth model example, cumulative MTBF.
Source: Breneman et al. (2022).

EXAMPLE 1.5

Duane and Crow-AMSAA Models

In the development of a single-use medical device, the reliability of a weld over repeated bending cycles was identified as a key requirement. During the development phase, prototypes were tested until failure, cycles to failure data were gathered, failures were rapidly traced to root causes, and improvements to the design of the weld join were implemented. How can the development team use the following cycles to failure data to model the improvement in the reliability of the weld joint?

Cycles to failure: 74, 235, 931, 1052, 1519, 1944, 2074, 2486, 2657, 2689, 4276, 5598, 5600, 6495, 6789

Applying the Duane model to this dataset illustrates the reliability growth of the cumulative MTBF from the earliest failure below 100 cycles to the longest-surviving units beyond 6000 cycles. The Duane plot in Figure 1.11 shows the straight-line Duane model when cumulative MTBF is plotted against cycles, both on log-transformed scales.

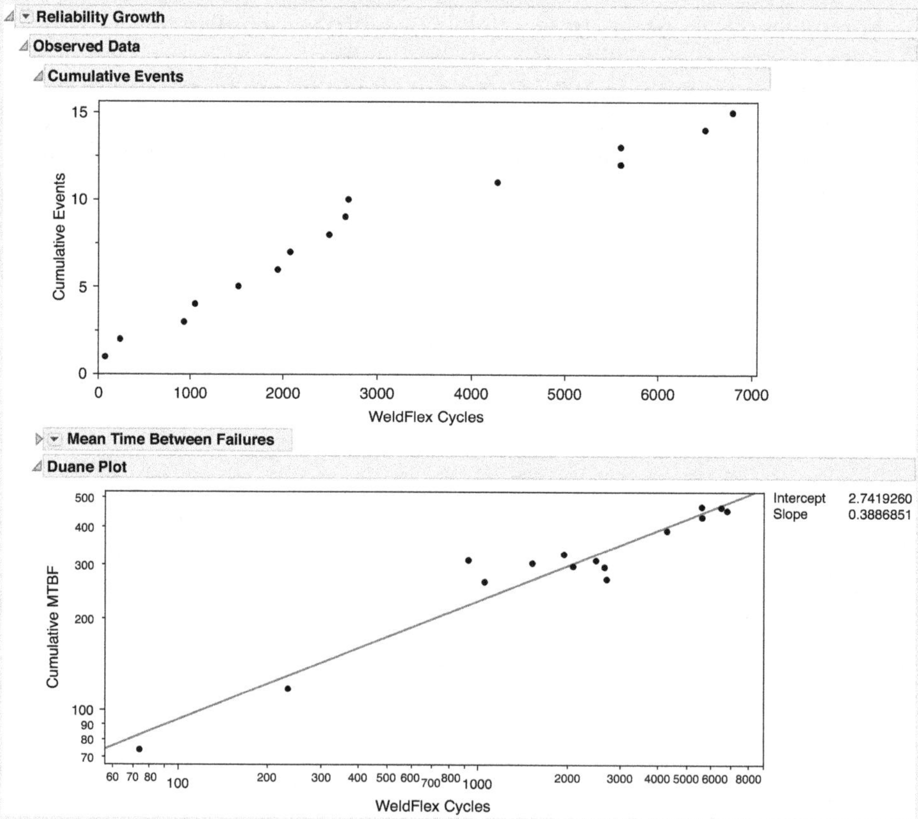

Figure 1.11 Cumulative event plot and Duane plot showing fitted line on log-log scales.

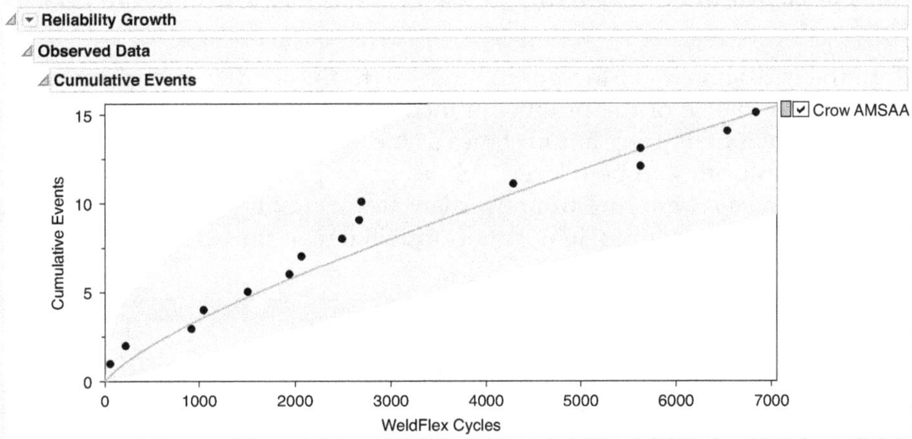

Figure 1.12 Cumulative event plot with Crow-AMSAA confidence bands.

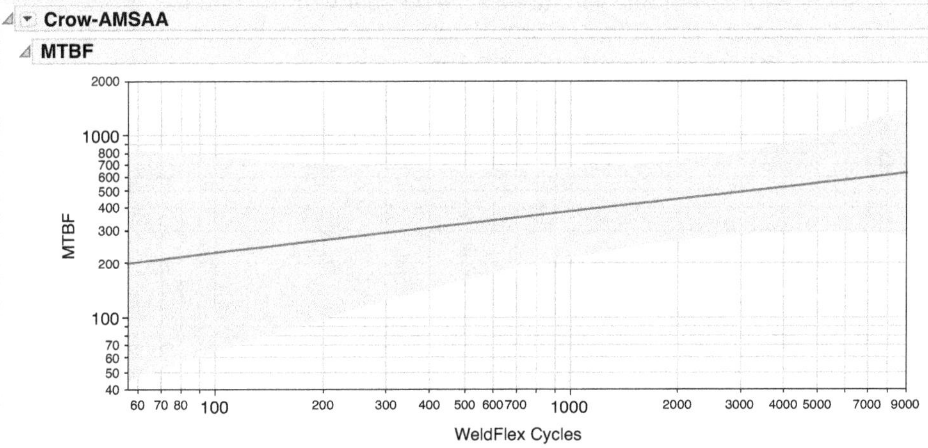

Estimates

Parameter	Estimate	Std Error	95% Confidence Interval (Weld) Lower	Upper
lambda	0.01650747	0.02934533	0.0005064	0.5380978
beta	0.77206697	0.19934684	0.4654528	1.2806613
Reliability Growth Slope	0.22793303	0.19934684	−0.2806613	0.5345472

Figure 1.13 Crow-AMSAA analysis of growth in MTBF.

The Crow-AMSAA model illustrated in Figures 1.12 and 1.13 provides confidence intervals for the cumulative event plot, the cumulative MTBF plot, the model parameters λ and β, and the reliability growth slope. The confidence interval of the reliability growth slope ranging from negative to positive shows that this data cannot yet prove that there is positive reliability growth.

Performance Monitoring During the Early Phases of a Product Released to the Field

In the early phases of the product release to the field, it is important to monitor the performance of the product immediately after manufacture, until such time that the manufacturing line becomes stable. Once the manufacturing line is stable, then monitoring can be relaxed. As design or process changes are implemented, performance monitoring from the manufacturing line should be implemented to ensure that the changes do not affect the final manufactured product.

EXAMPLE 1.6

Consider a product that has a warranty period of three months. During the later phases of the new product development process, products are tested that have been built under final manufacturing processes and verified to meet the warranty period. Once the manufacturing line is running at full capacity, a random sample is pulled and tested to ensure that the warranty reliability is still met. The random sampling and warranty reliability testing is implemented every three months until there is an acceptable variation in the reliability of the sampled products.

Another example of early field performance is warranty cost predictions versus actuals, as shown in Figure 1.14.

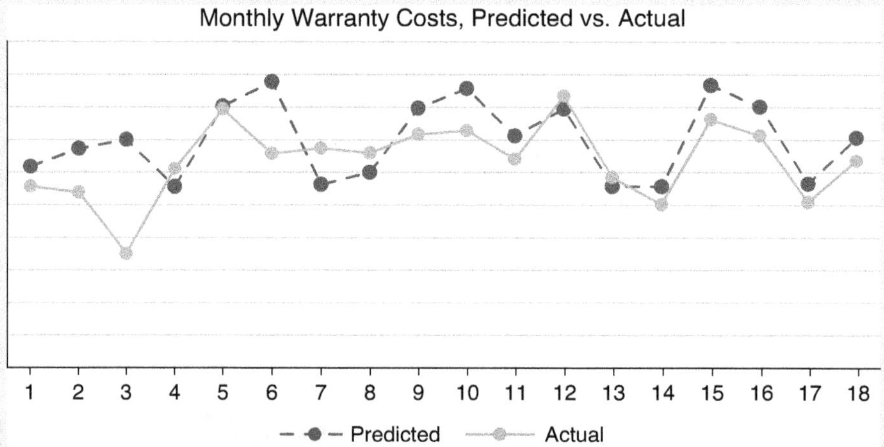

Figure 1.14 Monthly warranty costs (predicted versus actual).

Field Performance Monitoring

Field performance monitoring is a continuous process and feeds into many areas:

- Information feeds into new product development goals.
- Information feeds into design and process changes of current products in the field.
- Information feeds into customer satisfaction.

EXAMPLE 1.7

While reviewing complaints from the field, it is found that there is a trend around a specific area in the product. This prompted high visibility within the company, and a team was formed to understand the issue. This prompted a design change, which was implemented, and the trend was greatly reduced.

Overall Equipment Effectiveness Monitoring

Another metric that can be monitored is *overall equipment effectiveness* or OEE. The OEE can be used to evaluate processes and process improvement efforts. It is the product of three measures:

1. Availability, or percent uptime
2. Performance, or percent of published rate
3. Quality, or percent yield

To calculate OEE, we use:

$$\text{OEE} = \text{Availability} \times \text{Performance} \times \text{Quality}$$

where

$$Availability = \frac{Actual\ operating\ time}{Planned\ production\ time}$$

$$Performance = \frac{Total\ pieces\ /\ Actual\ operating\ time}{Designed\ (ideal)\ pieces\ /\ time\ period}$$

$$Quality = \frac{Number\ of\ good\ pieces\ produced}{Total\ pieces\ produced}$$

and

Planned production time = Shift length less time for breaks, lunch, and so on

Actual operating time = Planned production time less downtime for setup, maintenance, and so on.

These calculations are illustrated in Example 1.6, adapted from McShane-Vaughn (2022).

EXAMPLE 1.8

A machine is scheduled for a 10-hour shift, minus 80 minutes for breaks and a meal. This past shift, the machine was down for 70 minutes for maintenance. The designed production rate is 300 pieces per hour, or 5 pieces per minute. During the shift, the machine produced 1783 good parts and 91 bad parts. The OEE calculation follows (all times in minutes):

Planned production time (min) = 60 (10-hour shift) – 80 (breaks) = 520 minutes

Actual operating time = 520 – 70 (maintenance) = 450 minutes

Continued

Total pieces = 1783 (good parts) + 91 (bad parts) = 1874 parts

Ideal run time = 5 parts/min

Number of good pieces = 1783 parts

Total number of pieces = 1874 parts

$$\text{Availability} = \frac{\text{Actual operating time}}{\text{Planned production time}} = \frac{450\ \text{min}}{520\ \text{min}} = 0.865$$

$$\text{Performance} = \frac{\text{Total pieces / Actual operating time}}{\text{Designed (ideal) pieces / Time period}} = \frac{1874\ pc / 450\ \text{min}}{5\ pc / \text{min}} = 0.833$$

$$\text{Quality} = \frac{\text{Number of good pieces produced}}{\text{Total pieces produced}} = \frac{1783}{1874} = 0.951$$

OEE = (0.865)(0.833)(0.951) = 0.685 = 68.5%

OEE can be used to track efficiency over time. It also can be used to make comparisons between two identical machines making identical parts or to compare a machine or process before and after implementation of a quality improvement.

Chapter 2
Reliability Foundations

BASIC RELIABILITY TERMINOLOGY

> Clarify fundamental reliability terms and the corresponding metrics, including concepts related to system reliability (e.g., durability, failure rate, mean time to failure [MTTF], mean time between failures [MTBF], mean time between critical failure [MTBCF], and the bathtub curve), maintainability (e.g., service interval, mean time to repair [MTTR], mean time between maintenance [MTBM], and mean downtime [MDT]), and availability (e.g., operational availability, achieved availability, and inherent availability). (Apply)
>
> **Body of Knowledge 1.B.1**

Reliability is the probability that an item can perform its intended function for a specified interval under stated conditions. More formally, under stated conditions, $R(t)$ is a probability distribution.

When failure data have been fit to a statistical distribution, such as exponential or Weibull, reliability is found using the probability that a unit will function beyond time *t*, as shown in Equation 2.1.

Equation 2.1 Reliability at time *t*.

$$R(t) = Pr(X > t) = 1 - Pr(X \leq t)$$

$R(t)$ may be expressed as *BX life* or *B(X) life*, which is the time that has elapsed when $X\%$ of the population has failed. For example, $B(10) = 367$ hours means $R(367) = 0.90$.

Reliability at time *t*, or $R(t)$, can be estimated empirically by dividing the number of units functioning at the end of the time period by the number of units that were functioning at the start of the test, as shown in Equation 2.2.

Equation 2.2 Empirical reliability at time *t*.

$$R(t) = \frac{\text{Number of units functioning at the end of the time period}}{\text{Number of units that were functioning at the start of the test}}$$

Several measures of reliability and their definitions follow.

- *Durability* is described as a product's ability to continue to function for the duration of its expected useful life. The term *durability* addresses the life span of a product or system, which may be measured in units of time or cycles.
- The *mean life* of a product is the average time to failure of identical products operating under identical conditions. Mean life is also referred to as the expected time to failure.
- For nonrepairable products, mean life is expressed as *mean time to failure* (MTTF).
- For repairable products, the mean life is expressed as *mean time between failures* (MTBF).

  The reliability engineer should exercise care in using the terms MTTF and MTBF. These terms are generally used when the underlying failure distribution is exponential, and the failure rate is constant. The relationships given in the rest of this chapter are based on this assumption.

  The metrics MTTF and MTBF are often denoted with the Greek theta (θ). "Time," as used here, refers to some measure of life units for the product, which is not necessarily clock time. For example, with automotive products, the life units may be miles. In other equipment, life units may be cycles, rounds fired, and so forth.
- Some documents, for instance, replace MTBF with *mean cycles between failures* (MCBF). It is also appropriate in some cases to identify critical failures, or those that have a significant impact on the performance of a part or system.

  Mean time between critical failures (MTBCF) is the term used when analyzing critical failure patterns. MTBCF may be more commonly used in systems which demand high reliability, such as the aviation industry or healthcare systems. For a particular set of failure times, the mean life is obtained by averaging the failure times. This value serves as an estimate for θ.

The formulas for MTTF, MTBF, MCBF, and MTBCF are shown in Equation 2.3–Equation 2.6.

Equation 2.3 Mean time to failure.

$$MTTF = \theta = \frac{\Sigma t_i}{n}$$

where

n = Number of items

t_i = Failure times

Equation 2.4 Mean time between failures.

$$MTBF = \frac{nm}{r}$$

where

n = Number of items

m = Number of hours in the test

r = Number of failures

Equation 2.5 Mean cycles between failures.

$$MCBF = \frac{nc}{r}$$

where

n = Number of items

c = Number of cycles in the test

r = Number of failures

Equation 2.6 Mean time between critical failures.

$$MTBCF = \frac{nc}{C}$$

where

n = Number of items

c = Number of cycles in the test

C = Number of critical failures

- *Failure rate* is the reciprocal of the mean life and is defined as the number of failures per unit of time. Failure rate is usually denoted by the letter f or the Greek letter lambda (λ). For a constant failure rate, lambda can be expressed using Equation 2.7:

Equation 2.7 Constant failure rate.

$$\lambda = \frac{1}{MTBF}$$

$$\lambda = \frac{1}{MTTF}$$

- The *mean time to repair* (MTTR) is the average time to fix a repairable system. If the product is repairable and needs no preventative maintenance, and if repair can begin immediately when failure occurs, MTTR can be defined as in Equation 2.8:

Equation 2.8 Mean time to repair.

$$MTTR = \frac{\textit{Total time to repair the system}}{\textit{Number of repair operations}}$$

- *Dependability* is defined as the probability that a product will function at a particular point in time during a mission.

As component and product design decisions are made, the reliability engineer can aid in calculating the cost–benefit relationships by providing life expectancies for various design options.

- MTTF is defined as the average time elapsed until the product is no longer performing its function. If the item is repairable, MTBF is the metric used. For a constant failure rate λ, MTTF and MTBF are shown in Equation 2.9 and Equation 2.10.

Equation 2.9 MTBF, constant failure rate.

$$MTBF = \frac{1}{\lambda}$$

Equation 2.10 MTTF, constant failure rate.

$$MTTF = \frac{1}{\lambda}$$

- *Mean time between repairs* (MTBR) provides another measure of the reliability of a product. MTBR data should include information about the type of repair and resources required. These data are helpful in determining spare parts inventories, planning for resources, and scheduling preventative maintenance.
- *Mean downtime* (MDT) refers to the average of all downtime experienced by a machine or system. Sources of downtime may include planned maintenance, unplanned maintenance, or machine setup time, shown in Equation 2.11.

Equation 2.11 Mean downtime.

$$MDT = \frac{\textit{Total downtime experienced by a system}}{\textit{Number of occurrences of downtime}}$$

- *Availability* can be defined as the probability that a product is operable and in committable state when needed. In other words, it is the probability that an item has not failed or is not undergoing repair. This measure considers an item's reliability and its maintainability. Another way to express this is the proportion of time a system is in a functioning condition. This can be written as the fraction shown in Equation 2.12.

Equation 2.12 Availability.

$$Availability = \frac{\textit{Total time a functional unit is capable of being used during a given interval}}{\textit{The length of the interval}}$$

- A more general formula for availability can be written as the ratio of the average value of the uptime of a system to the sum of the average values of uptime and downtime as shown in Equation 2.13.

Equation 2.13 Availability, general formula.

$$Availability = \frac{\textit{Average uptime}}{\textit{Average uptime} + \textit{Average downtime}}$$

Availability can be broken down further into the following definitions:

- *Operational availability* is the percentage of time a system is working that it should be working, that is, not in a condition of unplanned downtime.
- *Achieved availability* is the percentage of time the system was actually running, including all sources of downtime, planned and unplanned.
- *Inherent availability* is the statistical likelihood that a system will be running as expected at any given time.

An important aspect of reliability engineering is understanding key terminology and knowing when to use which metrics and calculations.

Maintainability

It is expected that systems will break down and require maintenance. Effective maintenance programs account for this with robust planned maintenance tasks to reduce unplanned downtime, and maintenance must be considered in reliability engineering. Definitions specifically related to maintainability include:

- *Maintainability* is the probability that a failed product will be repaired within a given amount of time once it has failed. Thus, maintainability is a function of time. If there is a 95% probability that the product will be operable within three hours, then $M(3) = 0.95$. In defining maintainability, it is necessary to describe exactly what is included in the maintenance action. The following items are typical: diagnosis time, part procurement time, teardown time, rebuild time, and verification time.
- *Preventative maintenance* (PM) is planned maintenance, such as replacement of components that have not failed at specified intervals. It is frequently more cost-effective than reactive maintenance. Preventative maintenance reduces the diagnosis and part procurement times and thus may improve maintainability. Preventative maintenance also does not create unplanned downtime in a part or system.
- *Mean time between maintenance* (MTBM) refers to the average time between maintenance tasks on a piece of equipment. MTBM can be further specified by distinguishing between planned and unplanned maintenance activities.

Maintainability comprises two components: serviceability and repairability. *Serviceability* refers to the level of difficulty encountered when performing the maintenance activity, and *repairability* refers to the level of difficulty when returning a failed item to usefulness. The goal is to develop a *reliability-centered maintenance* (RCM) program. Steps in that direction include:

1. Develop a functional hierarchy of the machines and their components.
2. Make a complete list of failure modes for the machines.
3. Perform an evaluation of each PM activity to determine whether it really aids in preventing failure modes.
4. Reorganize PM activities to align with the results of data analysis.

Collecting and analyzing reliability metrics can be a critical step in the rollout of an RCM program. When designing or updating a maintenance program, the *service interval* refers to the time or cycle count between planned maintenance activities. Familiar examples include lubrication and filter change schedules. MTTR and MTBM are terms typically associated with the performance of an RCM program and would be optimized following the steps detailed in rolling out the program.

Product Lifecycle

Reliability engineers identify three stages in the lifecycle of a product:

1. The first stage is referred to variously as the *early failure* stage, the *infant mortality* stage, or the *decreasing failure rate* stage. The failures that occur during the early failure stage are usually associated with manufacturing rather than design. Examples of causes of failure include inadequate test or burn-in time, poor quality control, poor handling, weak materials or components, and human error in fabrication or assembly. Ideally, all these failures should occur in-house and be corrected before the customer takes possession.
2. The second stage is called the *constant failure rate* stage, the *random causes* stage, or the *useful life* stage. During the useful life stage, the failure rate is approximately constant. Note that the failure rate is not necessarily zero. During this stage, the failures have random causes and can't usually be assigned to production problems. Reducing the failure rate during this stage usually requires changes in product design.
3. The third stage is called the *wear-out* stage, *fatigue* stage, or *increasing failure rate* stage. The wear-out stage is characterized by an increasing failure rate over time. These failures are caused by product or component fatigue.

A graph of the failure rate is shown as the *bathtub curve* in Figure 2.1. Note that although the useful life stage is sometimes referred to as the random causes stage, random causes are generally present during all three stages.

DRIVERS OF RELIABILITY REQUIREMENTS AND TARGETS

> Describe how customer expectations, industry standards, safety considerations, liability issues, and regulatory concerns drive reliability requirements. Assess expected use conditions or use cases, align reliability requirements with environmental, social, and governance (ESG) policies, and draw valuable insights from lessons learned. (Understand)
>
> **Body of Knowledge I.B.2**

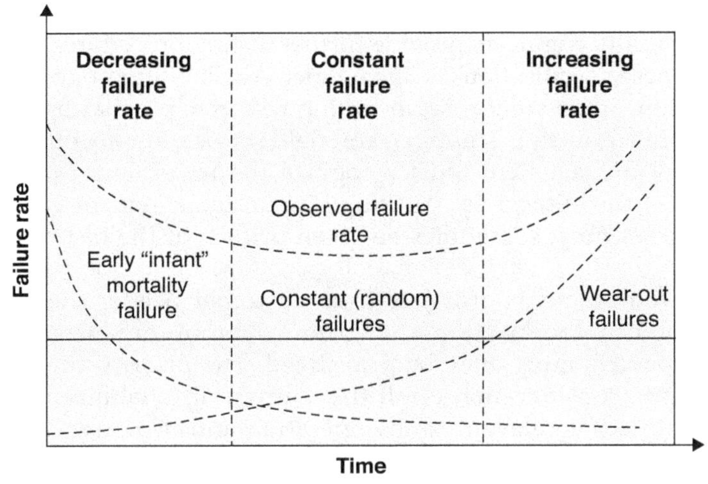

Figure 2.1 Reliability bathtub curve.
Source: Adapted from Durivage (2014).

Reliability analysis provides estimates of the probability of failure. The reliability engineer must go beyond these calculations and examine the consequences of failure. These consequences typically represent costs to the customer or the operational performance of the business. When the customer incurs costs, they often find ways of sharing these costs with the producer through the warranty system, loss of business, decrease in reputation, or civil litigation system. Therefore, an important reliability function is the anticipation of possible failures and the establishment of reliability acceptance goals that will limit their occurrence and consequent costs. The acceptance criteria associated with these goals generally fall into one of three categories:

1. Functional requirements (such as a minimum flow rate for a pump)
2. Environmental requirements (such as temperature, radiation, or pH)
3. Time requirements (such as failure rate during useful life, time elapsed before the wear-out phase begins)

Customers carry the expectation that product and component reliability will live up to their needs, but considerations for reliability go beyond just customer expectations. Regulatory bodies and industry standards must also be considered. Many industries operate under mandated requirements published by groups such as the Food and Drug Administration (FDA), the Federal Aviation Administration (FAA), and other governing bodies. Beyond the requirements to participate within certain industries, many companies voluntarily engage with industry standards such as ISO9001 or AS9100 through the process of external certification. Other expectations also come from safety and liability considerations as well. Companies must consider the expectations and requirements of component and product reliability derived from these sources at the beginning of the product lifecycle.

Once component, product, and system reliability goals have been set, a testing protocol should be implemented to provide validation that these goals will impact the failure rates and the associated consequences as planned. These reliability goals typically impose specifications on the product. In anticipation of the start of production, reliability engineers provide further testing procedures to provide verification that these specifications are being met. A reliability program should impact many functions in an enterprise, including research, purchasing, manufacturing, quality assurance, testing, shipping, and field service, among others.

Top-level management must recognize the importance of the program to the success of the enterprise. Working to gain management support for a reliability program must sometimes be given priority if the rest of the program is to succeed.

Design requirements originate from customer needs. These requirements, as determined by the marketplace, customer input, organizational objectives, and other sources, must often be prioritized. The prioritization of these objectives demands consideration of all the sources of reliability expectations and targets identified as relevant to the operation and use cases of the product or system.

Meeting reliability design requirements within time and resource constraints requires an efficient testing and documentation program. The tasks associated with the program must be accomplished in synchronization with other design, development, and manufacturing functions. These tasks must be given adequate priority at each stage of the design if unpleasant late-term surprises are to be avoided. When administering these tasks, the environmental, social, and governance (ESG) policies of the company should also be considered. While the standards, regulations, safety requirements, and customer expectations set the target for what the reliability goals are, the ESG policies should offer a level of ethical insight and guidance on how to go about achieving them. ESG policies are becoming more widespread within industry and often demand that consideration be given to how the work product will impact the people and environment surrounding the use cases, as well as how the company will govern its own policies and behaviors.

The reliability engineer should provide the project manager with time and resource needs during the project planning phase. The project manager is typically responsible for ensuring that these needs are fulfilled at the appropriate stages. After the lifecycle of the project has passed through each gating step, the reliability engineer should drive feedback from lessons learned back into the project, both for greater understanding of the project deliverables of the current project and also for insights into product performance and design, as well as process design for future projects.

CORRECTIVE AND PREVENTATIVE ACTION (CAPA)

> Evaluate specific situations that call for corrective and preventative actions. Explore the implementation of these actions and assess the effectiveness of the measures taken. (Evaluate)
>
> **Body of Knowledge 1.B.3**

Corrective and preventative action (CAPA) is a process designed to address failures. A typical CAPA workflow involves rapidly identifying a corrective action, usually a temporary solution such as heightened product inspection, to stop the failure from continuing. Corrective action is often implemented rapidly by an operations-focused team in order to contain a problem before it becomes unmanageable. In extreme cases, the corrective action may be to stop production while an investigation occurs. The next phase of the CAPA process is to develop preventative action. A preventative action is a permanent solution designed to prevent the problem from recurring and typically involves a more structured investigation to uncover root causes (root cause analysis is covered in the following section).

The ultimate test of the effectiveness of a CAPA is the ability to turn the action on or off and observe the corresponding effect on the process. Data should be collected before and after the installation of the corrective/preventative action. This section lists some tools for determining whether these data provide evidence that the action has been successful.

The histogram is probably the simplest tool to use. Suppose the proposed corrective/preventative action was designed to raise the mean effective lifetime of the item. Random samples from the process before and after the action should show a difference in their centers. If the two histograms look like those in Figure 2.2, one could conclude that the action was effective. If, however, the two histograms looked like those shown in Figure 2.3, the results would be inconclusive. Here, it would be necessary to subject the data to the hypothesis test for means of two populations. In a similar manner, if the corrective/preventative action was designed to reduce the variation in characteristic X, the histograms might look like those in Figure 2.4, in which case it would be safe to conclude that the action was effective. Again, if the results from the histograms were ambiguous, it would be possible to reach a more definitive conclusion by using a hypothesis test, in this case the test for two population standard deviations.

There is a caveat regarding the use of hypothesis tests for proportions: Consider avoiding the aggregation of defect types. It is sometimes best to delineate categories' defect types and study them individually.

One of the frustrating aspects of CAPA problem solving is the tendency for the same problem to recur. This may be because the installed solution did not really solve the problem, or it may be because the solution, although correcting the problem, was discontinued for some reason. Therefore, installing a system that monitors the process, to ensure that the problem doesn't recur, is a major CAPA activity.

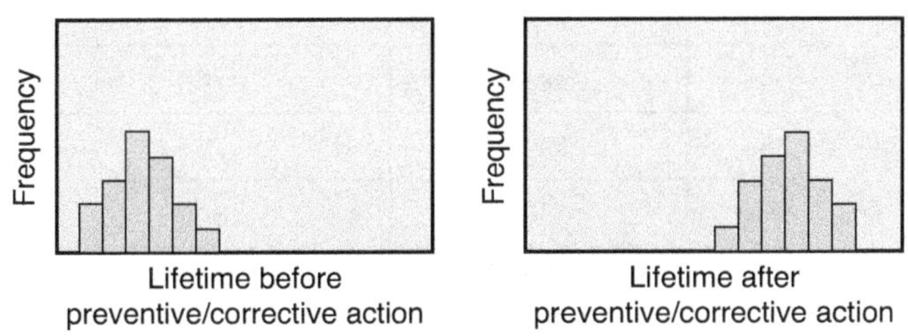

Figure 2.2 Histograms showing increase in mean lifetime.

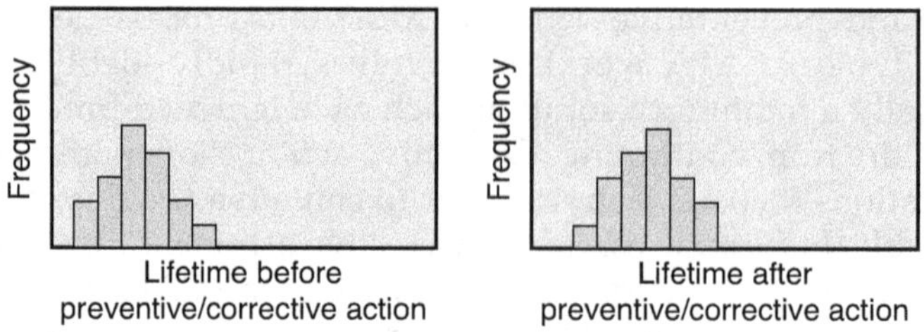

Figure 2.3 Inconclusive histograms.

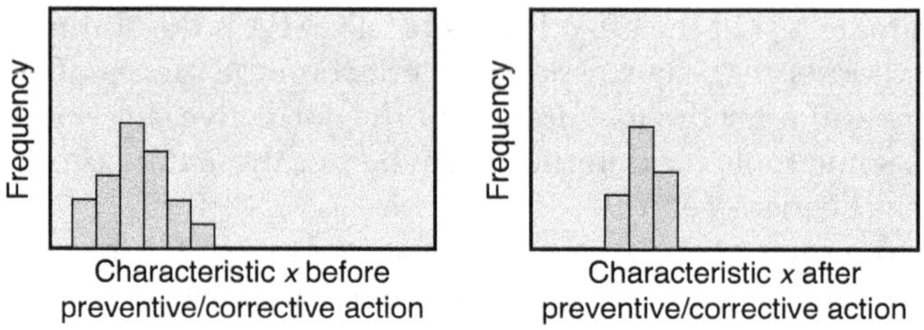

Figure 2.4 Histograms showing reduction of variation in characteristic *x*.

EXAMPLE 2.1

A reliability engineer is tasked with solving a problem identified by customer warranty claims. The company produces various configurations of polymer tubing for plumbing solutions, and customers have reported plugs in new spools of tubing. The engineer first identifies a test method to search for plugs in the spools in order to investigate.

It is discovered that spools occasionally have plugs in them; during a production run in which 10,000 meters of tubing is to be produced, a total of 12 plugs were discovered. The length of the tube between each plug was measured with data shown in Table 2.1:

Table 2.1 Tubing data.

Plug #	Meters produced	Meters between plugs
1	827	827
2	1578	751
3	2502	924
4	3242	740
5	3999	757
6	4848	849
7	5521	673

Table 2.1 Tubing data. (Continued)

Plug #	Meters produced	Meters between plugs
8	6307	786
9	7255	948
10	8118	863
11	8922	804
12	9714	792

The distribution of failures was also visualized in a histogram shown in Figure 2.5:

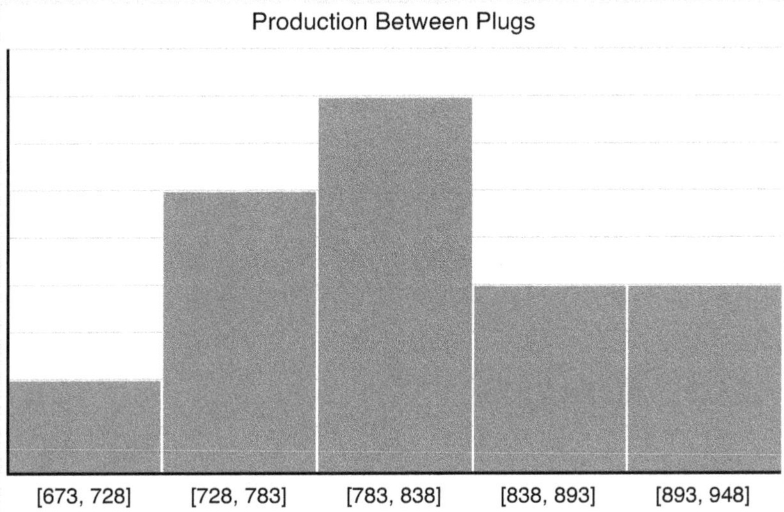

Figure 2.5 Histogram of tubing data, meters between plugs.

The meters between plugs data can be used to calculate MTTF = 809.5 m, which serves as an excellent example of an instance when MTTF may be represented in a unit other than clock time. The engineer reviewed the data and the histogram and initiated a CAPA.

The CAPA actions were as follows:

1. As a corrective action, the engineer implemented a 100% inspection plan to ensure that no tubing left the facility with a plug.
2. Samples of the plugs were submitted to the R&D departments to perform material analysis.

Continued

3. The maintenance team began cleaning the tooling after every 600 m of tubing production to prevent plugs from forming and analyze the behavior of the tools around the time interval in which the failure was occurring.
4. Historical data were collected on warranty claims surrounding plugs.

Step 1 provided the corrective action; Steps 2–4 were to investigate the root cause. The following discoveries were made:

- R&D identified the plug was made of a specific polymer, which was part of the blend they purchased from their suppliers.
- Maintenance discovered a gradual buildup of the polymer on the tooling during production.
- Historical data showed the warranty claims started around the same time as a supplier change to plastic resin feedstock into the process.

The data collected was used to develop the preventative action of switching back to the previous supplier of plastic resin. To prove the theory, a batch of the suspect material was saved, and after a run of good tubing, the suspect resin was used, and the issue recurred. The engineer issued a quality check of incoming goods to ensure that the resin supplier and recipe did not change again and closed the CAPA.

In situations where a problem rarely occurs, it is often best to establish a control chart to determine whether the desired change has occurred. Documentation regarding problem-solving activity should be maintained and available to team members. Properly cross-referenced, this documentation can aid team members by providing a history of current or similar problems.

ROOT CAUSE ANALYSIS

Assess root cause analysis by using specialized tools, such as fishbone diagrams, 5 Whys, and 8D to investigate the causes of degradation and failure. (Evaluate)

Body of Knowledge 1.B.4

Root cause analysis (RCA) is a critical phase of the problem-solving process. It is very difficult to prevent the recurrence of an issue without a comprehensive understanding of how that issue came to pass. RCA provides a structured and systematic approach to describing a problem in a way that makes it much easier to scope an appropriate permanent solution. RCA is also critical in preventing the identification of the wrong issue as a root cause. It may be tempting for professionals to trust their instincts and lock in on problems that they believe to be the root cause. However, without a structured approach to RCA, this poses a significant risk of misdiagnosing the issue.

There are many structured RCA tools that may be appropriate for different kinds of problems, including:

- *Ishikawa diagrams* are often referred to as *cause-and-effect* diagrams or *fishbone diagrams*, and they are used to understand root causes that may be varied, complex, or come from the interaction of several factors.
- *5 Why analyses* are deployed to explore the depth of a problem when it has been identified to ensure that the solution is not simply treating a symptom.
- *8D* is a problem-solving methodology that incorporates what are known as the Eight Disciplines. This is a structured problem-solving approach that breaks the process down into eight discrete, manageable deliverables.

It is critical to remember that not all tools are appropriate for all types of problems. A skilled facilitator can often discern which RCA tool is the most appropriate based on an initial description of a problem. Three common pitfalls to avoid when performing an RCA are the lack-of-solution cause, the proving-oneself-right cause, and people blaming.

Lack-of-solution may often be scoped as a cause of a problem when practitioners come into the RCA already having targeted a solution in their minds. These "solutions" often only treat the symptoms and not the causes themselves, and as such are better described as corrective actions. The goal of RCA is to identify preventative actions. Example 2.3 (the 5 Why example) demonstrates a lack-of-solution failure in the RCA process.

Another common pitfall, often related to the lack-of-solution, is practitioners trying to prove their pre-identified cause correct. The process of preventing this is to facilitate a healthy brainstorming session. The main risk is when a team member in the analysis articulates a compelling potential cause early in the discussion and the rest of the team gravitates toward that cause. It is natural for the people in the room to want to move past discussing causes and focus on crafting solutions, and locking in on a cause that is easy to get their arms around can lead to that trap. In order to drive healthy brainstorming, the facilitator must keep discussion going, keep the room focused on causes, and not allow the RCA to turn into solutions. Reminding the team that solving the problem comes after describing it may help. Tools such as catch ball—a brainstorming technique in which a ball is passed around a room and whoever catches it is expected to volunteer an idea—may be deployed to prevent a strong individual from dominating the brainstorming discussion, and with particularly headstrong groups, it may be effective to use the nominal group technique by prompting silence and having the group individually write down their theories about the cause. Proving themselves right is demonstrated in Example 2.2, the fishbone diagram.

The third pitfall to avoid when facilitating an RCA is people-blaming. It can be easy for a group to lock in on the idea that someone made a mistake. While human errors do occur, blaming people does not lead to a productive outcome for RCA. When articulating solutions, people-blaming only presents a limited path forward that can lead to the erosion of morale or loss of staff. If a person is the problem, the only potential solutions are to retrain that individual, to eliminate that person by rescoping their role, or to outright let them go. This closes the door to engineered solutions that provide more permanent prevention, and it also leaves the possibility that another individual could duplicate the same mistake.

Cause-and-Effect Diagram

The *cause-and-effect diagram,* also called the *fishbone* or *Ishikawa diagram,* helps a team identify, explore, and communicate all the possible causes of the problem. It does this by dividing potential causes into broad categories that help stimulate inquiry as successive steps delve deeper. The general structure of the diagram, shown in Figure 2.6, illustrates why it is also called the fishbone diagram.

The choice of categories or names of the main "bones" depends on the situation. Some alternatives might include policies, technology, tradition, legislation, and so on. A common categorical system for a cause-and-effect diagram is the 6Ms:

- Method—the processes used
- Machine—the equipment
- Materials—any material inputs or usage in the process
- Measurement—the ability to adequately detect the problem or other measurement inputs
- Mother Nature—environmental factors
- Manpower—the human element

A team may use a cause-and-effect diagram to generate a number of potential causes in each category by going around the room and asking each person to suggest one cause and its associated category. As each cause is identified, it is shown as a subtopic of the main category by attaching a smaller line to the main "bone" for that category. The activity continues until the team is satisfied that all possible causes have been listed. Individual team members can then be assigned to collect data on various branches or sub-branches for presentation at a future meeting. Ideally, the data should be derived by changing the nature of the cause being investigated and observing the result. For example, if voltage variation is a suspected cause, put in a voltage regulator and see if the number of defects changes. One advantage of this approach is that it forces the team to work on the causes and not focus on symptoms, personal feelings, history, and various other baggage. An alternative to the meeting format is to have an online fishbone diagram to which team members may post possible causes over a set period.

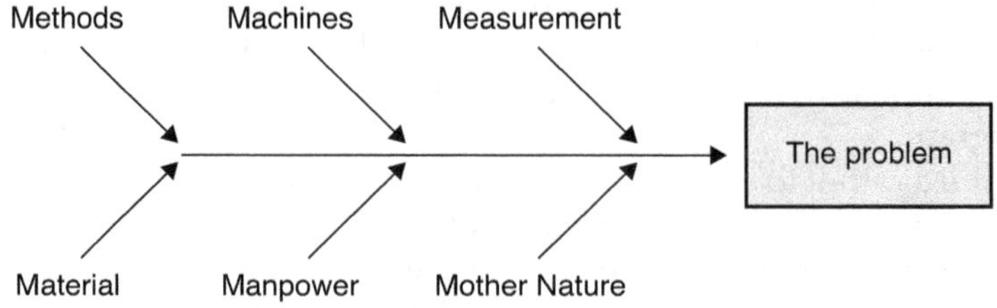

Figure 2.6 Traditional cause-and-effect diagram with the six Ms.

EXAMPLE 2.2

A batched product in a chemical facility has been contaminated. Many subject matter experts had theories on the sources of the contamination, so a reliability engineer was called to facilitate an RCA. The engineer chose to develop a cause-and-effect diagram to clarify the potential causes and choose likely culprits. The subject matter experts were assembled into the room, and several causes were identified. A snapshot of potential causes is highlighted in Figure 2.7.

After the diagram was generated, the team engaged in heavy debate about which causes to focus on. The operational team heavily suspected that the samples had been contaminated in the lab (measurement) and that the raw materials were already contaminated (materials). This cause was explored through historical data on incoming goods, which showed no contamination through historical incoming goods control sampling data. The lab investigated their sampling by sending duplicate samples to an external lab, which was able to validate their data. This round of investigation led to a multiweek delay while another production run was completed.

When the subsequent production run was also contaminated, the operations team was convinced to break down their equipment and clean it. It was discovered that a reservoir tank in the production process had no regular cleaning as part of a preventative maintenance program. The tank had been untouched for three years and had accumulated significant amounts of the contaminating agent. This tank was replaced, and the problem resolved.

The fishbone diagram is used when many potential causes are present. The most likely causes identified on a fishbone diagram merit further investigation and data collection. In this instance, the pitfall of "proving themselves right" caused the operations team to reject the notion that their equipment may be contaminated and compound the problem, but eventually the method did identify the true cause of the issue.

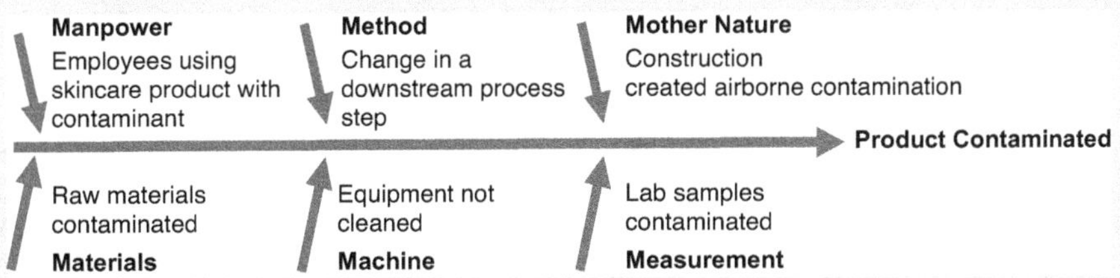

Figure 2.7 Fishbone diagram for a batched product in a chemical facility.

Five Whys

The 5 Whys is another technique that helps dig deeper into a problem as illustrated in Example 2.3. This tool consists of repeating the question, "Why does this happen?" as each answer surfaces. Performing an effective 5 Why analysis requires describing the factual basis of what has happened effectively. A well-articulated train of facts makes the answer to the question "why did this happen" much easier to identify. When training to facilitate these analyses, it is critical to develop the skill to describe *what* happened to lead to an effective *why*. When the 5 Why analysis is complete, an effective exercise is to read each "why" from the end of the analysis back to the original problem, replacing the word *why* with the word *therefore* to determine if the description of the problem follows a reasonable chain of logic and facts.

EXAMPLE 2.3

A facility has experienced an evacuation due to a gas leak. To prevent recurrence of the issue, a reliability engineer is called to perform a root cause analysis.

A group of scientists had already begun trying to assess the root cause and had come up with the following output by asking, "Why?" They first described what happened as, "What? The building was evacuated." This led to the question, "Why was the building evacuated?" to which the response was, "There was insufficient ventilation to remove toxic gases."

The reliability engineer diagnosed this as a lack-of-solution failure of the RCA process and determined that the scientists had already decided that a better ventilation system solved the problem. The engineer rejected the solution and restarted the analysis with the results in Table 2.2:

Table 2.2 Five Why analysis for building evacuation.

What happened?	The building had to be evacuated when the gas detectors alarmed.
Why did this happen?	Excessive hydrogen and carbon monoxide gases were discharged into the building.
Why did that happen?	A chemical reaction produced far more gases than had been expected to be present.
Why did that happen?	Organic materials were present within a raw material that was expected to be composed of inorganic compounds only.
Why did that happen?	The material suppliers' processes were not well understood by the company.

This example demonstrates that the 5 Why method is also not explicitly limited to asking "Why?" five times. Here, four was sufficient. However, problems also exist in which more than five repetitions of the question will be necessary.

We can check the flow of the RCA by replacing *why* with *therefore* and reading through the statements backward:

- A raw material supplier's processes were not well understood.
- Therefore, organic materials were present within a raw material that was expected to be comprised of inorganic compounds only.
- Therefore, a chemical reaction produced far more gases than had been expected to be present.
- Therefore, excessive hydrogen and carbon monoxide gases were discharged into the building.
- Therefore, the building had to be evacuated when the gas detectors signaled an alarm.

> This logical flow is sound and adequately described the problem, and the engineer was free to engage with a cross-functional team on developing solutions to prevent recurrence.
>
> A flowchart of the process will often aid in identifying root causes. Enhancements should include data on quality levels and potential causes at each step of the process.

8D

The 8D approach is a systematic method to define and solve a process or product problem. The steps are ordered from D0 to D8, as shown in Figure 2.8.

Actions associated with each of the 8D steps include:

- *D0: Plan.* Plan for solving the problem and determine the prerequisites.
- *D1: Use a team.* Select and establish a team of people with product/process knowledge.
- *D2: Define and describe the problem.* Specify the problem by identifying in quantifiable terms the who, what, where, when, why, how, and how many (5W2H) for the problem.
- *D3: Develop an interim containment plan; implement and verify interim actions.* Define and implement containment actions to isolate the problem from any customer.
- *D4: Determine, identify, and verify root causes and escape points.* Identify all applicable causes that could explain why the problem occurred. Also, identify why the problem was not noticed at the time it occurred. All causes shall be verified or proved, not determined by fuzzy brainstorming. One can use 5 Whys and cause-and-effect diagrams to map causes against the effect or problem identified.

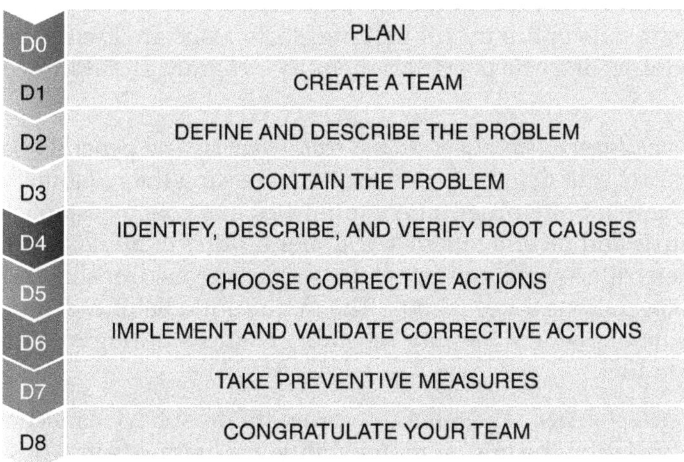

Figure 2.8. Steps in the 8D approach.

Source: https://asq.org/quality-resources/eight-disciplines-8d.

- *D5: Choose and verify permanent corrections (PCs) for the problem or nonconformity.* Through preproduction programs, quantitatively confirm that the selected correction will resolve the problem for the customer.
- *D6: Implement and validate corrective actions.* Define and implement the best corrective actions (CA).
- *D7: Take preventative measures.* Modify the management systems, operation systems, practices, and procedures to prevent a recurrence of this and all similar problems.
- *D8: Congratulate your team.* Recognize the collective efforts of the team. The team needs to be formally thanked by the organization.

PRODUCT LIFECYCLE ENGINEERING STAGES

> Examine how different lifecycle stages—concept / design, development / test, introduction, growth, maturity, and decline—affect reliability. Inspect the cost issues related to each of those stages, including aspects like product maintenance (both scheduled and unscheduled), life expectancy, and software defect phase containment. (Understand)
>
> **Body of Knowledge 1.B.5**

Reliability engineering techniques help quantify the "pay me now or pay me later" concept. The goal is to determine the reliability level that will minimize the total lifecycle cost of the product. The lifecycle cost of a product includes the cost of purchasing, operating, and maintaining the product during its useful lifetime. Sometimes, such as automotive products, where the customer seldom keeps the product for its entire useful lifetime, costs associated with depreciation may be factored into lifecycle costs. A reliability program should have the following elements:

1. *Established reliability goals and requirements.* The general goal of reliability efforts is to delight customers by increasing the reliability of products. The reliability program accomplishes this goal by establishing reliability goals and meeting them. Customer input and market analysis typically determine minimum reliability requirements. Consumers have rising expectations for reliability. The minimum reliability requirements are time-dependent because reliability changes throughout the life of the product.
2. *Product design.* The reliability program must have a mechanism for translating the minimum reliability requirements into design requirements. Reliability requirements should be documented for each stage of a product's design and for all subsystems and components.

3. *Process design.* As the product design firms up, attention can shift toward the design of the processes that will produce it. Reliability requirements must be finalized for components, whether produced in-house or purchased from suppliers. These requirements must be linked to manufacturing process parameters by determining what processes and what settings will produce components with the required reliability.
4. *Validation and verification.* As either prototypes or the first production pieces become available, the reliability program must facilitate tests that are conducted to validate that the reliability requirements do indeed produce the desired product reliability. When these requirements have been validated, it is necessary to verify that the production processes can produce products that meet these requirements.
5. *Post-production evaluation.* The reliability program must make provisions for collecting and analyzing data from products during their useful life:
    - Random samples from regular production should be collected and tested for reliability.
    - Customer feedback should be actively solicited and analyzed.
    - Field service and warranty records should be studied.
6. *Training and education.* Although listed last, this is certainly not the least important element of a reliability program. No reliability program can succeed without a basic understanding of its elementary concepts by people at all levels. Support from key managers is essential because their cooperation is needed for the testing and analysis process.

The real cost of failures is frequently underestimated. If a 90-cent natural gas valve component fails to function, the cost may far exceed the 90-cent replacement cost.

Reliability engineers take the long-term view and develop cost-effective ways to reduce lifecycle costs. These may range from design techniques, such as redundancy and derating, to specification of manufacturing parameters, such as burn-in time.

Increased reliability sometimes means increased manufacturing cost and selling price. Properly implemented, however, the result will be a decrease in lifecycle cost. Consider, for instance, a national truck line that discovered that its most frequent cause of vehicle downtime was loss of a headlight bulb. When the bulb went out, the driver had to stop the truck at the side of the road and summon a repair vehicle from the nearest company depot. This delay caused late deliveries and dissatisfied customers. The trucking company determined that a much more reliable bulb reduced lifecycle costs even though the new bulb had a considerably higher initial purchase price and required retrofitting a step-up transformer to obtain the required voltage. The company now specifies the more reliable bulb for new truck purchases. Here, the truck manufacturer can be faulted for producing a product that didn't have the lowest lifecycle cost.

As component and product design decisions are made, the reliability engineer can aid in calculating the cost–benefit relationships by providing life expectancies for various design options.

Knowing the locations of the three stages of the bathtub curve can help answer important questions, such as:

- Is the quality control system doing a good job?
- What is the optimum break-in/burn-in time?
- What is the optimum warranty period?
- What are the optimum replacement times for various components?
- What are the spare parts requirements?
- Is the preventative maintenance schedule optimized?
- Is the failure rate during the useful life stage low enough to meet customer expectations?

Each of these items relates to costs, so insight gained from the bathtub curve can have a direct impact on financial performance. Refer to the reliability bathtub curve in Figure 2.1.

Data are used to determine the shape of the bathtub curve, including locating the boundaries between the stages. Reliability engineers try to change the curve by improving the product, usually through one or more of the following actions:

1. *Improve the early failure stage by shortening its length and giving it a flatter slope.* This is usually accomplished by studying the processes used by the company and/or its suppliers to see where tighter control of process parameters is needed. Shortening the length of the early failure phase reduces costs by decreasing burn-in time and reducing the risk of early failure after installation.
2. *Improve the useful life stage by decreasing the constant failure rate.* Failure data are studied to determine the most frequent failure types. Reliability engineers work with product/process engineers to find changes that will decrease the failure rate. These activities reduce cost per unit of useful time by increasing life expectancy.
3. *Improve the wear-out stage by delaying its onset and flattening the curve.* The timing and steepness of the failure rate curve is generally a function of design. However, the wear-out phase can often be postponed somewhat and its slope reduced by more aggressive preventative maintenance and component replacement schedules. Delaying wear-out increases useful life and reduces cost per unit of useful time.

During the useful life (constant failure rate stage) of the product, the formula for calculating product reliability at time t is calculated using Equation 2.14.

Equation 2.14 Reliability at time *t*, constant failure rate.

$$R(t) = e^{-\lambda t}$$

where t is the time elapsed and λ is the constant failure rate.

EXAMPLE 2.4

A batch of 364 light bulbs has a constant failure rate of 0.000058 failures per hour. Find the reliability after 1500 hours of service. About how many bulbs have burned out at the end of the 1500-hour period?

$$R(t) = e^{-\lambda t}$$
$$R(1500) = e^{-0.000058(1500)} = 0.917$$

This result indicates that about 91.7% of the bulbs are still functioning, and 8.3% have burned out. 364(0.083) ≈ 30 of the bulbs have burned out.

Cost Considerations

The costs associated with reliability can vary greatly depending on where and how a company chooses to invest. The criticality of failure must be considered when creating an appropriate cost model to invest in reliability. For instance, the light bulbs in Example 2.4 have a much different cost consideration than a passenger jet. If a light bulb fails before it is expected, a customer may be upset. The customer may even try to return the lightbulb and recoup the costs.

If a commercial jet fails unexpectedly, the costs are likely to be immeasurable and go well beyond simple financial costs. For higher criticality products and systems, cost of reliability must be incurred in the design, development, and test phases of the product lifecycle. As criticality lessens, a portion of cost may transfer to later phases of a product lifecycle. These cost tradeoffs are also reflected in maintenance expectations and costs. A higher reliability system will demand more robust maintenance, whereas a product like the light bulb has no maintenance; it is simply replaced when it experiences failure. Later in this chapter, "Cost of Poor Reliability" explores the cost of poor reliability in greater depth, but a reliability engineer should consider the cost of reliability as it pertains to the product lifecycle.

Reliability Engineering for Software Products

The reliability effort for software products comprises three general phases:

1. Error prevention
2. Fault detection and removal
3. Fault containment through redundancy

The software package is developed and then tested against predetermined criteria. Typically, several faults are identified in the early stages of testing, and as this phase continues, the number of faults decreases. After the product is released to customers, faults continue to crop up, sometimes at a higher rate than before release. The software package may continue to be used until other considerations force its obsolescence. The typical reliability curve for a software product is shown in Figure 2.9.

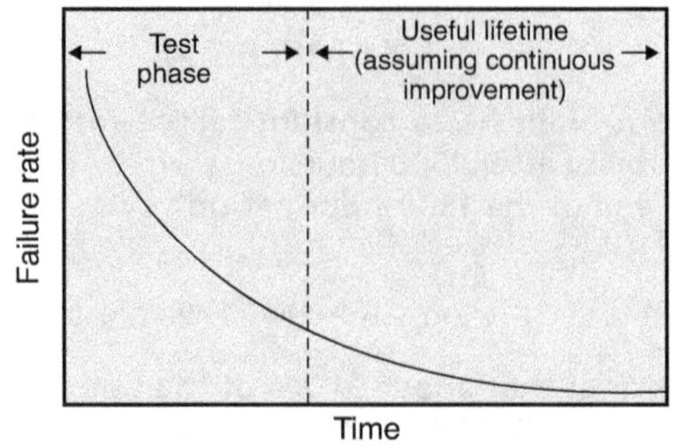

Figure 2.9 Reliability curve for a typical software product.

Software reliability engineering attempts to improve the reliability curve in Figure 2.9 by:

- Reducing the time and effort involved in the test phase (defect containment phase)
- Lowering the failure rate during the useful life of the product
- Employing redundant techniques to automatically detect and contain faults

Techniques that have been used include:

- Using alternate formulas to calculate a value and comparing results
- Using dual processors and comparing results
- Establishing known limits on a result and flagging values outside the limits

Of course, it is better to prevent errors than to detect faults. Many errors can be prevented by designing thorough requirements prior to code development. The requirements should be stated in terms of imperatives—statements that command that something must occur. Weak phrases that can be interpreted in more than one way should be eliminated from a requirements document. There should be no incomplete requirements of the "to be determined" (TBD) or "to be supplied" (TBS) type. In addition, the requirements should be structured much like a modular quality program. Developing a high-quality requirements document is worth the effort.

A software team that begins with a solid set of requirements can still introduce errors, of course. General rules for software code include:

1. Make it modular.
2. Keep it simple—excess complexity is difficult to debug and maintain. (Sometimes referred to as the Shirley Temple rule: "Don't be too cute.")
3. Provide copious documentation and comments.

The testing protocol must test every requirement at least once. Adequate resources must be provided to allow for complete testing.

Besides a product's service lifecycle, every product has a market lifecycle with these stages: concept/design, introduction, growth, maturity, decline. The reliability engineer can make useful contributions at each of these stages. The principal way that reliability engineers can influence the service lifecycle is to provide data during the concept/design stage to help ensure that components and assemblies will perform throughout their designed life. These data may be obtained through testing proposed designs, from standard tables, or by using components whose lifecycle information is available.

During the introduction and growth stages, the reliability engineer will rely more on warranty data from the field. In the product maturity and decline stages, the reliability engineer may be called on to study the use of the product in different environments or applications to extend its market life.

ECONOMICS OF PRODUCT MAINTAINABILITY AND AVAILABILITY

> Describe the cost tradeoffs associated with product maintainability strategies and availability. (Understand)
>
> **Body of Knowledge I.B.6**

System availability is determined by reliability (the probability of the system not failing) and maintainability (the ability to restore the system to service). Availability is the measure of the time the system is in an operating state compared to the total time. There are several commonly used measures of availability. The various availabilities depend on the downtimes that are included in the total time. In the following equations, the failure rate of the system is assumed to be constant.

Equation 2.15 Availability.

$$A = \frac{MTBF}{MTBF + MTTR}$$

The *mean time between maintenance actions* (MTBMA) includes both scheduled and unscheduled maintenance actions. MTBMA is a function of the failure rate (λ) and the preventative maintenance rate (μ). If the preventative maintenance rate is constant,

Equation 2.16 Mean time between maintenance actions.

$$MTBMA = \frac{1}{\lambda + \mu}$$

The *mean active maintenance time* (MAMT) is the average elapsed time needed to perform maintenance (preventative and corrective), excluding logistic and administrative delays. The achieved availability can be calculated as:

Equation 2.17 Availability.

$$A = \frac{MTBMA}{MTBMA + MAMT}$$

The effort during design should be to ensure a high probability that the equipment is ready to be used when the customer demands its use. This requires both reliability and the ability to maintain the equipment to be specified characteristics of the design. Design considerations from the customer factor into the equation. For some systems, it will be more appropriate to increase maintenance costs to keep design and product costs down. There is a tradeoff between the component level reliability and preventative maintenance demands of a system. Over-designed components and systems may achieve much greater operational performance with fewer maintenance demands, and for some systems, this will be what the customer demands. It is critical to understand the needs of the customer to appropriately design a system for its lifecycle, with stated tradeoffs being considered.

COST OF POOR RELIABILITY

> Describe the financial and non-financial expenses (e.g., availability, credibility, business operations, and reputation) incurred because of poor reliability. (Understand)
>
> **Body of Knowledge 1.B.7**

While there are many positives associated with having good reliability, a company can also experience negative effects of having poor reliability. Recent trends in consumer demand indicate a large-scale shift away from cost and toward performance. This means that when making purchases, more and more people prefer a higher performing option over the cheapest possible option. Reliability is one of the performance aspects that people care about.

The cost of poor reliability includes both financial and nonfinancial components. Financial costs have been broadly discussed in previous sections, including warranty costs, lost sales, increased maintenance, and others. These costs must be considered when designing a product for reliability. A reliability engineer should develop the skills required to calculate the return on investment associated with improving reliability at both the component and systems level in the product lifecycle.

The nonfinancial costs can be much harder to account for when targeting the early lifecycle steps for improvement. What is the most appropriate way to determine acceptable spending for considerations that do not have a clear cost associated? This can be very challenging, but the first step is understanding these costs. Some of these elements are more tangible and easier to quantify than others; some of them are very hard to quantify and must be analyzed more holistically.

Quantifiable Nonfinancial Costs

Many costs directly impact the business operations of a company. Take, for example, a product that has a high failure rate and warranty cost. The cost of the warranty claims is very clear and easy to calculate, but it is possible, and even likely, that the warranty claim also comes with a replacement shipment. Many aspects of this replacement should roll up into the financial cost of the warranty claim, but it may be harder to put a cost onto the system itself. Every unit produced as a replacement can be treated the same way that rework is treated within a production system. These components are not adding customer value, but they are consuming production capacity. In a high-complexity production environment, this impact may be exaggerated even further. In a production system in which a company has more demand than supply, every unit produced as a replacement for a warranty claim directly reduces the number of units sold, thus reducing the profitability of the company. This reduction can inadvertently impact the availability of a system as well. If the replacement product requires changeovers and retooling, there will be a trickle-down effect on availability and other production metrics.

These figures can all ultimately be quantified, so with enough detail and analysis, a financial cost can be calculated. However, even with quantifiable elements, it is not always easy to directly calculate a dollar figure of cost.

For reliability engineers, it may be worthwhile pursuing strong business-case development skills. Often there are diminishing returns that come from over-analyzing costs. When doing a cost justification, the analysis should only go to the level of detail required to confirm whether the cost is justified. Besides analytical skills, it is critical to develop the presentation and communication skills required to drive understanding of the implications of the costs of poor reliability. These skills are more valuable the farther removed from a direct dollar figure an analysis is.

Nonquantifiable Costs

The more challenging elements of the cost of poor reliability are the nonquantifiable costs. Many of these costs may be conceptually quantifiable; however, it is prohibitively difficult to collect the data needed to achieve a quantification. The credibility and reputation of a business are both significant to market success and can both be heavily impacted by poor reliability. The automotive industry provides many examples of the impact of reputation. During the second half of the 20th century, when books were being published on lean manufacturing, the phenomenon of wanting to go lean was heavily driven by Toyota's international success. Their cars achieved the reputation of being high quality and very reliable. In an era when many cars were not expected to last for over 100,000 miles, it was not uncommon for a Toyota to achieve double that number. This reliability success eventually led to that level of reliability being expected from a vehicle, and with less and less maintenance. Today, many vehicle manufacturers offer warranties that cover part or all of the vehicle for up to 100,000 miles. At the same time, the manufacturers only recommend maintenance every 5000 or even 10,000 miles compared to the historical expectation of an oil change every 3000 miles.

While the benefit of this reputation for high reliability is easy to understand, there is also a corresponding cost when that reliability is not achieved. In a marketplace where many vehicles could achieve mileages beyond 200,000, if a brand suddenly only produced cars that could achieve 40,000 miles before failure, that brand's reputation would be damaged, possibly irreparably. The reputation cost of poor reliability is likely impossible to quantify, but it is still important to understand and articulate the expected impact of poor reliability on a company and its potential performance.

QUALITY TRIANGLE

> Describe the relationship between cost, time, and quality with respect to reliability. (Understand)
>
> **Body of Knowledge 1.B.8**

Design requirements originate from customer needs. These requirements, as determined by the marketplace, customer input, organizational objectives, and other sources, must often be prioritized. A classic example is the motto NASA adopted in the 1990s, "Faster—Better—Cheaper," to which the engineers famously replied, "We can do any two." This *quality triangle* is illustrated in Figure 2.10.

Meeting reliability design requirements within time and resource constraints requires an efficient testing and documentation program. The tasks associated with the program must be accomplished in synchronization with other design, development, and manufacturing functions. These tasks must be given adequate priority at each stage of design if unpleasant late-term surprises are to be avoided. The reliability engineer should provide the project manager with time and resource needs during the project planning phase. The project manager is typically responsible for ensuring that these needs are fulfilled at the appropriate stages.

This approach to effective project management is the best course of action toward minimizing the tradeoff between cost, speed, and quality in project development. Effective project management and iterative project management strategies, such as agile, are designed to maximize project speed. With all other aspects being held consistent, the project management being effective may be able to achieve comparable quality and cost milestones in product development at the fastest possible duration.

The most important element of the quality triangle to consider is that investing in one element will come with a cost from another element. Priority setting on

Figure 2.10 The quality triangle.

new product development cycles should align with the company's value proposition and strategic approach to the market. A company known for being innovative and high quality while pivoting quickly to changing market demands will likely be unsuccessful if they begin compromising their strengths to prioritize lowering costs. Similarly, a company known for being the cost leader will likely damage their reputation with their customer base if they suddenly become very expensive to improve quality without compromising speed.

SIX SIGMA METHODOLOGIES

> Describe how six sigma principles serve as a supportive framework for enhancing reliability engineering practices. Explain how reliability engineering aligns with the principles of the DMAIC (Define, Measure, Analyze, Improve, Control) process, continuous improvement, and lean methodologies. (Understand)
>
> **Body of Knowledge 1.B.9**

This section is adapted from McShane Vaughn (2022), *The ASQ Certified Six Sigma Black Belt Handbook.*

Some implementations of reliability engineering have comprised testing products at the end of the manufacturing phase to determine their lifecycle parameters. At this point it is, of course, too late to have much impact on those parameters. Reliability engineering tools help the design engineer work more efficiently and effectively in various ways:

- MTBF values for existing products can be determined, and reasonable goals established.
- MTBF values for components and purchased parts can be determined.
- Failure types and times of occurrence can be anticipated.
- Optimal break-in/burn-in times can be determined.
- Recommendations for warranty times can be established.
- The impact of age and operating conditions on the life of the product can be studied.
- The effects of parallel or redundant design features can be determined.
- Accelerated life testing can be used to provide failure data.
- Field failure data can be analyzed to help evaluate product performance.
- Concurrent engineering can improve the efficiency and effectiveness of product development by scheduling design tasks in parallel rather than sequentially.

- Reliability engineering can provide information to individual teams about failure rates of their proposed components.
- Cost accounting estimates can be improved through the use of lifecycle cost analysis using reliability data.
- When management employs failure modes and effects analysis (FMEA)/failure modes, effects, and criticality analysis (FMECA) techniques, reliability engineering provides essential input.
- The Six Sigma problem-solving methodology can be used by reliability engineers to make data-based decisions and improve products and processes.

Kubiak and Benbow (2010) proffered this definition of Six Sigma:

> Six Sigma is a fact-based, data-driven philosophy of improvement that values defect prevention over defect detection. It drives customer satisfaction and bottom-line results by reducing variation and waste, thereby promoting a competitive advantage. It applies anywhere variation and waste exist, and every employee should be involved.

Linderman et al. (2003) define Six Sigma as "an organized and systematic method for strategic process improvement and new product and service development that relies on statistical methods and the scientific method to make dramatic reductions in customer defined defect rates."

Magnusson et al. (2003) define Six Sigma as "a business process that allows companies to drastically improve their bottom line by designing and monitoring everyday business activities in ways that minimize waste and resources while increasing customer satisfaction by some of its proponents."

Regardless of the exact definition, companies from a wide range of industries have found that when the Six Sigma approach is fully embraced, the enterprise thrives. What is this Six Sigma approach? Upon studying the various definitions, we notice the following common threads:

- Using teams that are assigned well-defined projects has a direct impact on the organization's bottom line.
- Training in statistical thinking at all levels with key people extensively trained in advanced statistics and project management. These key people are designated as Black Belts.
- An emphasis on the five-step problem solving approach of define, measure, analyze, improve, and control (DMAIC).
- A management environment that supports these initiatives as a business strategy.

The literature is replete with examples of projects that have returned high dollar amounts to the organizations involved. According to Harry and Schroeder (2000), in large organizations, each Black Belt brings saving of around $230,000, with many Black Belts completing four to six projects per year.

Six Sigma has been also described as a management philosophy, a problem-solving methodology, and a statistical standard of quality.

Philosophy

The philosophical perspective views all work as processes. Processes are a series of steps that take various inputs and produce outputs, such as a product or a service. Understanding the relationship between the inputs and outputs is a key concept in Six Sigma. By controlling the inputs, an enterprise can control the outputs. This notion can be expressed as the concept of $Y=f(X)$, where Y is the output, X is the input or input(s) and f is the function, or process, which transforms the inputs into outputs, as shown in Figure 2.11.

Methodology

The methodological view of Six Sigma describes the rigorous five-step project-solving approach of define, measure, analyze, improve, and control, or DMAIC, as shown in Figure 2.12. DMAIC defines the steps a Six Sigma practitioner is expected to follow, starting with identifying the problem and ending with implementing long-lasting solutions. While DMAIC is not the only Six Sigma methodology in use, it is certainly the most widely adopted and recognized.

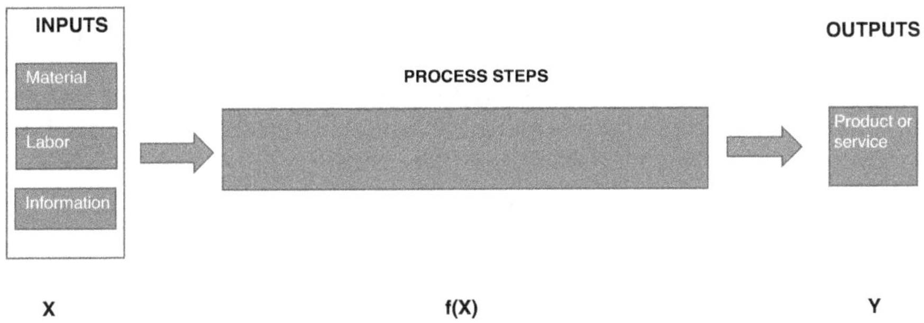

Figure 2.11 Inputs, process steps, and outputs.

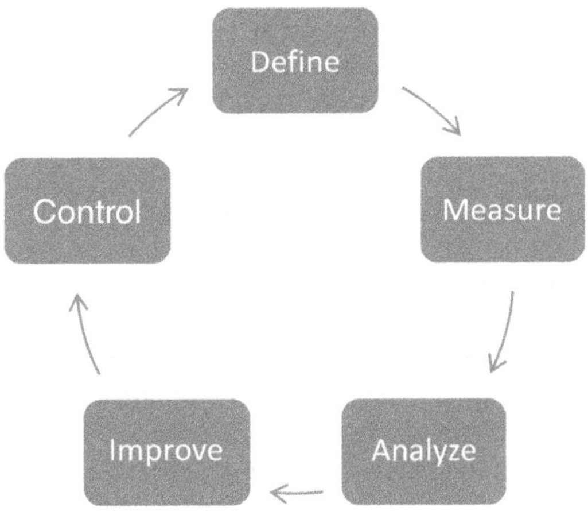

Figure 2.12 The DMAIC cycle.

Statistical Standard

In simple terms, a Six Sigma level of quality performance translates into 3.4 defects per million opportunities (accounting for a 1.5-sigma shift in the mean and assuming a normally distributed process output).

Lean

Lean is a separate methodology and philosophy from Six Sigma. However, Six Sigma has incorporated several lean tools into its toolbox. Lean can trace its origins to post–World War II Japan, in which the recovering industry was forced to do more with less. The National Institute of Standards and Technology (NIST) (2019) defines *lean* as "a systematic approach to identifying and eliminating waste through continuous improvement, flowing the product at the pull of the customer in pursuit of perfection."

At its heart, lean has five pillars, or principles:

1. *Specify value.* Define value from the customer's point of view.
2. *Identify the value stream.* Map the flow from customer order to delivery and payment, showing activities that add value to the customer and those that do not. ASQ defines the term *non-value-added* as a process step or function that is not required for the direct achievement of process output. These steps or functions are identified and examined for potential elimination.
3. *Create flow.* Concentrate on having pieces flow through the process without stopping or being put into inventory. Avoid creating large batches with the ultimate goal of one-piece flow.
4. *Pull from the customer.* Pulling from the customer means matching production output to customer demand. Here, organizations will only produce an item when it is required downstream in the flow.
5. *Seek perfection.* Unlike the Six Sigma goal of 3.4 defects per million opportunities, lean has zero defects as its goal. Organizations strive to continuously improve processes in pursuing perfection.

The term "lean thinking" expands the ideas originally used on the manufacturing floor into all areas of an organization. Lean is often thought of as a set of practices focused on the elimination of waste by concurrently reducing supplier, customer, and internal variability (Shah and Ward 2007).

While Six Sigma emphasizes reducing variation, lean concentrates on reducing waste. These wastes can be categorized by the Eight Wastes: transportation, inventory, motion, waiting, over-processing, over-production, defects and skills.

A lean organization is characterized by teamwork in which well-informed, cross-trained, and valued employees take part directly in improving processes. Work areas are clean and organized to allow for problems to be spotted more readily. Production systems are designed around flow instead of batch and queue, with the goal of one-piece flow. Work is centered on a pull system in which replenishment is based on what the customer has consumed. Finally, lead times are reduced due to concentrating on efficient processing, setups, and scheduling.

Lean and Six Sigma have the same general purpose of providing the customer with the best possible quality, cost, delivery, and a newer attribute, nimbleness.

There is a great deal of overlap, and disciples of both disagree about which techniques belong to which approach.

The most successful users of implementations have begun with the lean approach, making the workplace as efficient and effective as possible, reducing waste, and using value stream maps to improve understanding and throughput. When process problems remain, the more technical Six Sigma statistical tools may be applied. Both methodologies require strong management support to ensure that the quality approaches become the standard way of doing business.

Lean and Six Sigma

A combined definition of lean Six Sigma might read:

> Lean Six Sigma is a process-focused, data-driven philosophy of improvement that values defect prevention over defect detection. It drives customer satisfaction and bottom-line results by reducing variation, waste, and cycle time while promoting the use of work standardization and flow, thereby creating a competitive advantage. It applies anywhere variation and waste exist, and every employee should be involved.

Lean manufacturing, with its emphasis on workplace organization and standardized work procedures, can reduce product variation. This makes a more predictable lifecycle and helps the reliability engineer provide more useful data to the design function. The term "lean thinking" refers to the use of ideas originally employed in lean manufacturing to improve functions in all departments of an enterprise. NIST, through its Manufacturing Extension Partnership, defines *lean* as: "a systematic approach to identifying and eliminating waste (non-value-added activities) through continuous improvement by flowing the product at the pull of the customer in pursuit of perfection."

ASQ defines *non-value-added* as "A term that describes a process step or function that is not required for the direct achievement of process output. This step or function is identified and examined for potential elimination."

This represents a shift in focus for manufacturing engineering, which has traditionally studied ways to improve value-added functions and activities (for example, how can this process run faster and more precisely?). Lean thinking doesn't ignore the value-added activities, but it does shine the spotlight on waste. Lean manufacturing seeks to eliminate or reduce waste by use of:

- Teamwork with well-informed, cross-trained employees who take part in the decisions that impact their function
- Clean, organized, and well-marked workspaces
- Flow systems instead of batch and queue (that is, reduce batch size toward its ultimate ideal—one)
- Pull systems instead of push systems (that is, replenish what the customer has consumed)
- Reduced lead times through more efficient processing, setups, and scheduling

The history of lean thinking may be traced to Eli Whitney, who is credited with spreading the concept of part interchangeability. Henry Ford, who went to great lengths to reduce cycle times, furthered the ideas of lean thinking. Later, the Toyota Production System (TPS) packaged most of the tools and concepts now known as lean manufacturing.

SYSTEMS ENGINEERING AND INTEGRATION

> Comprehend the relationship between reliability engineering and systems engineering, with a focus on integrating components and their interactions within a system. (Understand)
>
> **Body of Knowledge 1.B.10**

Systems engineering may be defined as a branch of engineering that integrates the activities of other disciplines to produce a product or service. It is generally recognized as having six steps:

1. *State the problem.* Begin with an investigation of customer needs. This is different from customer requirements because there may be needs that are not understood or articulated by customers. Was there a huge outcry for iPads before they were produced? The problem should include a list of mandatory features and a list of "nice but not mandatory" features. Design requirements take shape during this phase.

   Some prioritization of these requirements may be possible as well. For example, "operates at 60 hertz" might be a higher-priority requirement than "switchable between 50 hertz and 60 hertz." The design team studies the customer needs in the market niche for which the product is intended as these requirements are formed and prioritized. The next two steps often uncover additional features that can be included in these requirements.

2. *Study alternatives.* Sometimes called divergent thinking, this step opens the box to innovative and creative options, which are then studied for feasibility. One or more of the alternatives is selected to proceed to Step 3. Steps 2 and 3 often are in a continuous loop as alternatives are studied and modeled. During these phases, it is necessary to solidify the design requirements and to prioritize them. Care must be taken to avoid "mission creep," or the tendency to add too many new features.

3. *Model or simulate the system.* Using physical or software approaches, the system is often operated in a primitive way, sometimes at a slower speed, to identify problems and opportunities.

4. *Integrate the various elements.* Interface the various elements of the system so the entire entity operates as one.

5. *Run the system.* Study the outputs to verify that they meet all the mandatory output requirements stated in Step 1.
6. *Study performance.* Evaluate the entire system, measuring its performance against alternatives and the requirements stated in Step 1.

Each of these steps is subject to constant reevaluation, so the series of steps is more like a series of interlocking and repeating loops. The objective is to build a system that not only has useful and productive elements but also blends these elements into a smoothly operating unit.

These systems engineering principles align well with the core concepts used in reliability engineering. Any system can be over-engineered to the point of performing well beyond its needed life, while also becoming such an inflated cost that it is infeasible to take to market and deploy tangibly. The ideas of identifying system parameters that limit or hinder performance, brainstorming, compromising where needed, and setting appropriate priorities for performance against cost are critical in both systems engineering and reliability engineering. Ultimately, the challenges faced by reliability engineers will present themselves as competing priorities and conflicting demands from various stakeholders and customers. Effective reliability engineers should be able to creatively diagnose and attack challenging issues while staying within the scope of business constraints to provide products that meet customer demands at prices customers can, and will, be happy to pay.

Part II

Risk Management

Chapter 3
Risk Identification

RISK MANAGEMENT TECHNIQUES

> Use risk management tools and processes to identify, document, and monitor environmental, liability, and security concerns. Incorporate methodologies such as p-diagrams to discern potential failure modes and use cases to conceptualize different real-world scenarios and interactions. Evaluate and prioritize risks related to safety, economics, performance, and customer satisfaction. Address these risks within an appropriate risk management framework. (Evaluate)
>
> **Body of Knowledge II.A.1**

Whether it's a new product or process, simple or complex, it is always advisable to perform risk identification and analysis. There are a variety of tools and methods that are available in support of performing a reasonable risk analysis. These various tools and methods have differing levels of complexity, so they can deliver different levels of risk coverage and depth of analysis. The full spectrum of risk exposure covers early concept development through the product's useful life and should include obsolescence and final disposition.

Historical records reveal that not performing a reasonable risk analysis can result in a collection of undesirable consequences. These undesirable consequences are not limited to product/process related issues but have business, legal, regulatory, environmental, liability, security, and social consequences as well. With system complexity growing due to the competitive application of innovative technologies coupled with applications of artificial intelligence (AI), adequate risk analysis is becoming more of a challenge. We will encounter functions that execute independently of the user and functions that require the user to be in the loop.

Functions that require the user's control are typically labeled as end user functions (EUFs). For example, riding an electric bike (e-bike) requires a user to operate all the functions of the e-bike. There is no automatic emergency braking, electronic stability control, adaptive cruise control, or lane-keeping support. The operator of the e-bike is the one who steers it, brakes it, and balances it.

Yet in the case of a new car, there are features like emergency braking, electronic stability control, and automatic transmission shifting that do not require that the operator be involved at all. These would be considered end-to-end (E2E) functions. There are types of E2E functions that can reconfigure the system without any input from the operator and, in many cases, without informing the operator of the changes being made to the system.

Another example of an E2E function would be the aftertreatment of vehicle exhaust gases. The aftertreatment system will monitor many variables and make necessary system control changes to adapt the system to the new environment. This is to ensure that the exhaust emissions are within regulatory limits. These functional changes may be in response to things like high humidity, driver demand variation, cargo load changes, elevation changes, and fuel quality. The risk analysis should strive to include all relevant aspects of the system.

This can be a challenging task considering that the traditional risk analysis tools and methods have not matured along with the ever-growing system complexity. As producers endeavor to create new products and/or processes in a highly competitive landscape, they look to include new and novel features and functions that are likely still in the test phases. There are some products and/or processes that are updated to add features and functions to gain competitive advantages. Then there are products and/or services that are already in use in some parts of the world and are in the process of being adapted for use in a different country, which likely requires numerous modifications to meet the host countries' applicable regulations, laws, and standards. These topics will be covered as well as the challenges that have already shown themselves to be imminent such as machine learning, AI, and neural networks.

Many enterprises choose to certify and/or show compliance with standards to demonstrate their organization's capability maturity. There are a variety of such standards within the International Standards Organization (ISO). ISO is an independent, nongovernmental international organization. ISO Guide 73:2009: Risk Management Vocabulary defines *risk* as "the effect of uncertainty on objectives." The guide also provides the definitions of generic terms related to risk management. It aims to encourage a mutual and consistent understanding of, and a coherent approach to, the description of activities relating to the management of risk, and the use of uniform risk management terminology in processes and frameworks dealing with the management of risk.

Risk assessment/risk management uses qualitative and quantitative data to provide estimates of the impact of various adverse events. The purpose is to identify and prioritize vulnerabilities so that appropriate corrective actions can be initiated. ISO 9001 is the international standard for creating a quality management system (QMS). ISO certification can enhance an enterprise's credibility by showcasing that its product or service meets or exceeds customers' expectations. For some industries, ISO certification is a legal or contractual requirement, and the lack of it would eliminate a company from consideration as partner, supplier, or developer.

For example, contracts in sectors like defense, aerospace, and healthcare typically demand higher standards of quality, security, and environmental management, making ISO certifications such as ISO 9001 (Quality Management), ISO 14001 (Environmental Management), and ISO/IEC 27001 (Information Security Management) more likely to be required.

Risk-based thinking is a crucial concept in ISO standards. Previously, ISO 9001 featured the concept of "preventive action" that focused on eliminating any potential causes that can lead to nonconformity. However, this term has been replaced in the latest release of the standard with "risk-based thinking," which offers a more systematic approach to dealing with problems and opportunities. It has now become a key part of managing regulatory compliance, as businesses need to demonstrate risk-based thinking through their systems, processes, strategies, and objectives across the organization to maintain compliance with ISO standards.

ISO 9001:2015 defines *risk-based thinking* as "the application of information, knowledge, and actions to determine uncertainties and potential opportunities." It is a proactive approach to handling the risks and opportunities that requires organizations to evaluate risk while establishing processes, controls, and improvements in a QMS. Enterprises that choose this path commit to the disciplines of the standard(s) and have a clear advantage over enterprises that do not.

New and/or Emerging Technologies

There are many risks associated with adopting a new technology or one that is still in the research stages. Large enterprises can afford to fund the investigation and development of these technologies and a risk analysis should be performed during that process. This approach can identify risks that include the adaptation of new technologies to a particular application as well as the risks that may arise during manufacturing or the implementation of the new process. Managing risk at this level includes a risk matrix that identifies risks related to schedule, cost, environment, and internal competencies. Such an approach can identify and therefore manage potential risks.

Smaller enterprises may not be able to afford to dedicate resources to developing new technologies or processes and as a result would need to accept the greater risk when adopting and integrating new technologies and/or processes. Early investigation would likely include an early failure mode and effects analysis (FMEA), design of experiments (DOE), and highly accelerated life testing (HALT) as tools to identify and better understand the variables associated with the introduction of a new approach.

P-diagrams

When performing a risk assessment, one of the preferred methods is to develop a *p*-diagram (parameter diagram). In general terms, the *p*-diagram is a way to visualize the overall process, including which factors are under control, and which are not controllable or not under control by choice, as shown in Figure 3.1.

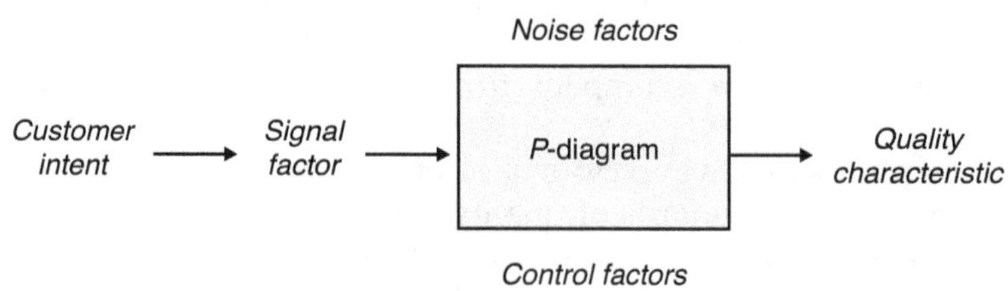

Figure 3.1 *P*-diagram (parameter diagram).

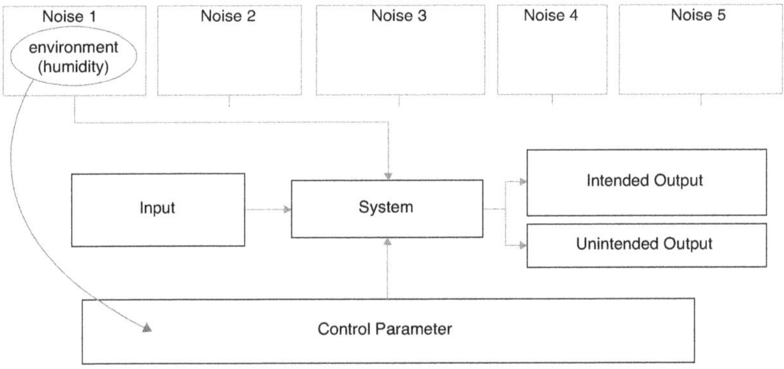

Figure 3.2 *P*-diagram showing uncontrolled variable shift to controlled.

For example, consider a manufacturing site that produces a product that is sensitive to humidity during assembly, like the on-board oxygen generating system (OBOGS) used in military cargo aircraft. This design is an individually-worn OBOGS that allows the crew members to move freely in the cargo area without having to manage an oxygen line tethered to them.

Testing of the devices showed that the catalyst (molecular sieve) suffered performance issues due to changing weather (humidity) in the manufacturing and assembly areas. This was initially considered to be a noise factor, and as such, uncontrolled. The assumption was that the existing manufacturing area was sufficient for the assembly without imposing any performance issues on the OBOG.

Given the performance issues due to humidity changes, a decision needed to be made. Based on the *p*-diagram, the humidity in the building would need to be controlled to an acceptable level. This would then add "humidity" to the control factors. The exceptionally high cost of such a change, having to control the humidity in an entire building, was prohibitive. The decision was to modify the design materials so that the OBOG catalyst was immune to changes in humidity. This was a preferred approach for numerous reasons that included cost, schedule, and design robustness.

The purpose of creating the *p*-diagram is to ensure that all uncontrolled (noise) factors are considered as well as showing any unintended output. This is in direct relationship with the development of the requirements. In the example of Figure 3.2, we can see that moving humidity from uncontrolled to controlled will add one or more requirements to the project.

Environmental, Liability, and Security Concerns

Environmental impacts on products or equipment are becoming increasingly concerning. Any materials being used to manufacture a product or equipment as well as any output from a product or equipment should be assessed for these impacts. Product lifecycle should be considered to include end-of-life disposal practices. The relevant environmental requirements need to be reviewed and analyzed to make sure that they comply with all applicable regulations, standards, and laws. This can be accomplished by using FMEA tools. Environmental requirements should be included for all FMEA types.

Liability of manufactured products and/or processes can differ significantly depending on the host country. International liability standards include the responsibility of a state under international law to compensate or make reparations for injuries. Product liability requires manufacturers and service providers to ensure their products and services are safe and free from danger. They must also provide warnings and instructions for emergency treatment if their products or services may be dangerous. Environmental liability includes any harm or injury that is caused by exposure to dangerous chemicals or hazardous materials. All relevant documents that would include the potential for liability should be reviewed. The appropriate requirements are added to the project and reviewed for compliance using FMEA tools. The liability requirements should be included in all FMEA types.

Security is a growing concern. Manufacturers and service providers need to be prepared to defend against the possibility of security attacks. The nature of the attacks can be attempts to extort money from producers by usurping control over their product or may simply be malicious in nature. Occasionally, system configurations originally designed to protect against security infringements became the security issue itself. The implementation of security measures needs to be aligned with a security strategy. Manufacturers need to consider a wide spectrum of potential threats. Automobile manufacturers need to ensure that control systems within the complexity of the vehicle are not susceptible to compromise.

There have been many instances where such systems have been breached. A large automobile manufacturer had suffered epidemic thefts of their vehicles due to a compromised software feature that was designed to thwart such thefts. Another example of such a loss was the devastating CrowdStrike (CrowdStrike is an American cybersecurity technology company) software glitch that crashed computers, canceled flights, and disrupted hospitals all around the globe. Security can be assessed in several different ways, depending on the nature of the product and/or service.

Using a fault tree analysis (FTA) or a FMEA to analyze potential security breaches can be effective if potential threats are thoroughly analyzed. The threats can be hard and/or soft, such as building access security versus cyber security. The nature of the business will dictate the approach used to analyze and fortify security. With the increasing integration of AI, machine learning (ML), and neural networks into all aspects of our economy, new tools will be developed to address potential security issues.

Use Cases

During the development and design stages of a project, there may be a need to explore customer use cases. This will help in identifying any foreseeable misuse and unintended use that is outside the scope of the design intended use. Some examples of this would be someone using a pickup truck to pull a tree down, or using a screwdriver as a chisel and hammering on the end of it. By developing the use case and reviewing it as a team, such examples of misuse can be identified and effectively managed. In some cases, it may require a redesign, and in others, it may simply require proper warning labels.

A use case can be structured as a brainstorming exercise or, more formally, using a method called lateral thinking. The term was first used in 1967 by Maltese psychologist Edward de Bono in his book *The Use of Lateral Thinking*. Using lateral thinking, the team consciously tries to generate ideas and scenarios that imagine uncertainties outside the information already available. The technique places importance on the ideas themselves and uses horizontal thinking to accomplish this. This type of thinking will allow teams to arrive at a more complete conclusion by looking at all sides of an argument instead of focusing too narrowly.

New Products and/or Services

A new product development process (NPDP) typically requires that risk analysis be performed from two different viewpoints: product and project. New product development would be achieved using a phase-gate approach (also referred to as a waterfall process). A phase-gate process is a project management technique in which a project is divided into distinct stages or phases, separated by decision points, defined as gates. At each gate, development continuation is decided by a manager, steering committee, or key stakeholders. These identified resources would be necessary signatories to allow passage through the relevant gate. One of the stakeholders should be the reliability engineer. This would ensure that the potential risks are being identified and managed. An example is shown in Figure 3.3.

The first viewpoint, that being the product, would be focused on the product and/or process that is being designed or developed. All the potential risks associated with the product and/or process would be assessed to determine if they have met the gate deliverables. For example, one deliverable for Concept Gate could be the review of evidence that a design alternative comparison was performed to determine that the best design option was chosen.

The approach would require the project team to have more than one design option, and the selection process would include a list of criteria by which to rank the competing designs. The design with the most favorable score, given all the criteria, would be the design chosen. It is an effective practice to semi-develop the second best design as well. This practice suggests that the alternative design is trailing the primary design significantly, but is developed to an extent where there is reasonable confidence to continue investment and development. The semi-design ensures that if for any reason, the primarily selected design is no longer possible to pursue, then the alternative design is the most practical alternative.

Each gate will have a list of various deliverables, such as design review, industrialization, budget review, schedule review, risk analysis of product and project, and so on.

The usual arrangement of phases and gates is:

1. Ideation
2. Scoping

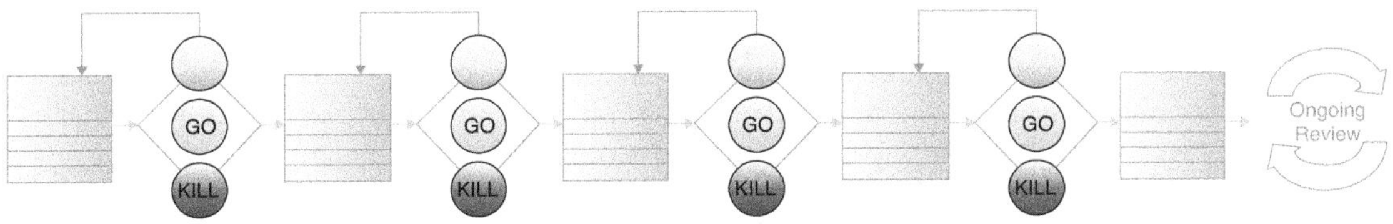

Figure 3.3 New product development process (NPDP) gates and phases.

3. Business case
4. Product design
5. Testing/validation
6. Product launch

The gates can be passed if all the required deliverables are satisfactorily completed. If there are shortfalls or missing deliverables, the decision makers can either allow the gate to be passed "with exceptions," meaning that the missed deliverables will need to be completed in the next phase, or the gate can be held closed until the established criteria are met. As some of the deliverables are likely prerequisites for the goals in the following phase, passing a gate with any exceptions or holding the gate closed will have negative effects on the project's ability to meet schedule.

The second viewpoint would be the project itself. This focuses on the capability of the project team to navigate the development challenges and stay within the projected schedule and budget. The responsibility for assurance of the project team's successful performance would fall to this assigned individual on the development team. The role is often referred to as the project assurance manager.

Carry-Forward and Carry-Across Products and/or Services

In our globalized economy, the development of new products and services can originate in different countries. Each country will have its own unique regulations and standards, which can differ dramatically from country to country. For products and services that are improved for use in the country of origin, we would identify them as "carry forward" improvements. This improved product or service would only have to comply with the unique regulations and standards of the country of origin. When a product is developed in one country and is then adapted to another country, the product will have to meet all the regulations and standards of the country it's being adapted to.

For example, automobiles from other countries that are to be sold in the Americas will need to meet all the specific regulations and standards of the Americas. This would be considered a "carry across" product improvement. In this case, a change point analysis would be necessary to make sure that all changes to the product and/or service are taken into consideration. The changes are not just meeting the regulations and standards but include other variables like environment and application.

The Challenges of Complexity

With the increasing application of AI comes new challenges. When systems include the ability to reconfigure autonomously, there is an increased risk that an undetected change can occur. Under these conditions, the change may render the system noncompliant with regulatory mandates or could impair the safety of the product and/or service. The system is functioning, all parts and routines are doing what they were designed to do, yet the system overall is now experiencing a loss of safety function or a violation of regulations. Given that no component failures or process failures have occurred even though the system is operating unsafely, then it can be said that "unsafety" is an emergent characteristic of the complex system.

Predicting such conditions in a complex system can be difficult to impossible, as the number of variants that the system can morph into may be unknown. In development of such a system, consideration of risk would need to include the potential for emergent characteristics such as unsafety. The traditional approaches for risk analysis and prevention, such as FMEA, are limited in their ability to identify such results.

It is essential to ensure the safety of products and services. Analysis of complex systems to ensure safe operability will require new and novel approaches. One such approach is systems theoretic process analysis (STPA). The STPA approach is used to open the context of the analysis and expose potential failure modes in addition to those failure modes that would be classified as emergent. STPA starts by rendering the hierarchical structure diagram of the system of interest, which can include an expanded representation of the system both in space and time. Using this approach, one can map failures to their historical roots such as policies, laws, and cultural influences, along with identifying all the conspicuous potential failure modes. This requires a unique set of skills to manage the investigation. The facilitator will develop the appropriate control structure with the team and lead them through the exploration of a brainstormed list of unsafe control actions.

STPA can and has been used to investigate system and process failures as well as is being used in product/process development to prevent potential failures that might arise due to unintended configuration scenarios and/or operational situations.

Prioritization of Risks

It is imperative to ensure that risks are properly prioritized. Risk prioritization is the process of identifying, evaluating, and ranking risks based on their potential impact and likelihood of occurrence. Prioritization will allow the most significant issues to be addressed early in the project phases, thus allowing time for solutions to be developed; in addition, early action minimizes expenses that would otherwise be incurred later in the project. Examples of risk prioritization schemes are shown in Figure 3.4.

Safety is a primary driver of priority. Risk priority matrices and FMEA ranking tables are familiar ways to prioritize risk. There are numerous other approaches, from a simple comparative analysis using an Excel spreadsheet to purchasable third-party risk priority tools. These approaches can be used to assess economics, performance, and customer satisfaction.

Economics includes the cost of developing the product and/or service along with the total cost of ownership by the customer. Performance and customer satisfaction can also be prioritized against selected criteria to ensure that the product features and functionality meet and/or exceed the customers' expectations.

The selected approach is adapted to the particular product and/or service that is being scrutinized, based on the established criteria. This allows the priority approach to be customized for the specific needs of the customer, product, and process.

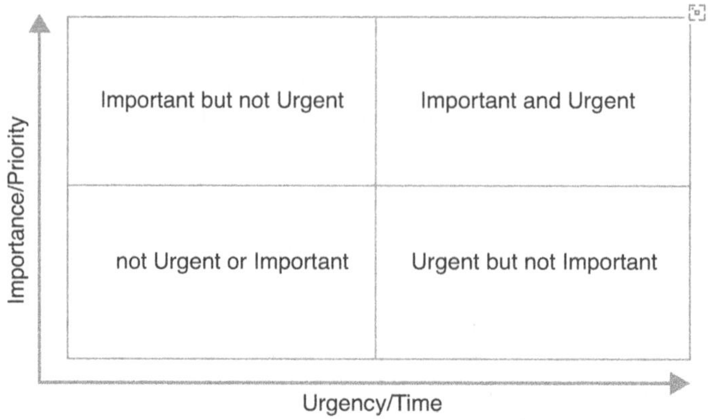

		Impact				
		Very Low	Low	Medium	High	Very High
Likelihood	Very High	Yellow	Yellow	Red	Red	Red
	High	Green	Yellow	Yellow	Red	Red
	Medium	Green	Yellow	Yellow	Red	Red
	Low	Green	Green	Yellow	Yellow	Red
	Very Low	Green	Green	Green	Green	Yellow

Figure 3.4 Examples of risk prioritization schemes.

RISK ASSESSMENT

> Employ different techniques for assessing and prioritizing risks including qualitative, quantitative, and semi-quantitative methods. Apply risk ranking methods to ascertain the potential impacts, considering likelihood and prioritization factors. Use probabilistic risk assessment (PRA) to categorize risk events and to understand the potential impacts. Perform vulnerability and threat-based assessments to analyze and gauge the susceptibility of the system to various threats. (Analyze)
>
> **Body of Knowledge II.A.2**

When performing risk assessment, one can use a qualitative and/or quantitative approach. A quantitative approach uses data to determine the risk level. An approach that uses historical data and/or surrogate data to determine risk level probabilities would be quantitative. Qualitative data can be either descriptive (nominal), or ordinal (e.g., ranks of risk levels denoted as low, medium, and high).

There are many approaches to investigate the potential risk of a system and/or process. These tools and methods come from a variety of global sources and are often aligned with particular disciplines, such as automotive, aerospace, medical, and defense. Some sources are:

- International Organization for Standardization ISO 26262, "Road vehicles—Functional safety," is an international standard for functional safety of electrical and/or electronic systems that are installed in serial production road vehicles, excluding mopeds.
- Aerospace Recommended Practices ARP4761, "Guidelines and methods for conducting the safety assessment process of civil airborne systems and equipment."
- ISO 14971 is a voluntary standard that provides a framework for risk analysis, evaluation, control, and review of medical devices.
- For defense, Military Standard (MIL-STD) 882E "Department of Defense Standard Practice System Safety" identifies the Department of Defense (DOD) approach for identifying hazards and assessing and mitigating associated risks encountered in developing, testing, production, using, and disposing of defense systems.

These are just a sampling of the many sources of risk analysis tools and methods available. Some of these approaches can be applied very early in the development process, such as the probabilistic risk analysis (PRA), preliminary system safety analysis/system safety analysis (PSSA/SSA), and functional hazards analysis (FHA). There is also a large community of vendors that supply their own proprietary version of risk analysis within the various disciplines.

The PRA strives to define the detrimental consequences of plausible failures in a complex system. The term *probabilistic risk analysis* is a categorical term and as such can be accomplished through a variety of means. As the title declares, PRA is accomplished by assigning probabilities to potential failures. It is worth noting that the acronym PRA, like many other acronyms, has other definitions, such as *particular risk analysis*, as found in ARP4761. There are differences in the way a PRA is structured, depending upon the industry, the discipline, or even the organization.

In a PRA, the consequences of a potential failure are typically characterized by two quantities:

1. The severity of the failure
2. The probability of the occurrence of the failure

Table 3.1 is representative of the Nuclear Regulatory Commission (NUREG) PRA approach as found in the NUREG/CR-2300 PRA procedures guide Vol. 1. The two volumes of the NUREG PRA guide include the use of event trees, fault trees, and FMEA. Assigning probabilities to the identified events, faults, and failures allows projections of the overall system failure rate under stated conditions. Table 3.1 is illustrative of the use of severity and probability in classifying a potential failure.

The error factor and severity of the event are coded and come from relevant tables specific to the nuclear industry. The use of tabulated error factors and severity risk evaluation matrices is a common approach tailored for specific industries and disciplines.

Table 3.1 Example of probabilistic risk analysis per NUREG/CR-2300.

Event	Description	Failure rate or failure on demand probability	Fault exposure time or mission time	Severity	Data source	Quantification model
EVLV12	Valve fails to open	Lognormal $1 * 10^{-3}$ per demand error factor 3	NA	I	Reactor safety study	Distribution: lognormal $1 * 10^{-3}$ (3) mean: $1 * 10^{-3}$
EPMI 2D	Pump discontinues running after start	Lognormal $3 * 10^{-5}$ per hour error factor 10	24 hrs	III	Reactor safety study	Distribution: lognormal $3 * 10^{-5}$ (10) mean: $3 * 10^{-5}$

A PSSA/SSA is a systematic evaluation of a proposed system architecture to determine how failures may lead to hazards. The purpose of the PSSA is to complete the list of safety requirements and determine how the design can satisfy the requirements. The derived safety requirements can then flow down to the lower-level items in the scope of the design. Experience informs us that if we fail in the requirements stage, the consequences can have an impact on the project schedule, budget, and ultimately on the safety of the system in the customer domain. Inputs to the PSSA are typically the FMEA, FTA, and the functional hazards analysis (FHA). Once the PSSA is sufficiently developed, the detailed design ensues. In the detailed design phase, the systems safety analysis (SSA) is initiated.

The SSA process is an inductive, bottom-up approach, accomplished using failure modes effects and criticality analysis (FMECA), the FMECA being the quantitative approach using probabilities from various sources, including test data, historical data, and surrogate data. This approach is used in the aerospace industry extensively. Aerospace Recommended Practice 4761 (ARP4761) defines a process for using the aforementioned common modeling techniques to assess the safety of a system under development. The first part of the document covers that process. Next is an overview of the modeling techniques and how they should be applied. The last section of the practice provides an example of the process in action.

Further development and examples of FTA, FMEA/FMECA, FHA, and PSSA/SSA are covered extensively in Chapter 4.

Vulnerability and Threat-Based Assessments

Product vulnerabilities and threat-based assessment are used to surface any weaknesses in the product or process being developed so that appropriate steps can be taken to prevent them. The terms *vulnerability* and *threats* are common terms in the cybersecurity world but also apply to other disciplines. For example, in the development of military systems, they need to be invulnerable to electromagnetic pulses. Medical devices used in an operating room need to be invulnerable to

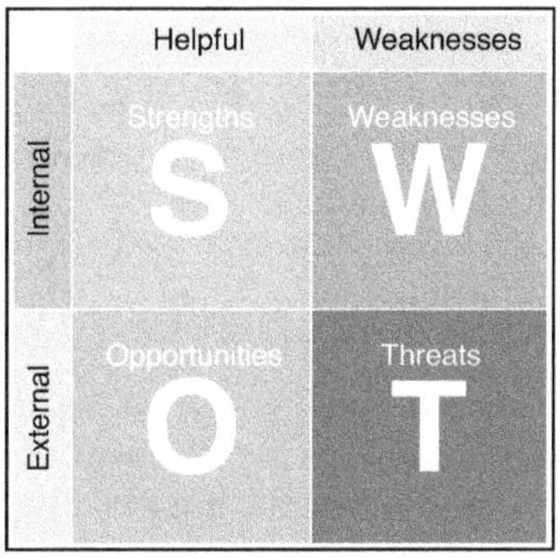

Figure 3.5 Strengths, weaknesses, opportunities, and threats (SWOT).

contamination. Self-driving cars need to be invulnerable to unauthorized access. A threat is an intention to inflict some type of loss. The types of loss include physical, economic, reputation, health, or other loss. In the present state of the economy, the need to investigate, identify, and take the appropriate countermeasures is imperative. Threats may also be part of the natural business cycle, where a threat is interpreted to mean that a competitor may be able to take your customers away by nature of their marketing and/or product offerings.

Strengths, weaknesses, opportunities, and threats (SWOT) analysis, illustrated in Figure 3.5, is one of a collection of tools that can be used to accomplish an analysis. Populating the SWOT table can be accomplished through brainstorming. The appropriate stakeholders need to be part of the process to ensure that all perspectives are covered. These different perspectives can be from a customer's point of view, a competitor's point of view, or even a malevolent actor's perspective.

TYPES OF RISK

> Examine the various types of operational risks (e.g., technical, scheduling, safety, and environmental), strategic risks (e.g., brand, reputation, stakeholder, and regulatory compliance), financial risks, cybersecurity risks, and analytical risks (e.g., informed and inherent) and their relationship to reliability. (Analyze)
>
> **Body of Knowledge II.A.3**

Reliability engineering is primarily focused on testing and evaluation of the intended operational functions of the products and/or processes. However, the reliability program must be cognizant of a broader view to include the various systems that bring the product to the customer.

The best lifecycle analysis may be made irrelevant if these systems fail. Examples of these issues follow.

Operational Risks

Technical. Are there foreseeable technological advances that can affect interfaces and/or integration of the product? Is there a possibility that advances in technology can render this product obsolete?

Scheduling. Are advance planning and provision for contingencies adequate? Will scheduling conflicts with other products or other competing priorities impact delivery?

Safety. Will various phases of the lifecycle cause health and safety hazards to the user or others? Will proposed changes in processes, installation, or operation impact safety? Does discard at the end of useful life pose health, environmental, or safety issues?

Environmental. Is testing sufficient to ensure advertised product performance in anticipated geographical and environmentally diverse customer markets? For example, exposure to extended sunlight, ozone, heat, humidity, altitude, and extended inventory holds. Have "foreseeable misuse" scenarios been explored to ensure safe use?

Strategic Risks

Strategic risks are considered in long-term plans identified specifically as strategic plans. A strategic plan term is typically 3–5 years or more. The strategic plan not only lays out the vision and mission of the enterprise for the future but includes the goals and objectives that are measurable over time. Included in the strategic plan are the potential strategic risks. Ongoing business decisions, marketing, product offerings, product features, and advertising all need to be aligned with the strategic plan. The strategic plan identifies inherent potential risks and should provide guidance on how to manage such risks if they arise. Some examples of strategic risks are:

- *Brand.* Plans to distinguish a brand in the marketplace could lead to unanticipated negative consequences. Brand damage can be irreversible. Has care been taken to ensure that brand visuals and text do not inadvertently create controversy? Branding promotions do not violate cultural or societal norms or practices.
- *Reputation.* With social media being a dominant element, an enterprise's reputation can easily be damaged. Have communications been reviewed and approved through all media channels prior to release to make sure that the messages are aligned and sanitized?
- *Stakeholder.* Loss of critical stakeholders can create undesirable disruption and could change the strategic direction of an enterprise. It's important to know who the stakeholders are and why they are stakeholders. Misaligned and/or conflicting priorities with major stakeholders need to be understood and managed.
- *Regulatory compliance.* As new products, processes, and practices become popular, there may be changes to regulations that need to be reviewed and adopted to avoid penalties, fines, and litigation. Dramatic changes like the introduction of AI into almost every element of life will require new regulations. Have all relevant regulations been reviewed, understood, and complied with?

- *Suppliers.* How would changes in supplier operations (for example, a buyout, process change, or adoption of lean manufacturing) impact the product's performance and/or delivery schedule? Will changes in the transportation infrastructure impact supplies? Can the supplier make changes that are not reportable under the form, fit, or function categories but that can result in the introduction of issues?
- *Materials.* How will changes in cost and availability of key materials and/or commodities impact the product? What effect will supply/demand and international relations have on the product?

Financial Risks

In the global economy, seemingly innocuous changes in one part of the world can have a significant impact in other areas of the world. Finances, which include funding for potential disruptions, need to be part of the strategic planning process. How will changes in exchange rates for international transactions impact the organization? Is the information regarding creditworthiness of customers adequate? Have tariffs been imposed? Can anticipated global events create financial strain? Have funds been reserved for potential natural disaster recovery activities?

Cybersecurity Risks

Protecting access to products, information, and systems is accomplished by implementing defensive tools along with real-time monitoring to make sure they are effective. As AI evolves, it is anticipated that the sophistication and frequency of cyber attacks will increase. Plans need to be in place to evolve and mature cybersecurity over time. This will require tracking trends in the capability of cybersecurity tools as well as awareness of cybersecurity threat trends. Have all mandatory compliance requirements been reviewed? Have industry specific requirements been reviewed for applicability, for example, HIPAA, Sarbanes-Oxley Act, GDPR, or CFR?

Analytical Risks

There have been many examples of analytical failures resulting in significant losses. Such losses can be due to faulty analysis of materials, incorrect mathematical computations, mistaken market trends, or incorrect unit conversions. To mitigate these risks, an organization should address the following questions:

- Are there adequate procedures to ensure correctness of computations?
- Is there guidance for decision making regarding the coverage and depth of analysis of analytical errors?
- Has the applicability of available analytical methods been reviewed for sufficiency?
- Have analytical results been reviewed objectively to ensure accuracy?
- Are the necessary analytical competencies inherent in the enterprise?
- Is information shared between the appropriate teams and stakeholders?

Chapter 4
Risk Analysis

Introduction to Risk Analysis

There are many tools and methods available for performing risk analysis early in the design and development cycle. Some of these approaches lose effectiveness when they are used at the wrong time or for the wrong purpose. Using the wrong tool, or using the right tool at the wrong time, will have undesirable consequences. The application of these tools and methods will be considered here, along with the recommended timing of initiation.

Products, equipment, and processes that have been in use for an extended period will require an approach different than what is used for new product development. The focus for existing equipment and processes will be day-to-day operational performance. The approach to improving and rectifying issues with existing products, equipment, and processes will rely on problem solving and root cause analysis (RCA). This approach will include the listing and ranking of historical failure modes and their frequencies so that the causes can be addressed. A system for managing existing day-to-day operations could be a work order processing system, a failure reporting and corrective action system, maintenance management system, or some other process in which problems are discovered and corrected.

As mentioned in Chapter 3 in the section on "New and/or Emerging Technologies," there may be research and development efforts to investigate new and emerging technologies and procedures to feed the ideation of new and/or improved products and/or processes. Typically, when researching and developing new technologies, there will be some establishment of what makes up readiness for adoption. There are several approaches to assess and integrate new technologies. However, some are more well-defined in practice than others are.

The Government Accounting Office (GAO) has a best practice entitled "Technology Readiness Assessment Guide" (GAO-20-48G). The Technology Readiness Assessment (TRA) is based on a scale that determines the Technical Readiness Level (TRL) of 1–9, with 9 being the most mature technology. This readiness is established using tools such as first principles testing, designed experiments, accelerated life testing, and prototyping and may not have been scrutinized using a risk analysis tool such as failure mode and effects analysis (FMEA). It is recommended that in scenarios where emerging or novel technologies are considered for inclusion into new products that all risk analysis documentation be reviewed, and if found to be insufficient, then an appropriate tool such as an FMEA should be implemented. Integrating new technologies, processes, and/or materials that have not been duly scrutinized into a project can negatively affect the project schedule and cost. In addition, it can also introduce unforeseen failure modes into the product and/or process lifecycle.

If a brand new "clean sheet" design is to be analyzed for its potential risk, then the most commonly used approaches are concept down-select (Pugh method); fault tree analysis (FTA); failure mode and effects analysis for functional, design, and process risk analysis (FFMEA, DFMEA, and PFMEA, respectively); and/or a risk matrix.

Concept Down-Select (Pugh Method)

The concept down-select, also known as the Pugh concept selection process or Pugh method, is a qualitative technique for selecting the best option from a list of potential concepts. An example of a Pugh matrix template is shown in Figure 4.1.

The Pugh analysis begins by selecting a baseline design. This can be the existing design that is now being modified or evolved, or it could be a selected design variation that is being promoted along with other competing designs. The criteria that will be used to select the preferred design are then established. These criteria should include relevant factors like cost, reliability, safety, and producibility. The team then assigns the rankings for the alternative design concepts, which in its simplest form assigns a "+", "-", or "S" to each criterion to show whether the

	A	B	C	D	E	F	G	H	I
			Design Concepts						
1	Pugh Concept Selection Matrix Comparison Criteria	Current Process (Baseline)	Concept 1	Concept 2	Concept 3	Concept 4	Concept 5	Concept 6	Concept 7
2	Criteria 1		+	S					
3	Criteria 2		S	–					
4									
5									
6									
7									
8									
9									
10									
11									
12									
13									
14									
15	Total +'s		1	0	0	0	0	0	0
16	Total –'s		0	1	0	0	0	0	0
17									
18	Compare current with	+ Better Alternative							
19	selected alternatives:	– Worse Alternative							
20		S Same Alternative							
21									
22		Focus on alternative with the most +'s and fewest –'s							

Figure 4.1 Concept comparison—Pugh method.

competing design is better, worse, or the same as the baseline design, respectively. The exercise produces two outcomes:

1. Ensuring that there is more than a single design concept under consideration
2. Assessing the design against a robust set of criteria to compare and minimize risk factors. It is common practice that certain features of one design be integrated into a competing design, reducing risk and improving reliability.

Note that there are many approaches that are used to facilitate competitive design selections, and the Pugh down-select methodology is just one.

FAULT TREE ANALYSIS (FTA)

> Use fault tree analysis (FTA) to evaluate the potential failures within a product or process. Employ action priority to categorize and prioritize the necessary corrective actions. (Analyze)
>
> **Body of Knowledge II.B.1**

FTA is a "top-down" approach that is used to identify the cause or the combinations of causes of failure. FTA utilizes a wide variety of symbols from the electronics and logic fields.

FTA can be performed without considering the likelihood of the occurrence of the event or cause, known as a qualitative fault tree analysis. FTA can be combined with probability techniques, failure rate analyses, and other relevant quantitative reliability data to better estimate the probability of failure and to prioritize actions, known as a quantitative fault tree analysis. A qualitative fault tree analysis focuses on identifying the logical sequence of events that could lead to a system failure, while the quantitative fault tree analysis can give a numerical probability of the occurrence of those events.

FTA can be used to analyze complex systems including electronic, mechanical, and software elements and the interaction and interface of subsystems independent of, or within, a larger product or process.

AND and OR Gates

The primary symbols used in FTA are the AND gate and the OR gate. Each of these has at least two inputs and a single output, as shown in Figure 4.2. The output for the AND gate occurs if and only if all inputs occur. The output for the OR gate occurs if and only if at least one input occurs. This is shown in the truth tables in Figure 4.2.

Rectangles are typically used for labeling inputs and outputs. The failure mode being studied is sometimes referred to as the "top" or "head" event. An FTA helps

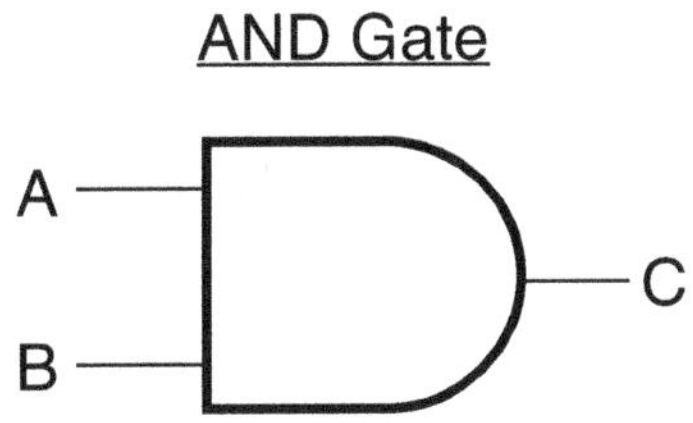

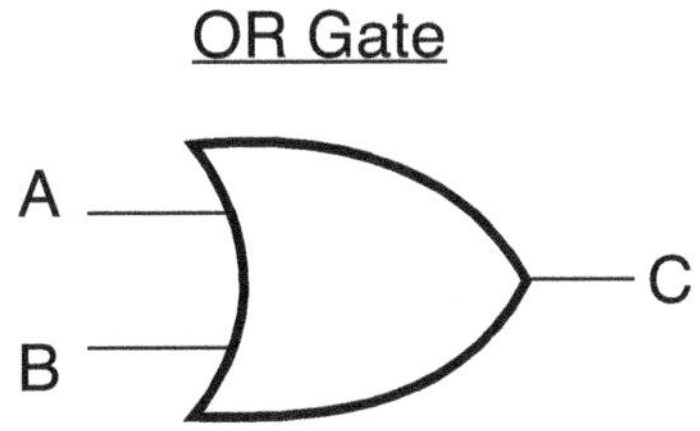

Truth Table		
A	B	C (A × B)
0	0	0
0	1	0
1	0	0
1	1	1

Truth Table		
A	B	C (A × B)
0	0	0
0	1	1
1	0	1
1	1	1

Figure 4.2 Comparison of AND and OR logical gates.

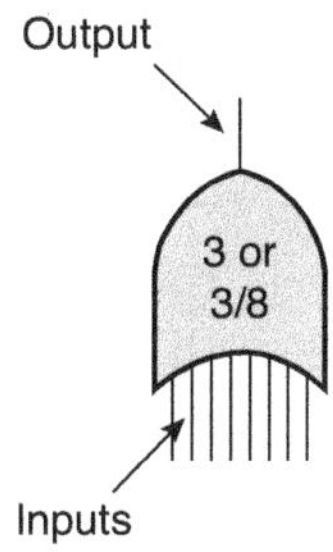

Figure 4.3 Voting OR gate.

the user to consider underlying causes of a failure mode and to study relationships between various failures.

Voting OR Gates

A voting OR gate is used to model fault tolerant systems. In a voting OR gate, the output occurs if and only if *k* or more of the input events occur, where *k* is specified, usually on the gate symbol, shown in Figure 4.3.

FTA is very useful in the early stages of design, especially in situations where the product being designed is complex and/or has interdependencies with other components or systems. FTA provides a diagram of the complexities and helps users visualize possible corrective/preventive actions. In typical use, the analysis begins with a fault or failure that has been singled out for further study and asks, "What condition(s) could cause this failure?"

Example 4.1 demonstrates FTA for the agitator function of a mixing tank.

EXAMPLE 4.1

The failure mode being studied is the stoppage of agitation in a tank before mixing is complete. This becomes the top event. Further team study shows that this stoppage will occur if any of the following events occur:

- Power loss
- Timer shuts off too soon
- Agitator motor failure
- Agitator power train failure

Power loss will occur if the external power source fails and the backup generator fails. The timer shuts off too soon if it is set incorrectly or has a mechanical failure. The agitator motor fails if overheated or if a fuse or capacitor fails. The agitator power train fails if both belts A and B break or if a clutch or transmission fails. This is symbolized by the FTA diagram shown in Figure 4.4.

Further analysis and assignment of probabilities can be done using metrological records and insulator testing results. At this point the design team could determine whether the wire strength, insulator design, amount of sag, and so on are appropriate.

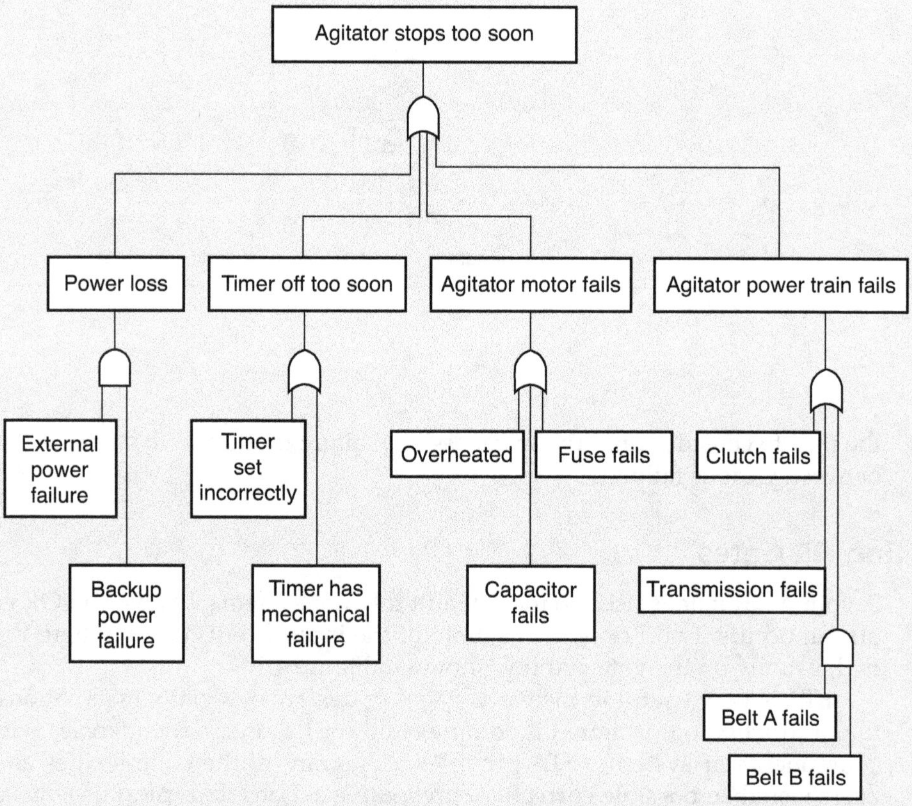

Figure 4.4 Fault tree analysis (FTA) diagram of agitation failure.

Example 4.2 shows an FTA for the loss of continuity of a power transmission line.

EXAMPLE 4.2

A loss of continuity between points A and B on a power transmission line is being studied. Suppose the team has determined three events, any one of which would cause the outage:

1. Wind more than 130 miles per hour (mph)
2. Wind more than 70 mph with more than one inch of ice buildup on the line
3. Insulator failure

The resulting fault tree is shown in Figure 4.5.

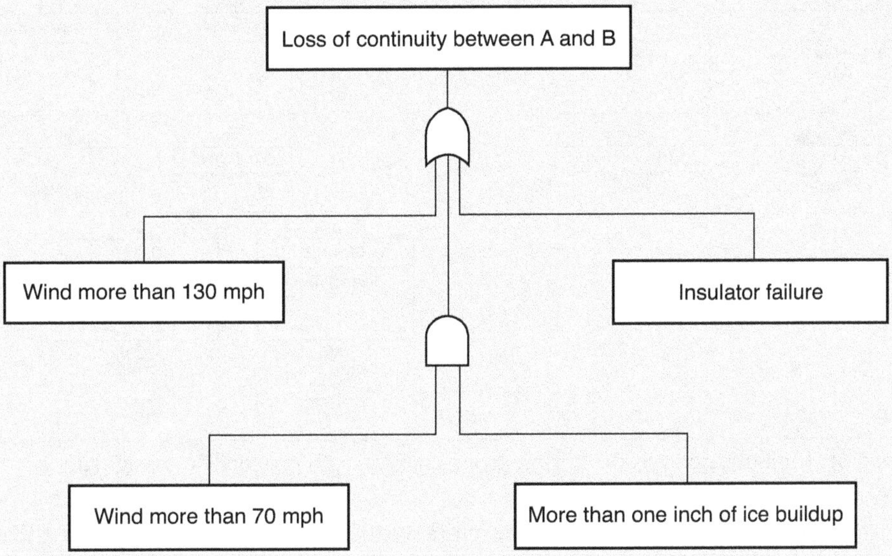

Figure 4.5 Fault tree analysis (FTA) of a power outage.

Action Priority

In FTA, there can be more than one way to determine what actions are to be prioritized. Using probability of failure is one approach. Requirements will often dictate what the expected life of a component, assembly, subsystem, or system needs to be. The useful life of a product can be based on numerous relative factors, such as the useful life of the model's predecessor, material enhancements, or design changes that reduce parts count. For example, consider the case of a small private jet. The expected useful life of a jet is established to determine quantities such as the warranty period and spares stocking. The oxygen system of the jet can then be specified to exceed the useful lifetime of the jet, giving it an additional factor of safety over what is required. This would also take into consideration any assigned safety factors to ensure sufficient safety margin. An example of an FTA for an aircraft drone collision is shown in Figure 4.6.

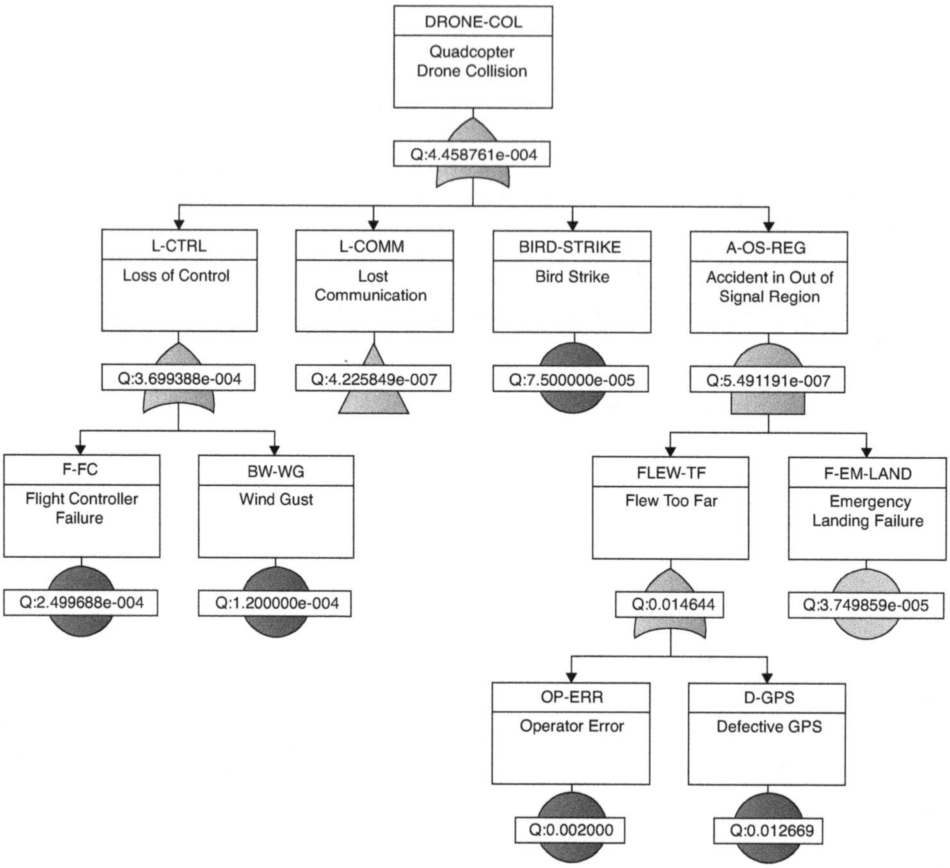

Figure 4.6 Fault tree analysis (FTA) for a drone collision with probabilities assigned.

Once the FTA has been completed with assigned probabilities for the various events in the FTA trees/branches, severity levels can be set. Figure 4.7 shows an action priority (AP) table using severity versus probability of occurrence to determine what zone is applicable when prioritizing a risk. The AP table is tailored to the needs of an industry, discipline, culture, and application, giving it contextual significance. It is recommended that a guidance document be developed to provide decision rules to resolve situations where the AP rankings are the same. The scales used in establishing the priority levels are flexible and can be tailored to include all relevant regulations, standards, and laws.

Action priority can be set using a wide variety of approaches. The high, medium, and low categories in Figure 4.7 are not a prioritization of the risk but rather a prioritization of the need to take action to eliminate or reduce the risk.

Success Tree Analysis (STA)

Success tree analysis (STA) approaches a system more positively than FTA, asking "What must occur for the system to function successfully?"

Example 4.3 shows an STA for the agitator function first presented in Example 4.1.

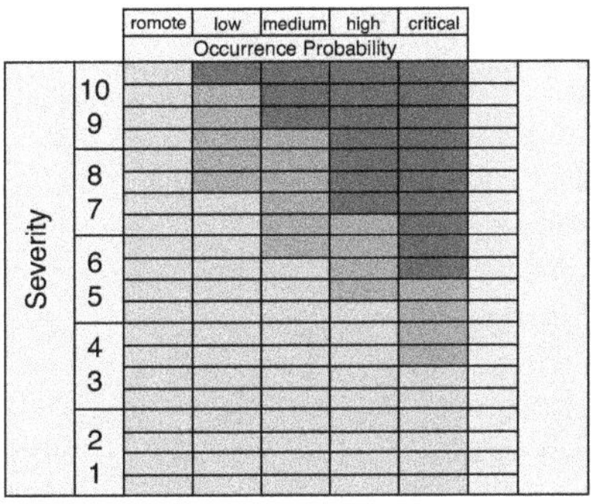

HIGH	Highest level of priority for review and action. Prevention is considered first. May require design change. If not preventable, then probability of occurrence is reduced.
MEDIUM	Medium level of priority for review and action. Recommend actions to improve prevention. May require added controls. If not preventable, reduce probability of occurrence.
LOW	Low level of priority for review and action. Recommendations for improvement are encouraged. Document the decision/justification for no actions taken.

Figure 4.7 Action priority example.

EXAMPLE 4.3

An STA diagram for the agitator function is shown in Figure 4.8.

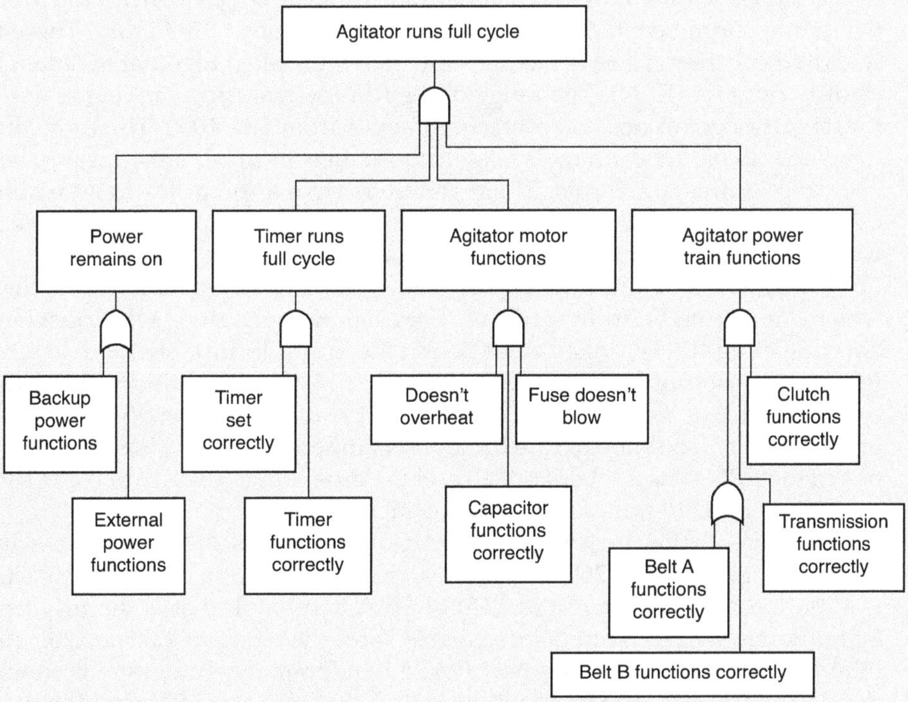

Figure 4.8 Success tree analysis (STA) diagram for the agitator function.

Refer to IEC 61025 FTA for additional information.

FAILURE MODE AND EFFECTS ANALYSIS (FMEA)

> Define and distinguish between failure mode and effects analysis (FMEA), failure mode, effects, and criticality analysis (FMECA), functional failure mode and effects analysis (FFMEA), and use failure mode and effects analysis (UFMEA). Evaluate these techniques for use on systems, products, processes, and designs. (Evaluate)
>
> **Body of Knowledge II.B.2**

FMEA

FMEA is one of the most widely used risk analysis tools across all disciplines for identifying potential failures and generating actions to prevent them. The essence of an FMEA is to list all the potential failure modes that could occur in the system design and then rank the severity of the failure consequences, the probability the failure could occur, and finally, the ability to detect the failure. These three—severity, occurrence, and detection—are then multiplied together to obtain the risk priority number (RPN). The range of the RPN when using ranking scales of 1–10 for severity, occurrence, and detection ranges from 1 to 1000. The RPN allows for the potential failures and their related consequences to be ranked using the Pareto principle, or the 80/20 rule. There are alternative approaches to assigning RPN. One such approach would be to take the face value of each rating. For example, with Severity = 8, Occurrence = 6, Detection = 3, assign the RPN of 863.

When RPNs in an FMEA are the same, there needs to be a way to determine which one should be addressed first. For example, consider a scenario where there are competing RPNs ranked at 300 using the multiplication method, in which one RPN is the result of Severity = 10, Occurrence = 3, and Detection = 10 and the other is the product of Severity = 6, Occurrence = 10, and Detection = 5. One way to prioritize actions would be to rate these RPNs independently by severity. Given that one of the RPNs has a severity of 10 and the other has a severity of 6, the RPN with the higher severity rank would be prioritized for action.

At present, the most popular FMEA approach is the Automotive Industry Action Group (AIAG) (2008) handbook or some variation of that approach. There is a more recent update to the AIAG FMEA handbook that is the result of a collaboration and agreement between AIAG and the German Verband der Automobilindustrie (VDA) to create a new FMEA handbook that integrates both the AIAG and VDA methods. The first edition of the AIAG/VDA FMEA handbook is dated 2019. The new AIAG/VDA handbook is not a mandated replacement of the AIAG (2008) handbook and as such has been accepted as an optional approach by the automobile manufacturers. Based on this standing, the AIAG (2008) methodology is still recognized as the dominant approach and the AIAG/VDA approach only finds mention here.

FMEA Templates

There are a wide variety of FMEA templates in use. The one shared in Figure 4.9 is from the AIAG (2008) handbook with a minor modification. Namely, the template has been changed from the AIAG (2008) format only to provide the ability to distinguish the type of FMEA being developed, as indicated by the checkbox identifiers. It is not unusual to see mixed FMEA types in one document, where design, process, and even software failure modes are intermingled together. Such a document with mixed modes is difficult to populate and even more difficult to interpret. It is desirable to stay within the context of the particular FMEA type and not mix failure modes. For example, process (manufacturing) failure modes and/or manufacturing process prevention and detection methods are covered in the PFMEA and should not be found in a DFMEA.

Types of FMEAs

This section will cover system, use, equipment, software, functional, design, and process FMEAs. Additional detail is provided for design and process FMEAs, as they are the most often used.

System FMEA

The AIAG (2008) handbook includes system FMEA (SFMEA). However, the coverage of the SFMEA topic is minimal and only indicates that the system FMEA is an FMEA made up of various subsystems. The text notes that the SFMEA is to be focused on interfaces of the subsystems and the interactions among them. In practice, if the design FMEA (DFMEA) is scoped and performed correctly, the subsystem interfaces and interactions will be covered in the DFMEA.

In the AIAG/VDA FMEA handbook, the SFMEA is not mentioned, and another type of FMEA is identified as an FMEA-MSR. This is described as FMEA for monitoring and system response (MSR) and is focused on the failure cases that might occur under the customer operating conditions. Reliability practitioners have also identified an SFMEA and have used it to define failures very early in the development process, at the functional level. Many organizations strive to identify, develop, and integrate new technologies into their products and processes to gain competitive advantages. Novel approaches and materials to be integrated into their products and processes are investigated, tested, and developed using FMEAs.

POTENTIAL
FAILURE MODE AND EFFECTS ANALYSIS
☐ System ☐ Design ☐ Process ☐ Software

____ System
____ Subsystem
____ Component ____________
Model Year(s)/Program(s) ____________
Core Team ____________

Design Responsibility ____________
Key Date ____________

FMEA Number ____________
Page ______ of ______
Prepared By: ____________
FMEA Date (Orig.) ____________

Item / Function	Requirement	Potential Failure Mode	Potential Effect(s) of Failure	Severity	Classification	Potential Cause(s) of Failure	Current Design				RPN	Recommended Action	Responsibility & Target Completion Date	Action Results				
							Controls Prevention	Occurrence	Controls Detection	Detection				Actions Taken Completion Date	Severity	Occurrence	Detection	RPN

Figure 4.9 FMEA template.

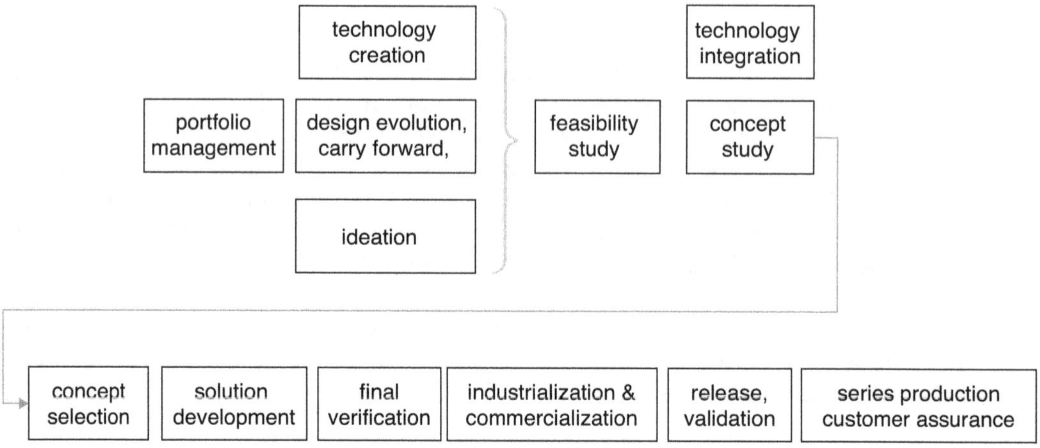

Figure 4.10 New product development process flow.

The SFMEA is often focused on the high-level functional requirements, performed early in the stage gate process where concepts are considered for possible integration into a design. This interpretation of SFMEA is often identified as a functional FMEA (FFMEA) and is synonymous with the SFMEA. When exploring concepts, high-level functions can be identified prior to any design efforts and the conspicuous failure modes of the high-level functional requirement failures can be assessed. This approach can also be very useful in the early development of requirements and will have a significant influence on the choices made when moving into the concept selection phase of the project.

The stages shown in Figure 4.10 illustrate a typical idea-to-product development cycle.

The initiation point, content, and cautions for typical types of FMEAs are denoted in Table 4.1.

Use FMEA

Use FMEA (UFMEA) and equipment FMEA (EFMEA) are not commonly used in practice due to the availability of other tools and methods that are more efficient and effective. The UFMEA is performed from the perspective of the user of the system, product, or process. For example, a failure mode could be "User accidentally moves automotive gear shift from DRIVE to REVERSE while at highway speed." This is intended to capture the consequences that the driver experiences in such a scenario. The UFMEA would be analogous to the FMEA-MSR in the AIAG/VDA handbook but due to scope and preference in the field, it will not be covered in any detail here.

The UFMEA is essentially a process FMEA from the point of view of the owner, user, or operator. The UFMEA is focused on the operation of a system. Instructions on the operation of the system along with maintenance schedules and any cautions (operational) are developed during the design of the system. Take for example the operation of an automobile. The manufacturer provides all the necessary instructions on the correct operation (use) of the automobile along with published maintenance programs and safety cautions. The operational instructions

Table 4.1 FMEA types and start times.

FMEA type	Initiation	Content	Cautions
Function/ System	Ideation, concept analysis/feasibility	High level requirements	Prevention and detection indeterminable.
Design	Upon readiness of the bill of materials	Design process	Mixed modes—for example, PFMEA entries. Attempt to FMEA after design freeze.
Process	Closely follows DFMEA start	Process flow diagram, process steps, and process step requirements	Mixed modes—for example, DFMEA, SWFMEA entries. Failure to include lean in the context of the FMEA.
Use	After product/process realization—in the customer domain	User experience with product, direct data feedback (warranty)	Failure to assemble the appropriate stakeholders, meaning a diverse set of users of the product.
Software	Software methodology decided (agile, waterfall, static, or dynamic). Software flowchart, language selected, architecture established	Breadth and depth of analysis, subject to the chosen methodology (agile, waterfall, static, or dynamic)	Design, process (manufacturing), or UFMEA entries.

and limits of the automobile are specified, and all regulatory and legal obligations are met and/or identified with recommendations to ensure lawful and regulatory compliance. Upon the sale of the automobile, the system risk transfers to the new owner. The requirements that populate the UFMEA comprise the operating instructions, maintenance schedules, and safety notes, as provided by the manufacturer (designer). For example, one entry might address starting the engine with a keyless (push button) switch, as shown in Figure 4.11.

The process requirements are:

1. Check that the parking brake is set.
2. Check that the shift lever is in P.
3. Firmly depress the brake pedal.
4. Check display for key indicator.
5. Press the Engine Start Stop switch shortly and firmly.

The failure modes from the user point of view are then all of the type "the failure to meet the listed requirements." The first failure mode is the failure to check if the parking brake is set, shown in Figure 4.12.

The facilitator would continue populating the template in this fashion to complete the UFMEA. The PFMEA tables for severity, occurrence, and detection are

Starting the engine

1 Check that the parking brake is set.

2 Check that the shift lever is set in P.

3 Firmly depress the brake pedal.

 will be displayed on the multi-information display.

If it is not displayed, the engine cannot be started.

4 Press the engine switch shortly and firmly.

When operating the engine switch, one short, firm press is enough. It is not necessary to press and hold the switch.

The engine will crank until it starts or for up to 30 seconds, whichever is less.

Continue depressing the brake pedal until the engine is completely started.

The engine can be started from any engine switch mode.

Figure 4.11 Keyless ignition starting instructions.

Process Step	Requirements	Potential Failure Mode	Potential Effect(s) of Failure	Severity	Classification	Potential Cause(s) of Failure	Operational				RPN
							Controls Prevention	Occurrence	Controls Detection	Detection	
Start the engine	Operator checks that the parking brake is set	Operator fails to check and/or set the parking brake	No discernable effects on start function	1		Operator intentionally skips step	Operators manual - start engine instructions	8	Operator diligence	9	72
						Parking brake is not operable - stuck - frozen actuator					

Figure 4.12 UFMEA entry example.

easily adaptable to represent the operational domain. This approach provides unique insight with respect to the operation of the system by the user. The importance regarding foreseeable misuse, unintended use, and unanticipated consequences also needs to be in context.

For example, when big SUV trucks became popular with all drivers, it was noticed that steering column issues started to appear in warranty requests. This was a direct consequence of drivers using the steering wheel to hoist themselves into the driver's seat. Another example that surfaced was in the early days of catalytic converters. It was discovered that if a user was parked in high grass, the converter could ignite the grass and destroy the vehicle.

These examples suggest that the participants in a UFMEA would need to be a cross-section of the general population of owner/users. These UFMEA team members would need to have experience using the product so that these additional potential failure modes can be discovered. It is good practice to stage user studios during the design phase so that random users have a chance to interact with the product prior to design freeze. Those random users need to represent the expected purchasers/users. It would be ineffective to do a UFMEA on the "starting the engine" analysis in Figure 4.12 if the FMEA team members are not vehicle owners or drivers.

Equipment FMEA

For an EFMEA, the focus is on improving the performance of an existing capital asset. For example, a failure mode might be "Vertical mill drifts out of alignment during long runs." A PFMEA or a root cause analysis (RCA) would be more effective in addressing these issues. The EFMEA is, for all intents and purposes a DFMEA. If there is equipment that has been in use for an extended period of time, then the EFMEA approach is undesirable. Given that the equipment is being addressed for failure modes and has a performance history, the reliability centered maintenance (RCM) or total productive maintenance (TPM) approach would be two of the preferred approaches.

Software FMEA

In software development there are additional challenges. For example, when an agile code development approach is used, a software FMEA (SWFMEA) is not practical. Agile is based upon a set of four values and 12 principles laid out in the Manifesto for Agile Software Development. Agile is flexible and iterative, emphasizing continuous feedback and adaptation. On the other hand, a waterfall software development approach is sequential and rigid, with distinct phases completed in order. Agile promotes ongoing collaboration, while waterfall focuses on thorough planning and execution in predetermined stages. The differences in coding approaches are shown in Figure 4.13 and are discussed in more detail in Chapter 11.

A SWFMEA is a viable tool for development under the waterfall model. However, different approaches for risk assessment are taken when agile is the software development process. In an FMEA we need the defined requirements to begin the FMEA assessment, and requirements can be more fluid in the agile approach than in the waterfall approach.

Software development is unique, especially with the current advances in artificial intelligence (AI), and includes additional elements like security, atomicity, coupling, and cohesion. The language, terminology, and additional considerations

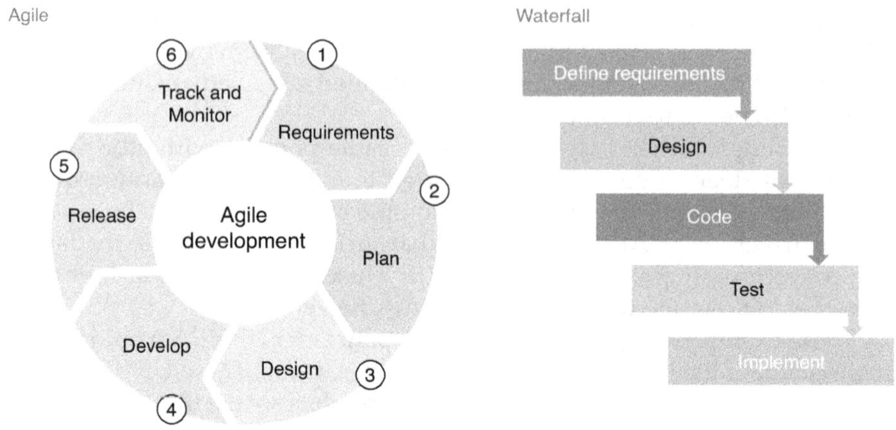

Figure 4.13 Agile versus waterfall software coding.

depend on the selection of the source code (for example, C++, Python, or Java), which will dictate the available approaches. Because of the diversity of software coding languages and applications, software FMEAs and agile-focused risk analysis approaches are beyond the scope of this handbook.

Functional (System) FMEAs

Functional (system) FMEAs or FFMEAs can be helpful in establishing and analyzing the high-level requirements of a new concept. For example, consider the addition of a hot tub into the architecture of a private jet. Examples of the high-level requirements include:

- Onboard pumping ability to load/unload the water
- Water weight distribution shall not change the center of gravity of the aircraft
- Ensure no sloshing of water
- Temperature control of the water

To begin populating the FFMEA, the facilitator will ask the team what the requirements are for the concept realization. The listed requirements are at the functional level, and no design approaches have been defined yet. The high-level requirements can drive decisions about the design path. For example, the requirement "Onboard ability to load/unload the water" will increase the parts count and weight of the aircraft and will require additional maintenance activities of the aircraft as well. With the increased parts count comes an increase in potential failures. Early consideration of this requirement could result in the onboard hardware being separated from the system and the requirement changed to an "External ability to load/unload the water," thus designing out the potential failure modes and related issues of an onboard load/unload subsystem.

Once a requirement is listed in the template, the conspicuous failure mode is simply the failure to be able to satisfactorily fulfill the requirement. For example, the requirement that states, "Temperature control of the water" has a failure mode of

Item / Function	Requirement	Potential Failure Mode	Potential Effect(s) of Failure	Severity	Classification	Potential Cause(s) of Failure	Current Design Controls Prevention	Occurrence	Controls Detection	Detection
	Temperature control of the water	Failure to adequately control the water temperature								

Figure 4.14 Functional FMEA potential failure mode.

System of interest Item and function	Function/ Requirements	Potential Failure Mode	Potential Cause(s) of Failure	Potential Effect(s) of Failure	Severity	Classification	Controls Prevention	Occurrence	Controls Detection	Detection
Electro-mechanical automotive brake system	r1: System shall ensure that necessary power is available for brake initiation under all conditions.	Brake system fails due to loss of power to system	Loss or interruption of electrical connection to the wheel end actuator	Loss of braking capability, Potential for accident, no warning of lost connection						

Figure 4.15 Functional FMEA entry for brake-by-wire system concept.

"Failure to adequately control the water temperature," as shown in Figure 4.14. It is important to move through the template row by row so that the information flows logically and so that any actionable issues can be identified and addressed early in the process.

Continue moving through the template by next listing the potential effects of the failure. The facilitator solicits from the participants the potential effects of the failure and enters that information into the template. Continuing in a stepwise manner, the team will work together to rank the severity of the potential effects. The team strives to complete the row entries but may not be able to because in the FFMEA there is no design yet. The team is developing a list of high-level requirements. Without a design it is impossible to make any entries about failure causes. The design may use any one of a number of ways to control the water temperature, and until a design decision is made as to how the water temperature will be controlled, we don't know what it is that can potentially fail. The severity, occurrence, and detection are then tabulated, most often on a scale of 1–10 where 1 is low and 10 is high.

The severity ranking scale can be the same for the system/functional FMEA as well as the design FMEA. The process FMEA will have a different ranking scale because it is a process step that is being scrutinized, and there is likely a human operator involved with the process step. The difference in the severity table is that it needs to include potential issues related to the operator. These issues include operator safety issues, human cognition errors, part/equipment damage, and contamination.

The FFMEA entry shown in Figure 4.15 does not show the design methodology prevention or detection as this is a concept, and no physical design has yet

Table 4.2 Suggested FMEA severity ranking for system/functional and design FMEAs.

Effect	Criteria: Severity of effect on product (customer effect)	Rank
Failure to meet safety and/or regulatory requirements	Potential failure mode affects safe vehicle operation and/or involves noncompliance with government regulation without warning.	10
	Potential failure mode affects safe vehicle operation and/or involves noncompliance with government regulation with warning.	9
Loss or degradation of primary function	Loss of primary function (vehicle inoperable, does not affect safe vehicle operation).	8
	Degradation of primary function (vehicle operable, but at reduced level of performance).	7
Loss or degradation of secondary function	Loss of secondary function (vehicle operable, but comfort/convenience functions inoperable).	6
	Degradation of secondary function (vehicle operable, but comfort/convenience functions at reduced level of performance).	5
Annoyance	Appearance or audible noise, vehicle operable, item does not conform and noticed by most customers (>75%).	4
	Appearance or audible noise, vehicle operable, item does not conform and noticed by many customers (50%).	3
	Appearance or audible noise, vehicle operable, item does not conform and noticed by discriminating customers (<25%).	2
No effect	No discernible effect.	1

been defined. Without knowing what the detailed design looks like, it is not yet possible to say definitively how the design methodology can prevent and detect failures. In the foregoing illustration, it is not determined as to what type of connections will be used, what the routing of the power delivery medium will be, or what physical/environmental protections will be applied. For example, the design team can elect to implement some features in hardware and others in firmware or software. Given these different design realizations, the design risk methodology will be different. For example, design features in hardware use a different investigative approach than implementing a design feature in software.

The FMEA severity table verbiage shown in Table 4.2 is automotive specific, so appropriate changes need to be made to reflect the discipline or industry being analyzed. The table shows severity level 10 is "failure to meet safety or regulatory requirements without warning" and severity level 9 is "failure to meet safety or regulatory requirements with warning." These entries at severity 10 and 9 should always include safety and regulatory failures. Given the stage of this project, we cannot pursue the potential causes of these potential failures because, at this level in the design stage, we are not decided on how a requirement will or can be satisfied. As noted earlier, the design solution can be hardware, firmware, software, or some undetermined combination of these solutions.

For example, some early automatic garage door openers were completely physical, all wires, relays, and switches. They used an infrared optical pair that established a beam across the driveway. When the beam was broken, the signal triggered logic to start the motor to drive the gears that were used to open the garage door. Although the approach was never a good residential offering, it was popular with parking garages and warehouses. This allowed the garage door to be kept closed for the benefit of temperature control, deterring pests, shielding rain, and security, and only opening when there was a vehicle coming into the driveway. In contrast, modern garage door openers are software centric and have eliminated the need for driveway optical sensors, the wiring necessary to connect them, and the hardware and firmware logic that was used to switch the motor on/off (circuit boards, relays).

It is easy to imagine that we can elect to implement hardware or software to whatever degree we prefer in our designs. Without knowing what that design decision is specifically, we are not able to explore the potential causes in the FFMEA. At this point we are in the requirements generation phase, not cause prevention. Along with the obvious requirements being listed, consideration needs to be given to standards, regulations, cultural conventions, foreseeable misuse, and nonfunctional requirements.

In an FFMEA we are unable to make any determination of potential causes or the probability of such causes. There will be cases where the project team, even at this early stage of feasibility, will offer a particular design solution. Given this solution, the team can rank the occurrence and if possible, the detection. This is accomplished using the DFMEA occurrence and detection tables. Table 4.3 shows the AIAG (2008) documentation to rank occurrence.

Table 4.4 shows the suggested DFMEA detection rankings.

Design FMEA

The timing of the design and process FMEAs is important. The DFMEA should be started once the design has been through a comparative down select process, in which the design team has a bill of materials (BOM) and they are talking to suppliers, getting prototypes and cost/time estimates. Then it is time to start the DFMEA. If the DFMEA is started too soon, this could cause some design churn. The design team may keep making changes based on the challenges in the DFMEA, and as a result the schedule will be impacted. If the facilitator waits for the design to reach a maturity level where the team has a high level of confidence (as witnessed by supplier engagements and prototype testing) then the DFMEA can proceed, knowing the amount of rework needed will be minimized.

The PFMEA should start shortly after the DFMEA has started. This allows the manufacturing side of the house to give real-time feedback on the practicality of building the design. This then allows for the design to be challenged by the manufacturing floor if necessary. The design can be changed to avoid undesirable issues in operations. For example, if it is possible to assemble part A and part B backward, and bad things happen when this is the case, then the design should be changed with new or modified design features to ensure that it can only be assembled in the proper configuration. It is also highly recommended that there be a manufacturing

Table 4.3 Suggested FMEA occurrence ranking for design FMEAs.

Likelihood of failure	Criteria: Occurrence of cause—DFMEA (design life/reliability of item/vehicle)	Criteria: Occurrence of cause—DFMEA (incidents per items/vehicles)	Rank
Very High	New technology/new design with no history.	≥100 per thousand ≥1 in 10	**10**
High	Failure is inevitable with new design, new application, or change in duty cycle/operating conditions.	50 per thousand 1 in 20	9
	Failure is likely with new design, new application, or change in duty cycle/operating conditions.	20 per thousand 1 in 50	8
	Failure is uncertain with new design, new application, or change in duty cycle/operating conditions.	10 per thousand 1 in 100	7
Moderate	Frequent failures associated with similar designs or in design simulation and testing.	2 per thousand 1 in 500	6
	Occasional failures associated with similar designs or in design simulation and testing.	0.5 per thousand 1 in 2,000	5
	Isolated failures associated with similar design or in design simulation and testing.	0.1 per thousand 1 in 10,000	4
Low	Only isolated failures associated with almost identical design or in design simulation and testing.	0.01 per thousand 1 in 100,000	3
	No observed failures associated with almost identical design or in design simulation and testing.	≤0.001 per thousand 1 in 1,000,000	2
Very Low	Failure is eliminated through preventive control.	Failure is eliminated through preventive control.	1

representative on the DFMEA team. This allows the operations side of the house to begin development of work instructions based on the design requirements.

The DFMEA is an inductive, bottom-up approach, alternatively referred to as the many-to-one approach. The DFMEA process starts with the BOM. It is good practice to prioritize what parts to start with, as all parts are not equal in their failure impact on the system. It is recommended that the areas of greatest concern be addressed first. This will allow remedial actions to be undertaken should the failure consequences be unacceptable, as determined by the FMEA team. As noted in the ranking tables, changes such as new designs, new technologies, and new materials are likely to be accompanied by uncertainties that will likely require added levels of scrutiny such as experiments, accelerated testing, and stress testing.

Table 4.4 Suggested FMEA detection ranking for design FMEAs.

Opportunity for detection	Criteria: Likelihood of detection by design control	Rank	Likelihood of detection
No detection opportunity	No current design control; cannot detect or is not analyzed.	10	**Almost impossible**
Not likely to detect at any stage	Design analysis/detection controls have a weak detection capability; virtual analysis (e.g., CAE, FEA, etc.) is *not correlated* to expected actual operating conditions.	9	**Very remote**
Post design freeze and prior to launch	Product verification/validation after design freeze and prior to launch with *pass/fail* testing (subsystem or system testing with acceptance criteria such as ride and handling, shipping evaluation, etc.).	8	**Remote**
	Product verification/validation after design freeze and prior to launch with *test to failure* testing (subsystem or system testing until failure occurs, testing of system interactions, etc.).	7	**Very low**
	Product verification/validation after design freeze and prior to launch with *degradation* testing (subsystem or system testing after durability test, e.g., function check).	6	**Low**
Prior to design freeze	Product validation (reliability testing, development or validation tests) prior to design freeze using *pass/fail* testing (e.g., acceptance criteria for performance, function checks, etc.).	5	**Moderate**
	Product validation (reliability testing, development or validation tests) prior to design freeze using *test to failure* (e.g., until leaks, yields, cracks, etc.).	4	**Moderately high**
	Product validation (reliability testing, development or validation tests) prior to design freeze using *degradation* testing (e.g., data trends, before/after values, etc.).	3	**High**
Virtual analysis correlated	Design analysis/detection controls have a strong detection capability. Virtual analysis (e.g., CAE, FEA, etc.) is *highly correlated* with actual or expected operating conditions prior to design freeze.	2	**Very high**
Detection not applicable; failure prevention	Failure cause or failure mode cannot occur because it is fully prevented through design solutions (e.g., proven design standard, best practice or common material, etc.).	1	**Almost certain**

For example, consider that a new technology vehicle transmission has been selected for a new automotive power train design. In such a case, it would be wise to prioritize the transmission FMEA entries so that any undesirable failure consequences can be addressed early in the analysis. If, for any reason, the failure consequences of the transmission cannot be reduced to an acceptable level of risk, it may require a radical design change. Early detection of this issue will allow a design change with minimal impact on schedule. See Figure 4.16.

In this example, shown in Figure 4.17, additional investigations need to be conducted. Although normal transmission temperature range is typically 180°F–250°F, the design team determines that the 300°F failure onset does not provide an adequate safe operational temperature margin. This early recognition of potential failures should minimize impact on the schedule. The recommended action is to eliminate the failure by finding materials that will provide a greater temperature margin. If this testing does not find the right combination of materials, then the design alternative transmission will need to replace this design option. The sequence of filling out the template is row by row, with some sense of priority in the order of analysis.

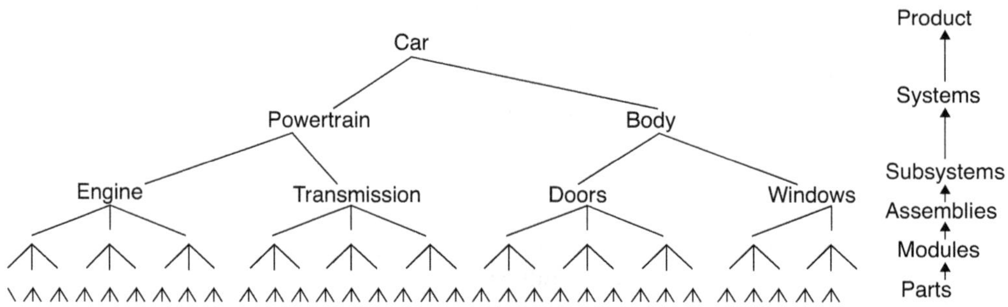

Figure 4.16 Inductive approach (bottom-up) for DFMEA.

Item / Function	Requirements	Potential Failure Mode	Potential Effect(s) of Failure	Severity	Classification	Potential Cause(s) of Failure	Current Design Controls Prevention	Occurrence	Current Design Controls Detection	Detection	RPN	Recommended Action	Response
Transmission: provide power transfer from the prime mover to wheels and provides gear ratio changes	Shall be immune to environmental stresses	Susceptible to environmental stress - heat sensitivity	Internal wet clutches fail - slipping as heat increases above 300°F. potential safety issue	#		Clutch material selection not adequate for the design thermal stresses	None	9	Early performance testing	2	180	Investigate alternative clutch material	
			Internal wet clutches fail - slipping as heat increases above 300°F. potential safety issue			Transmission fluid cooling not adequate for design thermal stresses	None	9	Early performance testing	2	180	Investigate alternative transmission fluid cooling. e.g. trans oil cooler	

Figure 4.17 Transmission DFMEA example.

If the proposed design includes new technology, for example, the order of analysis in the FMEA may prioritize this part of the design first, as it is the most likely to generate significant actions.

We know that requirements development is critical to being able to complete a design project successfully. These requirements provide the "item/function" entries, along with any nonfunctional and derived requirements. There are many lists of nonfunctional requirements, such as color, accessibility, serviceability, and security.

With respect to the derived requirements, these are not explicitly stated in the designer–customer requirements agreement. Often in the design cycle, there are some general requirements such as "must be fast," but they may not specify what "fast" is, as it may not have been agreed upon with the stakeholders yet. The designers will not stop designing; they will establish what "fast" is by whatever logical means are available. That could simply be "10% faster than the last version of this product." This requirement then is internally derived by the design team to keep the team progressing and needs to eventually be formalized between the designer and the customer. Until then, such requirements need to be tracked and resolved, typically under the requirements development process. It also means that the actual customer requirement may be different than the derived requirement and could result in additional time and expenses to fulfill when reviewed for acceptance with the customer. The FMEA facilitator should take this into account as it also impacts the timeline of the FMEA. The sooner the requirement is finalized, the less risk there is to the project timeline.

In the FMEA template, we will list the requirements and the failures we input will be the failure to meet each stated requirement. These will be entered as potential failure modes. The "potential effects of failure" entries are developed by the team, with the facilitator providing guidance.

Requirements: A comprehensive collection of requirements, including any requirements needed to fulfill adherence to standards, regulations, and laws. Include nonfunctional requirements and any derived requirements.

Potential failure mode: The conspicuous failure modes are the failure to fulfill a requirement. This is simply the negation of the requirement. For example, if the requirement is "must be waterproof," then the failure mode is "fails to be waterproof."

Potential effects of failure: List the effects that this failure will have on the product's performance and/or its compliance with any standards, regulations, and laws. The effects include health, environmental, and safety consequences. Extenuating consequences should be included, for example, vehicle steering assist failure, potential for unsafe operation, accident, traffic/pedestrian safety, and lawsuits.

Severity ranking: Once the team has agreed on the potential effects of the failure, refer to the severity table to determine the ranking. In a group, it is typical to get a range of numbers, some people being overconfident and some being risk averse. Taking the conservative approach would be to enter the highest number offered. There will undoubtedly be some pushback on that because high severity numbers will require additional work. It is the conservative approach to ensure that consequences, especially safety-related consequences, are addressed. If testing shows that the consequences of failure were overestimated, then the severity can be reduced, with comment, on the second ranking of the RPN. Otherwise, the severity ranking can only be changed because of a design change. Severity ranking of 9 or 10 always requires a recommended action, as these are safety or regulatory violations. Some organizations, upon adapting the severity rankings, include the assignments of critical characteristics (CCs) and special characteristics (SCs), requiring any severity of 7 or higher to require action. Although organizations have the capability to define for themselves what severity rankings to use, caution needs to be exercised so that there is consistency among projects, suppliers, and design partners.

Potential causes of failure: There may be multiple causes of a single potential failure mode. These different causes need to be entered on separate lines in the DFMEA template to allow the relative ranking of the probability of occurrence. Different causes may have different occurrences and thus different relative risks. Separating the causes allows the specific preventive design controls to be assessed and, when recommended actions are necessary, they can be specific to the listed cause.

Design methodology prevention controls: This question is specific for the DFMEA, asking, "In the design methodology, what preventive measures were taken?" The question is specific to the applied design methodology. For example, if the failure mode is "interference in fit up," then the preventive measure in the design methodology could be "tolerance stack-up analysis performed."

Caution: Prevention is not to be misconstrued as something that can be done during the assembly process and relegated to the PFMEA. Prevention specifically refers to modifying the design to minimize failure in assembly. This is especially true when a CC or SC is being addressed. If it is critical to safe operation, it needs to be mistake-proofed, or poka-yoked, in the design.

Occurrence ranking: This is the estimated and/or projected probability that the failure cause and its related consequences will occur. Analysis of testing data, historical data, and surrogate data can be used to project an estimate of the probability of the cause. If the design is new and no data is available, then the "low, medium, high" qualitative scale in the occurrence ranking table can be used. Consideration of the prevention controls needs to be part of the analysis. If there have been many entries in the prevention column, then this can have a favorable influence on the occurrence ranking entry. If the entry in the controls prevention column is "none," then it is likely that the team will assume the worst case and enter a potentially higher occurrence ranking. In the AIAG (2008) FMEA table found in Figure 4.2, the occurrence entry follows the controls prevention entry, ensuring that any information entered is reviewed prior to the occurrence ranking consideration.

Design methodology detection controls: Here, a specific question for the DFMEA is posed: "In the design methodology, what detection measures were taken?" The question is specific to the applied design methodology. For example, in the case of the stack-up analysis, the detection method could be to have someone verify the calculations used in the stack-up analysis. This answers the question, "What detection method is employed to verify the accuracy of the stack-up analysis?"

Caution: Detection is not to be misconstrued as something that can be addressed during the assembly process and relegated to the PFMEA.

Detection ranking: This is the inherent ability of the design methodology to be able to detect potential failures. The detection ranking does not include any controls implemented on the operations (manufacturing) floor. The entries in the detection column, as witnessed in the detection ranking table, are listed for their effectiveness in surfacing any potential issues related to the design process. Entries support establishing which recommended actions might be appropriate. Do not suppose or rely on inspections, operator diligence, or line testing in the manufacture of the product to detect a potential failure that results from a lack of robustness in the design methodology.

Risk priority number (RPN) ranking: The risk priority number is the product of severity, occurrence, and detection. There are numerous arguments regarding the use of the RPN in making decisions with respect to recommended actions. There are some advocates that recommend setting an RPN action threshold; others propose using the Pareto analysis (that is, the 80/20 rule), and others advocate using the numbers at face value (for example, with $S=6$, $O=5$, $D=9$ the RPN would be 659) rather than the product of $S \times O \times D$. The adopted method of using the RPN needs to be an agreed-upon method and acceptable by the FMEA team. The decision rule needs to be stated in the notes of the FMEA or risk plan so that anyone reading it understands how the actions were triggered.

Recommended actions: The recommended action(s) should not be limited due to schedule, cost, and resources. If there are reasons that the recommended actions cannot be fulfilled, then the reasons for this would be stated in the actions taken column. In the example of Figure 4.9, the recommended action is to "investigate alternative clutch material." This would require numerous steps.

1. Collaborating with the clutch supplier to see if they can provide an alternative material, are willing to sponsor the search for the new materials, or if they can provide an alternative solution.
2. Investigating the market to see what research has been done and if anyone has already developed a higher temperature resistant material.
3. Estimating the time and cost for acceptance and entry into the project schedule and budget.

Responsible: It is highly recommended that the person assigned to the responsibility for an action generated in an FMEA be a member of the FMEA team. In the example of Figure 4.9, the person assigned to "investigate alternative materials" does not need to be a materials scientist or be involved with suppliers. It is the job of the person who is delegated responsibility for an action to engage the necessary resources, guide those resources toward the stated goal, and report the ongoing results to the project team. If an action is assigned to someone who is not on the team, there may be delays and inconclusive results due to misaligned organizational reporting structure, overloaded resources, and competing priorities. Such issues have the potential to negatively impact the overall project.

Target date/completion date: It is necessary to ensure that timely results are shared with the project team, and setting milestones to accomplish these actions is essential to the success of the project. If there are significant delays in the established target completion dates as evidenced by the extended actual completion dates, the project manager and/or the project assurance manager may need to take additional actions to address the issues.

Actions taken: Not all recommended actions are achievable. Given the numerous variables involved, such as cost, time, technology limits, and materials availability, the actual actions taken may differ from the recommended actions and will need to be scrutinized by the project team for efficacy. If the project team determines that the "taken actions" are insufficient, then further actions or alternatives will need to be negotiated.

Updated RPN ranking: The severity can only be changed because of a design change. Any actions that reduce the occurrence (probability of failure) need to show clear evidence of what was done to reduce it and the rationale used to justify the lowered occurrence ranking. The detection ranking changes need to be supported by evidence that clearly shows the actions justifying the change. There is a caution that improving the detection can lower the RPN yet not change the severity or the occurrence of the failure mode. Ranking the product of severity and occurrence, independent of detection, is one way to ensure that a low detection ranking does not reduce an RPN to a level below the action threshold.

Process FMEA

The PFMEA is focused on the steps for manufacturing, service, or software processes such as the manufacture of a newly designed product, the service of establishing a new account, or the software for a self-service gasoline pump. These process steps are identified and laid out in a sequential, stepwise manner. For example, in the design of a new vehicle transmission, the process flow diagram (PFD) will be required to identify all the process steps necessary to assemble and test the transmission's functionality, as shown in Figure 4.18. One of the assembly steps is to attach the transmission bell housing to the transmission body.

For each step in the process, there is a collection of requirements that need to be met. Using a PFD allows the sequential assembly step requirements to be listed for each step, as shown in Figure 4.19.

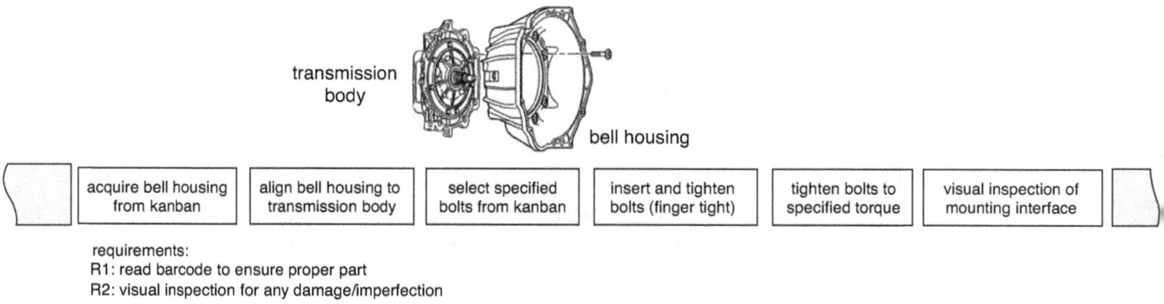

Figure 4.18 Process flow diagram (PFD) for bell housing mounting.

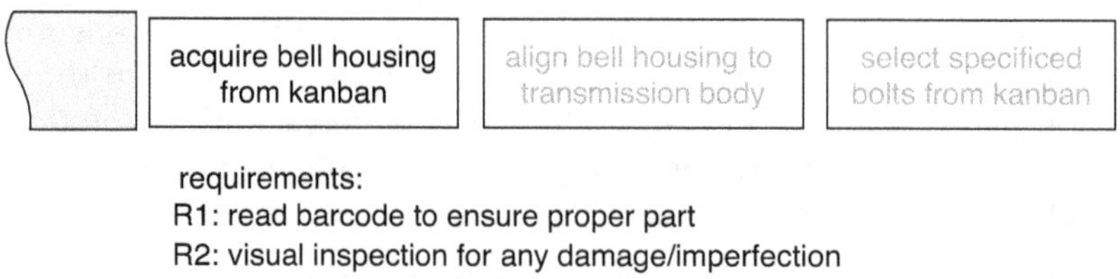

Figure 4.19 Process step requirements for assembly.

The failure modes will be the failures in the manufacturing and assembly process to meet the requirements. The tables used to determine severity, occurrence, and detection are specific to the PFMEA. The process of manufacturing and assembling the system can have failures that affect the performance of the system should the defect not be detected during manufacturing and the assembly end up being delivered to the customer. Another impact can be on the process line itself, resulting in stoppage of the manufacturing line. Stopping a production line can have a major impact on the operations, which is especially true for manufacturers who are using the lean manufacturing methodology.

The PFMEA tables can be customized to reflect the needs of the organization. The tables include both a numerical ranking and a qualitative scale (low, medium, high). This is intended to provide flexibility in determining the appropriate entry. The severity ranking can be found in Table 4.5.

The occurrence and detection scales are shown in Table 4.6 and Table 4.7, respectively.

The FMEA template is modified to reflect that this is a PFMEA. The process steps are now under scrutiny and the controls are now process controls.

In Figure 4.20, one of the process step requirements is to ensure that the right part is selected. In this example, the wrong part cannot be attached because of its different geometry, but selection of the wrong part will result in a line stoppage to investigate and correct any parts selection issues. There are many instances when the wrong part can still fit up to the assembly and, as a result, will move to the next processing station. It is conceivable that if the wrong part fits up and no other issues arise, then the flawed final assembly may end up in the customer's possession. This will then have a consequence in the field. Those consequences would be listed here in the PFMEA and will affect the severity ranking.

Requirements: A comprehensive collection of requirements, including any requirements needed to fulfill adherence to standards, regulations, and laws.

Potential failure mode: The conspicuous failure modes are the failures to fulfill a requirement. This is simply the negation of the requirement. For example, the PFMEA requirement stated is that the "assembly parts shall be clean and undamaged." The corresponding failure mode is "assembly parts are contaminated and/or damaged."

Potential effects of failure: List the effects that this failure will have on the final product if the product gets released with the failure mode in place. Include any effects at this process step that may involve the operator or may impact the flow of the materials in the process line. In high-volume manufacturing, any line stoppage is costly, and even though the line stop did not involve the potential release of nonconforming product, it is costly and can create a host of other issues.

Severity ranking: Once the team has agreed on the potential effects of the failure, refer to the PFMEA severity table to determine the ranking. In a group it is typical to get a range of numbers, some people being overconfident and some being risk averse. Taking the conservative approach would be to enter the highest number offered. There will undoubtedly be some pushback on that because high severity numbers will require additional work. It is the conservative approach to ensure that consequences, especially safety-related consequences, are addressed. Severity ranking of 9 or 10 always requires a recommended action.

Some organizations, upon adapting the severity rankings, include the assignments of CCs or SCs, requiring any severity of 7 or higher to require action. In a

Table 4.5 Process FMEA severity ranking table.

Effect	Criteria: Severity of effect on product (customer effect)	Rank	Effect	Criteria: Severity of effect on process (manufacturing/assembly effect)
Failure to meet safety and/or regulatory requirements	Potential failure mode affects safe vehicle operation and/or involves noncompliance with government regulation without warning.	10	**Failure to meet safety and/or regulatory requirements**	May endanger operator (machine or assembly) without warning.
	Potential failure mode affects safe vehicle operation and/or involves noncompliance with government regulation with warning.	9		May endanger operator (machine or assembly) with warning.
Loss or degradation of primary function	Loss of primary function (vehicle inoperable, does not affect safe vehicle operation).	8	**Major disruption**	100% of product may have to be scrapped. Line shutdown or stop ship.
	Degradation of primary function (vehicle operable, but at reduced level of performance).	7	**Significant disruption**	A portion of the production run may have to be scrapped. Deviation from primary process including decreased line speed or added manpower.
Loss or degradation of secondary function	Loss of secondary function (vehicle operable, but comfort/convenience functions inoperable).	6	**Moderate disruption**	100% of production run may have to be reworked off line and accepted.
	Degradation of secondary function (vehicle operable, but comfort/convenience functions at reduced level of performance).	5		A portion of the production run may have to be reworked off line and accepted.
Annoyance	Appearance or audible noise, vehicle operable, item does not conform and noticed by most customers (>75%).	4	**Moderate disruption**	100% of production run may have to be reworked in station before it is processed.
	Appearance or audible noise, vehicle operable, item does not conform and noticed by many customers (50%).	3		A portion of the production run may have to be reworked in-station before it is processed.
	Appearance or audible noise, vehicle operable, item does not conform and noticed by discriminating customers (<25%)	2	**Minor disruption**	Slight inconvenience to process, operation, or operator.
No effect	No discernible effect.	1	**No effect**	No discernible effect.

Table 4.6 Process FMEA occurrence ranking table.

Likelihood of failure	Criteria: Occurrence of cause—PFMEA (incidents per items/vehicles)	Rank
Very high	≥100 per thousand ≥1 in 10	10
High	50 per thousand/1 in 20	9
	20 per thousand/1 in 50	8
	10 per thousand/1 in 100	7
Moderate	2 per thousand/1 in 500	6
	0.5 per thousand/1 in 2,000	5
	0.1 per thousand/1 in 10,000	4
Low	0.01 per thousand/1 in 100,000	3
	≤0.001 per thousand/1 in 1,000,000	2
Very low	Failure is eliminated through preventive control.	1

manufacturing process, it is undesirable to have a CC or an SC at a process step, especially if the satisfactory accomplishment of the CC/SC is dependent on a human operator to successfully complete it. See the section on "Human Error and Mistake Proofing" for additional information.

Potential causes of failure: There may be multiple causes of a single potential failure mode. These different causes need to be entered on separate lines to allow the relative ranking of probability of occurrence. Separating the causes allows the specific preventive process controls to be assessed, and when recommended actions are necessary, they can be specific to the listed cause, as illustrated in Figure 4.20.

Process prevention controls: This question is specific to the PFMEA, asking, "In the manufacturing process, what preventive measures are being taken?" In the case of a CC, or SC, the actions taken to guarantee prevention need to be mistake-proofed using poka-yoke methods to ensure that human error is minimized as a potential contributing factor.

Occurrence ranking: This is the estimated and/or projected probability that the failure cause and its related consequences will occur. Analysis of testing data, historical data, and surrogate data can be used to project an estimate of the probability of the cause. If the process is new and no data is available, then the "low, moderate, high" scale in the occurrence ranking table can be used. Consideration of the prevention controls needs to be part of the analysis. If there have been many entries in the prevention column, then this can have a favorable influence on the occurrence ranking entry. If the entry in the controls prevention column is "none," then it is likely that the team will assume the worst case and enter a potentially higher occurrence ranking. In the AIAG (2008) FMEA table presented earlier in Figure 4.9, the occurrence entry follows the controls prevention entry, ensuring that any information entered as preventative is reviewed prior to the occurrence ranking consideration.

Table 4.7 Process FMEA detection scale.

Opportunity for detection	Criteria: Likelihood of detection by process control	Rank	Likelihood of detection
No detection opportunity	No current process control; Cannot detect or is not analyzed.	10	**Almost impossible**
Not likely to detect at any stage	Failure mode and/or error (cause) is not easily detected (e.g., random audits).	9	**Very remote**
Problem detection post processing	Failure mode detection post-processing by operator through visual/tactile/audible means.	8	**Remote**
Problem detection at source	Failure mode detection in-station by operator through visual/tactile/audible means or post-processing through use of attribute gauging (go/no-go, manual torque check/clicker wrench, etc.).	7	**Very low**
Problem detection post processing	Failure mode detection post-processing by operator through use of variable gauging or in-station by operator through use of attribute gauging (go/no-go, manual torque check/clicker wrench, etc).	6	**Low**
Problem detection at source	Failure mode or error (cause) detection in-station by operator through use of variable gauging or by automated controls in-station that will detect discrepant part and notify operator (light, buzzer, etc.). Gauging performed on setup and first-piece check (for set-up causes only).	5	**Moderate**
Problem detection post processing	Failure mode detection post-processing by automated controls that will detect discrepant part and lock part to prevent further processing.	4	**Moderately high**
Problem detection at source	Failure mode detection in-station by automated controls that will detect discrepant part and automatically lock part in-station to prevent further processing.	3	**High**
Error detection and/or problem prevention	Error (cause) detection in-station by automated controls that will detect error and prevent discrepant part from being made.	2	**Very high**
Detection not applicable; error prevention	Error (cause) prevention as a result of fixture design, machine design, or part design. Discrepant parts cannot be made because item has been error-proofed by process/product design.	1	**Almost certain**

Process detection controls: This question is specific to the PFMEA, asking, "At this particular step in the process, what failure detection measures can be applied?" The question is specific to the process step under scrutiny. It is unacceptable to propose or accept that a failure at one processing step will be detected at a later process step. Detection of a failure mode at a later step may mean that the assembly is not reworkable and has to be discarded. Detection by a human (observe, measure, perceive, or reason), especially when CC/SC process steps are being established, will not be

Process step	Requirements	Potential Failure Mode	Potential Effect(s) of Failure	Severity	Classification	Potential Cause(s) of Failure	Process Controls: Controls Prevention	Occurence	Process Controls: Controls Detection	Detection	RPN	Recom-mended Action	Response
Select bell housing from kanban	R1: Read barcode to ensure proper part selected	Failure to read barcode, wrong part selected	line stoppage, rework	8		Process step skipped	Operator diligence	5	None	10	400	Modify process station controls to force bar code reading	
				8		Bar code reader failed	Calibration and maintenance. Operator work instructions to check calibration sticker.	1	Calibration and maitenance schedule. Operator work instructions to check calibration sticker.	2	16	None	

Figure 4.20 PFMEA for transmission bell housing.

acceptable for a detection method. Human error failures are covered in the later section titled "Human Error and Mistake Proofing."

Detection ranking: This is the inherent ability to be able to detect potential failures. The entries in the detection column, as witnessed in the detection ranking table, are listed for their effectiveness in surfacing any potential issues related to the manufacturing and assembly process. The detection ranking is improved by adopting improved detection methods.

Risk priority number (RPN) ranking: The product of severity, occurrence, and detection yields the RPN. There are numerous arguments regarding the use of the RPN in making decisions with respect to recommended actions. There are some advocates that recommend setting an RPN action threshold, others propose using the Pareto analysis (80/20 rule), and others advocate using the numbers at face value (for example, if $S=6$, $O=5$, $D=9$, then the RPN would be 659) rather than the product $S\times O\times D$. The adopted method of using the RPN needs to be an agreeable method and acceptable to the FMEA team. The decision rule needs to be stated in the notes of the FMEA so that anyone reading it understands how the actions were triggered.

Recommended actions: The recommended action(s) should not be limited due to schedule, cost, or resources. If there are reasons that the recommended actions cannot be fulfilled, then the reasons for this would be stated in the actions taken column. In the example of Figure 4.20, the recommended action is to "modify process station controls to force bar code reading." If there is an exception to this recommendation that eliminates it as a viable option, then an alternative recommended action needs to be proposed. The team will determine if the alternate actions are acceptable. If the PFMEA is related to a new design (DFMEA), then recommended actions can propose basic design changes to eliminate potential failures in manufacturing and assembly. Changing a design can reduce a severity ranking or even eliminate a failure mode completely. This reinforces the need to make sure that the PFMEA is following the DFMEA closely enough to allow for a design change.

Responsible: It is highly recommended that the person assigned to the responsibility for an action generated in an FMEA be a member of the FMEA team. The responsible team member is not necessarily the subject matter expert or the process engineer but is the person who will identify the necessary resources, initiate the change request, and follow up with the PFMEA team to report progress. If an action is assigned to someone who is not on the team, there may be delays and inconclusive results due to misaligned organizational reporting structure, overloaded resources, and competing priorities. Such issues have the potential to negatively impact the overall project.

Target date/completion date: It is necessary to ensure that timely results are shared with the project team, and setting milestones to accomplish these actions is essential to the success of the project. If there are significant delays in the established target completion dates as evidenced by the extended actual completion dates, the project manager and/or the project assurance manager may need to take additional actions to address the issues.

Actions taken: Not all recommended actions are achievable. Given the many variables involved, such as cost, time, technology limits, and materials availability, the actual actions taken may differ from the recommended actions and will need to be scrutinized by the project team for efficacy. If the project team determines that the "taken actions" are insufficient, then further actions or alternatives will need to be negotiated.

Updated risk priority number (RPN) ranking: The severity can only be changed as the result of a process step change or a design change. Any actions that reduce the occurrence (probability of failure) need to show clear evidence of what was done to reduce it and the rationale used to justify the lowered occurrence ranking. The detection ranking changes need to be supported by evidence that clearly shows the actions justifying the change. There is a caution that improving the detection can lower the RPN yet not change the severity or the occurrence of the failure mode. Ranking the product of severity and occurrence independent of detection is one way to ensure that a low detection ranking does not reduce an RPN to a level below the action threshold.

FMEA Preparation

Regardless of which type of FMEA is being used, there are some cardinal rules to be honored, as well as a collection of "effective practices" to ensure the intent of the FMEA is being met. One of the cardinal rules is that FMEAs are *team events.* All the required associates representing key elements of the new design and/or process need to be in the same meeting at the same time. Attempts to complete an FMEA by passing the template around to the various design contributors does not meet the intent of an FMEA.

Another cardinal rule is to ensure that a nonbiased and independent individual facilitates the FMEA. It would be a conflict of interest to have a designer of the system or any of the major contributors of the FMEA take on the role of the facilitator. These contributors need to be participants in the team analysis and cannot effectively perform both the roles of facilitator and participant at the same time. In preparing for an FMEA, the facilitator needs to perform numerous tasks to ensure the efficiency and effectiveness of the FMEA process. The facilitator owns the process and is responsible for the journey of the team. In a new product development scenario, the FMEA facilitator would be involved with the project leader to make sure that the appropriate schedule estimates are loaded into the project.

One of the biggest mistakes that can be made with respect to performing an FMEA for a new product or process is to incorrectly estimate the time needed to complete the FMEA or to neglect adding the time estimate to the project timeline altogether. The time estimates depend on the facilitator's competence and experience with FMEAs. An experienced FMEA facilitator will base their estimates on past performance. A novice FMEA facilitator will need to include additional time as a precaution.

It is advised that the FMEA facilitator, to the best of their ability, adhere to the following proposed tasks. As noted, FMEAs are team events, and these suggestions are made in order to make sure that impact on team member schedules is minimized and that their participation in the FMEA is as productive as possible.

- Prepare the FMEA Excel template (pre-populate information where practical).
- Collect necessary documentation, for example, drawings, bill of materials, and design requirements.
- Catalog all assumptions, for example, all supplier parts are received without defect.
- Identify needed participants, for example, design, manufacturing, warranty, and supplier.
- Acquire team participant commitments to the FMEA schedule.
- Determine the naming convention and document control management (revision control).
- Establish the meeting cadence and reserve the meeting room (recurring calendar meeting).
- Keep an action item list for offline and unanticipated activity (for example, document acquisition).

The facilitator should designate time before the FMEA meetings to prepare for the facilitation. The facilitator should review the template, review the action items from the last meeting, and arrive early to ensure the meeting room is available and appropriately arranged. This allows the facilitator to start the meeting on time and begin making progress on the FMEA without delays such as rearranging furniture, clearing out any lingering participants from the last meeting, cleaning up leftover materials, removing any distractions, and making sure the AV equipment is functioning.

It is also advisable that the facilitator acquaint the participants with the FMEA process, the objectives of the FMEA, the ranking scales to be used, and the expectations of the participants. This could be a simple "orientation for FMEA teams" presentation at the first meeting. The participants may come from organizations that have performed FMEAs in the past using a radically different approach. There may also be some participants that have never seen an FMEA before and are unfamiliar with the terminology. An orientation meeting will establish the rules for the process and reduce the potential for interruptions during the facilitation such as, "That's not the way we did it at my last company."

It is important to know that FMEAs are legal documents and, in the event of litigation, are searchable and admissible evidentiary documents. As such, the FMEA documents need to be under revision control. It is too often the case that there is a legal action and the most recent FMEA is not readily identifiable or cannot be found at all. It is advisable that a repository be created for storing and retrieving FMEAs to ensure that they are properly stored and under change management. ISO 9001 is the international standard that specifies requirements for a quality management system (QMS). ISO 9001—Document Control (4.2.3), is summarized here.

The mandatory document control procedure must define:

- How documents are approved, such as procedures, flow charts and process maps prior to use. For example, is there a signed-off paper copy, or are they added to the computer network via a password protected system?
- How amended documents are updated and reapproved.
- How changes are identified such as date or issue number, or by using different fonts or colors.
- How documents are made available where they are needed.
- How documents of external origin are controlled.
- How the inadvertent use of obsolete documents is prevented. Note that obsolete-but-still-in use is the single most common noncompliance.

The standard tells us that documents such as specifications or procedures are considered as information and their supporting medium (for example, paper or electronic) must be controlled and readily available. The standard implies that over time, these documents will evolve as new information supersedes old and that change must be managed. Documents (FMEAs) are active and dynamic by nature and should be reviewed and updated when necessary.

FMECA

Failure mode, effects, and criticality analysis (FMECA) bases the prioritization scheme of addressing failure modes on severity of the failure modes and probability of occurrence. These two variables can be used to construct a two-dimensional graph displaying each failure mode as a point. Severity can be plotted on the horizontal axis and occurrence on the vertical axis. More severe failure modes will plot farther to the right, and more probable ones farther up from the horizontal axis. If a set of failure modes is grouped in the upper right corner, those should be addressed with the highest priority.

EXAMPLE 4.4

A team has identified 10 failure modes denoted 1 through 10 for a product and has determined the following severity and probability of occurrence for each. These ratings are plotted in Figure 4.21. Failure modes 8, 5, and 3 appear to be the place to start.

Failure mode	S	O
①	5	5
②	2	1
③	9	5
④	1	3
⑤	9	7
⑥	4	1
⑦	6	5
⑧	8	8
⑨	1	2
⑩	1	1

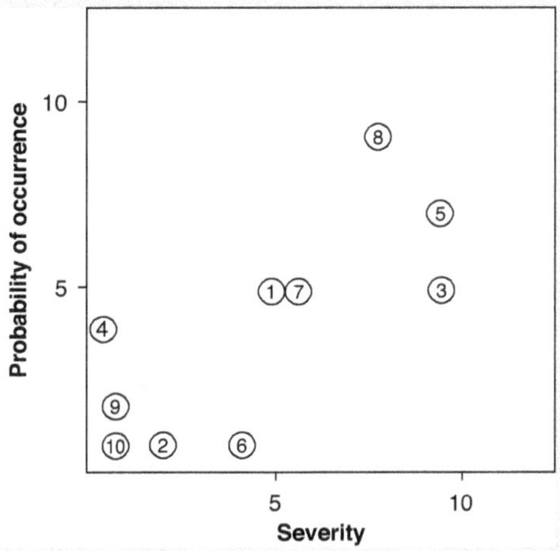

Figure 4.21 Severity and probability of occurrence rankings for 10 failure modes.

Other tools for assessing criticality include flowcharts and block diagrams. These graphical tools display the relationships and dependencies of various subsystems and components. They are especially useful for complex systems. Users of FMECA have the option of including additional concerns such as safety, downtime, preventive maintenance, and stocks of spare parts.

FMEA and FMECA in Design

Reliability improvement will occur when design changes are incorporated into the system to eliminate a failure, reduce the probability the failure will occur, or reduce the effect the failure has on the operation of the system. A most important factor for the success of the technique is that the FMEA/FMECA is performed in time to economically change the design using the results of the FMEA/FMECA effort. The FMEA/FMECA is meant to be a before-the-fact action, not an after-the-fact exercise.

The FMEA/FMECA is changed from a qualitative analysis to a quantitative analysis by assigning values to the probability of failure occurring, to the severity of the effect of the failure on the operation of the system, and to the probability that the system controls will detect and eliminate the failure before the design is complete. The FMEA team should represent a broad cross-section of technical and non-technical expertise. The team should include, but not be limited to, representatives from product design, product service, manufacturing engineering, quality engineering, reliability engineering, purchasing (to represent suppliers), and marketing (to represent the customer), as well as experts in materials, thermodynamics, dynamics, and other specializations. This breadth and depth of experience for the design FMEA team is critical.

Addressing critical failure modes at the earliest possible stage of design is most cost-effective, as seen in Example 4.5.

> **EXAMPLE 4.5**
>
> A group is charged with producing a design for an operator cab for an existing agricultural tractor. The FMECA team identifies ultraviolet (UV) deterioration of the seal where the steering column intersects the cab wall as one of the critical failure modes. Possible preventive actions include:
>
> - Protecting the seal from UV exposure
> - Using UV-resistant seal material
> - Designing the steering linkage so the seal isn't needed
>
> It is likely that none of these solutions would be available once the design has been finalized.

COMMON MODE FAILURE ANALYSIS

> Describe common mode failure (also known as common cause failure) and how it affects risk. (Understand)
>
> **Body of Knowledge II.B.3**

To avoid confusion, it is necessary to distinguish common cause failure from common cause variation as applied to control charts. A *common cause failure* (CCF) occurs when two or more components fail due to a shared cause.

Recognizing and mitigating common cause failure is a vital part of designing for reliability, since eliminating or mitigating the causes of failure is a primary aim of reliability efforts. When performing risk analysis (for example, FTA or FMEA), consider whether there are common causes that could result in multiple faults, hazards, or failures, and place special emphasis on eliminating these common causes, since their elimination will often produce the greatest improvement in reliability for the effort.

Fault tree analysis is especially susceptible to missing the total impact of a common cause, since FTA is often performed with a very narrow focus. For example, if the fault being studied is a sudden rise in engine compartment temperature, an FTA that only considers the impact of this event on the braking system would be too narrow. Other functions could be impaired by the common cause, including the engine control computer, electronic controls, wiring, and power steering system. So, a single cause might generate multiple faults and failures.

Another caution in considering common mode failures is with redundant systems, whereby one cause can lead to the failure of otherwise redundant elements, leading to system failure. For example, a sudden rise in engine compartment temperature could cause the failure of redundant circuits and components in the engine control computer. Or a bent or damaged conduit could cause the failure of redundant wiring that is routed through the same conduit.

HAZARD ANALYSIS

> Explain how hazard analysis informs the development process and how the information obtained from it is used by reliability engineers. (Understand)
>
> **Body of Knowledge II.B.4**

A *hazard* is a condition, event, or circumstance that could lead to or contribute to an unplanned or undesirable event. The risk of the hazard is often classified as high risk, moderate risk, or low risk.

Hazard analysis is a method to assess risk based on the severity of the risk and the likelihood of the occurrence of the risk. Hazard analysis is often used early in development to begin to assess risk. This includes products and processes.

Three fundamental questions are asked in a hazard analysis:

- What might go wrong? The risk.
- What is the probability that it will go wrong? The likelihood.
- What are the consequences? The severity.

The hazard is then the combination of the likelihood and severity of the risk. Figure 4.22 shows a typical hazard analysis risk matrix, with risk scoring and color coding.

Figure 4.24 shows an example of a functional hazards analysis (FHA) used to develop and refine the functional requirements of a new system concept for an electric wheel braking system (brake-by-wire), shown in Figure 4.23.

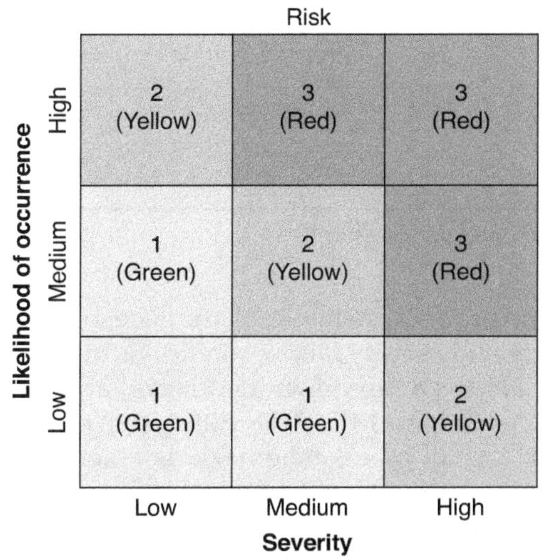

Figure 4.22 Hazards analysis risk matrix.

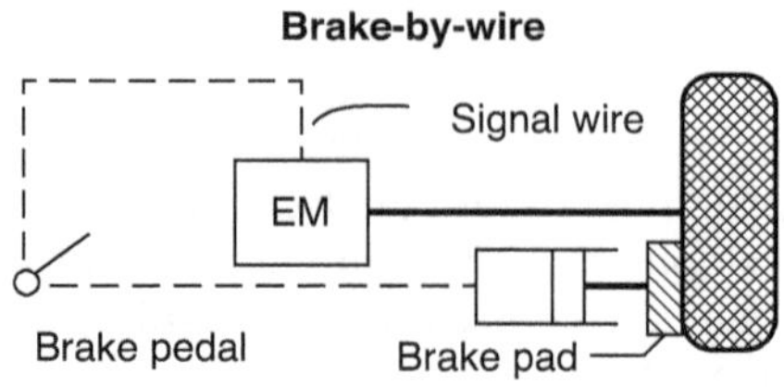

Figure 4.23 High level functional diagram.

Aircraft FHA						
Function	Failure Condition	Flight Phase	Effects of Failure Condition	Classification	likelihood of occurrence	Verification
Deceelerate aricraft on ground	Loss of adequate deceleration for Flight Phase	Rejected take off (RTO)	Crew is unable to stop aircraft on runway	Catastrophic	moderate	Fault Tree Analysis (FTA)
	Loss of adequate deceleration for Flight Phase	Landing	Loss of deceleration function without warning - crew is unable to stop aircraft on runway	Catastrophic	moderate	Fault Tree Analysis (FTA)

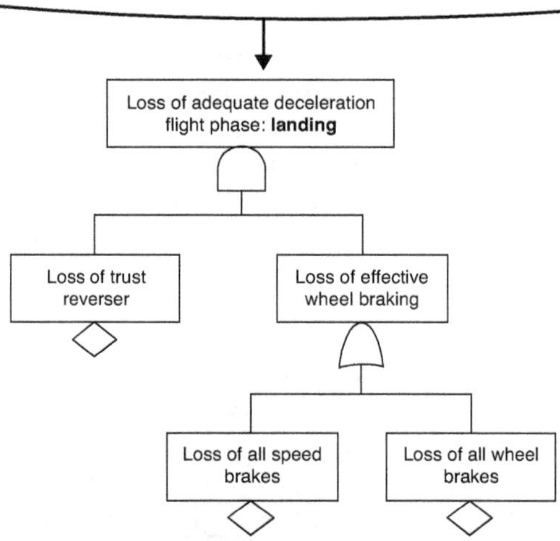

Figure 4.24 Example of FHA relationship to FTA.

An FHA is defined as a systematic, comprehensive examination of the system functions to identify and classify failure conditions of those functions according to their severity. The FHA is a top-down (deductive) approach examining functions to identify failure conditions. MIL-STD-882E task 208 states that the facilitator in performing the FHA shall assess "the hazards associated with loss of function, degraded function or malfunction, or functioning out of time or out of sequence for the subsystems, components, and interfaces. The list of hazards should consider the next effect in a possible mishap sequence and the final mishap outcome." The following example shows the relationship of the FHA to other tools used to further develop the causes of failures so that actions to manage the risk can be established.

	Importance	Risk types																						
		Technical			Scheduling				Safety		Financial			Regulatory				Supplier			Materials			Etc.
		A	B	C	A	B	C	D	A	B	A	B	C	A	B	C	D	A	B	C	A	B	C	
Product 1																								
Product 2																								
Product 3																								
Service 1																								
Service 2																								

Figure 4.25 Example of a risk assessment matrix.

Note that hazard analysis is like FMEA except that hazard analysis typically does not include detection in the analysis. Hazard analysis may also be used as an input to the FMEA.

Several hazard analysis models exist for various organizations and industries; therefore, using a hazard analysis, its format, and its scoring for a particular effort should be determined beforehand. A notable application of hazard analysis is the Food and Drug Administration (FDA) 21 CFR Part 120 Hazard Analysis and Critical Control Point (HACCP) System.

RISK MATRIX

> Explain the application of risk matrices in evaluating risk, focusing on how they aid in assessing the likelihood and the severity of potential adverse events or outcomes. (Understand)
>
> **Body of Knowledge II.B.5**

The risk matrix approach is common in reliability engineering and other efforts that involve the identification, classification, and prioritization of risk. Several reliability-specific techniques such as hazard analysis, FMEA, and FMECA are customized techniques in the use of a risk matrix.

A common risk matrix approach is to list potential risks and evaluate their severity and their likelihood to then determine the risk. "Risk Matrix" provides more detail on this approach. Another common risk matrix approach is to create a matrix listing products and/or services in a vertical column, with an estimate of the importance of each in another vertical column. Across the top of the matrix a list of risk types can be listed. A matrix such as that shown in Figure 4.25 typically provides for multiplying an estimated importance score of an adverse event by an estimate of the probability that the event will occur.

Reliability data can improve the risk assessment process by providing more accurate values for the probability of failure at given points in the lifecycle. Changes in reliability obviously impact the resulting risk assessment scores. The risk matrix helps to prioritize risk vulnerabilities.

Many risk matrix models exist for various organizations and industries; therefore, using a risk matrix, its format, and its scoring for a particular effort should be determined beforehand.

SYSTEM SAFETY

> Identify safety-related issues by analyzing multiple sources, including customer feedback, design data, field data, and other pertinent information. Prioritize safety concerns, focusing on those that possess the highest risk of occurrence and impact. Determine steps to minimize the improper or unintended use of equipment, products, or processes, aligning with safety system. Employ tradeoff analysis to determine the optimal risk control measures. (Evaluate)
>
> **Body of Knowledge II.B.6**

A typical system safety program has eight elements, as shown in Figure 4.26.

A system safety program aims to fulfill three key functions:

1. *Identification of safety hazards.* The reliability engineer, working with the safety organization to achieve a safe product, must be innovative and diligent in the discovery of all possible ways that any failure, combination of failures, or other combination of circumstances would present a safety hazard to personnel. Hazard analysis can be used to achieve this

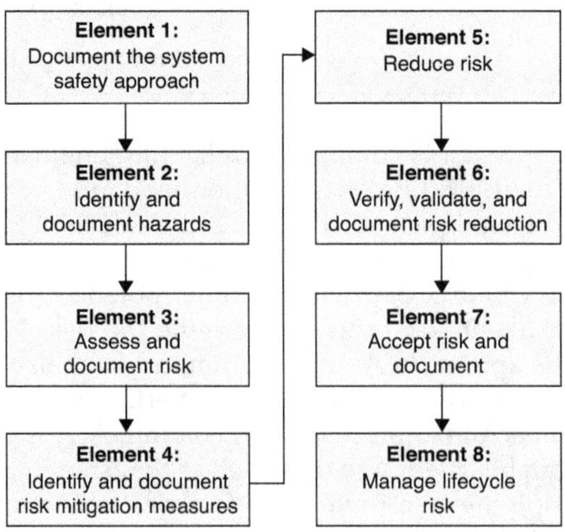

Figure 4.26 Typical system safety program.

objective. Warranty data and other forms of customer feedback should also be analyzed. Product testing reports should include safety issues that may have emerged. All available product data should be searched for the occurrence of safety hazards. In addition, if databases for similar products are available, these should also be studied.

2. *Risk analysis*. The standard analysis techniques such as risk matrix, FTA, hazard analysis, and FMEA/FMECA can be employed in risk analysis for system safety. These techniques can be used to estimate the risk associated with various events. It then becomes possible to establish a prioritized list that will provide guidance as the root causes are attacked and resolved.

3. *Corrective and preventive action* should be employed for any risk analysis technique in general. In the case of safety hazards, there is the additional urgency to avoid harm to personnel. In cases in which flaws in products or processes permit hazardous conditions to occur, design and process changes should be initiated.

As Low as Reasonably Practicable (ALARP)

As low as reasonably practicable (ALARP) is a principle used in system safety whereby risk is reduced as far as reasonably practicable. For a risk to be determined to be ALARP, it must be demonstrated that the cost involved in reducing the risk further would be grossly disproportionate to the benefit gained. More information on ALARP is found in Chapter 5.

Human Error and Mistake-Proofing

It is always necessary to consider human error. From one perspective, human error occurs because no system is in place to prevent it. In other words, the onus is on the product/process design community to reduce or eliminate errors. Procedures for preventing or mitigating human error are variously called mistake-proofing, poka-yoke, and zero quality control (ZQC). Human error falls into the following categories: misunderstanding, misidentification, inexperience, inattention, and lack of standards. The ideal way to deal with human mistakes is to incorporate design elements that will prevent them from occurring or prevent their occurrence from causing defects.

Often, the best way to prevent a reoccurrence of human error is to ask the person who made the error, "How can we design a system that will make that error impossible to occur?"

Principal types of mistake-proofing techniques include:

- *Physical barriers to errors*. The round shaft won't fit through the square hole.
- *Visual reminders*. A photograph of correct and incorrect results is better than a note written on a print or a paragraph of text.
- *Use of automated equipment*. The conveyor will stop if the microswitch detects an error.
- *Standardizing*. The operator takes the same action on a family of parts.

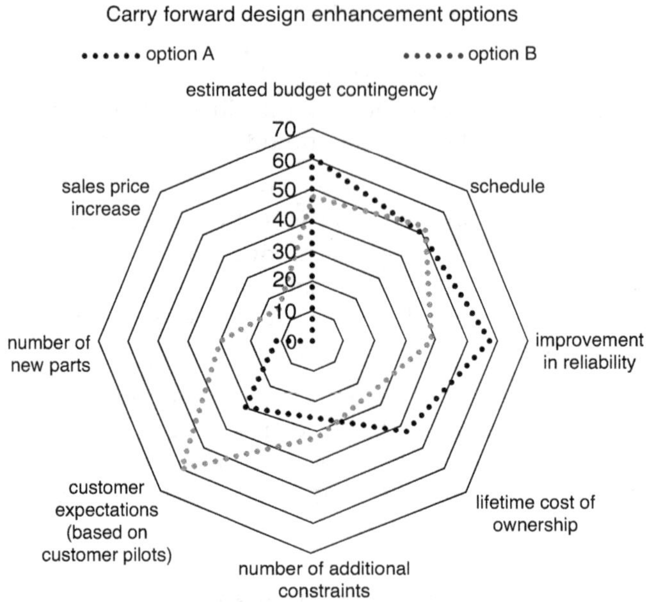

Figure 4.27 Radar graph displaying improvement option differences.

		A	B	C	D	E
	criteria	cost	schedule	reliability	parts count	etc.
A	cost					
B	schedule	B				
C	reliability	C	B			
D	parts count	A	B	C		
E	etc.	A	B	C	D	

COUNT	A	B	C	D	E
	2	4	3	1	0

PRIORITY	
B(4)	schedule
C(3)	reliability
A(2)	cost
D(1)	parts count
E(0)	etc.

Figure 4.28 Simple comparative criterion prioritization.

Tradeoff Analysis

Tradeoff analysis is a method used at the systems engineering level that provides guidance on the selection of product functions, features, and properties using a comparative analysis method. There are many ways to accomplish a tradeoff analysis. One approach would be to use the previously discussed Pugh down-select method, as shown in Figure 4.1. Multiple criteria are identified by which the various proposed solutions are compared. This approach also allows for the adoption and integration of particularly attractive functions, features, and properties that are inherent in one design option but are missing from another.

A radar graph, shown in Figure 4.27, can also be instrumental in comparing various aspects of a project, process, or product. Although the radar graph does not specify how the differences are achieved, it does show where the greatest opportunities lie.

Tradeoff analysis requires that decisions be made regarding what criteria are used, the priority (weighting) of each criterion, and the process by which decisions are made in the analysis. To avoid possible contention and disagreement regarding the criterion and how it is applied, the decision can be supported by a prioritization matrix. The various elements are compared to each other in an exhaustive manner by the team. Consensus on the prioritization is necessary to ensure that all team members agree.

In Figure 4.28, the elements are compared, resulting in an agreeable prioritization. This table should be included in the project documentation so that it is clear who established the criteria, how it was analyzed, and what the final priority list looks like. Activity must be continuous in the elements of the system safety program. As the higher-priority hazards are corrected/prevented, action can focus on lower-priority items.

Several system safety approaches exist for various organizations and industries; therefore, the approach used for system safety for a particular effort should be determined beforehand.

Chapter 5
Risk Mitigation

> Identify appropriate risk mitigation plans by integrating the guidelines and principles outlined in standards such as ISO 31000 and ISO 55000. Assess risk control plans that incorporate effective controls, aiming to minimize both inherent risks and their subsequent impacts, focusing on safety, liability, and regulatory compliance. Distinguish between the reliability goals of As Low As Reasonably Practical (ALARP), As Low As Reasonably Achievable (ALARA), and As Low As Possible (ALAP). (Evaluate)
>
> **Body of Knowledge II.C**

Reliability analysis provides estimates of the probability of failure. The reliability engineer must go beyond simple calculations and examine the consequences of failure, particularly safety. These consequences typically represent costs to the customer. The customer finds ways of sharing these costs with the producer through the warranty system, loss of business, decrease in reputation, or the civil litigation system. Therefore, an important reliability function is the anticipation of possible failures and the establishment of reliability acceptance goals that will limit their occurrence and consequent costs.

The acceptance criteria or specifications associated with these goals generally fall into three categories:

1. Functional requirements (such as flow rate, material of construction, physical dimensions)
2. Environmental requirements (such as temperature, radiation, pH)
3. Time requirements (such as failure rate during useful life, time elapsed before the wear-out phase begins)

Once component, product, and system reliability goals have been set, a testing protocol should be implemented to provide verification that these goals will influence the failure rates and the associated consequences as planned. These reliability goals typically impose specifications on the product. In anticipation of the start of production, reliability engineers use further testing procedures to provide *verification* that these specifications are being met.

As well as verification, the product should undergo *validation*, or objective evidence that the intended use or function has been fulfilled by the product. This difference is commonly stated as, "Verification confirms we are building the product right; validation confirms we are building the right product."

For physical products, there is an additional step in *process validation*. This shows by repeated testing that the specific process can consistently produce products that can fulfill their requirements. Reliability goal setting is key to validating a process, because the determination of whether a process passes or fails is determined by its ability to meet or exceed the reliability requirements. Reliability metrics should also be used over the lifetime of the process to confirm that the process is still meeting its goals over time.

STRATEGIES TO MINIMIZE RISK

Once reliability goals have been set and the risks identified, the reliability engineer must consider the risk–benefit analysis and determine the best strategies to minimize the associated risks to an acceptable level. The four major categories of risk mitigation, known as the 4T's, are:

1. *Tolerate the risk.* For low priority risks, the organization accepts the risk, allowing it to direct resources to higher priority issues.
2. *Terminate the risk.* This action completely removes the risk by redesigning the product, changing suppliers or mistake-proofing the system, for example.
3. *Treat the risk.* Actions are taken to reduce the risk based on feasibility and cost.
4. *Transfer the risk.* Here the organization shifts the risk to a different entity, for example, by outsourcing.

Applications of these risk mitigation categories are illustrated in Example 5.1.

EXAMPLE 5.1

A manufacturer produces a part in-house that is part of the final product assembly. It finds that the latest batches of parts are consistently 1.0 mm out of tolerance. This manufacturer may be able to:

1. *Tolerate the risk.* Verification testing shows that the final product assembly functions within specification with a part that is up to 1.0 mm wider than the current specification. The current specification is amended to allow for an additional 1.0 mm in width.

2: *Terminate the risk.* The final product assembly is redesigned to tolerate a larger part or to eliminate the need for that part.

3. *Treat the risk.* The part production process is investigated to determine the root cause of the out-of-tolerance parts, and the system is remediated to produce in-tolerance parts.
4. *Transfer the risk.* The manufacturer outsources the production of the part to another facility.

RISK TREATMENT

Once the overall level of risk has been confirmed via testing or other means, the risks must be evaluated for overall acceptability. If risks are found to be unacceptable, these risks should be addressed through an iterative process to implement options to reduce the overall risk. Per ISO 31000, risk treatment involves:

- formulating and selecting risk treatment options
- planning and implementing risk treatment
- assessing the effectiveness of that treatment
- deciding whether the remaining risk is acceptable
- if not acceptable, taking further treatment

The selection of risk treatment methods will depend not only on the risk level itself but on the type of industry, the effect on all stakeholders, and the overall tolerance for risks. Risk treatment should be conducted on design, usability, and manufacturing risks to provide comprehensive risk control.

REQUIREMENTS TO REDUCE RISK: ALARP, ALAP, AND ALARA

The degree to which a risk needs to be reduced to be at an acceptable level heavily depends upon the context of the risk and its severity.

The commonly used approach for risk reduction is as low as reasonably practicable (ALARP). This approach balances the need to reduce risks with the cost and practicality of undertaking risk reduction measures. ALARP is typically used in situations where the overall risk and impact on safety are low, and it ensures risks are controlled to a level where the cost of further risk reduction would be disproportionate to the benefit gained. This point where the benefit of mitigating risk is outweighed by the resources required is subjective, and rationale should justify the selected risk mitigation activities.

Regulations, standards, and other relevant guidance must also be considered when undertaking risk reduction. Additionally, per ISO 31000, the justification of the risk mitigation options should not consider solely external requirements and economic considerations, but "should take into account all of the organization's obligations, voluntary commitments and stakeholder views" (p. 13). Other considerations that may factor into risk mitigation decisions include customer satisfaction, company reputation, and environmental impact.

The more conservative approach, where risks must be reduced as far as possible, regardless of cost or feasibility, is referred to as ALAP, as low as possible, or AFAP, as far as possible. Reducing a risk as low as possible is the standard for situations with a high likelihood of occurrence and/or catastrophic consequences of failure.

Industries with a potential of high risks generally use ALAP for risk mitigation, which may be required by regulations or standards. For example, the ISO standards for medical device and aerospace risk management both require risks to be reduced ALAP.

Other industries may use different philosophies for risk reduction. For example, ALARA, as low as reasonably achievable, covers the radiation safety industry and aims to minimize a person's exposure to radiation that does not directly benefit them. Dental x-rays and cancer treatment benefit a person while exposure to radiation from a faulty gamma irradiator does not and so must be minimized. According to the US Centers for Disease Control (CDC), this reduction is covered under three factors: time, shielding, and distance. Increasing shielding and distance and reducing time spent in the vicinity of a radiation source help those who work around radiation reduce risks as low as reasonably achievable.

These risk reduction approaches are illustrated in Example 5.2.

EXAMPLE 5.2

1. *ALARP.* A contractor was aware of the possibility of falls on the worksite that may lead to injury. Ladders are often used to perform work at height but are unsafe if used for a long period or for tasks that require both hands. The contractor decided to limit the use of ladders to simple tasks under 30 minutes and provide a scaffold with fall protection for demanding tasks.

2: *ALAP.* An electrocautery tool was meant to be energized only in the closed position for sealing blood vessels during surgery. The information for use stated that the device was not to be operated while the jaws were open. During testing, users disregarded the instructions and used the open jaw as a cautery pen. The design team for this tool added software to allow for safe activation while the jaws were open, tested the new configuration, and removed the caution in the instructions for use.

3. *ALARA.* A cancer treatment center installed a new particle accelerator to better treat its patients. The accelerator needed to be positioned so it exposes the operator to minimal radiation while still being able to treat the patient. A series of hallways leading to the treatment room were built of shielding material that reduces the level of radiation, the control chamber was built on the opposite side of the hallway system from the accelerator, and a series of interlocks were put into place that only allows accelerator activation from the control chamber.

All these approaches require a process to determine how best to reduce risk. Manuele (2003) provides one example, a "safety decision hierarchy" that is readily applied to ensure safety, liability, and regulatory compliance. Manuele suggests using the following activities in the order presented:

1. Eliminate hazards and risks through system design and redesign.
2. Reduce risks by substituting less hazardous methods or materials.
3. Incorporate safety devices (fixed guards, interlocks).
4. Provide warning systems (lights, buzzers, labels).
5. Apply administrative controls (work methods, training, etc.).
6. Provide personal protective equipment.

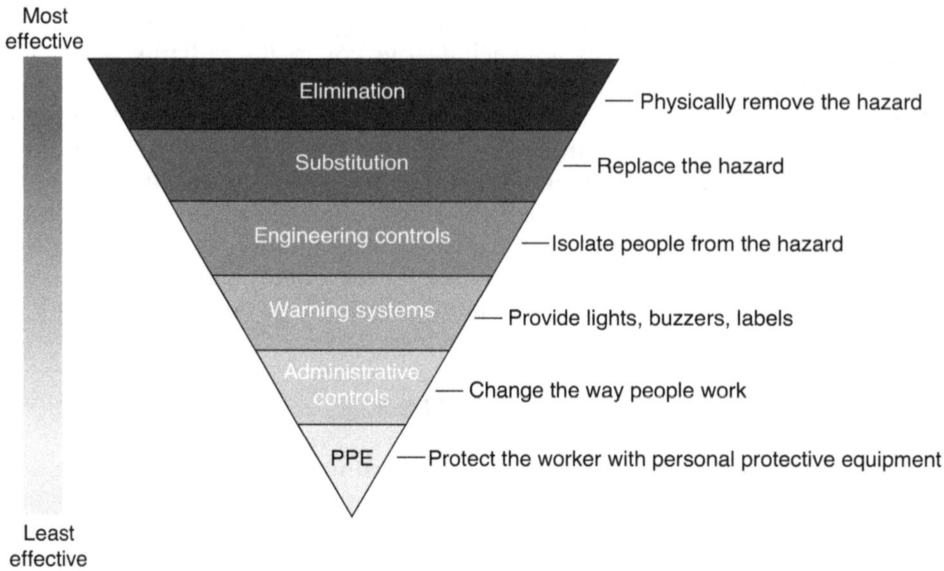

Figure 5.1 Hierarchy of controls.

Source: Adapted from https://wwwcdc.gov/niosh/topics/noise/images/HOC.png.

A graphic representation of the risk reduction activities suggested by the US National Institute for Occupational Safety and Health (NIOSH) is presented in Figure 5.1.

SECONDARY RISK

When risk mitigations are put into place, there is a possibility that these mitigations will create new risks that could cause failure in the product or process. These secondary, or introduced, risks must be evaluated and mitigated themselves until all risks are at an acceptable level. These risks can be reviewed using the same mechanisms used to identify and control the original risks.

If using a failure mode and effects analysis (FMEA), any failure modes of the new risk control should be listed and their impact assessed. Adding complexity via risk controls may impact parts of the system that had not been previously considered, so the secondary risk assessment should not be limited to the area where the control was implemented. Using a fault tree analysis (FTA) can help determine how the proposed control could impact other parts of the system.

Example 5.3 illustrates this point.

EXAMPLE 5.3

A backup generator is installed to keep equipment running in safe mode during a power failure. When power fails, the critical systems that require uninterrupted power need to be switched over to the generator power. If there is a fault in the mechanical switch or the monitoring software, then the equipment will not switch over to the generator power and critical function is lost during the power outage.

There should also be checks in place to make sure users will be able to continue to use the system without issue. This includes determining availability, how likely the control is to cause human error and whether the user will be able to bypass the control.

For example, putting in a new patch to improve cybersecurity might make the software incompatible with older hardware or older versions of operating systems. Users who are using the updated software on these older systems will be forced to upgrade their systems or stop using the software altogether. If a significant number of users are on outdated systems, then additional mitigations will need to be put in place to make the software accessible.

Ultimately, all risks—primary, secondary, and so on—must be reduced to a level that is deemed acceptable by the organization before risk mitigation is considered complete.

RESIDUAL RISK

What Is Residual Risk?

Residual risk is the risk that remains after all risk control measures have been implemented and verified to function as intended. In any risk management process, the goal is to reduce risks to an acceptable level by applying various controls, such as safety features, design changes, protective measures, or user warning and instructions. However, no product or process is completely risk-free, and after mitigation efforts are completed, some risks inevitably remain. These are the residual risks, which must be evaluated to ensure they do not exceed the thresholds considered acceptable by the organization. Residual risk can exist in many forms, such as a small chance of equipment failure, a minimal risk of human error, or a low-probability occurrence with high severity.

Residual risk differs from the initial, or inherent, risk (the risk before any controls are applied) and represents the state after the risk has been reduced to the lowest level possible. Even after applying risk controls, the remaining risks may still result in harm or negative outcomes, though they should be much less likely or less severe than before any mitigation was put in place. In industries like healthcare, aviation, and automotive, the practice of determining the amount of residual risk is crucial to ensuring safety while maintaining functionality and performance.

Determining the Acceptability of Residual Risk

To determine the acceptability of residual risk, organizations must compare it against predefined criteria, which are often informed by industry standards, regulations, and company policies. The acceptability of residual risk is a decision-making process based on factors like the probability of occurrence, the severity of potential harm, the environment in which the product or process operates, and the types of users who will be exposed to the residual risks. There are multiple factors that should be considered when deeming residual risk acceptable, including:

- *Risk acceptance criteria.* Many industries use formal guidelines or standards to define acceptable risk levels. For example, the medical device industry uses ISO 14971 to establish risk thresholds that must be met to ensure that residual risks are acceptable. The risk must be ALAP, meaning that further reduction efforts are either impossible or are outside of state of the art for that particular device. Other fields, like aerospace or pharmaceuticals, also follow stringent risk acceptance guidelines based on the specific context of use and the potential for harm.

- *Benefit–risk analysis.* A key factor in determining the acceptability of residual risk is a *benefit–risk analysis*, where the benefits of a product, system, or process are compared to the post-mitigation risks. If the benefits, such as improved patient health, increased efficiency, or better performance, clearly outweigh the residual risks, then those risks may be deemed acceptable. This is especially important in sectors like healthcare, where the inherent risk of using medical treatments or devices must be balanced against the potential benefits to patients. A device that is literally lifesaving will be able to tolerate a higher risk level than a device that is meant to improve the quality of life but not save it.
- *Stakeholder input and risk tolerance.* Acceptability of residual risk also depends on the risk tolerance of stakeholders, including regulatory bodies and their designees, customers, and users. Different organizations and industries have different thresholds for what constitutes acceptable risk, based on the nature of the product, the consequences of failure, and the potential for harm. For example, in consumer products, residual risk must be extremely low because users expect safety and may not have the knowledge to manage risk themselves. In contrast, highly skilled operators in industrial settings may be trained to handle a higher degree of residual risk.

Overall Residual Risk Assessment

Before finalizing a product or process and releasing it into the wild, the question must be answered, "Will this cause more problems than it solves?" Just because all the residual risks are considered acceptable on their own does not mean that the cumulative risk will not cause issues when taken as a whole. Multiple small risks that are acceptable on their own could combine into a more likely or more harmful risk. Also, if there are many low probability risks, it will be more likely that issues will occur, especially if the product or process is used over a long period of time.

Risk interaction can lead to major unforeseen issues, whether the risks are dependent on each other to occur, cause a chain of failures to initiate, or are able to occur at the same time. In any of these situations, one risk may cause or amplify the effect of another risk, leading to greater harm than expected. Tools such as bow-tie diagrams, shown in Figure 5.2, and FTA can help to determine if risks are predisposed to interact.

The final decision on risk acceptability is determined by a combination of regulations, industry standards, and the corporation's risk management strategy. Depending on the product or industry, a higher overall risk level or number of risks may be acceptable than a product or industry that is more tightly controlled. This threshold of acceptability should be determined and documented prior to finalizing risks.

If the overall residual risk is considered unacceptable, action needs to be taken. Individual risks can be reduced or decoupled, and system-wide mitigations can be put in place to stop simultaneous failures or failure cascades.

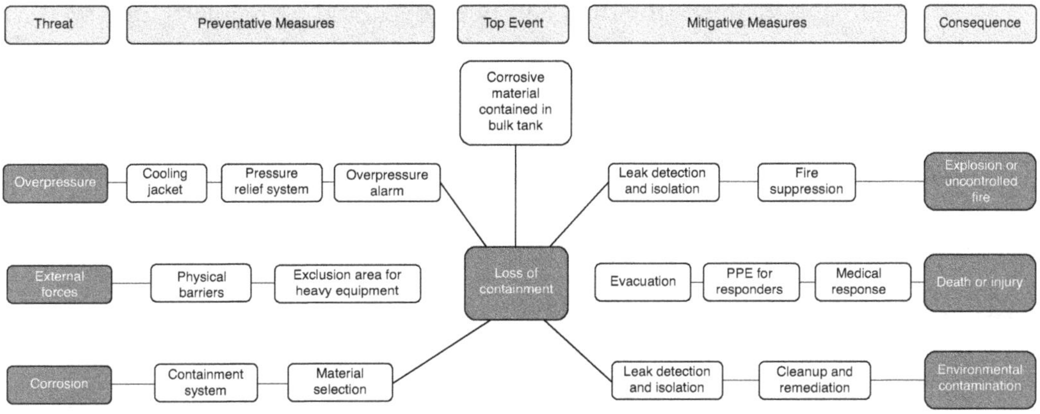

Figure 5.2 Bow-tie diagram illustrating preventative and mitigative measures for loss of containment in a pressure vessel holding volatile liquids.

Example 5.4 details a failed overall risk analysis of a medical device.

EXAMPLE 5.4

The **Therac-25** was a radiation therapy machine developed in the 1980s. The device could deliver two types of radiation depending on the type of treatment:

1. Low-energy X-rays (electron beam)
2. High-energy X-rays (photon beam)

Operators used software to switch between high and low settings. The machine's hardware included mechanisms to position a "target" in the path of the beam to attenuate its power when it was set to deliver a photon beam. However, due to critical flaws in its hardware and software, the machine was able to deliver lethal doses of radiation in certain conditions.

Previous models of Therac had physical hardware interlocks in place that made it physically impossible to deliver a photon beam unless the target was in the correct location to mitigate the energy from the beam. In the Therac-25 model, the physical interlock was replaced by a software interlock.

The software also contained a bug called a "race condition" where the operator could enter commands faster than the software could check it. This could lead to a situation where the software would position the hardware into the low energy configuration when it was actually set to deliver high energy. Additionally, the error codes were confusing, and dismissing them did not cause any change in machine functionality, so they were routinely dismissed by the operators.

Taken as individual risks, the lack of physical interlocks, race condition, and poorly written error codes were all low. However, when the operator entered commands too quickly and did not understand that the system was in a fault state, all of them interacted to fire a lethal beam of energy that ultimately resulted in the death of six patients.

It was later determined that those testing the software did not activate the race condition during testing because they were not able to enter data into the system as fast as the operators and were unaware of the effect these risks could have when occurring simultaneously.

Residual risk, both individual and overall, should be regularly reviewed as part of ongoing risk management efforts. Conditions can change over time, such as new technology availability, updated standards, or feedback and complaints from real-world use. This can change both the inherent level of the risk and also change the perception of the level of risk. If there are better alternatives available, what was once considered an acceptable level of residual risk may need to be reevaluated in the context of the current state of the industry.

RISK REVIEW AND REPORTING

Prior to release, the risk management process should be evaluated to ensure that all elements of the risk plan have been executed, the overall risk is acceptable, and an appropriate system is in place to gather, analyze, and act upon risk-related data generated after commercial release. If all of these have been completed and deemed acceptable, the decision to commercialize and proof of success execution of the risk management process should be documented in a risk management report and approved by the relevant stakeholders. The risk management report is a living document and should be updated over the lifetime of the product or process if information becomes available that impacts the acceptability of the risks.

ONGOING RISK CONTROL

Risk control is not just a checkbox. Risks must be controlled throughout the life of the product. As circumstances change, new risks may be introduced, controls may be rendered less effective, or user feedback may indicate a higher instance of issues than anticipated. Regardless of the reason, continuing risk control is needed to maintain a good product.

Key standards for risk management, including ISO 31000, ISO/IEC 27001, ISO 14971, and ISO 550001, all emphasize that ongoing monitoring and review of risks is crucial to ensuring that risk controls remain effective over the lifetime of the product or process. Individual systems will have different frequencies and thresholds for monitoring and action, depending on the inherent risk of the device, the industry, and the organization's tolerance for risk. This risk system should include key risk indicators (KRIs) to monitor high-priority risks, how and when data is gathered, how often risks are assessed, and what actions should be taken after a threshold is reached.

The KRIs to be monitored should be related to the safety of the user, stakeholder requirements, and other high-risk items documented in the risk report.

The system should dictate what types of data to monitor and how often reviews of that data take place. For comprehensive results, data should be acquired passively (e.g., complaints and internal nonconformances) and actively (e.g., user surveys, competitor benchmarking). The review frequency will depend on the product and industry, but it is recommended to be completed at least annually.

Scheduled activities include reviewing risk assessments using the new data to determine if the original classifications are still valid or comparing KRI calculations to defined thresholds. How often this data should be analyzed should be specified in the risk monitoring system.

Some industries will have specific risk monitoring activities that need to be performed on a regular basis. For example, ISO/IEC 27001 recommends periodic testing of security controls and ISO 55001 suggests using asset performance data, predictive maintenance systems, or condition monitoring technologies to assess risk. ISO 14971 also requires an assessment of whether the state of the art has changed and, if so, what impact that has on the benefit–risk assessment.

The conclusions drawn from these assessments should be added to the existing risk management report. Updates should include any trends that have been observed, changes to documented risk thresholds, and any corrective actions that have been discovered as part of the assessment. As with data gathering and review, the results should be reported on a regular basis as determined by the risk monitoring system. If no updates to the KRIs or risk levels are required, the data and analysis should still be documented as part of the report, as it shows the requirements of periodic assessment set forth by the system have been completed.

The role or function of who is responsible for completing each of the tasks should be documented before the product or process is released. This ensures no break in the ongoing assessment of risk due to confusion over who should be doing the work.

Auditing the risk management system to ensure that all the inputs to the assessment are valid and effective should also be scheduled as part of the regular internal audit of the quality system. Updates should be described as part of the document charge and as part of the internal audit report.

RISK CONTROL PLANS

For organizations that have multiple products or operate across multiple industries, a single system may not be adequate to cover all the different risk monitoring needs. In this case, the organization can create a generic system covering how to set thresholds, gather data, and act upon nonconformities. Individualized plans will cover the specific parameters for their product or process.

The system will specify what needs to be done, specifically:

- Selecting which KRIs need to be monitored and measured over time.
- Determining the alert and threshold for the KRIs that will be monitored over time.
- Designing which methods are at minimum needed to gather data for assessment over time.
- Setting a minimum frequency for analysis of the data.
- Setting a minimum frequency for reporting the results of the data and where they should be reported.

The details of how these tasks will be accomplished can be documented in the system, only leaving the product-specific details in the plan, or the system can dictate only the steps that need to be taken, leaving most of the details in the product-specific plan. If the what, who, how, and when are part of either the system or the plan, the specifics can be adapted to meet the organization's needs.

At a minimum, the plan should include:

- A list of KRIs and their thresholds. If there is no rubric for determining KRIs and their thresholds in system procedures, that information is part of the plan.
- The frequency of data collection, analysis, and reporting, if different from the system requirements.
- What actions need to be taken if the KRI breaches a threshold or is trending in that direction.
- Which people or roles are responsible for all the necessary activities.

PRODUCT IMPROVEMENT

Some products with long lifecycles will undergo multiple product redesigns over the lifetime of the device to improve some aspect of the product. Product improvement often means redesigning components, and the integration of new components requires an orderly procedure if genuine improvement is to occur. If the changes impact the form or function of the device, additional risk mitigation and controls must be undertaken to ensure secondary risks will not cause an increased level of risk in the redesigned product.

Conversely, the relatively short product lifetimes in areas such as electronics, software, and some consumer goods have caused increased emphasis on obsolescence management and end-of-life consideration, in addition to design changes.

END-OF-LIFE CONSIDERATIONS

It is inevitable that components or entire systems will eventually become obsolete due to the unavailability of replacement parts or changes in technology. Components and systems should document end-of-life considerations during the design and development phase to ensure there is a documented plan to address these considerations.

OBSOLESCENCE FROM THE PRODUCER'S VIEWPOINT

When any component of a system is changed, the result is a new system, which must be tested to verify that it performs as specified. Many times, qualifying the component independently from the system will miss undesirable interactions with other components. Judgment is required, of course, to decide how much of the new system must be requalified. Redesign of a pivot pin in a vehicle door latch linkage does not require requalification of an entire vehicle, for example. It is necessary, however, that the new latch system be tested not only on a lab bench but also on vehicles in their designed environment.

Once the decision has been made to make the change, the disposition of obsolete parts must be managed. In some cases, the change occurs when the supply of obsolete parts has been exhausted. If not, the inventory of obsolete parts must be segregated to prevent its use in the production process.

The repair and maintenance of parts in the field is another consideration. Will the new design be installed, or are the obsolete parts to be used?

OBSOLESCENCE FROM THE CUSTOMER'S VIEWPOINT

What strategies can a customer use when an environment of continuous improvement means component changes? The warranty documents the responsibility that the supplier has to the customer. Some provisions that are useful include:

- Commitment to support the product with replacement parts for a stated period of time.
- Lifetime buy, which is the practice of providing the buyer with enough replacement parts to cover the needs for the lifetime of the system. This option requires a careful forecasting methodology.
- Backward compatibility, guaranteeing that new or redesigned components will function within the existing system.

Part III

Probability and Statistics for Reliability

Chapter 6
Probability and Statistics for Reliability, Basic Concepts

This chapter will cover basic statistical terms and fundamental concepts, such as hypothesis testing and the central limit theorem. Probability concepts and rules are addressed, as well as probability distributions, probability functions, sampling plans for statistics and reliability testing, statistical process control (SPC) and capability, and confidence and tolerance intervals.

The material in this chapter has been adapted from probability and statistics content in the third edition of this handbook, from McShane-Vaughn, *The Probability Handbook* (2015), and McShane-Vaughn, *The ASQ Certified Six Sigma Black Belt Handbook* (2022). These materials have been reworked to present the material with a reliability engineering focus. New material has been developed, and several new reliability-specific examples have been added.

Why Statistics?

Statistics is a broad field dedicated to understanding, organizing, and communicating information through the collection, analysis, interpretation, and presentation of data. Statistical methods extend beyond basics like summarization and descriptive methods. More advanced methods include processes for data collection, test planning for design of experiments (DOE), and inferential analysis methods for estimation and prediction. Statistical tools enable us to facilitate decision-making by quantifying the uncertainty in a given scenario. Focus areas include:

- Data collection and preparation, including sampling and experiment or test study design
- Summarization, estimation, analysis, and prediction/forecasting, including model-building
- Interpretation, visualization, and effective communication of results

Recall the steps of the **scientific method**: (1) observe, (2) research, (3) hypothesize, (4) design and perform experiment, (5) test hypotheses, (6) draw conclusions, and (7) report results. Statistical tools and techniques provide structured and quantitative approaches in every step of this process and are integral to establishing cause and effect. In industry, statistical applications make use of a robust set of tools designed for monitoring and managing product and process quality, as well as advanced analytical modeling required for reliability engineering.

"Basic Statistics" in this chapter provides an overview of basic univariate statistical methods, akin to material in a first semester undergraduate statistics course.

Note that statistical tools and techniques are specific to data type and the statistical question under consideration. Subject matter expertise is always important to understand context. In addition, this section provides an overview of basic bivariate and multivariate modeling methods.

It may be helpful to first review probability foundations, in "Basic Probability Concepts," which also covers probability trees. "Probability Distributions" provides a detailed reference to different probability distributions, both discrete and continuous. "Probability Functions" walks through various probability function types, their meaning, interpretation, and how they are used, from probability distribution function to cumulative hazard rate function. "Sampling Plans for Statistics and Reliability Testing" discusses data collection, specifically sample size calculation, in a reliability context where standard sample size methods may not be valid and shows how these calculations are linked to hypothesis testing and interval estimation. More in-depth discussion of design of experiments is included in "Design of Experiments (DOE)," and reliability test planning in "Testing." "Statistical Process Control (SPC) and Process Capability Studies (C_p, C_{pk}, P_p, and P_{pk})," in this chapter, covers basic control chart methods and the calculation, appropriate usage, and interpretation of various process capability and performance measures. "Confidence and Tolerance Intervals" discusses the various forms of interval estimation and its connection to hypothesis testing.

BASIC STATISTICS

> Define basic statistical terms such as population, parameter, sample, and statistic. Understand statistical concepts such as sampling and the central limit theorem. Differentiate between parametric and non-parametric (e.g., Kaplan-Meier Analysis) statistical methods. Use appropriate methods to estimate and interpret statistical values. (Analyze)
>
> **Body of Knowledge III.A.1**

Basic Concepts and Definitions

Statistical studies are done with the intent of making some **inference** about a **population** or **universe** about which there is limited information. *Inference* means to draw conclusions; statistical inference is drawing conclusions with a statistical foundation.

The *population* is defined as the collection of all individuals, items, or data under consideration, which is often impractical to fully evaluate and can rarely be fully characterized. A reliability engineer might define the population for a study as the products currently in use in the field. Alternatively, they might consider the population to be the parts produced by production lines A and B last week. Researchers study a subset of the *population* from which information can be collected, called the **sample**, from which numeric measures called **statistics** are calculated, which are then used to make **estimates** of various population **parameters**. A *statistic* is therefore some **measure** or quantity that describes a sample.

Data *collection* practices include data gathering and preparation for analysis. Statistical studies are typically done to address a specific question or problem, usually to make a decision. They are classified by the nature of the data collected: **random** versus **nonrandom**, **prospective** versus **retrospective**, **observational** versus **experimental**.

For example, if 1500 citizens are randomly selected from the United States and their heights are measured, the population would be all US citizens, and the sample, in this case a random sample. A *random* sample is chosen by selecting units for a sample such that all combinations of units under consideration have an equal or ascertainable chance of being selected. Note that **sampling** should be done in a way to minimize potential sources of **bias**. It might be more convenient to sample the local basketball team, for example, but this would not provide a representative sample. **Randomization** is used as insurance to avoid inadvertently oversampling nonrepresentative groups. *Nonrandom* sampling methods should be used with caution since they can result in incorrect conclusions.

A *prospective* study is one where data is collected for the purpose of that study. In this scenario, those implementing the study control the process. A *retrospective* study is one where historical or legacy data are used. Due to availability and convenience, retrospective data is often used to first identify a hypothesis; however, recommendations may be misleading and should be confirmed and validated in a prospective or experimental study. Similarly, an *observational* study is based on data where the conditions are only able to be watched, or observed, instead of specific interventions being applied to assess the impact. Observational studies may be either prospective or retrospective. Conclusions from observational studies should likewise be confirmed and validated with additional experimental testing.

The highest quality data are considered those provided by an **experiment**, in which factor levels are deliberately manipulated to identify causality. The experimental test process is an integral part of the **scientific method** and is the foundation of repeatable research. However, not all test plans are equal. A family of statistically designed experimental test practices known as the **Design of Experiments** (**DOE**), discussed in Chapter 11, provides a rigorous and structured approach to obtaining experimental data that can help maximize the information gained from the testing, given robust test practices.

Statistical studies are interested in determining the characteristics of the population under consideration. When it is not feasible to measure a characteristic of each item in a defined population, a statistical study will randomly select a sample from the population, measure each item in the sample, and analyze the resulting data. The analysis of the sample data produces statistics. Examples of sample statistics include the sample mean, the sample median, and the sample standard deviation. These sample statistics are then used to estimate the corresponding population parameters and may be used for other purposes of estimation or prediction. For example, if the mean of those 1500 heights is 64.29 inches, the conclusion is that the sample mean is 64.29 inches. The value 64.29 is called a statistic, which is defined as a descriptive measure of a sample. The next step is to infer the mean height of the population, which is likely to be around 64.29. The actual population mean is called a parameter, which is defined as a descriptive measure of a population. So, it can also be said that a statistic is an estimated value of a parameter.

Different symbols are used to differentiate between parameters and statistics. Parameters are usually designated by Greek letters, and statistics are usually denoted by Latin letters. For instance, the Greek letter mu (μ) is used for the population mean and the Latin letter X with a bar above it ($\overline{X}$), pronounced "x-bar," is used for the sample mean.

The aim of **inferential statistics** is to make statements about a target **population** of interest whose parameters are unknown, and unknowable, to us. To do this, we take **samples** from that population and analyze them to make judgments, or **inferences**, about the population with a certain level of **confidence**.

Inferential statistics has two key functions: estimating parameters and testing hypotheses about parameters. These functions are handled using methods of **estimation**, specifically through the development of **confidence intervals** and **hypothesis testing**, at a designated level of **confidence**, which represents a level of assurance desired for the analysis.

A **hypothesis test** addresses a specific question about a parameter. **Confidence intervals** can provide a range of plausible values for the parameter of interest. The analyses are inherently paired and represent different formulations of the same problem. For the same statistical question, confidence intervals and hypothesis tests conducted on the same parameter and at the same level of confidence will result in the same conclusion. Higher confidence means more certainty but also requires a more stringent threshold: Therefore, the analysis will yield a correspondingly wider interval. Both approaches may be used on a stand-alone basis or integrated into a larger analysis. These two techniques provide the underlying framework for many statistical methods. Methods for determining sample size also use the same framework.

Statistical methods may be classified as **parametric** or **nonparametric**. Parametric methods leverage the use of models, and so make assumptions about model parameters, whereas nonparametric methods minimize assumptions and consider only the empirical data.

Types of Data

Statistical methods are designed for use with specific types of **data**. It is the responsibility of the user to ensure appropriate tools are chosen, given the context of the problem. If a statistical method is used with the wrong type of data, mathematical calculations can be generated, but the resulting model or analysis can provide suboptimal or meaningless information. A quick summary is provided in Figure 6.1 for reference.

Quantitative, or **numeric** data, may be either **discrete** or measured on a **continuous** scale. Discreteness means that the values may only take on certain values. This is also known as **attribute** data, of which there are several forms. The simplest and least informative is **binary**, taking on only two values, such as 0 or 1, and often called pass/fail. **Classification** data include two or more levels, which may be ordered (**ordinal**) or not ordered, but with names (**nominal** or **categorical**). Integer **count** data is also considered to be *attribute* data. Data measured on a *continuous* scale is also known as **variable** data, not to be confused with factors or variable names.

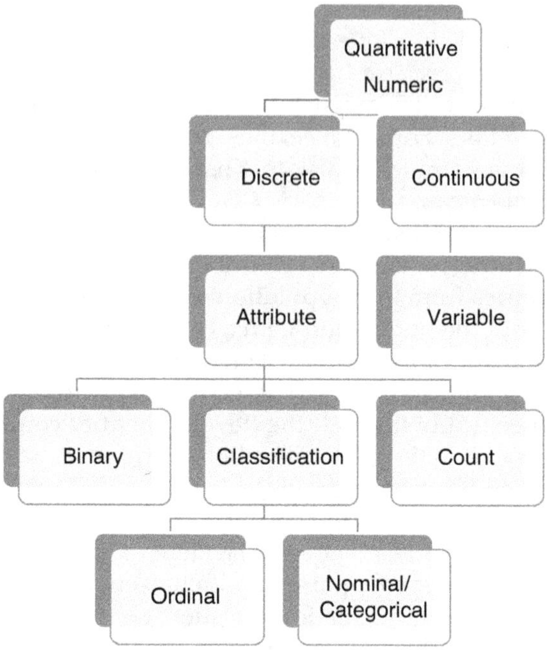

Figure 6.1 Data types flowchart.

Reliability lifecycle data may be simplified to pass/fall (binary), but this comes with a loss of information. Variable or count data may be subject to **censoring** or suspensions, where data exact magnitudes are not known precisely. Not all censored data are censored in the same way. Data may be **right-censored**, **left-censored**, or **interval-censored**. Right- and left-censored data are data where the true time (or other measure) is one-sided, known to be above or below a certain value, respectively. Interval-censored data are those known to be within an interval, and thus have two limits, an upper and lower bound for each data point. This is also known as arbitrary censoring, but censoring is not done in an arbitrary way, per the conventional meaning. All right and left censored data may be written in interval-censored format, so interval censoring is considered the more generalized form.

Other special considerations may include but are not limited to constraints, time-ordering or "time series," spatial information for locations along a surface, or repeated measurements on the same unit over time, known as longitudinal data. In many cases, the true output of interest is in fact itself a profile or function. Standard approaches here are to identify and analyze key features or "landmarks." For these cases, advanced methods are available to conduct appropriate analyses. These are outside the scope of this handbook but are mentioned here for awareness.

One final note; the physical mechanism of the failure may also point to the most appropriate type of data that should be analyzed. In general, there are two broad categories of structural failures, called *low cycle fatigue* (LCF) and *high cycle fatigue* (HCF). Physical hardware usually experiences multiple stressors that lead to degradation, which may be driven by either (or both) LCF and HCF. For example, consider a light bulb. The action of turning it on and off would be a more countable discrete LCF failure mode, whereas total usage time could drive more of an HCF style mode, measured in terms of time. Any item that is turned on and off, then is subject to some operational exposure time, would behave similarly. Modeling an ignition system for example, therefore, would require more discrete countable data on starts. For a system experiencing wear due to contact stress, it would be most relevant to look at operational time when the unit is experiencing contact.

Measurement

The user should also always have some understanding of how data are obtained and measured.

Measured values will differ from a true value, and the size of that difference will depend on the accuracy and precision of the measurement process, as illustrated in Figure 6.2. The accuracy of a measurement system refers to how close the measured values are to the true value, on average. A systematic difference between the measured and true values is called bias. A measurement that is unbiased is centered on the true value, or on target.

Precision reflects the variation in the measurements produced by a measurement process. Accuracy and precision are independent of each other, as shown in Figure 6.3. Data may be both accurate and precise, either accurate or precise, or neither.

Measurement systems analyses (MSA), more commonly known as **Gage R&R**, that is, Gage repeatability and reproducibility studies, are used to help

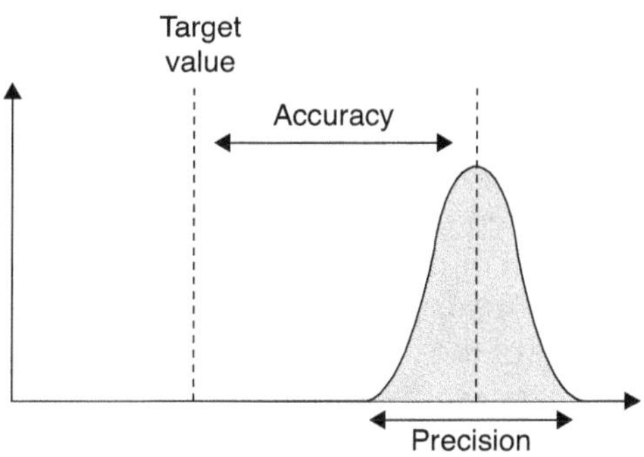

Figure 6.2 Control chart accuracy and precision.
Source: Durivage (2014).

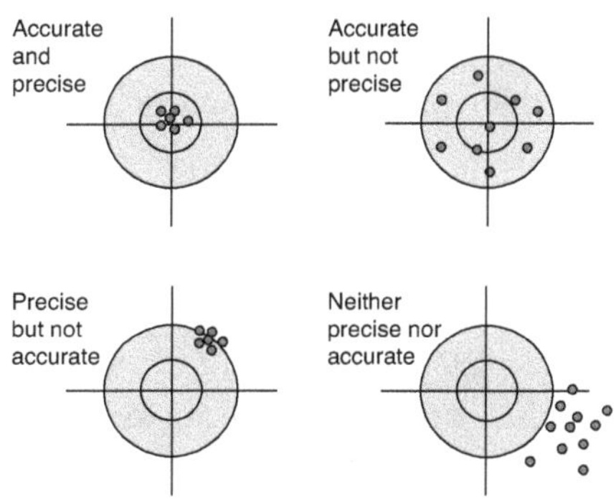

Figure 6.3 Accuracy versus precision.
Source: Durivage (2014).

understand and characterize the measurement process and are important for both inputs and outputs. Users of measurement tools and devices often rely purely on calibration processes, which tend to primarily focus on maintaining system accuracy. However, issues of precision are frequently underestimated. Gage R&R studies are designed to quantify how much of the variability in your data is attributable to the measurement system itself. Ideally, we want to be able to differentiate between **repeatability** and **reproducibility**.

Repeatability refers to the variability inherent to the system itself: What happens if you simply repeat the measurement? *Reproducibility* refers to variability associated with how the system is used, typically involving human factors. Excessive variability in data could be attributable to differences in set up or how the system is used, which may itself be due to differences in training and standard work.

Descriptive Statistics

Raw data itself provides little insight into underlying patterns and limits our ability to draw meaningful conclusions. This is where descriptive statistics become useful. **Descriptive statistics** are the collection of tools and techniques for displaying and summarizing data. There are many ways in which to display and summarize data. Therefore, it is important that one selects the best way to facilitate an understanding of the data. In some cases, multiple statistics and presentations or views are required to gain the most insight from the data.

The simplest statistical summary measures are well known and should be generally familiar. In technical statistics language, these are classified as **univariate point estimates**, since each calculation is performed for a single variable (*univariate*) and generates a single point value (known as a *point estimate or statistic*). Point estimates are calculated to provide an estimate of an unknown parameter of the underlying distribution of the data.

Point estimates can be **measures of central tendency**, whose purpose is to estimate the middle or center of the data, or **measures of dispersion**, which characterize the level of variability or scatter seen in the data. Data observations themselves are typically denoted with a variable identifier and subscript ($x_1, x_2, \ldots x_n$), or as a data vector **X**. Any letter may be used, but a common convention is to use Xs for explanatory data and Ys for response or dependent variables. Subscripts may indicate time or production order of hardware and therefore be associated with time, or data may be randomly ordered. A variable with data generated by a random process is referred to as a **random variable**.

The **mean**, **median**, and **mode** are all *measures of central tendency*. The *mean* of a sample is a more technical term for the arithmetic average, calculated for $x_1, x_2, \ldots x_n$ using Equation 6.1.

Equation 6.1 Sample mean.

$$\bar{X} = \frac{1}{n}\sum_{i=1}^{n} X_i$$

The *median* is the middle number of the sorted data. The calculation depends on whether the dataset contains an even or odd number of observations. For example, the median of a sorted set of 11 values is the sixth from the end. If there are an even number of values, the median is the mean of the two middle values.

The *mode* is the number in the data set that appears most frequently. Data that is **bimodal** or **multimodal** will have two or more peaks and is likely produced by a mixture of overlapping subpopulations.

These measures can be calculated for any type of numeric data, including *binary* (0,1), counts, or variable quantitative data. For binary data, averaging the X_is provides an estimate of the **proportion** $\hat{p} = k/n$, where k is the total number of occurrences out of a total of n instances.

In theoretical statistics, the mean is also referred to interchangeably as the **expectation** or **expected value**, denoted $E(X)$, and indicates the nominal or most likely value associated with that random variable. For example, $E(X) = E(\overline{X}) = \mu$. Also, $E(X, Y) = E(X) \cdot E(Y)$ if X and Y are two independent random variables. A distribution's expectation or expected value refers to its theoretical mean (a parameter), not an empirical mean calculated from a sample. Intuitively, it makes sense that sample means should cluster near the theoretical mean. The theoretical rule that governs this is known as the **central limit theorem** (CLT), described later in this section.

Measures of dispersion include the **range**, and the **variance** and **standard deviation**, which are functions of each other. The range is the simplest to calculate. It is the difference between the largest and smallest values, that is, the **maximum** minus the **minimum**:

$$R = X_{Max} - X_{Min}$$

The *population standard deviation* σ is the square root of the *population variance* σ^2. Estimates may be generally denoted $\hat{\sigma}$ and $\hat{\sigma}^2$, with the hat indicating an estimate. Estimates obtained from the *sample standard deviation*, denoted s, are the square root of the *sample variance*, s^2. There are several versions of the same calculation that differ only in the divisor. It is important to understand the distinction between them. First, the *population variance* is obtained by calculating each observation's deviation, or difference from the mean, squaring that deviation, then summing the squared values and dividing by the population size N. This assumes the entire population is available, which is not typically true when doing estimation. The purpose of the *sample variance* is to estimate the population variance.

There are two versions of the "variance" that you might encounter on a calculator or in software, one dividing by n and the other by $n-1$. *The version dividing by n is intended to be used for populations,* if available, and has a bias if used for a sample. The version with the $n-1$ in the divisor should be used for samples, as shown in Table 6.1

Table 6.1 Variance and standard deviation equations.

	Population Assumes finite population of size N with known mean μ	**Sample** Assumes sample of size n with estimate $\overline{x}$ for unknown mean μ
Variance (Var)	$\sigma^2 = \frac{1}{N}\sum_{i=1}^{N}(x_i - \mu)^2$	$s^2 = \frac{1}{n-1}\sum_{i=1}^{n}(x_i - \overline{x})^2$
Standard Deviation (Std Dev)	$\sigma = \sqrt{\frac{1}{N}\sum_{i=1}^{N}(x_i - \mu)^2}$	$s = \sqrt{\frac{1}{n-1}\sum_{i=1}^{n}(x_i - \overline{x})^2}$

A few notes: Variance calculations of this form are the foundation for many statistical methods, such as Analysis of Variance (ANOVA), which uses this structure to assess and quantify how much variation is being produced from different sources. Clearly the range is computationally much simpler, which has advantages, and is/was historically more easily implemented for manual calculations. It is also the foundation of methods used to produce estimates of short-term variation for small subgroup control charts. However, the standard deviation is a more robust estimator of variability. The range will also tend to increase as the sample size increases, unlike the standard deviation.

In practice, it is generally inadvisable for computations to be done by hand due to the risk of errors. Real-time SPC or other statistical software are widely available. Excel provides the formulas for samples, VAR.S and STDEV.S, as well as formulas for the population parameters, VAR.P and STDEV.P. For manual computations, calculators provide shortcut functions usually with options to differentiate the population from the sample version. Users should default to the sample version for conservatism since the true mean is hardly ever known in practice, and we rarely have access to all the data in a population.

The variance has its own operator $Var(X)$, which has its own rules. For example, if $Var(X) = \sigma^2$ then $Var(2X) = 4\sigma^2$. Other rules are provided here for reference.

1. $Var(cX) = c \cdot Var(X) = c^2 \cdot \sigma^2$, where c is any constant
2. $Var(X + Y) = Var(X) + Var(Y) = \sigma_X^2 + \sigma_Y^2$ for X and Y independent

When X and Y are not independent,

3. $Var(X + Y) = Var(X) + Var(Y) + 2 \cdot Cov(X, Y) = \sigma_X^2 + \sigma_Y^2 + 2 \cdot \rho_{XY} \cdot \sigma_X \cdot \sigma_Y$

where ρ_{XY} is the correlation (rho), and $Cov(X,Y)$ is the covariance between X and Y. Note that $Cov(X, Y) = E[(X - E(X)) \cdot (Y - E(Y))]$.

Expectation and variance operators can be used to obtain the distribution of new functions of random variables, for instance, for **linear combinations**. A linear combination is a function where elements are combined only by adding and subtracting with constant multipliers, such as the average.

EXAMPLE 6.1

Estimate Measures of Central Tendency: Mean, Median, Mode

Consider the following data for a sample of hole diameters taken from a drilling operation. For convenience, the data in the table have been sorted, but the observations were collected in a random fashion. Compute all three measures of central tendency.

From Table 6.2, we see that the sum of the diameters is 2.468. Using Equation 6.1, we have:

$$\text{Sample Mean} = \bar{X} = \frac{1}{n} \cdot \sum_{i=1}^{n} x_i = \frac{2.468}{20} = 0.1234$$

There are an even number of data points in the ordered set of 20. Thus, the median is the average of points 10 and 11, with the result that the median = 0.123.

We also see that the most frequently found value in the dataset is 0.123: Mode = 0.123.

Table 6.2 Data for Examples 6.1 and 6.2.

Number	Diameter	Number	Diameter
1	0.121	11	0.123
2	0.121	12	0.124
3	0.121	13	0.124
4	0.122	14	0.124
5	0.122	15	0.125
6	0.122	16	0.125
7	0.123	17	0.125
8	0.123	18	0.125
9	0.123	19	0.126
10	0.123	20	0.126
		TOTAL	2.468

EXAMPLE 6.2

Estimate Measures of Dispersion—Range, Variance, and Standard Deviation

For the hole diameter data in Example 6.1, calculate all measures of dispersion. We can use the formulas provided in Table 6.1 to solve.

The range is the simplest to calculate:

$$R = x_{Max} - x_{Min} = 0.126 - 0.121 = 0.005$$

The sample variance and standard deviation are related, so we will calculate the variance first, then take the square root to get the standard deviation. There is a computational version of the equation that helps simplify calculations.

$$s^2 = \frac{1}{n-1} \cdot \sum_{i=1}^{n} (x_i - \bar{X})^2 = \frac{1}{19} \cdot \sum_{i=1}^{27} (x_i - 0.1234)^2$$

$$= \frac{1}{n-1} \cdot \left(\sum_{i=1}^{n} x_i^2 - n \cdot \bar{X}^2 \right) = \frac{1}{19} \cdot (0.3046 - 20 \cdot 0.1234^2) = 0.00000257$$

$$s = \sqrt{s^2} = \sqrt{0.00000257} = 0.00160$$

EXAMPLE 6.3

Estimate Proportion

A total of $n=30$ sensors are tested to determine if they will detect a given force. Two of the sensors failed to detect the force, denoted as $r=2$. What is the estimate of the reliability ($\hat{R}$) of the sensor to detect the force?

$$\hat{p} = \frac{x}{n} = \frac{n-r}{n} = \frac{30-2}{30} = 0.933 = \hat{R}$$

EXAMPLE 6.4

Use Expectation and Variance Operators on a Linear Combination

We can calculate the mean and variance of new variables that are linear combinations of each other. If X_1, X_2, and X_3 are independent random variables with means μ_i and variances σ_i^2, we can find the mean and variance of any linear combination using expectation and variance operators.

We will find the mean and variance of a new variable y that is a linear combination of X_1, X_2, and X_3:

$$y = x_1 + 2x_2 + 3x_3$$

Take the expectation to find the mean, that is, the expected value of the linear combination:

$$\begin{aligned} E(Y) &= E(x_1 + 2x_2 + 3x_3) \\ &= E(x_1) + E(2x_2) + E(3x_3) \\ &= E(x_1) + 2E(x_2) + 3E(x_3) \\ &= \mu_1 + 2\mu_2 + 3\mu_3 \end{aligned}$$

To obtain the variance:

$$\begin{aligned} Var(Y) &= Var(x_1 + 2x_2 + 3x_3) \\ &= Var(x_1) + Var(2x_2) + Var(3x_3) \text{ since } x_i \textit{ are independent} \\ &= Var(x_1) + 4Var(x_2) + 9Var(x_3) \\ &= \sigma_1^2 + 4\sigma_2^2 + 9\sigma_3^2 \end{aligned}$$

And standard deviation:

$$StdDev(Y) = \sqrt{Var(Y)} = \sqrt{\sigma_1^2 + 4\sigma_2^2 + 9\sigma_3^2}$$

If the X_i random variables were also *iid, independently and identically distributed*, then $\mu_i = \mu$ and $\sigma_i^2 = \sigma^2$ for all random variables i, and

$$E(Y) = \mu_1 + 2\mu_2 + 3\mu_3 = \mu + 2\mu + 3\mu = 6\mu$$
$$Var(Y) = \sigma_1^2 + 4\sigma_2^2 + 9\sigma_3^2 = \sigma^2 + 4\sigma^2 + 9\sigma^2 = 14\sigma^2$$
$$StdDev(Y) = \sqrt{Var(Y)} = \sqrt{14\sigma^2} = \sigma\sqrt{14} = 3.74\sigma$$

Central Limit Theorem (CLT)

One of the most important and useful theorems in statistics is the central limit theorem, or CLT. The foundation of many statistical methods, the CLT is invoked when we perform hypothesis testing, linear regression, analysis of variance, design of experiments, and control charting. While there are several forms, a commonly used version is this:

If the random variable X has a mean μ and variance σ^2, and we take n independent samples from the distribution, then the sum of the Xs will approximately follow a normal distribution with mean equal to $n\mu$ and variance equal to $n\sigma^2$.

A few comments are in order. First, the reason why the CLT is so useful is that the underlying distribution of X can be *any* distribution that has a defined mean and variance. This relationship provides the underlying theoretical basis for hypothesis tests constructed around any sum or mean, including proportions. Given enough independent samples, the distribution of the sum will become approximately normal with a known mean and variance. Note that this applies equivalently to any mean, which is just a sum divided by a constant. Second, as n increases, this approximation will improve. For more skewed or asymmetric distributions, it takes more samples to look normal, but the distribution can be theoretically derived. If the underlying distribution of X is fairly symmetric, the approximation looks better faster as n increases. The CLT can also be seen in action as the mechanism underpinning control charting applications. Control charts generally operate under the assumption that the region bounded by the three sigma limits will contain 99.73% of the points, as would be the case if the data were generated by a normally distributed process. A frequent question that arises is the validity of $\bar{X}$ control charts when the population is not normally distributed. However, the CLT comes to the rescue in non-normal cases. Because the $\bar{X}$ chart plots averages, the normality assumption is (approximately) met. This approximation is of course better for larger subgroup sizes, which should be kept in mind when setting up a chart.

Parametric Versus Nonparametric Methods

Parametric statistical methods involve the use and application of mathematical models defined by a small number of terms known as **parameters**. The normal distribution, for example, is a parametric statistical model with two parameters, μ and σ. A simple linear regression would have three parameters: intercept and slope terms β_0 and β_1, and error variance σ^2. The proper application of these models would depend on the context of the data in a particular scenario and whether the assumptions for the model are met in that case. Most conventional statistical methods involve parametric models of some form. However, using parametric models is not always desirable or appropriate, particularly in cases where users want to rely on fewer assumptions or want to rely purely on the data themselves without the use of a model. Nonparametric statistical methods make very limited assumptions about the underlying data distribution and may be used to address situations where models could not or should not be used.

Parametric statistical methods are more powerful in that they provide more information about the population than nonparametric statistical methods. When the assumptions are correct, parametric methods will produce more accurate and precise estimates than nonparametric methods. However, nonparametric methods

Table 6.3 Parametric tests with nonparametric analogs.

Test for	Parametric test	Nonparametric alternative(s)
Location	Z-test or Student's t-test for one mean, or mean difference for paired data	Sign test for median Wilcoxon signed-ranks test for median
	Z-test or Student's t-test for two means	Wilcoxon rank-sum test (also known as the Mann-Whitney U test) for equal medians Kruskal-Wallis test for equal medians Mood's median test
	Analysis of variance (three or more means)	Kruskal-Wallis test for equal medians Mood's median test
Variation	Chi-square test for one standard deviation or variance	Bonett's test*
	F-test for two standard deviations or variances	Bonett's test* Levene's test*
	Bartlett's test for three or more standard deviations or variances	Levene's test*
Proportions	Z-test for one population proportion	Exact test (assumes binomial conditions hold)
	Z-test for two population proportions	Fisher's exact test (assumes binomial conditions hold)

*Note that Bonett and Levene tests are both performed using medians.

are generally considered more robust since they are applicable to any dataset. However, they will usually provide a more conservative result. Methods often involve sorting the data and using order statistics, which are the sequenced values. For example, the nonparametric version of correlation is performed on ranks instead of the raw data.

Standard parametric tests and their nonparametric analogues are listed in Table 6.3. Readers can refer to a statistics text such as Montgomery and Runger (2010) for more information on nonparametric testing. Those who want a more in-depth resource on nonparametrics may refer to Hollander et al. (2001). Note that basic methods listed here generally do not apply to censored data.

For reliability applications, one nonparametric method in general use is the **Kaplan-Meier** (K-M) **estimator**, used to estimate the reliability function of a population with censored information based on only the empirical data and making no assumptions about the underlying distribution. Some additional preliminaries are required, so this method is discussed in more detail in Chapter 6, "Probability Functions," which covers the mathematics and usage of statistical distributions in more depth. Note that while the K-M method computations are not inherently difficult, they require tabulation and counting that become increasingly more difficult to generate manually as the size of the dataset increases, especially with censored data. It is recommended that users obtain software with K-M capability instead of relying on either hand calculations or a spreadsheet.

HYPOTHESIS TESTING AND INTERVAL ESTIMATION METHODOLOGY OVERVIEW

The purpose of a **statistical hypothesis test** is to evaluate a claim about an unknown population parameter using the available data in a structured way that takes into account both dispersion in the data as well as the nominal estimate. The hypotheses are presented in pairs: a **null hypothesis** (H_0) and an **alternative hypothesis** (H_1 or H_A, or H_a). The term H_0 is often read as "H sub zero" or "H naught," and the alternative hypothesis is read as "H one" or "H sub one," "H A," or "H sub A."

The *null hypothesis* can be thought of as the baseline or status quo condition that is assumed to be true. The *alternative hypothesis* is the statement that the analyst is evaluating against the null. Hypotheses are written in terms of the unknown population parameters, such as this two-sided test shown here:

$$H_0: \theta=0$$
$$H_A: \theta\neq 0$$

The parameter *theta* here is generic and represents any population parameter. The value zero is used here for simplicity, but any relevant value could be used, denoted by Θ_0. This hypothesis test will be looking for evidence that the null hypothesis H_0 is not true by calculating an appropriate **test statistic** and evaluating it against a **reference distribution.** The comparison will check to see how unusual it is relative to the baseline. Values "near" the hypothesized value do not provide evidence against the null.

Two-sided hypothesis tests are used when there is no prior knowledge about what direction the difference might be, that is, in this case where the value could be either positive or negative. One-sided tests would be used if the direction of the change was important, so may be written in two ways:

Left-tailed	Right-tailed
$H_0: \theta\geq 0$	$H_0: \theta\leq 0$
$H_A: \theta<0$	$H_A: \theta>0$

Regardless of the form used, the hypotheses should represent all the possible values for the parameter being tested. Ideally, the alternative reflects the condition for which the consequences of making a wrong decision are most severe.

Hypothesis tests can be constructed for any population parameter, such as μ, σ, or ρ, and are widely used as the foundation for more complex statistical assessments. Hypothesis tests are also directly related to confidence intervals, which simply present the test outcome in a more generalized way. While there are a wide variety of different hypothesis tests, each with their own specific equations, the general process for performing hypothesis tests is the same, as shown in Figure 6.4.

When conducting a statistical hypothesis test, we take random representative samples from the population of interest and, using the information from the sample, we calculate a test statistic. We compare the value of the test statistic to a critical value in a corresponding distribution. Based on the value of the test statistic, we decide whether we have enough evidence to reject the null hypothesis.

Table 6.4 Four outcomes associated with statistical hypothesis tests.

		Decision based on statistical test	
		Do not reject H_0	**Reject H_0**
Truth	**H_0 is true**	**Correct Decision** *(True Negative)* **Confidence** $1-\alpha$	**Type I Error** *(False Positive)* **Producer's Risk** α
	H_A is true	**Type II Error** *(False Negative)* **Consumer's Risk** β	**Correct Decision** *(True Positive)* **Power** $1-\beta$

Alternatively, and more commonly, we can use the confidence interval method or *p*-value method to make the decision. Both methods are presented in detail later in this chapter.

Since the hypothesized parameters are unknown, we can never be certain that our decision is correct. However, we can be right in different ways, and we can also be wrong in different ways. The two ways that we can be correct in the context of a hypothesis test are described as a **true negative**, correctly selecting H_0 when H_0 is true, and a **true positive**, correctly selecting H_A when H_A is true. The probability of a *true negative* is known as the **confidence**, $1-\alpha$, which plays into numerous calculations that are presented later in this section. The probability of a *true positive* is also known as the **power**, $1-\beta$. It is often as described as the probability of detecting an effect if it truly exists.

For incorrect decisions, there is a possibility that our decision causes us to commit either a **Type I** or a **Type II error**, which correspond with the **producer's risk** and **consumer's risk** respectively, as shown in Table 6.4. These concepts can easily be extended to describe the ability of an automated inspection system to distinguish between conforming and nonconforming items.

A Type I error occurs when we reject the null hypothesis when it is true. More formally,

$$Pr(\textit{Type I error}) = Pr(\textit{Reject } \mathrm{H_0} \mid \mathrm{H_0} \textit{ true}) = \alpha$$

A Type I error is also known as a **false positive**, or an α error. One way to remember this connection is to note that α is the *first* (1) letter of the Greek alphabet. Another is to remember that an alpha error is an error of *A*ction, in which we incorrectly reject H_0 and leap over to the alternative hypothesis. The Pr(Type I error) is also known as the α-value, producer's risk, and the significance level.

A Type II error occurs when we do not reject the null hypothesis when the alternative H_a is true. More formally,

$$Pr(\textit{Type II error}) = Pr(\textit{Do not Reject } \mathrm{H_0} \mid \mathrm{H_A} \textit{ true}) = \beta.$$

A Type II error is also known as a **false negative**, or a β error. The term β is the second letter of the Greek alphabet, which is a way to remember that it is used to represent Type II error. The Pr(Type II error) is also known as a false negative, β-error, or consumer's risk. The Type II error is also referred to as the "do-nothing error" since we stay with the null hypothesis, or the status quo, when we should have rejected it.

As mentioned previously, the **power** of a statistical test is the probability of correctly rejecting H_0. More formally, it is the conditional probability of rejecting the null hypothesis, given that the alternative hypothesis is true,

$$Power = Pr(Reject\ H_0 \mid H_A\ true) = 1 - \beta$$

The power of a test depends upon the variance of the data, the confidence level $(1-\alpha)$, the size of the difference we are trying to detect E, and the sample size n. The higher the power of a hypothesis test, the more sensitive it is in detecting small differences in the values of parameters. For a given difference and alpha level, we can increase the power of a test by increasing the sample size. This concept will be explored further in the sampling plan section of this chapter.

And finally, the **confidence** level $(1-\alpha)$, the probability that we correctly select H_0, is written more formally as:

$$Confidence = Pr(Do\ not\ Reject\ H_0 \mid H_0\ true) = 1 - \alpha$$

The desired confidence level is prespecified by the user, conventional levels of which are 90%, 95%, or 99%, i.e., 0.90, 0.95, or 0.99, typically 95% or 1/20 risk. The higher the specified confidence, the more stringent the threshold. The consequence of making an incorrect decision should be higher for higher confidence levels. Another situation where higher confidence levels are advised is when many comparisons are being performed where each one has the same probability and the risk of making an incorrect decision compounds due to sheer volume. This is called the problem of **multiple comparisons**, or **multiplicity** issue. Multiple comparison tests that control the overall Type I error are shown in Chapter 11, "Design of Experiments."

The procedure for hypothesis testing is shown in Figure 6.4 and explained as follows:

- **Check Test Assumptions**

Each type of test makes different assumptions about the population. For results to be valid, these conditions should be reasonable.

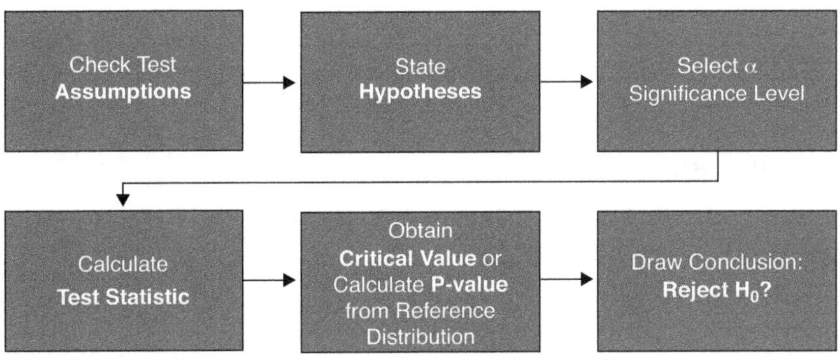

Figure 6.4 The hypothesis testing process.

- **State the Hypotheses**

Specify the statistical parameter being evaluated by the test, e.g., μ, σ, ρ, π, and the particular comparison value of interest. Is this a two-sided or one-sided test? What direction? Write out the hypotheses clearly. What is the consequence of a mistake?

- **Select the Level of Significance, Alpha (α)**

The alpha (α) value represents the significance level chosen for a given test that represents the probability that H_A was chosen in error when H_0 was true. Conventional values are considered to be 0.01, 0.05, or 0.10 (corresponding with 99%, 95%, and 90% confidence, respectively). An alpha value of 0.10 is not often used in practice due to the high probability of false positives when doing multiple tests. Note that this value should be specified *in advance* of conducting the test.

The confidence level of the test is $C = 1 - \alpha$ and is often referred to in terms of percent.

- **Calculate the Test Statistic**

Test statistics usually take one of several generic forms, such as, Student's *t*-test, for comparing a shift in location to an expected level of dispersion, calculated as:

$$t^* = \frac{\theta - \theta_0}{std\ error(\theta)} \sim t_{df = v = n - 1}$$

where Θ_0 represents the hypothesized value under the null hypothesis H_0.

The test statistic for an *F*-test for the ratio of two variances is calculated using:

$$F^* = \frac{\chi_1^2 / v_1}{\chi_2^2 / v_2} \sim F_{v_1, v_2}$$

The test statistic for the chi-squared test of association is calculated using:

$$X^{2*} = \sum_{i=1}^{k} \frac{(O_i - E_i)^2}{E_i} \sim \chi^2_{df = k - 1}$$

where O_i and E_i refer to the **Observed** and **Expected** counts, respectively, for one factor with *k* levels.

The calculated test statistic, often designated with a star or asterisk, e.g., t^*, F^*, or χ^{2*}, and read "*t*-star," "*F*-star," and so on, can be shown theoretically to follow a particular reference distribution, such as normal or *t*, or *F*, or chi-squared χ^2, as shown here, with some number of degrees of freedom, depending on the type and nature of the test.

- **Obtain the Critical Value (or *P*-Value) from the Reference Distribution**

The critical value is defined by the distribution, the degrees of freedom, and which tail, namely, upper or lower, is being referenced. The necessary distribution is associated with the type of test statistic, the degrees of freedom is a function of the sample size, and the choice of tail is dependent on whether the test is one-sided or two-sided. The appendices in this handbook include distribution tables for the binomial, Poisson, standard normal, Student's t, F, and chi-squared distributions.

- **Draw a Conclusion: Reject H_0?**

Three related approaches can be used to equivalently assess statistical significance.

1. *Critical value.* Provides a binary (yes/no) result. First, we identify the critical value from the appropriate reference distribution with given degrees of freedom and desired alpha level. The table value is then compared to the calculated value of the test statistic. If the absolute value of the test statistic is more extreme than the specified critical value, then H_0 is rejected at that given alpha level.
2. *P-value.* Provides a probability value between zero and one. P-values are obtained by using the appropriate reference distribution and degrees of freedom and calculating the probability of obtaining a test statistic equal to or more extreme than the one calculated, assuming the null hypothesis is true. These values are given by statistical software and are more precise than the critical value method. The p-value is then compared to the desired alpha to assess statistical significance.
3. *Confidence interval.* A generalized form that combines elements of the test statistic with the critical value from the reference distribution to obtain a nominal estimate ± limits calculated to reflect the uncertainty associated with the estimate. The confidence interval can then be checked to see if it contains the hypothesized value to assess statistical significance. Refer to "Confidence and Tolerance Intervals" for more information on this method.

Introductory statistical texts begin with the critical value approach. However, this method is not often used in practice, because it requires the alpha to be known in advance. Statistical software and other analysis packages provide p-values or confidence intervals so that the user can specify their own desired alpha level for assessing significance. Users should be able to recognize and interpret results provided in different formats.

It is critical to understand that the null hypothesis is never accepted or assumed to be true, only that there is insufficient evidence against it. One might "accept" the alternative hypothesis, but never the null. Similarly, however, even accepting the alternative hypothesis does not guarantee that it is true; it is accepted at a certain level of confidence. Hypothesis tests are constructed in such a way that the null hypothesis cannot be proven to be true.

A few comments on p-values:

P-values are defined as the probability of obtaining a test statistic of the value seen or something more extreme assuming that the null hypothesis is true, defined formally as the conditional probability:

$$p = \Pr(|\textit{calculated test statistic}| > |\textit{critical value}| \mid H_0 \textit{ true})$$

If p is less than our chosen α value, we reject H_0. If the p-value is greater than the α value, we fail to reject H_0. A small p-value is an indication that the validity of the null hypothesis assumption is suspect.

Using the *p*-value method gives us more information about the confidence we have in our conclusion compared to the critical-value method. If the α value for a study is changed, a new critical value must be determined from the statistical tables compared to the test statistic. In contrast, the *p*-value method is designed to tell us the lowest level of α at which the null hypothesis will be rejected. Note that we should not be relying solely on a single occurrence of a small *p*-value for making important decisions. Rather, the ability to **replicate** the study results will go far to ensure that our original conclusion was correct.

A reliability engineer should be able to state the interpretation correctly and precisely: The *p*-value is the probability of obtaining the current value of the test statistic, or one that is more extreme, assuming that the null hypothesis is correct.

A final note on statistical versus practical significance, which has become particularly important since big data has given us access to extremely large sample sizes: When the sample size is very large, it is possible to detect a statistically significant difference between, say, two population parameters even when the size of the difference has no practical importance. Just because a *p*-value is less than our chosen α, it does not immediately follow that the result has any practical utility. Note that the magnitude of the *p*-value does not indicate the size of the effect.

HYPOTHESIS TEST SELECTION

The next several sections summarize methods and equations with examples used for hypothesis tests for single proportions, means, and variances, as well as differences between (or the ratio of) two proportions, means, and variances (including standard deviation as the square root of the variance). Full equations are provided instead of simplified forms so that users can understand the underlying structure of the tests. Note that hypothesized values are denoted with the "naught" subscript, such as p_0. The decision tree in Figure 6.5 walks through key determining factors for hypothesis test selection.

Hypothesis Tests for Proportions

Hypothesis tests for proportions are of particular interest for reliability applications, since the analyst is often estimating the probability of success or failure. To model a proportion, data are typically assumed to be an independent random sample from a binomial distribution, implying a fixed number of trials, two mutually exclusive outcomes, with constant probability of success for all trials. Equations for one and two-sided hypothesis tests for p and $p_1 - p_2$ are organized in Table 6.5 and Table 6.6. **Note that two-sided tests allocate $\alpha/2$ to each side, whereas one-sided limits retain the full α on that side.**

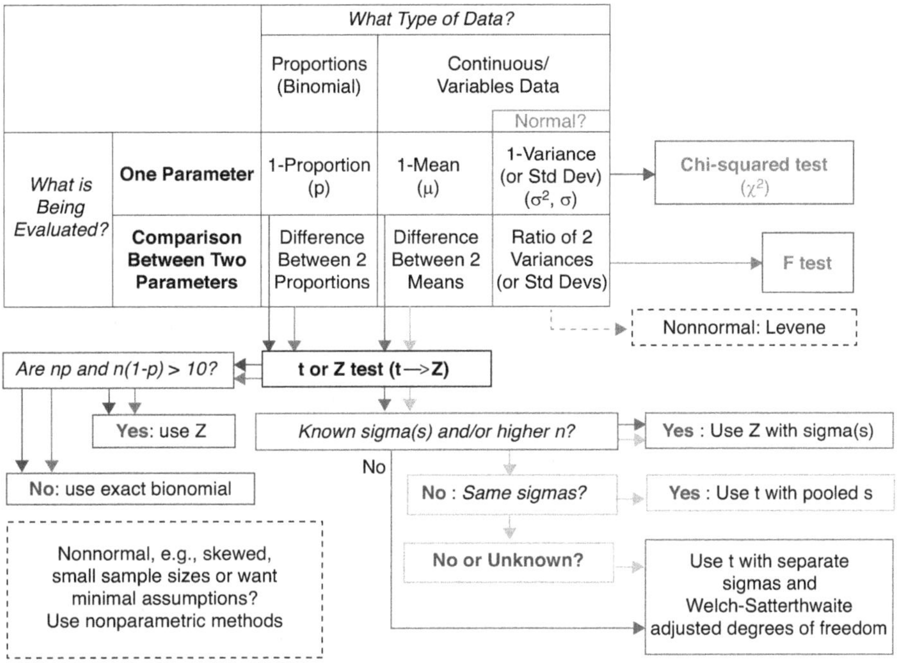

Figure 6.5 Decision tree for hypothesis test selection.

Table 6.5 Equations for hypothesis tests for proportions.

Parameter	Method	Equations for hypothesis tests at 100 $(1-\alpha)$% confidence	
p	*Normal approximation* use when $x \geq 10$ and $(n-x) \geq 10$	Two-sided hypotheses: $H_0: p = p_0$ $H_A: p \neq p_0$	One-sided hypotheses: $H_0: p \geq p_0$ $\quad H_0: p \leq p_0$ $H_A: p < p_0$ $\quad H_A: p > p_0$
		Point estimate, test statistic, and reference distribution: $\hat{p} = \frac{x}{n} \qquad z^* = \frac{\hat{p} - p_0}{\sqrt{\frac{\hat{p}\cdot(1-\hat{p})}{n}}} \sim Z$ The Agresti-Caffo method gets added robustness by using a variation on the point estimate: $\hat{p}_{AC} = \frac{x+1}{n+2}$ for smaller sample sizes.	
		Rejection rules: Two-sided: Reject if $\lvert z^* \rvert > Z_{\alpha/2}$ One-sided: Reject if $\lvert z^* \rvert > Z_{\alpha}$ For one-sided, use α instead of $\alpha/2$. *Refer to Appendix D.	

Table 6.6 Equations for hypothesis tests for the difference between two proportions.

Parameter	Method	Equations for 100 $(1-\alpha)$% Confidence Limits	
p_1-p_2	*Normal approximation* use when $x \geq 10$ and $(n-x) \geq 10$	Two-sided hypotheses: Zero difference $H_0: p_1=p_2$ $H_A: p_1 \neq p_2$ $H_0: p_1-p_2=0$ $H_A: p_1-p_2 \neq 0$ Nonzero difference $H_0: p_1-p_2=p_0$ $H_A: p_1-p_2 \neq p_0$	One-sided hypotheses: Zero difference $H_0: p_1 \geq p_2$ $H_0: p_1 \leq p_2$ $H_A: p_1 < p_2$ $H_A: p_1 > p_2$ $H_0: p_1-p_2 \geq 0$ $H_0: p_1-p_2 \leq 0$ $H_A: p_1-p_2 < 0$ $H_A: p_1-p_2 > 0$ Nonzero difference $H_0: p_1-p_2 \geq p_0$ $H_0: p_1-p_2 \leq p_0$ $H_A: p_1-p_2 < p_0$ $H_A: p_1-p_2 > p_0$
		Point estimate, test statistic, and reference distribution: $\hat{p}_1-\hat{p}_2=\frac{\lambda_1}{n_1}-\frac{x_2}{n_2}$ where $\hat{p}_1=\frac{x_1}{n_1}$ and $\hat{p}_2=\frac{x_2}{n_2}$	$z^*=\frac{(\hat{p}_1-\hat{p}_2)-p_0}{\sqrt{\frac{\hat{p}_1 \cdot \hat{q}_1}{n_1}+\frac{\hat{p}_2 \cdot \hat{q}_2}{n_2}}}$
		Rejection rules: Two-sided: Reject if $\lvert z^* \rvert > Z_{\alpha/2}$ One-sided: Reject if $\lvert z^* \rvert > Z_{\alpha}$ For one -sided, use α instead of $\alpha/2$. *Refer to Appendix D.	
	Exact Fisher's exact test Assumes hyper-geometric	This scenario is analogous to a 2×2 contingency table, for which there are several options for analysis depending on the specific comparison of interest. Fisher's exact test is one option, but it is computationally intense. Fisher assumes the hypergeometric distribution (refer to "Probability Distributions") and calculates the probability of seeing a table as or more extreme than the table observed.	

EXAMPLE 6.5

Hypothesis test for one proportion: A vendor claims that at most 2% of a shipment of parts is defective. Receiving inspection chooses a random sample of 500 and finds 15 defectives. At the 0.05 significance level, do these data indicate that the vendor is wrong?

First calculate the estimated proportions: $\hat{p}=\frac{15}{500}=0.03$. To conduct the hypothesis test: Check assumptions and define hypotheses—this is a one-sided test to be conducted at $\alpha=0.05$.

$$H_0: p \le 0.02$$
$$H_A: p > 0.02$$

Calculate test statistic:

$$z^* = \frac{\hat{p} - p_0}{\sqrt{\frac{\hat{p} \cdot (1 - \hat{p})}{n}}} = \frac{0.03 - 0.02}{\sqrt{\frac{0.03 \cdot 0.97}{500}}} = \frac{0.01}{0.00763} = 1.31$$

Calculate critical value
Here, we use Z obtained from Appendix D.

$$Z_\alpha = 1.645$$

Define rejection rule: We reject the null hypothesis if our test statistic $|z^*| \ge Z_{crit} = 1.645$.

Make comparison and draw conclusion: $|z^*| = 1.31 < Z_\alpha = 1.645$, so we fail to reject H_0. No, there is insufficient evidence to suggest that the lot has a higher mean than the 2% claimed by the vendor. The lot should not be rejected.

EXAMPLE 6.6

Hypothesis Test for Difference Between Two Proportions

Two machines produce the same parts. A random sample of 1500 parts from machine 1 has 36 defectives, and a random sample of 1680 parts from machine 2 has 39 defectives. Does machine 2 have a lower defective rate? Test at 0.01 significance level.

First calculate the estimated proportion:

$$\hat{p}_1 = \frac{36}{1500} = 0.024 \text{ and } \hat{p}_2 = \frac{39}{1680} = 0.0232$$

To conduct the hypothesis test: Check assumptions and define hypotheses—this is a one-sided test to be conducted at $\alpha = 0.01$.

$$H_0: p_1 \ge p_2 \qquad H_0: p_1 - p_2 \ge 0$$
$$H_A: p_1 < p_2 \quad \Leftrightarrow \quad H_A: p_1 - p_2 < 0$$

Calculate test statistic:

$$z^* = \frac{(\hat{p}_1 - \hat{p}_2) - p_0}{\sqrt{\frac{\hat{p}_1 \cdot \hat{q}_1}{n_1} + \frac{\hat{p}_2 \cdot \hat{q}_2}{n_2}}}$$

$$= \frac{(0.024 - 0.0232) - 0}{\sqrt{\frac{0.024 \cdot 0.976}{1500} + \frac{0.0232 \cdot 0.9768}{1680}}} = \frac{0.0008}{\sqrt{0.000015616 + 0.0000134891}} = \frac{0.0008}{0.00539} = 0.148$$

Note: $q = 1 - p$

Calculate critical value: Here, we use Z obtained from Appendix D.

$$Z_\alpha = 2.326$$

Continued

Define rejection rule:

We reject the null hypothesis if our test statistic $z^* \geq z_{crit} = 2.326$.

Make comparison and draw conclusion: $z^* = 0.148 < Z_\alpha = 2.326$, so we fail to reject H_0. No, there is insufficient evidence to suggest that the lot has a higher mean than the 2% claimed by the vendor. The lot should not be rejected.

Note also that we would fail to reject for higher significance levels as well, above 0.10.

Hypothesis Tests Centered Around Means

Table 6.7 shows generic equations for hypothesis tests that apply broadly to estimates for means, regardless of underlying distribution. Data are assumed to be an *iid* random sample. Here, the critical value is shown as t, since t converges to Z as n increases.

The standard deviation is genericized as $\hat{\sigma}$, which may be estimated in different ways. If σ is known, then $\hat{\sigma} = \sigma$, otherwise use $\hat{\sigma} = S$, the sample standard deviation. If σ is known, then Z may be used instead of t. **Note that 2-sided limits allocate $\alpha/2$ to each side, whereas one-sided limits retain the full α on that side.**

The practice of assuming that unknown variances are equal and calculating the pooled standard deviation for the Student's t test has fallen out of favor. Moser and Stevens (1992) note that we rarely know that $\sigma_1 = \sigma_2$, and instead, we should calculate the Student's t test statistic using the Welch-Satterthwaite method, which does not require the equal variance assumption.

The calculations for the Welch-Satterthwaite method are illustrated in Table 6.7. The hand calculations are more onerous than those for the pooled standard deviation method, but this is not an issue when using statistical software. In fact, the unequal variance assumption and Welch-Satterthwaite method is the default setting for two-sample t-tests in many statistical packages.

Table 6.7 Equations for hypothesis tests for a mean and the difference δ between two means.

Parameter	Method	Equations for hypothesis tests at 100 $(1-\alpha)$% confidence	
μ	<u>*Student's t-test*</u> If σ is known, then use $\hat{\sigma} = \sigma$ and Z, otherwise use sample standard deviation $\hat{\sigma} = s$ and t	Two-sided hypotheses: $H_0: \mu = \mu_0$ $H_A: \mu \neq \mu_0$	One-sided hypotheses: $H_0: \mu \geq \mu_0$ $H_0: \mu \leq \mu_0$ $H_A: \mu < \mu_0$ $H_A: \mu > \mu_0$
		Point estimate, test statistic, and reference distribution: $\hat{\mu} = \bar{X} = \frac{1}{n} \cdot \sum_{i=1}^{n} x_i \qquad t^* = \frac{\bar{X} - \mu_0}{\sqrt{\frac{\hat{\sigma}^2}{n}}} \sim t_{df=n-1}$	
		Rejection rules: Two-sided: Reject if $\lvert t^* \rvert \geq t_{\alpha/2}, df = v = n-1$ One-sided: Reject if $\lvert t^* \rvert > t_{\alpha, df=v=n-1}$ For one-sided, use α instead of $\alpha/2$. Refer to Appendix E.	

Table 6.7 Equations for hypothesis tests for a mean and the difference δ between two means. (Continued)

<table>
<tr><th>Parameter</th><th>Method</th><th colspan="2">Equations for hypothesis tests at 100 $(1-\alpha)$ % confidence</th></tr>
<tr><td rowspan="4">$\delta = \mu_1 - \mu_2$</td><td rowspan="4"><u>Student's t-test</u>
If σ_i is known, then use $\hat{\sigma}_i = \sigma_i$ and Z, otherwise use sample standard deviations $\hat{\sigma}_i = s_i$.
Pool variations if unknown, but assumed equal, $\hat{\sigma} = S_p$.
Otherwise, use $s_1^2 \neq s_2^2$ with Welch-Satterthwaite adjustment to degrees of freedom below.
$$v = \frac{\left(\frac{s_1^2}{n_1} + \frac{s_2^2}{n_2}\right)^2}{\frac{(s_1^2/n_1)^2}{n_1 - 1} + \frac{(s_2^2/n_2)^2}{n_2 - 1}}$$</td><td>Two-sided hypotheses:
Zero difference
$H_0 : \mu_1 = \mu_2$
$H_A : \mu_1 \neq \mu_2$

$H_0 : \mu_1 - \mu_2 = 0$
$H_1 : \mu_1 - \mu_2 \neq 0$

Nonzero difference
$H_0 : \mu_1 - \mu_2 = \mu_0$
$H_A : \mu_1 - \mu_2 \neq \mu_0$</td><td>One-sided hypotheses:
Zero difference
$H_0 : \mu_1 \geq \mu_2$ $H_0 : \mu_1 \leq \mu_2$
$H_A : \mu_1 < \mu_2$ $H_A : \mu_1 > \mu_2$

$H_0 : \mu_1 - \mu_2 \geq 0$ $H_0 : \mu_1 - \mu_2 \leq 0$
$H_A : \mu_1 - \mu_2 < 0$ $H_A : \mu_1 - \mu_2 > 0$

Nonzero difference
$H_0 : \mu_1 - \mu_2 \geq \mu_0$ $H_0 : \mu_1 - \mu_2 \leq \mu_0$
$H_A : \mu_1 - \mu_2 < \mu_0$ $H_A : \mu_1 - \mu_2 > \mu_0$</td></tr>
<tr><td colspan="2">Point estimate, test statistic, and reference distribution:
$$(\hat{\mu}_1 - \hat{\mu}_2) = (\bar{X}_1 - \bar{X}_2) = \frac{1}{n_1} \cdot \sum_{i=1}^{n_1} x_{1i} - \frac{1}{n_2} \cdot \sum_{i=1}^{n_2} x_{2i}.$$
$$t^* = \frac{(\bar{X}_1 - \bar{X}_2) - \mu_0}{\sqrt{\frac{\hat{\sigma}_1^2}{n_1} + \frac{\hat{\sigma}_2^2}{n_2}}} \sim t_{df = n_1 + n_2 - 2}$$
Note that these equations simplify if $\sigma_1 = \sigma_2 = \sigma$, meaning that both groups have the same variation. If standard deviations are unknown but assumed equal, then the sample standard deviations may be pooled to obtain an aggregate estimate, which may be used for $\hat{\sigma}$:
$$\hat{\sigma} = s_p = \sqrt{\frac{(n_1 - 1) \cdot s_1^2 + (n_2 - 1) \cdot s_2^2}{n_1 + n_2 - 2}}$$
If $\sigma_1 \neq \sigma_2$, then separate $\hat{\sigma}_1$ and $\hat{\sigma}_2$ estimates should be used with the Welch-Satterthwaite adjustment for degrees of freedom.</td></tr>
<tr><td colspan="2">Rejection rules:
Two-sided: Reject if $|t^*| \geq t_{\alpha/2}$, $df = v = n_1 + n_2 - 2$
One-sided: Reject if $|t^*| > t_{\alpha, df = v = n_1 + n_2 - 2}$
For one-sided, use α instead of $\alpha/2$.
Refer to Appendix E.</td></tr>
</table>

EXAMPLE 6.7

Hypothesis Test for One Mean, Variance Known

A vendor claims that the average weight of a shipment of parts is 1.84. The customer randomly chooses 64 parts and finds that the sample has an average weight of 1.88, higher than the requirement. Suppose that the standard deviation of the population is known to be 0.03. Should the customer reject the lot? Assume the customer wants to be 95% confident that the supplier's claim is incorrect before he rejects.

To conduct the hypothesis test, check assumptions and define hypotheses. This is a two-sided test to be conducted at $\alpha = 0.05$:

$$H_0 : \mu = 1.84$$
$$H_A : \mu \neq 1.84$$

Calculate test statistic:

$$z^* = \frac{\bar{X} - \mu_0}{\hat{\sigma} / \sqrt{n}} = \frac{1.88 - 1.84}{0.03 / \sqrt{64}} = \frac{0.04}{0.03 / 8} = 10.67$$

Calculate critical value: We can use Z for two reasons. Sigma is considered known, and there is also a higher n ($n = 64$), meaning that the central limit theorem allows us to assume that X-bar is normally distributed. Here we use $Z_{0.025} = 1.96$

Define rejection rule: We reject the null hypothesis if our test statistic $|z^*| \geq Z_{crit} = 1.96$, obtained from Appendix D.

Make comparison and draw conclusion: $|z^*| = 10.67 \gg Z_{\alpha/2} = 1.96$. Reject H_0. Yes, there is sufficient evidence to suggest that the lot has a higher mean than claimed by the vendor. The lot should be rejected.

How would this analysis have changed if the sigma was not assumed to be known? The hypotheses would have been unchanged. The test statistic would have been t^* instead of a z^*; however, the value would have been unchanged.

$$t^* = \frac{\bar{X} - \mu_0}{\hat{\sigma} / \sqrt{n}} = \frac{1.88 - 1.84}{0.03 / \sqrt{64}} = \frac{0.04}{0.03 / 8} = 10.67$$

The critical value would therefore need to be obtained from a t-table instead, obtained from Appendix E, which would have $v_1 = n - 1 = 63$ degrees of freedom:

$$t_{0.025,63} = 1.999 \text{ (slight interpolation required)}$$

However, this is negligibly different from the previous Z value obtained. The prior conclusion is not impacted since, $|t^*| = 10.67 \gg t_{\alpha/2} \approx 2$. Calculated p-values would have decreased very slightly.

EXAMPLE 6.8

Hypothesis Test for One Mean, Variance Unknown (One-Sided)

A cut-off saw has been producing parts with a mean length of 4.125. A new blade is installed, and we want to know whether the mean length has decreased. We select a random sample of 20, measure the length of each part, and find that the average length is 4.123 and the sample standard deviation is 0.008. The population is reasonably normally distributed and does not appear to be trending. Use a significance level of 0.10 to determine whether the mean length has decreased.

To conduct the hypothesis test:

Check assumptions and define hypotheses. This is a one-sided test to be conducted at $\alpha = 0.10$.

$$H_0 : \mu \geq 4.125$$
$$H_A : \mu < 4.125$$

Calculate test statistic:

$$t^* = \frac{\bar{X} - \mu_0}{\hat{\sigma} / \sqrt{n}} = \frac{4.123 - 4.125}{0.008 / \sqrt{20}} = \frac{-0.002}{0.00179} = -1.118$$

Calculate critical value: We must use the t distribution for two reasons. Sigma is considered unknown, and there is also a lower n ($n = 20$), meaning that the $t \neq Z$ for this sample size. The critical value needs to be obtained from a t-table, Appendix E, with $v = n - 1 = 19$ degrees of freedom.

$$t_{\alpha = 0.10, v = 19} = 1.328$$

Define rejection rule: We reject the null hypothesis if our test statistic $|t^*| > t_{crit} = 1.328$.

Make comparison and draw conclusion: $|t^*| = |1.118| < t_{\alpha = 0.10, 19} = 1.328$. Here, we fail to reject H_0 since there is insufficient evidence at the 90% level of confidence to suggest that the new blade is producing parts with a decreased mean.

EXAMPLE 6.9

Hypothesis Test for Difference Between Two Independent Means (One-Sided)

Two vendors of a valve diaphragm present significantly different cost quotations. The wall thickness is the critical quality characteristic. Use the following data to determine whether the average thickness of the products from Vendor 1 is greater than that from Vendor 2. Test at the 0.10 significance level. Assume the populations are normally distributed and that the samples are independent.

The wall thickness measurements are as follows:

- Vendor 1: 86, 82, 91, 88, 89, 85, 88, 90, 84, 87, 88, 83, 84, 89
- Vendor 2: 79, 78, 82, 85, 77, 86, 84, 78, 80, 82, 79, 76

Summary statistics for the two groups are:

$$\bar{X}_1 = 86.71 \qquad \bar{X}_2 = 80.50$$
$$s_1 = 2.76 \qquad s_2 = 3.26$$
$$n_1 = 14 \qquad n_2 = 12$$

To conduct the hypothesis test: Check assumptions and define hypotheses—this is a one-sided test to be conducted at $\alpha = 0.10$.

$$\begin{matrix} H_0 : \mu_1 \leq \mu_2 \\ H_A : \mu_1 > \mu_2 \end{matrix} \Leftrightarrow \begin{matrix} H_0 : \mu_1 - \mu_2 \leq 0 \\ H_A : \mu_1 - \mu_2 > 0 \end{matrix}$$

Continued

Calculate test statistic:

$$t^* = \frac{(\bar{X}_1 - \bar{X}_2) - \mu_0}{\sqrt{\frac{s_1^2}{n_1} + \frac{s_2^2}{n_2}}} = \frac{(86.7 - 80.5) - 0}{\sqrt{\frac{(2.76)^2}{14} + \frac{(3.26)^2}{12}}} = \frac{6.21}{\sqrt{0.54 + 0.89}} = \frac{6.21}{1.195} = 5.2$$

Calculate critical value: We must use the t distribution for two reasons. Sigma is considered unknown and sample sizes for both groups are low, meaning that the $t \neq Z$. For unequal standard deviations, we use the Welch-Satterthwaite equation for degrees of freedom. This method should be the default approach for unknown sigmas.

$$v = \frac{\left(\frac{s_1^2}{n_1} + \frac{s_2^2}{n_2}\right)^2}{\frac{(s^2_1 / n_1)^2}{n_1 - 1} + \frac{(s^2 / n_2)^2}{n_2 - 1}} = \frac{(0.54 + 0.89)^2}{\frac{0.54^2}{13} + \frac{0.89^2}{11}} = \frac{2.04}{0.023 + 0.071} = 21.7 \rightarrow 22$$

The critical value is obtained from a t-table, Appendix E, with $v = 22$ degrees of freedom.

$$t_{\alpha = 0.1, v = 22} = 1.321$$

Define rejection rule: We reject the null hypothesis if our test statistic $|t^*| > t_{crit} = 1.321$.

Make comparison and draw conclusion: $|t^*| = 5.2 > t_{\alpha = 0,1, v = 22} = 1.321$. Here we do reject H_0 since there is sufficient evidence at the 90% level of confidence to indicate that the first vendor is producing parts with a thicker mean wall thickness.

EXAMPLE 6.10

Hypothesis Test for Difference Between Two Independent Means (One-Sided)

A pilot study was conducted to determine whether a new catalyst resulted in increased yields, measuring in ounces. Yields have historically followed a normal distribution with similar levels of variation. Test batches were produced in the lab using both catalysts with the results shown in Table 6.8. Perform a test at the 99% confidence level to determine if the new catalyst results in an increase in mean yield.

Table 6.8 Yield study results for Example 6.10.

	Yield study	
	Current catalyst	**New catalyst**
Average	28.6 oz	31.9 oz
Std. deviation	3.5 oz	3.6 oz
Sample size	15	20

To conduct the hypothesis test: Check assumptions and define hypotheses—this is a one-sided test to be conducted at $\alpha=0.01$.

$$\begin{matrix} H_0: \mu_1 \geq \mu_2 \\ H_A: \mu_1 < \mu_2 \end{matrix} \Leftrightarrow \begin{matrix} H_0: \mu_1 - \mu_2 \geq 0 \\ H_A: \mu_1 - \mu_2 < 0 \end{matrix}$$

Calculate test statistic: We first need to calculate the pooled estimate of sigma, which is a type of weighted average of the two values.

$$\hat{\sigma} = s_p = \sqrt{\frac{(n_1-1)\,s_1^2 + (n_2-1)\,s_2^2}{n_1+n_2-2}} = \sqrt{\frac{(14)\,(3.5)^2 + (19)\,(3.6)^2}{15+20-2}}$$

$$= \sqrt{\frac{171.5+246.24}{33}} = \sqrt{12.6} = 3.56$$

$$t^* = \frac{(\bar{X}_1 - \bar{X}_2) - \mu_0}{s_p \cdot \sqrt{\frac{1}{n_1} + \frac{1}{n_2}}} = \frac{(28.6-31.9)-0}{3.558 \cdot \sqrt{\frac{1}{15} + \frac{1}{20}}} = \frac{-3.3}{3.558 \cdot 0.342} = \frac{-3.3}{1.215} = -2.715$$

Calculate critical value: The critical value is obtained from a *t*-table with $v=n_1+n_2-2=33$ degrees of freedom.

$$t_{\alpha=0.01, v=33} = 2.445$$

Define rejection rule: We reject the null hypothesis if our test statistic $|t^*| > t_{crit} = 2.445$.

Make comparison and draw conclusion: $|t^*| = |-2.715| > t_{\alpha=0.01,33} = 2.445$. Here we reject H_0 since there is sufficient evidence at the 99% level of confidence to indicate that the new catalyst is associated with increased yields.

Alternatively, we can use the *p*-value method to be able to give a more precise statement. Statistical software calculates $p=0.005$ for this test. Since this value is less than our chosen $\alpha=0.01$, we reject H_0.

Analysis of Paired Data

When making a comparison between two groups, we can produce a much more powerful analysis if we recognize **paired data**, if it is present, in which there are two observations associated with each measured item, for example, before/after, left/right, top/bottom. One indication of paired data is that the sample sizes for the two groups should be the same, that is, $n_1 = n_2$. (A scatter plot of the data would also suggest a relationship is present.) Situations in which paired data arise include two readings on the same item or subject at different points in time, such as:

- Cholesterol readings for patients before and after beginning a new drug protocol
- Test scores for learners before and after a training program
- Hardness readings for metal parts before and after an annealing process.

In each of these examples, the two sets of measurements collected are not independent groups because they each measure the same items, just at different points in time. Often what is seen here is that those units that test high to start with would be expected to test high later, while lower units may see the effect, but still stay low. A scatter plot of the data (detailed later in this chapter) would show a relationship present. In this case, conducting a standard two-sample hypothesis test is less likely to be able to detect an overall difference between the two groups if the data are considered all together, because a small effect may be masked by the overall scatter in the data. However, first calculating a difference between the measurements on each unit allows us to leverage this dependency to do a more robust analysis.

To conduct a paired analysis, we must first calculate the differences (also known as deltas) between each unit's measurements:

$$X_{D,i}=X_{1,i}-X_{2,i}$$

If the two values are exactly equal for a given unit, the $X_D=0$. If one method has a consistent effect, the differences will tend to be more positive or more negative. Once the differences have been calculated, standard analysis methods can be used for the population parameter of interest, for instance, a hypothesis test or confidence interval for the true mean equal to zero. If it fails to reject, there is insufficient evidence of a difference. This method enables greater sensitivity to detect smaller effects.

For example, suppose that we want to determine if a gasoline additive increases average mileage in the population consisting of several hundred company cars of various types and vintages. One approach would be to randomly select 10 vehicles and record mileage using gasoline without the additive and randomly select another 10 vehicles and record mileages using gasoline with the additive. These two samples could be tested using the hypothesis test for comparing two independent means. If the additive causes a large increase in average mileage, this procedure will likely reject the null hypothesis that the averages are equal. If the additive causes a small increase in average mileage, the test might not detect it because of the large variation between cars. Thus, the test might fail to reject the null hypothesis even though it is false. Statisticians would say the test lacks **sensitivity**.

An alternate approach has probably already occurred to you: Choose 10 cars at random, record their mileages using gas without the additive, then use gas with the additive in those same 10 cars. This arrangement reduces the sampling variation encountered when two samples of 10 are used. The approach is called the paired sample method and provides a very powerful test when it can be used. Sometimes it is not viable or practical to use. Suppose we need to know whether the average effect of a particular drug is different for people with type A blood than for people with type B. If we attempted to do a paired approach, we would select 10 type A people and measure the effects of the drug on them, then drain their blood and refill them with type B and again measure the effects of the drug. One might conclude that the drug was fatal for the population having type B blood. Obviously, this is not a sensible approach to evaluating this particular question.

Example 6.11 presents an example of the paired data showing the effect of a dependency, using both the hypothesis test for two means and the paired analysis for the single mean of the paired data on the same dataset. A scatter plot of the data is shown in Figure 6.6 as an illustration.

Further discussion of paired analyses is omitted from this book since the formulas used are identical to the single mean case. Note that unequal sample sizes will preclude the paired analysis from being performed; however, having equal sample sizes could still indicate either is appropriate. Pairing data that should not be paired will typically not result in a significant outcome since there would be no dependency. However, the results are unlikely to be clearly interpretable.

EXAMPLE 6.11

Paired Data Versus Two Independent Means

For the gasoline additive problem discussed in this section, suppose the data are as shown in Table 6.9. Assuming that the differences are normally distributed, do the data indicate that the additive increases average gas mileage at the 0.05 significance level?

Table 6.9 Gasoline additive data for Example 6.11.

Vehicle #:	**1**	**2**	**3**	**4**	**5**	**6**	**7**	**8**	**9**	**10**
mpg with additive:	21	23	20	20	27	18	22	19	36	25
mpg without additive:	20	20	21	18	24	17	22	18	37	20
Difference *d*:	1	3	−1	2	3	1	0	1	−1	5

We will conduct the analysis first on the raw data, then on the differences. Summary statistics for the two groups and the differences are:

$$\begin{array}{lll} \bar{X}_1 = 23.1 & \bar{X}_2 = 21.7 & \bar{X}_D = 1.4 \\ s_1 = 5.3 & s_2 = 5.8 & s_D = 1.9 \\ n_1 = 10 & n_2 = 10 & n_D = 10 \end{array}$$

To conduct the hypothesis test by (incorrectly) comparing **two independent means**: Check assumptions and define hypotheses—this is a one-sided test to be conducted at $\alpha = 0.05$. Variances are similar and are pooled here.

$$\begin{array}{l} H_0 : \mu_1 \le \mu_2 \\ H_A : \mu_1 > \mu_2 \end{array} \Leftrightarrow \begin{array}{l} H_0 : \mu_1 - \mu_2 \le 0 \\ H_A : \mu_1 - \mu_2 > 0 \end{array}$$

Calculate test statistic:

$$t^* = \frac{(\bar{X}_1 - \bar{X}_2) - \mu_0}{s_p\sqrt{\frac{1}{n_1} + \frac{1}{n_2}}} = \frac{(23.1 - 21.7) - 0}{5.56\sqrt{\frac{1}{10} + \frac{1}{10}}} = \frac{1.4}{5.56 \cdot 0.447} = \frac{1.4}{2.49} = 0.56$$

where

$$\hat{\sigma} = s_p = \sqrt{\frac{(n_1 - 1)s_1^2 + (n_2 - 1)s_2^2}{n_1 + n_2 - 2}} = \sqrt{\frac{(9)(5.3)^2 + (9)(5.8)^2}{10 + 10 - 2}}$$

$$= \sqrt{\frac{252.81 + 302.76}{18}} = \sqrt{30.865} = 5.56$$

Note that this test statistic value is so extremely low that we should know without even doing the formal test that it cannot possibly be statistically significant. Walking through the math, calculate the critical value: We must use the *t* distribution for two reasons. Sigma is considered unknown and sample sizes for both groups are low, meaning that the $t \neq Z$. The critical value is obtained from a *t*-table, Appendix E, with $\nu = 18$ degrees of freedom.

$$t_{\alpha=0.05, \nu=18} = 1.734$$

Continued

Define rejection rule: We reject the null hypothesis if our test statistic $|t^*| > t_{crit} = 1.734$.

Make comparison and draw conclusion: $|t^*| = 0.56 < t_{\alpha=0.05,\nu=18} = 1.734$. Here we fail to reject H_0 since there is insufficient evidence at 95% confidence to indicate that the first vendor is producing parts with a thicker wall thickness.

Next, we will perform the correct analysis. To conduct the hypothesis test on **paired differences**: Check assumptions and define hypotheses—this is a one-sided test to be conducted at $\alpha = 0.05$.

$$H_0 : \mu_d \leq 0$$
$$H_A : \mu_d > 0$$

The appropriate test is therefore a Student's *t* test on the mean difference.

Calculate test statistic:

$$t^* = \frac{\bar{X} - \mu_0}{\hat{\sigma} / \sqrt{n}} = \frac{1.4 - 0}{1.9 / \sqrt{10}} = \frac{1.4}{0.6} = 2.33$$

Calculate critical value: The critical value is obtained from a *t*-table, Appendix E, with $\nu = 9$ degrees of freedom.

$$t_{\alpha=0.05,\nu=9} = 1.833$$

Define rejection rule: We reject the null hypothesis if our test statistic $|t^*| > t_{crit} = 1.833$

Make comparison and draw conclusion: $|t^*| = 2.33 > t_{\alpha=0.05,9} = 1.833$. Here we **do** reject H_0 despite having fewer degrees of freedom and a higher threshold. There is sufficient evidence at 95% confidence to indicate that average miles per gallon (mpg) is increased by using the additive.

The reason for the conflict in results is because the variation associated with the differences is much less than that of the raw data. Vehicle-to-vehicle variation is very high relative to the size of the effect with $s_p = 5.56$ versus $s_d = 1.9$ relative to $\bar{X}_D = 1.4$.

Data are plotted here. The answer is not obvious, but a clue that there is an effect is that there are more points to the left of the $y = x$ line shown on the chart. Variation between measurements on the same vehicle is much less than variation between vehicles. Also note that data appear positively correlated. Vehicles with high mileage will tend to have high mileage whether or not the additive is present. Similarly, vehicles with low mileage will tend to have low mileage regardless.

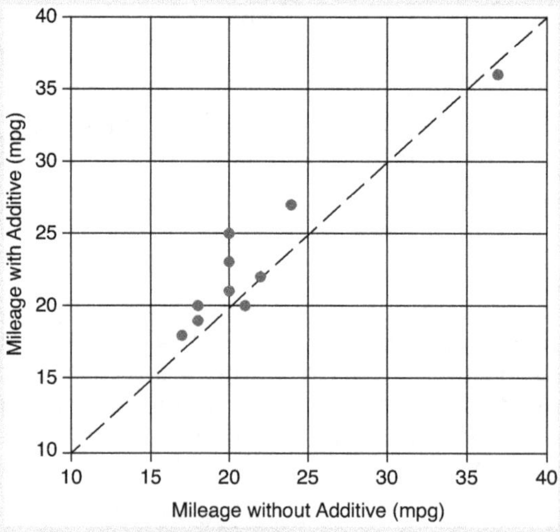

Figure 6.6 Plot of paired gas mileage data for Example 6.11.

EXAMPLE 6.12

Hypothesis Test for Paired Data

The grip strength of 35 male patients was measured before and after a rehabilitation program that included a weight training protocol. Because each patient was sampled before and after the protocol, this is a paired-data scenario. Summary statistics on the (After–Before) readings are summarized in Table 6.10. Test whether the weight training program significantly increased mean grip strength using $\alpha = 0.05$.

Table 6.10 Grip strength study summarized results for Example 6.12.

Grip strength results for weight training protocol	
Average difference in grip strength (After–Before)	2.229 lbs
Sample standard deviation of the difference	2.850 lbs
Sample size	35

To conduct the hypothesis test: Check assumptions and define hypotheses—this is a one-sided test to be conducted at $\alpha = 0.05$.

Here, we will be performing a test on the mean difference, not the difference in means. The difference was calculated as (After–Before), so if the weight training protocol had been successful, the mean difference would be significantly greater than zero.

$$H_0 : \mu_d \leq 0$$
$$H_A : \mu_d > 0$$

The appropriate test is therefore a Student's *t* test on the mean difference.

Calculate test statistic:

$$t^* = \frac{\bar{X} - \mu_0}{\hat{\sigma} / \sqrt{n}} = \frac{2.229 - 0}{2.850 / \sqrt{35}} = \frac{2.229}{0.482} = 4.63$$

Calculate critical value: The critical value is obtained from a *t*-table, Appendix E, with $v = n - 1 = 34$ degrees of freedom.

$$t_{\alpha = 0.05, v = 34} = 1.691$$

Define rejection rule: We reject the null hypothesis if our test statistic $|t^*| > t_{crit} = 1.691$, that is, in the more positive direction.

Make comparison and draw conclusion: $|t^*| = 4.63 > t_{\alpha = 0.05, 34} = 1.691$. Here we reject H_0 since there is sufficient evidence at 90% confidence to indicate that the weight training protocol increased the mean difference in grip strength.

Alternatively, we can use the *p*-value method to be able to give a more precise statement. Statistical software calculates $p = 0.000$ for this test. Since this value is less than our chosen $\alpha = 0.05$, we reject H_0.

Hypothesis Tests for Variability

Table 6.11 summarizes the equations for hypothesis tests for variance-related estimators. Equations are provided for both variance and standard deviation. However, the standard deviation is the square root of the variance, so these tests are functionally equivalent and will result in identical conclusions.

Hypothesis tests may be conducted for a single parameter, either σ^2 or $\sigma = \sqrt{\sigma^2}$, or for the ratio, used when comparing variances. If the ratio of variances is significantly greater than 1, then the source in the numerator is associated with greater variation. If the ratio is significantly less than 1, then the source in the denominator has more. If the effect is not statistically significant, then the result is inconclusive.

Data are assumed to be an *iid* random sample from a normal distribution. Note that the assumption of normality for these hypothesis tests is quite strict since both the chi-square and *F* distributions are sensitive to departures from normality. There are nonparametric methods that can be used to calculate confidence intervals when the normal assumption is violated. Although beyond the scope of this handbook, these methods are listed in Table 6.3 in the "Nonparametric Tests" section of this chapter.

The notation used here for the estimate of the population standard deviation σ is the lowercase letter *s*. Since the purpose of this analysis is to estimate σ, it can't be considered known. However, the ratio estimators could be done using both known or estimated sigma values, if known values are available. **Note that two-sided limits allocate $\alpha/2$ to each side, whereas one-sided limits retain the full α on that side.**

Table 6.11 Equations for hypothesis tests for variances and standard deviations.

Parameter	Method	Equations for hypothesis tests at 100 $(1-\alpha)$% confidence	
Population variance or standard deviation σ^2 $\sigma = \sqrt{\sigma^2}$	χ^2-*test* (chi-squared) Data are *iid* **Normal** (μ, σ^2)	Two-sided hypotheses: $H_0 : \sigma = \sigma_0$ $H_A : \sigma \neq \sigma_0$	One-sided hypotheses: $H_0 : \sigma \geq \sigma_0$ $H_0 : \sigma \leq \sigma_0$ $H_A : \sigma < \sigma_0$ $H_A : \sigma > \sigma_0$
		Point estimate, test statistic, and reference distribution: *Variance*: $\hat{\sigma}^2 = s^2 = \frac{1}{n-1}\sum_{i=1}^{n}(x_i - \bar{x})^2$ *Std Dev*: $\hat{\sigma} = s = \sqrt{\frac{1}{n-1}\sum_{i=1}^{n}(x_i - \bar{x})^2}$ $\chi^{2*} = \frac{(n-1)s^2}{\sigma_0^2} \sim \chi^2_{v=n-1}$	

Table 6.11 Equations for hypothesis tests for variances and standard deviations. (Continued)

Parameter	Method	Equations for hypothesis tests at 100 (1 – α)% confidence
		Rejection rules: Two-sided: Reject if $\chi^{2*} \ge \chi^2_{\alpha/2, v=n-1}$ or $\chi^{2*} < \chi^2_{1-\alpha/2, v=n-1}$ For one-sided, use α instead of $\alpha/2$. One-sided lower: Reject if $\chi^{2*} < \chi^2_{1-\alpha, v=n-1}$ One-sided upper: Reject if $\chi^{2*} \ge \chi^2_{\alpha, v=n-1}$ Refer to Appendix G.
Ratio of two population variances or standard deviations $\frac{\sigma_1^2}{\sigma_2^2}$ $\frac{\sigma_1}{\sigma_2} = \sqrt{\frac{\sigma_1^2}{\sigma_2^2}}$	*F-test* Data are *iid* X_1 ~**normal** (μ_1, σ_1^2) and X_2 ~**normal** (μ_2, σ_2^2)	Two-sided hypotheses: $H_0: \sigma_1^2 = \sigma_2^2$, $H_A: \sigma_1^2 \ne \sigma_2^2$ $\Leftrightarrow$ $H_0: \frac{\sigma_1^2}{\sigma_2^2} = 1$, $H_A: \frac{\sigma_1^2}{\sigma_2^2} \ne 1$ One-sided $H_0: \sigma_1^2 \le \sigma_2^2$, $H_A: \sigma_1^2 > \sigma_2^2$; $H_0: \frac{\sigma_1^2}{\sigma_2^2} \le 1$, $H_A: \frac{\sigma_1^2}{\sigma_2^2} > 1$ $H_0: \sigma_1^2 \ge \sigma_2^2$, $H_A: \sigma_1^2 < \sigma_2^2$; $H_0: \frac{\sigma_1^2}{\sigma_2^2} \ge 1$, $H_A: \frac{\sigma_1^2}{\sigma_2^2} < 1$ Point estimate, test statistic, and reference distribution: *Variance:* $\frac{\hat{\sigma}_1^2}{\hat{\sigma}_2^2} = \frac{s_1^2}{s_2^2}$ *Std Dev:* $\frac{\hat{\sigma}_1}{\hat{\sigma}_2} = \frac{s_1}{s_2}$ $F^* = \frac{s_1^2}{s_2^2} \sim F_{v_1=n_1-1, v_2=n_2-1}$
		Rejection rules: Two-sided: Reject if For one-sided, use α instead of $\alpha/2$. $F^* \ge F_{\alpha/2, v_1=n_1-1, v_2=n_2-1}$ *or* $F^* < F_{1-\alpha/2, v_1=n_1-1, v_2=n_2-1}$ One-sided lower: Reject if $F^* < F_{1-\alpha, v_1=n_1-1, v_2=n_2-1}$ One-sided upper: Reject if $F^* \ge F_{\alpha, v_1=n_1-1, v_2=n_2-1}$ Refer to Appendices F.1–F.8. Note that this handbook provides F table values for left-tail probabilities, but not all references do. If we do not have a table for the left tail, we can use the following relationships to calculate it: One-sided: $F_{1-\alpha, v_2, v_1} = \frac{1}{F_{\alpha, v_1, v_2}}$ Two-sided: $F_{1-\alpha/2, v_2, v_1} = \frac{1}{F_{\alpha/2, v_1, v_2}}$

EXAMPLE 6.13

Hypothesis Test for One Variance

A test engineering team would like to estimate the repeatability in a new test process and evaluate whether the true system standard deviation is lower than a downstream system requirement of 1 psig. Ten pressure samples were obtained with a standard deviation of 0.450 psig. Will the new process meet the new requirement? Test at the 0.01 level.

To conduct the hypothesis test: Check assumptions and define hypotheses. This is a one-sided chi-squared test to be conducted at $\alpha=0.01$. This test is appropriate for normal data. Here, we will be conducting a test on whether the new test process has a lower level of variability than the requirement, with the baseline condition being assumed to be equal or higher.

$$H_0 : \sigma^2 \geq 1 \quad H_A : \sigma^2 < 1$$

Note that this is equivalent to testing:

$$H_0 : \sigma \geq 1 \quad H_A : \sigma < 1$$

Calculate test statistic:

$$\chi^{2*} = \frac{(n-1)s^2}{\sigma_0^2} = \frac{(10-1)(0.45)^2}{1^2} = 1.83$$

Calculate critical value: Refer to χ^2 table in Appendix G. Obtain χ^2 value for *lower* limit only since we need to demonstrate that variability is *lower* than the target.

$$\chi^2_{1-\alpha=0.99, v=9} = 2.088$$

Define rejection rule: We reject the null hypothesis if test statistic $\chi^{2*} < \chi^2_{0.99,9} = 2.088$.

Make comparison and draw conclusion: Since the test statistic $\chi^{2*} = 1.83$ is less than the critical value $\chi^2_{0.99,9} = 2.088$, we reject H_0 and conclude that there is sufficient evidence to indicate that the variance, and correspondingly the standard deviation, of the new process is less than that of the requirement.

Alternatively, we can use the *p*-value method to be able to give a more precise statement. Statistical software calculates $p=0.173$ for this test. Since this value is less than our chosen $a=0.01$, we fail to reject H_0.

EXAMPLE 6.14

Hypothesis Test for Ratio of Two Variances

An investigation team would like to test whether a new fixture has more precision in locating drilled holes compared to the original fixture. The location readings are known to follow a normal distribution. The original fixture will be denoted as Sample 1, and the new fixture will be Sample 2. Sample data gives $s_1=0.250$, $s_2=0.195$, and $n_1=20$ and $n_2=15$. We will test at the $\alpha=0.10$ level.

To conduct the hypothesis test: Check assumptions and define hypotheses—this is a one-sided test to be conducted at $\alpha=0.10$. This test is appropriate for normal data. Here, we will be conducting a test on whether the new fixture has a smaller variance than the original fixture. This will occur if the ratio $\frac{\sigma_1^2}{\sigma_2^2}$ is greater than one.

$$H_0: \frac{\sigma_1^2}{\sigma_2^2} \leq 1 \quad H_A: \frac{\sigma_1^2}{\sigma_2^2} > 1$$

Calculate test statistic:

$$F^* = \frac{s_1^2}{s_2^2} = \frac{0.25^2}{0.195^2} = 1.644$$

Calculate critical value: Refer to F table in Appendix F.4: $F_{\alpha=0.10,\nu_1=19,\nu_2=14} = 1.97$

Define rejection rule: We reject the null hypothesis if our test statistic $F^* > F_{0.10,19,14}$.

Make comparison and draw conclusion: Since the test statistic $F^* = 1.644$ is less than the critical value $F_{0.10,19,14} = 1.97$, we fail to reject H_0 and conclude that there is insufficient evidence to claim the variance of the new fixture is less than that of the original fixture.

Alternatively, we can use the p-value method to be able to give a more precise statement. Statistical software calculates $p = 0.173$ for this test. Since this value is greater than our chosen $a = 0.10$, we fail to reject H_0.

Correlation, Regression, and ANOVA

Correlation and linear regression describe the nature of a linear relationship between two variables. Correlation is used to describe the general degree and direction of the association, while linear regression uses values of an independent explanatory variable to predict those of a dependent response variable to arrive at a predictive equation.

Graphical Methods: Scatter (X-Y) and Matrix Plots

When exploring relationships between variables, our first instinct should always be to graph the data on an **X-Y plot**, or **scatter plot**. Plotting the (X, Y) pairs will show what type of relationship exists between the two variables, as well as showing any unusual or outlying observations. Typically, in this type of plot, the variable that we want to predict is on the vertical or y-axis, and the input variable is placed on the x-axis. The **Y-variable** is also called the **output variable, response variable**, or **dependent variable**. The **X-variable** is referred to variously as the **input variable**, the **predictor variable**, or the **independent variable**. Another colloquial term for an X-variable is **driver**.

Data must be organized in *columns* for plotting, with invalid characters or extraneous textual information removed.

Not every pair of variables shown on a scatter plot will always be a predictor–response pair. They could both be predictor variables, or they could both be responses. Scatter plots are used widely for exploratory data analysis and data visualization, so they may include either or both. **Matrix plots** allow us to look at many scatter plots at the same time, and these may be a mix of variable types.

Temperature	Seal Strength
225	123.1
235	125.0
238	123.8
240	119.4
241	125.8
241	122.5
245	127.8
250	126.5
252	127.0
250	126.0
255	129.5
260	130.8
270	127.5
269	128.7
265	124.8
275	125.3
292	131.5
321	132.2
315	138.2
309	137.4

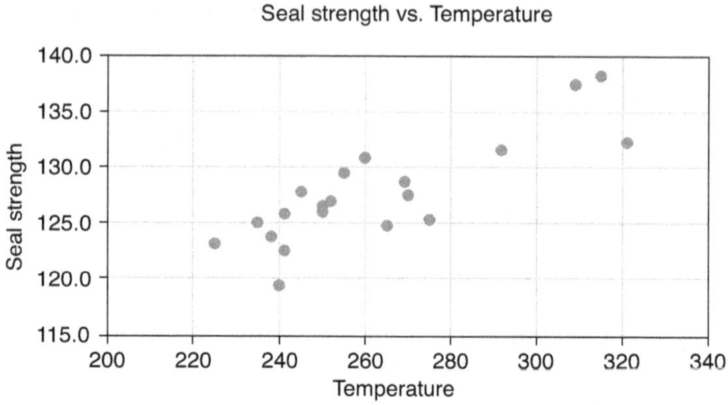

Figure 6.7 Example (*X*, *Y*) data organization and scatterplot.

Note that in order to generate a scatterplot or do any of the subsequent analyses, data must be related pairs, in that each X value should have a corresponding value for Y. In most cases, X and Y will be variables that are both measured on each sampled item or experimental unit. Data points may be missing, but if so, they cannot be included in the analysis using standard methods and will not show up on plots unless specially indicated. Figure 6.7 gives an example of the data input and resulting scatterplot for seal strength versus temperature data.

Correlation

Correlation is a mathematical tool used to measure the strength of a linear association between two variables X and Y. It provides a conceptual and intuitive way to describe simple relationships between two variables. Correlation may be positive or negative, referring to the direction of the slope seen when the two variables are plotted against each other. Two variables with no linear relationship would be said to have no correlation, indicated mathematically as zero, while perfect functional relationships have values of ±1, depending on the direction, as shown in Figure 6.8. Values closer to ±1 have stronger correlations; lower values have weaker correlations. The true population correlation is denoted by the Greek letter rho, ρ, estimated by r. Percentages may also be used.

Graphically, the correlation coefficient of −1 has (x, y) pairs falling along a straight line with a negative slope, and the correlation coefficient $r=+1$ has pairs falling along a straight line with positive slope. Figure 6.8 shows examples of scatter plots with their corresponding correlation coefficients. Notice that as the plotted points become more scattered around the best fit line, the correlation coefficient gets closer to zero.

Also note that correlation looks only for *linear* relationships—other relationships may be present but may fail to be detected, for example, curvilinear (parabolic), exponential and logarithmic curves, and S-shaped logistic curves, as shown in Figure 6.9.

Two factors can be correlated for many reasons. The order in which the correlation is calculated doesn't matter—the variables x and y in the formula can be interchanged without affecting the calculation:

$$r(x, y) = Corr(x, y) = Corr(y, x) = r(y, x) = \hat{\rho}$$

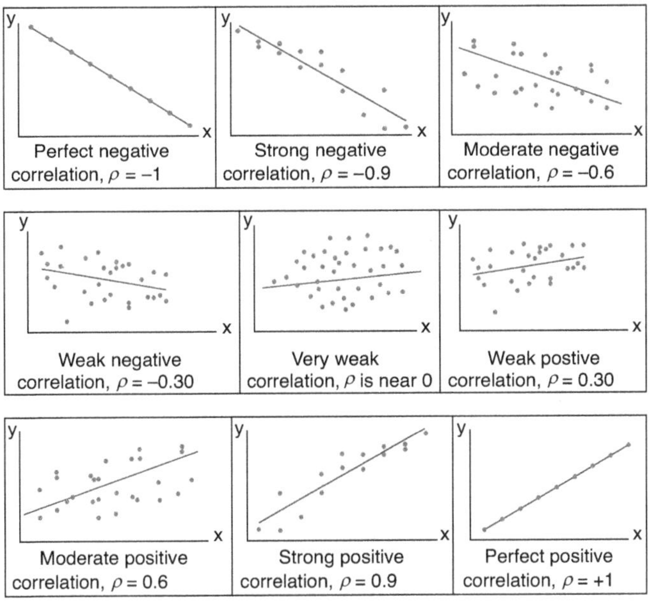

Figure 6.8 Correlation examples from perfect negative to perfect positive.

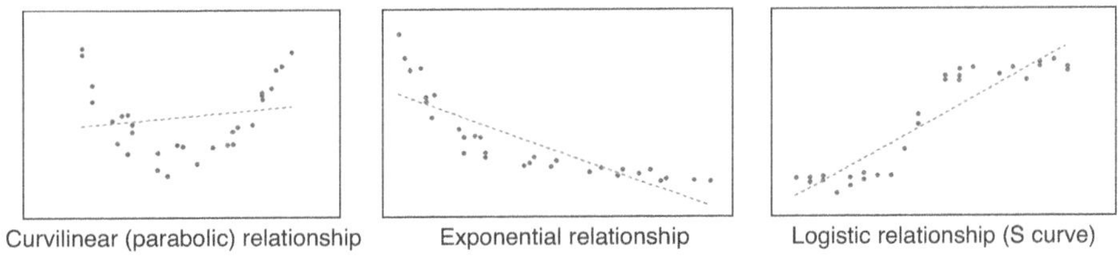

Figure 6.9 Scatterplot example for nonlinear relationships.

The correlation coefficient is also location and scale invariant, remaining unchanged if a constant c is added to or multiplied by the values of one or both variables. For example, the correlation calculated between a group's height and weight will remain constant whether the height is measured in inches or centimeters, or weight is measured in pounds or kilograms.

However, just because a correlation seems to be present does not indicate that it is meaningful or practically significant. Likewise, no discussion of correlation is complete without the requisite caution that *correlation does not imply causation!* If the variables X and Y are correlated, it is incorrect to then claim that X causes Y. For example, consider a study that shows that there is a significant positive correlation between the monthly amount of ice cream sold (X) and number of shark attacks that occur per month (Y). Will outlawing ice cream save swimmers from these dangerous attacks? Clearly, there is something else afoot. But what? Each of the variables, ice cream sales and shark attacks, is itself correlated to another, **latent**, variable—in this case, season of the year, or temperature. Shark attacks are correlated with ambient temperature because when it is hotter, more people go swimming in the ocean. Ice cream sales are also correlated with temperature because we eat ice cream to cool off from the heat. If we remove ambient temperature from the picture, shark attacks and ice cream sales are correlated themselves by means of their mutual relationship to temperature.

There are two forms of the standard measure of correlation, one parametric and one nonparametric, but they use the same equation. The parametric version is known formally as **Pearson's product–moment correlation coefficient**, less formally as the *Pearson correlation*, or just as *correlation*, when applied to numeric data. The nonparametric version uses the identical formula but is applied to **ranked data** instead and is known as **Spearman's rank correlation coefficient** or **Spearman's** ρ. Ranked data are obtained by sorting data from low to high then relabeling, such as 1, 2, 3 . . . n, based on the ordered value. Ordered data are denoted $X_{(i)}$ to indicate the ith ranked datapoint. The ranked values can then be used in analyses to lessen the effects of outliers or other unusual data features. There are a number of different forms of the equation, of which several are given here:

Equation 6.2 Pearson sample correlation coefficient *r*.

$$
\begin{aligned}
r &= \frac{Cov(X,Y)}{s_x s_y} \\
&= \frac{\sum(x_i-\bar{x})(y_i-\bar{y})/(n-1)}{\sqrt{\frac{\sum(x_i-\bar{x})^2}{(n-1)}}\sqrt{\frac{\sum(y_i-\bar{y})^2}{(n-1)}}} \\
&= \frac{\sum x_i y_i - n\overline{xy}}{\sqrt{\left(\sum x_i^2 - n\bar{x}^2\right)\left(\sum y_i^2 - n\bar{y}^2\right)}} \\
&= \frac{SS_{xy}}{\sqrt{SS_{xx}SS_{yy}}}
\end{aligned}
$$

where

$$
\begin{aligned}
SS_{xy} &= \sum x_i y_i - n\overline{xy} \\
SS_{xx} &= \sum x_i^2 - n\bar{x}^2 \\
SS_{yy} &= \sum y_i^2 - n\bar{y}^2
\end{aligned}
$$

The sample correlation coefficient can also be calculated in Excel using the = CORREL(,) function, providing two paired ranges as inputs.

EXAMPLE 6.15

Pearson Correlation

Consider the chemical process data presented in Figure 6.10, which shows the reaction times X in minutes and the yield of a chemical Y in grams. Calculate the sample Pearson correlation coefficient r between the reaction time and the yield of the chemical process. (The number of data points in this example has been kept small by design so that the reader can practice performing the calculations manually.)

Recall from Equation 6.2 that:

$$
r = \frac{Cov(X,Y)}{s_x s_y} = \frac{\sum x_i y_i - n\overline{xy}}{\sqrt{\left(\sum x_i^2 - n\bar{x}^2\right)\left(\sum y_i^2 - n\bar{y}^2\right)}} = \frac{SS_{xy}}{\sqrt{SS_{xx}SS_{yy}}}
$$

X(min)	Y(g)
7	9
8	10
9	13
11	14
12	15
14	12
16	17
17	18

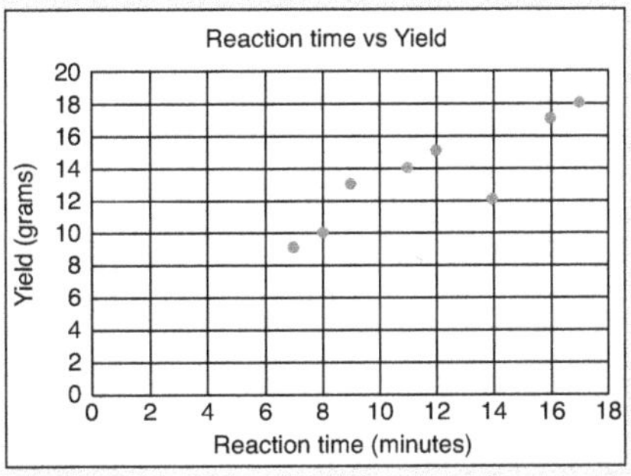

Figure 6.10 Reaction time and yield data.

We first determine the values of SS_{xx}, SS_{yy}, and SS_{xy}. Using the given values x and y, we obtain $n = 8$, $\bar{x} = 11.75$, and $\bar{y} = 13.5$ and use to calculate:

$$SS_{xy} = \sum x_i y_i - n\bar{x}\bar{y} = 1340 - 8(11.75)(13.5) = 71$$
$$SS_{xx} = \sum x_i^2 - n\bar{x}^2 = 1200 - 8(11.75)^2 = 95.5$$
$$SS_{yy} = \sum y_i^2 - n\bar{y}^2 = 1528 - 8(13.5)^2 = 70$$

Now, substituting in the values of SS_{xx}, SS_{yy}, and SS_{xy}, yields

$$r = \frac{SS_{xy}}{\sqrt{SS_{xx}}\sqrt{SS_{yy}}} = \frac{71}{\sqrt{95.5}\sqrt{70}} = 0.868$$

This is a moderately strong correlation, but it should be treated with caution due to small sample size of $n=8$. Additional data could be heavily influential.

EXAMPLE 6.16

Expanded Spearman Correlation

Consider the chemical process data presented in Example 6.15. Now use the ranked data to calculate the sample Spearman correlation coefficient r between the reaction time and the yield of the chemical process shown in Figure 6.11. (The number of data points in this example has been kept small by design so that the reader can practice performing the calculations manually.)

Applying Equation 6.2, we first determine the values of SS_{xx}, SS_{yy}, and SS_{xy}. Using the given values x and y, we obtain $n=8$ and $\bar{x} = \bar{y} = 4.5$ and use to calculate:

$$SS_{xy} = \sum x_i y_i - n\bar{x}\bar{y} = 198 - 8(4.5)(4.5) = 36$$
$$SS_{xx} = \sum x_i^2 - n\bar{x}^2 = 204 - 8(4.5)^2 = 42$$
$$SS_{yy} = \sum y_i^2 - n\bar{y}^2 = 204 - 8(4.5)^2 = 42$$

Now, substituting in the values of SS_{xx}, SS_{yy}, and SS_{xy} yields

$$r = \frac{SS_{xy}}{\sqrt{SS_{xx}}\sqrt{SS_{yy}}} = \frac{36}{(\sqrt{42})^2} = 0.857$$

For these data there is not a big difference between the parametric and non-parametric methods.

Continued

RankX	RankY
1	1
2	2
3	4
4	5
5	6
6	3
7	7
8	8

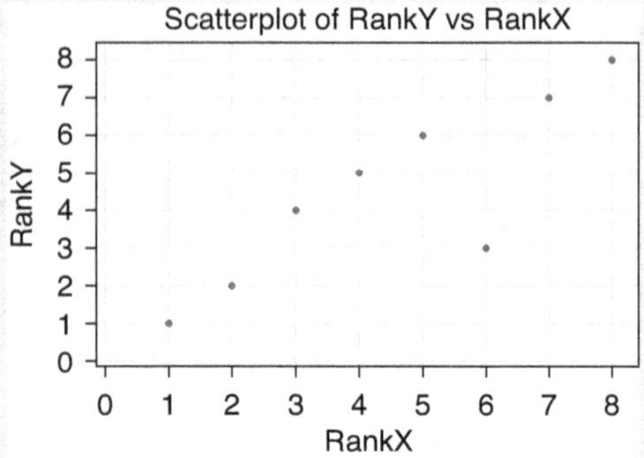

Figure 6.11 Ranked reaction time and yield data.

Simple Linear Regression (SLR)

Imagine that we are interested in predicting the diameter of a drilled hole. If the only data we had access to was historical diameter data, our best prediction of the diameter would be equal to the average diameter, or $\overline{y}$. Now imagine that we also have access to the age of the drill bit for each hole diameter. By incorporating this age data, we could potentially improve upon our prediction. In regression, we in effect partition the total variability of Y into two: the part explained by X, and the part not explained by X, which we call the **residual**, or random, **error**.

Regression analysis is comprised of a large family of related parametric methods used to construct prediction models by quantifying the effects of one or more predictors on a dependent variable, in the form of a **regression model**. The method of least squares is used to determine the values of the regression coefficients for the corresponding model. **Simple linear regression**, or SLR, the simplest regression model, produces a straight line through the data, using only a single X and a single Y.

We can use the concept of partitioning to develop a mathematical model to explain the relationship between the response variable Y and the input, or independent, variable X. SLR allows us to predict an expected value y from an input x. The goal in SLR is to quantify the total variability of the response variable, Y, and explain as much of that variability as possible using a single input variable, X, assuming it takes the specific form of a straight line. The term "simple" refers to the fact that we are using only one input variable in the prediction equation, in contrast to multiple linear regression (MLR) that uses more than one input variable. The term "linear" refers to the fact that the model parameters are estimated using linear combinations of the response variable, Y.

The general form of a *simple linear regression* model is shown in Equation 6.3:

Equation 6.3 General form of the simple linear regression model.

$$y = \beta_0 + \beta_1 x + \varepsilon$$

The variable Y is referred to as the response, or dependent, variable. The terms β_0 and β_1, the "betas", are model coefficients, unknown parameters that represent the y-intercept and slope, respectively. The variable X is the independent, or predictor, variable, and the term ε, "epsilon," is a random variable referred to as the residual, or random, error, which is the difference between the actual and the predicted values. This is a parametric model with three parameters, β_0, β_1, and ε. Because β_0 and β_1 are unknown parameters, they must be estimated from the data. We will denote these estimates as b_0 and b_1, respectively. Some texts use $\hat{\beta}_0$ and $\hat{\beta}_1$, respectively. The random error term ε is also unknown but considered to have a mean of zero, which intuitively makes sense since we want our errors to average out to zero. Putting the estimates in for the parameters, we can write the fitted regression equation or fitted trend line shown in Equation 6.4.

Equation 6.4 Fitted regression equation.

$$\hat{y} = b_0 + b_1 x$$

Note that the term $\hat{y}$ is pronounced "y hat." Statisticians often use the ^, or hat, symbol to denote an estimated value, as mentioned previously. The average value of the response variable y at a value of the predictor variable x can be estimated by Equation 6.4. The interpretation of the estimate b_0, or y intercept, is the value of y when it crosses the y-axis at $x=0$. The term b_1 is the estimated slope of the line. To interpret the slope, we can say that as the variable x increases by one unit, the predicted y will change by b_1 units.

The formulas for b_0 and b_1 are derived using the method of ordinary least squares. We choose the values of b_0 and b_1 that will minimize the sum of the squared residuals (SSE) from the model:

$$\min SS_E = \min \sum_{i=1}^{n} \left(y_i - \hat{y}_i\right)^2 = \min \sum_{i=1}^{n} \left(y_i - \left(b_0 + b_1 x_i\right)\right)^2$$

With a bit of calculus, we arrive at formulas for calculating the estimated intercept and slope of the regression equation. Note that b_1 is used in the calculation of b_0.

Equation 6.5 Estimated intercept and slope.

$$b_0 = \bar{y} - b_1 \bar{x}$$

$$b_1 = \frac{SS_{xy}}{SS_{xx}} = \frac{\sum_{i=1}^{n} x_i y_i - n\bar{x}\,\bar{y}}{\sum_{i=1}^{n} x_i^2 - n\bar{x}^2}$$

The estimated residuals e_i are calculated using Equation 6.6.

Equation 6.6 Estimated residuals.

$$e_i = \left(y_i - \hat{y}_i\right)$$

A visual depiction of the intercept, slope, observed values y_i, and residuals is shown in Figure 6.12. It is important to note that the linear regression model should only be used for **interpolation**, to predict responses from x values that are within the original range of the data. Using values outside this x range is called **extrapolation** and is not recommended. Since we have no information about the relationship between x and y outside the range of our original inputs, we should exercise great caution in predicting responses for values outside this input range.

Once we calculate the estimated model parameters for the SLR, we need to determine whether this model explains enough of the variability to be useful. How do we know this is better than noise? These questions can be answered by

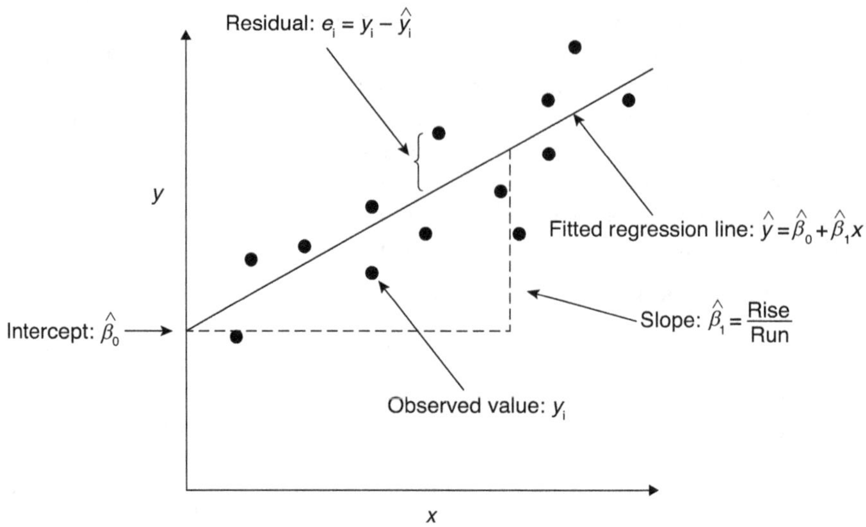

Figure 6.12 Graphical depiction of regression concepts.

conducting hypothesis tests to evaluate whether the model coefficients are significantly different from zero. A slope β_1 that is equal to zero implies that there is no relationship between the input $\boldsymbol{X}$ and the response variable $\boldsymbol{Y}$. We test the following hypotheses:

$$\mathrm{H}_0 : \beta_1 = 0$$
$$\mathrm{H}_1 : \beta_1 \neq 0$$

Equation 6.7 Two-sided test statistic for slope with decision rule.

$$t^* = \frac{b_1 - \beta_1}{s_{b_1}}$$

where

$$s_{b_1} = \frac{s_e}{\sqrt{SS_{xx}}} \text{ and } s_e = \sqrt{\frac{\sum_{i=1}^{n}(y_i - \hat{y}_i)^2}{(n-2)}} = \sqrt{\frac{SS_E}{(n-2)}}$$

Decision rule: Reject H_0 if $|t^*| > t_{\alpha/2,\, n-2}$

An analogous test is defined for other model terms, such as β_0.

When evaluating a model fit, hypothesis tests are conducted on each model term, to assess the hypothesis that a given β_i is statistically different from zero. Output from these tests will be provided automatically in statistical software. Output also typically provides residual plots for the model fit so that assumptions may be verified graphically. A model that fits well should have errors that are normally distributed with constant variance, also known as homoskedasticity. It should look generally random. The presence of unnatural patterns suggests a failure of assumptions then may need to be addressed with a more complex model in which case you should contact a statistician for additional support.

The slope b_1 is mathematically related to the correlation coefficient r discussed previously, using the relationships here:

$$r = \frac{b_1\sqrt{SS_{xy}}}{\sqrt{SS_{yy}}} \text{ and } b_1 = \frac{r\sqrt{SS_{yy}}}{\sqrt{SS_{xx}}}$$

where $SS_{xy} = \sum x_i y_i - n\bar{x}\bar{y}$, $SS_{xx} = \sum x_i^2 - n\bar{x}^2$, and $SS_{yy} = \sum y_i^2 - n\bar{y}^2$

Analysis of Variance (ANOVA)/ Covariance (ANCOVA)

Analysis of variance (ANOVA) methods are frequently paired with regression model results and are used with more complicated models to assess which factors are contributing to variability and how. They are commonly used to provide an alternative quantification of model inputs with multiple levels relative to baseline variability. ANOVA is covered in more depth in Chapter 11, "Design of Experiments (DOE)."

Models may be **fixed-effects** models, **random-effects** models, or **mixed-effects** models. *Fixed-effects* models contain factors with a limited and fixed discrete set of levels. For example, suppliers ABC, BCD, and CDE would be considered the only vendors available. *Random-effects* model factors have levels that are not fixed, but that are obtained from a larger population, namely, suppliers ABC, BCD, and CDE would be considered only three of many vendors available, of which these are a sample. *Mixed-effect* models contain both fixed and random factors. An ANOVA model with one or more continuous variables, or covariates, is called ANCOVA, or **analysis of covariance**.

Quantified results are summarized in an ANOVA table, which breaks out how variability is allocated in the dataset by source. Table 6.12 shows a notional table for two fixed factors, A and B, with k_A and k_B levels.

SS terms are sums of squares, as previously introduced, MS terms are mean squares calculated by dividing the SS by the degrees of freedom, which is a measure of information available about each item. For example, if there are three levels for each *A* and 2 for B, then *A* has two degrees of freedom and *B* has one. *F* statistics are used to assess the significance of the effect seen against the baseline level of unexplained variability as measured by the mean square for error, MSE.

Table 6.12 Calculations for a two-way ANOVA table.

Source	Degrees of freedom (df)	Sums of squares (SS)	Mean squares (MS)	F* *(test statistics)*	*p*-value *(Assesses significance)*
Factor A	$k_A - 1$	SSA	MSA = SSA/df_A	F_A^* = MSA/MSE	$\Pr(F_A^* > F_{crit,\, dfA,\, dfError})$
Factor B	$k_B - 1$	SSB	MSB = SSB/df_B	F_B^* = MSB/MSE	$\Pr(F_B^* > F_{crit,\, dfB,\, dfError})$
A*B Interaction	$(k_A - 1)(k_B - 1)$	SSAB	MSAB = SSAB/df_{AB}	F_{AB}^* = MSAB/MSE	$\Pr(F_{AB}^* > F_{crit,\, dfAB,\, dfError})$
Error	$(n_T - 1) - (k_A - 1) - (k_B - 1) - (k_A - 1)(k_B - 1)$	SSE	MSE		
Total	$n_T - 1$	SST			

Assessing Model Quality

Models should be judged based on how well they predict outcomes. Statistically significant p-values should be considered necessary but not sufficient. A number of different measures are typically provided in statistical output, of which a brief overview is provided here. A simple graphical depiction is to create a scatterplot of the actual Ys versus the predicted Ys. Better results will show a stronger correlation, with perfect results having a correlation of 1.0.

The **R-squared** (R^2) value given in the output is the **coefficient of determination**. This statistic is interpreted as the percentage of the variation in the response Y that is explained by the model. It takes on values between 0 and 100%. In the case of simple linear regression, R^2 is the square of the sample correlation coefficient r. When additional factors are added to the model, R^2 will always increase, but this does not necessarily reflect a benefit provided by the additional terms. *R^2 should not be used for model selection.*

$$R^2 = 1 - \frac{SSE}{SST}$$

Adjusted R-squared (R_a^2), the ***adjusted* coefficient of determination**, makes an adjustment for additional terms in the model and will decrease if unexplained (residual) variation does not decrease. It also takes on values between 0 and 100%. *R_a^2 is recommended for model selection*, since it is effective for comparing models of different sizes, that is, different numbers of parameters:

$$R_a^2 = 1 - \frac{\frac{SSE}{n-p}}{\frac{SST}{n-1}} = 1 - \left(\frac{n-1}{n-p}\right) \cdot \frac{SSE}{SST}$$

A good quality model will minimize **residual variation** left unexplained by the model, as measured by "root MSE," the square root of mean squared error (MSE), also referred to as RMSE. This value is the simplest metric but unscaled, so most users prefer to use R_a^2 instead. It takes on nonnegative values, so has no upper bound. R_a^2 will improve if RMSE decreases and decline if it increases.

$$RMSE = \sqrt{MSE} = \sqrt{\frac{SSE}{n-p}}$$

Akaike's information criterion (AIC) and the **Bayesian information criterion** (BIC) are also commonly provided in statistical output. These are both likelihood-based methods, so beyond the scope of this handbook, but they have similar intent to minimize residual variation with different adjustments to penalize for the model size. Lower values indicate a better model, but outcomes are not scaled.

EXAMPLE 6.17

Statistical Versus Practical Significance

Suppose that a statistical test with $n = 10{,}000$ data points randomly drawn from a database has shown that adding an additional step to a chemical process results in a statistically significant ($p = 0.001$) improvement of mean yield, from

560.40 g to 560.49 g. If the specifications for the yield are [550, 570], does this statistically significant result translate into practical significance? In this case, the large sample size increased the power of the test to a degree that was not necessary.

BASIC PROBABILITY CONCEPTS

Apply basic probability concepts such as independence, mutually exclusive events, and conditional probability. Calculate probabilities using tools and techniques (e.g., probability trees and expected frequency trees) and interpret their meaning. (Analyze)

Body of Knowledge III.A.2

The mathematical foundation of statistical methods is probability, which defines a way to consider the likelihood of uncertain events.

Definitions and Notation

The **probability** that a particular **event** occurs is a number between 0 and 1 inclusive, expressed mathematically as $p \in [0, 1]$. An event with a probability of exactly 0 means that the **outcome** is impossible and cannot happen, while a probability of exactly 1 means that the outcome is definite and will certainly happen.

The basic building blocks of probability are **events**, **trials**, and **outcomes**. Each unique occurrence that is possible for a particular scenario is known as a **simple event**. These simple events each have a specific probability of occurring and are collected into a **sample space**, which can be given a label, such as Ω (omega), often used to refer to a **universe**. The *sample space* contains all the possible simple events that can occur for a particular experiment or procedure. For example, if we toss a standard single six-sided die, the sample space for this procedure is composed of six simple events {1, 2, 3, 4, 5, 6}. If we are tossing two coins with outcomes either heads, H, or tails, T, the sample space would contain the simple events {HH, HT, TH, TT}. A **compound event** can be any combination of simple events that are of interest. For example, we could define compound event A as rolling a die and getting a number less than 3, resulting in the *event set* {1, 2}. Or, for our coin toss, we could define compound event R as having at least one head in two tosses. We see that Event R *contains* three simple events: {HH, HT, and TH}, also referred to as being **mapped** to the event or space.

By throwing the die, or tossing the coins, we are performing a *trial* of the experiment or procedure. The result of the trial is the *outcome*. Any outcome can be *mapped* to a simple event in the sample space. In summary, simple events are occurrences that are possible, and outcomes are actual occurrences, or results, of a trial. Compound events are a collection of two or more simple events.

There is a **classical definition of probability** that is predicated on the fact that the probability of each event in the sample space is known and equal. This will be true for cases such as die or coin tosses or selecting cards from a standard poker deck. Thus, the probability of event A, denoted Pr(A), is calculated as shown here:

$$Pr(A) = \frac{number\ of\ ways\ A\ can\ occur}{number\ of\ total\ simple\ events} = \frac{A}{N}$$

For situations in which the probabilities of the events are not known, we can perform an experiment with several trials and then count the number of favorable outcomes and divide by the number of trials. This is referred to as **empirical probability** or **relative frequency** shown in Equation 6.8.

Equation 6.8 Relative frequency.

$$Pr(A) = \frac{number\ of\ times\ A\ occurred}{total\ number\ of\ trials} = \frac{a}{n}$$

For example, if a lot consisting of 100 parts has four defective units, we would say that the probability that any unit in the lot may be defective is 0.04, or 4%. The empirical probability is an **estimate** of the true probability that A will occur. This true probability might be denoted Pr(A), p_A, or π_A, with the estimate denoted $\hat{p}_A$ or $\hat{\pi}_A$ (with "hats"). As the number of trials increases, this estimate will get closer and closer to the true probability of Event A due to the *law of large numbers* (LLN). The empirical probability equation is more useful for most real-world situations that cannot be modeled using simple events and probabilities are not generally known *a priori* (Latin phrase meaning "ahead of time").

Empirical probabilities can also be estimated computationally by using **indicator** variables, which can be helpful in summarizing results efficiently using standard equations. An Event A can be defined as a 0 or 1 for presence/absence or pass/fail or red/green or other *binary* (two-level) definition. Then a nominal estimate of the probability can be calculated from the *n* trials, indexed by *i*, using the formula for the mean to obtain the relative frequency, also known as the **proportion**:

$$\hat{p}_A = \frac{1}{n}\sum_{i=1}^{n} X_i,\ \text{where}\ X_i = \begin{cases} 1 & \text{if A present} \\ 0 & \text{if A not present} \end{cases}$$

The *proportion* thus reflects the share of the sample that has condition A. It can be used to estimate the probability of A but should *not* be expected to be exactly equal to the true probability of A.

In **reliability engineering**, the study of how and why things fail, we are often concerned with probabilities near 0 or 1. **Reliability** is defined to be the probability that a unit or system or process will operate successfully for the desired lifetime. High reliability and, therefore, low probabilities of failure are clearly desirable. When dealing with reliability and/or probabilities of failure near 1 or 0, it is important *not to round* estimates. Reliabilities of 0.99, 0.999, 0.9999, or 0.99999, are not equivalent to a reliability of exactly 1.0 but are instead associated with nonzero failure probabilities of 1/100, 1/1000, 1/10K, and 1/100K, any of which may or may not be materially relevant depending on the context of the situation. Thresholds of concern will vary depending on unique industry and company requirements and organizational practices. Some companies will define thresholds quantitatively, such as 1/10K, for hazard risk matrices used in risk assessment and management activities. Finally, one common convention when speaking of reliability estimates is to refer to them descriptively, to add clarity as to exactly what value is being discussed. For example, 0.9999 is referred to a "four 9s," 0.999995 as "five 9s 5," 0.991 as "two 9s 1," and so on.

Rules of Probability

Probability math has its own special rules, which can be more easily visualized using **Venn diagrams** (Figure 6.13). This graphical tool helps make clear the relationships between sets.

The **complement** of Event A includes all the simple events in which A does not occur. The total probability of an event and its complement *must always equal exactly 1*, representing the *universe* for that set. The notation for the complement of A is $\bar{A}$, read as "not A" or "A complement," also written A^c or $\sim A$, or –A, depending on the text.

The probability of $\bar{A}$ is equal to total probability 1 – Pr(A), formally stated in Equation 6.9:

Equation 6.9 Complement rule.

$$Pr(\bar{A}) = 1 - Pr(A)$$

In addition, Pr(A) can be found using $1 - Pr(\bar{A})$ since $Pr(A) + Pr(\bar{A}) = 1$. This complement rule comes in handy when we need to find the probability of Event A, but the calculation is laborious. We can instead find the probability of $\bar{A}$ and subtract the result from 1.

EXAMPLE 6.18

Complements

To find the probability of rolling a die and getting a number greater than or equal to 2, we would need to solve:

$$Pr(X \geq 2) = Pr(X = 2) + Pr(X = 3) + Pr(X = 4) + Pr(X = 5) + Pr(X = 6)$$

Using the complement, we instead can calculate the simpler calculation to arrive at the same answer:

$$Pr(X \geq 2) = 1 - Pr(X = 1)$$

The probability of two or more events occurring can be calculated by using the ideas of **union** and **intersection**. The *union* of two events A and B contains the

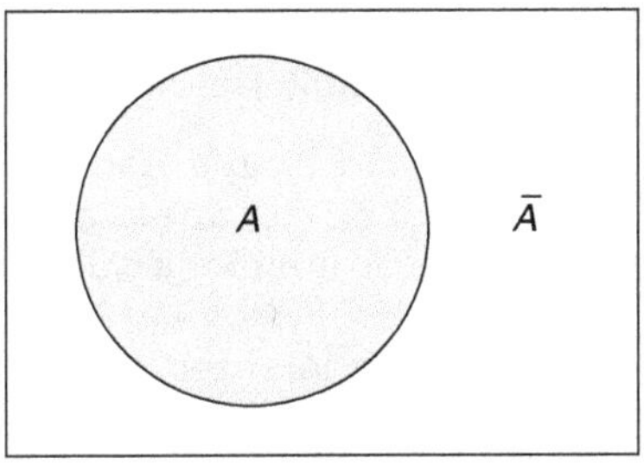

Figure 6.13 Venn diagram illustrating universe with Event A and its complement $\bar{A}$.

total number of unique simple events that comprise each event. For example, for a single die toss, let Event A be defined as rolling either a 1 or a 3, and Event B as rolling either a 1 or a 5. The union of these two events is {1, 3, 5}. We can write this by using the symbol for a union, ∪:

$$A \cup B = \{1, 3\} \cup \{1, 5\} = \{1, 3, 5\}$$

Note that A∪B is equivalent to B∪A. We can think of the union set as the numbers that will satisfy A **OR** B. An *intersection* between two events A and B contains only the simple events they have in common. The intersection of Events A and B, then, is {1}. We can use the symbol for intersection, ∩:

$$A \cap B = \{1, 3\} \cap \{1, 5\} = \{1\}$$

Note that A∩B is equivalent to B∩A. The intersection is the set of simple events that satisfy A **AND** B. Events A and B can be represented in a Venn diagram, as shown in Figure 6.14, which depicts both the union and intersection. Both AND and OR operations may be referred to as "gates," namely, the "∩ gate" or the "∪ gate."

Two events A and B are said to be **mutually exclusive** or **disjoint** if they cannot occur at the same time. For example, if we toss a coin one time, we cannot simultaneously flip a head (H) and a tail (T). As soon as the coin lands on heads, we know with certainty that the coin has not landed on tails. In quality, when we check the dimensions of a part with a go/no-go gage, the part will either pass or fail, not both. In this situation, the intersection is the empty, or null, set, denoted { } or ∅, also shown in Figure 6.14. When two events are mutually exclusive, calculating the probability of either event occurring requires only that we add the probabilities of the events together, known as the **special addition rule**:

$$Pr(A \cup B) = Pr(A) + Pr(B)$$

The special addition rule is *not* recommended for general use, since events may not always be mutually exclusive. Instead, the **general addition rule** should be used.

Equation 6.10 General Addition Rule.

$$Pr(A \cup B) = Pr(A) + Pr(B) - Pr(A \cup B)$$

This equation adjusts for the fact that the intersection area has been counted twice when adding together the probabilities of both Events A and B, so subtracts out one of them. The special addition rule leaves out this adjustment since it is unnecessary. The general formula can be safely used in *all* situations. If the events do not intersect, then the value for the intersection is zero.

EXAMPLE 6.19

Simple Unions

Suppose a card is randomly selected from a standard 52-card deck. What is the probability that the card is a club? Since there are 13 clubs, P(♣) = 13/52 = 0.25. What is the probability that the card is either a club or a spade? Since there are 26 cards that are either clubs or spades, P(♣ or ♠) = 26/52 = 0.5. Therefore, it appears that P(♣ or ♠) = P(♣) + P(♠), which, generalized, becomes the special addition rule: P(A or B) = P(A) + P(B). Caveat: Use only if events A and B are mutually exclusive and cannot occur simultaneously! The general addition rule described next is safer to use because it is always valid.

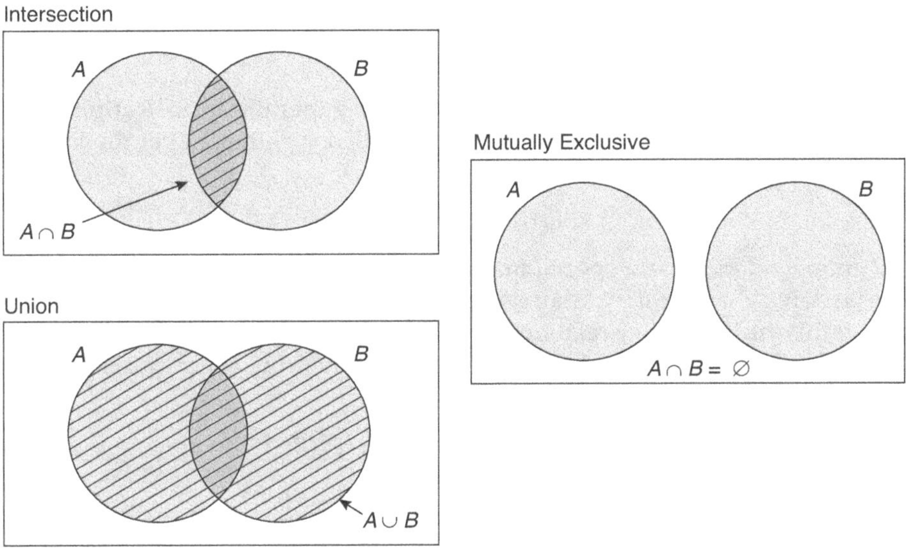

Figure 6.14 Venn diagrams depicting intersection, union, and mutually exclusive events.

EXAMPLE 6.20

General Unions and Intersection

In a standard 52-card deck, what is the probability of selecting either a king or a club? Using the special addition rule:

$$P(\text{K or } \clubsuit) = P(K) + P(\clubsuit) = \frac{4}{52} + \frac{13}{52} = \frac{17}{52}$$

This is incorrect, because there are only 16 cards that are either kings or clubs (the 13 clubs plus K♠, K♦, and K♥). The reason that the special addition rule doesn't work here is that the two events (drawing a king and drawing a club) can occur simultaneously.

The correct answer can be obtained by using the general addition rule:

$$P(K \cup \clubsuit) = P(K) + P(\clubsuit) - P(K \cap \clubsuit) = \frac{4}{52} + \frac{13}{52} - \frac{1}{52} = \frac{16}{52}$$

The intersection probability is obtained by recognizing that only one card is both a king and a club.

Two events, A and B, are said to be **independent** if the occurrence of one event does not influence the occurrence of the other. For example, let Event A be the outcome of a single toss of a coin, and Event B the outcome of a second coin toss. Events A and B are independent because the outcome of the first toss does not influence the outcome of the second. Regardless of what transpires on the first toss, the simple events for B are still {H, T}.

We can also think of independence relative to the transfer of information from trial to trial. If Events A and B are independent, knowing if Event A has occurred does not give us any additional information on the possible outcomes of Event B. Events that happen sequentially, such as **drawing (sampling) without replacement**, can induce dependencies.

If Events A and B are independent, then the **general multiplication rule** holds. In this case, their probabilities can be multiplied together to get the intersection, the probability that they will occur together:

$$Pr(A \cap B) = Pr(A) \cdot Pr(B)$$

Independence has a special mathematical property, in that this multiplication can be done *"if and only if"* (abbreviated *iff*) the two events are independent. Therefore, if this multiplicative relationship is true, then the events must be independent. Similarly, if Events A and B are independent, then the following statements are also true about A, B, $\bar{A}$, and $\bar{B}$:

$$Pr(A \cap \bar{B}) = Pr(A) \cdot Pr(\bar{B})$$

$$Pr(\bar{A} \cap B) = Pr(\bar{A}) \cdot Pr(B)$$

$$Pr(\bar{A} \cap \bar{B}) = Pr(\bar{A}) \cdot Pr(\bar{B})$$

EXAMPLE 6.21

Independence

A company determines that 25% of its customers are located in-state, 68% of customers are members of the reward program, and 17% of customers are both in-state and reward members. We can prove that Event A = (in-state location) and Event B = (reward member) are independent by showing that $Pr(A \cap B) = Pr(A) \cdot Pr(B)$:

Pr(In-state and Reward member) = 0.17
Pr(In-state) · Pr(Reward member) = (0.25) · (0.68) = 0.17

What is the probability that customers are either in-state or reward program members?

$$Pr(A \cup B) = Pr(A) + Pr(B) - Pr(A \cap B) = 0.25 + 0.68 - 0.17 = 0.76$$

In this example, independence of the events implies that customer location does not influence whether a customer joins the reward program, and vice versa. Knowing that a customer is in-state does not give us any information about whether they are a reward member.

For two events A and B, the **conditional probability** of Event A *given that Event B has occurred* is denoted as Pr(A | B), read as "A given B." In this case, the order of the events does matter: $Pr(A \mid B) \neq Pr(B \mid A)$, unless A and B are independent. Conditional probability can be explained graphically. Consider Event A residing in a given universe denoted Ω, that is divided into two parts, B and $\bar{B}$, as shown in the lefthand plot in Figure 6.15. By definition, the area of the universe is equal to one.

Once Event B has occurred, we can divide the universe to form Ω′ as shown to the right in Figure 6.15. Here, the probability of Event A given Event B is the ratio of the area of Event A divided by the area of Ω′, or Event B. The conditional probability Pr(A | B) is computed as shown here. Note that $Pr(A \cap B) \leq Pr(B)$.

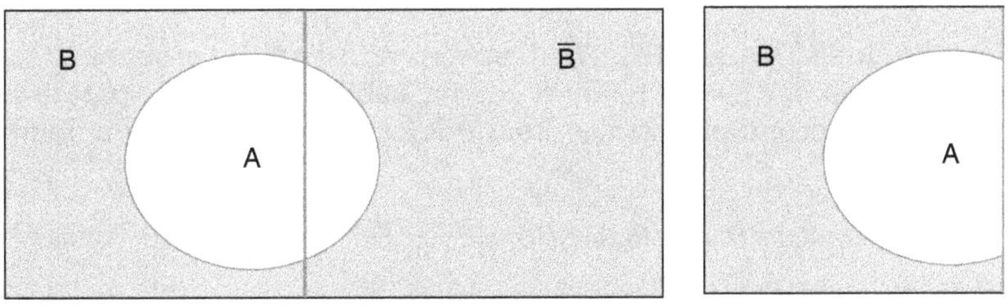

Figure 6.15 Full universe with Events A and B (left), conditional universe with A given B (right).

Equation 6.11 Conditional probability.

$$Pr(A \mid B) = \frac{Pr(A \cap B)}{Pr(B)}$$

This equation also leads to two different alternative expressions for the probability of two independent events, based on conditional probability:

Equation 6.12 Probability of an intersection.

$$Pr(A \cap B) = Pr(B \mid A) \cdot Pr(A) = Pr(A \mid B) \cdot Pr(B)$$

Thus, Events A and B are statistically independent when the probability of event A occurring does not depend on the level of event B:

$$Pr(A \mid B) = Pr(A)$$

EXAMPLE 6.22

Conditional Probability

If $Pr(A \cap B) = 0.25$ and $Pr(B) = 0.40$, find $Pr(A \mid B)$.

$$Pr(A \mid B) = \frac{Pr(A \cap B)}{Pr(B)} = \frac{0.25}{0.40} = 0.625$$

EXAMPLE 6.23

Dependence

A box holds 129 parts, of which six are defective. A part is randomly drawn from the box and placed in a fixture. A second part is then drawn from the box. What is the probability that the second part is defective? This is referred to as *sampling without replacement*. In other words, the probabilities associated with successive draws depend on the outcome of previous draws. Use the symbol D_1 to denote the event that the first part is defective and G_1 to denote the event that the first part is good, and so on. There are two mutually exclusive events that can result in a defective part on the second draw: good on first draw and defective on second, or else defective on first and defective on second. Symbolically, these two events are

Continued

(G_1 and D_2) or else (D_1 and D_2). The first step is to find the probability for each of these events. So, if we wish to know the probability of the first part drawn being good and the second part drawn being defective, by the general multiplication rule:

$$Pr(G_1 \cap D_2) = Pr(G_1) \cdot Pr(D_2 \mid G_1) = \frac{(129-6)}{129} \cdot \frac{6}{128} = \frac{123}{129} \cdot \frac{6}{128} \approx 0.045$$

Also, by the general multiplication rule:

$$Pr(D_1 \cap D_2) = Pr(D_1) \cdot Pr(D_2 \mid D_1) = \frac{6}{129} \cdot \frac{5}{128} \approx 0.002$$

Since the two events (G_1 and D_2) and (D_1 and D_2) are mutually exclusive, the special addition rule may be used:

$$Pr(D_2) \approx 0.045 + 0.002 = 0.047$$

When drawing two parts, what is the probability that one will be good and one defective? Drawing one good and one defective can occur in two mutually exclusive ways:

$$\begin{aligned} &Pr(\text{one good and one defective}) \\ &= Pr((G_1 \cap D_2) \cup (G_2 \cap D_1)) \\ &= Pr(G_1 \cap D_2) + Pr(G_2 \cap D_1) \end{aligned}$$

$$Pr(G_1 \cap D_2) = Pr(G_1) \cdot Pr(D_2 \mid G_1) = \frac{123}{129} \cdot \frac{6}{128} \approx 0.045$$

$$Pr(G_2 \cap D_1) = Pr(D_1) \cdot Pr(G_2 \mid D_1) = \frac{6}{129} \cdot \frac{123}{128} \approx 0.045$$

So,

$$Pr(\text{one good and one defective}) \approx 0.045 + 0.045 = 0.090$$

EXAMPLE 6.24

Dependence Using Complements

Consider a box with 10 electrical switches of which 6 are functional and 4 are defective. If we select two switches from the box, one after the other, what is the probability that we select exactly one defective switch?

Let Event A be defined as the condition of the switch on the first trial, and Event B be the condition of the switch on the second trial. When we first open the box, we have a probability of 6/10 of choosing a functional switch and 4/10 of selecting a defective switch.

What is the probability of obtaining a particular condition on the second trial? It depends. If we choose a functional switch on the first trial, then the box is left with 5 functional switches and 4 defectives. If we choose a defective switch first, we would be left with 6 functional switches and 3 defectives. Since the sample space for Event B depends on the outcome of Event A, we say that Events A and B are not independent.

The probability of choosing functional switches on both draws is $6/10 \cdot 5/9 = 1/3$. The probability they are both defective will be $4/10 \cdot 3/9 = 2/15$. The probability of selecting exactly one defective switch is thus

$$1 - \frac{1}{3} - \frac{2}{15} = 1 - \frac{7}{15} = \frac{8}{15}.$$

Discrete data are often tabulated in the form of a **contingency table**, where the frequency of events is indexed by rows and columns for different factors. Contingency tables provide an efficient way to organize data and enable simple statistical comparisons. However, it is important to be aware that effects of hidden factors may be masked by summaries that oversimplify results.

EXAMPLE 6.25

Conditional Probability with Contingency Table—Independence

Consider the count data in Table 6.13:

Table 6.13 Process count data XYZ versus FGH.

	X	Y	Z	Totals
F	17	18	14	49
G	18	11	16	45
H	25	13	18	56
Totals	60	42	48	150

Are events G and X independent?

$$Pr(G) = \frac{45}{150} = 0.30$$

$$Pr(G \mid X) = \frac{18}{60} = 0.30$$

The events G and X are statistically independent.

$$Pr(G) = Pr(G \mid X)$$

Knowing that a part is of type X does not affect the probability that it is of type G. Intuitively, two events are called independent if the occurrence of one does not affect the probability that the other occurs.

EXAMPLE 6.26

Conditional Probability with Contingency Table

The data in Table 6.14 shows the number of defective units found in samples from four different production lines.

Using this table, we can find the probability that a unit came from Line 2, given that it is defective. Define Event A as Line 2, and Event B as a defective unit. The probability of a unit being both defective and from Line 2 is 10/500. The probability of a defective unit is 45/500. Putting the information into the conditional formula yields:

$$Pr(A \mid B) = \frac{Pr(A \text{ and } B)}{Pr(B)} = \frac{10/500}{45/500} = 0.222$$

Continued

Table 6.14 Defects found per production line.

	Defective	Nondefective	Total
Line 1	3	87	90
Line 2	10	75	85
Line 3	25	230	255
Line 4	7	63	70
Total	45	455	500

We can also solve the problem by dividing the universe and concentrating only on the defective column as shown in Table 6.15.

Table 6.15 Reduced table, conditional on defectives.

	Defective
Line 1	3
Line 2	10
Line 3	25
Line 4	7
Total	45

The proportion of Line 2 units in the defective column is 10/45 = 0.2222.

Now, given that a unit is from Line 2, what is the probability that unit is defective? Here, we define Event A as a defective unit and Event B as Line 2.

$$Pr(A \mid B) = \frac{Pr(A \text{ and } B)}{Pr(B)} = \frac{10/500}{85/500} = 0.118$$

Counting Techniques

Statistical mathematics includes some special notation and mechanisms for correctly counting the number of ways different events could happen. Obtaining accurate probabilities is contingent on correctly counting these possibilities.

A more convenient way to determine the number of possible orderings of items is to use a **factorial**. A *factorial* of a positive integer (meaning 1, 2, 3, etc.) is the product of all the positive integers from 1 up to that number. For example, the factorial of 3 is equal to $1 \cdot 2 \cdot 3 = 6$. Note that due to the commutative property of multiplication we can also write this as $6 = 3 \cdot 2 \cdot 1$. The mathematical operator for factorial is an exclamation point (!). The factorial of 3 is denoted as 3! and is read as "three factorial." Note here that 4! can be written as $4 \cdot 3!$, $5! = 5 \cdot 4!$, $6! = 6 \cdot 5!$, and $7! = 7 \cdot 6!$. We can say in general that n items can be arranged in $n \cdot (n-1)!$ ways. We can also denote the factorial formula using a multiplication or product operator, the capital letter pi, Π. This is used analogously to the summation operator, Σ, but elements are multiplied together instead of added:

$$\prod_{i=1}^{n} i = n!$$

The multiplication operator, denoted by the Greek capital letter pi (Π), tells us to multiply an integer variable i letting i equal 1, all the way up to some number n. For $n=3$, we can find 3! by using the formula:

$$\prod_{i=1}^{3} i = 1 \cdot 2 \cdot 3 = 3!$$

This is an elegant way to succinctly represent the factorial formula. Note that there is only one way for zero things to be arranged in order. *Therefore, as a special case, zero factorial, or 0!, by definition, equals one.* This fact will become quite useful for permutations and combinations.

EXAMPLE 6.27

Factorial—Counting Unique Orderings

A reliability engineer selects four products and lines them up on a workbench for test. How many ways can the products A, B, C, and D be arranged on the bench? There are four choices for the first spot in the line. Once we have chosen a product for the first space, we have three products to choose from for the second spot, then two for the third spot, and then the remaining product has to be in the last space. Hence, we have $4 \cdot 3 \cdot 2 \cdot 1 = 4! = 24$ ways to arrange products A, B, C, and D. If product E is added to the sample, we now have $5! = 5 \cdot 4 \cdot 3 \cdot 2 \cdot 1 = 120$ arrangements. If product F is then added as well, we have $6! = 720$ unique orderings.

In the foregoing scenario, each product could be distinguished from the others. However, if we want to order n items where some items are identical to others, we need to modify our approach. Note that explicitly enumerating options may be viable for very small cases, but it is not recommended as a general rule.

The number of **permutations** gives us the number of ways r items out of n items can be arranged in a certain order. If we aren't concerned with the order of the selections, when we choose r items from a total of n items, we need the number of **combinations**. The number of *permutations* will always be higher than the number of *combinations* for the same values of n and r.

Total permutations can be obtained using the permutation formula:

Equation 6.13 Permutation Formula.

$$nPr = \frac{n!}{(n-r)!}, \ \textit{where } r \leq n$$

Total combinations can be obtained using the combination formula:

Equation 6.14 Combination Formula.

$$nCr = \binom{n}{r} = \frac{n!}{r!\,(n-r)!}, \ \textit{where } r \leq n$$

We read the term nCr as "n choose r." These equations are very similar, but the number of combinations has an extra factor $r!$ in the divisor that removes the ordered cases. Note that for $n=r$, both the permutation combination formulas reduce to the factorial since $(n-r)=0$ and $0!=1$.

EXAMPLE 6.28

Permutations

A box has 20 parts labeled A through T. Two parts are randomly selected. What is the probability that the two parts are A and T in that order? Note that selecting A and then T is different from selecting T and then A.

$$P = \frac{\#\ \textit{of ways an event can occur}}{\#\ \textit{of possible outcomes}}$$

The bottom term of the fraction is the number of orderings or permutations of two objects from a collection of 20 objects.

$$20P2 = \frac{20!}{(20-2)!} = \frac{20!}{18!} = \frac{20\cdot 19\cdot 18!}{18!} = 20\cdot 19 = 380$$

Of the 380 possible permutations, only one is AT, so the top term in the fraction is 1. The probability that the parts are selected in the specified order is $1/380 \approx 0.0026$.

EXAMPLE 6.29

Combinations

A box of 20 parts has two defectives. The quality technician inspects the box by randomly selecting two parts. What is the probability that both parts selected are defective?

$$P = \frac{\#\ \textit{of ways an event can occur}}{\#\ \textit{of possible outcomes}}$$

The "event" in this case is selecting two defectives, so "number of ways an event can occur" refers to the number of ways two defective parts could be selected. There is only one way to do this because there are only two defective parts; therefore, the numerator in the fraction is 1. The denominator in the fraction is the number of possible outcomes. This refers to the number of different ways of selecting two parts from the box, in other words, the *number of combinations of* two objects from a collection of 20 objects, calculated using the combination formula:

$$20C2 = \binom{20}{2} = \frac{20!}{2!\cdot(20-2)!} = \frac{20!}{2!\cdot(18)!} = \frac{20\cdot 19\cdot 18!}{2\cdot 1\cdot(18!)} = \frac{20\cdot 19}{2\cdot 1} = 10\cdot 19 = 190$$

The probability that both parts selected are defective is therefore $1/190 \approx 0.005$.

Also see "Binomial and Hypergeometric Distributions" in this chapter "Probability Distributions," where combinations are used in a statistical distribution.

Probability and Event Frequency Trees

Probability tree diagrams are used to model sequences of independent events, using the *general multiplication rule* discussed in the previous section. Both a graphical and quantitative tool, they consist of branches connecting **event nodes** to **outcomes** with an associated *probability* and can be easily sketched out as well as set up in a spreadsheet program for implementation.

In its simplest form, a tree diagram delineates all possible outcomes and their associated probabilities, which must sum to 1.0 for each node. Each node can have as many branches as necessary to describe the scenario, with probabilities p_i that quantify the probability of that set of outcomes occurring. The diagram is then extended as many steps as required. Probabilities along each branch are multiplied together to obtain the probability of that branch occurring, with final probabilities totaling to 1.0, so intended to capture all possible conditions. Figure 6.16 shows a simple tree diagram with an initial node, and the tree in Figure 6.17 has two nodes.

Consider a series of three independent coin tosses, where each outcome has $p=0.5$. There are $2^3=8$ possible outcomes, each branch with probability:

$$\prod_{j=1}^{3} p_{ij} = 0.5 \cdot 0.5 \cdot 0.5 = 0.5^3 = 0.125,$$

where i represents branches 1 to 8. The probabilities of the eight branches sum to 1.0, as shown in Figure 6.18 and Table 6.16.

Tree diagrams are very flexible in their structure and can be used to model a wide variety of scenarios. Some popular variations include an **event frequency tree**, where results are in terms of number of events instead of probability, using the binomial distribution for the conversion. Another type of **decision tree** is used in financial applications, using the probability at each node to weight an

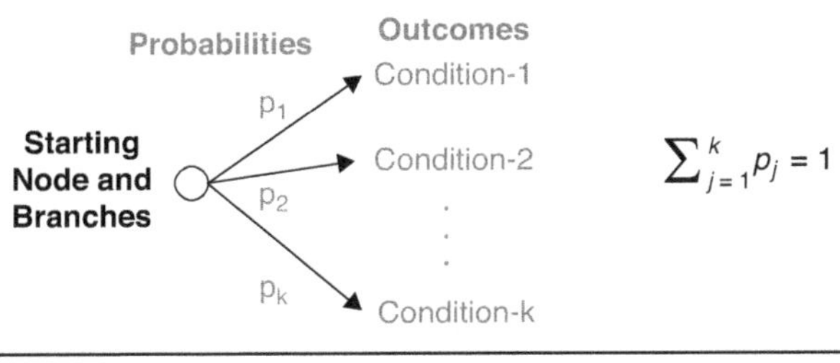

Figure 6.16 Simple tree diagram with initial node.

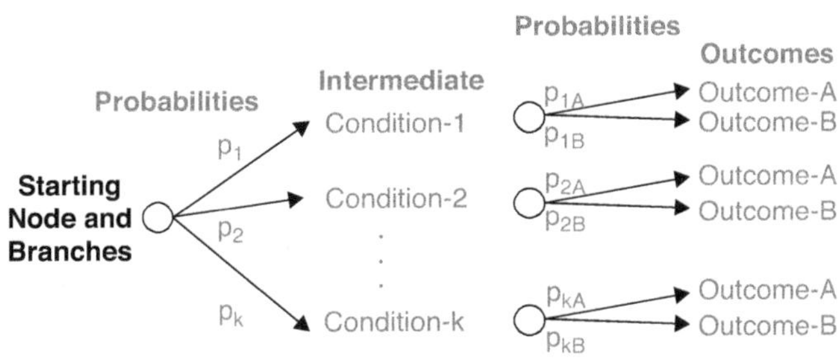

Figure 6.17 Simple tree diagram with two nodes.

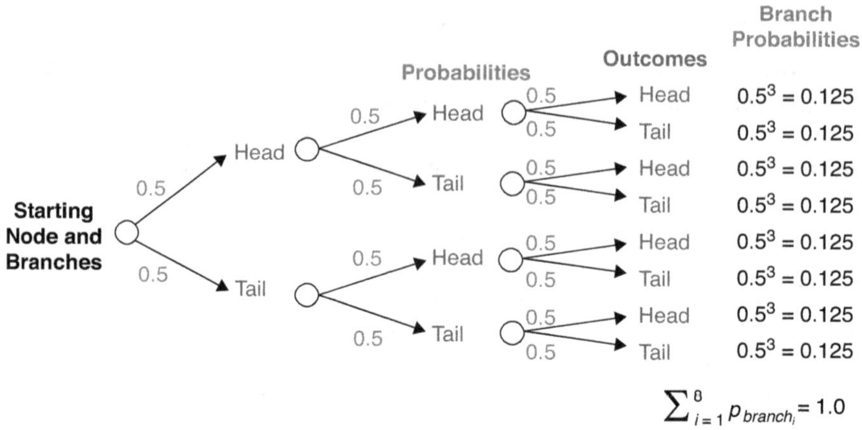

Figure 6.18 Tree diagram for the three-coin-toss scenario.

Table 6.16 Spreadsheet layout for three-coin-toss tree diagram.

Event1	Prob1	Event2	Prob2	Event3	Prob3	Sequence	Branch probability
H	0.5	H	0.5	H	0.5	HHH	0.125
				T	0.5	HHT	0.125
		T	0.5	H	0.5	HTH	0.125
				T	0.5	HTT	0.125
T	0.5	H	0.5	H	0.5	THH	0.125
				T	0.5	THT	0.125
		T	0.5	H	0.5	TTH	0.125
				T	0.5	TTT	0.125
						Total	1.000

expected gain or loss, for instance, a cost assessment such as net present value (NPV), to make investment decisions. The structure also lends itself well toward incorporating uncertainty in estimates, which can be done via simulation, considering the information available for each node probability and/or value for other incorporated factors.

In reliability engineering, an **event sequence diagram** (ESD) is used to lay out the sequence of events leading to failure, or other undesired events, to understand and assess specific risks. When quantified, with a probability assigned to each step, these are known as a **quantitative risk analysis** (QRA). ESDs focus only on the branches that lead to undesirable outcomes, so are incomplete compared with a true probability tree diagram but are simpler and easier to use. An **event tree** is an alternative version that includes all possible paths, usually to ensure that all paths in a complex series of events have been identified and appropriately quantified, so is a true probability tree diagram. **Fault trees** are another tree structure most often used in investigations or to model very complex systems, but they are not often fully quantified due to the resources required. Fully quantified fault trees are widely used in the nuclear industry, where this is known as **probabilistic risk assessment** (PRA) and used to manage system risks associated with operating and managing nuclear facilities.

EXAMPLE 6.30

Coin Flip Tree Diagram with Indicator Variable

Refer to Figure 6.18 and calculate the probability that the series of three coin flips has at least two heads, in any order.

Four events have at least two heads, resulting in an estimate for at least two out of three heads, leading to an answer of 4/8 = 0.5. We can obtain this by inspection in this case, but an alternative approach useful for larger problems is to set up an indicator variable testing for the condition of concern. The indicator variable can be used to flag ("indicate") when the condition is present, by assigning a value of 1 if present and 0 if not. These can then be multiplied with the branch probability and then totaled to obtain the estimate for the case of interest (see Table 6.17).

Table 6.17 Three-coin-toss tree diagram with indicator variables.

Event1	Prob1	Event2	Prob2	Event3	Prob3	Sequence	Branch Probability	Indicator: 1 if 2H+	Indicator probability
H	0.5	H	0.5	H	0.5	HHH	0.125	1	0.125
				T	0.5	HHT	0.125	1	0.125
		T	0.5	H	0.5	HTH	0.125	1	0.125
				T	0.5	HTT	0.125	0	0
T	0.5	H	0.5	H	0.5	THH	0.125	1	0.125
				T	0.5	THT	0.125	0	0
		T	0.5	H	0.5	TTH	0.125	0	0
				T	0.5	TTT	0.125	0	0
						Total	1.000		0.500

EXAMPLE 6.31

Coin Flip Tree Diagram with Event Frequencies

Refer to Example 6.30 and estimate the number of expected events in $n=550$ total coin flips where we see at least two heads (see Table 6.18).

Table 6.18 Three-coin-toss tree diagram with expected events.

Event1	Prob1	Event2	Prob2	Event3	Prob3	Sequence	Branch Probability	Indicator: 1 if 2H+	Indicator Probability	Expected events out of Total 550
H	0.5	H	0.5	H	0.5	HHH	0.125	1	0.125	68.75
				T	0.5	HHT	0.125	1	0.125	68.75
		T	0.5	H	0.5	HTH	0.125	1	0.125	68.75
				T	0.5	HTT	0.125	0	0	0
T	0.5	H	0.5	H	0.5	THH	0.125	1	0.125	68.75
				T	0.5	THT	0.125	0	0	0
		T	0.5	H	0.5	TTH	0.125	0	0	0
				T	0.5	TTT	0.125	0	0	0
						Total	1.000		0.500	275

Results are obtained by multiplying the branch probabilities by the total number of coin flips and totaling. The answer is 275, which is 50% of 550 planned flips.

EXAMPLE 6.32

Engineering Analysis Event Tree Diagram

The following event tree shows the sequence of events that could lead to failure for an issue. For conservatism, assume that initiating event (IE) definitely occurs with probability equal to one. Then calculate the probability of failure if pivotal events (PE) PE1, PE2, and PE3 are assumed to have values of 0.01, 0.00135, and 0.5.

The probability of failure is calculated as follows:

$$\prod_{i=0}^{3} p_{\text{event}_i} = p_{IE} \cdot p_{PE1} \cdot p_{PE2} \cdot p_{PE3} = (1)(0.01)(0.0135)(0.5) = \frac{1}{148200}$$

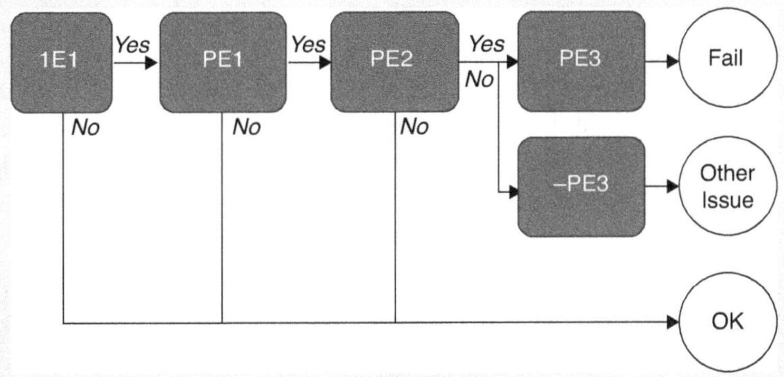

Figure 6.19 Event tree diagram for pivotal events.

PROBABILITY DISTRIBUTIONS

> Analyze reliability data using appropriate statistical distributions such as binomial, Poisson, exponential, Weibull, normal, log-normal, chi-square, and Student's t distribution. Interpret the associated probability plots for each continuous distribution. Assess goodness-of-fit to a distribution. (Analyze)
>
> **Body of Knowledge III.A.3**

Statistical distribution functions are the building blocks of statistical methods, providing concise yet flexible mathematical expressions used in many modeling scenarios to represent data or other assumptions.

Statistical Distribution Functions

Statistical distributions are used to mathematically express the nature of a random variable in a concise manner. However, not every distribution can be used in all circumstances. Different distributional forms must be used for different types of data. These are broadly classified into models used for **discrete** data and those for **continuous** data. There is a difference in terminology as well. When discussing *discrete* data models, functions are described in terms of **mass**, where *continuous* data models are described in terms of **density**. These are all *parametric* models, as discussed in this chapter, "Basic Statistics."

Each statistical distribution model can be expressed in multiple forms, related through calculus. **Probability functions**, for either *mass* or *density*, referred to as PMFs or PDFs, can be visualized as histograms or curves that reflect the probability at that value of the random variable. **Cumulative functions** are used to calculate the area under the curve to the left of a point X and always progress from 0 to 1.

They are referred to as CMFs or CDFs: cumulative mass functions or cumulative distribution functions. Probability functions are always scaled so that the total area under the curve is exactly 1.0.

Reliability engineering is concerned with the probability that a system fails after a specified time, t. Statistical CDFs are used to determine the probability of failure prior to time t, then subtracted from 1 to obtain the probability that it is still operating afterward, known as the **reliability**:

$$R(t) = Pr(X > t) = 1 - Pr(X \leq t)$$

Note that this definition of "success" is arbitrary and can be better described as the "outcome of interest." For example, we could just as well be counting the number of defectives in a sample as the number of good parts. However, the general practice in industry is to focus on the parts that are *not* working, not the ones that are, so that convention is typically used here. The next section provides an overview of commonly used statistical distributions—first discrete, then continuous. This list is by no means exhaustive, but it summarizes models used most often in reliability analyses. A summary table of the models for these probability functions appear at the end of this subsection in Table 6.17 for easy reference. Model forms that are particularly complex have been omitted. Models may have different parametrizations, but the one most common in reliability applications is shown here. Tables of critical values are provided in the appendices, but in general, computations are done in software using preprogrammed functions. The focus of this section is the different model forms, typical uses, as well as how to assess **goodness-of-fit** (GOF), that is, whether a given model is providing a good quality fit to the data. "Probability Functions," in this chapter goes into more depth on the relationship between different functional forms and discusses how to perform a distributional comparison across groups.

Distributions differ by where they are located or centered, the amount of spread or scatter present, whether the distribution is symmetric or skewed, and how peaked the mode is. Theoreticians refer to these as the first four moments, but for practical purposes it is important to recognize these general concepts: **location**, **scale** (spread), and **shape** (skewness and kurtosis**)**. Each distribution has one or more parameters that, either individually or in combination, are used to change the location, scale, and shape of the distribution. For variables that have a constraint, such as lifetimes that cannot be less than zero, there may also be a **threshold** parameter to shift this lower bound.

The commonly used parametric statistical distribution models are all unimodal by default, meaning that they only have one peak. Distributions with more than one peak are known as **bimodal** if they have two modes and otherwise **multimodal**. Multiple modes will generally indicate the presence of more than one subgroup, or a process that is moving, and therefore not stable or predictable. *Skewness* can indicate the same issue, so it is critical to **always look at data in time order** before attempting to fit distributions. **Mixture distributions** can be obtained that fit multiple distributions simultaneously, for instance, a mixture of two normal distributions; however, these should be used with caution since they may not be truly representative. More data are required to fit more complex models, so more parsimonious models are preferred. In reliability engineering this may be even more of a challenge due to the presence of censored data, which is data known incompletely in some way.

The probabilities for the PMF and CMF of the binomial distribution are tabulated in Appendices B.1 and B.2, respectively. The Poisson PMF and CMF probabilities can be found in Appendices C.1 and C.2, respectively.

Discrete Distributions

Bernoulli, Binomial, Negative Binomial, and Geometric

Binary **Bernoulli trials** are the foundation for the binomial, negative binomial, and geometric distributions. These are closely related and so discussed together here.

In a *Bernoulli trial*, there are only two possible mutually exclusive outcomes, such as heads or tails, pass or fail, success or failure, or defective or nondefective. This model has a single shape parameter p, representing the true proportion of the target outcome in the population. The probability of this outcome of interest is equal to p, and by the complement rule, the probability of the second outcome is equal to $(1-p)$, also denoted q. Trials must be independent with the probability of "success" p constant from trial to trial. Trials are designated "*iid*," for "*independent and identically distributed*," read as "I-I-D."

A Bernoulli random variable X is denoted $X \sim$ *iid Bernoulli*(p). It would have expected value $E(X)=p$, with standard deviation $\sqrt{p\cdot(1-p)}=\sqrt{(p\cdot q)}$, with p estimated using the equation for the mean:

$$E(X)=\hat{p}=\frac{x}{n}=\frac{1}{n}\sum_{i=1}^{n}X_i$$

where $X_i=0$ *or* 1 and x is total count of events of interest.

The **binomial distribution** is obtained by counting (summing) the number of outcomes of interest from a series of independent Bernoulli trials. This results in a discrete, finite set of possible values for the total sum. For example, if we toss a coin three times, each toss is an *iid* Bernoulli trial. If we count the number of heads in the three trials, this count, X, will then follow a binomial distribution, with possible values of 0, 1, 2, or 3, assuming the conditions stated previously are met,

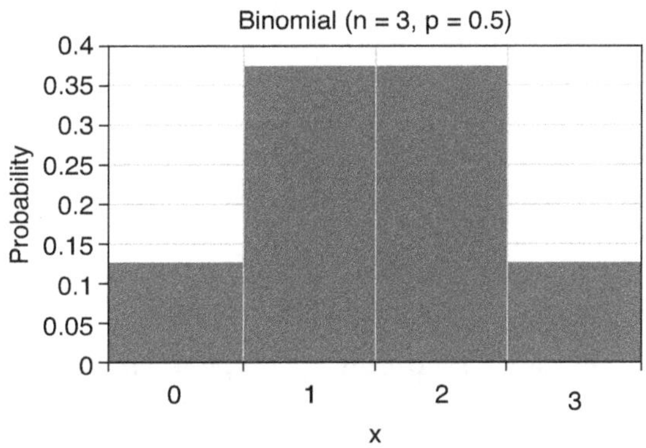

Figure 6.20 Probability mass function of binomial ($n=3, p=0.5$).

meaning that there is a certain known calculable probability associated with each possible outcome, as shown here for this example.

A binomial random variable X is denoted $X \sim Binomial(n, p)$ or further abbreviated $B(n, p)$, $Bin(n, p)$, or $Binom(n, p)$ as needed for clarity. It would have expected value $E(X) = n \cdot p$, with standard deviation $\sqrt{np(1-p)} = \sqrt{npq}$, with p estimated as previously, for n known. The *binomial probability mass function* (Figure 6.20) gives the probability of exactly x "success" outcomes out of n possible outcomes, assuming the probability p of success remains the same for each trial:

$$Pr(X = x \mid n, p) = \binom{n}{x} p^x (1-p)^{n-x}$$

The *binomial cumulative mass function* gives the probability of x or fewer "successes" out of n, for a given p:

$$Pr(X \le x \mid n, p) = \sum_{k=0}^{x} \binom{n}{k} p^k (1-p)^{n-k}$$

These values traditionally were obtained from tables for specific combinations of n and p, but now spreadsheet programs or statistical software provide exact calculations for computational purposes. Example 6.33 and 6.34 demonstrate how to interpret binomial tables, found in Appendices B.1 and B.2. Equations are summarized in Table 6.21.

The binomial and Bernoulli distributions are the foundation of many methods in reliability and quality engineering. For example, the binomial is used when counting the number of defective units in a sample. It is the basis for the np and p statistical process control charts. When the binomial is used to model reliability data, x could be the number of units that fail, and is denoted as r. If x becomes the number of units that do not fail, it is denoted as $n-r$.

EXAMPLE 6.33

Binomial Calculations

Suppose 25% of a large population of parts has failed. If six parts are selected at random, find the probability that none of the six has failed.

In this case, $n=6$, $p=0.25$, $x=0$. Substituting into the binomial formula,

$$Pr(X = 0) = \binom{6}{0} (0.25)^0 (1-0.25)^6 = \frac{6!}{6! \cdot 0!} (1)(0.75)^6 = (0.75)^6 \approx 0.18$$

EXAMPLE 6.34

Binomial Distribution Complete Enumeration

Find the binomial distribution for $p=0.25$ and $n=6$, and draw the associated histogram. What is the probability that the sample has one or fewer nonconforming units?

Following are binomial calculations for $X=0$ to 6. See Figure 6.21.

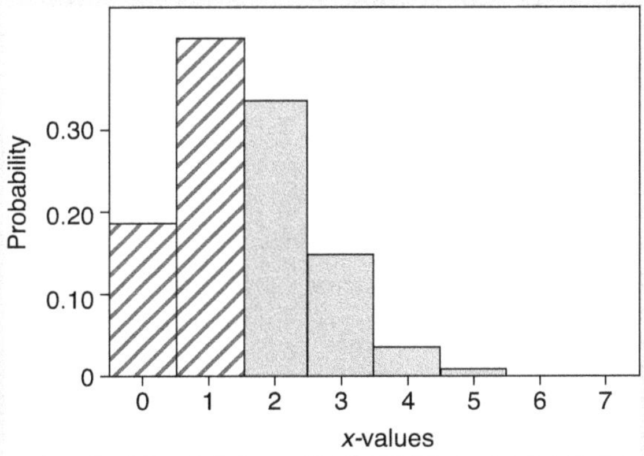

Figure 6.21 Binomial probability distribution and histogram for $p=0.25$ and $n=6$, with striped area designating $X \le 1$.

$$Pr(X=0)=\binom{6}{0}(0.25)^0(1-0.25)^6=\frac{6!}{(6!)(0!)}(0.25)^0(0.75)^6=(1)(1)(0.75)^6=0.1780$$

$$Pr(X=1)=\binom{6}{1}(0.25)^1(1-0.25)^5=\frac{6!}{(5!)(1!)}(0.25)^1(0.75)^5=(6)(0.25)(0.75)^5=0.3560$$

$$Pr(X=2)=\binom{6}{2}(0.25)^2(1-0.25)^4=\frac{6!}{(4!)(2!)}(0.25)^2(0.75)^4=\frac{6\cdot 5}{2}(0.25)^2(0.75)^4=0.2966$$

$$Pr(X=3)=\binom{6}{3}(0.25)^3(1-0.25)^3=\frac{6!}{(3!)(3!)}(0.25)^3(0.75)^3=\frac{6\cdot 5\cdot 4}{3\cdot 2}(0.25)^3(0.75)^3=0.1318$$

$$Pr(X=4)=\binom{6}{4}(0.25)^4(1-0.25)^2=\frac{6!}{(2!)(4!)}(0.25)^4(0.75)^2=\frac{6\cdot 5}{2}(0.25)^4(0.75)^2=0.0330$$

$$Pr(X=5)=\binom{6}{5}(0.25)^5(1-0.25)^1=\frac{6!}{(1!)(5!)}(0.25)^5(0.75)^1=(6)(0.25)^5(0.75)=0.0044$$

$$Pr(X=6)=\binom{6}{0}(0.25)^6(1-0.25)^0=\frac{6!}{(0!)(5!)}(0.25)^6(0.75)^0=(1)(0.25)^6(1)\approx 0.0002$$

We can then calculate $Pr(X \le 1)$ as $0.1780+0.3560=0.5339$.

We can also use a table of binomial probabilities, provided in the Appendix B.1 (PMF) and B.2 (CMF). Based on n and p, we can find $Pr(X=x)$ in the interior of the table. These are available for common probabilities. Note that p and $1-p$ must be interchanged to obtain values from 0 to 1. These calculations can also be performed in Excel using the BINOM.DIST function, using False to obtain the PDF and True for the CDF.

The **negative binomial distribution** similarly uses a sequence of *iid* Bernoulli trials to model a related scenario, the number of trials to achieve r events. Note that n here is not known, but is the random variable being modeled. There are multiple formulations of this in different texts that vary based on how the number of events r are assigned to the "success" and "failure" probabilities p and $1-p$. The version used here is from a statistical text and assigns the r events to success probability p.

A negative binomial random variable X is denoted $1 \cdot X \sim NegBin(r, p)$ or further abbreviated $NB(r, p)$ as needed for clarity. It would have expected value

$E(X) = r/p$, with standard deviation $\sqrt{r(1-p)}\,/\,p$, with p estimated as previously. The *negative binomial probability mass function* gives the probability of exactly r "success" outcomes, assuming the probability p of success remains the same for each trial:

$$Pr(X = n \mid r, p) = \binom{n-1}{r-1} p^r (1-p)^{n-r}, \text{ where } n \geq r$$

The *negative binomial cumulative mass function* gives the probability of n or fewer trials, for a given p and number of events r:

$$Pr(X \leq n \mid r, p) = \sum_{k=0}^{n} \binom{k-1}{r-1} p^r (1-p)^{k-r}$$

The **geometric distribution** is a special case of the negative binomial, with $r = 1$. It models the number of trials to the first event. Equations are summarized in Table 6.21 later in the chapter.

A geometric random variable X is denoted $X \sim Geometric(p)$ or further abbreviated $G(p)$, $Geo(p)$, or $Geom(p)$ as needed for clarity, or equivalently $NB(1, p)$ or $NegBin(1, p)$. It would have expected value $E(X) = 1/p$, with standard deviation $\sqrt{(1-p)}\,/\,p$, with p estimated as above.

The geometric distribution is the only discrete distribution with the "**memoryless" property**, meaning that the length of time spent already waiting for a given event does not provide any information about time remaining until it occurs. The probability of it occurring is the same s. This is referred to as having a **constant hazard rate**, analogous to the exponential distribution, to be discussed in the next section, which is the only memoryless continuous distribution.

EXAMPLE 6.35

Negative Binomial Calculations and Interpretation

Approximately 5% percent of engines sold in the last year are thought to have been manufactured with a dangerous suspect condition, but inspection requires an expensive teardown. The engineering team would like to identify three sample engines with this condition as soon as possible for an investigation. What is the mean number of total number of teardowns expected? What is the probability that the company might need to tear down more than 100 units? Where is the 80th percentile?

The expected number of teardowns can be obtained through the expectation formula for the negative binomial distribution:

$$E(X) = \frac{r}{p} = \frac{3}{0.05} = 60$$

An analyst produces Table 6.19 using the negative binomial CMF.

The probability of this effort requiring over 100 teardowns is:

$$\begin{aligned} Pr(X > 100) &= \sum_{101}^{\infty} \binom{n-1}{r-1} p^r (1-p)^{n-r} \\ &= 1 - Pr(X \leq 100) = 1 - 0.882 = 0.118 \text{ } or \text{ } 11.8\% \end{aligned}$$

Per the table, the 80th percentile can be seen to be between 80 and 100. Additional calculation results in a value of 85 teardowns. The median, the 50th percentile, is 54 teardowns.

Table 6.19 Probability table for number of engine teardowns until three defectives found.

Total teardowns required	Cumulative probability	Probability in interval
Less than 20	7.5%	7.5%
20 to 40	32.3%	24.8%
40 to 60	58.3%	25.9%
60 to 80	76.9%	18.7%
80 to 100	88.2%	11.2%
100 to 120	94.2%	6.1%
120 to 140	97.3%	3.1%
140 to 160	98.8%	1.5%
160 to 180	99.5%	0.7%
180 to 200	99.8%	0.3%

Negative binomial calculations can be done in Excel using the NEGBIN.DIST function, using False to obtain the PDF and True for the CDF (Figure 6.22). This is a different parametrization than is quoted here. The number of successes is 3, failures is $n-3$, and success probability is 0.05.

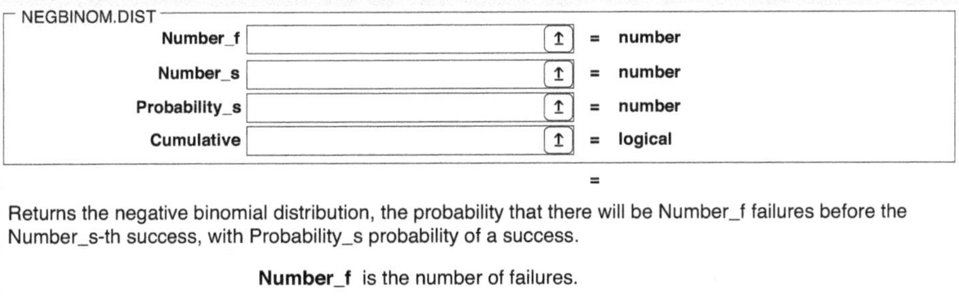

Figure 6.22 Excel formula dialog box for negative binomial distribution.

EXAMPLE 6.36

Geometric Calculations and Interpretation

For the previous problem, how many teardowns are expected until the first defective unit is found? What is the probability it will require over 40 teardowns to find the first unit needed for the investigation?

Continued

$$E(X) = \frac{1}{p} = \frac{1}{0.05} = 20$$

See Table 6.20.

Table 6.20 Probability table for number of teardowns until first defective found.

Total teardowns required	Cumulative probability	Probability in interval
Less than 20	64.2%	64.2%
20 to 40	87.1%	23.0%
40 to 60	95.4%	8.2%
60 to 80	98.3%	3.0%
80 to 100	99.4%	1.1%
100 to 120	99.8%	0.4%
120 to 140	99.9%	0.1%
140 to 160	100.0%	0.0%
160 to 180	100.0%	0.0%
180 to 200	100.0%	0.0%

Using the geometric probability distribution, the probability of this effort requiring more than 40 teardowns is:

$$\Pr(X > 40) = \sum_{n=41}^{\infty} p(1-p)^{n-1}$$
$$= 1 - \Pr(X \leq 40)$$

Using Table 6.20, we have $\Pr(X>40) = 1-0.871 = 0.129$.

Per the table, the 80th percentile can be seen to be below 40. Additional calculation results in a value of 45 teardowns. The median, the 50th percentile, is 18 teardowns.

Geometric calculations can be done in Excel using the NEGBIN.DIST function, using False to obtain the PDF and True for the CDF. This is a different parametrization than is quoted here. The number of successes is 1, failures is $n-1$, and success probability is 0.05.

EXAMPLE 6.37

Geometric Memoryless Property

The probability of a major system malfunction on any day is assumed to be 0.02. What is the probability that a major system malfunction will occur tomorrow? How many days do we expect to wait until the next major system malfunction?

The probability that a major system malfunction will occur tomorrow is $p = 0.02$.

$$Pr(X = 1) = p(1-p)^{n-1} = (0.02)(1-0.02)^{1-1}$$
$$= 0.02$$

The expected number of days until a major system malfunction is 50:

$$E(X) = \frac{1}{p} = \frac{1}{0.02} = 50$$

What is the probability of experiencing a major system malfunction in the next 3 days?

$$\begin{aligned} Pr(X \le 3) &= \sum_{n=1}^{3} p(1-p)^{n-1} = \sum_{n=1}^{3} (0.02)(1-0.02)^{n-1} \\ &= 0.02 \cdot 0.98^0 + 0.02 \cdot 0.98^1 + 0.02 \cdot 0.98^2 \\ &= 0.02 + 0.0196 + 0.0192 \\ &= 0.0588 \end{aligned}$$

If we make it through the next 50 days without a major system malfunction, what is the probability that we will have one on the 51st day?

The geometric distribution is memoryless, with constant probability of failure, so the probability that a major system malfunction will occur on the 51st day is $p = 0.02$. This can be verified by calculating the instantaneous hazard function, see Chapter 6, "Probability Functions."

$$h(t) = \frac{f(t)}{R(t)} = \frac{Pr(X=51)}{1-Pr(X \le 50)} = \frac{(0.02)(0.98)^{50}}{1-0.6358} = \frac{0.00728}{0.3642} = 0.0204$$

This value is the same regardless of the waiting time. Geometric calculations can be done in Excel using the NEGBINOM.DIST function with the number of failures equal to 1, using False to obtain the PDF and True for the CDF.

Hypergeometric

The **hypergeometric distribution** models a scenario known as the classic urn problem, in which a sample of size n is drawn without replacement from a hypothetical urn containing N total marbles, K of one color and $N-K$ the other, that is, red and green. In quality or reliability engineering, we can consider the red units analogous to defectives and the green units acceptable. In this situation, we are interested in the probability that our sample contains defective units. This model addresses finite population sampling, in which a sample is both countable and small. It has no connection to the geometric distribution discussed previously.

A hypergeometric random variable X is denoted $X \sim Hypergeometric\ (N, n, K)$ or further abbreviated $Hyp(N, n, K)$ or $HypGeo(N, n, K)$, as needed for clarity. It would have expected value $E(X) = n \cdot K/N$, with standard deviation $\sqrt{n\left(\frac{K}{N}\right)\left(1-\frac{K}{N}\right)\left(\frac{N-n}{N-1}\right)}$. The *hypergeometric probability mass function* uses combinations to calculate the exact probability of obtaining exactly k items from the target subgroup (*"defective"* outcomes in our example), out of a sample of n, from a finite population known to have N total units, of which K are defective:

$$\Pr(X = k \mid N, K, n) = \frac{\binom{K}{k}\binom{N-K}{n-k}}{\binom{N}{n}}$$

This equation comprises three elements, all calculated from combinations. The numerator separates the population into defective parts on the left and nondefective parts on the right, and for each calculates the number of ways a sample of the designated type can be obtained from the total of each available. These are multiplied together to obtain the total available number of combinations, which is then divided by the total number of ways a sample of size n can be formed out of N total units. The *hypergeometric cumulative mass function* uses combinations to calculate the exact probability of obtaining k or fewer items from the target subgroup (*"defective"* outcomes in our example), out of a sample of n, from a finite population known to have N total units, of which K are defective:

$$\Pr(X \leq n \mid r, p) = \sum_{k=0}^{n} \binom{k-1}{r-1} (p^r)(1-p)^{k-r}$$

For more information on combinations, see "Counting Techniques." Equations are provided in Table 6.21 later in the chapter. **Note that large factorials produce very large numbers!** Two methods are shown in the example for doing computations, if manual calculations are necessary. One demonstrates a cancelation process that helps to simplify expressions; the other uses tabulated values of logarithms of factorials.

EXAMPLE 6.38

Hypergeometric Calculations

Consider a lot of 50 units, of which two are nonconforming. A sample of 3 is selected from the lot at random. Find the probabilities associated with finding exactly zero, one, or both nonconforming units in the sample of 3.

Substituting into the hypergeometric formula for $N=50$, $K=2$, and sample size $n=3$, we obtain:

$$Pr(X=0) = \frac{\binom{2}{0}\binom{48}{3}}{\binom{50}{3}} = \frac{\left(\frac{2!}{2!\cdot 0!}\right)\left(\frac{48!}{45!\cdot 3!}\right)}{\left(\frac{50!}{47!\cdot 3!}\right)} = \frac{(1)\left(\frac{48\cdot 47\cdot 46}{3\cdot 2}\right)}{\left(\frac{50\cdot 49\cdot 48}{3\cdot 2}\right)} = 0.8825$$

$$Pr(X=1) = \frac{\binom{2}{1}\binom{48}{2}}{\binom{50}{3}} = \frac{\left(\frac{2!}{1!\cdot 1!}\right)\left(\frac{48!}{46!\cdot 2!}\right)}{\left(\frac{50!}{47!\cdot 3!}\right)} = \frac{(2)\left(\frac{48\cdot 47}{2}\right)}{\left(\frac{50\cdot 49\cdot 48}{3\cdot 2}\right)} = 0.115$$

$$Pr(X=2) = \frac{\binom{2}{2}\binom{48}{1}}{\binom{50}{3}} = \frac{\left(\frac{2!}{0!\cdot 2!}\right)\left(\frac{48!}{47!\cdot 1!}\right)}{\left(\frac{50!}{47!\cdot 3!}\right)} = \frac{(1)\left(\frac{48}{1}\right)}{\left(\frac{50\cdot 49\cdot 48}{3\cdot 2}\right)} = 0.0025$$

Hypergeometric calculations can be performed in Excel using the HYPGEOM. DIST function, using False to obtain the PDF and True for the CDF. It is important to ensure that each element in the calculation is identified properly.

Poisson

The **Poisson distribution** is used for **counts,** to find the probability that an event will occur a specified number of times in some standard interval of time or space, that is, per sampling unit. A Poisson random variable can take on the value of any whole number, so the probability distribution is unbounded, unlike the binomial, which has a max value. Assumptions are that occurrences are random, independent of each other, and the probability of occurrence is constant across the interval of interest, or, in other words, *iid*. Applications of Poisson include *queuing*, when counting the number of arrivals in a certain period of time or counting the number of *defects* found in a sampling unit of product.

Before continuing, it is important to distinguish here between the terms **defect** and **defective.** A *defective unit* will have one or more *defects*. However, a unit with a *defect* is not necessarily classified as *defective.* For example, an automobile manufacturer may sample three car hoods and count the total number of paint defects found on those hoods. These defects could include any number of occurrences of dirt, runs, sags, overspray, underspray, or orange-peel defects, which may or may not be sufficiently major to justify a rejection. Defects differ from the defectives considered previously, in which units classified as pass/fail, either defective or nondefective, were modeled with the *binomial.* Analysis of defects, or other count data, is typically handled with the *Poisson.*

One final note on terminology: Usage of the term "defect" versus "defective" in this context is intended *solely* to allow differentiation between the type of data available, to ensure appropriate distribution selection, that is, are the data countable, as per convention. It is not intended to implicate hardware as being noncompliant per definitions in ANSI Z1.4 2003. Defects identified here may or may not be sufficiently severe to force classification as a defective. See "Statistical Process Control (SPC)" in this chapter for additional discussion.

A Poisson random variable X is denoted $X \sim Poisson\ (\lambda)$ or further abbreviated $P(\lambda)$ or $Poi(\lambda)$ as needed for clarity. It would have expected value $E(X) = \lambda$, with standard deviation $\sqrt{\lambda}$, with single shape parameter λ estimated by taking the mean of the data. Lambda represents a rate and must take on a positive value. The *Poisson probability mass function* gives the probability of exactly x countable events, assuming the probability of occurrence remains the same for each sample:

$$Pr(X = x \mid \lambda) = \frac{e^{-\lambda}\lambda^{x}}{x!},\ where\ x \in \mathbb{Z}$$

The *Poisson cumulative mass function* gives the probability of x or fewer countable events, assuming the probability of occurrence remains the same for each sample:

$$Pr(X \le x \mid \lambda) = \sum_{k=0}^{x} \frac{e^{-\lambda}\lambda^{k}}{k!}$$

These values traditionally were obtained from tables for specific combinations of λ and x, but now spreadsheet programs or statistical software provide exact calculations for computational purposes.

Example 6.39 and Example 6.40 demonstrate how to interpret Poisson tables, found in Appendices C.1 (PMF) and C.2 (CMF). Equations are summarized in Table 6.21 later in the chapter.

EXAMPLE 6.39

Poisson Calculations

Records indicate that the number of customers that arrive at a bank drive-up window between 1:00 P.M. and 2:00 P.M. has a Poisson distribution with $\lambda = 5.4$. Find the probability that exactly six people arrive. Substituting into the Poisson formula for $x = 6$ and $\lambda = 5.4$,

$$Pr(X = 6) = \frac{e^{-5.4}(5.4)^6}{6!} \approx 0.16$$

Poisson calculations can be done in Excel using the POISSON.DIST function, using False to obtain the PDF and True for the CDF.

EXAMPLE 6.40

Poisson Calculations

The number of defects per shift has a Poisson distribution with $\lambda = 4.2$. A plot of the probability mass function is shown in Figure 6.23 with the striped area representing the probabilities of $X = 0$ and $X = 1$. Find the probability that the second shift produces fewer than two defects.

$$\begin{aligned} Pr(X < 2) = Pr(X \le 1) = \sum_{x=0}^{1} \frac{e^{-\lambda}\lambda^x}{x!} &= \frac{e^{-4.2}(4.2)^0}{0!} + \frac{e^{-4.2}(4.2)^1}{1!} \\ &= e^{-4.2} + (4.2)e^{-4.2} \\ &= 0.015 + 0.063 \\ &= 0.078 \end{aligned}$$

We can also find the answer using a table of cumulative Poisson probabilities, provided in Appendix C.2. Based on the value of λ we can find $Pr(X \le x)$ in the interior of the table.

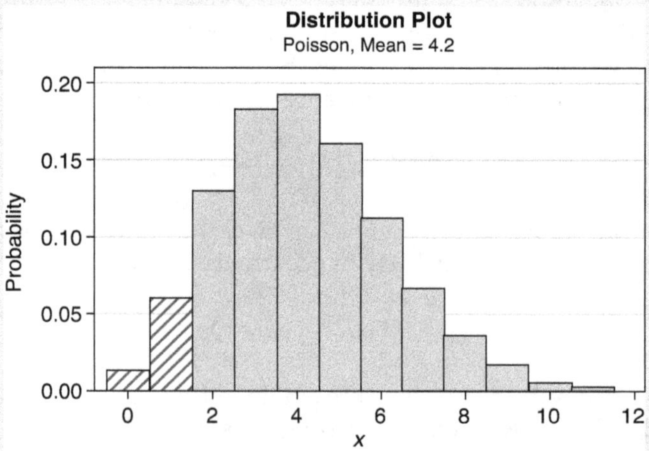

Figure 6.23 Poisson distribution plot.

Discrete Distribution Reference Table

Table 6.21 lists the mass functions, expectation formulas, and standard deviation formulas of each of the discrete distributions discussed in this section.

Continuous Distributions

Weibull and Exponential

The **Weibull** and **exponential distributions** are both used extensively in reliability to model time to failure in products, among other applications. They are discussed here together because they are closely related, since the exponential is a special case of the Weibull. The exponential distribution is also special in that it is the only continuous distribution with the **memoryless** property, meaning that it has a **constant hazard rate.** It therefore has the same behavior as the geometric, discussed previously in the section on discrete distributions. There are multiple parametrizations popular in different arenas. The forms included here are most common for reliability applications. Note that X must be positive and nonzero. For the purposes of this discussion, we are ignoring threshold parameters used to translate or shift the distribution, which can be applied to any model, as mentioned previously.

An **exponential** random variable X has one scale parameter λ, the failure rate, and is denoted $X \sim$ *Exponential* (λ) or abbreviated $Exp(\lambda)$. It has expected value $E(X)=1/\lambda$, with standard deviation also equal to $1/\lambda$. Its shape is fixed and known. Figure 6.24. shows an example of an exponential probability density function.

The *exponential probability density function* gives the probability of an event at exactly time t:

$$Pr(X=t \mid \lambda)=\lambda exp\{-\lambda t\}$$

The *exponential cumulative density function* gives the probability of an event at or before time t:

$$Pr(X \leq t \mid \lambda)=1-exp\{-\lambda t\}$$

Thus, the *exponential reliability distribution* takes on a simple form:

$$R(t)=Pr(X>t \mid \lambda)=exp\{-\lambda t\}$$

The exponential is also often alternatively written in textbooks using the notation $\beta=1/\lambda$, inversing the scale parameter, transforming to units of time. This is denoted $X \sim$ *Exponential* (β) or abbreviated $Exp(\beta)$. It has expected value $E(X)=\beta$, with standard deviation also equal to β. However, this specific notation (too many betas!) can be a source of confusion since the form of the Weibull distribution

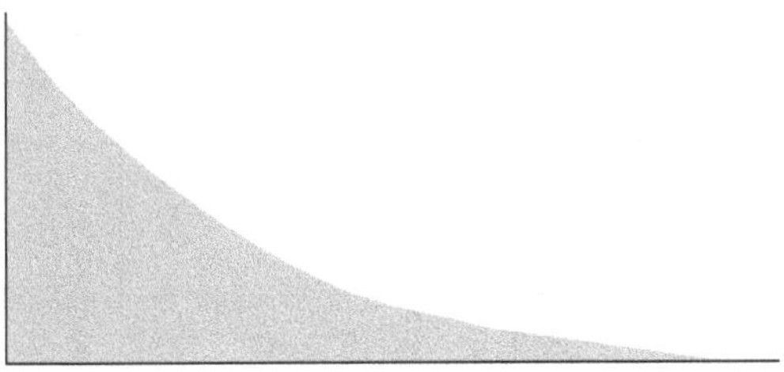

Figure 6.24 Exponential distribution.

Table 6.21 Reference table for discrete distributions.

Distribution	Description	Mass functions: PMFs and CMFs	Expectation & Standard Deviation ($Variance = StdDev^2$)
Bernoulli (p)	Used to model binary scenarios with $x=0, 1$ for probability p. Special case of Binomial with $n=1$. Foundation of binomial-related methods.	$PMF: Pr(X=x \mid p) = \begin{cases} p & if\ x=1 \\ q=1-p & if\ x=0 \end{cases}$ $CMF: \begin{cases} 0 & if\ x<0 \\ 1-p & if\ 0 \le x < 1 \\ 1 & if\ x \ge 1 \end{cases}$	$E(X)=p$ $StdDev(X)=\sqrt{p(1-p)}$ $=\sqrt{pq}$
Binomial (n, p)	Models sum of n Bernoulli trials, each with probability p where the number of occurrences is $x=0, 1, 2, \ldots, n$.	$PMF: Pr(X=x \mid n, p) = \binom{n}{x} p^x (1-p)^{1-x}$ PMF table values in Appendix B.1 CDF table values in Appendix B.2	$E(X)=np$ $StdDev(X)=\sqrt{np(1-p)}$ $=\sqrt{npq}$
Negative Binomial (r, p)	Version of binomial used to model the number of trials before r events occur. $x=r, r+1, r+2, \ldots$	$PMF: Pr(X=n \mid r, p) = \binom{n-1}{r-1} p^r (1-p)^{n-r}$ CMF: Complex form, not covered here. Note parametrization has multiple variations.	$E(X)=\frac{r}{p}$ $StdDev(X)=\sqrt{\frac{r(1-p)}{p^2}}$

Geometric (p) =NegBin$(1, p)$	Version of binomial used to model the number of trials before first event occurs. $x=1, 2, \ldots,$	$PMF: Pr(X=n \mid p)=p(1-p)^{n-1}$ CMF: Complex form, not covered here.	$E(X)=\frac{1}{p}$ $StdDev(X)=\sqrt{\frac{1-p}{p^2}}$
Hypergeometric (N, n, K, k)	Models the probability of obtaining k successes out of n samples, without replacement, from a finite population of size N containing K objects with the target feature.	$PMF: Pr(X=k \mid N, K, n)=\frac{\binom{K}{k}\binom{N-K}{n-k}}{\binom{N}{n}}$ CMF: Complex form, not covered here.	$E(X)=\frac{K}{N}n$ $StdDev(X)=\sqrt{n\left(\frac{K}{N}\right)\left(1-\frac{K}{N}\right)\left(\frac{N-n}{N-1}\right)}$
Poisson (λ) Pronounced "lambda"	Models the count of events occurring in a fixed interval. $x=0, 1, 2 \ldots,$ (unbounded)	$Pr(X=x \mid \lambda)=\frac{e^{-\lambda}\lambda^x}{x!}, \lambda \geq 0$ PMF table values in Appendix C.1 CMF table values in Appendix C.2	$E(X)=\lambda$ $StdDev(X)=\sqrt{\lambda}$

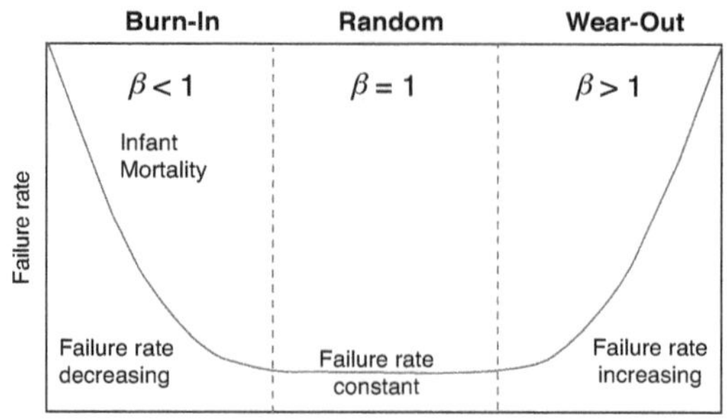

Figure 6.25 Reliability bathtub curve relative to Weibull beta parameter range.

conventionally used in reliability also is written with two parameters β and η, "beta" and "eta." However, the Weibull β parameter is *not* the same parameter as the exponential β in the parametrization just discussed. Some additional clarification is required.

The Weibull distribution was developed by a Swedish engineer, Waloddi Weibull, who popularized its use in modeling metallurgic failures. The **Weibull** has a relatively simple distributional form. It has two parameters, a shape parameter β and scale parameter η, both of which must be positive. This distribution is most famously applied in the failure rate **bathtub curve**, discussed previously in Chapter 2, "Benefits of Reliability Engineering," (Figure 6.25). The bathtub curve is made up of three separate Weibull functions visualized over time. The first period of the curve is called **burn-in** or **infant mortality**, in which the failure rate is high and then falls off quickly. The middle region represents a period of **constant failure rate**, or random failures, which is then followed by the **wear-out** period of rapidly increasing failures. Mathematically, these curves are obtained by changing the Weibull shape parameter β. The η parameter, also known as the **characteristic life**, adjusts the location of the distribution, has units of time, and always identifies the 63.2nd percentile.

Low Weibull β parameter values, with $0<\beta<1$, generate models with decreasing failure rates, reflect phenomena described as infant mortality. A process called burn-in may be used to remove these from the population. High Weibull β parameter values, with $\beta>1$, generate model with increasing failure rates, those where condition is degrading with age. The center region with Weibull $\beta=1$ is equivalent to the exponential distribution with $\lambda = \frac{1}{\eta}$. This may also be appropriate for situations that include a mix of failure modes, for example, a system level analysis would tend toward random. Note that a single Weibull distribution does not simultaneously represent the entire bathtub curve! But the Weibull distribution has the mathematical flexibility to take on the full range of shapes, hence why the model is so popular (Figure 6.26).

Returning to formal descriptions, a *Weibull* random variable X is denoted $X \sim$ *Weibull*(β, η) or abbreviated $X \sim$ *Weib*(β, η). The *Weibull probability density function* gives the probability of an event at exactly time t, assuming $t>0$ (positive):

$$Pr(X=t\,|\,\beta,\eta)=\frac{\beta}{\eta^{\beta}}\left(t^{\beta-1}\right)exp\left\{-\left(\frac{t}{\eta}\right)^{\beta}\right\}$$

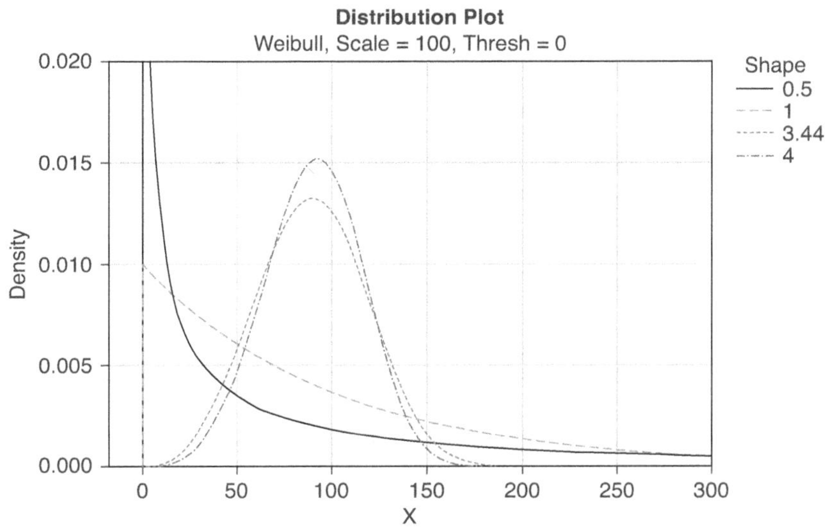

Figure 6.26 Plot of various members of the Weibull distribution.

The *Weibull cumulative density function* gives the probability of an event at or before time t, for $t > 0$:

$$Pr(X \leq t \mid \beta, \eta) = 1 - exp\left\{-\left(\frac{t}{\eta}\right)^{\beta}\right\}$$

This means that the Weibull reliability function takes on the form:

$$R(t) = Pr(X > t \mid \beta, \eta) = exp\left\{-\left(\frac{t}{\eta}\right)^{\beta}\right\}$$

One oddity of the Weibull distribution is that the functional forms for the mean, that is, expected value, and standard deviation are particularly complex. These equations are included in Table 6.21 but are not typically computed by hand. Instead, users reference the parameters. Summary metrics are provided in software. Calculations can also be obtained via spreadsheet programs. Excel has WEIBULL.DIST where alpha = beta and beta = eta, causing even further confusion, and EXPON.DIST, both with True for CDF and False for PDF. Statistical software provides exact calculations for computational purposes. When using a software program, it is as always critical to verify the nature of the parametrization of the distribution.

Calculations for the exponential and Weibull distributions are demonstrated in Example 6.41 through Example 6.44.

EXAMPLE 6.41

Exponential Calculations

Find the value of the PDF at 1000 hours given that the failure rate is 0.00053 failures per hour. Substituting $t = 1000$ and $\lambda = 0.00053$ into the exponential probability density function, we get:

$$f(1000) = 0.00053e^{-0.00053(1000)} \approx 0.00031$$

This says that the probability of failure at 1000 hours is about 0.00031. The mean and standard deviation are both calculated as $\frac{1}{\lambda} = \frac{1}{0.00053} = 1887$ *hours*. A sketch of the PDF in this example is shown in Figure 6.27. Find the reliability at 1000 hours if $\lambda = 0.00053$ failures/hour.

$$R(1000) = e^{-0.00053(1000)} \approx 0.59$$

which indicates that approximately 59% of the units are still operating after 1000 hours, or the probability that a particular unit is still operating after 1000 hours is 0.59.

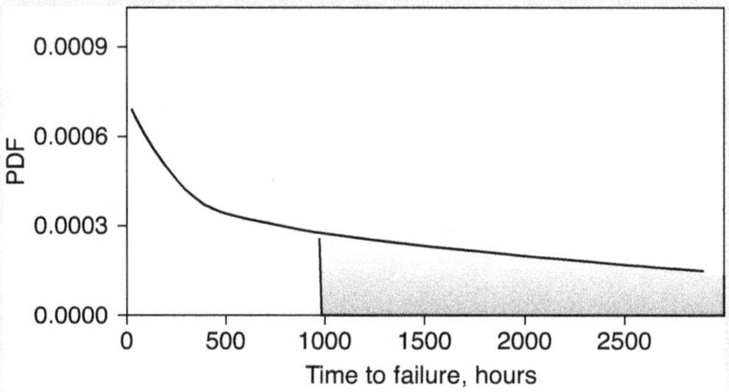

Figure 6.27 Probability density function for Example 6.41.

EXAMPLE 6.42

Weibull Calculations

The time to failure of a belt on an engine is distributed as a Weibull random variable with $\beta = 0.5$ and $\eta = 10{,}000$ hours. Determine the probability that the belt lasts up to 15,000 hours.

$$F(x) = Pr(X \leq x) = 1 - e^{-\left(\frac{x}{10{,}000}\right)^{0.72}} = 1 - e^{-\left(\frac{15{,}000}{10{,}000}\right)^{0.72}}$$
$$= 1 - e^{-(1.5)^{0.72}} = 1 - 0.2938 = 0.7062$$

What is the reliability of the belt at 15,000 hours?

$$R(15{,}000) = 1 - F(15{,}000) = 1 - 0.7062 = 0.2938$$

EXAMPLE 6.43

Weibull Calculations R(t)

A product's time to failure has a Weibull distribution with $\beta = 0.72$ and $\eta = 10{,}000$. Find reliability at 200 hours:

$$R(200) = e^{-\left(\frac{200}{10{,}000}\right)^{0.72}} \approx 0.94$$

indicating that about 94% of the products are operating after 200 hours.

EXAMPLE 6.44

Weibull Eta Proportion Calculation

What is the probability of failure before $\eta = 5$ for a Weibull distribution with $\beta = 2$ and $\eta = 5$? What is the probability of failure after $\eta = 5$?

$$Pr(X \le t) = 1 - exp\left\{-\left(\frac{t}{\eta}\right)^{\beta}\right\}$$

$$Pr(X \le \eta) = 1 - exp\left\{-\left(\frac{\eta}{\eta}\right)^{\beta}\right\}$$

$$Pr(X \le \eta) = 1 - exp\{-(1)^{\beta}\} = 1 - e^{-1} = 1 - 0.368 = 0.632$$

$$R(t = \eta) = 1 - Pr(X \le \eta) = Pr(X > \eta) = exp\{-(1)^{\beta}\} = e^{-1} = 0.368$$

Figure 6.28 corroborates this result.

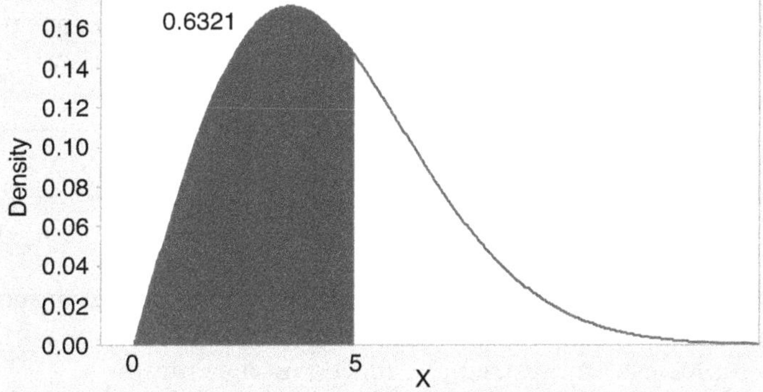

Figure 6.28 Weibull ($\beta = 2$, $\eta = 5$) probability distribution function with $Pr(T < t)$ at $x = \eta$.

Normal, Standard Normal, Lognormal

The **normal distribution**, also known as the **Gaussian**, may be the best known of all statistical distributions. It is generally considered the most important distribution in both the theory and practice of statistics. It has certain properties that make it both easy to work with and broadly useful since often representative of real data. It is also used as the foundation for many theoretical methods. It is rarely

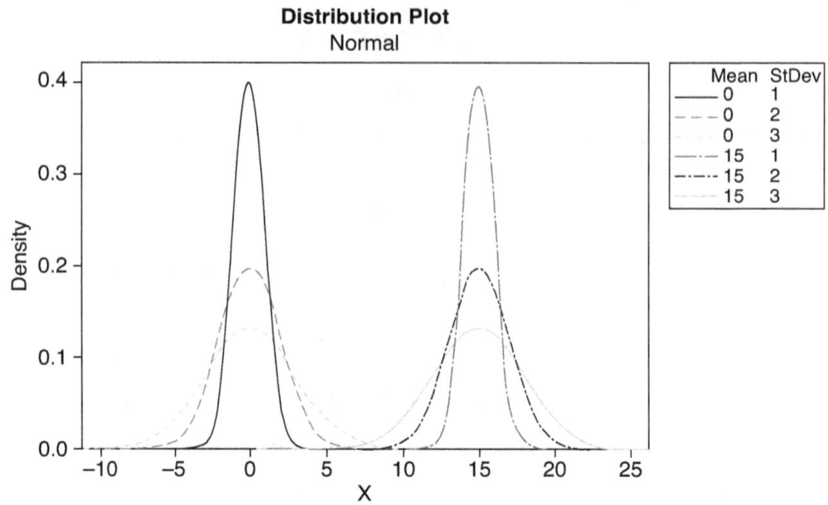

Figure 6.29 Normal distribution with various values for μ and σ.

used for modeling lifetime data, which is not reliably symmetric and prone to skew. The normal is **symmetric** and **bell-shaped**, centered at mean μ, "mu", with standard deviation σ, "sigma," with zero skew. **Mu** is the shape parameter, which controls the location, and **sigma** is the scale parameter, which controls the spread. Figure 6.29 shows that changing μ moves the center of the distribution along the x-axis, while increasing σ makes the distribution more spread out.

A **normal** random variable X is denoted $X \sim Normal(\mu, \sigma^2)$ or abbreviated $Norm(\mu, \sigma^2)$, or $N(\mu, \sigma^2)$, which has expected value $E(X) = \mu$, with standard deviation also equal to σ. The parameter μ can take on any value, but σ must be positive. Estimates for μ and σ are estimated using standard mean and standard deviation equations:

$$\hat{\mu} = \bar{X} = \frac{1}{n}\sum_{i=1}^{n} X_i \text{ and } s = \sqrt{\frac{1}{n-1}\sum_{i=1}^{n}\left(X_i - \bar{X}\right)^2} = \sqrt{\frac{\sum_{i=1}^{n} X_i^2 - n\bar{X}^2}{n-1}}$$

Notice that the estimated standard deviation shown above has $(n-1)$ in its denominator. This quantity is called the **degrees of freedom**, and it reflects the size of the sample, less any restrictions placed on that sample.

The *normal probability density function* gives the probability of an event at exactly $x = t$:

$$Pr(X = t \mid \mu, \sigma^2) = \frac{1}{\sqrt{2\pi}(\sigma)} exp\left\{-\frac{1}{2}\left(\frac{t-\mu}{\sigma}\right)^2\right\}$$

The *normal cumulative density function* gives the probability of an event at or less than $x = t$:

$$Pr(X \le t \mid \mu, \sigma^2) = \int_{-\infty}^{t} \frac{1}{\sqrt{2\pi}(\sigma)} exp\left\{-\frac{1}{2}\left(\frac{x-\mu}{\sigma}\right)^2\right\} dx = \Phi\left(\frac{x-\mu}{\sigma}\right)$$

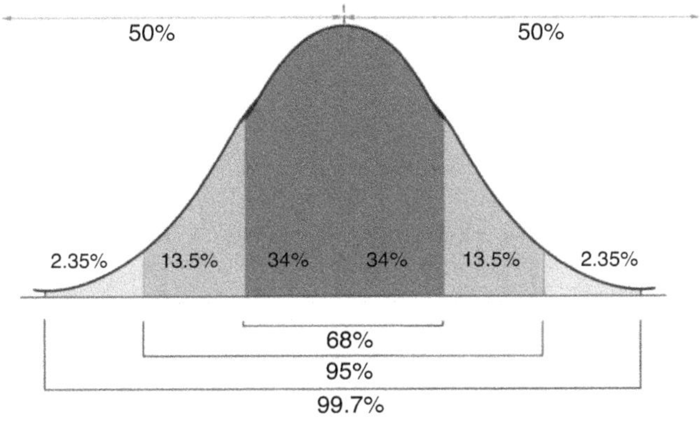

Figure 6.30 Normal distribution with empirical rule.

Data that are normally distributed follow what is known as the **empirical rule**, where approximately 68% of the distribution falls within $\mu \pm 1\sigma$, about 95% within $\mu \pm 2\sigma$, and ~ 99.7% within $\mu \pm 3\sigma$, as shown in Figure 6.30.

The transformation shown here is known as **normalization** and is used to obtain a special metric called a **Z-score**:

$$Z = \frac{X - \mu}{\sigma}$$

By using the Z transformation, we are given a relative location of a data point x from a normal distribution: The point is $|z|$ standard deviations away from the mean. A negative Z indicates that the point is below the mean, while a positive Z tells us the point is above the mean. A Z equal to zero indicates that the data point is equal to the mean. This general structure is used in many methods for standardization purposes, since it results in data from the **standard normal distribution**, denoted Z, which is centered at $\mu = 0$ with $\sigma = 1$, that is,

$$Z \sim Normal(\mu = 0, \sigma = 1)$$

Standard normal Z-values may be referenced to evaluate the significance of test statistics for certain hypothesis tests. A Z-table can be found in Appendix D.

Another common use of the normal distribution in the quality arena is to evaluate how close we are to process **requirements**, as defined by **upper and lower specification limits**, abbreviated USL and LSL. These limits designate the target range specified in the design to meet objectives for system performance or operability considerations. Ideally a process produces products entirely within the specification, but natural process variability may result in a nonzero risk of being outside the limits and therefore nonconforming or possibly requiring *rework and/or repair* (R&R). More on these topics of **statistical process control** (SPC) and **process capability** is found in Chapter 6.

The **lognormal distribution** is a related variant of the *normal*, obtained using a transformation. A random variable X is said to be distributed lognormally if its natural log is normally distributed, as:

$$X \sim LogNormal(\mu, \sigma^2) \Leftrightarrow Ln(X) \sim Normal(\mu, \sigma^2)$$

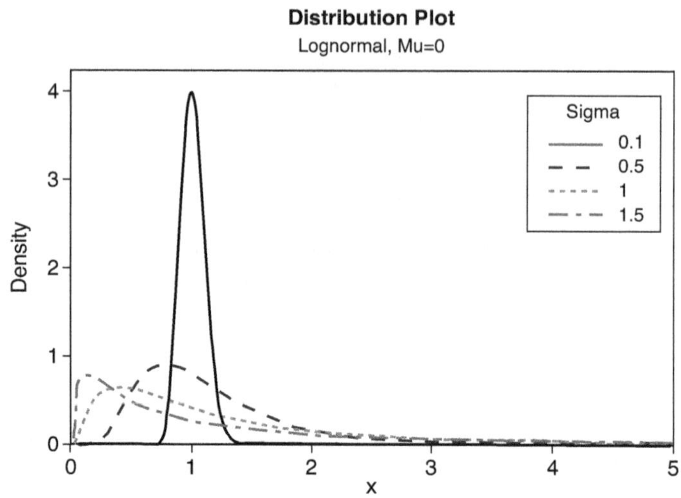

Figure 6.31 Shape of lognormal distribution, various σ.

Alternatively, we can write:

$$X \sim Normal(\mu, \sigma^2) \Leftrightarrow exp(X) \sim LogNormal(\mu, \sigma^2)$$

It is important when working with the lognormal to always keep in mind which scale you are using.

Figure 6.31 shows the shape of the lognormal distribution for various values of σ. As the value of sigma nears zero, the log normal distribution has an increasingly higher peak near e^{μ}.

A **lognormal** random variable X is denoted $X \sim LogNormal(\mu, \sigma^2)$ or abbreviated $LogNorm(\mu, \sigma^2)$, or $LogN(\mu, \sigma^2)$. Note that μ and σ here are obtained on log scale and will not have an intuitive meaning. This is why the equations for mean and variance are more complicated as well, since changing back to the original scale:

$$E(X) = exp\left\{\mu + \frac{\sigma^2}{2}\right\}$$

$$SD(X) = \sqrt{Var(X)} = \sqrt{exp\{2(\mu + \sigma^2)\} - exp\{2\mu + \sigma^2\}}$$
$$= \sqrt{exp\{2\mu + \sigma^2\} \cdot (exp\{\sigma^2\} - 1)}$$

On log scale, the parameter μ can take on any value, but σ must be positive. Compared to other right-skewed distributions, it tends to have heavier tails. The log-normal distribution has been found to be a good mathematical model for times to failure for some electronic and mechanical products, including transistors, bearings, and electrical insulation. It is sometimes a good model for times to repair a unit after a failure.

The *lognormal probability density function* gives the probability of an event at exactly $x = t$:

$$Pr(X = t \mid \mu, \sigma^2) = \frac{1}{t\sqrt{2\pi \cdot \sigma^2}} exp\left\{-\frac{1}{2}\left(\frac{\ln(t) - \mu}{\sigma}\right)^2\right\}$$

The *lognormal cumulative density function* gives the probability of an event at or less than $x=t$:

$$Pr(X \le t \mid \mu, \sigma^2) = \int_{-\infty}^{t} \frac{1}{x\sqrt{2\pi \cdot \sigma^2}} exp\left\{-\frac{1}{2}\left(\frac{\ln(x)-\mu}{\sigma}\right)^2\right\} dx = \Phi\left(\frac{\ln(t)-\mu}{\sigma}\right)$$

EXAMPLE 6.45

Simple Normal *Z*-Scores with Table Calculations

A system performance measurement is normally distributed with $\mu=200$ and $\sigma=2$. What is the probability of an outcome with a value between 200 and 202, as shown in Figure 6.32?

From the empirical rule, we know $\mu \pm 1\sigma$ has probability 68%. The shaded area shows the upper side only, so we know from symmetry that the probability of being between 200 and 202 must be 34%. This can be confirmed using a standard normal table, Appendix D.

The *Z*-scores are for 200 and 202 $A=\pi r^2$ calculated as:

$$Z = \frac{X-\mu}{\sigma} = \frac{200-200}{2} = \frac{0}{2} = 0 \text{ and } Z = \frac{202-200}{2} = \frac{2}{2} = 1$$

Appendix D shows the area to the left of *Z* under the standard normal curve.

Therefore, to calculate the probability of being between 200 and 202, we look first for the probability of being to the left of 202, which aligns with $Z=1$. To read the value from the chart, we look down the column on the left for the tenths digit, then across the top row to match the hundredths digit. Negative values of *Z* begin on the first page and continue on the second page, followed by positive values. To locate $Z=1$, a positive value, we look on the second page to find the row designated 1.0, then move across the row to find the cell the matches the hundredths digit in the first column (X.X0). The resulting value is 0.8413. For $Z=0$ we are in the upper left corner, which gives us a value of 0.5000. To obtain the probability of being between 200 and 202, we subtract 0.5 from 0.8413, since we are using the CDF:

$$\begin{aligned} Pr(200 \le X \le 202) &= Pr(X \le 202) - Pr(X \le 200) \\ &= 0.8413 - 0.5000 = 0.3413 \end{aligned}$$

The probability that a data value will be between 200 and 202 is therefore 0.88413–0.5000=0.3413 or ~34%, as calculated previously using the empirical rule.

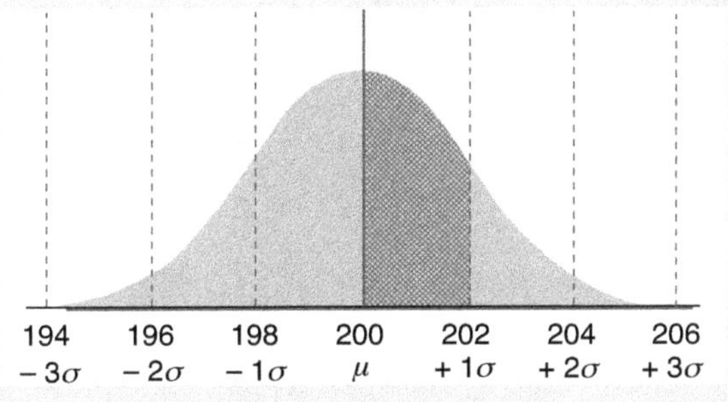

Figure 6.32 Normal distribution with mean = 200 and standard deviation = 2.

Continued

Normal calculations can also be performed in Excel using the NORM.S.DIST function for the standard normal and the NORM.DIST for any normal. Cumulative = True yields the CDF, with False the PDF. Note that this is the area to the *left,* not the right as shown here, but the identical principle applies. Tables may be constructed in different ways.

EXAMPLE 6.46

Normal Z-Scores and Table Calculations

The diameters of a batch of turned shafts are normally distributed with $\mu = 2.015$ and $\sigma = 0.0053$ cm. Specification limits are 2.000 to 2.025 cm.

(a) What percentage of the batch is above the upper specification limit (USL)?

(b) What percentage of the batch is below the lower specification limit (LSL)?

Find the *z*-score for each side of the specification limit and obtain probabilities from Appendix D.

$$Z_{USL} = \frac{X_{USL} - \mu}{\sigma} = \frac{2.025 - 2.015}{0.0053} = \frac{0.010}{0.0053} = 1.89$$

First find the *Z* value in the *Z* table. From the standard normal table in Appendix D in the row labeled 1.8 and the column labeled 0.09, the value is 0.9706. However, this is the area to the left of 1.89. We need the area to the right, which is obtained by subtraction: $1 - 0.9706 = 0.0294$, which indicates that 2.94% of the batch exceeds the upper specification.

The *Z* score for the LSL is calculated:

$$Z_{LSL} = \frac{X_{LSL} - \mu}{\sigma} = \frac{2.000 - 2.015}{0.0053} = \frac{-0.015}{0.0053} = -2.83$$

The area to the left of −2.83 in the standard normal table is 0.0023. Therefore, 0.23% of the batch violates the lower specification.

EXAMPLE 6.47

Lognormal Calculations

Suppose that the random variable *X*, the lifetime of a plasma television in hours, is distributed as lognormal with parameters $\mu = 10$ hours and $\sigma = 2$ hours. What is the probability that the lifetime of a plasma television exceeds 1000 hours? Based on the information given, it follows that $\ln(X) \sim N(10, 4)$.

$$\begin{aligned}
Pr(X > 1000) &= 1 - Pr(X < 1000) \\
&= 1 - Pr(\ln X < \ln 1000) \\
&= 1 - Pr\left(Z < \frac{(\ln 1000 - 10)}{2}\right) \\
&= 1 - Pr\left(Z < \frac{6.9078 - 10}{2}\right) \\
&= 1 - Pr(Z < -1.55) \\
&= 1 - 0.0606 \\
&= 0.9394
\end{aligned}$$

Therefore, the probability that the lifetime of the plasma television exceeds 1000 hours is 0.9394.

EXAMPLE 6.48

Lognormal Calculations, Uncensored Data

The times to failure for a product are known to be lognormally distributed. The times to failure for a sample of six parts are 850, 925, 1250, 1550, 1800, and 2750 cycles. Estimate the mean and sigma parameters and use to calculate the probability of failing before 2000 cycles.

Begin by taking the natural log of each data point as shown in Table 6.22.

Table 6.22 Times to failure and natural log conversion for Example 6.48.

x	$x' = \ln x$
850	6.75
925	6.83
1250	7.13
1550	7.35
1800	7.50
2750	7.92

Calculate the mean and standard deviation of the values in the second column.

$$\bar{X}' = 7.25$$
$$\sigma_{X'} = 0.44$$

We can now write the distribution of the transformed data like this:

$$Ln(X) = X' \sim Normal(\hat{\mu}' = 7.25, \hat{\sigma}' = 0.44)$$

Then back out to the original random variable X:

$$X = EXP(Ln(X)) \sim LogNormal(\hat{\mu}' = 7.25, \hat{\sigma}' = 0.44)$$

The parameters are still on log-scale.

This distribution can now be used for calculations. We can calculate the probability of failing by 2000 cycles several ways. One method is to stay in log scale and do calculations assuming a normal. If we do this, we need to convert the threshold of interest to the log scale by taking the natural log of 2000, which results in 7.6. We can then calculate:

$$\begin{aligned} Pr(X \le 2000) &= Pr(X' \le 7.6) \\ &= \Phi\left(\frac{X' - \mu'}{\hat{\sigma}'}\right) \\ &= \Phi\left(\frac{7.6 - 7.2445}{0.4388}\right) \\ &= 0.792 \end{aligned}$$

This can also be done using the NORM.S.DIST function in logscale, using the CDF.

Alternatively, we can convert back to original scale and do calculations for lognormal, using spreadsheet functions for lognormal, LOGNORM.DIST, with parameters in log-scale, which gets the same result.

Chi-Squared

The **chi-squared distribution** (pronounced "kai," not "chai") or *-square* distribution is a workhorse in statistical analysis. The chi refers to the Greek letter χ, and so the distribution is also written χ^2 using the Greek letter. It is derived as the distribution of the sum of squared standard normal random variables Z_i, which were introduced in the previous section.

$$Q = \sum_{i=1}^{k} Z_i^2 \sim \chi^2_{df=k-1}, \text{ where } Z_i \text{ are iid.}$$

Common usages include performing parametric tests on the variance of a normal distribution, and for nonparametric tests of independence for contingency tables, and goodness-of-fit (GOF) tests.

A χ^2 random variable X is denoted $X \sim \chi^2(\nu)$ if using the Greek (note degrees of freedom are the Greek letter "nu," not a v), or chi-sq(ν), which has expected value $E(X) = \nu$, with standard deviation equal to $\sqrt{2\nu}$. The parameter ν is positive and a natural number ($\nu \in \mathrm{N}$) since derived from a count of the number of *iid* random variables that make up X. Note that X must be positive and nonzero since obtained as a sum of squared values. The chi-squared distribution is always skewed to the right, with shape determined by its single parameter, ν. It becomes more symmetric as the degrees of freedom increase, as shown in Figure 6.33.

The χ^2 distribution is what allows us to make inferences about the variance for a normal population, which has wide-ranging implications and uses. The chi-square random variable itself is a function of the sample variance of the normal distribution, s^2:

$$\frac{(n-1)s^2}{\sigma^2} \sim \chi^2_{df=n-1}$$

The χ^2 is used in the construction of both t-tests and F-tests, used to relate the magnitude of an effect to the underlying variability. Thus, it is the linchpin that ties other foundational methods together. For example, hypothesis testing, regression, and analysis of variance (ANOVA), previously discussed in "Basic Statistics," are all connected by this distribution.

Another key application of the χ^2 distribution is for comparing sets when data are summarized in categories. One commonly used approach is a **Pearson test of association**, first published in the year 1900, for which the test statistic takes the following general form:

$$X^2 = \sum_{i=1}^{k} \frac{(O_i - E_i)^2}{E_i} \sim \chi^2_{df=k-1}$$

where O_i and E_i refer to the **Observed** and **Expected** counts, respectively, for one factor with k levels. The divisor puts each element on a common scale, so that when totaled, large observations won't bias the results. Observed data are what actually occurred. Expected values are estimated values obtained assuming no effects, which is the null hypothesis here. If the null hypothesis is true and there is no effect, we would expect the observed to generally align with the calculated expected values.

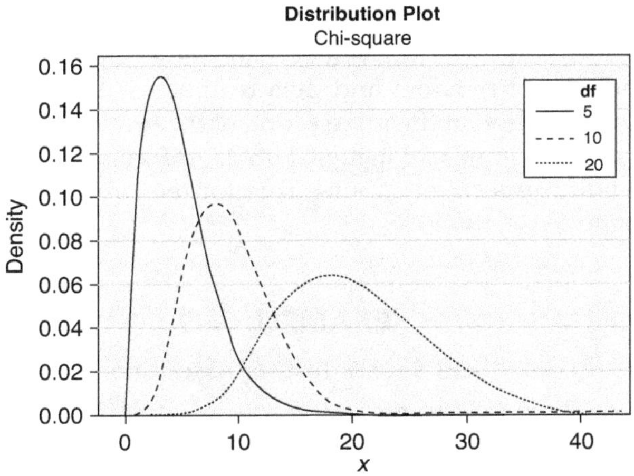

Figure 6.33 Example of chi-squared distributions with various degrees of freedom.

For the one-factor case, expected values are calculated as N/k, where N is the total of the counts and k is the number of levels.

This approach is easily generalized, e.g., for 2 and 3 factors:

$$X^2 = \sum_{i=1}^{r}\sum_{j=1}^{c}\frac{\left(O_{ij} - E_{ij}\right)^2}{E_{ij}} \quad \text{and} \quad X^2 = \sum_{i=1}^{r}\sum_{j=1}^{c}\sum_{k=1}^{p}\frac{\left(O_{ijk} - E_{ijk}\right)^2}{E_{ijk}},$$

which have $(r-1)$ $(c-1)$ and $(r-1)$ $(c-1)$ $(p-1)$ degrees of freedom, respectively.

For the 2-factor case, expected values are calculated in the following way:

$$E_{ij} = \frac{n_{i+} \cdot n_{+j}}{n_{++}}, \text{ where}$$

n_{i+} is the total number of observations in the ith row, n_{+j} is the total number of observations in the jth column, and n_{++} is the overall total number of observations.

This formulation is broadly used when we need to test if the same behavior is present in different circumstances, such as *independence* as discussed in "Basic Probability Concepts," *homogeneity* in sampling frames, to be discussed in "Sampling Plans for Statistics and Reliability Testing," and *goodness-of-fit*, discussed later in this section.

The χ^2 probability density function gives the probability of an event at exactly $x = t$:

$$Pr(X = t \mid \nu) = \frac{1}{\Gamma\left(\frac{\nu}{2}\right)2^{\nu/2}}\left(t^{\frac{\nu}{2}-1}\right)exp\left\{-\frac{t}{2}\right\}$$

The term Γ is the *gamma function*, a complex function out of scope of this handbook. The gamma function is an integral, but for any positive integer n, the function simplifies to $\Gamma(n) = (n-1)!$.

The χ^2 cumulative density function gives the probability of an event at or less than $x = t$:

$$Pr(X \le t \mid \nu) = \int_0^t \frac{1}{\Gamma\left(\frac{\nu}{2}\right)2^{\nu/2}}\left(x^{\frac{\nu}{2}-1}\right)exp\left\{-\frac{x}{2}\right\}dx$$

The chi-square table in Appendix G gives the chi-square values x associated with $\Pr(X>x)$. To find a critical value, read down the first column to find the appropriate degrees of freedom and then read across to the desired alpha level. The critical value is found at the intersection of the row and column. Alternatively, these can be calculated in Excel using CHISQ.DIST and CHISQ.DIST.RT to obtain left and right tails, respectively. Using True for the cumulative parameter generates the CDF, and False, the PDF.

EXAMPLE 6.49

Chi-Square Test for Association

The data shown in Table 6.23 were previously discussed in "Basic Probability Concepts." Example 6.26 as showing evidence of difference in quality between four production lines:

Table 6.23 Defectives found per production line.

Source	Defective	Nondefective	Total	Percent defective
Line1	3	87	90	3.3%
Line2	10	75	85	11.8%
Line3	25	230	255	9.8%
Line4	7	63	70	10.0%
Total	**45**	**455**	**500**	**9.0%**

A chi-square test for association was conducted for these data to look at effects of production line, for which the following output was obtained. Do these results support or undermine the initial characterization of independence? How are the degrees of freedom calculated for this hypothesis test?

The row factor Line has $k=4$ levels, for which there are $k-1=3$ degrees of freedom. Likewise, the category factor defective also has 2 levels and therefore 1 degree of freedom. The degrees of freedom for the combination are obtained by multiplying the degrees of freedom for all the factors together, resulting in the value of $3\times1=3$ shown previously.

The statistical results summarized in Figure 6.34 can be obtained by calculating the associated chi-square test to obtain the value in the column labeled "chi-square." The table shows two test results, a Pearson chi-square and the likelihood ratio test, which are different methods of doing the test of association, as described previously. *P*-values for both are in a similar range, greater than 0.1, above conventional thresholds for statistical significance. The null hypothesis here is not explicitly stated but assumes no associations. From this test we would conclude that there is insufficient evidence to show a lack of independence here for the *full* table.

Alternatively, we can use the chi-square table in Appendix G to determine if there is a significant association between production line and defectives.

Rows: Line Columns: Defective

	Defective	Non-Defective	All
1	3 8.10	87 81.90	90
2	10 7.65	75 77.35	85
3	25 22.95	230 232.05	255
4	7 6.30	63 63.70	70
All	45	455	500

Cell Contents
Count
Expected count

Chi-Square Test

	Chi-Square	DF	P-Value
Person	4.609	3	0.203
Likelihood Ratio	5.560	3	0.135

Figure 6.34 Output for chi-square test of association for Example 6.49.

EXAMPLE 6.50

Chi-Square Test for Association

The contingency data shown in Table 6.24 were previously discussed in Example 6.25 as showing evidence of independence:

Table 6.24 Process count data for XYZ versus FGH.

	X	*Y*	*Z*	Totals
F	17	18	14	49
G	18	11	16	45
H	25	13	18	56
Totals	60	42	48	150

A chi-square test for association was conducted for these data to look at effects of both Group FGH and Category XYZ, for which the output in Figure 6.35 was obtained. Do these results support or undermine the initial characterization of independence? How are the degrees of freedom calculated for this hypothesis test?

The table shows two test results, a Pearson chi-square and the likelihood ratio test, which are different methods of doing the test of association, as described previously. *P*-values for both are in a similar range, greater than 0.5, well above any conventional thresholds for statistical significance. The null hypothesis here is not explicitly stated but assumes no associations. We would conclude that there is no evidence for rejecting the independence null hypothesis in favor of the alternative.

Continued

CHISQCOUNTS

Chi-Square Test for Association: Group, Category

Rows: Group Columns: Category

	X	Y	Z	All
F	17 19.60	18 13.72	14 15.68	49
G	18 18.00	11 12.60	16 14.40	45
H	25 22.40	13 15.68	18 17.92	56
All	60	42	48	150

Cell Contents
Count
Expected count

Chi-Square Test

	Chi-Square	DF	P-Value
Person	3,001	4	0.558
Likelihood Ratio	2.924	4	0.571

Figure 6.35 Output for chi-square test of association for Example 6.50.

The row factor FGH has $k=3$ levels, for which there are $k-1=2$ degrees of freedom. Likewise, category factor XYZ also has 3 levels and therefore 2 degrees of freedom. The degrees of freedom for the combination are obtained by multiplying the degrees of freedom for all the factors together, resulting in the value of $2\times2=4$.

The statistical results summarized here can be obtained by calculating the associated chi-square test to obtain the value in the column labeled "Chi-Square." That value is then referenced to the chi-square table in Appendix G or other calculator using the appropriate degrees of freedom.

EXAMPLE 6.51

Chi-Square Goodness of Fit Test

Table 6.25 provides historical data for IT helpline requests. Percentages of each request type have been computed from the previous year's data. Additionally, the total number of calls last month has been categorized by request type. Is the distribution of last month's calls the same as last year's distribution? Use $\alpha=0.05$.

Table 6.25 IT Helpline request data for Example 6.51.

IT Helpline request type	Last year's percentage of requests	Observed number of calls last month
File recovery	16%	40
Log-in issue	28%	35
Slow Internet	42%	80
Printer issue	14%	28
Total	**100%**	**183**

Step 1. *State the null hypothesis and the alternative hypothesis.*

H_0: The data follow last year's distribution
H_1: The data do not follow last year's distribution

Alternatively, we could state
H_0: The distribution of request types has not changed
H_1: The distribution of request types has changed

Step 2. *Assign an appropriate value to the level of significance.*

$$\alpha = 0.05$$

Step 3. *Determine the probability distribution of the appropriate test statistic.*

We then determine that the appropriate test is a chi-square goodness of fit test.

Step 4. *Calculate the test statistic.*

We can calculate the expected number of calls last month using the last year's percentage. There are $k=4$ request types. Here, we take the total number of observed calls for the month and multiply it by last year's percentage for each category. Once these expected values are calculated, we calculate the chi-square component for each request type and sum the values to find the chi-square test statistics. These calculations are shown in Table 6.26.

Table 6.26 Observed and expected numbers of calls per month and chi-square contribution.

IT helpline request type	Observed number of calls last month	Last year's percentage of requests	Expected number of calls last month	Chi-square $=(O-E)^2/E$
File recovery	40	16%	$183 \times 0.16 = 29.28$	$\frac{(O_1-E_1)^2}{E_1} = \frac{(40-29.28)^2}{29.28} = 3.925$
Log-in issue	35	28%	$183 \times 0.28 = 51.24$	$\frac{(O_2-E_2)^2}{E_2} = \frac{(35-51.24)^2}{51.24} = 5.147$
Slow Internet	80	42%	$183 \times 0.42 = 76.86$	$\frac{(O_3-E_3)^2}{E_3} = \frac{(80-76.86)^2}{76.86} = 0.128$
Printer issue	28	14%	$183 \times 0.14 = 25.62$	$\frac{(O_4-E_4)^2}{E_4} = \frac{(28-25.62)^2}{25.62} = 0.221$
Total	**183**	**100%**	**183**	$\chi_0^2 = 9.421$

Step 5. *Determine the rejection region(s) and locate the critical point(s) and compare these to the calculated test statistic. Alternatively, calculate the p-value for the test and compare it to alpha.*

The degrees of freedom for the critical value are equal to $k-1=(4-1)=3$. From Appendix G, we find the critical value $\chi^2_{0.05,3} = 7.815$. The rejection criterion for this test is:

$$\text{Reject } H_0 \text{ if } \chi_0^2 > \chi^2_{\alpha,k-p-1}$$

Continued

Step 6. *State the statistical conclusion in business terms.*

Since the test statistic $\chi_0^2 = 9.421$ is greater than the critical value $\chi^2_{0.05,3} = 7.815$, we reject H_0. The distribution of the IT helpline requests last month do not follow the historical distribution.

Alternatively, we can use the *p*-value method to be able to give a more precise statement. Statistical software calculates $p = 0.024$ for this test. Since this value is smaller than our chosen α of 0.05, we reject H_0.

Some additional analysis can give us more insight into what is driving the change in the call types. Figure 6.36 shows the differences between the expected and observed call counts for each category. We see that file recovery calls were more frequent than expected, and log-in issues were less frequent. Figure 6.37 is a Pareto chart of contributions to the chi-square test statistic. We see that the changes in log-in issues and file recovery were the top contributing categories to the change in the call mix.

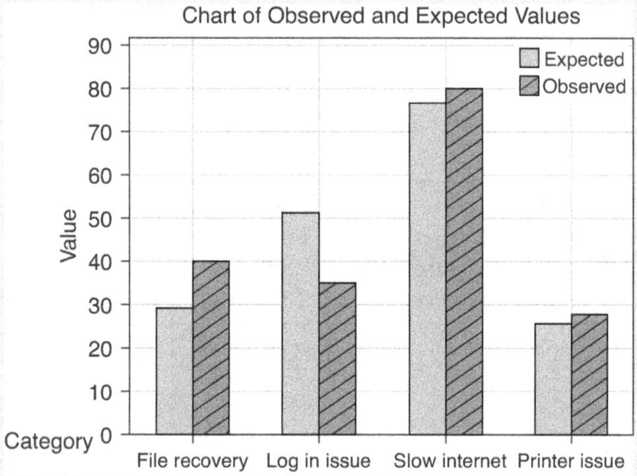

Figure 6.36 Expected versus observed requests for Example 6.51.

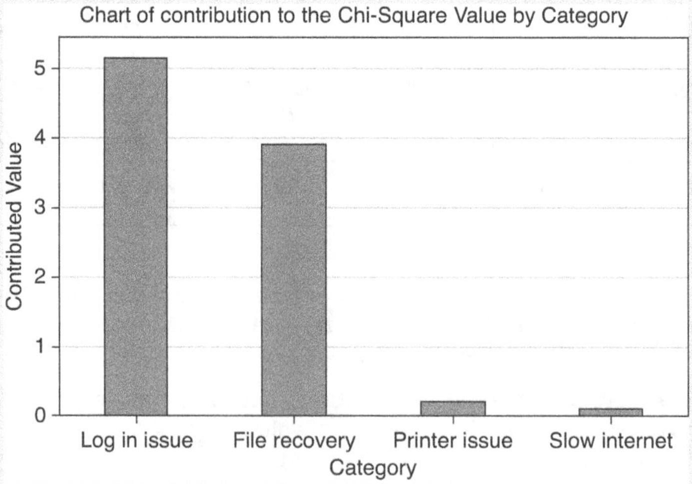

Figure 6.37 Contribution to the chi-square value by request category for Example 6.51.

Student's t *and* F

The **Student's *t* and *F* distributions** are related, as well as derived from, the other distributions presented earlier. Their usage is more like that of the χ^2, part of the underlying methodology used in statistical assessments related to hypothesis testing. Neither distribution is ever used for reliability calculations like the distributions discussed at the beginning of this section. For a given comparison, there are parallel sets of results.

The F test statistic with $(1, \nu)$ degrees of freedom is the square of the Student's t statistic, and they will have the same p-value:

$$F(1, \nu) = t^2(\nu)$$

This phenomenon is present in simple line regression analysis: The F statistic in the ANOVA table is the square of the Student's t static for the independent variable. In this case, the independent variable has 1 degree of freedom, and the v is equal to the error degrees of freedom.

The t distribution is formed by combining two distributions discussed previously in a ratio:

$$t_\nu = \frac{Z}{\sqrt{\chi^2 / \nu}}$$

The Z in the numerator is a standard normal, the divisor the square root of the χ^2 with ν degrees of freedom. The square root puts the two measures on a common scale. This ratio creates a scaled measure that can be used to evaluate whether the difference seen in the numerator is "large" relative to the variation in the population, as quantified in the denominator.

An F statistic is formed analogously as the ratio of two χ^2 random variables. This is a ratio of variances, a test case over a baseline to evaluate whether the variability in the test group is statistically different. This is the foundation of the F tests conducted in analysis of variance (ANOVA) methods highlighted in "Basic Statistics."

$$F = \frac{\chi_1^2 / \nu_1}{\chi_2^2 / \nu_2}$$

If we take n independent samples from a normal distribution with mean μ and standard deviation σ, then the sample will follow a **Student's *t* distribution** with ν (nu) **degrees of freedom**. The value of ν is based on the sample size n, where $\nu = (n-1)$. Note the random variable's similarity to the normal Z score for an average:

$$t = \frac{(\bar{X} - \mu)}{s / \sqrt{n}} \sim t_{df = n-1}$$

The Student's t distribution has a single scale parameter ν, referred to as its **degrees of freedom**. It is centered at $\mu = 0$ and is symmetric around its mean. Its location is fixed. The distribution resembles the bell shape of the normal distribution, but with heavier tails, known as leptokurtic. The distribution becomes more and more like a normal distribution as the degrees of freedom increase. Figure 6.38 shows the difference in shape for the Student's t distribution with various degrees of freedom.

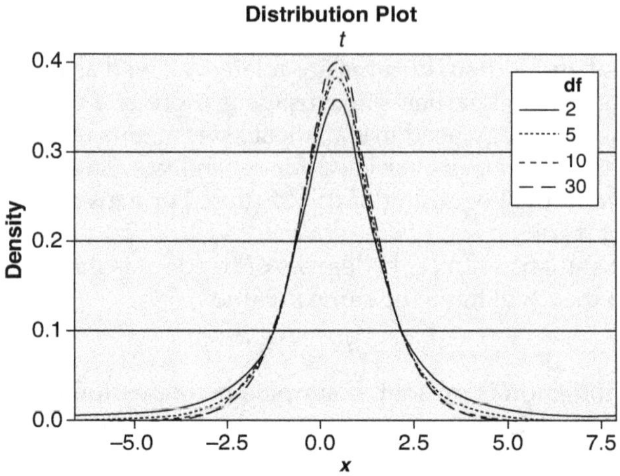

Figure 6.38 Example of a Student's t distribution with various degrees of freedom.

A **Student's *t*** random variable t^* is denoted $t^* \sim t(v)$. It has expected value $E(X)=0$ and standard deviation:

$$SD(X)=\sqrt{Var(X)}=\sqrt{\frac{v}{v-2}},\ for\ v>2$$

The parameter v is positive and a natural number ($v\in$ N). A t-table is provided in Appendix E. In Excel there are both one-tailed and two-tailed t functions: T.DIST and T.DIST.RT are one-sided, while T.DIST.2T is two-sided. Whether a hypothesis test is one- or two-sided needs to align with whether a one- or two-sided calculation is required.

EXAMPLE 6.52

Obtaining Student's *t* Critical Values from a Table

We will find the $(1-\alpha)$ critical value for a Student's t distribution with $(n-1)=9$ degrees of freedom for $\alpha=0.05$.

Critical values of Student's *t* distribution with v degrees of freedom

Student's *t* Cumulative Distribution Table					
Degrees of freedom	α				
v	0.10	0.05	0.025	0.01	0.005
1	3.078	6.314	12.706	31.821	63.657
2	1.886	2.920	4.303	6.965	9.925
3	1.638	2.353	3.182	4.541	5.841
4	1.533	2.132	2.776	3.747	4.604
5	1.476	2.015	2.571	3.365	4.032
6	1.440	1.943	2.447	3.143	3.707
7	1.415	1.895	2.365	2.998	3.499
8	1.397	1.860	2.306	2.896	3.355
9	1.383	1.833	2.262	2.821	3.250
10	1.372	1.812	2.228	2.764	3.169

Figure 6.39 Excerpt from the Student's t table.

Probability less than the critical value ($t_{1-\alpha,v}$)

Using the excerpt of the Student's t table in Figure 6.39, we find the row with $n=9$ degrees of freedom and read across to the 0.05 column. The critical value is 1.833, meaning that $\Pr(t>1.833)=0.05$. The full Student's t table is in Appendix E.

The F distribution takes on various shapes depending on its degrees of freedom, as shown here. The F distribution changes shape based on two sets of degrees of freedom, as shown in Figure 6.40. The F tables presented in statistics textbooks generally give separate tables for the often-used, right-tail probabilities $\Pr(F>F_0)$ of 0.10, 0.05, 0.025, and 0.01, which map to the 90th, 95th, 97.5th, and 99th percentiles, respectively. For convenience, this handbook provides the cumulative F tables for the lower tails as well, which map to $\Pr(F<F_0)$, or the right tail probabilities 0.90, 0.95, 0.975, and 0.99.

To use the F tables, match the percentile of interest, find the column and row associated with the numerator and denominator degrees of freedom, respectively, and read off the value of the F statistic. An excerpt of a 95th percentile table is shown in Figure 6.41. The full set of F tables is found in Appendix F.1–F.8. In Excel, the F functions are F.DIST and F.DIST.RT.

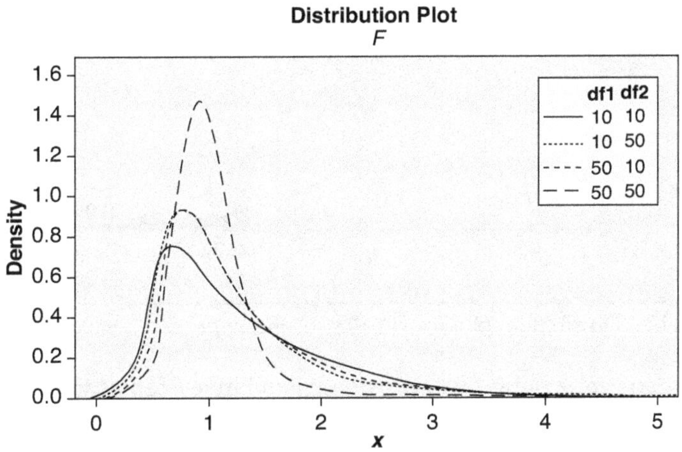

Figure 6.40 Example of F distributions with various degrees of freedom.

EXAMPLE 6.53

Obtaining *F*-table Critical Values

To find the value associated with the 95th percentile of an $F(6,9)$ distribution, we can use the table excerpt in Figure 6.41 to find $x=3.374$. This value would solve the statement $\Pr(F<x)=0.95$, as shown in Figure 6.42. Note that the tables in Appendices F.1–F.6 give the upper tail probability of the distribution. If we are testing at the 95% level, for example, then we would refer to the $F(0.05)$ table in Appendix F.3.

Continued

F distribution 95th percentile (5 percent in the right tail) df numerator									
df denominator	1	2	3	4	5	6	7	8	9
1	161.448	199.500	215.707	224.583	230.162	233.986	236.768	238.883	240.543
2	18.513	19.000	19.164	19.247	19.296	19.330	19.353	19.371	19.385
3	10.128	9.552	9.277	9.117	9.013	8.941	8.887	8.845	8.812
4	7.709	6.944	6.591	6.388	6.256	6.163	6.094	6.041	5.999
5	6.608	5.786	5.409	5.192	5.050	4.950	4.876	4.818	4.772
6	5.987	5.143	4.757	4.534	4.387	4.284	4.207	4.147	4.099
7	5.591	4.737	4.347	4.120	3.972	3.866	3.787	3.726	3.677
8	5.318	4.459	4.066	3.838	3.687	3.581	3.500	3.438	3.388
9	5.117	4.256	3.863	3.633	3.482	3.374	3.293	3.230	3.179
10	4.965	4.103	3.708	3.478	3.326	3.217	3.135	3.072	3.020

Figure 6.41 Excerpt of $\alpha = 0.05$, 95th percentile *F* table, denoted as *F*(0.05) in Appendix F.3.

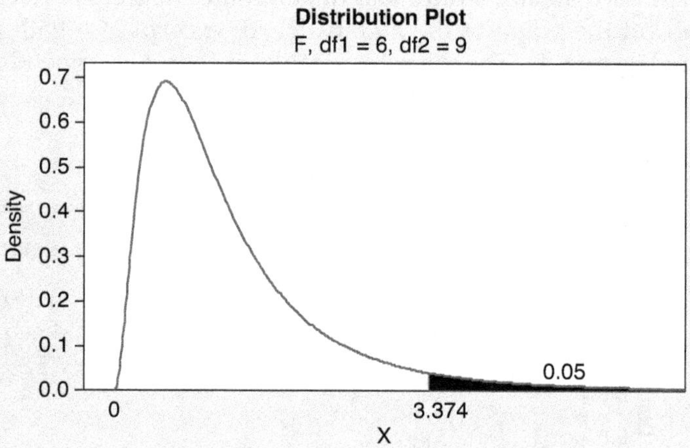

Figure 6.42 Distribution plot for *F*(6, 9) showing upper 5th percentile.

This handbook generously provides cumulative *F* tables for left-tail and right-tail probabilities, but many textbooks do not.

Continuous Distribution Reference Table

Pertinent information about each of the continuous distributions discussed in this section is summarized in Table 6.27.

Table 6.27 Reference table for continuous distributions.

Distribution	Description	Distribution functions: PDFs and CDFs	Expectation & Std Dev (*Variance* = *StdDev*²)
Weibull(β, η) Pronounced "beta" and "eta"	Widely used for lifetime data X, $0<x<\infty$; Shape parameter β; Scale parameter η=63.2%ile	$PDF: Pr(X=x\mid\beta,\eta)=\frac{\beta}{\eta}\left(\frac{x}{\eta}\right)^{\beta-1} exp\left\{-\left(\frac{x}{\eta}\right)^{\beta}\right\}$ $CDF: Pr(X\le x\mid\beta,\eta)=1-exp\left\{-\left(\frac{x}{\eta}\right)^{\beta}\right\};\beta,\eta>0$ Note parametrization has multiple different conventions.	$E(X)=\eta\Gamma\left(1+\frac{1}{\beta}\right)$ $StdDev(X)=\eta\cdot\sqrt{\Gamma\left(\frac{2}{\beta}+1\right)-\Gamma\left(\frac{1}{\beta}+1\right)^{2}}$ Gamma function $\Gamma(\cdot)$ has complex form, if k is an integer, $\Gamma(k)=(k-1)!$
Exponential (λ) =Weibull $\left(\beta=1,\eta=\frac{1}{\lambda}\right)$ Pronounced "lambda," "beta," and "eta"	Used to model time to failure for products when the failure rate is constant. Special case of the Weibull. $0<t<\infty$. Shape fixed; λ is scale parameter.	$PDF: Pr(X=t\mid\lambda)=\lambda\cdot exp\{-\lambda\cdot t\}$ $CDF: Pr(X\le t\mid\lambda)=1-exp\{-\lambda\cdot t\}$ Note parametrization may be inversed where λ is $\frac{1}{\lambda}$.	$E(X)=\frac{1}{\lambda}$ $StdDev(X)=\sqrt{\frac{1}{\lambda^2}}=\frac{1}{\lambda}$
Normal (μ, σ^2) = Gaussian Pronounced "mu" and "sigma-squared"	Symmetric and bell-shaped, with the center of the distribution defined by μ, the spread by σ. Foundation of many statistical methods.	$PDF: Pr(X=x\mid\mu,\sigma^2)=\frac{1}{\sqrt{2\pi\sigma^2}}exp\left(-\frac{(x-\mu)^2}{2\sigma^2}\right)$ $CDF: Pr(X\le x\mid\mu,\sigma^2)=\Phi\left(\frac{x-\mu}{\sigma}\right)$	**For X:** $E(X)=\mu$ $StdDev(X)=\sigma$
	X and μ may take on values of any real number x, $\mu\in\Re$, $\sigma>0$.	CDF table values Appendix D. **Empirical "68-95-99.7" Rule:** Approximately 68% of the data falls within $\pm1\sigma$, 95% within $\pm2\sigma$, and 99.7% within $\pm3\sigma$.	**For Standard Normal, Z:** $E(X)=0$ $StdDev(X)=1$

Continued

Table 6.27 Reference table for continuous distributions. (Continued)

Distribution	Description	Distribution functions: PDFs and CDFs	Expectation & Std Dev (*Variance* = *StdDev*²)
Lognormal (μ, σ^2) Note: μ and σ are on natural log scale	A random variable X is said to be distributed lognormally if its natural log is distributed normally. Or, exp(x) ~ normal if x is lognormal.	$PDF: Pr(X = x \mid \mu, \sigma^2) = \frac{1}{x\sqrt{2\pi\sigma^2}} exp\left(-\frac{(\ln(x)-\mu)^2}{2\sigma^2}\right)$ $CDF: Pr(X \le x \mid \mu, \sigma^2) = \Phi\left(\frac{\ln(x)-\mu}{\sigma}\right).$	$E(X) = \exp\left\{\mu + \frac{\sigma^2}{2}\right\}$ $StdDev(X) = \sqrt{\exp\{2\mu+\sigma^2\}(\exp\{\sigma^2\}-1)}$
The following distributions are part of the theoretical foundations of statistical testing methodology and are not utilized for reliability analysis.			
Chi-Squared(v) χ_v^2 Pronounced "kai," not "chai," and "nu."	Foundation of many statistical methods, since used to model the sum of squared, independent standard normal: $\sum Z_i^2$. Used in common statistical tests. $0 \le x < \infty$	$PDF: Pr(X = x \mid v) = \frac{1}{2^{\frac{v}{2}}\Gamma\left(\frac{v}{2}\right)} x^{\frac{v}{2}-1} exp\left(-\frac{x}{2}\right)$ PDF & CDF: Complex form, not intended for hand calculations. CDF table values in Appendix G.	$E(X) = v$ $StdDev(X) = \sqrt{2v}$
Student's $t(v)$ t_v.	Basis of hypothesis tests of mean effects, using t as test statistic. Constructed as $t_v = \frac{Z}{\sqrt{\chi^2 / v}}$, with v degrees of freedom (scale parameter).	PDF & CDF: Complex forms. CDF table values in Appendix E.	$E(X) = 0, v > 1$ $StdDev(X) = \sqrt{\frac{v}{v-2}}, v > 2$
F(v_1, v_2)	Basis of hypothesis tests of variance, using F as test statistic. Constructed as $F = \frac{\chi_1^2 / v_1}{\chi_2^2 / v_2}$, with v_1 degrees of freedom for the numerator and v_2 for the denominator.	PDF & CDF: Complex forms. CDF table values in Appendices F.1–F.8.	$E(X) = \frac{v_2}{v_2-2}, v_2 > 2$ $StdDev(X) = \sqrt{2\left(\frac{v_2}{v_2-2}\right)\left(\frac{v_1+v_2-2}{v_1 \cdot (v_2-4)}\right)}$ $v_2 > 4$

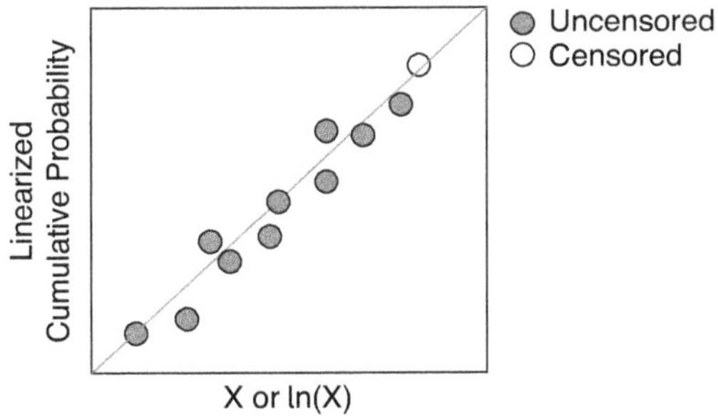

Figure 6.43 Example of probability plot.

Probability Plots and Assessing Goodness-of-Fit

Probability Plot Basics

Probability plots and **goodness-of-fit** (GOF) **tests** are used in conjunction to assess whether a given theoretical distribution is in fact a good representation of the actual data in question. A *probability plot* provides a graphical visualization of how well the data align with the hypothesized distribution. The *goodness-of-fit* statistics and tests quantify the strength of the relationship. Using an inappropriate or poorly fitting distribution will result in incorrect analyses, conclusions, and recommendations. Correct interpretation of probability plots is particularly critical in reliability applications, where data may be sparse, censored, and there are major business implications relying on your conclusions.

Probability plots are generally constructed as a plot of the ranked data on the x-axis, plotted against the sample cumulative frequency on a special linearized scale, through which a regression line could be fit. This special scale for the y-axis is uniquely defined for each distribution in such a way that the resulting plot will present as a straight line, if that distribution provides a reasonable representation for that data set. Data are assumed to be *iid*. An example plot is shown in Figure 6.43.

Typically, the x-axis is the x-ordinate or the natural log. Other variations exist that plot a quantile on the x-axis instead, but here we will focus on more standard convention for statistical software for use in reliability. Censored data may be included but must be handled in a way that recognizes that these points are observed incompletely.

Probability Plot Construction

Statistical software packages efficiently generate a number of distributional options needed to identify the distribution that provides the best fit. They are strongly recommended for general use over manual construction. A brief overview will be provided here of construction methods, but readers are advised for more detail to review reliability textbooks, such as Hahn and Shapiro (1994), Meeker and Escobar (2022), and O'Connor and Kleyner (2025), all of which dedicate an entire chapter to the mechanics of probability plots. Historically, special graph paper was used for these analyses, available for different distributions. Practices have evolved over time, resulting in a variety of different methods being developed over the years for defining the plotted points. However, the focus of this discussion will be on interpretation rather than mechanics.

To create a probability plot, data are first sorted from lowest to highest, then assigned to a percentile value by calculating the proportion of points that are lower. This is similar to the process outlined in "Probability Functions," in the construction of empirical CDFs. The linearization process is different for each distribution. For example, for the normal distribution, y-ordinates are obtained using the standard normal CDF function on the percentile p, $\Phi^{-1}(p)$ and plots against x. The lognormal uses the same CDF, but plots against $\ln(x)$. The exponential is $-(\ln(1-p))$ against x, and so forth.

A special mention is due here for the linearization of the Weibull as being of particular interest. Weibull's probability paper is derived using the following relationship:

$$\ln\left(\ln\left(\frac{1}{1-F(t)}\right)\right) = \beta\ln(t) - \beta\ln(\eta)$$

which has the form of a straight line with $x = \ln(t)$, with slope β and intercept $-\beta\ln(\eta)$. This allows for parameter estimation directly using regression parameter estimates. See "Basic Statistics" in this chapter for an overview of simple linear regression.

Probability Plot Interpretation

Figure 6.44 shows a normal probability plot for 30 observations simulated from a normal distribution with $\mu = 1.5$ and $\sigma = 3$. Since these data were obtained from a normal distribution, we should be able to verify that the normal is representative using this normal probability plot.

The probability plot in Figure 6.44 is a scatter plot with the data tracking a straight line. The x-axis shows the full range of the data, which, from the empirical rule, we expect to cover the range $1.5 \pm 3.3 \Rightarrow 1.5 \pm 9 \Rightarrow (-7.5, 10.5)$, which looks reasonable. The sample mean and standard deviation are not exactly the same as the source distribution, but they are close, and we wouldn't expect them to replicate the true mean and standard deviation anyway due to sampling variation.

Next consider the y-axis. It ranges from 0 to 100%, that is, probability of 0 to 1, but with an unusual scale. If you look closely, you can see percentiles are closer

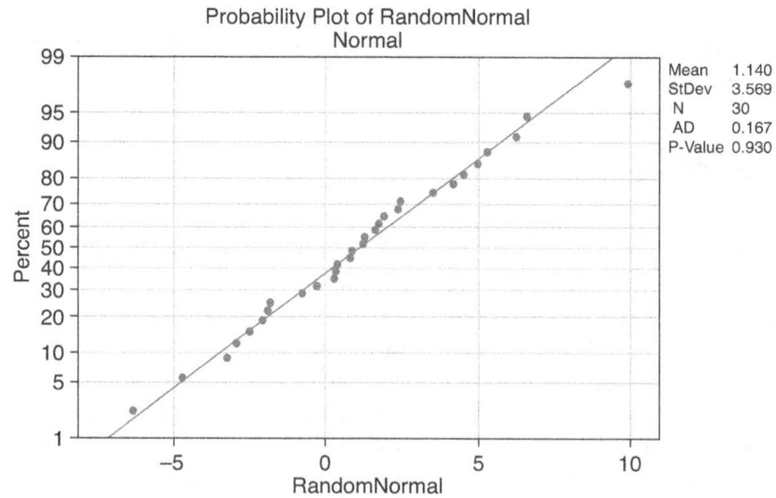

Figure 6.44 Normal probability plot based on simulated normal ($\mu = 1.5$, $\sigma = 3$) data.

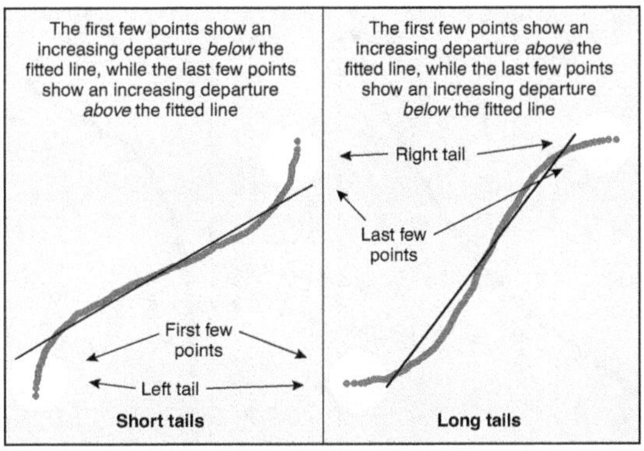

Figure 6.45 Direction of the tail bend on the normal probability plot tails for short- and long-tailed distributions.

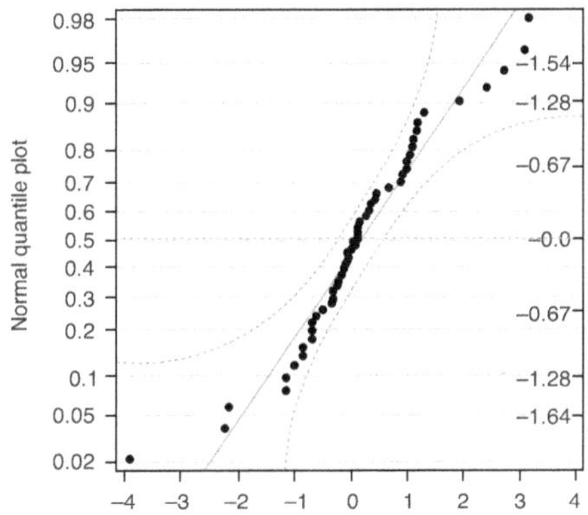

Figure 6.46 Heavy-tailed, or long-tailed, probability plot.

together in the center of the distribution and stretch farther apart as we approach the extremes. This shows the normal probability scale and reflects exactly what is going on with the symmetric bell-shaped distribution.

Finally, we check if the data align well with the normal. These data follow the straight line, so the allocation of points in the data distribution matches what we would expect to see. This is what a "good fit" looks like in a probability plot, regardless of distribution, for a distribution with a healthy amount of data. Plots will be more ambiguous for smaller datasets and those that include censored observations. Interpretation can be somewhat subjective, so it is important to keep in mind what type of error would be more conservative for a given scenario.

What does a "bad fit" look like? Data that do not track the straight line will be poorly represented by the candidate distribution. Note that certain deviations are associated with particular patterns and can provide insight into a more appropriate model. For example, in Figure 6.45, the direction of the tail bend for the plot on the left indicates a contraction in the data relative to expected for a normal, that is, "short tails." The tail bend for the plot on the right shows an expansion, or "long tails." Note the presence of the same curve on both sides indicates a symmetric distribution.

The probability plot in Figure 6.46 shows a **heavy-tailed (leptokurtic) distribution** in which more points are seen in the tails than expected if this were a good

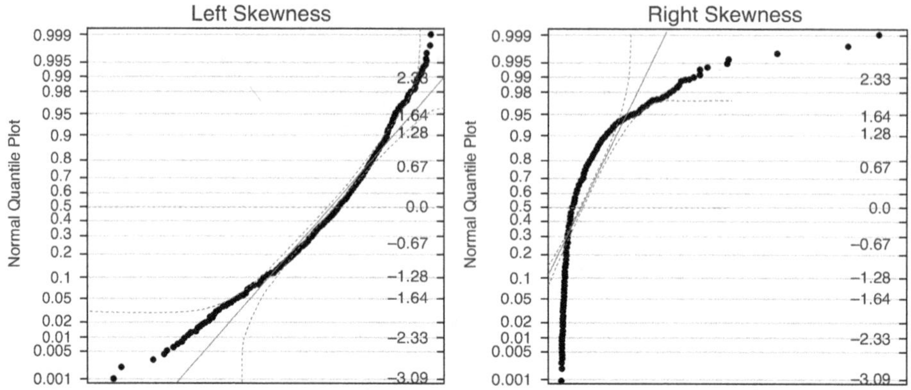

Figure 6.47 An illustration of left and right skewness in a normal probability plot.

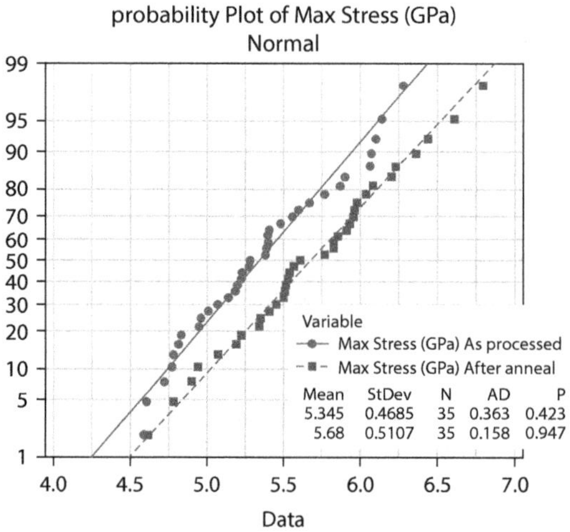

Figure 6.48 Probability plot showing data from max stress as processed versus after annealing.

fit. The slope of the straight line is steeper than the fitted line, meaning that the data in the center of the distribution have less scatter than if it were truly normal, and there are more points from the tails than expected. Typically, we would *not* want to assume the fitted normal in this case, since we would underestimate the probability of having extreme values.

Figure 6.47 illustrates normal probability plots exhibiting skewness. The plot on the left shows the first points, representing the left tail, falling above the line, indicating that the distribution has a left, or negative, skew. Conversely, the plot on the right shows the right-most points falling below the line, indicting a right, or positive, skew.

Probability plots can also be used to compare different datasets or different groups within a dataset. The probability plot in Figure 6.48 shows two lines, each representing a different group. These data were taken from Tong et al. (2015) and show 35 samples of each material, where one sample was tensile tested as processed, and the other sample after annealing.

Summary statistics are shown for the fitted normal distribution (mean $= \hat{\mu}$, standard deviation $= \hat{\sigma}$) for both groups, which indicate similar standard deviations around 0.5, but a shift in the mean from 5.345 to 5.68, a difference of 0.335 associated with the annealing. The statistical output accompanying the plot should include different comparisons and formal statistical tests for various aspects of this distribution. For example, the results shown in Figure 6.48 include the Anderson-Darling (AD) goodness-of-fit measure, with p-values, to be discussed next.

Goodness-of-Fit (GoF)

Goodness-of-fit test results are usually provided along with a probability plot. Popular measures are **Anderson-Darling** (AD), and **Shapiro-Wilk** (SW) which are reported with a p-value, and the **Pearson correlation coefficient** (r), discussed in "Basic Statistics," calculated from the linearized variables that comprise the probability plot.

Goodness-of-fit measures are providing a quantitative characterization of how well the data track that straight line, using different approaches. The same general null and alternative hypotheses apply to every test scenario.

H_0: Data follow the candidate distribution

H_A: Data do not follow the candidate distribution

In contrast to other hypothesis tests in which we want to prove the alternative hypothesis, in the goodness of fit tests, rejecting H_0 tells us that our candidate distribution does not fit the data. If we fail to reject the null hypothesis, it does not prove that the candidate distribution fits; it simply tells us that we have insufficient evidence to reject H_0.

Aside from the correlation, GOF measures themselves tend to be very complex functions that we will not replicate here. Statistical software provides p-values for the GOF test, so they can now be interpreted as any other formal statistical test.

Distribution Identification and Selection

Distribution identification and selection methods do not result in definitive conclusions. The process is one of elimination, of identifying a "best" option that provides a representative model. The list discussed here is not exhaustive but provides a starting point and a toolbox. Some distributions may be clearly unfit for a given dataset, while there may be multiple other options. It is important to remember that goodness-of-fit hypothesis tests only allow us to reject options but do not confirm them.

Distribution selection must consider the nature and constraints of the situation. What information is available about the process? Also, what is the nature of the problem? The intended usage of the estimate to be developed may provide guidance. Several examples will be discussed here more thoroughly.

EXAMPLE 6.54

Distribution Identification Uncensored ($n=6$)

The data for this example were first presented in Example 6.48 in the lognormal distribution section. Note that all data are uncensored. The times to failure for a product are known to be lognormally distributed. The times to failure for a sample of six parts are 850, 925, 1250, 1550, 1800, and 2750 cycles. Estimate the mean and sigma parameters and use to calculate the probability of failing before 2000 cycles.

Continued

Previously the estimates were done manually and assumed to be lognormal. However, we should also assess if lognormal is a good assumption. How sensitive are our results to this assumption?

Four different distributional fits are shown in Figure 6.49 for the normal, lognormal, exponential, and Weibull. All are bounded below by zero except the normal. Looking at the order of the AD statistics, the quality of the fits is quantitatively ordered from best to worst as lognormal, Weibull, normal, and exponential. Having only six points means that conclusions are going to have high uncertainty. We would drop out the normal and exponential for technical reasons, in that using the normal would allow us to generate negative data. Similarly, the exponential is poorly behaved relative to the others. We could move forward and consider the Weibull and lognormal more closely.

More detailed results were run for the Weibull and lognormal distributions, shown in Figure 6.50.

The two models produce different results for probability of failure by $t = 2000$ cycles. The Weibull estimates 77%, the lognormal, 81%. This probability is in the same ballpark but cannot be estimated very precisely with this amount of data. Estimates should be used with caution. Note that extrapolating further beyond the range of the data will allow estimates to diverge more because of the difference in the heaviness of the tails.

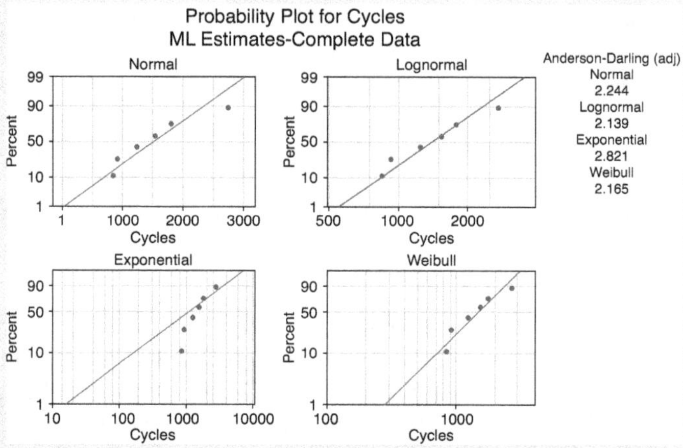

Figure 6.49 Normal, lognormal, exponential, and Weibull probability plots.

Distribution: Lognormal

Parameter Estimates

Parameter	Estimate	Standard Error	95.0% Normal Upper Bound
Location	7.24447	0.163546	7.51348
Scale	0.400605	0.115645	0.644069

Goodness-of-Fit

Anderson-Darling (Adjusted)
2.139

Table of Cumulative Failure Probabilities

Time	Probability	95.0% Normal Upper Bound
2000	0.813194	0.953820

Distribution: Weibull

Parameter Estimates

Parameter	Estimate	Standard Error	95.0% Normal Upper Bound
Shape	2.54006	0.775883	4.19806
Scale	1720.23	293.592	2277.75

Goodness-of-Fit

Anderson-Darling (Adjusted)
2.165

Table of Cumulative Failure Probabilities

Time	Probability	95.0% Normal Upper Bound
2000	0.769226	0.943566

Figure 6.50 Lognormal and Weibull detailed analysis.

EXAMPLE 6.55

Distribution Identification with Uncensored Groups

As presented in Figure 6.48, Tong et al. (2015) report tensile strength data for two samples of 35 units each of nanocrystalline α-iron. One sample was tensile tested as processed, and the other was tested after the units went through an annealing process. The maximum stress during tensile deformation in GPa was recorded as "processed" and "after annealing" for each paired sample. No points were censored. What can we learn from these data? Are these populations different from each other, and if so, how? Is there a difference in their lower tails?

Four different distribution fits are shown in Figure 6.51 for the normal, lognormal, exponential, and Weibull. All are bounded below by zero except the normal. Looking at the order of the AD statistics, the quality of the fits is quantitatively ordered, from best to worst, as normal and lognormal, Weibull, and exponential. All the distributions look plausible except for the exponential. We could decide to drop the normal for technical reasons, in that using the normal could allow us to generate negative data. We could then move forward and consider the lognormal more closely. The Weibull appears to fall off the line on the lower side, which is more of an issue with stress distributions, since we care about worst case performance. This fit might benefit from incorporating a threshold parameter.

The detailed results for the lognormal distribution are shown in Figure 6.52 for as processed versus after annealing. The lognormal and normal distributions would provide similar fits to the center of the data, but the lognormal allows for additional skew in the upper tail.

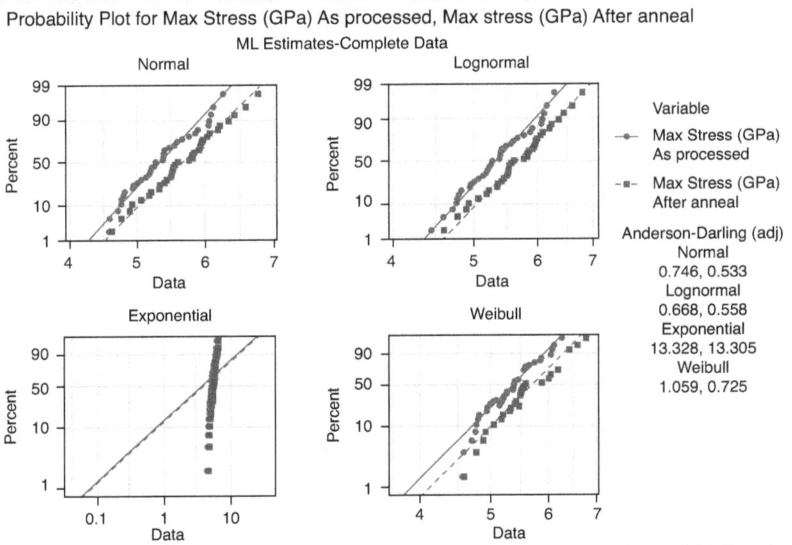

Figure 6.51 Probability plot for max stress (GPa) for annealing data.

Continued

Distribution Analysis: Max Stress (GPa) As processed

Distribution: Lognormal

Parameter Estimates

Parameter	Estimate	Standard Error	95.0% Normal CI Lower	Upper
Location	1.67254	0.0145117	1.64410	1.70099
Scale	0.0858524	0.0102613	0.0679226	0.108515

Table of Percentiles

Percent	Percentile	Standard Error	95.0% Normal CI Lower	Upper
1	4.36152	0.121844	4.12913	4.60699
2	4.46480	0.114242	4.24641	4.69442
3	4.53159	0.109422	4.32212	4.75121
4	4.58249	0.105825	4.37970	4.79467
5	4.62431	0.102933	4.42691	4.83053
6	4.66022	0.100505	4.46733	4.86143
7	4.69192	0.0984093	4.50296	4.88882
8	4.72050	0.0965645	4.53498	4.91361
9	4.74663	0.0949172	4.56420	4.93636
10	4.77082	0.0934305	4.59117	4.95750
20	4.95446	0.0836662	4.79316	5.12118

Distribution Analysis: Max stress (GPa) After anneal

Distribution: Lognormal

Parameter Estimates

Parameter	Estimate	Standard Error	95.0% Normal CI Lower	Upper
Location	1.73298	0.0151036	1.70338	1.76259
Scale	0.0893541	0.0106799	0.0706930	0.112941

Table of Percentiles

Percent	Percentile	Standard Error	95.0% Normal CI Lower	Upper
1	4.59568	0.133622	4.34111	4.86518
2	4.70899	0.125405	4.46951	4.96131
3	4.78233	0.120187	4.55248	5.02379
4	4.83825	0.116289	4.61562	5.07163
5	4.88422	0.113153	4.66741	5.11111
6	4.92370	0.110519	4.71178	5.14514
7	4.95857	0.108244	4.75089	5.17533
8	4.99000	0.106241	4.78606	5.20263
9	5.01876	0.104452	4.81816	5.22772
10	5.04538	0.102838	4.84780	5.25102
20	5.24766	0.0922320	5.06997	5.43158

Figure 6.52 Detailed output for lognormal distribution fit.

EXAMPLE 6.56

Distribution Identification with Censoring

Twenty products are tested for 1000 hours. Fourteen products fail at the following times, respectively: 70, 128, 204, 291, 312, 377, 473, 549, 591, 663, 748, 827, 903, 955 hours.

Estimate the B10 predicted life, which is the 10th percentile of the cumulative distribution function. Four different distributional fits are shown in Figure 6.53 for normal, lognormal, exponential, and Weibull. All are bounded below by zero except the normal. Looking at the order of the AD statistics, the quality of the fits is quantitatively ordered, from best to worst, as Weibull, normal, lognormal, and exponential. All look plausible. This dataset contains 20 points, of which 14 ran to failure. Despite this, our conclusions are going to have high uncertainty. We drop the normal distribution from consideration for technical reasons, namely, using the normal would allow us to generate negative data. Similarly, the exponential is the worst fit relative to the others. Of these four options, the Weibull appears the best, as it does a better job of tracking the line.

The detailed Weibull results are shown in Figure 6.54. The best fit model for this distribution is a Weibull (β= 1.44, η=887). The B10, or 10th percentile of the CDF, is nominally estimated at 186.5 hours.

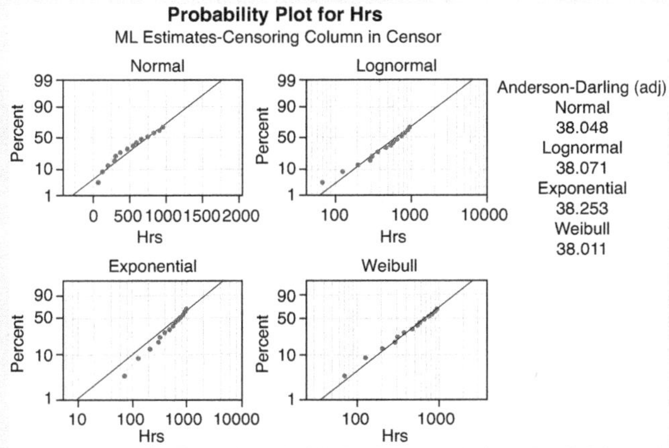

Figure 6.53 Distribution fits for censored data.

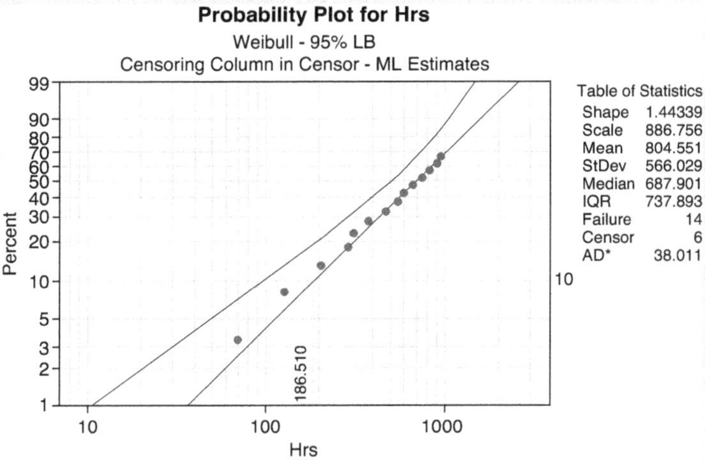

Figure 6.54 Weibull distribution fit results.

EXAMPLE 6.57

Distribution Identification with Heavy Censoring ($n=10$)

Ten units are on test. The units are not replaced when they fail (nonreplacement). One unit fails at $t_1=685$ hours, and a second unit fails at $t_2=1690$ hours. The test is ended at $t=2500$ hours with no additional failures. We need to estimate the B10 predicted life, the 10th percentile of the cumulative distribution function.

Four different distributional fits are shown in Figure 6.55 for normal, lognormal, exponential, and Weibull. All are bounded below by zero except the normal. Looking at the plots and the order of the AD statistics, the distributions are indistinguishable. **Fitting a distribution will not provide a robust approach. Estimates can be obtained mathematically but are not generally recommended for use.** Results are shown in Figure 6.55 for comparison purposes.

Continued

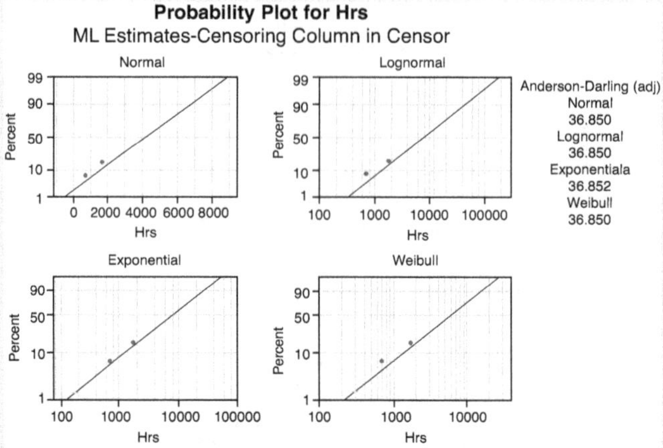

Figure 6.55 Distribution plots for heavily censored data.

Estimates were obtained for the B10 from each distribution, shown in Table 6.28 and Figure 6.56.

Table 6.28 B10 estimates for heavily censored data.

Distribution	B10 (hours)	Comments
Normal	1539	Highest value, has nonnegligible probability below zero so reject
Lognormal	1331	Mid-range value, but heavy upper tail, usage not recommended without rationale that heavy upper tail is plausible
Exponential	1178	Lowest estimate
Weibull	1364	Mid-range value, with more moderate upper tail

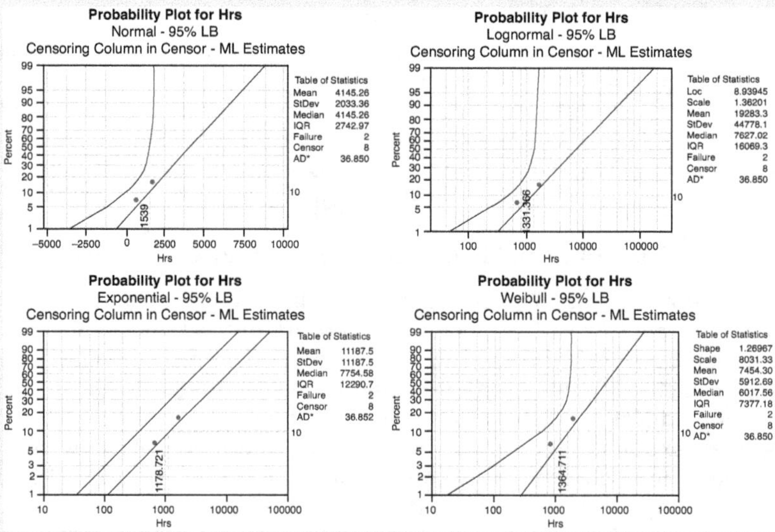

Figure 6.56 B10 estimates using normal, lognormal, exponential, and Weibull fits.

PROBABILITY FUNCTIONS

> Assess various probability functions (e.g., cumulative distribution functions [CDFs], probability density functions [pdfs], and hazard functions) and recognize their application in various situations. (Evaluate)
>
> **Body of Knowledge III.A.4**

Statistical distribution functions are written equivalently in multiple forms, which are used to answer different questions.

Probability Functions

As presented in "Probability Distributions," a **probability density function** (PDF) is used to represent the statistical distribution for a *continuous* random variable in terms of instantaneous probability for any given value. The distribution is a continuous curve that begins and ends with zero values. The area under the curve is defined to be exactly 1.0 to reflect that total probability is exactly one, which is mathematically enforced through scaling factors. The height of the curve reflects the instantaneous probability at any given point. A **probability mass function** (PMF) is used to represent the statistical distribution for a *discrete* random variable in terms of instantaneous probability. PDFs appear to be curves, while PMFs use point mass and look like bar charts. The PMF and PDF of a Poisson and normal distribution are shown in Figure 6.57. PDFs and PMFs are typically denoted as a function f evaluated at time t, given some parameter set "theta," Θ, abbreviated $f(t)$ for generality:

$$f(t \mid \Theta) = f(t)$$

The examples used in this section are arbitrary distributions that are not necessarily associated with reliability analyses. They have been chosen to show how the different types of statistical functions can be calculated for any distribution, and that the same concepts hold for both discrete and continuous families.

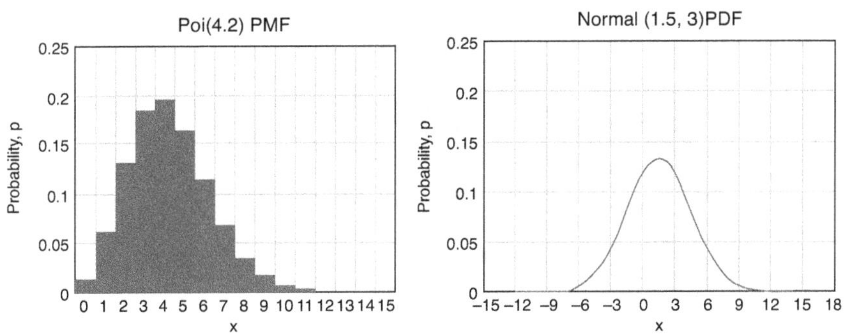

Figure 6.57 Example of discrete and continuous mass and density probability functions for Poisson ($\lambda = 4.2$) and normal ($\mu = 1.5$, $\sigma = 3$), respectively.

Cumulative Functions

Cumulative functions are used to calculate the probability of being less than a certain point by calculating the area under the curve from left to right. They always progress from 0 to 1 and take on an S-shape. They are referred to as CMFs or CDFs: cumulative mass functions or cumulative distribution functions. The cumulative mass function for a discrete distribution can be constructed through standard summation. However, integral calculus must be used to obtain cumulative functions for continuous distributions. A cumulative distribution sums up the probability to the left of X = t. CDFs and CMFs are typically denoted as a function F evaluated at time t, given some parameter set Θ, abbreviated $F(t)$ for generality:

$$F(t \mid \Theta) = F(t) = Pr(X \leq t)$$

In calculus terms, cumulative functions are defined as:

$$F(t) = \Pr(X \leq t) = \int_0^t f(x)\mathrm{d}x$$

The cumulative distributions for a Poisson ($\lambda = 4.2$) and normal ($\mu = 1.5$, $\sigma = 3$) distribution are shown in Figure 6.58.

Note that the standard normal distribution uses special notation for its CDF: Φ. The value x must be normalized first:

$$F(x \mid \mu, \sigma^2) = \Phi\left(\frac{x-\mu}{\sigma}\right) = \Phi(Z) = Pr(Z < z)$$

Reliability Functions

A **reliability function**, denoted $R(t)$, is used to calculate the probability that a system fails *after* a specified time, t. Statistical CMFs or CDFs are therefore used to determine the probability of failure (or the event of concern) *before* time t, then is subtracted from 1 to obtain the probability that it is still operating afterward:

$$R(t) = Pr(X > t) = 1 - Pr(X \leq t)$$

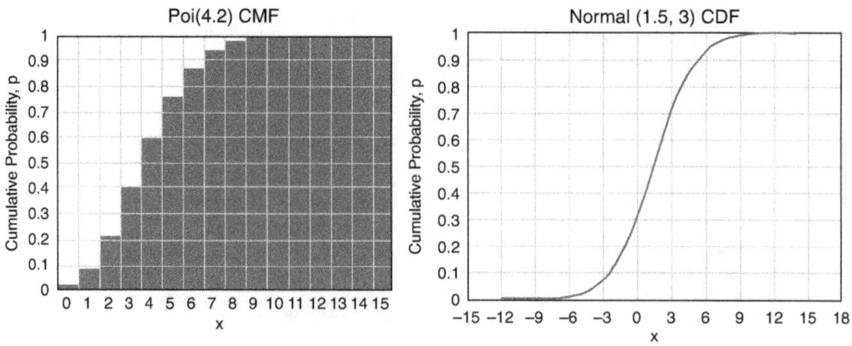

Figure 6.58 Example of discrete and continuous cumulative mass and distribution functions for Poisson ($\lambda = 4.2$) and normal ($\mu = 1.5$, $\sigma = 3$), respectively.

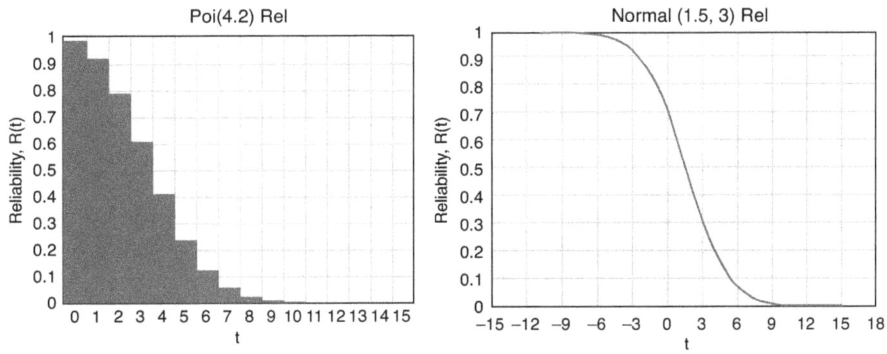

Figure 6.59 Example of discrete and continuous reliability functions *R*(*t*).

In calculus terms, reliability functions are defined as:

$$R(t) = Pr(X > t) = \int_t^{\infty} f(x)\mathrm{d}x = 1 - \int_0^t f(x)\mathrm{d}x$$

In other contexts, the reliability function may be referred to as the **survival function** S(t). Both refer to the same function, and therefore the notation and methods are interchangeable, although terminology and usage may differ.

$$S(t) = R(t)$$

Examples of discrete and continuous reliability functions $R(t)$ are shown in Figure 6.59.

Some organizations may reference **BX life**, which is defined as the time at which $X\%$ of the units in a population will have failed. For example, if an item has a B10 life of 100 hours, that means that 10% of the population will have failed by 100 hours of operation. These values are obtained by extracting the appropriate quantile from the reliability function $R(t)$, that is, by identifying for what t $R(t) = 0.9$.

The **conditional reliability function** is used to calculate the probability that a system does not fail during the next mission, or other interval of concern of length M, given that it has survived to time t.

$$R(t + M \mid t) = \frac{Pr(X > t + M)}{Pr(X > t)} = \frac{1 - F(t + M)}{1 - F(t)} = \frac{R(t + M)}{R(t)}$$

Longer missions mean longer exposure to failure opportunities and thus are higher risk, which can be seen in lower calculated conditional reliability values for higher times M. Figure 6.60 shows conditional reliability for $M = 1$, 3, and 5, using arbitrary plausible values for comparison purposes.

Hazard Functions

Hazards are frequently referenced but are a source of much confusion and easily misinterpreted. The following is an attempt to disentangle the web by going back to the underlying math. As with probability functions, there are two versions: the **instantaneous hazard rate function** $h(t)$ and the *cumulative* version $H(t)$, which is the integral of $h(t)$ and so calculates area under the $h(t)$ curve. Interpretation will be discussed at the end.

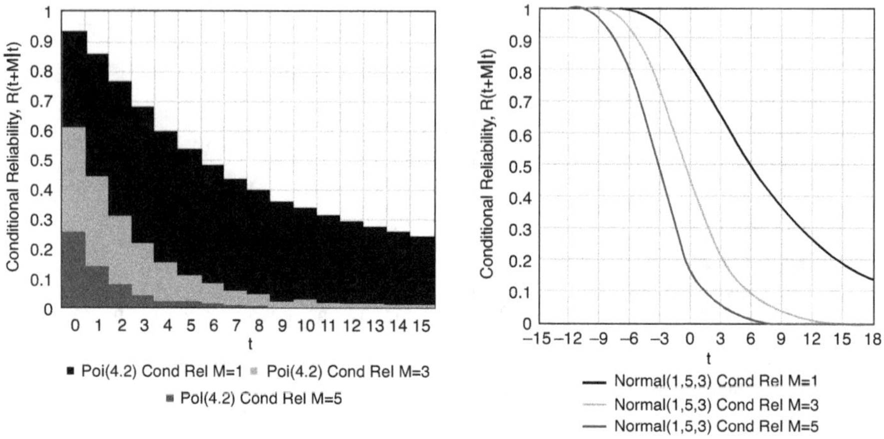

Figure 6.60 Example of discrete and continuous conditional reliability functions $R(t+M \mid t)$ for Poisson ($\lambda = 4.2$) and normal ($\mu = 1.5$, $\sigma = 3$) for missions of length $M = 1$, 3, and 5.

The derivation of the hazard rate function $h(t)$ itself involves calculus and so is beyond the scope of this book, but it is provided subsequently for reference since it is helpful for understanding its meaning. In formal mathematical terms, $h(t)$ is defined in terms of a limit over the width of the next subsequent time interval Δt:

$$h(t) = \lim_{\Delta t \to 0} \frac{Pr(t < X \le t + \Delta t \mid X > t)}{\Delta t}$$

This expression describes the probability of failing in the next time interval Δt, given that the unit has not failed prior to time t. Since this is defined using a limit function, this measure involves the instantaneous rate of change at time t, as Δt becomes vanishingly small. A derivative is therefore involved. This also explains why this is referred to as the *instantaneous* hazard rate function. Another name is the conditional failure density function. It is more casually referred to as the hazard rate or just the hazard.

We can rearrange this expression by expanding the conditional probability and rewriting in terms of cumulative functions $F(\cdot)$ and extracting the divisor $R(t)$:

$$h(t) = \lim_{\Delta t \to 0} \frac{1}{\Delta t} \frac{Pr(t < X \le t + \Delta t)}{Pr(X > t)} = \lim_{\Delta t \to 0} \frac{1}{\Delta t} \frac{Pr(X \le t + \Delta t) - Pr(X \le t)}{1 - Pr(X \le t)}$$

$$h(t) = \lim_{\Delta t \to 0} \frac{1}{\Delta t} \frac{F(t + \Delta t) - F(t)}{1 - F(t)} = \lim_{\Delta t \to 0} \frac{1}{1 - F(t)} \frac{F(t + \Delta t) - F(t)}{\Delta t}$$

At this point we are left with the definition of the derivative for $F(t)$, with $R(t)$ in the numerator:

$$h(t) = \frac{1}{R(t)} \lim_{\Delta t \to 0} \frac{[F(t + \Delta t) - F(t)]}{\Delta t} = \frac{F'(t)}{R(t)} = \frac{f(t)}{R(t)}$$

The derivative of the cumulative function $F(t)$ is the probability function $f(t)$, by definition, so we are back to the PDF or PMF. So, what are we looking at here? We arrived at the standard definition of $h(t)$, the ratio of the density to the reliability. It is not obvious in this form that this calculation produces a rate, but the fact

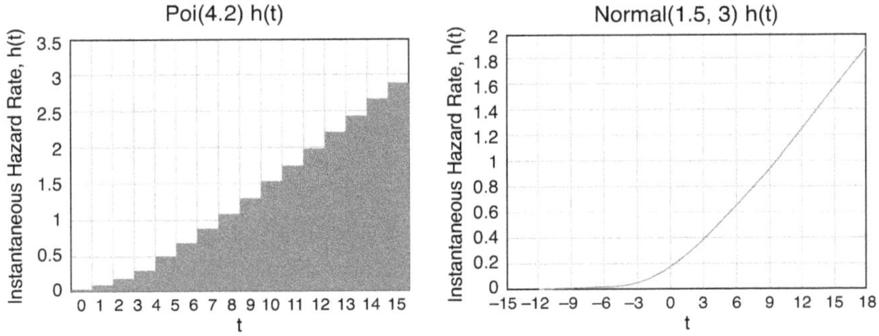

Figure 6.61 Example of discrete and continuous *instantaneous* hazard functions $h(t)$ for Poisson ($\lambda = 4.2$) and normal ($\mu = 1.5$, $\sigma = 3$).

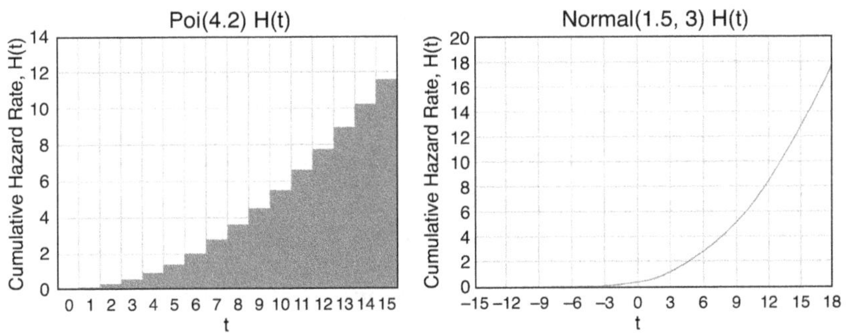

Figure 6.62 Example of discrete and continuous *cumulative* hazard functions $H(t)$ for Poisson ($\lambda = 4.2$) and normal ($\mu = 1.5$, $\sigma = 3$).

that we are dividing here in the derivation by Δt, infinitesimal though it may be, converts the units to a rate. Dividing by the reliability $R(t)$ accounts for the conditional success up until time t.

Thus, we can interpret the **hazard rate** as the frequency that a failure or other event of interest is expected to occur per unit time, given success up until that time. Hazards are *not* probabilities: Values can exceed 1, although they do not take on negative values. Hazard functions can increase, decrease, or stay constant, depending on the shape of the original distribution. As discussed in "Probability Distributions," the geometric and exponential distributions have constant failure rate, that is, constant hazards, meaning a flat line for the entire range of X. The hazard rates shown in Figure 6.61 are both increasing.

The **cumulative hazard function** $H(t)$ is the integral of the instantaneous hazard rate function $h(t)$ from 0 to t. It effectively accumulates the events of interest over time and can be thought of as the total potential exposure that a population would experience:

$$H(t) = \int_0^t h(x)\,dx = -\ln(R(t))$$

Since $H(t)$ is an integral, its value cannot decrease—it will always increase or stay the same, regardless of the behavior of the hazard function $h(t)$ itself (Figure 6.62).

Empirical CDFs

Thus far this section has focused solely on theoretical aspects of probability functions. The next step is to discuss plotting of actual data to create an empirical cumulative mass or distribution function (CMF or CDF). The empirical CDF is a step function that progresses from 0 to 1 (i.e., 0 to 100%) and follows a general S-shape, as did the theoretical cumulative functions described previously. Empirical CDFs enable us to get a quick visual assessment of the entire distribution, allowing us to get a sense of the many features all at once, that is, the center location, amount of scatter, and shape. The term "empirical" here refers to the chart being generated from real data.

To create an empirical CDF, data are sorted from lowest to highest, then assigned to a percentile value by calculating the proportion of points that are lower. Each step in the plot represents the next data point, beginning from zero. Higher steps represent more points; flat lines are no data. The median is easily identified as the point at which the data line crosses the 50th percentile. Other percentiles of interest may also be readily identified.

EXAMPLE 6.58

Empirical Distribution Comparison Between Poisson and Normal

The plot in Figure 6.63 was created from 30 observations simulated from each of the Poisson (λ=4.2) and normal (μ=1.5, σ=3) distributions used in the previous comparison plots. The Poisson data are shown as a solid and normal as a dashed line, showing distinctly different curve shapes that overlap only up at the upper end of the *x*-range.

By inspecting the plot, we see that the normal data show more scatter than the Poisson data, ranging from about −6 to +10 with a median of ~1.5, while the Poisson data are more compact, beginning at +2 and going up to +9, with a median of ~4 to 5. This result is consistent with values we would expect from theory: Per the normal distribution's empirical rule, we would expect 95% of the data to be in the range 1.5±2.3 or the interval (−4.5, +7.5). The S-shape looks consistent with a symmetric bell curve.

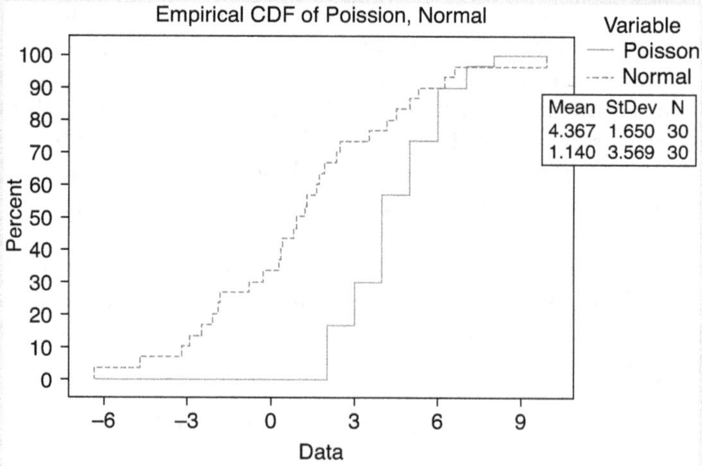

Figure 6.63 Example of discrete and continuous empirical cumulative functions for n=30, simulated observations from Poisson (λ=4.2) and normal (μ=1.5, σ=3).

The Poisson data are more compact, with mean around 4 and standard deviation ~2, which is also theoretically consistent, since $\sqrt{4.2} \approx 2.05$. The actual summary statistics for this hypothetical case are shown to the side in Figure 6.63.

In Example 6.58, each product could be distinguished from the others. However, when there is more overlap, a key question of interest might be to determine if the distributions are different from each other.

Reliability Distribution Estimation Using the Kaplan-Meier Method

The **Kaplan-Meier (K-M)** or **product limit estimator** is a popular *nonparametric* method commonly used to estimate the reliability (survival) function $R(t)$ from empirical lifetime data, as shown in Figure 6.64. The two distributions cover different ranges but taper out in the same vicinity. These are step functions and resemble the empirical CDF discussed in the previous section, except inversed, beginning with a probability of one declining to zero. More data will create a smoother curve.

To create the K-M chart, sort the failure times from smallest to largest, record how many failures occurred at each time, and indicate if there was any censoring. Each step-down on the K-M plot represents the next time-point, beginning from zero. Higher steps represent more points; flat lines are no data (or suspensions), as with the empirical CDF previously described. The median and other percentiles of interest are readily identifiable, and it is straightforward to make comparisons between distributions. Note that these methods assume all censoring is noninformative; else biases may be induced in the results.

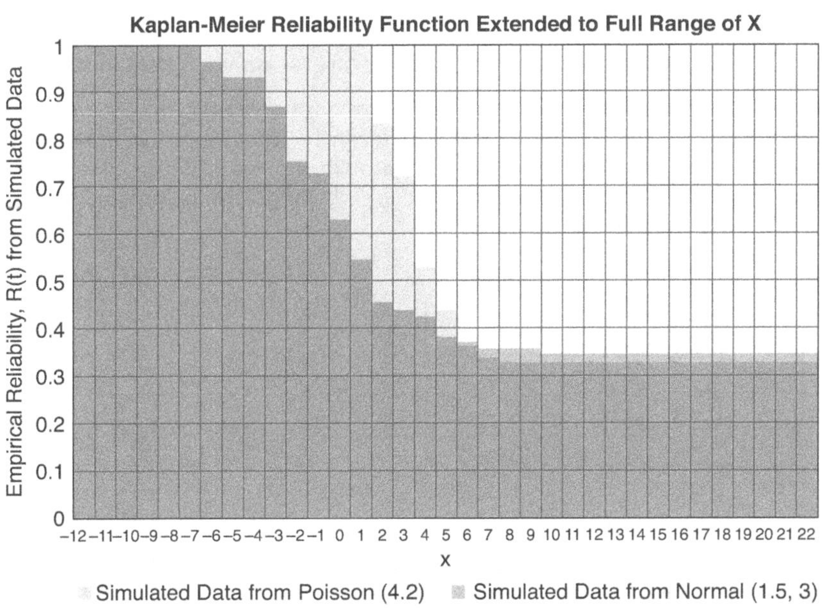

Figure 6.64 Example of discrete and continuous reliability functions obtained using Kaplan-Meier estimator for $n = 30$ simulated observations from Poisson ($\lambda = 4.2$) and normal ($\mu = 1.5$, $\sigma = 3$).

The K-M can be obtained using the following equations:

$$\hat{R}(t) = \hat{S}(t) = \prod_{i;t_i \le t}^{k} \left(1 - \frac{r_i}{n_i}\right) = \prod_{i;t_i \le t}^{k} \left(1 - \hat{h}_i\right)$$

where t_i is the time (or time interval) where at least one failure occurred, for k times where data are available. For each time window, we tabulate r_i and n_i, the number of failures and total number of uncensored surviving units present at time t_i, and use to calculate the proportion of surviving units, which are then multiplied together up through time to estimate the overall reliability function $R(t)$. The term

$$\hat{h}_i = \frac{r_i}{n_i}$$

can be thought of as a discrete estimate of the hazard at time t_i, obtained simply through the fraction of failed units of those with known condition, not suspensions. The "hats" indicate estimates, as always. In survival analysis literature the model is defined in terms of deaths, d_i. This method is generally preferred due to the lack of assumptions needed for the analysis, since it does not require a distributional fit. The computations are also relatively straightforward.

Comparisons can be also made between distributions. Formal statistical tests have been defined for lifetime (nonnegative) data. Statistical software may include the log-rank and Wilcoxon tests, for which low p-values would provide evidence of a difference between the distributions that were sampled. The log-rank test is a type of chi-squared test that compares estimates of the reliability functions between groups at each observed event time. The Wilcoxon signed-rank test compares just the medians of the distributions using paired averaged data.

EXAMPLE 6.59

Kaplan-Meier Mechanics and Interpretation with No Censoring

A reliability engineer collects time-to-failure data on two processes, ABC and XYZ, to determine if they are performing differently. No observations were censored. Raw data are shown in the left-hand plot in Figure 6.65, grouped data in the center, and the Kaplan-Meier plot on the right, with the associated test statistics. What should she conclude?

Unit	ABC	XYZ
1	100	120
2	100	120
3	110	120
4	110	120
5	110	120
6	90	110
7	110	120
8	110	110
9	100	120
10	100	100

	ABC	XYZ
Time	r1	r2
70	0	0
80	0	0
90	1	0
100	4	1
110	5	2
120	0	7
130	0	0

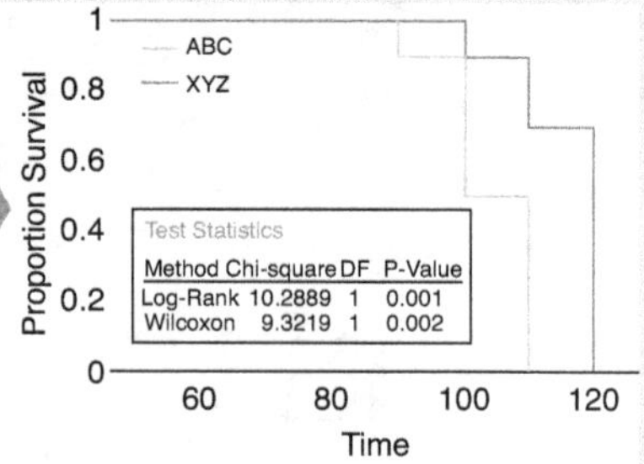

Figure 6.65 Raw data and resulting Kaplan-Meier plot for Example 6.59.

The Wilcoxon test compares the medians of these two processes, treating the data as paired. The log-rank test compares the reliability curves $R(t)$. Both tests are statistically significant at conventional levels. There is therefore evidence that these processes are not equivalent. The XYZ process is producing higher results than ABC. There is a clear shift upward in the data.

EXAMPLE 6.60

Kaplan-Meier Interpretation with No Censoring

Recall from Example 6.55, thirty-five samples of nanocrystalline α-iron were cut into two equal pieces. One half of the sample was tensile tested as processed, and the other half was tested after the sample went through an annealing process. The maximum stress during tensile deformation in GPa was recorded as processed and after annealing for each paired sample. No points were censored. What can we learn from these data? Are these populations different from each other, and if so, how?

Plotted data in Figure 6.66 show no correlation between as processed or annealed, despite being paired.

The analyst generated Kaplan-Meier curves to compare the distributions and looked at the results (Figure 6.67). The curves are similar in shape, suggesting a similar level of variation, but show a shift. Is it statistically significant?

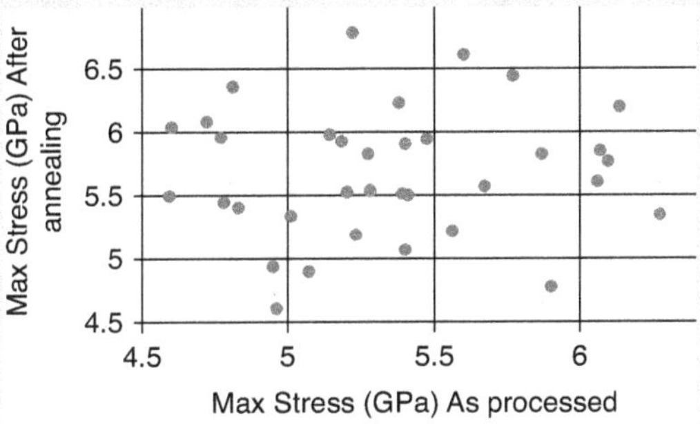

Figure 6.66 Max stress (GPa) for samples as processed and after annealing.

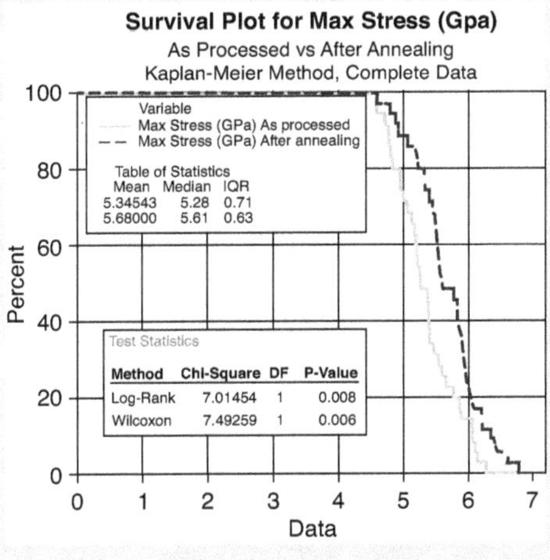

Figure 6.67 Survival plot for max stress (GPa) as processed and after annealing.

Continued

The Wilcoxon test compares the medians of these two processes, treating the data as paired. The log-rank test compares the reliability curves $R(t)$. Both tests are statistically significant at conventional levels. There is therefore evidence that these processes are not equivalent. The annealed process produces higher results than the baseline process. There is a clear shift upward in the data. The difference in both mean and median is 0.33.

EXAMPLE 6.61

Kaplan-Meier Interpretation with Censoring

The data for this example was also used in Example 6.56. Twenty products are tested for 1000 hours. Fourteen products fail at the following times, respectively: 70, 128, 204, 291, 312, 377, 473, 549, 591, 663, 748, 827, 903, and 955 hours. Estimate the B10 predicted life, the 90th percentile of the reliability function, using a nonparametric approach. The analyst generated the Kaplan-Meier curve in Figure 6.68 for the reliability distribution.

The straight line produced by the failures in this plot indicates a constant failure rate over the time frame tested, which may not be desirable (Figure 6.69).

The B10 life can be obtained from the detailed output, which indicates a 90% survival probability at 128 hours.

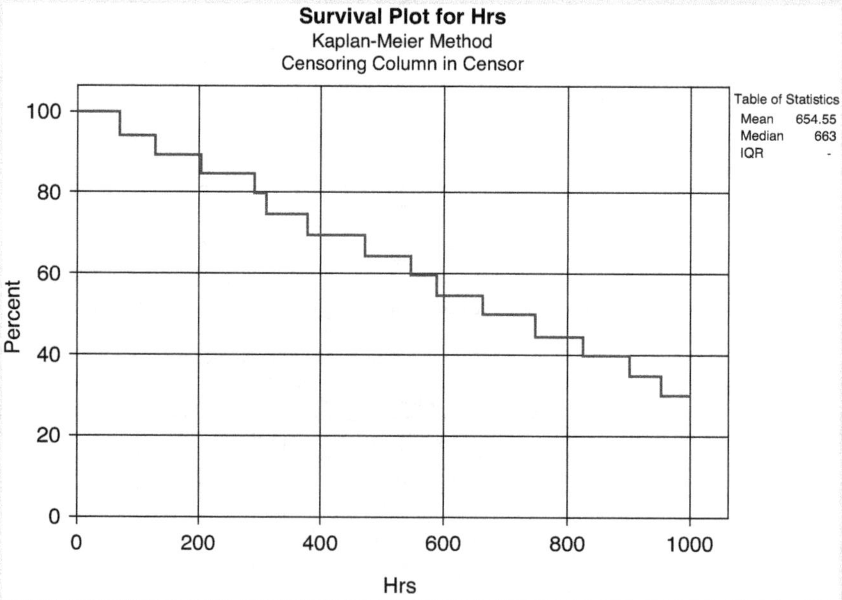

Figure 6.68 Survival plot for censored data.

Characteristics of Variable

Mean(MTTF)	Standard Error	95.0% Normal CI Lower	Upper	Q1	Median	Q3	IQR
654.55	75.0569	507.441	801.659	312	663	*	*

Kaplan-Meier Estimates

Time	Number at Risk	Number Failed	Survival Probability	Standard Error	95.0% Normal CI Lower	Upper
70	20	1	0.95	0.048734	0.854483	1.00000
128	19	1	0.90	0.067082	0.768522	1.00000
204	18	1	0.85	0.079844	0.693509	1.00000
291	17	1	0.80	0.089443	0.624695	0.97530
312	16	1	0.75	0.096825	0.560227	0.93977
377	15	1	0.70	0.102470	0.499163	0.90084
473	14	1	0.65	0.106654	0.440963	0.85904
549	13	1	0.60	0.109545	0.385297	0.81470
591	12	1	0.55	0.111243	0.331968	0.76803
663	11	1	0.50	0.111803	0.280869	0.71913
748	10	1	0.45	0.111243	0.231968	0.66803
827	9	1	0.40	0.109545	0.185297	0.61470
903	8	1	0.35	0.106654	0.140963	0.55904
955	7	1	0.30	0.102470	0.099163	0.50084

Figure 6.69 Kaplan-Meier detailed output.

EXAMPLE 6.62

Kaplan-Meier Interpretation with Heavy Censoring

This example uses the same data as Example 6.57. Ten units are on test. The units are not replaced when they fail (nonreplacement). One unit fails at $t_1 = 685$ hours, and a second unit fails at $t_2 = 1690$ hours. The test is ended at $t = 2500$ hours with no additional failures. We need to estimate the B10 predicted life, the 90th percentile of the reliability function, using a nonparametric approach.

The analyst generated the Kaplan-Meier curve for the reliability distribution (Figure 6.70).

The B10 life is obtainable from the detailed output, which indicates a 90% survival probability at 685 hours.

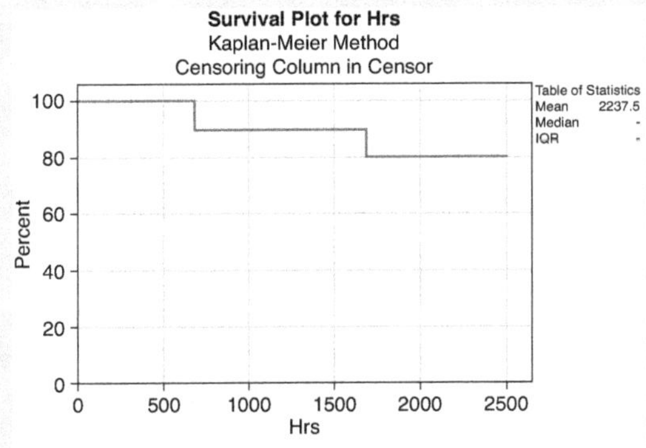

Characteristics of Variable

Mean(MTTF)	Standard Error	95.0% Normal CI Lower	Upper	Q1	Median	Q3	IQR
2237.5	221.176	1804.00	2671.00	*	*	*	*

Kaplan-Meier Estimates

Time	Number at Risk	Number Failed	Survival Probability	Standard Error	95.0% Normal CI Lower	Upper
685	10	1	0.9	0.094868	0.714061	1
1690	9	1	0.8	0.126491	0.552082	1

Figure 6.70 Survival plot output for data with heavy censoring.

EXAMPLE 6.63

Kaplan-Meier Interpretation Compare Two Populations with Censoring ($n = 10$ versus 20)

Two designs need to be compared to assess whether improvements were effective. Design Seven was analyzed in Example 6.62; Design Nine was analyzed in Example 6.63. The analyst generated Kaplan-Meier curves in Figure 6.71 to compare the reliability distributions. What should they conclude?

The newer Device Nine appears to be an improvement over Device Seven. The B10 went from 128 to 685 hours, approximately a 4× increase. Despite the limited data in these studies, a difference is clearly apparent, provided there weren't any changes in operating test conditions affecting exposure severity or differences in material quality that could have caused this difference. Refer to Chapter 11 for information on reliability test planning and the design of experiments (DOE).

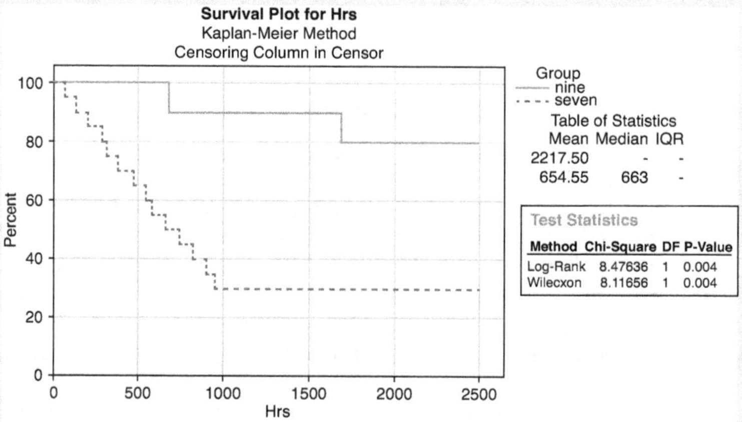

Figure 6.71 Survival plot comparison of designs.

SAMPLING PLANS FOR STATISTICS AND RELIABILITY TESTING

> Use various theories, tables, and formulas to determine appropriate sample sizes or testing time needed for statistical and reliability testing. Incorporate representative and randomized sampling techniques. (Apply)
>
> **Body of Knowledge III.A.5**

Data must first be collected before analysis can be done. Good sampling practices can reduce biases and improve the interpretability of results.

Sampling Concepts

In statistical analysis, our aim is to make statements about a target **population** of interest whose parameters are unknown, and unknowable, to us. To do this, we take **samples** from that population and analyze them to make judgments, or **inference**, about the population with a certain level of **confidence**, $C = 1 - \alpha$. However, not just any sample will do this effectively. If done improperly, **sampling bias** will be induced, resulting in inaccurate results. Sampling is often considered in the context of surveys, so some of that language is used here, but sampling itself is a broad topic that may be conducted in any area.

Complete knowledge would theoretically be obtained through **100%** or **full sampling**, since every unit in the population would be evaluated. Sampling methods where only a subset of the population is evaluated fall under one of two umbrellas: **probability sampling**, and **nonprobability sampling**. Probability sampling methods provide the foundation for effective data gathering that leads to meaningful statistical analyses. Each unit in the sample has a known, nonzero chance of being selected. Nonprobability sampling methods use selection criteria that are undefined, unspecified, or even changeable, where the selection probability is unknown, so no real inferences can be made. For studies to have **statistical validity**, probability sampling methods are required (Figure 6.72).

More detail is provided later in this section on specific types of sampling plans. Regardless of type, however, they have the same objectives: to ensure that (1) the sample is **representative** of the target population, and (2) the **homogeneity** of the sample is controlled. These attributes help limit the presence of biases and improve both the **accuracy** and **precision** of the sample. As discussed in Chapter 6, "Basic Statistics," *accuracy* refers to how close the sample statistic is to the true population parameter, while *precision* is the repeatability, that is, how close estimates from different samples are to each other, or how close different measurements of the same sample are to each other.

In sampling, the quality of the estimate to be obtained can be improved through good test conduct, whether the sample is conducted as a survey or a torque check.

Representative selection results in a sample that is *representative of the target population*. From a statistical point of view, a representative sample is a random sample that is drawn from a population so that observed values of the samples have the same distribution as the population. When samples are not representative of the population, valid statistical inferences cannot be made. True representative samples are sometimes difficult to accomplish. Consider a group of 32 suppliers, of

100% Sampling (Full Inspection)	
Probability Sampling Methods	**Nonprobability Sampling Methods**
1. Simple Random Sampling (SRS) 2. Systematic (Interval) Sampling 3. Stratified Sampling 4. Cluster Sampling 5. Multistage Sampling	1. Convenience Sampling 2. Judgment or Purposive Sampling 3. Quota Sampling 4. Voluntary Response Sampling 5. Snowball Sampling
0% Sampling (No Inspection)	

Figure 6.72 Probability versus nonprobability sampling methods.

which 16 are located in the Midwest and 16 in the South. A representative sample of size 8 based on location alone would comprise 4 each in the Midwest and South. A nonrepresentative sample, for example, would be 8 Midwestern suppliers. If we added company size in terms of sales to the characteristics, finding a representative sample would become more difficult.

Homogeneity refers to the similarity or sameness of specified characteristic(s) of the members of the targeted population. Different sampling methods will make use of homogeneity between and within subgroups and use it strategically to improve the quality of the estimate. For example, simple random sampling assumes the population is homogeneous; however, in stratified sampling, strata are optimally defined to be as similar as possible within each one but different between.

Sampling bias is clearly something we want to avoid in order to develop an understanding of the truth. Biased samples are not representative of the population. **Bias can occur when the sample is selected in a nonrandom way**, resulting in the sample statistics underestimating or overestimating the population parameter. Some types of sampling bias include:

- **Under** (or **over**) **coverage bias** occurs when some members of the population are not proportionally represented in the sample. For example, an employee survey may be sent only to the workers on first shift and omit those working on the second shift.
- **Self-selection bias** occurs when individuals themselves decide whether to participate in the study. Those who choose to participate are often not representative of the population. For example, only those with strong opinions may respond to a survey.
- **Survivorship bias** occurs when the sample is comprised of units that pass the selection criteria while others are ignored. For example, a sample composed solely of assemblies that have survived 10,000 hours of use in the field will ignore the assemblies that have failed.
- **Recall bias** occurs when respondents do not remember events or information accurately. This type of bias can be reduced by collecting data in a timely manner.
- **Exclusion bias** occurs when specific groups are excluded from the sample if they have recently moved into the study or have left the study area. An analogous situation could occur in sampling manufacturing hardware if certain lots are available or more accessible than others.
- **Convenience bias** occurs when the sample collected is based on expediency. For example, a quality inspector who collects samples only from the production line closest to their office would be gathering a convenience sample.

Understanding and controlling sample homogeneity enables an appropriately representative sample to be taken. This reduces **uncertainty** and improves both accuracy and precision of the sample. When considering risks associated with reliability, it is important to consider what additional factors may be contributing that could be used to refine the assessment. Statistical tools address one type of uncertainty, known as **aleatory uncertainty**, that variability that can be quantified and modeled. However, many situations are subject to **epistemic uncertainty** due to lack of knowledge. Computing standard errors quantifies only aleatory uncertainty. We rely on good test practices, such as randomization and leveraging known subgroups, for insurance against epistemic uncertainty affecting our conclusions.

Finally, we will discuss **sample size**, an often-critical question, but one that also often seems to require more information than we have (at least in the test planning phase), so some assumptions are required. Sample size questions are addressed through specifying the maximum desired confidence interval width as the **margin of error**, variously referred to as E, ME, or B for bound, or given as a probability for exact binomial computations. Statistical intervals are discussed in all of their various forms in "Confidence and Tolerance Intervals."

Sampling Terminology

First, a few additional sampling terms:

- *Sampling unit.* Smallest unit of observation in a study, the item available to be sampled from a target population, may be a single item or a group.
- *Sampling frame.* Complete list or index of the items in the target population from which a sample is drawn.
- *Random sample.* Sample chosen by selecting units for a sample such that all combinations of units under consideration are equally likely, that is, that they have an equal or ascertainable chance of being selected as the sample.
- *Representative sample.* Random sample drawn from the target population such that observed values of the samples have the same distribution as the population. When samples are not representative of the population, valid statistical inferences cannot be made.
- *Sample size.* The number of samples to be obtained as the sample.
- *Strata.* Mutually exclusive population subgroups defined by known characteristic, for example, population of subassemblies separated by production line and/or shift, which are then randomly sampled within the strata.
- *Cluster.* Mutually exclusive population subgroups defined to be similar, then sampled at the cluster level and then subjected to 100% sampling, for example, evaluating a training program by randomly sampling classes, but including every member of the class.
- *Confidence level (or coefficient).* In hypothesis testing terms, **confidence**, $C = 1 - \alpha$, is the probability of not rejecting the null hypothesis, given that the null hypothesis is true, (that is, a "true negative"). It is an attribute of the testing procedure. It is specified by the user, not calculated.
- *Margin of error (ME, E).* Maximum desired difference between the sample estimate of the population parameter and the true population parameter for a given confidence level. The margin of error, *E*, is equal

to one-half of the width of a two-sided confidence interval used to estimate a population parameter.

- *Power* $(1-\beta)$. Probability that a statistical test will detect an effect that is in fact there (that is, a "true positive").

Sample Size Considerations

Several general factors influence what sample size is necessary to detect an effect of a particular size.

First, **what is the statistical question of interest** and **what kind of data are available?** More data are required for binary data versus other attribute data versus continuous data. Estimating a mean, sigma, or proportion are all very different calculations. Similarly, characterizing variation takes more effort than estimating a mean, and more data would be required to develop a tolerance interval than a confidence interval. The answer is rarely if ever 30; however, many people have heard this rule of thumb, which originates with the central limit theorem and the law of large numbers. This is the point at which a t distribution is closely converging to the Z-distribution and at which some say you have enough data to consider a population "known." However, having fewer than 30 datapoints does not prevent an analysis from being conducted.

Choosing the correct sample size calculation method depends on the type of data available and the analysis planned for the data. There are a number of statistical sample size equations available for use in different scenarios, which largely parallel the selection of hypothesis tests. For every way we can ask if some effect is statistically significant, there is a related question that can be asked regarding how much data would be required to generate an effect of a particular size, in other words, what sample size is needed. The nature of the study also contributes here. What sampling design was adopted? Are there relevant subgroups? How should the sampling plan be laid out?

Next, **how much variation or scatter is present in the data?** More scatter means more samples. Precision can be thought of as the noise factor, against which we are trying to detect a signal. If the noise is high, then we will need more data to detect the same signal. Note that precision is inversely related to the standard error of the estimate, which is defined as the sample standard deviation divided by the square root of the sample size n. Precision is high when the standard error is small, and vice versa. There may be a benefit to allocating more samples to more variable strata, for example, a process known as **optimum allocation**. The baseline analyses discussed here can be modified with weights.

How small an effect are you trying to detect? Needing to detect smaller effects will also mean more samples. A larger signal is easier to detect than a smaller signal, given the same amount of noise.

Also, **how certain do you need to be?** Importance plays a role, for instance, in defining the confidence and desired power of the test. What kind of resolution is necessary to address the problem? Is this one step of several? Needs in research may be different from those of operations. Every organization will have its own norms and needs.

And last but not least, there are a number of **practical logistical considerations**. How long would it take to collect each unit? For sampling hardware, how big are the parts? Will they need to be moved and stored and labeled? How will you maintain **traceability**? How difficult will samples be to obtain and access and evaluate? What special work instructions are being provided to the people doing the actual evaluations?

How much time and other resources are available? What are the estimated costs? What is the budget? If an observation takes 2 weeks and costs $10,000, it is not always going to be a viable solution, especially in a reliability environment. If relative costs vary, these factors could be incorporated into the test design to obtain more data where it is less expensive.

Types of Sampling Plans

The next page depicts the four varieties of sampling plans: **simple random sampling** (SRS), **systematic**, **stratified**, and **cluster**. The fifth type, **multistage**, involves at least two of those in combination. A brief overview of each type is given here. More details are provided in the summary in Table 6.29.

1. *Simple random sampling* (SRS) involves drawing a sample from a targeted population in which each member of the population has an equal probability of being chosen for the sample. Each member of the sample is selected independently (Figure 6.73).
2. *Systematic (interval) sampling* takes a sample from an ordered population at a specified sampling interval, *i*. Systematic sampling can be fast and efficient, but issues arise if, for some reason, the sampling interval synchronizes with a natural periodicity in the ordered population (Figure 6.74).
3. *Stratified sampling* first segregates the population into *k* mutually exclusive and exhaustive strata, which are then sampled. These strata are dissimilar to each other, but members in each stratum are similar based on one or more characteristics (Figure 6.75).
4. *Cluster sampling* first separates the population into *k* clusters using some criterion such as proximity. The clusters are considered similar to each other. Then, a sample of *n* clusters is chosen, and 100% of the members of the chosen clusters are measured (Figure 6.76).
5. *Multistage sampling* involves some combination of these types in order to accommodate more complex hierarchical structures.

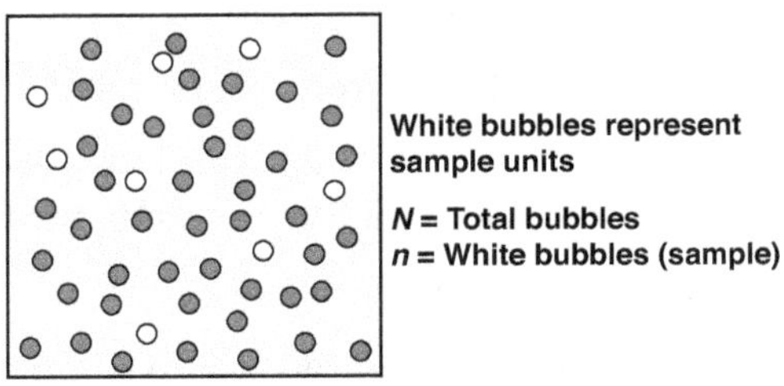

Figure 6.73 Simple random sampling.

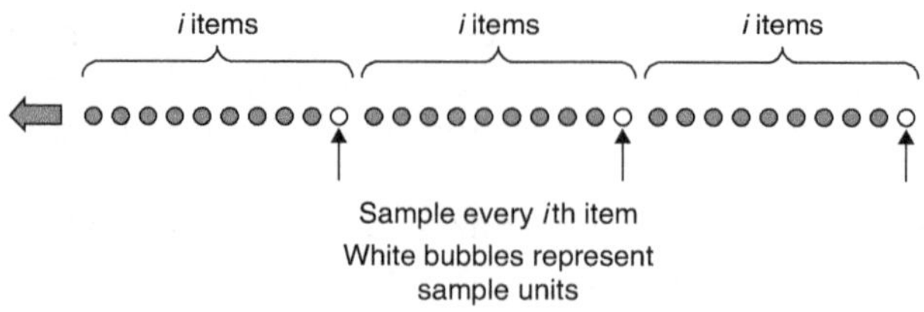

Figure 6.74 Systematic sampling.

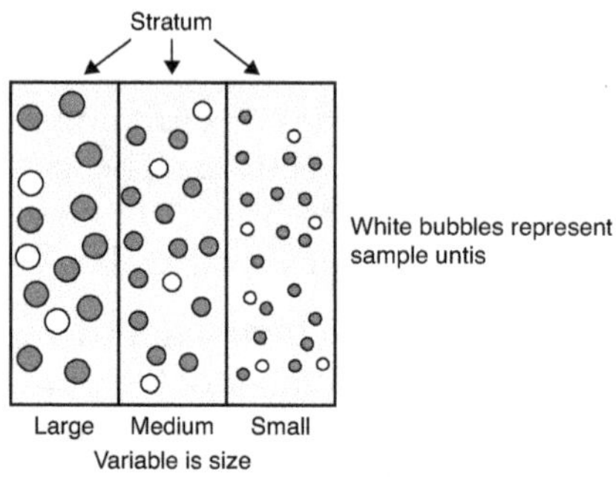

Figure 6.75 Stratified sampling.

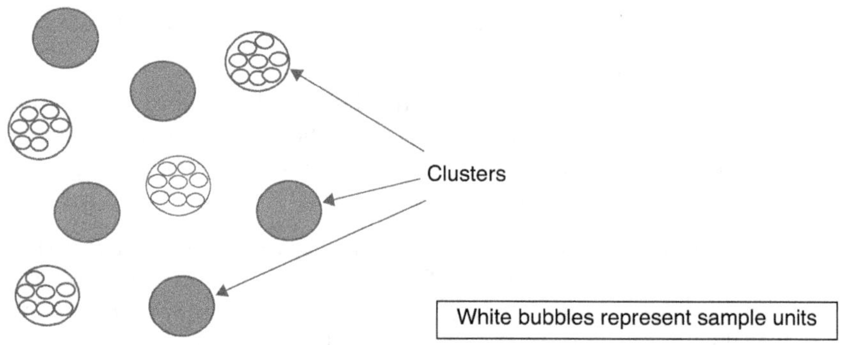

Figure 6.76 Cluster sampling.

Assessing the Quality of a Sampling Plan

A well-designed sampling plan should include the following aspects:

- Samples are drawn randomly from the targeted population as a whole or within strata.
- The sample is representative of the population from which it is drawn.
- The design is economical and efficient.
- The sampling design is sufficiently easy to conduct.
- The standard error is minimal.

- Bias is minimal.
- Statistical assumptions are not violated.
- Inferences can be made from the sample about the population with the desired level of confidence.
- The selection criteria are objective, not subject to a data collector's bias.

These characteristics will be affected by sampling design. Table 6.29 summarizes the advantages and disadvantages of the four sampling methods.

Nonprobability methods to be aware of include shortcuts of different kinds: A *convenience* sample contains readily available items, a *judgment* sample contains those that are deliberately selected by the researcher, a *quota* sample has predefined groups, *voluntary response* means participants are self-selected, and *snowball* means that subjects recruit others to join (see Table 6.72). In some cases, there may be reasons to justify an approach, for example, to identify hard-to-find participants. However, the results must be handled with care, and some adjustment for known biases may need to occur during analysis. A two-step approach may identify candidates, then sample to obtain statistical validity.

Sample Size Determination

If we perform a hypothesis test and fail to reject the null hypothesis, it does not mean that the null hypothesis is necessarily true. It might be the case that the null hypothesis is false, but we do not have enough evidence to reject it. By increasing the sample size of a given test, we can increase the probability of rejecting H_0 when it is false, thus increasing the **power** of the test, the probability an effect is detected if it is there. In this section we present sample size calculations for achieving a specific **margin of error** E in a confidence interval and for detecting a difference δ in a hypothesis test. Statistical intervals are discussed in depth in "Confidence and Tolerance Intervals."

With a little algebra we can determine *a priori* what sample size will be necessary to be able to achieve a specific margin of error E or detect a difference δ in a population parameter with a certain level of confidence $(1-\alpha)$. Sample sizes are obtained through calculation by inverting a statistical interval of some form to achieve the targeted confidence level with the targeted bound. An overview of how hypothesis tests, confidence intervals, and sample size estimation are all connected is shown in Figure 6.77 for the specific case of t-tests.

This method applies to asymptotically constructed tests. An analogous approach applies to methods constructed with exact equations.

Tables 6.31 and 6.32 later in the chapter provide sample size equations using two different methods, one involving the margin of error method E, and the other incorporating desired power to detect an effect.

Finite Population Correction (FPC)

The **finite population correction** (FPC) is an adjustment factor that may be applied in sample size calculations to account for the fact that a population is not infinite, as is the standard assumption, and the sample size may take up a large proportion of the available population. Applying the FPC in appropriate

Table 6.29 Comparison of probability sampling methods.

Method	Advantages	Disadvantages	Comments
Simple Random	• Easy to conduct • Requires minimum knowledge about the population • Human bias eliminated	• Identifying all members of the population can be difficult • Could be costly	• Can be performed with or without replacement • Population must be homogeneous
Systematic	• Cost effective • Less time-consuming • Easy to conduct • If the order of the units is related to the characteristic of interest, representativeness of the sample is increased	• Not all members of the population have an equal chance of being included once sampling begins • If the order of the units produces a cyclic pattern, may reduce representativeness • Creates a risk of bias is there is a natural periodicity in the data, especially if it coincides with sampling interval	• Useful in survey sampling • Sample size typically known in advance to determine sampling interval • Frequently used with a random start
Stratified	• Higher sample accuracy • Higher sample precision • Assures that each stratum is represented	• Identifying all members of the population can be difficult • Identifying all members of a stratum may be difficult • May be difficult to select stratification variables	• There are both proportional and nonproportional stratified sampling methods • Random sampling is applied within each stratum • Each stratum is different from the others, but units within each stratum are similar
Cluster	• Effective for comparing programs • Cost effective • Less time-consuming • High reliability • Reduces logistical problems for program rollout	• Clusters may be difficult to define • Sample can be very large if cluster membership is large • May be difficult to implement • Sampling error can be high	• Clusters are considered similar groups • Each member of a chosen cluster is measured

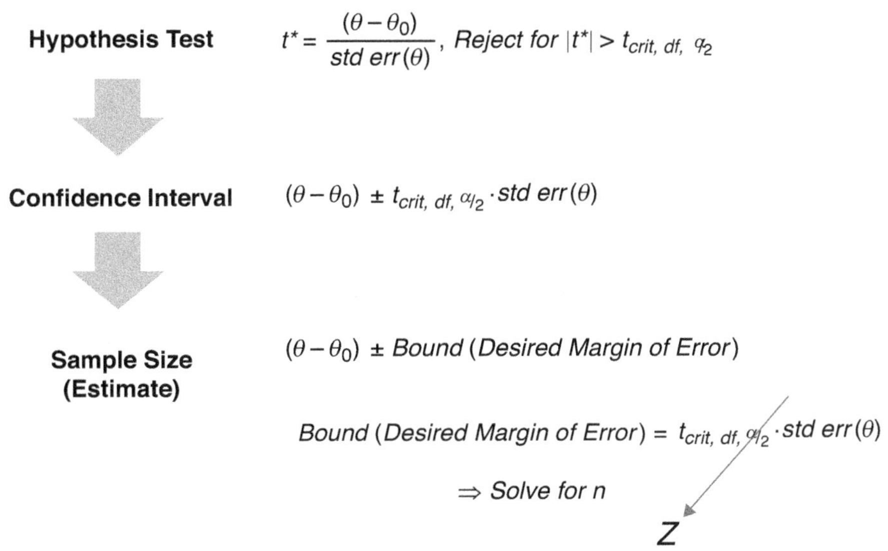

Figure 6.77 Formulaic connections between hypothesis testing, intervals, and sample size estimation.

circumstances can reduce the sample size and increase statistical power. The formula for the FPC is this:

$$fpc = \sqrt{\frac{N-n}{N-1}},$$

where N = the population size and n is the sample size. It is implemented as an additional multiplier applied to the standard error that contributes to the margin of error. Sample size calculations with the FPC do not have a closed-form solution and must be solved iteratively, but this can be done with Solver in Excel or another simple optimizer.

Reliability and Low-Probability Event Sampling (Zero Failures)

In reliability applications, there is a particular interest in characterizing low probability events, those where p (or $1-p$) is near zero. As we move farther from $p=0.5$, the standard asymptotic methods presented previously for calculating sample size begin to fail since they become less applicable due to assumption violations, so other methods are used. This is discussed in more detail in "Confidence and Tolerance Intervals."

Also, in these cases, we often only care about estimating the parameter's worst case condition, so we only look to place an upper bound on a probability near zero (or inversely a lower bound on a probability near 1).

Reliability Test Planning

A frequently asked question is, "What sample size is required to obtain a given reliability with a given confidence?" If the test can be run with no failures, the required sample size can be found from the binomial equation as shown:

$$\Pr(X=0\mid n,p)=(1-p)^n=\alpha$$
$$n\ln(1-p)=\ln(\alpha)$$
$$n=\frac{\ln(\alpha)}{\ln(1-p)}=\frac{\ln(\alpha)}{\ln(R)}$$

EXAMPLE 6.64

Sample Size for Zero-Failures Reliability Test

How many no-failure tests are required to demonstrate a reliability of 0.95 with a confidence of 0.90?

Use the previous equation:

$$n=\frac{\ln(\alpha)}{\ln(R)}=\frac{\ln(0.10)}{\ln(0.95)}=44.9\Rightarrow 45$$

"The Rule of 3": Binomial for Zero Failures

A special case of the preceding formula at 95% confidence produces the calculation equation known as the Rule of 3. To show that a proportion is less than p given zero failures at 95% confidence, a quick estimate can be obtained by taking 3 divided by p. The derivation of the rule follows:

$$\Pr(X=0\mid n,p)=(1-p)^n=0.05$$
$$n\ln(1-p)=\ln(0.05)$$
$$n\ln(1-p)=-3$$
$$n=\frac{-3}{\ln(1-p)}=\frac{3}{-\ln(1-p)}\approx\frac{3}{p}\ \text{since}\ -\ln(1-p)\approx p\ \text{for small}\ p$$

The approximation yields a slight overestimate, typically of 2 units, which may be subtracted to obtain the exact answer.

EXAMPLE 6.65

Zero-Failures Rule of 3

A team thinks that they have implemented sufficient corrective action to preclude or at least significantly prevent an issue from reoccurring. They want a quick estimate of how many samples would be required to show that the problem has probability below 0.01 at 95% confidence.

Use the Rule of 3:

$$n\approx\frac{3}{p}=\frac{3}{0.01}=300$$

Using the Rule of 3, exact formula: $n=\dfrac{3}{-\ln(1-p)}=\dfrac{3}{0.01005}=298.5\Rightarrow 299$

Sample Sizes Obtained from the Clopper-Pearson Exact Upper Bound Beta-Binomial Equation

The following tables were developed from what is known as the Clopper-Pearson equation, used to compute "exact" bounds on proportions, meaning that it is obtained directly without additional assumptions. This is discussed in more detail in "Confidence and Tolerance Intervals." Sample sizes were generated to show what would be necessary to demonstrate that the true proportion p is less than 0.10, 0.05, and 0.01 at a wide range of confidence levels. In these cases, p is the width of the interval E. For one-sided bounds, 50% confidence values can be used to produce a nominal estimate, since with zero failures $\hat{p}=0$ always (Table 6.30).

Results would be equivalent for $q=1-p$ as well, if thinking in terms of reliability. Note that sample size increases with higher desired confidence, an increase in

Table 6.30 Sample sizes required to demonstrate $p \leq 0.01$, 0.05, 0.10 at various confidence levels using exact Clopper-Pearson beta-binomial equation assuming zero or very low failures, $x \leq 4$.

p=0.10		*Number of failures x of n*				
		x=0	x=1	x=2	x=3	x=4
Confidence	99%	44	64	81	97	113
	95%	29	46	61	76	89
	90%	22	38	52	65	78
	80%	16	29	42	54	66
	50%	7	17	27	37	47
p=0.05		*Number of failures x of n*				
		x=0	x=1	x=2	x=3	x=4
Confidence	99%	90	130	165	198	229
	95%	59	93	124	153	181
	90%	45	77	105	132	158
	80%	32	59	85	110	134
	50%	14	34	54	74	94
p=0.01		*Number of failures x of n*				
		x=0	x=1	x=2	x=3	x=4
Confidence	99%	459	662	838	1001	1157
	95%	299	473	628	773	913
	90%	230	388	531	667	798
	80%	161	299	427	551	671
	50%	69	168	268	367	467

Computed using equation 6.1 in Meeker, Hahn, and Escobar (2017) and the qbeta function in the R package 'stats', version 4.4.3, downloaded 2025-03-01.

events, and with the level of resolution in the estimate desired, that is, how small we want the estimate to be.

Sample Size and Testing Time to Demonstrate Reliability

If test units can be run without failure for multiples of their specified lifetime, the number of samples required to demonstrate a given reliability level at that lifetime can be reduced. Doganoksoy, Meeker, and Hahn (2021) provide the formula for sample size, given zero-failure testing to k times the lifetime, with an assumption of a Weibull lifetime distribution with assumed β parameters:

$$n = \frac{1}{k^{\beta}} \times \frac{\log(\alpha)}{\log(R)}$$

EXAMPLE 6.66

Sample Size and Test Time to Demonstrate Reliability

Reliability demonstration testing is being planned for an electronic module that has an expected lifetime of 1000 hours. Relevant life data analysis has shown that a Weibull distribution with $\beta = 2$ is an appropriate model. If the engineering team expects the module to run without failure for 4000 hours, what sample size would be required to demonstrate reliability of 0.99 for this lifetime of 1000 hours with 95% confidence?

$$k = \frac{4000}{1000} = 4$$

$$n = \frac{1}{4^2} \times \frac{\log(.05)}{\log(.99)} = 19$$

Thus, the team can demonstrate 0.99 reliability with 0.95 confidence at 1000 hours by testing 19 samples to 4000 hours without failure. This planning can also be done with software, as shown in the output in Figure 6.78.

WORKSHEET 1

Demonstration Test Plans

Reliability Test Plan
Distribution: Weibull, Shape = 2
Reliability Goal = 0.99. Target Confidence Level = 95%

Test Plans

Failure Test	Testing Time	Sample Size	Actual Confidence Level
0	1000	298	94.9963
0	2000	75	95.0959
0	3000	33	94.9458
0	4000	19	95.2895

Figure 6.78 Test time and sample size to demonstrate reliability.

Tables of Equations Sample Size Equations to Achieve a Specified Margin of Error

Table 6.31 provides sample size formulas for assuring a certain margin of error E in a confidence interval on a mean or proportion or on the difference between two means or proportions.

Sample Size Equations to Detect a Difference with a Specified Probability

An alternative approach incorporates the desired power level $(1-\beta)$ instead of the target margin of error, E. Table 6.32 shows equations for the comparison for both estimation of population means and proportions, as well as those for detecting a difference δ between two population means and proportions. The equations include a second z value to control the type II error as well.

Table 6.31 Sample size calculations to assure margin of error, E, in a confidence interval.

Population parameter	Purpose	Sample size formula for $(1-\alpha)\%$ two-sided confidence interval	Sample size formula for $(1-\alpha)\%$ one-sided confidence interval
Population mean, μ	Assuring a $(1-\alpha)\%$ confidence interval on μ has a margin of error of at most E.	$n=\left(\frac{z_{\alpha/2}\sigma}{E}\right)^2$	$n=\left(\frac{z_{\alpha}\sigma}{E}\right)^2$
Difference between two population means, $\mu_1-\mu_2$	Assuring a $(1-\alpha)\%$ confidence interval on $\mu_1-\mu_2$ has a margin of error of at most E.	$n=n_1=n_2=\frac{\left(z_{\alpha/2}\right)^2\left(\sigma_1^2+\sigma_2^2\right)}{E^2}$	$n=n_1=n_2=\frac{\left(z_{\alpha}\right)^2\left(\sigma_1^2+\sigma_2^2\right)}{E^2}$
Population proportion, p	Assuring a $(1-\alpha)\%$ confidence interval on p has a margin of error of at most E.	$n=\left(\frac{z_{\alpha/2}}{E}\right)^2\hat{p}(1-\hat{p})=\frac{1}{4}\left(\frac{z_{\alpha/2}}{E}\right)^2$ Use $p=0.5$ to maximize the estimate of n if unknown, since the most conservative	$n=\left(\frac{z_{\alpha}}{E}\right)^2\hat{p}(1-\hat{p})=\frac{1}{4}\left(\frac{z_{\alpha}}{E}\right)^2$ Use $p=0.5$ to maximize the estimate of n if unknown, since the most conservative
Difference between two population proportions, p_1-p_2	Assuring a $(1-\alpha)\%$ confidence interval on p_1-p_2 has a margin of error of at most E.	$n=n_1=n_2=\left(\frac{z_{\alpha/2}}{E}\right)^2 \times(\hat{p}_1(1-\hat{p}_1)+\hat{p}_2(1-\hat{p}_2)) =\frac{1}{2}\left(\frac{z_{\alpha/2}}{E}\right)^2$ when $p_i=0.5$	$n=n_1=n_2=\left(\frac{z_{\alpha}}{E}\right)^2 \times(\hat{p}_1(1-\hat{p}_1)+\hat{p}_2(1-\hat{p}_2)) =\frac{1}{2}\left(\frac{z_{\alpha}}{E}\right)^2$ when $p_i=0.5$

Table 6.32 Sample size calculations for detecting a difference with a specified power $(1-\beta)$.

Population parameter	Purpose	Sample size formula for $(1-\alpha)$% two-sided confidence interval	Sample size formula for $(1-\alpha)$% one-sided confidence interval
Population mean, μ	Detecting a difference of δ with a probability $(1-\beta)$	$n=\frac{\left(z_{\alpha/2}+z_\beta\right)^2\sigma^2}{\delta^2}$	$n=\frac{\left(z_\alpha+z_\beta\right)^2\sigma^2}{\delta^2}$
Difference between two population means, $\mu_1-\mu_2$	Detecting a difference of δ with a probability $(1-\beta)$	$n=n_1=n_2=\frac{\left(z_{\alpha/2}+z_\beta\right)^2\left(\sigma_1^2+\sigma_2^2\right)}{\delta^2}$	$n=n_1=n_2=\frac{\left(z_\alpha+z_\beta\right)^2\left(\sigma_1^2+\sigma_2^2\right)}{\delta^2}$
Population proportion, p	Detecting a difference of δ with a probability $(1-\beta)$	$n=\left(\frac{z_{\alpha/2}\sqrt{\hat{p}_0\left(1-\hat{p}_0\right)}+z_\beta\sqrt{\hat{p}(1-\hat{p})}}{\delta}\right)^2$	$n=\left(\frac{z_\alpha\sqrt{\hat{p}_0\left(1-\hat{p}_0\right)}+z_\beta\sqrt{\hat{p}(1-\hat{p})}}{\delta}\right)^2$
Difference in population proportions, p_1-p_2	Detecting a difference of δ with a probability $(1-\beta)$	$n=\left(\frac{z_{\alpha/2}\sqrt{(p_1+p_2)(q_1+q_2)/2}+z_\beta\sqrt{p_1q_1+p_2q_2}}{\delta}\right)^2$ *where* $q_1=(1-p_1),\ q_2=(1-p_2)$	$n=\left(\frac{z_\alpha\sqrt{(p_1+p_2)(q_1+q_2)/2}+z_\beta\sqrt{p_1q_1+p_2q_2}}{\delta}\right)^2$ *where* $q_1=(1-p_1),\ q_2=(1-p_2)$

EXAMPLE 6.67

Sample Size for Mean with Known Sigma, *E* Method

A reliability engineer wants to estimate the mean diameter of fiberoptic cable. How many samples must be randomly selected if they want 95% confidence that the sample mean is within 0.5 mm of the population mean? From past experience, it is known that the population standard deviation, σ, is equal to 2 mm.

Here, $\alpha = 0.05$, and from Appendix D, we find that $z_{0.025} = 1.96$. The population standard deviation is $\sigma = 2$ and the desired margin of error is $E = 0.5$.

We will use the following formula from Table 6.31:

$$n = \left(\frac{z_{\alpha/2}\sigma}{E} \right)^2$$

Substituting in the values, we obtain

$$n = \left(\frac{z_{\alpha/2}\sigma}{E} \right)^2 = \left(\frac{1.96 \cdot 2}{0.5} \right)^2 = 61.47$$

$$\lceil n = 62$$

Based on our confidence level and the inherent variation in the data, we will need a sample size of at least $n = 62$ to obtain 95% confidence with a margin of error of at most 0.5 mm.

Note that the calculation of the sample size n is a *ceiling function*, denoted by the symbol $\lceil$, in which any noninteger result is rounded up to the next whole number. For example, if the result of the calculation were 61.07, we would round this up to 62 as well. Sample size results are always rounded up.

EXAMPLE 6.68

Sample Size for Difference Between 2 Means, Known Sigma

A vendor claims that product A has an average thickness at least 2 mm larger than that of product B. How many of each product type should be sampled to construct a 95% confidence interval to check the vendor's claim? Assume the two populations have standard deviations of 3.3 mm and 3.8 mm, respectively.

Here, $\alpha = 0.05$, and from Appendix D, we find that $z_{0.025} = 1.96$. The desired margin of error is $E = 2$ and sigma values are provided.

$$n = \frac{(z_{\alpha/2})^2(\sigma_1^2 + \sigma_2^2)}{E^2} = \frac{(1.96)^2(3.3^2 + 3.8^2)}{(2)^2} = 24.3 \Rightarrow 25$$

EXAMPLE 6.69

Sample Size for one Proportion

A polling firm would like to report the estimated proportion of respondents who are in favor of a proposed sales tax increase that would benefit the local public schools. How many respondents must be randomly selected to ensure a margin of error equal to three percentage points with 95% confidence?

Continued

Here, $\alpha=0.05$, and from Appendix D, we find that $z_{0.025}=1.96$. The desired margin of error is $E=0.03$. We do not have an estimate of the proportion in favor at this time, so we will use the most conservative value of $\hat{p}=0.50$. Using this value for the estimated proportion will maximize the estimate of n. Using the equation in Table 6.31,

$$n=\left(\frac{z_{\alpha/2}}{E}\right)^2 \hat{p}(1-\hat{p})$$
$$n=\left(\frac{1.96}{0.03}\right)^2 (0.50)(1-0.50)$$
$$n=1067.11$$
$$\lceil n=1068$$

To achieve a margin of error of +/−3 percentage points, the polling firm will need to randomly sample 1068 respondents for the survey.

EXAMPLE 6.70

Sample Size for Difference Between two Proportions

A reliability engineer has two competing designs that are stress-tested for 100 hours. Let p_A=the proportion of design A that fails, and p_B=the proportion of design B that fails. It is necessary to calculate the 95% confidence interval with width=0.4 for the difference p_A-p_B. Using the equation in Table 6.32,

$$n_1=n_2=\frac{1}{2}\left(\frac{z_{\alpha/2}}{E}\right)^2=\frac{1}{2}\left(\frac{1.96}{0.2}\right)^2=48.02\Rightarrow 49.$$

EXAMPLE 6.71

Sample Size for Mean with Known Sigma, Power Method

A reliability engineer plans a test at the 95% confidence level to determine whether a new drilling station has resulted in a change in the mean diameter of a metal flange. How many samples should be randomly collected to be able to detect a change of 1 mm in the mean diameter with a probability of 90%? The standard deviation of the diameters is known and equal to 1.7 mm, and the chosen α is equal to 0.05.

Here, $\alpha=0.05$, and from Appendix D, we find that $z_{0.025}=1.96$ and $z_\beta=z_{0.10}=1.28$. (Note that even for a two-sided test, the β error is never split.) The desired detectable change is $\delta=1$. Using the equation in Table 6.32,

$$n=\frac{\left(z_{\alpha/2}+z_\beta\right)^2\sigma^2}{\delta^2}$$
$$n=\frac{(1.96+1.28)^2 1.7^2}{1^2}$$
$$n=30.34$$
$$\lceil n=31$$

Using computer software, we can also generate a power curve that gives the probability of detection at $a=0.05$ and $n=31$ for various differences in the mean, shown in Figure 6.79. From this curve, we can see that the sample size equal to 31 achieves the required power for a one-millimeter change in the mean.

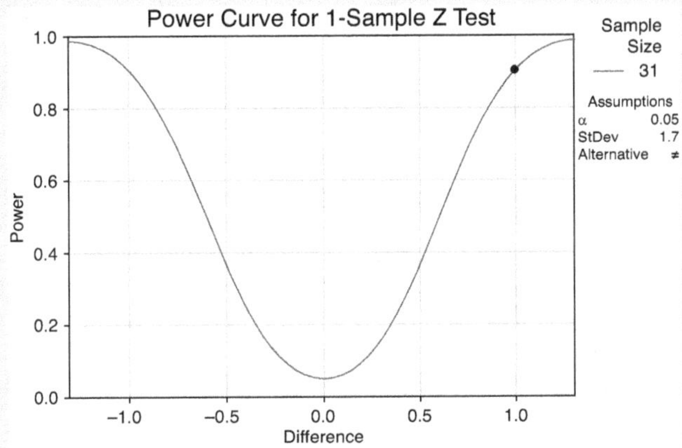

Figure 6.79 Power curve for Example 6.71.

EXAMPLE 6.72

Sample Size for Difference Between Two Means with Known Sigma, Power Method

A chemical engineer is studying the yields of an existing process and a newly designed process that both produce the same product. They plan to conduct a test on the difference between means at the 99% confidence level. They would like to be able to detect whether the newly designed process increases the mean yield by at least 275 grams with a probability of 95%. What sample sizes should be used to collect yield data from the existing process (μ_1) and the new process (μ_2)? The variance of the new process is equal to 110,000 g, and the variance of the existing process is 125,000 g.

Here we have a one-sided test on the difference between means. We are trying to show that the new process mean yield is at least 275g larger than the existing process mean. The corresponding hypotheses can be written as:

$$H_0 : \mu_2 - \mu_1 \leq 275$$
$$H_A : \mu_2 - \mu_1 > 275$$

Here, we have $a=0.01$, and from Appendix D, we find that $z_\alpha=z_{0.01}=2.33$ and $z_\beta=z_{0.05}=1.645$. The desired detectable change is $\delta=275$. Substituting these values in the equation from Table 6.72, we have:

$$n = n_1 = n_2 = \frac{(z_\alpha + z_\beta)^2(\sigma_1^2+\sigma_2^2)}{\delta^2}$$
$$= n_1 = n_2 = \frac{(2.33+1.645)^2(110{,}000+125{,}000)}{275^2}$$
$$= n_1 = n_2 = \lceil 49.10 = 50$$

Quality Engineering and Reliability Engineering Applications of Sampling

100% Quality Control Inspection

Certain quality engineering applications may require **100% inspection** of products. This involves inspecting every unit or every feature on a given component. This is the most thorough sampling method, but it is naturally time-consuming and expensive and so is typically applied only when necessary. It is not practical for low-value, high-volume parts and may not always be viable, for instance, for destructive inspections. When putting together a sampling plan, 100% inspection may be used as a baseline for comparison.

Subgroup Sampling for Statistical Process Control (SPC)

This method of sampling is performed in the context of statistical process control charts in which samples of size n are chosen at set time intervals. The samples are selected using a technique known as **rational subgrouping**. Here, the samples are selected to make the units with subgroups as homogeneous as possible.

Lot Acceptance Tests (LAT)

Lot acceptance testing (LAT) is a special type of sampling process that considers lots or batches as units, instead of individual parts. LAT is designed for processes where acceptability decisions can be made at the lot level. The process works as follows: (1) units are manufactured in a series of lots, (2) each lot is randomly sampled per its LAT plan requirements, (3) the sampled units are tested and results determined, and (4) the lot is accepted or rejected based on the criteria in the designated LAT plan. The intention here is that the agreed-upon plan appropriately balances producer and consumer risk. In practice, the producer and consumer may want to work out an alternative plan to reassess the units in the rejected lot, if so desired, but that is not part of a standard LAT process.

Some organizations may be contractually required to implement a certain type of LAT sampling plan and may be audited that they are in compliance. Standards documents ASQ/ANSI Z1.4 and Z1.9 provide guidance on processes and implementation. Z1.4 contains *Sampling Procedures and Tables for Inspection by Attributes*, while Z1.9 has *Sampling Procedures and Tables for Inspection by Variables for Percent Nonconforming*. The term "attributes" refers to inspection outcomes that yield attribute data, specifically pass/fail (binary) outcomes. Inspection by variables covers processes with test results that produce a quantitatively measured value. Often these may be applied when inspection requires destructive testing, i.e., when the unit must be sacrificed in order to prove that the product works, often a pass/fail outcome.

As a general note, inspection by attributes will always require more samples than inspection by attributes, but variable data may not always be available or cost-effective. Z1.4 and Z1.9 contain a variety of different sampling plans for a range of batch sizes, their underlying assumed quality level in terms of percent defective, and the desired sensitivity level. Plans may involve a single set of tests or sequential multistep testing. For additional details, please reference ANSI/ASQ

Z1.4-2003 (R2018) and ANSI/ASQ Z1.9-2003 (R2018). However, note that these tables are specific in application to processes where the *entire* lot or batch will be rejected if the units fail the LAT. If your process is not an LAT process, you should not be using LAT. These tables should *not* be used for general sampling applications.

STATISTICAL PROCESS CONTROL (SPC) AND CAPABILITY STUDIES (C_P, C_{PK}, P_P, P_{PK})

> Apply statistical process control (SPC), capability studies, and their associated indices (e.g., C_p, C_{pk}, P_p, and P_{pk}) and describe how they relate to reliability. Considering differing sample sizes, select control charts to represent variability over time. (Apply)
>
> **Body of Knowledge III.A.6**

SPC methods allow users to monitor process behavior over time and differentiate between common and special causes of variation. Processes in a state of statistical control are stable and predictable.

For processes under statistical control, process capability indices compare requirements to natural process variation, providing information as to how well the process is performing.

Special Versus Common Cause Variation

Variation is everywhere. The sources of **process variation** can be divided into two categories: special and common.

Common causes, or **natural sources of variation**, are those that are inherent to the process and generally not controllable by process operators. Examples of common causes include variation in raw materials, variation in ambient temperature and humidity, variation in electrical or pneumatic sources, and variation within equipment, such as worn bearings. In the case of service processes, common causes typically include variation in the input data, variation in customer load, and variation in computer operations. A process that has only common causes of variation is considered to be in a **state of statistical control**, where outcomes are **stable** and **predictable** (Figure 6.80).

Special, or **assignable**, **causes of variation** include unusual events that the operator (or if needed, the maintenance crew or process engineer), when properly alerted, can often remove or adjust. Examples include large changes in raw materials and broken equipment.

A principal problem in process management is distinguishing between special and common causes. If the process operator tries to adjust a process in response to common cause variation, the result is usually *more* variation rather than less. This

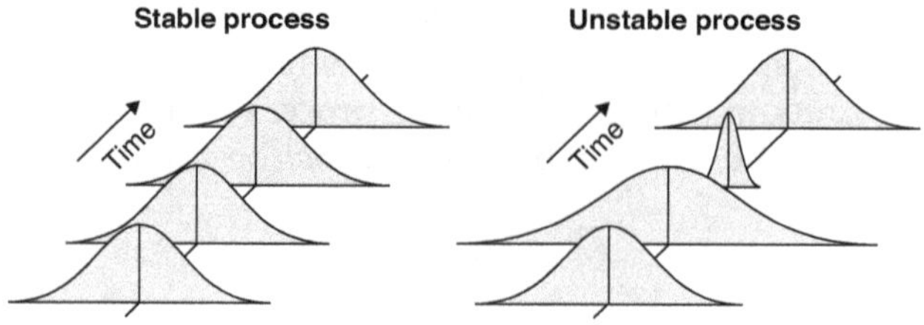

Figure 6.80 Stable and unstable variation.

is sometimes called **overadjustment** or **overcontrol**. Similarly, if a process operator fails to respond to the presence of a special cause of variation, this is similarly likely to produce additional process variation. This is referred to as **under adjustment** or **under control**.

Statistical process control (SPC) charts are the key tool used to help differentiate between special and common cause variation, and they are essential to satisfying continuous process monitoring requirements for validated processes. They are real-time decision-making tools that provide information for timely decisions concerning recently produced products, to help the process operator recognize the presence of special causes so that appropriate action can be taken in a timely manner. They are also problem-solving tools that help locate and investigate the causes of poor or marginal quality. The data from control charts can provide useful information when performing revalidation or retrospective validation activities. SPC charts may also be used in numerous applications to monitor the behavior of any kind of process. After all, everything is a process!

Control Chart Basics

Constructing control charts is straightforward and, more often than not, aided by computer software designed specifically for this purpose. Figure 6.81 illustrates the general form for a control chart. Its critical components are:

1. *X-axis*. This axis represents the time order of subgroups. Subgroups represent samples of data taken from a process. It is critical that the integrity of the time dimension be maintained when plotting control charts.

2. *Y-axis*. This axis represents the measured value of the quality characteristic under consideration when using variables charts. When attributes charts are used, this axis is used to quantify defectives or defects.

3. *Centerline*. The centerline represents the process average.

4. *Control limits*. Control limits typically appear at ±3 standard errors (of the characteristic being plotted) from the centerline. The width of the control limits is sometimes referred to as the *voice of the process (VOP)*.

5. *Zones*. The zones represent the distance between each standard error and are useful when discussing specific out-of-control rules.

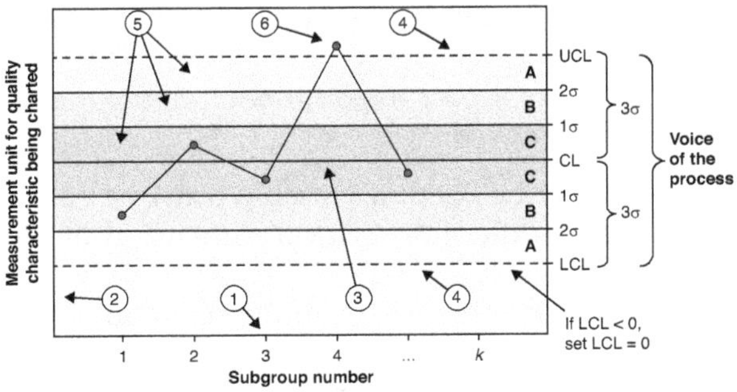

Figure 6.81 Structure of a typical control chart.

6. *Rational subgroups*. The variation within subgroups should be only common-cause variation so it's easier to detect special causes in subgroup-to-subgroup variation. It follows that units in a subgroup should be sampled close together in time.

Notice that there are no specification limits present on the control chart in Figure 6.81. This is by design, not by accident. The presence of specification limits on control charts could easily lead to inaction, particularly when a process is out of control but within specification. In addition, specifications are meant to be applied to individuals, and in the case of the *X-bar* chart, we are plotting averages.

A single chart is shown in Figure 6.81, but charts for variables data are often paired, with the top chart focusing on the process mean and the lower chart on process variation. Measures of short-term or within subgroup variation in the lower chart are used in the calculations of the control limits for both charts.

Because control limits are calculated based on data from the process, they represent the **voice of the process** (VOP). The difference between the UCL and the LCL constitutes a 6σ spread and should capture 99.73% of the plotted points from a normal distribution, as we know from the empirical rule discussed in "Probability Distributions." This spread represents the natural process variation and is important in evaluating **process capability**, to be discussed later in this section, which brings together the VOP with the **voice of the customer** (VOC).

There are two general classifications of univariate control charts: **variables charts** and **attributes charts** (see Table 6.33 later in the chapter). Charts fall into the variables category when the data to be plotted are the result of measurement on a continuous scale, for example, length, temperature, pressure or weight. Attributes charts are used for discrete data, for example, counts of dents, scratches, defects, days, cycles, or yes/no decisions. This aligns with the breakdown of different types of data into continuous versus discrete, as discussed in Chapter 6, "Basic Statistics." There are also several other varieties of specialty charts. **Cumulative sum (CUSUM) charts** are designed to monitor deviations from target. **Exponentially weighted moving average (EWMA) charts** give older data less weight and so are more sensitive to process shifts. ***Z*** and **Zed short run charts** allow combination of data from related but different sources by calculating deviations from average or normalizing the data. **Multivariate charts** are a way to monitor multiple variables simultaneously, using an aggregate measure. The focus of this discussion here is the standard univariate methods that are the foundation of SPC methodologies.

Selection of Variables to Be Tracked

When a control chart is to be used, a variable must be selected for monitoring. Sometimes that variable is the most critical dimension of the product. In some cases, the variable of choice is a "leading indicator" of special causes—one that detects special causes before others do. Contractual requirements with a customer sometimes specify the variable(s) to be monitored via control charts. If the root cause of the special variation is known, an input variable may be monitored. Often, the variable to be monitored is the one that is the most difficult to hold, as determined by capability analyses, to be discussed later in this section. It is possible to monitor several variables on separate control charts, especially if computerized charting is employed. In the end, the selection of the variable to be charted depends on experience and judgment.

Chart Selection: Attribute Versus Variable

Variables charts are preferred over attributes charts because the data contain more information and are typically more sensitive to detecting process shifts than attributes charts. It should be noted that measurement data collected for a variables control chart can be categorized and applied to an attributes control chart. For example, consider five temperature readings in which each reading can be classified as being within specification (for example, good) or out of specification (for example, bad). Had each reading been classified as attribute data (for example, either within or out of specification) at the time of collection and the actual measurement value not recorded, the attribute data could not be transformed for use in a variables control chart.

Chart Selection: Individuals and Moving Range (X and mR)

Individuals and moving range charts are appropriate in the increasingly common situation that every unit produced is measured. Caution should be exercised when using these charts, since normality is usually assumed. Scenarios where it may be necessary and useful are those where rational subgrouping is difficult, for example, when automated inspection and measurement technology are used; when the production rate is very slow and waiting to accumulate a subgroup would be inefficient; repeated measurements on the process output vary only due to laboratory error, such as in chemical processing industry; and if the measurement involves a destructive test of an expensive product for large samples are not desirable. See Montgomery (2019).

Attribute Charts: Defects Versus Defectives

It is important to distinguish here between the terms **defect** and **defective.** A *defective unit* may have one or more *defects* that are serious enough to make the unit unacceptable. However, a unit with a *defect* is not necessarily classified as a *defective, since the defects may be minor.*

Also note that usage of the terms "defect" and "defective" in this context is intended *solely* to allow differentiation between the type of data available in order to ensure appropriate chart selection. It is not intended to implicate hardware as being noncompliant per definitions in ANSI Z1.4 2003. "Defects" identified here may or may not be sufficiently severe to force classification as "defective." There is additional discussion of this topic later in this section.

Rational Subgrouping

Many types of charts use subgrouping, in which data are obtained from subsequent subgroups to monitor process health. That means that special causes of variation occur between successive subgroups while the variation within the subgroup is kept low. This serves the same purpose as blocking in the DOE. Subgroups may consist of parts that are produced successively by the same process, producing natural clusters. However, care should be taken to make sure the same process is producing each item. One way to minimize within-sample variation is to space out samples so that any process shifts that have occurred will be displayed on the chart as between-subgroup variation.

Constant Versus Varying Subgroup Sizes

An SPC monitoring plan may be designed with constant subgroup sizes in mind. However, in reality, there may be variation in subgroup size since some units may be lost or damaged or otherwise unavailable. Software can typically handle varying subgroup sizes, which will be readily apparent since the limits will have different widths for each subgroup. Smaller subgroups will result in wider control limits for that subgroup, while larger subgroups will result in tighter limits. Some charts are unable to accommodate varying subgroup sizes, the np and c attribute charts. However, this can be handled easily by switching to the p and u attribute charts, which take subgroup size into account.

Getting Started: Phase I Versus Phase II

SPC activities can be divided into two phases. **Phase I** is more exploratory, where the primary objective is to establish process stability and control, and often involves smaller datasets. Control charts are used here as a tool for understanding process behavior, then diagnosing and resolving issues. Once a process has stabilized, special causes of variation are eliminated, and we are looking primarily at common cause variation, we can move into **Phase II**, where stable processes are subject to ongoing monitoring. Note that the inclusion of new data here may still be influential, depending on the amount of data available. Estimates of the control limits have their own uncertainty. Limits at this point usually should not be recalculated automatically with the addition of new data until that point has been evaluated and shown to be consistent with the baseline population, since the inclusion of outliers, for example, can impact existing process limits. Software allows data to be evaluated against predetermined limits.

When first assessing whether a process is in a state of statistical control, it is prudent to collect as much data as practical. Many authors suggest at least 25 to 30 subgroups. If out-of-control conditions are found in the initial assessment, root causes should be investigated, and the associated data removed, if appropriate. A revised set of control limits may then be computed. If it is not possible to determine root causes, the associated data should remain in the calculation for the limits. In this situation, it is possible the out-of-control condition was a rare event in an otherwise stable process. This is often an iterative process as new information becomes available and the team's understanding of process behavior matures.

Negative Limits

It is important to be aware that calculating the lower control limit on the process variation chart may sometimes result in a negative value. If this should occur, the LCL is set to zero because it is not possible to have negative ranges or standard deviations. Likewise, when an LCL computes to a negative value on attributes charts, the LCL is set to zero since it is not possible to have a negative percentage defective or negative defect counts. This can also be an indication that something is wrong with either the dataset or the underlying assumptions.

Handling Out-of-Control Signals

Points falling outside the limits or that meet any of the out-of-control conditions described subsequently are attributed to special cause variation. Such points, regardless of whether they constitute "good" or "bad" occurrences, should be investigated immediately while the cause-and-effect relationships and individual memories are fresh and access to documentation for process changes is readily available. As the time between the out-of-control event and the beginning of the investigation increases, the likelihood of determining root causes diminishes greatly. Hence, the motto "time is of the essence" is most appropriate.

Finding root causes for out-of-control conditions may be frustrating and time-consuming, but the results are worthwhile. Ideally, root causes for good out-of-control conditions are incorporated into the process, while root causes for bad out-of-control conditions are removed. Note, however, that each signal has a non-negligible probability of occurring through chance alone, thus leading to some events that may be naturally occurring and *not* an indication of special cause variation. Signals tend to occur together. It is important to use your judgement regarding chasing down obscure signals.

A word of caution: Adjusting a process when it is not warranted by out-of-control conditions constitutes process tampering. This usually results in destabilizing a process, causing even greater variation.

Defects Versus Nonconformities

Finally, readers should understand the differing terminology that surrounds these charts. Often, quality and reliability professionals tend to use the terms "defects" and "nonconformities" interchangeably. Consequently, we have used them interchangeably in this handbook. However, there is a technical distinction between these two terms that is worth noting. ANSI Z1.4-2003 (R2013), which is in accordance with ISO 3534–2:2006(E), provides the following definitions:

- *Defect*. A departure of a quality characteristic from its intended level or state that occurs with a severity sufficient to cause an associated product or service not to satisfy intended normal, or foreseeable, usage requirements.
- *Nonconformity*. A departure of a quality characteristic from its intended level or state that occurs with severity sufficient to cause an associated product or service not to meet a specification requirement.

ISO 3534–2:2006(E) adds two important notes to the definition of defects:

- *Note 1*. The distinction between the concepts *defect* and *nonconformity* is important as it has legal connotations, particularly those associated with product liability issues. Consequently, the term "defect" should be used with extreme caution.
- *Note 2*. The intended use by the customer can be affected by the nature of the information, such as operating or maintenance instructions provided by the customer.

Based on these definitions and supporting notes, one might argue that we can generally align a nonconformity to occurring within the organization and a defect to occurring outside the organization (that is, external customer).

Control Chart Types and Selection

Figure 6.82 and Table 6.33 can be used to identify the best options for basic control chart selection.

Charts for Variables Data (X-mR, X-bar-R/S)

Three common charts used to track the output of continuous, or measurement, data are the $\bar{X}$ **and** $\boldsymbol{R}$ ($\bar{X}-R$), $\bar{X}$ **and** $\boldsymbol{S}$ ($\bar{X}-S$), and individuals and moving range (variously referred to as ImR, I-MR, **X-mR**) charts. Usage here is typically determined by subgroup size, as shown in the flowchart in Figure 6.83.

If only one observation is available at a time, the only option is the individuals (X) chart, which plots each point individually. It is paired with the moving range (mR) chart, which plots the difference between each consecutive pair of points to quantify the short-term variation used in the limit calculations. If using an $\bar{X}$ chart, it could theoretically be paired with either a range (R) or standard

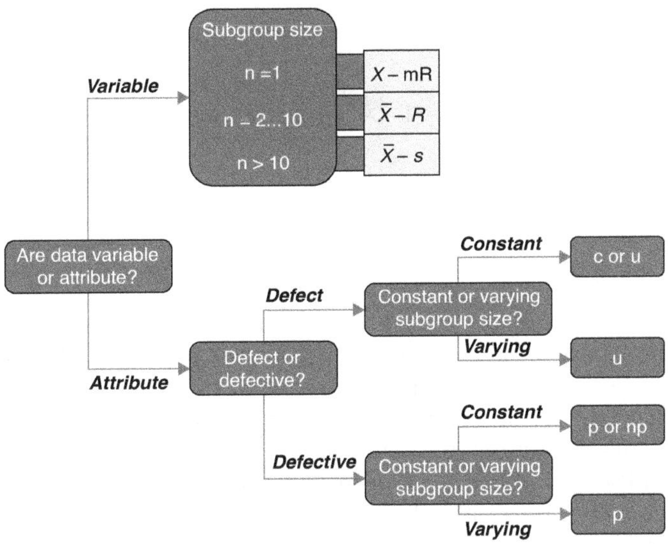

Figure 6.82 Control chart selection flowchart.

Table 6.33 Variables and attributes control charts.

Variables control charts			
Type	**Distribution**	**Sample**	**Application**
$\bar{X}$ and R	Normal	2 to 10	Measurement subgroups
$\bar{X}$ and s	Normal	>10	Measurement subgroups
Attributes control charts			
Type	**Distribution**	**Sample**	**Application**
c	Poisson	Constant	Count number of defects per item
u	Poisson	Varies	Count number of defects per item
np	Binomial	Constant	Count of defective items
p	Binomial	Varies	Count of defective items
g	Geometric	Individual	Interval between rare events
Variables or attributes control charts			
Type	**Distribution**	**Sample**	**Application**
X and mR	Normal	1	Individual counts or measurements

Source: Mark Durivage. *Practical Engineering, Process, and Reliability Statistics* (ASQ Quality Press, 2014).

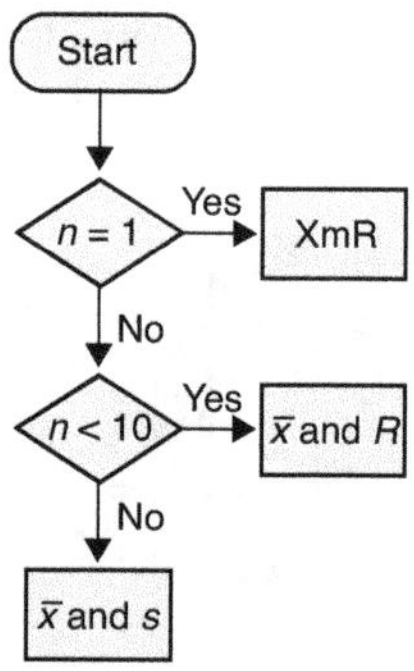

Figure 6.83 Selecting a variables control chart based on the subgroup size.

deviation (S) chart to assess variation. General practice is to use the range for smaller subgroups, then switch over to standard deviation for larger subgroups, since standard deviation is a more robust measure of variability. End users may also have a specific interest in quantifying either the range or standard deviation explicitly on their chart, such as for monitoring process repeatability.

Other charts of interest include the g chart, based on the geometric distribution to model time between rare events, and the **X-mR-R/S** chart, a three-level chart, which allows multiple sources of variability to be shown in a hierarchy. This is particularly useful when monitoring processes where multiple tests are done on the same unit, for example.

About the Moving Range

The **moving range** (MR) provides a mathematically simple way to estimate **short-term** process variation. It is calculated by taking the absolute value of the difference between each measurement and the previous one. For this reason, the moving range chart has one fewer point plotted than the corresponding individuals chart. Short-term or within subgroup variation is (typically) less than long-term or overall variation.

Table 6.34 gives the formulas for the center line and upper and lower control limits for variables SPC charts, and Table 6.35 provides the constant values needed for the calculations, based on sample size n.

Charts for Attribute Data (p, np, c, *and* u)

There are four types of attribute charts (Figure 6.84). Attributes charts are used for count data in which either defects or defectives are counted. The ***p*-charts** and ***np*-charts** are based on the binomial distribution and used to plot proportion defective and number of **defectives**, respectively. The ***c*-charts** and ***u*-charts** are based on the Poisson and are used to plot counts of **defects** and defects per unit, respectively. Defects and defectives were previously discussed in "Probability Distributions."

For p-charts and np-charts, it is assumed that items are categorized using two outcomes (for example, defective/non-defective, success/failure, pass/fail, go/no go), and the probability of a defective remains constant from subgroup to subgroup. The number of defectives found in a subgroup is also bounded by the subgroup size n. For c-charts and u-charts it is assumed that the probability of the occurrence of a defect is assumed constant from subgroup to subgroup. There is also no upper limit to the number of occurrences that can be found in a sample. When constructing attributes charts, it is important to keep these assumptions in mind to ensure statistical validity of the results.

Additionally, users of attributes charts should note that the np-chart and the c-chart require constant sample sizes. The p-chart and u-chart, however, permit

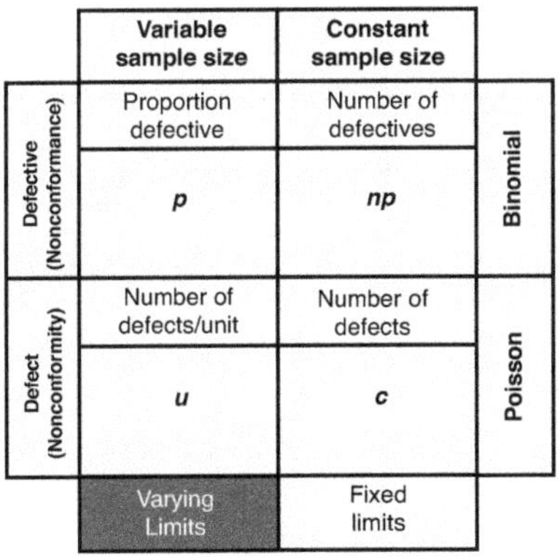

Figure 6.84 Characteristics of attributes charts.

Table 6.34 Formulas for calculating the centerline and control limits for variables charts.

Average chart				Individuals chart			
Chart	**Center-line**	**LCL**	**UCL**	**Chart**	**Center-line**	**LCL**	**UCL**
$\bar{X}$	$\bar{\bar{X}}$	$\mathrm{LCL}_{\bar{X}} = \bar{\bar{X}} - A_2\bar{R}$	$\mathrm{UCL}_{\bar{X}} = \bar{\bar{X}} + A_2\bar{R}$	X	$\bar{X}$	$LCL_X = \bar{X} - E_2\overline{mR}$	$UCL_X = \bar{X} + E_2\overline{mR}$
R	$\bar{R}$	$\mathrm{LCL}_R = D_3\bar{R}$	$\mathrm{UCL}_R = D_4\bar{R}$	mR	$\overline{mR}$	$LCL_{reR} = D_3\overline{mR}$	$UCL_{mR} = D_4\overline{mR}$
$\bar{X}$	$\bar{\bar{X}}$	$\mathrm{LCL}_X = \bar{\bar{X}} - A_3\bar{s}$	$\mathrm{UCL}_{\bar{R}} = \bar{\bar{X}} + A_3\bar{s}$				
s	$\bar{s}$	$\mathrm{LCL}_s = B_3\bar{s}$	$\mathrm{UCL}_s = B_4\bar{s}$				

Average chart	Individuals chart
$\bar{X}_i = \frac{\sum_{j=1}^{n} x_{ij}}{n}$ = Average of the ith subgroup; plot points	$\bar{X} = \frac{\sum_{i=1}^{n} x_i}{n}$ = Centerline of the X chart
$\bar{\bar{X}} = \frac{\sum_{i=1}^{k} \bar{X}_i}{k}$ = Centerline of the $\bar{X}$ chart	$mR_i = \lvert x_1 - x_{i-1} \rvert$ for $i = 1, 2, 3, \ldots, n$ plot points
R_i = Range of the ith subgroup; plot points	$\overline{mR} = \frac{\sum_{i=2}^{n} mR_j}{n-1}$ = Centerline for the mR chart
$\bar{R} = \frac{\sum_{i=1}^{k} R_i}{k}$ = Centerline of the R chart	x_i = Individual data measurements; plot point
s_i = Standard deviation of the ith subgroup; plot point	n = Number of individual data measurements
$\bar{s} = \frac{\sum_{i=1}^{k} s_i}{k}$ = Centerline of the s chart	$x_{ij} = j$ individual data measurements within i subgroups
k = Number of subgroups	n = Sample size of each subgroup

Table 6.35 Constants for variables control charts.

Sub-group size										A_2 for median charts			
N	A_2	d_2	D_3	D_4	A_3	C_4	B_3	B_4	E_2		A_4	D_5	D_6
2	1.880	1.128	-	3.267	2.659	0.798	-	3.267	2.660	1.880	2.224	-	3.865
3	1.023	1.693	-	2.574	1.954	0.886	-	2.568	1.772	1.187	1.091	-	2.745
4	0.729	2.059	-	2.282	1.628	0.921	-	2.266	1.457	0.796	0.758	-	2.375
5	0.577	2.326	-	2.114	1.427	0.940	-	2.089	1.290	0.691	0.594	-	2.179
6	0.483	2.534	-	2.004	1.287	0.952	0.030	1.970	1.184	0.548	0.495	-	2.055
7	0.419	2.704	0.076	1.924	1.182	0.959	0.118	1.882	1.109	0.508	0.429	0.078	1.967
8	0.373	2.847	0.136	1.864	1.099	0.965	0.185	1.815	1.054	0.433	0.380	0.139	1.901
9	0.337	2.970	0.184	1.816	1.032	0.969	0.239	1.761	1.010	0.412	0.343	0.187	1.850
10	0.308	3.078	0.223	1.777	0.975	0.973	0.284	1.716	0.975	0.362	0.314	0.227	1.809

variable sample sizes. As a result, the control limits for p-charts and u-charts may vary since their calculation is dependent on n. There will be no difference in the appearance of the p- and np-charts except for the y-axis scale if the sample size is constant; likewise, the c- and u-charts will be identical except for scale in this case. The p- and u-charts are more generally applicable since they are utilized regardless of whether sample size is constant or not. Narrower limits indicate more data were available for that subgroup, wider limits mean less.

Table of Equations for Attribute Charts

Limit calculation equations for attribute charts are summarized in Table 6.36.

Control Chart Interpretation

Improper control chart interpretation can lead to several problems, including blaming people for problems they cannot control, spending time and money looking for problems that do not exist, spending time and money on process adjustments or new equipment that are not necessary, taking action where no action is warranted, and asking for worker-related improvements where process or equipment improvements need to be made first.

Control Chart Rules

Rules have been developed for assessing whether data for a process is in a state of statistical control. Initially implemented in 1956 as a set of four rules and known variously as the Western Electric rules, Nelson rules, and AIAG rules, they have evolved and expanded over time by different user communities. Some form of

Table 6.36 Formulas for calculating the centerline and control limits for attributes charts.

Chart	Centerline	LCL	UCL	Plot point
p	$\overline{p}=\dfrac{\sum_{i=1}^{k} D_i}{\sum_{i=1}^{k} n_i}$	$\mathrm{LCL}_p=\overline{p}-3\sqrt{\dfrac{\overline{p}(1-\overline{p})}{n_i}}$	$\mathrm{UCL}_p=\overline{p}+3\sqrt{\dfrac{\overline{p}(1-\overline{p})}{n_i}}$	$p_i=\dfrac{D_i}{n_i}$
np	$n\overline{p}=n\dfrac{\sum_{i=1}^{k} D_i}{\sum_{i=1}^{k} n_i}$	$\mathrm{LCL}_{np}=n\overline{p}-3\sqrt{n\overline{p}(1-\overline{p})}$	$\mathrm{UCL}_{np}=n\overline{p}+3\sqrt{n\overline{p}(1-\overline{p})}$	D_i
c	$\overline{c}=\dfrac{\sum_{i=1}^{k} c_i}{k}$	$\mathrm{LCL}_c=\overline{c}-3\sqrt{\overline{c}}$	$\mathrm{UCL}_c=\overline{c}+3\sqrt{\overline{c}}$	c_i
u	$\overline{u}=\dfrac{\sum_{i=1}^{k} c_i}{\sum_{i=1}^{k} n_i}$	$\mathrm{LCL}_u=\overline{u}-3\sqrt{\dfrac{\overline{u}}{n_i}}$	$\mathrm{UCL}_w=\overline{u}+3\sqrt{\dfrac{\overline{u}}{n_i}}$	$u_i=\dfrac{c_i}{n_i}$

Notes:
1. n = Constant sample size of each subgroup
2. n_i = Sample size of the ith subgroup
3. k = Number of subgroups
4. D_i = Number of *defective* units in the ith subgroup
5. c_i = Number of *defects* in the ith subgroup
6. u_i = Number of *defects*/unit in the ith subgroup

these rules is typically available in statistical software, which allows users to select which specific rules they want to apply.

The probability of occurrence for each rule is individually very low, namely approximately 0.0027 or 1/370, the probability of being beyond $\pm 3\sigma$ for normally distributed data. When all rules are enabled together, however, the combination can significantly increase the rate of false alarms. Users should always consider the context of the data. A recommended approach is to just use the 3-sigma rule or use that plus one or two specific rules that match the types of special causes typical of the process being monitored.

The rules in Table 6.37 can be applied to formally assess whether a process is out of statistical control. Illustration of the patterns represented by these rules can be found in Figure 6.85.

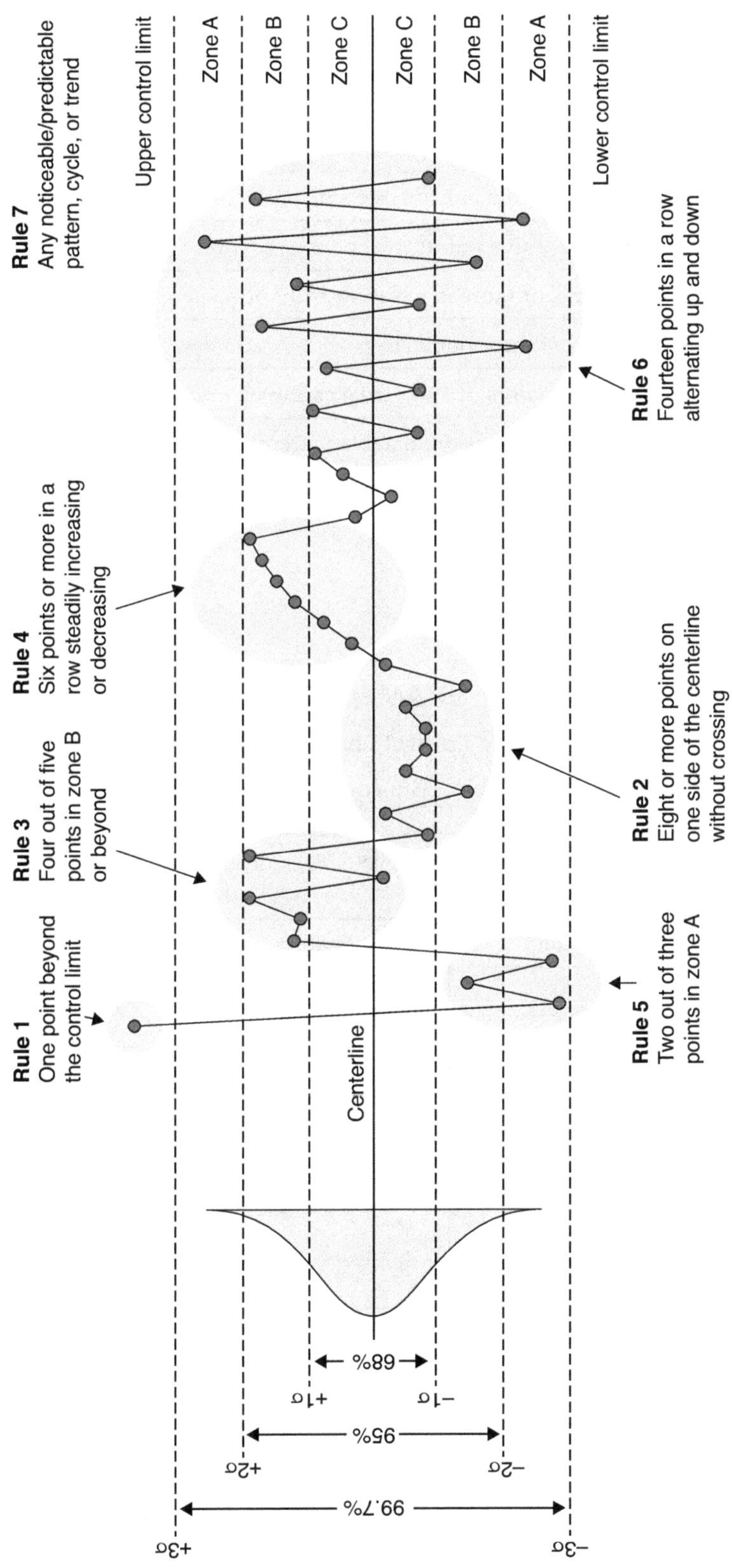

Figure 6.85 Control chart interpretation rules.

Table 6.37 Control chart runs rules.

Rule number	Out of control condition
1	One point beyond the 3-sigma control limit
2	Eight or more points on one side of the centerline without crossing
3	Four out of five points in zone B or beyond
4	Six points or more in a row steadily increasing or decreasing
5	Two out of three points in zone A
6	Fourteen points in a row alternating up and down
7	Any noticeable or predictable pattern, cycle, or trend

Control Chart Examples

EXAMPLE 6.73

Variable Control Chart: *X*-MR Charts

Using the data provided in Table 6.38, construct an *X*-mR chart.

Table 6.38 Data for Example 6.73—individuals and moving range chart.

Reading	Individual data element
1	290
2	288
3	285
4	290
5	291
6	287
7	284
8	290
9	290
10	288

Several of the intermediate calculations for this example are shown in Table 6.39. The values of E_2, D_3, and D_4 are found in Table 6.35 for $n=2$.

Table 6.39 Calculations for Example 6.73—individuals and moving range chart.

Reading	Individual data element	Difference	Absolute value of the difference
1	290		
2	288	−2	2
3	285	−3	3
4	290	5	5
5	291	1	1
6	287	−4	4
7	284	−3	3
8	290	6	6
9	290	0	0
10	288	−2	2
Average	**288.3**	**Moving range**	**2.89**

The completed individuals and moving range chart is shown in Figure 6.86. Both control charts demonstrate statistical control.

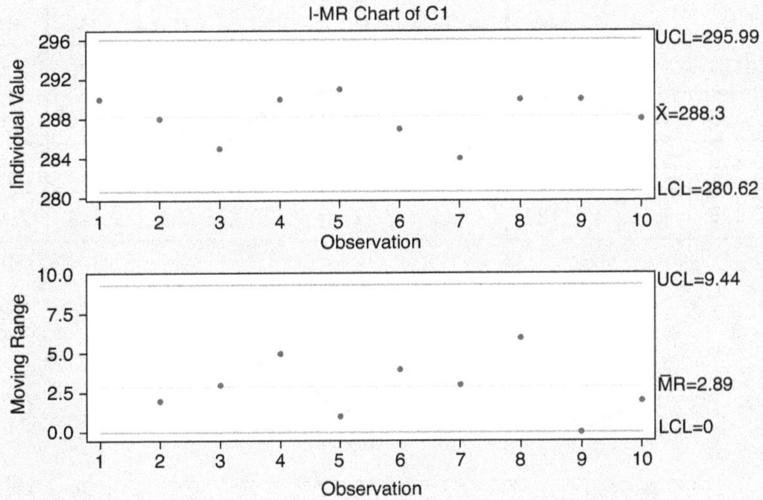

Figure 6.86 *X*-mR chart for Example 6.73.

EXAMPLE 6.74

Variable Control Chart: *X*-bar-*R* Chart

A face-and-plunge operation is performed on a lathe. The dimension of interest is the groove inside diameter (ID). Four parts are measured every hour, as shown in Table 6.40. Determine the centerlines and control limits for the $\bar{X}$ and R chart.

Using Tables 6.34 and 6.35, we have

$$\bar{\bar{X}} = 7.12565$$
$$\bar{R} = 0.0037$$

$$LCL_{\bar{X}} = \bar{\bar{X}} - A_2\bar{R}$$
$$= 7.12565 - (0.729)(0.0037) = 7.1230$$

$$UCL_{\bar{X}} = \bar{\bar{X}} + A_2\bar{R}$$
$$= 7.12565 + (0.729)(0.0037) = 7.1283$$

$$LCL_R = D_3\bar{R} = (0)(0.0037) = 0$$
$$UCL_R = D_4\bar{R} = (2.282)(0.0037) = 0.0084$$

The values of A_2, D_3, and D_4 are found in Table 6.35 for the value of $n=4$. The completed $\bar{X}$ and R chart is given in Figure 6.87. Both charts are in statistical control.

Table 6.40 Data for Example 6.74 and Example 6.75.

7:00	8:00	9:00	10:00	11:00	12:00	1:00	2:00	3:00	4:00
A.M.	A.M.	A.M.	A.M.	A.M.	P.M.	P.M.	P.M.	P.M.	P.M.
7.127	7.125	7.123	7.127	7.128	7.125	7.126	7.126	7.127	7.128
7.123	7.126	7.129	7.127	7.125	7.125	7.123	7.126	7.129	7.123
7.123	7.121	7.129	7.124	7.126	7.127	7.123	7.127	7.128	7.122
7.126	7.122	7.124	7.125	7.127	7.128	7.125	7.128	7.129	7.124

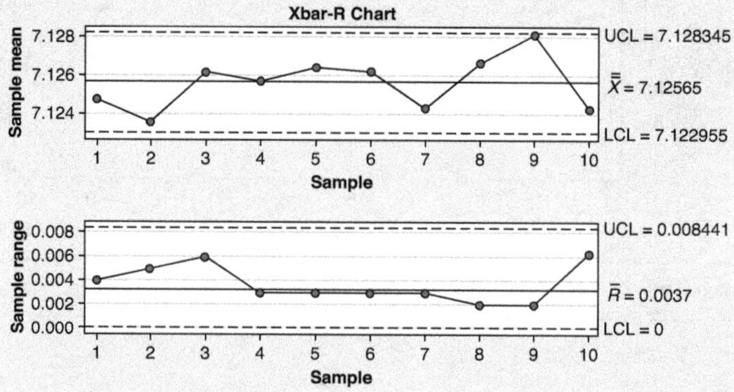

Figure 6.87 $\bar{X}$ and R chart for Example 6.74.

EXAMPLE 6.75

Variable Control Chart: X-bar-S Chart

Using the same data provided in Table 6.40, construct an $\bar{X}$ and S chart. Using Tables 6.34 and 6.35, we have

$$\bar{\bar{X}} = 7.12565$$
$$\bar{s} = 0.001798$$
$$LCL_{\bar{X}} = \bar{\bar{X}} - A_s\bar{s}$$
$$= 7.12565 - (1.6278)(0.001798) = 7.1227$$
$$UCL_{\bar{X}} = \bar{\bar{X}} + A_3\bar{s}$$
$$= 7.12565 + (1.6278)(0.001798) = 7.1286$$
$$LCL_S = B_3\bar{s} = (0)(0.001798) = 0$$
$$UCL_{\bar{S}} = B_4\bar{s} = (2.266)(0.001798) = 0.0041$$

The values of A_3, B_3, and B_4 were found in Table 6.35 for the value of $n=4$. The completed $\bar{X}$ and S chart is given in Figure 6.88. Both charts are in statistical control.

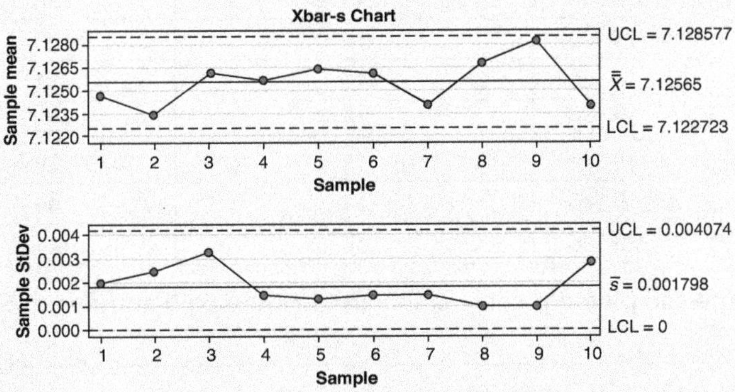

Figure 6.88 $\bar{X}$ and S chart for Example 6.75.

EXAMPLE 6.76

Attribute Control Chart: *p*-Chart

Packages containing 1000 light bulbs are randomly selected, and varying number of bulbs are light-tested each day. The results of the last 13 days of testing are shown in Table 6.41. Construct a *p*-chart.

We obtain the results shown in Table 6.42 using Table 6.36. Note that the control limits will change for each value of n_i.

$$\bar{p} = \frac{189}{1625} = 0.1163$$
$$LCL_p = \bar{p} - 3\sqrt{\frac{\bar{p}(1-\bar{p})}{n_i}}$$
$$= 0.1163 - 3\sqrt{\frac{0.1163(0.8837)}{n_i}}$$

Continued

$$UCL_p = \bar{p} + 3\sqrt{\frac{\bar{p}(1-\bar{p})}{n_i}}$$

$$= 0.1163 + 3\sqrt{\frac{0.1163(0.8837)}{n_i}}$$

Table 6.41 Data for Example 6.76—*p*-chart.

Date	Number of defectives (*D*)	Sample size (n_i)
8-Sep	12	116
9-Sep	14	126
10-Sep	18	112
11-Sep	13	134
12-Sep	17	121
13-Sep	15	119
14-Sep	15	138
15-Sep	16	109
16-Sep	11	111
17-Sep	14	125
18-Sep	13	129
19-Sep	14	144
20-Sep	17	141

The completed *p*-chart is shown in Figure 6.89 and demonstrates statistical control.

Table 6.42 Calculations for Example 6.76—*p*-chart.

Date	Number of defectives (D_i)	Sample size (n_i)	Plot points (p_i)	Centerline ($\bar{p}$)	LCL	UCL
8-Sep	12	116	0.1034	0.1163	0.0270	0.2056
9-Sep	14	126	0.1111	0.1163	0.0306	0.2020
10-Sep	18	112	0.1607	0.1163	0.0254	0.2072
11-Sep	13	134	0.0970	0.1163	0.0332	0.1994
12-Sep	17	121	0.1405	0.1163	0.0289	0.2037
13-Sep	15	119	0.1261	0.1163	0.0281	0.2045
14-Sep	15	138	0.1087	0.1163	0.0344	0.1982
15-Sep	16	109	0.1468	0.1163	0.0242	0.2084
16-Sep	11	111	0.0991	0.1163	0.0250	0.2076
17-Sep	14	125	0.1120	0.1163	0.0303	0.2023
18-Sep	13	129	0.1008	0.1163	0.0316	0.2010
19-Sep	14	144	0.0972	0.1163	0.0362	0.1964
20-Sep	17	141	0.1206	0.1163	**0.0353**	**0.1973**
Total	**189**	**1625**				
***p*-bar**	**0.1163**					

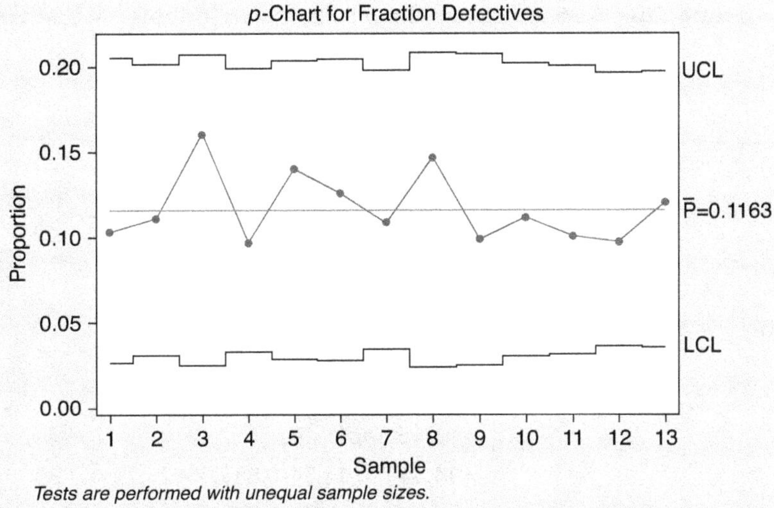

Figure 6.89 *p*-chart for Example 6.76.

EXAMPLE 6.77

Attribute Control Chart: *np*-Chart

Packages containing 1000 light bulbs are randomly selected, and $n=125$ bulbs from the package that are light-tested. The results of the tests are given in Table 6.43. Construct an *np*-chart.

Table 6.43 Data for Example 6.77—*np*-chart.

Date	Number of defectives (D_i)	Sample size (n)
8-Sep	12	125
9-Sep	14	125
10-Sep	18	125
11-Sep	13	125
12-Sep	17	125
13-Sep	15	125
14-Sep	15	125
15-Sep	16	125
16-Sep	11	125
17-Sep	14	125
18-Sep	13	125
19-Sep	14	125
20-Sep	17	125

Continued

Using Table 6.36 we obtain the results shown in Table 6.44 and as follows:

$$\bar{p} = \frac{\sum_{i=1}^{k} D_i}{nk} = \frac{189}{1625} = 0.1163$$

$$n\bar{p} = \frac{\sum_{i=1}^{k} D_i}{k} = \frac{189}{13} = 14.54$$

$$\begin{aligned} LCL_{np} &= n\bar{p} - 3\sqrt{n\bar{p}(1-\bar{p})} \\ &= 14.54 - 3\sqrt{14.54(0.8837)} = 3.78 \end{aligned}$$

$$\begin{aligned} UCL_{np} &= n\bar{p} + 3\sqrt{n\bar{p}(1-\bar{p})} \\ &= 14.54 + 3\sqrt{14.54(0.8837)} = 25.29 \end{aligned}$$

Table 6.44 Calculations for Example 6.77—*np*-chart.

Date	Number of defectives (D_i)	Sample size (n)	Centerline (n*p-bar)	LCL	UCL
8-Sep	12	125	14.54	3.78	25.29
9-Sep	14	125	14.54	3.78	25.29
10-Sep	18	125	14.54	3.78	25.29
11-Sep	13	125	14.54	3.78	25.29
12-Sep	17	125	14.54	3.78	25.29
13-Sep	15	125	14.54	3.78	25.29
14-Sep	15	125	14.54	3.78	25.29
15-Sep	16	125	14.54	3.78	25.29
16-Sep	11	125	14.54	3.78	25.29
17-Sep	14	125	14.54	3.78	25.29
18-Sep	13	125	14.54	3.78	25.29
19-Sep	14	125	14.54	3.78	25.29
20-Sep	17	125	14.54	3.78	25.29
Total	**189**	**1625**			
$\bar{p}$	**0.1163**				
$n\bar{p}$	**14.54**				

The completed *np*-chart is given in Figure 6.90 and demonstrates statistical control.

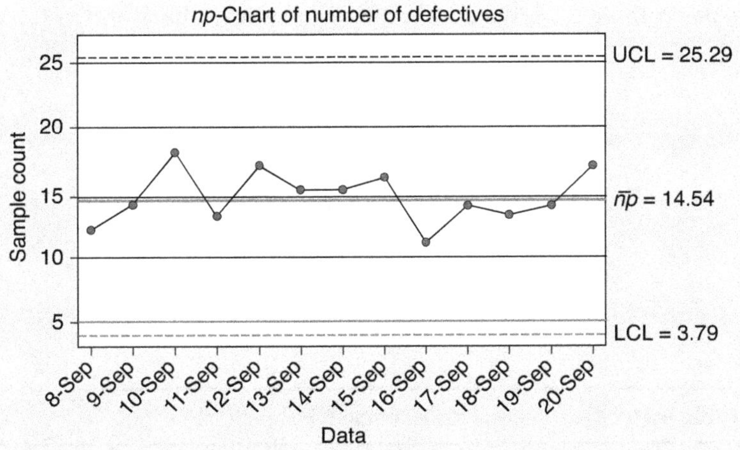

Figure 6.90 *np*-chart for Example 6.77.

EXAMPLE 6.78

Attribute Control Chart: *c*-Chart

Panes of glass of uniform area are inspected for defects such as bubbles, scratches, chips, inclusions, waves, and dips with a constant sample size equal to 125. The results of the inspection are given in Table 6.45. Construct a *c*-chart.

Table 6.45 Data for Example 6.78–*c*–chart

Date	Number of defects (c_i)	Sample size (n)
8-Sep	12	125
9-Sep	14	125
10-Sep	18	125
11-Sep	13	125
12-Sep	17	125
13-Sep	15	125
14-Sep	15	125
15-Sep	16	125
16-Sep	11	125
17-Sep	14	125
18-Sep	13	125
19-Sep	14	125
20-Sep	17	125

Continued

Using Table 6.36, we obtain the results shown here and in Table 6.46:

$$\bar{c} = \frac{\sum_{i=1}^{k} c_i}{k} = \frac{189}{13} = 14.54$$

$$\begin{aligned} LCL_c &= \bar{c} - 3\sqrt{\bar{c}} \\ &= 14.54 - 3\sqrt{14.54} = 3.10 \end{aligned}$$

$$\begin{aligned} UCL_c &= \bar{c} + 3\sqrt{\bar{c}} \\ &= 14.54 + 3\sqrt{14.54} = 25.98 \end{aligned}$$

Table 6.46 Calculations for Example 6.78—*c*-chart.

Date	Number of defects—plot points (*c*)	Sample size (*n*)	Centerline ($\bar{c}$)	LCL	UCL
8-Sep	12	125	14.54	3.10	25.98
9-Sep	14	125	14.54	3.10	25.98
10-Sep	18	125	14.54	3.10	25.98
11-Sep	13	125	14.54	3.10	25.98
12-Sep	17	125	14.54	3.10	25.98
13-Sep	15	125	14.54	3.10	25.98
14-Sep	15	125	14.54	3.10	25.98
15-Sep	16	125	14.54	3.10	25.98
16-Sep	11	125	14.54	3.10	25.98
17-Sep	14	125	14.54	3.10	25.98
18-Sep	13	125	14.54	3.10	25.98
19-Sep	14	125	14.54	3.10	25.98
20-Sep	17	125	14.54	3.10	25.98
Total	**189**				
$\bar{c}$	**14.54**				

The completed *c*-chart is given in Figure 6.91 and demonstrates statistical control.

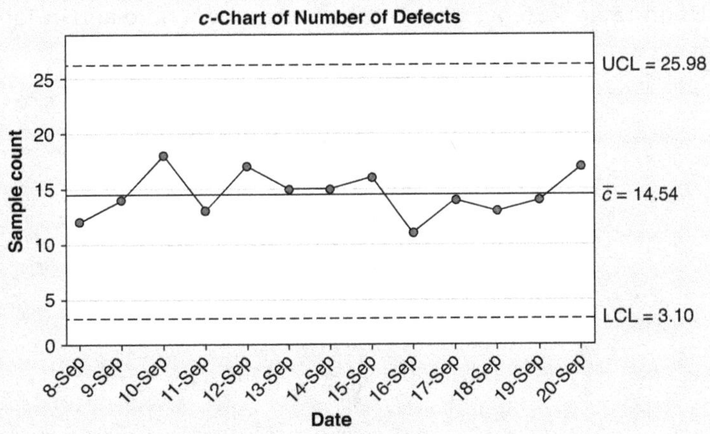

Figure 6.91 *c*-chart for Example 6.78.

EXAMPLE 6.79

Attribute Control Chart: *u*-Chart

Panes of glass of uniform area are inspected for defects such as bubbles, scratches, chips, inclusions, waves, and dips, with sample size varying with the production count. The results of the inspection are given in Table 6.47. Construct a *u*-chart.

Table 6.47 Data for Example 6.79—*u*-chart.

Date	Number of defects (c_i)	Sample size (n_i)
8-Sep	12	116
9-Sep	14	126
10-Sep	18	112
11-Sep	13	134
12-Sep	17	121
13-Sep	15	119
14-Sep	15	138
15-Sep	16	109
16-Sep	11	111
17-Sep	14	125
18-Sep	13	129
19-Sep	14	144
20-Sep	17	141

Continued

Using Table 6.36, we obtain the results shown here and in Table 6.48:

$$\bar{u} = \frac{\sum_{i=1}^{k} c_i}{\sum_{i=1}^{k} n_i} = \frac{189}{1625} = 0.1163$$

$$LCL_u = \bar{u} - 3\sqrt{\frac{\bar{u}}{n_i}} = 0.1163 - 3\sqrt{\frac{0.1163}{n_i}}$$

$$UCL_u = \bar{u} - 3\sqrt{\frac{\bar{u}}{n_i}} = 0.1163 + 3\sqrt{\frac{0.1163}{n_i}}$$

Table 6.48 Calculations for Example 6.79—*u*-chart

Date	Number of defectives (c_i)	Sample size (n_i)	Plot points (c_i/n_i)	Centerline $\bar{u}$	LCL	UCL
8-Sep	12	116	0.1034	0.1163	0.0213	0.2113
9-Sep	14	126	0.1111	0.1163	0.0252	0.2074
10-Sep	18	112	0.1607	0.1163	0.0196	0.2130
11-Sep	13	134	0.0970	0.1163	0.0279	0.2047
12-Sep	17	121	0.1405	0.1163	0.0233	0.2093
13-Sep	15	119	0.1261	0.1163	0.0225	0.2101
14-Sep	15	138	0.1087	0.1163	0.0292	0.2034
15-Sep	16	109	0.1468	0.1163	0.0183	0.2143
16-Sep	11	111	0.0991	0.1163	0.0192	0.2134
17-Sep	14	125	0.1120	0.1163	0.0248	0.2078
18-Sep	13	129	0.1008	0.1163	0.0262	0.2064
19-Sep	14	144	0.0972	0.1163	0.0310	0.2016
20-Sep	17	141	0.1206	0.1163	**0.0301**	**0.2025**
Total	**189**	**1625**				
$\bar{u}$	**0.1163**					

The completed *u*-chart is given in Figure 6.92 and demonstrates statistical control.

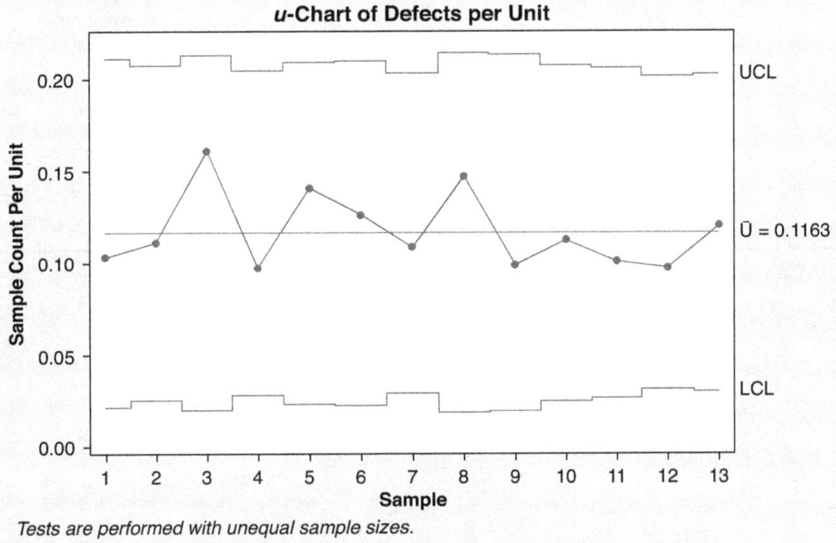

Figure 6.92 *u*-chart for Example 6.79.

Process Capability and Performance

Process capability and performance indices provide a quantitative method for characterizing how well a process is meeting specifications, and by extension, customer requirements. The performance to specifications is measured over a period of stable operations. A process is said to be **capable** when the output conforms to the process specifications to a high degree (Figure 6.93).

However, process capability measures can only predict future performance for stable and predictable processes, those under statistical control. They provide a standardized quantification that can be used to compare process behavior. **Short term or within subgroup capability indices**, denoted $\mathbf{C}_x$, are calculated using short term or within subgroup variation measures as used in control charts. **Long-term or overall performance indices**, denoted $\mathbf{P}_x$, are calculated using the overall estimate of standard deviation *s*. For succinctness, we will refer to these as "C/P" when referring generally to either process capability or performance. There are several versions of both capability and performance that provide insight into different aspects of potential process performance. Higher values are always better. Negative values indicate that the process mean is outside of the specification limits. In general, these indices are obtained by dividing the estimates of the required engineering specification or tolerance range by the full range of natural process variation:

$$\frac{\textit{Engineering Tolerance Range}}{\textit{Natural Tolerance for Process Variation}} = \frac{ET}{NT}$$

Note that "tolerance limits" described here are the engineering specifications and *not* as the statistical tolerance intervals discussed in "Confidence and Tolerance Intervals."

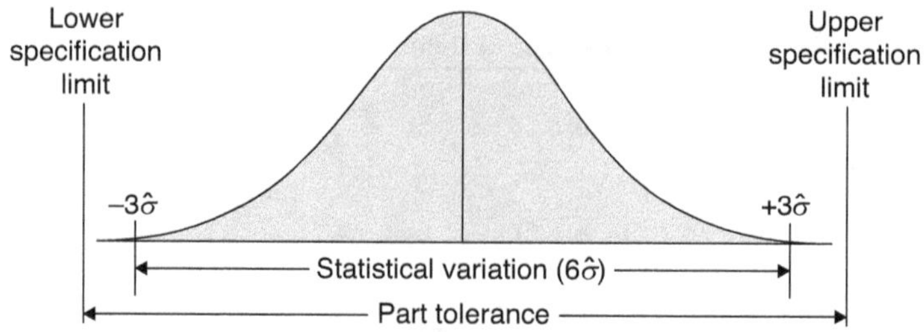

Figure 6.93 Illustration of variation, specification limits, and tolerance range.
Source: Durivage (2014).

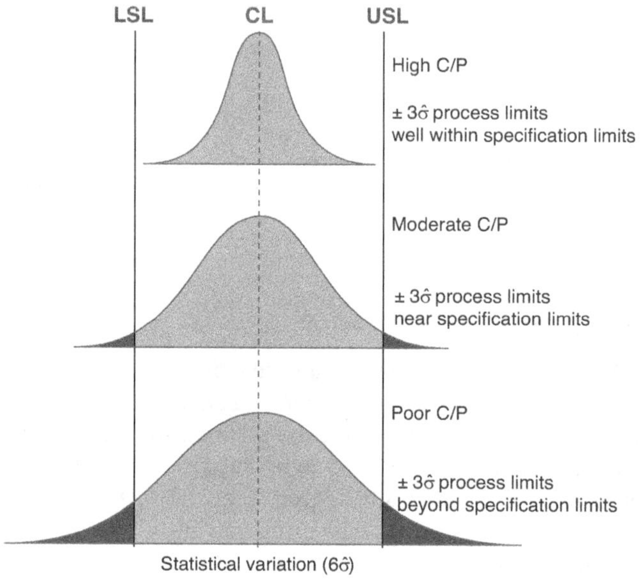

Figure 6.94 Process capability and performance relative to specification limits for centered process.

The chart in Figure 6.94 shows how differences in variation affect the process behavior for a centered process. Centered processes with lower variation are much less likely to generate products that are out of specification.

Consider what happens if processes move off center, which they do (Figure 6.95). An uncentered process has a much higher probability of yielding out-of-specification product.

Capability and Performance (C/P) Indices

Indices are calculated for each individual set of data being monitored. Equations for normally distributed data are shown in Table 6.49 with short-term or within-subgroup sigma estimates provided for control chart formulas. Long-term variation for process performance measures is obtained using the usual standard deviation estimate for the overall dataset. Alternative distributions can be handled by identifying the appropriate percentiles from that distribution to obtain the range for natural process variation.

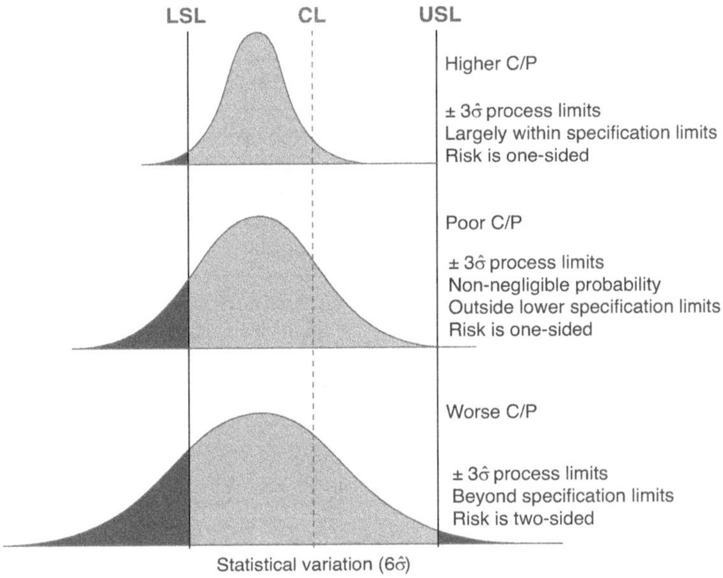

Figure 6.95 Process capability and performance relative to specification limits for uncentered process.

Results for many processes can be aggregated together using a **goal capability chart**, which can also be used to isolate targeting issues from those involving variation. Some organizations use capability metrics to manage their process quality, while others will use sigma.

Note short-term variation can be estimated using $\hat{\sigma} = \dfrac{\overline{mR}}{1.128}$ if using an X-mR chart, $\hat{\sigma} = \dfrac{\overline{R}}{d_2}$ if using $\overline{X} - R$, and $\hat{\sigma} = \dfrac{\overline{s}}{c_4}$ for $\overline{X} - S$.

Interpretation

Returning to the previous examples, we can see how the indices relate to each other for these three general cases (Figure 6.96). The upper side is not shown since would have analogous relationships, just opposite. These relationships would also hold for process performance measures *Px*.

A process with C_{pk} or $P_{pk} = 0$ means that the process is centered on one of the specification limits. A process with negative C_{pk} or P_{pk} means that the process is centered completely outside the limits.

What Is Process Sigma?

Having a "# sigma process" is a statement about process capability or performance. Inverting the capability ratio tells you how many standard deviations, or sigmas, your specification limits are from the engineering limits (Table 6.50). This is also called process sigma.

Some organizations use capability metrics to manage their process quality, while others will use process sigma.

Table 6.49 Equations for process capability and performance indices.

Type		Equation	Description	Notes
Total Potential Process C/P		$\hat{C}_p\left(or\ \hat{P}_p\right)=\frac{USL-LSL}{6\cdot\hat{\sigma}}$	Most optimistic measure, shows what the process is capable of if perfectly centered.	Compares full ET to full NT, does NOT account for centering.
One-Sided	Process C/P Lower Limit	$\hat{C}_{pL}\left(\text{or } \hat{P}_{pL}\right)=\frac{\overline{X}-LSL}{3\cdot\hat{\sigma}}$	Considers lower (left) side only, used as intermediate calculation in overall C/P, also appropriate for processes with only a lower limit.	Accounts for centering, compares half ET to half NT, lower side only.
One-Sided	Process C/P Upper Limit	$\hat{C}_{pU}\left(or\ \hat{P}_{pU}\right)=\frac{USL-\overline{X}}{3\cdot\hat{\sigma}}$	Considers upper (right) side only, used as intermediate calculation in overall C/P, also appropriate for processes with only an upper limit.	Accounts for centering, compares half ET to half NT, lower side only.
Two-Sided Overall Process C/P *Worst Case*		$\hat{C}_{pK}\left(or\ \hat{P}_{pK}\right)=min\left\{\hat{C}_{pL},\hat{C}_{pU}\right\}$ or $\hat{P}_{pK}=min\left\{\hat{P}_{pL},\hat{P}_{pU}\right\}$	Overall C/P is the worst case of the upper and lower C/P, use minimum.	Shows the capability to the specification closest to the mean.

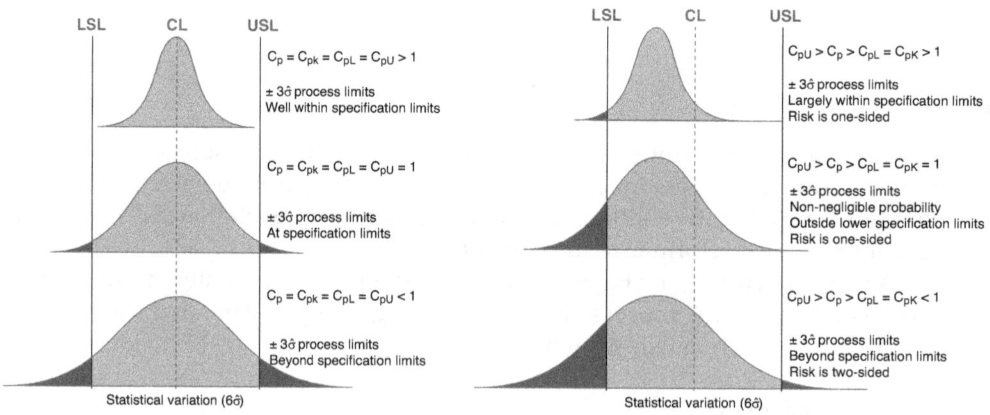

Figure 6.96 Process capability and performance relationships for centered and uncentered processes.

Table 6.50 Process capability relative to process sigma.

C_P	Process sigma
0.25	0.75
0.50	1.50
0.75	2.25
1.00	3.00
1.25	3.75
1.50	4.50
1.75	5.25
2.00	6.00

Process Capability and Reliability

Process capability and performance measures directly link process to quality and to reliability. Although implemented and evaluated at the part and component level, poor process capability has systemic effects on downstream processes, resulting in increased **cost of poor quality** (COPQ), with likely impacts on performance. Higher process capability directly translates to fewer nonconformities, reducing time and resources spent on fire-fighting, and higher potential long-term performance and reliability.

Capability Examples

EXAMPLE 6.80

Process Capability Measures: C_p

Note that natural tolerance and C_p measures both concentrate on the variability of the process output. However, the central tendency of the process output, in addition to its dispersion, can affect how well the output meets specifications. Consider the two process outputs shown in Figure 6.97, each with a known short-term standard deviation equal to ½, and lower specification limit equal to −3 and upper specification limit equal to +3. The specification target, also known as specification nominal, is equal to 0.

We can estimate the C_p value for the process shown in the image on top as:

$$
\begin{aligned}
C_{p1} &= \frac{(UCL-LCL)}{6\sigma} \\
&= \frac{(3-(-3))}{3} \\
&= 2.0
\end{aligned}
$$

Continued

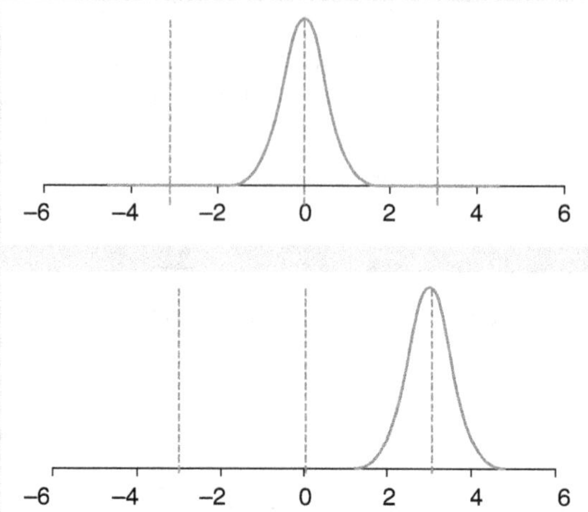

Figure 6.97 Two processes with equal C_p values.
Source: Used with permission from McShane-Vaughn (2021, p. 256).

This process is comparable to a "6 sigma" level process, in which it would produce only 1.9 parts-per-billion defective in the short term. The C_p for the second process is calculated similarly:

$$\begin{aligned}\hat{C}_{p2} &= \frac{(UCL-LCL)}{6\hat{\sigma}} \\ &= \frac{(3-(-3))}{6} \\ &= 2.0\end{aligned}$$

From the diagram, we see that although the second process has a tight standard deviation relative to the specification width, the mean of the process is not centered between specification limits. Rather, it is centered on the upper specification limit, and 50% of its output is nonconforming. Clearly, both the mean and the variation in the process must be considered when assessing capability of a process.

We can calculate the C_{pk} values for the two processes:

$$\begin{aligned}C_{pk1} &= \min(C_{pl1}, C_{pu1}) \\ &= \min\left(\frac{(\mu_1 - LSL)}{3\sigma}, \frac{(USL-\mu_1)}{3\sigma}\right) \\ &= \min\left(\frac{(0-(-3))}{1.5}, \frac{(3-0)}{1.5}\right) \\ &= \min(2.0, 2.0) \\ &= 2.0\end{aligned}$$

For a process that is centered at the specification target, the C_{pk} value will be equal to the C_p.

For the second process, we have:

$$\begin{aligned}C_{pk2} &= \min(C_{pl2}, C_{pu2}) \\ &= \min\left(\frac{(\mu_2 - LSL)}{3\sigma}, \frac{(USL-\mu_2)}{3\sigma}\right)\end{aligned}$$

$$= \min\left(\frac{(3-(-3))}{1.5}, \frac{(3-3)}{1.5}\right)$$
$$= \min(2.0, 0)$$
$$= 0$$

When the process mean is equal to one of the specification limits, the C_{pk} value will equal zero.

CONFIDENCE AND TOLERANCE INTERVALS

Compute confidence intervals and tolerance intervals to analyze reliability with chosen confidence. Assess the use of confidence and tolerance intervals for reliability analysis using Weibull, normal, and lognormal distributions. Describe how point estimates are used to determine the interval. (Evaluate)

Body of Knowledge III.A.7

Interval estimates are widely used throughout statistics as estimation tools, to designate a range of plausible values at the desired confidence level. Intervals are constructed from point estimates and may be one- or two-sided with either or both lower and upper bounds.

Types of Intervals

Classical statistical methods make use of three different types of intervals for purposes of estimation: **confidence intervals**, **prediction intervals**, and **tolerance intervals**. They may appear similar in form and in nature, but they estimate different things and so are not interchangeable. These interval forms are all constructed at a given level of **confidence**, $C = 1 - \alpha$. Bayesian statistical methods also generate outcomes in the form of intervals, but they are produced in a completely different way using computational methods—these are known as **credible intervals** and are beyond the scope of this book. Intervals are an alternative way to view the results of a hypothesis test, where the user looks to see if the interval contains the value of interest, as discussed previously. Intervals can also be generated using computational approaches that involve resampling, such as bootstrapping.

Intervals take on the general form:

$$L < \theta < U$$

where the parameter of interest, θ (whatever it may be), is expected to fall between some lower endpoint, L, and upper endpoint, U, often referred to as "bounds" or "limits." An interval with both lower and upper limits is described as a two-sided interval. Some are symmetric, others are not. One-sided intervals are those with only one bound, for instance, $L < \theta$ and $\theta < U$.

Confidence intervals are constructed around **point estimates** of population parameters, or functions thereof. Examples are the population mean μ or variance σ^2, any proportion p, regression coefficients β_i, or distributional parameters as for the Weibull, defined by β and η. Possible functions include the mean prediction lines, shown here for simple linear regression (SLR) $\hat{Y} = \hat{\beta}_0 + \hat{\beta}_1 \cdot X$, discussed elsewhere in this chapter, or reliability estimates calculated using estimated population parameters. Confidence intervals are a very versatile tool used extensively throughout classical parametric statistics to construct estimates that incorporate the underlying mathematical structure, the sample size, and the empirical variability seen in the data. However, it is important to recognize that a confidence interval around a mean only provides information about the center of the distribution. This interval is not designed to capture the full range of variation seen in the data.

Prediction and *tolerance intervals* are different methods used to account for the greater variability seen in the full sample.

Prediction intervals provide an interval in which a future predicted observation is expected to be. They are constructed similarly to confidence intervals but incorporate an additional source of variability, which results in a wider interval than the confidence interval. Prediction intervals are commonly used in regression analysis but can also be constructed for univariate data.

Tolerance intervals provide an interval that is designed to capture a designated proportion of the population at a given level of confidence. Note that limits widen as both coverage and confidence increase. For example, a team might ideally want coverage of 99% at 99% confidence for maximum protection, but this may produce an impractically wide interval unless the database available is sufficiently large to support estimates that far out in the tail of the distribution.

In reliability engineering, cases are frequently encountered where only one limit may be of interest, but there is still a constraint on the other end that produces an additional bound. This occurs when estimating proportions near zero or one, such as reliability, R, since usually only the lower endpoint is of interest for estimating R, but there is still an effective upper bound of 1.0 since mathematically reliability is constrained and cannot exceed that value. Conversely, when estimating probabilities of rare events, such as failures $(1 - R)$, only the upper limit may be of interest, but the lower bound cannot mathematically go below zero.

Two engineering applications where confidence and tolerance intervals are embedded are in the development of probability of detection (POD) curves in the field of nondestructive evaluation/test (NDE/T), and in the definition of material design allowables for metallic material properties, known as "A-basis" and "B-basis."

In NDE/T, for example for fluorescent penetrant inspection (FPI), individual inspectors are qualified and their capabilities certified that they can identify flaws down to a given size. This is administered by having them evaluate a set of test specimens of different sizes, the results of which are then analyzed and used to generate a POD curve. This POD curve is a one-sided confidence interval around a logistic regression, a method used to analyze binary data, in this case pass/fail for each individual flaw. Their score is referred to as an "a90/95" (read as "a 90 95," denoting the flaw length a), indicating the indication length they can detect with 90% coverage at 95% confidence.

A-basis and B-basis values are one-sided tolerance intervals defined in the reference text *Metallic Materials Properties Development and Standardization* (MMPDS), formerly known as MIL-HDBK-5J, a product of the U.S. government (USG) Department of Defense (DOD) entitled *Metallic Materials and Elements for Aerospace Vehicle Structures*. These are strength values below which 1% and 10% of the population are expected to fall at 95% confidence. Upper limits are not of particular interest for strength since the primary concern is failure occurring due to strength being too low, or insufficient.

Understanding Confidence

Confidence is a critical concept in statistics, and it applies to the construction of confidence, prediction, and tolerance intervals. We will explain the concept first in the context of confidence intervals.

In hypothesis testing terms, *confidence*, $C = 1 - \alpha$, is the probability of accepting the null hypothesis, given that the null hypothesis is true. The risk denoted as α is known as the significance. See "Basic Statistics" for prior discussion. Common values of C are 90%, 95%, and 99%, with 50% and 80% occasionally quoted in special cases. Confidence is preselected for use in an analysis.

Confidence intervals are used to mathematically characterize the uncertainty associated with the method used to obtain the estimate. A statistical confidence interval incorporating a given risk, α, can be placed about the point estimate of any population parameter made from a sample taken from the population. The statement can then be made that the true parameter of the population is within that interval with a confidence of $1 - \alpha$.

Higher confidence values are analogous to using a wider bracket to capture a target, as in the game of horseshoes. Imagining that the true mean is the stick, using a lower confidence value equates to using a narrower horseshoe, with higher confidence values increasing the width of the horseshoe. To achieve higher confidence, a wider interval is needed to be more certain that we can capture the truth. The width of the interval is also affected by sample size and the variability in the data. Larger samples will decrease the width of the interval (tighten) while smaller samples will be wider. All other things equal, a sample with lower variability will be tighter than one with higher variability.

Laman (2022) describes the mechanism for how confidence intervals work:

> The probabilistic interpretation of the confidence level is the proportion of all confidence intervals constructed on that parameter (under repeated sampling and identical conditions) that would contain the true parameter. For example, a 95% confidence interval on μ would be interpreted as the percentage of all confidence intervals constructed (under repeated sampling and identical conditions) that would contain the population mean μ.

Consider a normal population with a mean $\mu = 25$ and variance $\sigma^2 = 9$ from which 20 samples of size $n = 40$ are selected. Figure 6.98 shows the twenty 95% confidence intervals constructed for μ. We see that 19 of the 20 (or 95%) of the

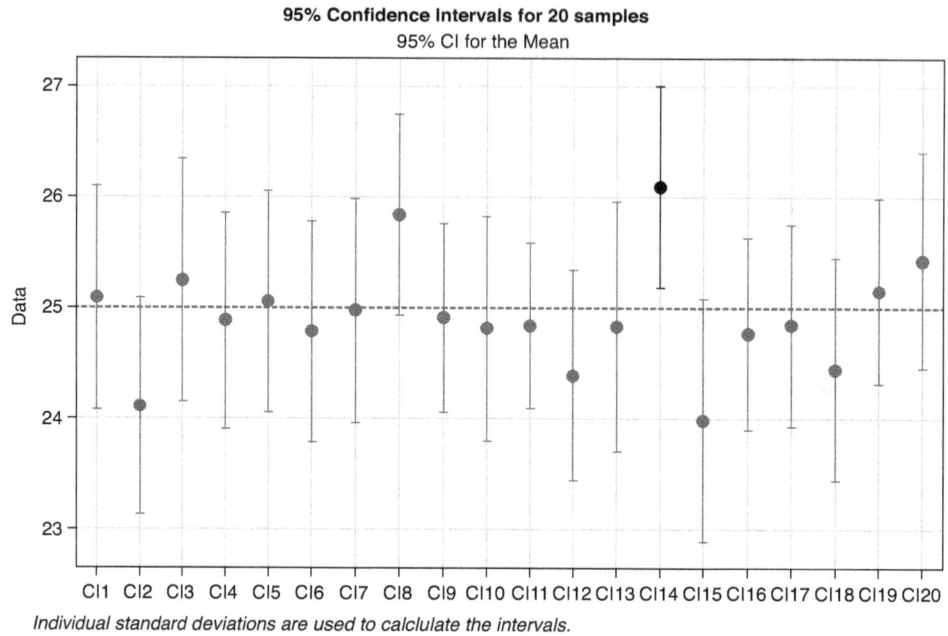

Figure 6.98 Plot of 95% confidence intervals on the mean for each of 20 samples from the same distribution.

confidence intervals contain the true population mean $\mu = 25$, which aligns with expectations.

The empirical rule for the normal distribution, refer to "Probability Distributions," states that 99.73% of this population would be expected to be between 16 and 34 since $\mu \pm 3 \cdot \sigma \rightarrow 25 \pm 3 \cdot \sqrt{9} \rightarrow 25 \pm 3 \cdot 3 \rightarrow 25 \pm 9 \rightarrow [16, 34]$. Note the difference in scale on the y-axis, which only ranges from 23 to 27, or 25 ± 2. The confidence intervals around the mean are narrow compared with the overall population. The width of the confidence interval is driven by the sample size: More data results in a narrower confidence interval.

Interval Construction Concepts

Confidence Intervals

Suppose an estimate is needed for the mean coating thickness of a population of 1000 circuit boards received from a supplier. Rather than measuring all 1000 boards, a random sample of 40 is selected for measurement. The sample mean of the 40 values is 0.003 inches. Based on this sample, the estimate for the mean coating thickness for the entire lot of 1000 boards is 0.003 inches. This value is called the *point estimate.*

A *point estimate* is a *statistic* used to estimate a *parameter* of a population, a concept that was introduced earlier in this chapter. Examples of point estimates include the sample mean, the sample variance, and sample proportion defective. Point estimates are just as the name implies—single points on the number line. Consider the average dimension discussed previously: $\bar{X} = 0.003\ in.$ While it is highly unlikely that this value is exactly equal to the true mean, due to

Hypothesis Test $t^* = \dfrac{(\theta - \theta_0)}{Std\ err\,(\theta)}$, *Reject for* $|t^*| > t_{crit,\ df,\ \alpha/2}$

Confidence Interval $(\theta - \theta_0) \pm t_{crit,\ df,\ \alpha/2}\ std\ err(\theta)$

Figure 6.99 Formulaic connections between hypothesis testing and confidence intervals.

sampling error, we have no information on how close this point estimate actually is to μ.

We can provide more information beyond the point estimate by calculating a *confidence interval* on the population parameter μ. A confidence interval is therefore a range of values that combines information from a point estimate, the variability of the data, and the sample size to give a more complete estimate of a population parameter.

Confidence intervals are constructed from the identical information used to produce hypothesis tests, except that the information is combined in a way that makes a more general statement about the range of plausible values (Figure 6.99).

In hypothesis testing, a test statistic in the form of a standardized ratio is calculated that is compared against a critical value to determine significance, in which case a binary decision is made. Confidence intervals take the same elements but combine them in a different way. The numerator of the test statistic provides the point estimate for the parameter, the nominal or best guess based on available data. The standard error in the denominator is then multiplied by the critical value indicating the threshold associated with the appropriate confidence level, which is then added to and/or subtracted from the point estimate. This yields upper and lower limits where the values represent exactly that point at which the hypothesis test conclusion would flip from accepting to rejecting the null hypothesis. An equivalent test can be done by looking at the confidence interval and determining whether it contains the hypothesized value θ_0, which is often but not necessarily zero.

The type of hypothesis test and confidence interval described previously is known as a Wald test, one of several families of what are known as **asymptotic** tests that provide approximations obtained using parametric solutions and perform best with larger datasets. **Exact** tests are those obtained using the underlying distribution, typically referring to nonparametric tests. *Exact intervals* may be obtained in some cases, such as for binomial proportions, and will be discussed subsequently.

Tolerance Intervals

A tolerance interval is designed to capture a designated proportion of the population at a given level of confidence. So, in addition to specifying a confidence level, we must also specify a *coverage proportion*, which is the proportion of the population that we would like the interval to contain. The coverage proportion is often specified as a percentage. For example, we may want to cover 99% of the population with 95% confidence.

For tolerance intervals, both parametric and nonparametric methods are discussed here for obtaining the upper and lower tolerance limits (UTL and LTL).

Note that estimates should be treated with caution unless the distribution is well characterized due to the extrapolation required to achieve high coverage and high confidence.

Tolerance Intervals—Nonparametric

Nonparametric methods for estimating tolerance intervals use order statistics. Depending on the confidence, coverage proportion, and sample size, the nonparametric LTL and UTL take on values of the minimum and maximum values, that is, the first-order statistics from either end. With larger sample sizes, the bounds may be set at the second-order statistics, or third, and so on. Higher confidence and higher coverage require more data. Smaller datasets are unlikely to be capable of achieving the desired confidence levels.

Tolerance Intervals—Normal and Lognormal

The tolerance interval for a normally distributed variable is symmetric and looks similar to a confidence interval for the mean but is calculated differently. The UTL and LTL are given by:

$$\bar{X} \pm K \cdot s$$

where $\bar{X}$ and s are the sample mean and standard deviation, respectively. The "K-factor" is defined by sample size n, confidence level $C = 1 - \alpha$, and the desired coverage proportion. Tabulated values for the normal distribution are found in Appendix H.1-H.2.

For lognormal data, ln(X) is distributed normally, and tolerance intervals can be obtained by using the process described subsequently for normal data after being transformed. Note different software and codes have different language and conventions for natural log, base e, versus log base 10. The lognormal distribution is well known as having particularly heavy tails, which results in particularly high upper tolerance limits. Estimates should be used with caution or with a large database to provide the rationale that the distributional fit is appropriate.

Tolerance Intervals—Other Nonnormal (Weibull, Exponential)

For Weibull data, ln(X) is distributed according to the smallest extreme value (SEV) distribution. Tolerance intervals can be obtained by using the process described subsequently for SEV data after being transformed (data after applying natural log, ln). The SEV distribution is outside the scope of this text. Tolerance intervals for SEV and Weibull data may be obtained in statistical software. The exponential is a special case of the Weibull, with $\beta = 1$.

Intervals for Proportions (p and $p_1 - p_2$)

Intervals for proportions are of particular interest for reliability applications, since we often are interested in estimating the probability of success or failure. Both conventional asymptotic methods are shown here, as well as the exact methods needed for low probability events and handling zero failure outcomes. Recall that standard asymptotic methods are not applicable for extreme p (near either 0 or 1) since limits may go above 1 or below 0.

Table 6.51 Equations for confidence intervals for proportions.

<table>
<tr><th>Parameter</th><th>Method</th><th colspan="2">Equations for 100 $(1-\alpha)$% confidence limits</th></tr>
<tr><td rowspan="4">p</td><td rowspan="2">Normal approximation
use when $X \ge 10$ and $(n-X) \ge 10$</td><td>Point Estimate:
$\hat{p}=\frac{x}{n}$</td><td>Two-Sided Limits:
$\hat{p} \pm z_{\alpha/2} \cdot \sqrt{\frac{\hat{p}\cdot(1-\hat{p})}{n}}$</td></tr>
<tr><td>One-Sided Lower Limit:
$LL: \hat{p}-z_{\alpha} \cdot \sqrt{\frac{\hat{p}\cdot(1-\hat{p})}{n}}$
Note upper boundary at 1, intervals may be written (LL, 1]</td><td>One-Sided Upper Limit:
$UL: \hat{p}+z_{\alpha} \cdot \sqrt{\frac{\hat{p}\cdot(1-\hat{p})}{n}}$
Note lower boundary at 0, intervals may be written [0, UL)</td></tr>
<tr><td rowspan="2">Exact
Clopper-Pearson Beta-Binomial Equation</td><td>Point Estimate:
$\hat{p}=\frac{x}{n}$ Note: The exact intervals are not centered at $\hat{p}$.</td><td>Two-Sided Limits:
Use both equations below, but with $\alpha/2$ instead of α.
Note: UL and LL have different degrees of freedom.</td></tr>
<tr><td colspan="2">One-Sided Upper and Lower Limits:
$UL \,\&\, LL: \frac{v_1 \cdot F_{\alpha,v_1,v_2}}{v_2+v_1 \cdot F_{\alpha,v_1,v_2}}, \textit{ where}$
$UL: v_1=2\cdot(x+1) \textit{ and } v_2=2\cdot(n-x)$
$LL: v_1=2\cdot x \textit{ and } v_2=2\cdot(n-x+1)$
Note: Upper boundary at 1, intervals may be written (LL, 1]
Note: Lower boundary at 0, intervals may be written [0, UL)</td></tr>
</table>

To model a proportion, data are typically assumed to be an independent random sample from a binomial distribution, implying a fixed number of trials, two mutually exclusive outcomes, with constant probability of success for all trials. One- and two-sided confidence interval equations for p and p_1-p_2 are organized in Tables 6.51 and 6.52 for both conventional and exact methods. **Note that two-sided limits allocate $\alpha/2$ to each side, whereas one-sided limits retain the full α on that side.**

Exact Binomial for Zero Failures: The Rule of 3 and Obtaining Lower Limits on Reliability

In this chapter, "Sampling Plans for Statistics and Reliability Testing," introduced the Rule of 3 to estimate sample size when zero events were expected, which assumes the binomial. That equation has inverse forms that may be used to provide an approximate confidence interval for p:

$$\left(0,\frac{3}{n}\right) \text{ for small } p\text{, and } \left(1-\frac{3}{n},1\right) \text{ for high } p.$$

A more general form of this equation can be obtained and is used for success testing in reliability. If there are no failures during a test, a point estimate of the

Table 6.52 Equations for confidence intervals for the difference between two proportions.

<table>
<tr><th>Parameter</th><th>Method</th><th colspan="3">Equations for 100 (1 – α)% confidence limits</th></tr>
<tr><td rowspan="4">$p_1 - p_2$</td><td rowspan="3">Normal approximation use when $X \geq 10$ and $(n - X) \geq 10$</td><td rowspan="2">Point estimate:
$\hat{p}_1 - \hat{p}_2 = \frac{x_1}{n_1} - \frac{x_2}{n_2},$
where
$\hat{p}_1 = \frac{x_1}{n_1}$ *and* $\hat{p}_2 = \frac{x_2}{n_2}$</td><td colspan="2">Two-sided limits:
$(\hat{p}_1 - \hat{p}_2) \pm z_{\alpha/2} \cdot \sqrt{\frac{\hat{p}_1 \cdot (1 - \hat{p}_1)}{n_1} + \frac{\hat{p}_2 \cdot (1 - \hat{p}_2)}{n_2}}$</td></tr>
<tr><td>The Agresti-Caffo method gets added robustness by using a variation on the point estimate:</td><td>$\hat{p}_{AC} = \frac{x + 1}{n + 2}$</td></tr>
<tr><td colspan="3">One-sided upper and lower limits:
$UCL: (\hat{p}_1 - \hat{p}_2) + z_\alpha \cdot \sqrt{\frac{\hat{p}_1 \cdot (1 - \hat{p}_1)}{n_1} + \frac{\hat{p}_2 \cdot (1 - \hat{p}_2)}{n_2}}$
$LCL: (\hat{p}_1 - \hat{p}_2) - z_\alpha \cdot \sqrt{\frac{\hat{p}_1 \cdot (1 - \hat{p}_1)}{n_1} + \frac{\hat{p}_2 \cdot (1 - \hat{p}_2)}{n_2}}$
Note: Upper boundary at 1, intervals may be written (LL, 1]
Note: Lower boundary at 0, intervals may be written [0, UL)</td></tr>
<tr><td>Fisher's exact test assumes hypergeometric</td><td colspan="3">This scenario is analogous to a 2×2 contingency table, for which there are several options for analysis depending on the specific comparison of interest. Fisher's exact test is one option, but it is computationally intensive, and results are not often converted into a confidence interval form. Fisher assumes the hypergeometric (refer to Chapter 6, "Probability Distributions") and calculates the probability of seeing a table as or more extreme than the table observed.</td></tr>
</table>

proportion $\hat{p}$ cannot be made. However, a lower confidence bound can be calculated. The lower $1 - \alpha$ confidence bound for a no-failure test is given as:

$$R_L = \alpha^{1/n}$$

Exact Binomial for Obtaining Lower Limits on Reliability

The following equation can be used to obtain a lower confidence bound on reliability:

$$R_L = \frac{(n - r)}{(n - r) + (r + 1) \cdot F_{\alpha, v_1, v_2}}$$

where $v_1 = 2 \cdot (r + 1)$ and $v_2 = 2 \cdot (n - r)$. This is a simplified form obtained by calculating an upper bound on the estimate for probability of failure, then taking the complement. *F* tables are in Appendix F. There are alternative ways to write this equation, but it derives from the Clopper-Pearson exact equation shown in Table 6.51.

EXAMPLE 6.81

Confidence Interval for One Proportion

A quality inspector tests a random sample of 50 assemblies and finds that 10 are defective. Calculate a 95% confidence interval for the true proportion defective. Here, we have $\alpha = 0.05$, $x = 10$, and $n = 50$. The estimated proportion is calculated to be

$$\hat{p} = \frac{x}{n} = \frac{10}{50} = 0.20$$

Checking assumptions, we note that we have a random sample, there is a fixed number of trials, each assembly is classified as defective or nondefective, the trials are independent, and there is a constant probability of finding a defective on each trial. Furthermore, we see that both X and $n-X$ are greater than or equal to 10.

$$n \cdot p = 50 \cdot (0.20) = 10$$
$$n \cdot (1-p) = 50 \cdot (0.80) = 40$$

We will thus use the normal approximation method formula from Table 6.51. From Appendix D, we find $z_{\alpha/2} = 1.96$.

$$\hat{p} \pm z_{\alpha/2} \cdot \sqrt{\frac{\hat{p} \cdot (1-\hat{p})}{n}}$$
$$0.20 \pm 1.96 \cdot \sqrt{\frac{(0.20) \cdot (0.80)}{50}}$$
$$0.20 \pm 0.111$$
$$[0.089, 0.311]$$

Thus, we are 95% confident that the interval [0.089, 0.311] covers the true proportion defective.

EXAMPLE 6.82

Confidence Interval for the Difference in Two Proportions

A heavy equipment manufacturer tracks the percentage of warranty claims for its dump trucks each quarter. Last quarter, a random sample of 225 vehicle records showed that sales Region A had 25 warranty claims. In Region B, 200 records were sampled, and 10 had claims. Construct a 95% confidence interval to determine if there is a significant difference in warranty claims between the two sales regions.

Here, we have $\alpha = 0.05$, $X_1 = 25$, $X_2 = 10$, $n_1 = 255$, and $n_2 = 200$. The estimated proportions are calculated to be

$$\hat{p}_1 = \frac{X_1}{n_1} = \frac{25}{255} = 0.098 \quad and \quad \hat{p}_2 = \frac{X_2}{n_2} = \frac{10}{200} = 0.050$$

Checking assumptions, we note that we have random samples, there is a fixed number of trials, and each truck either does or doesn't have a warranty claim. We assume the status of each vehicle is independent, and that there is a constant probability of a warranty claim for each vehicle. Furthermore, we see that both

Continued

X and $n-X$ are greater or equal to 10 for each region. Therefore, we can use the normal approximation method. From Appendix D, we know that $z_{0.025}=1.96$.

We will use the following formula from Table 6.52:

$$(\hat{p}_1-\hat{p}_2)\pm z_{\alpha/2}\cdot\sqrt{\frac{\hat{p}_1\cdot(1-\hat{p}_1)}{n_1}+\frac{\hat{p}_2\cdot(1-\hat{p}_2)}{n_2}}$$

$$(0.098-0.050)\pm 1.96\cdot\sqrt{\frac{0.098\cdot(0.902)}{255}+\frac{0.05\cdot(0.95)}{200}}$$

$$[0.00063, 0.09537]$$

We are 95% confident that the difference in the population proportions is covered by the interval [0.00063, 0.09537]. Because this interval does not include the value 0, we conclude that Region A had a higher rate of warranty claims.

EXAMPLE 6.83

LCL for Reliability with Zero Failures

Thirty sensors were tested with no failures. Calculate the lower 90% confidence limit for reliability:

$$R_L=\alpha^{1/n}=(0.10)^{1/30}=0.926$$

The reliability of the sensor is greater than 0.926 at 90% confidence.

EXAMPLE 6.84

LCL for Reliability with Nonzero Failures

Thirty-two sensors were tested with *two* failures. Calculate the reliability point estimate and the lower 90% confidence limit for the reliability of the sensors:

Inputs are $n=32$, $r=2$, $n-r=30$, and $\alpha=0.10$.

$v_1=2\cdot(2+1)=6$ and $v_2=2\cdot(32-2)=60$.

$$F_{\alpha,v_1,v_2}=F_{0.1,6,60}=1.87$$

$$R_L=\frac{n-r}{(n-r)+(r+1)}=\frac{30}{(30+3\cdot 1.87)}=0.842$$

The reliability of the sensor is therefore greater than 0.842 at 90% confidence.

Intervals Centered on Means (μ and $\mu_1-\mu_2$)

Table 6.53 shows generic equations for confidence intervals that apply broadly to estimates for means, regardless of the underlying distribution of the data. Data are assumed to be an independent identically distributed (*iid*) random sample. The standard deviation is genericized as $\hat{\sigma}$, which may take on different values. If σ is known, then $\hat{\sigma}=\sigma$, otherwise use $\hat{\sigma}=s$, the sample standard deviation.

Table 6.53 Equations for confidence intervals for a mean and the difference δ between two means.

Parameter	Method	Equations for 100 $(1-\alpha)$% confidence limits	
μ	*Wald t-test* If σ is known, then use $\hat{\sigma}=\sigma$ and $t=Z$, otherwise use sample standard deviation $\hat{\sigma}=s$.	Point estimate: $\hat{\mu}=\bar{X}=\frac{1}{n}\cdot\sum_{i=1}^{n}x_i$	Two-sided limits: $\bar{X}\pm t_{\alpha/2,n-1}\cdot\sqrt{\frac{\hat{\sigma}^2}{n}}$
		One-sided lower limit: $LCL:\bar{X}-t_{\alpha,n-1}\cdot\sqrt{\frac{\hat{\sigma}^2}{n}}$	One-sided upper limit: $UCL:\bar{X}+t_{\alpha,n-1}\cdot\sqrt{\frac{\hat{\sigma}^2}{n}}$
$\delta=\mu_1-\mu_2$	*Wald t-test* If σ_i is known, then use $\hat{\sigma}=\sigma_i$ and $t=Z$, otherwise use sample standard deviations $\hat{\sigma}=s_i$.	Point estimate: $(\hat{\mu}_1-\hat{\mu}_2)=(\bar{X}_1-\bar{X}_2)$ $=\frac{1}{n_1}\cdot\sum_{i=1}^{n_1}x_{1i}-\frac{1}{n_2}\cdot\sum_{i=1}^{n_2}x_{2i}$	Two-sided limits: Use both equations below, but with $\alpha/2$ instead of α.
	Note $t\rightarrow Z$ as n increases to 30+ Pool variations if unknown, $\hat{\sigma}=s_p$. *Welch-Satterthwaite* $\nu=\frac{\left(\frac{s_1^2}{n_1}+\frac{s_2^2}{n_2}\right)^2}{\frac{\left(s_1^2/n_1\right)^2}{n_1-1}+\frac{\left(s_1^2/n_2\right)^2}{n_2-1}}$	One-sided upper and lower limits: $UCL:\ (\bar{X}_1-\bar{X}_2)+t_{\alpha,n_1+n_2-2}\cdot\sqrt{\frac{\hat{\sigma}_1^2}{n_1}+\frac{\hat{\sigma}_2^2}{n_2}}\Rightarrow +t\cdot\hat{\sigma}\sqrt{\frac{1}{n_1}+\frac{1}{n_2}}$ $LCL:(\bar{X}_1-\bar{X}_2)-t_{\alpha,n_1+n_2-2}\cdot\sqrt{\frac{\hat{\sigma}_1^2}{n_1}+\frac{\hat{\sigma}_2^2}{n_2}}\Rightarrow -t\cdot\hat{\sigma}\sqrt{\frac{1}{n_1}+\frac{1}{n_2}}$ Note that these equations simplify if $\sigma_1=\sigma_2=\sigma$, meaning that both groups have the same variation. If standard deviations are unknown but assumed equal, then the sample standard deviations may be pooled to obtain an aggregate estimate which may be used for $\hat{\sigma}$: $\hat{\sigma}=s_p=\sqrt{\frac{(n_1-1)\cdot s_1^2+(n_2-1)\cdot s_2^2}{n_1+n_2-2}}$ If $\sigma_1\neq\sigma_2$, then separate $\hat{\sigma}_1$ and $\hat{\sigma}_2$ estimates should be used with the Welch-Satterthwaite adjustment for degrees of freedom.	

If σ is known, then Z may be used instead of t. However, generally speaking, σ should not be assumed to be known, and users should default to t and $\hat{\sigma}=s$ for general use. **Note that two-sided limits allocate $\alpha/2$ to each side, whereas one-sided limits retain the full α on that side.**

What follows are examples that illustrate intervals centered on the mean, as well as examples that demonstrate the use of tolerance intervals in other situations.

EXAMPLE 6.85

Confidence Interval for a Mean

Suppose an estimate is needed for the average coating thickness for a population of 1000 circuit boards received from a supplier. Rather than measure the coating thickness on all 1000 boards, one might randomly select a sample of 40 for measurement. Suppose that the average coating thickness on these 40 boards is 0.003, and the standard deviation estimated from the sample is 0.0005. Calculate a 90% confidence interval for the true mean.

We have $\bar{X} = 0.003$, $s = 0.0005$, and $n = 40$. From Appendix E, we know that $t_{39,\ 0.10/2} = 1.685$.

We use the following formula from Table 6.53 to calculate the confidence interval:

$$\bar{X}_{\pm t_{\alpha/2,n-1}} \cdot \frac{s}{\sqrt{n}}$$

$$0.003 \pm 1.685 \cdot \frac{0.0005}{\sqrt{40}}$$

$$0.003 \pm 0.00013$$

$$0.0029 \leq \mu < 0.0031$$

Thus, we are 95% confident that the true mean coating thickness is covered by the interval [0.0029, 0.0031].

EXAMPLE 6.86

Confidence Interval for the Difference Between two Means, Known Sigmas

A reliability engineer conducts a study to determine if there is a difference in the mean weight of packages being produced on two different assembly lines. The weights from each line have been shown to follow normal distributions with known variances of 85.5 and 69.2, respectively. Twenty samples are randomly selected from Line 1, yielding a sample mean equal to 105.2 oz. Fifteen samples are randomly selected from Line 2, resulting in a sample mean equal to 99.7 oz.

Construct a 95% confidence interval to determine if there is a significant difference in mean weights. Here we have $\bar{X}_1 = 105.2$, $\bar{X}_2 = 99.7$, $\sigma_1^2 = 85.5$, $\sigma_2^2 = 69.2$, $n_1 = 20$, $n_2 = 15$, and $\alpha = 0.05$. Because we can assume normal distributions with known but unequal variances, we can use the following formula from Table 6.53. Using $\alpha/2 = 0.025$, we find $z_{0.025} = 1.96$ from Appendix D. Substituting in the values, we obtain:

$$\left(\bar{X}_1 - \bar{X}_2\right) \pm z_{\alpha/2}\sqrt{\frac{\sigma_1^2}{n_1} + \frac{\sigma_2^2}{n_2}}$$

$$(105.2 - 99.7) \pm 1.96\sqrt{\frac{85.5}{20} + \frac{69.2}{15}}$$

$$5.5 \pm 5.843$$

We are 95% confident that the true difference in mean weights is covered by the interval [−0.343, 11.343]. Because the interval includes the value zero, we can conclude there is insufficient evidence to claim that the mean package weights produced on the two lines are different. Remember that this does not prove that the null is true.

EXAMPLE 6.87

Confidence Interval for Difference Between two Means, Unknown Sigmas

The training department at a tooling firm designed and delivered a new program to improve the technical skills of its customer service representatives. To determine the efficacy of the training, the department will test whether the mean wait time for customers calling in to the help line has decreased. Before the training program, sample data showed that customers waited an average of 10.9 minutes before being helped. After training, the customer mean wait time was 6.8 minutes. Details on the samples are shown in Table 6.54. Calculate a 95% one-sided confidence interval on the difference in the mean wait times to determine if the training was effective.

Table 6.54 Wait time data for Example 6.87.

	Baseline	After training
n	40	35
$\bar{X}$	10.9	6.8
S	10.5	6.4

Even though wait times often follow a skewed distribution such as an exponential, the central limit theorem allows us to make comparisons between means, especially with the larger sample sizes available here, using the *t* distribution equations in Table 6.53.

Here, we will be calculating a one-sided confidence interval because the training department wants to know whether the mean wait time after the training is less than the mean time at baseline. We will calculate a lower bound on the difference, and if the confidence interval does not include the value 0, we will conclude that the training did significantly decrease the mean wait times.

For this problem, we will *not* assume that the population variances are equal. From Table 6.53 we will use the following formulas. We first calculate the degrees of freedom, *n*, for the *t* statistic:

$$LB: (\bar{X}_1 - \bar{X}_2) - t_{\alpha,v}\sqrt{\frac{s_1^2}{n_1} + \frac{s_2^2}{n_2}}$$

where

$$v = \frac{\left(\frac{s_1^2}{n_1} + \frac{s_2^2}{n_2}\right)^2}{\frac{(s_1^2 / n_1)^2}{n_1 - 1} + \frac{(s_2^2 / n_2)^2}{n_2 - 1}}$$

$$v = \frac{\left(\frac{10.5^2}{40} + \frac{6.4^2}{35}\right)^2}{\frac{\left(10.5^2/40\right)^2}{39} + \frac{\left(6.4^2/35\right)^2}{34}}$$

$$= \lfloor 65.59 = 65$$

Continued

We use a floor function for the degrees of freedom, meaning that we round any noninteger result down to the next lowest integer. Here, we obtain $n=65$.

The α value for our one-sided interval is 0.05. Using Appendix E, we find $t_{0.05,65}=1.669$. The one-sided confidence interval for the difference in means is calculated:

$$\left(\bar{X}_1-\bar{X}_2\right)-t_{\alpha,v}\sqrt{\frac{s_1^2}{n_1}+\frac{s_2^2}{n_2}}$$

$$(10.9-6.8)-1.669\sqrt{\frac{10.5^2}{40}+\frac{6.4^2}{35}}$$

$$[0.793,\infty)$$

We are 95% confident that the reduction in mean wait times is at least 0.793 minutes, or about 48 seconds. We conclude that the training program was effective in reducing mean wait times for customers.

EXAMPLE 6.88

Confidence Interval Versus Tolerance Interval for 99% Coverage at 95% Confidence—Normal

A process characteristic follows a normal distribution. Measurements taken on a 15-piece random sample result in a sample mean equal to 10.821 and sample standard deviation of 0.027. Find the two-sided confidence interval for the mean and the two-sided tolerance interval for 99% coverage probability, both at 95% confidence.

Here, $\bar{X}=10.821$, $s=0.027$, and $n=15$. The confidence interval for the mean is calculated as:

$$\bar{X}\pm t_{\alpha/2,n-1}\cdot\frac{s}{\sqrt{n}}$$

$$10.821\pm 2.145\cdot\frac{0.027}{\sqrt{15}}$$

$$10.821\pm 0.015$$

$$10.806\le\mu\le 10.836$$

For the tolerance interval, the confidence level is 95%, and the coverage proportion is 99%, and the sample size is 15. Thus, the tolerance factor K is 3.885 from the two-sided tolerance factor table in Appendix H.2. Using the formulas given, we calculate:

$$\begin{aligned}[LTL,UTL]&\Rightarrow\bar{X}\pm K\cdot s\\&\Rightarrow 10.821\pm 3.885\cdot 0.027\\&\Rightarrow 10.821\pm 0.105\\&\Rightarrow[10.716,10.926]\end{aligned}$$

Given these two calculations, we are 95% confident that the interval [10.806, 10.836] will cover the true mean of the population, and we are 95% confident that the interval [10.716, 10.926] will cover 99% of the population. Note that the confidence interval is much narrower than the tolerance interval. The width of the confidence interval for the mean is $2\cdot 0.015=0.030$ and the width of the tolerance interval is $2\cdot 0.105=0.210$, which is 7 times as wide.

EXAMPLE 6.89

Nonparametric Tolerance Interval for 99% Coverage at 95% Confidence

The difference in maximum stress for the 35 observations in the annealed versus as-processed data discussed in several other examples are shown here:

1.56, 1.55, 0.47, −0.33, 0.13, 1.44, 1.02, 0.26, −0.03, 0.33, 0.52, −0.93, 0.58, 1.36, −0.01, −0.16, 0.67, −0.34, −0.35, −0.45, −0.10, 0.10, −0.04, 0.33, 0.85, 0.75, 1.19, −0.33, 0.84, 0.56, 0.91, −0.22, 0.67, 0.06, −1.12

Find a nonparametric tolerance interval for 99% coverage proportion at 95% confidence, and find a nonparametric tolerance interval for 90% coverage proportion at 95% confidence.

Data range from a minimum of −1.12 to maximum of 1.56. A nonparametric tolerance interval for these data will be the minimum to the maximum and LTL=−1,12, UTL=1.56 (Figures 6.100 through 6.102). Note that for the 99% coverage proportion, this amount of data achieves only 4.8% confidence, which is much less than the 95% confidence desired. For the 90% coverage proportion, this same interval achieves 87.8% confidence, still missing the target of 95%. If we then try the one-sided nonparametric upper bound on the difference with coverage probability of 90%, UTL=1.56 achieves 97.5% confidence, which exceeds the target.

95% Tolerance Interval

Variable	Nonparametric Method	Achieved Confidence
Difference	(−1.120, 1.560)	4.79%

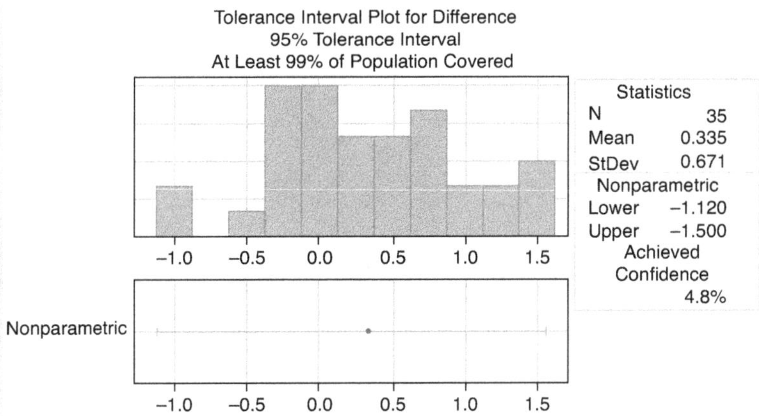

Figure 6.100 Nonparametric two-sided tolerance interval for annealing data (99% coverage).

Continued

95% Tolerance Interval

Variable	Nonparametric Method	Achieved Confidence
Difference	(–1.120, 1.560)	87.76%

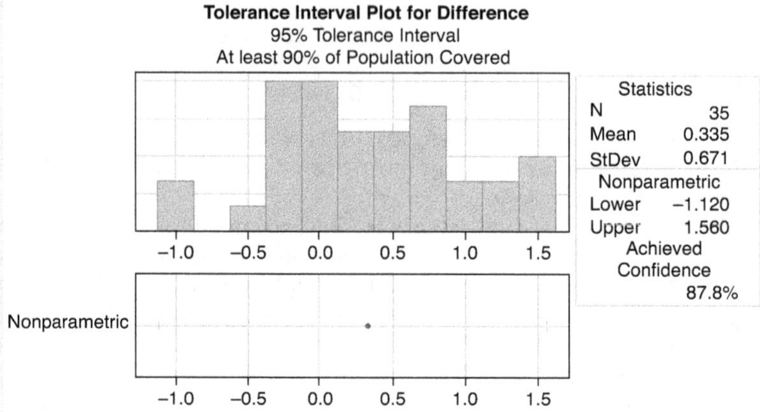

Figure 6.101 Nonparametric two-sided tolerance interval for annealing data (90% coverage).

95% Upper Tolerance Bound

Variable	Nonparametric Method	Achieved Confidence
Difference	1.560	97.50%

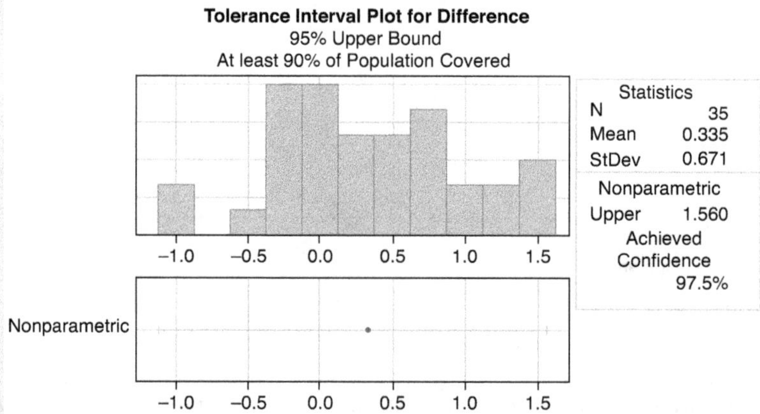

Figure 6.102 Nonparametric one-sided upper tolerance bound for annealing data (90% coverage).

EXAMPLE 6.90

Lognormal Tolerance Interval for 95% Coverage at 95% Confidence—Complete Data

The cycles to failure for a product are known to be lognormally distributed. The times to failure for a sample of six parts are 850, 925, 1250, 1550, 1800, and 2750 cycles (Figure 6.103). Estimate the mean and sigma parameters and use to obtain a tolerance interval for 95% of the population at 95% confidence.

Mean is 1521 cycles and standard deviation 702.6 cycles. Notice that the non-parametric tolerance interval is minimum to maximum: 850 to 2750 but achieves only 3.3% confidence due to very limited data. The lognormal tolerance interval [201, 9751] is very wide, also due to the small sample size. This interval is not very useful as an estimate of the width of the population but could be used to demonstrate that at least 95% of the products will survive at least 201 cycles. A one-sided lognormal tolerance interval could also be used for this purpose and is shown in Figure 6.104.

95% Tolerance Interval

Variable	Lognormal Method	Nonparametric Method	Achieved confidence
Cycles	(201.110, 9750.702)	(850.000, 2750.000)	3.28%

Achieved confidence level applies only to nonparametric method

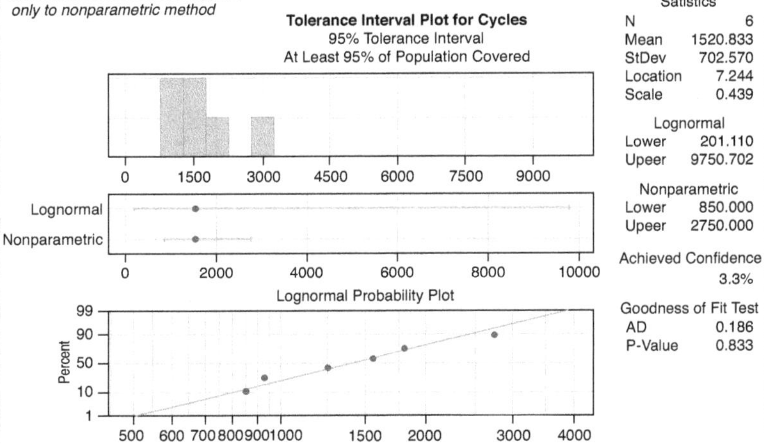

Figure 6.103 Lognormal two-sided tolerance interval for cycles data.

95% Lower Tolerance Bound

Variable	Lognormal Method	Nonparametric Method	Achieved Confidence
Cycles	275.170	850.000	26.4%

Achieved confidence level applies only to nonparametric method

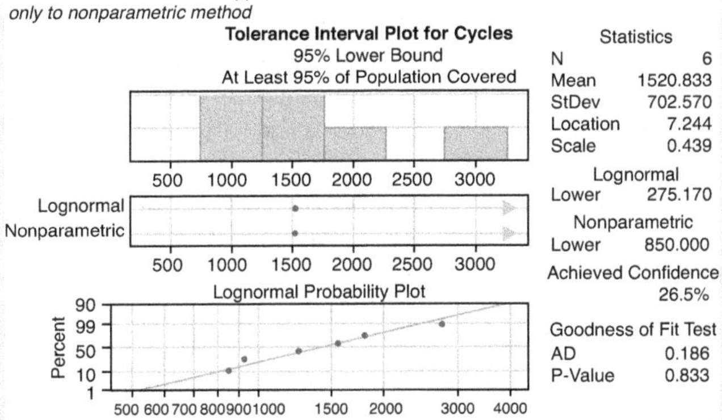

Figure 6.104 Lognormal one-sided lower tolerance bound for cycles data.

EXAMPLE 6.91

Weibull Tolerance Interval for 99% Coverage at 95% Confidence—Censored Data

Twenty products are tested for 1000 hours. Fourteen products fail at the following times, respectively: 70, 128, 204, 291, 312, 377, 473, 549, 591, 663, 748, 827, 903, and 955 hours. These data were previously evaluated using probability plots showing that a Weibull (β=1.44, η=887) appears to fit best (Figure 6.105). The team needs to estimate the number of hours with 95% reliability at 95% confidence.

Two lines are shown on this plot. The center line represents the nominal estimate, equivalent to median probability, so 50% confidence, or the best guess as to the percentiles. The line to the left is the one-sided lower bound at 95% confidence. Here the 5th percentile corresponds to 95% reliability, which has nominal value of 113 hours. The lower 95% confidence bound for this point can be obtained from the table of probabilities, in the box, 49.5 hours.

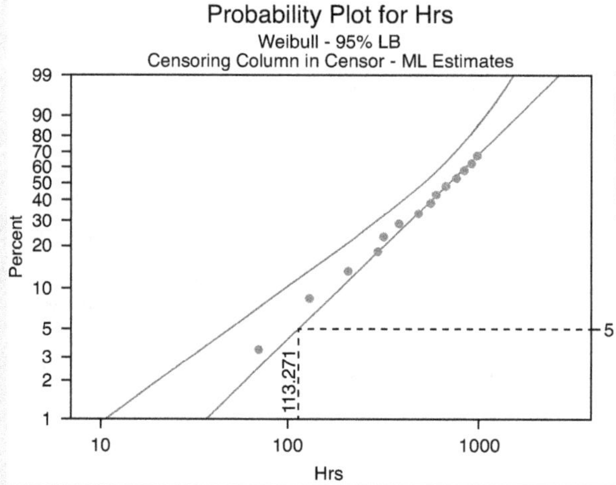

Table of Statistics	
Shape	1.44339
Scale	886.756
Mean	804.551
StDev	566.029
Median	687.901
IQR	737.893
Failure	14
Censor	6
AD*	38.011

Table of Percentiles

Percent	Percentile	Standard Error	95.0% Normal Lower Bound
1	36.6176	27.7727	10.5169
2	59.3978	38.4769	20.4657
3	78.9415	46.0577	30.2360
4	96.6972	52.0249	39.9098
5	113.271	56.9696	49.5262
6	128.989	61.1957	59.1080
7	144.054	64.8843	68.6704
8	158.602	68.1534	78.2242
9	172.730	71.0852	87.7774
10	186.510	73.7394	97.3363
20	313.686	91.6467	193.995
30	434.123	103.153	293.680
40	556.791	114.600	396.883
50	687.901	130.270	503.787
60	834.642	154.629	615.399
70	1008.46	193.688	735.284
80	1233.08	258.748	873.157
90	1580.35	384.253	1059.40
91	1630.10	404.222	1084.11
92	1684.93	426.731	1110.87
93	1746.16	452.453	1140.21
94	1815.68	482.378	1172.88
95	1896.41	518.040	1210.02
96	1993.19	562.006	1253.51
97	2114.97	619.094	1306.77
98	2281.54	700.109	1377.28
99	2554.52	839.505	1487.81

Figure 6.105 Weibull lower tolerance bound with 95% confidence and 95% coverage.

Intervals for Variability Estimators (σ, σ^2 and σ_2/σ_1, σ_2^2 / σ_1^2)

Table 6.55 summarizes equations for confidence intervals for variance-related estimators. Equations are provided for both variance and standard deviation, but recall that standard deviation is just square root of the variance, so these intervals are functionally equivalent and will result in identical conclusions.

Intervals may be obtained for a single parameter, either σ^2 or $\sigma = \sqrt{\sigma^2}$, or for the ratio, used when comparing variations. If the ratio is significantly greater than 1, that is, the interval does not contain 1, then the source in the numerator is associated with greater variation. If the ratio is significantly less than 1, that is, the interval does not contain 1, then the source in the denominator has more. If the effect is not statistically significant, that is, the interval does contain 1, then the result is inconclusive.

Data are assumed to be an independent and identically distributed *(iid)* random sample from a normal distribution. Note that the assumption of normality for these intervals is quite strict since both the chi-square and *F* distributions are sensitive to departures from normality. There are nonparametric methods that can be used to calculate confidence intervals when the normal assumption is violated.

Table 6.55 Equations for confidence intervals for variability estimators.

<table>
<tr><th>Parameter</th><th>Method</th><th colspan="2">Equations for 100 (1 – α)% confidence limits</th></tr>
<tr><td rowspan="2">Population variance or Standard deviation σ^2
$\sigma = \sqrt{\sigma^2}$</td><td rowspan="2">χ^2-test
(chi-squared) Data are iid
Normal (μ, σ^2)</td><td>Point Estimate:
$Variance: \hat{\sigma}^2 = s^2 = \frac{1}{n-1} \cdot \sum_{i=1}^{n} (x_i - \bar{x})^2$
$Std\ Dev: \hat{\sigma} = s = \sqrt{\frac{1}{n-1} \cdot \sum_{i=1}^{n} (x_i - \bar{x})^2}$</td><td>Two-Sided Limits: Use both equations below, but with $\alpha/2$ instead of α.</td></tr>
<tr><td colspan="2">One-Sided Upper and Lower Limits:
$Variance\ LCL(\sigma^2): \frac{(n-1) \cdot s^2}{\chi^2_{\alpha,n-1}}$ $UCL(\sigma^2): \frac{(n-1) \cdot s^2}{\chi^2_{1-\alpha,n-1}}$
$Std\ Dev\ LCL(\sigma): s \cdot \sqrt{\frac{(n-1)}{\chi^2_{\alpha,n-1}}}$ $UCL(\sigma): s \cdot \sqrt{\frac{(n-1)}{\chi^2_{1-\alpha,n-1}}}$
One-sided confidence intervals: [LCL, ∞) or (–∞, UCL]</td></tr>
<tr><td rowspan="2">Ratio of two population variances or standard deviations $\frac{\sigma_1^2}{\sigma^2}$
$\frac{\sigma_1}{\sigma_2} = \sqrt{\frac{\sigma_1^2}{\sigma_2^2}}$</td><td rowspan="2"><u>F-test</u>
Data are iid
X_1 ~Normal (μ_1, σ_1^2) and X_2 ~Normal (μ_2, σ_2^2)</td><td>Point Estimates:
$Variance: \frac{\hat{\sigma}_1^2}{\hat{\sigma}_2^2} = \frac{s_1^2}{s_2^2}$
$Std\ Dev: \frac{\hat{\sigma}_1}{\hat{\sigma}_2} = \frac{s_1}{s_2}$</td><td>Two-Sided Limits: Use both equations below, but with $\alpha/2$ instead of α.</td></tr>
<tr><td colspan="2">One-Sided Upper and Lower Limits:
$Variance\ LCL\left(\frac{\sigma_1^2}{\sigma_2^2}\right): \frac{s_1^2}{s_2^2} \cdot F_{1-\alpha,\nu_2,\nu_1}$ $UCL\left(\frac{\sigma_1^2}{\sigma_2^2}\right): \frac{s_1^2}{s_2^2} \cdot F_{\alpha,\nu_2,\nu_1}$
$Std\ Dev\ LCL\left(\frac{\sigma_1}{\sigma_2}\right): \frac{s_1}{s_2} \cdot F_{1-\alpha,\nu_2,\nu_1}$ $UCL\left(\frac{\sigma_1}{\sigma_2}\right): \frac{s_1}{s_2} \cdot F_{\alpha,\nu_2,\nu_1}$
$\nu_1 = n_1 - 1$
$\nu_2 = n_2 - 1$
One-sided confidence intervals: [LCL, ∞) or (–∞, UCL]
Note that this handbook provides F table values for left-tail probabilities, but not all references do. If we do not have a table for the left-tail, we can use the following relationships to calculate it:
One-sided: $F_{1-\alpha,\nu_2,\nu_1} = \frac{1}{F_{\alpha,\nu_1,\nu_2}}$
Two-sided: $F_{1-\alpha/2,\nu_2,\nu_1} = \frac{1}{F_{\alpha/2,\nu_1,\nu_2}}$</td></tr>
</table>

Although beyond the scope of the handbook, these methods are listed in Table 6.3 in the "Nonparametric Tests" section of this chapter.

The notation used here for the standard deviation is s to estimate the true population σ. Since the purpose of this analysis is to estimate σ, it can't be considered known. However, the ratio estimators could be done using both known or estimated sigma values if known values are available. **Note that two-sided limits allocate $\alpha/2$ to each side, whereas one-sided limits retain the full α on that side.**

The ANOVA is based on *F*-tests conducted using residual variation as measured by MSE, which is always in the denominator. The effects of other possible sources of variation in the analyses as measured by other mean squares are always compared to residual variation. In that framework, *F*-tests are set up by default to look for greater variation in the numerator relative to the denominator.

EXAMPLE 6.92

Confidence Intervals for Variance and Standard Deviation

Calculate the 0.90 confidence interval for the population standard deviation using the data here, in which a sample of five units tested to failure: 162, 157, 146, 173, 155.

$$\bar{X} = \frac{1}{n}\sum X_i = \frac{793}{5} = 158.6; \hat{\sigma} = s = 9.915\ hrs$$

From the chi-squared tables in Appendix G, $\chi^2_{0.05,4} = 9.488$ and $\chi^2_{0.95,4} = 0.711$.

$$[LCL, UCL] \Rightarrow LCL: \sqrt{\frac{4}{9.488}} \cdot 9.915 = 6.44$$

$$UCL: \sqrt{\frac{4}{0.711}} \cdot 9.915 = 23.5$$

The population standard deviation σ is therefore estimated to be between 6.44 and 23.5 at 90% confidence.

EXAMPLE 6.93

Confidence Interval for Standard Deviation

Let's revisit the coating thickness data first introduced in Example 6.81. Past experience shows that these thickness readings are normally distributed. Alternatively, we could conduct a formal test for normality, such as an Anderson-Darling test. We would like to calculate 95% confidence intervals on the population standard deviation. The α value is 0.05. Recall that $\bar{X} = 0.003$, $s = 0.0005$, $n = 25$, and that variance is the square of the standard deviation, so $s^2 = 0.00000025$.

We will use the formula from Table 6.55. From Appendix G, we find $\chi^2_{0.025,24} = 39.364$ and $\chi^2_{0.975,24} = 12.401$. Substituting the values, we obtain

$$\frac{(n-1)s^2}{\chi^2_{a/2,n-1}} \le \sigma^2 \le \frac{(n-1)s^2}{\chi^2_{1-a/2,n-1}}$$

$$\frac{(24)0.0005^2}{39.364} \le \sigma^2 \le \frac{(24)0.0005^2}{12.401}$$

$$0.0000001524 \le \sigma^2 \le 0.0000004838$$

$$0.00039 \le \sigma \le 0.00070$$

Thus, the 95% confidence interval for the population standard deviation is (0.00039, 0.00070). Notice that the 95% confidence interval for the population standard deviation is simply the square root of the endpoints found for the population variance. Alternatively, the formula can be rewritten in root form:

$$s\sqrt{\frac{(n-1)}{\chi^2_{a/2,n-1}}} \le \sigma \le s\sqrt{\frac{(n-1)}{\chi^2_{1-a/2,n-1}}}$$

$$0.0005\sqrt{\frac{(24)}{39.364}} \le \sigma \le 0.0005\sqrt{\frac{(24)}{12.401}}$$

$$0.00039 \le \sigma \le 0.00070$$

Note that these confidence intervals are not symmetric around the point estimates for σ and σ^2 because the chi-square distribution is not symmetric.

EXAMPLE 6.94

Confidence Intervals for Ratios of Variances and Standard Deviations

A reliability engineering team is considering implementing a new design feature to reduce the failure rate of a mechanical switch. The team would also like to test whether the new design has less variability in a key performance feature compared to the original fixture. The data for this performance feature are known to follow a normal distribution. A random sample of 20 switches are produced using the original design, and the sample standard deviation was equal to 0.250 hrs. A sample of 15 parts was produced using the new design, and these parts have a standard deviation equal to 0.195. Construct a 90% one-sided lower confidence bounds on the ratio of the two population variances and two population standard deviations to determine if the variability of the new design is less than the original design.

Here the original design will be denoted as Sample 1, and the new design will be Sample 2. It follows that $s_1 = 0.250$, $s_2 = 0.195$, $n_1 = 20$, and $n_2 = 15$. If the standard deviation from the new design is less than that of the original design, the ratios $\frac{\sigma_1^2}{\sigma_2^2}$ and $\frac{\sigma_1}{\sigma_2}$ will be greater than one. We will use the formula from Table 6.55 to calculate a lower bound for the ratio of the population variances. Using $\alpha = 0.10$, from Appendix F.5, we find $F_{0.90,14,19} = 0.51$. *Note that when we look up the F value in the table for the confidence interval, we switch the degrees of freedom of the numerator and denominator.* Substituting the values, we obtain

$$\frac{s_1^2}{s_2^2} F_{1-\alpha,v_2,v_1} \le \frac{\sigma_1^2}{\sigma_2^2} < \infty$$

$$\frac{0.250^2}{0.195^2}(0.51) \le \frac{\sigma_1^2}{\sigma_2^2} < \infty$$

$$0.838 \le \frac{\sigma_1^2}{\sigma_2^2} < \infty$$

We are 90% confident that the ratio of the population variances is covered by the interval [0.838, ∞). Since the value of one is included in the interval, we do not have sufficient evidence to claim the variance of the failure rates of the new design is less than that of the original design. We will obtain an equivalent conclusion if using standard deviation.

Continued

The 90% lower confidence bound for the ratio of population standard deviations is simply the square root of the endpoints found for the ratio of population variances, or alternatively, the formula can be rewritten in root form:

$$\frac{s_1}{s_2}\sqrt{F_{1-\alpha,v_2,v_1}} \le \frac{\sigma_1}{\sigma_2} < \infty$$

$$\frac{0.250}{0.195}\sqrt{0.51} \le \frac{\sigma_1}{\sigma_2} < \infty$$

$$0.916 \le \frac{\sigma_1}{\sigma_2} < \infty$$

It is mathematically impossible for these tests to result in different conclusions. Software may provide one or both.

Intervals Using the Exponential Distribution

The **exponential** distribution is usually expressed in one of two alternate forms:

1. (1) $f(t) = \lambda \cdot e^{-t\cdot\lambda}$
2. (2) $f(t) = \frac{1}{\theta} \cdot e^{\frac{-t}{\theta}}$

The second form shown here is parametrized using the mean of the exponential distribution as equal to θ and is referred to as the mean time to failure if the units are not repairable and the mean time between failures if the units are repairable.

The exponential distribution is unique in that its failure rate (λ) is constant. The failure rate is equal to the reciprocal of the mean:

$$\lambda = \frac{1}{\theta}$$

Because the failure rate is constant, the probability of failure is not dependent on the age of the unit but only on the length of the mission time. Also, because of the constant failure rate property, the mean can be estimated using censored data. It is not necessary for each unit in the sample to fail in order to make an estimate. A test can be terminated on a given failure, referred to as failure censored, or at a given time, referred to as time censored. To make an estimate of the mean, it is only necessary to know the total time (T) accumulated on the units and the number of failures (r). The total time is accumulated using the units that fail as well as the units that do not fail. The estimate of the mean is given as

$$\hat{\theta} = \frac{T}{r}$$

The χ^2 distribution is used to calculate the confidence limits on the mean of the exponential distribution. For the two-sided $1-\alpha$ confidence interval, the lower limit is given by

$$\theta_L = \frac{2 \cdot T}{\chi^2_{\frac{\alpha}{2}, v=2r+2}}$$

if the test is time censored (ended at a given time), and

$$\theta_L = \frac{2 \cdot T}{\chi^2_{\frac{\alpha}{2}, v=2r}}$$

if the test is failure censored (ended on a given failure). The difference in the equations is in the number of degrees of freedom for the χ^2 term. If the test is time censored, the number of degrees of freedom is $2r+2$, and if the test is failure censored, the number of degrees of freedom is $2r$. The upper limit for both time-censored and failure-censored tests is:

$$\theta_U = \frac{2 \cdot T}{\chi^2_{1-\frac{\alpha}{2}, v=2r}}$$

For one-sided confidence limits, α replaces $\alpha/2$.

EXAMPLE 6.95

Confidence Interval for Exponential Mean

Ten units from a distribution assumed to be exponential are placed on a life test. Four of the units failed at the times shown in Table 6.56. The units were not replaced when they failed. The test ended at 5000 hours (time censored) with six units still operating. Estimate the population mean and failure rate, then calculate the 90% confidence interval for the mean.

Table 6.56 Life test results for Example 6.95.

Unit	Time to Fail (hours)	Outcome
1	1150	Test failed
2	2150	Test failed
3	3350	Test failed
4	4950	Test failed
5	5000	Test terminated without failure
6	5000	Test terminated without failure
7	5000	Test terminated without failure
8	5000	Test terminated without failure
9	5000	Test terminated without failure
10	5000	Test terminated without failure

Results are $r=4$ from $n=10$, with $n-r=6$ suspensions. Total time on test is $T = \sum_{i=1}^{n} t_i = 41{,}600$ *hours*. Total time accumulated on the four failed units

Continued

is $T = \sum_{i=1}^{r} t_i = 11{,}600$ *hours*. We estimate mean time to failure as $\hat{\theta} = \frac{T}{r} = \frac{41{,}600}{4} =$ $10{,}400$ *hours*, which also provides us an estimate of the failure rate:

$$\hat{\lambda} = \frac{1}{\hat{\theta}} = \frac{1}{10{,}400} = 9.6 \times 10^{-5} = 96 \times 10^{-6} \textit{ failures per hour } (96 \textit{ failures per million hours})$$

Confidence limits for the mean are found using the equations:

$$\theta_L = \frac{2 \cdot T}{\chi^2_{\frac{\alpha}{2}, \nu = 2r+2}} \quad \text{and} \quad \theta_U = \frac{2 \cdot T}{\chi^2_{1-\frac{\alpha}{2}, \nu = 2r}}$$

From the χ^2 tables in Appendix G,

$$\chi^2_{\frac{\alpha}{2}, \nu = 2r+2} = \chi^2_{0.05, \nu = 10} = 18.307 \quad \text{and} \quad \chi^2_{1-\frac{\alpha}{2}, \nu = 2r} = \chi^2_{0.95, \nu = 8} = 2.733.$$

Therefore, we obtain:

$$\begin{aligned}[\theta_L, \theta_U] &\Rightarrow \left[\frac{2 \cdot T}{\chi^2_{\frac{\alpha}{2}, \nu = 2r+2}}, \frac{2 \cdot T}{\chi^2_{1-\frac{\alpha}{2}, \nu = 2r}}\right] \\ &\Rightarrow \left[\frac{2 \cdot (41{,}600)}{18.307}, \frac{2 \cdot (41{,}600)}{2.733}\right] \\ &\Rightarrow [4545 \textit{ hours}, 30{,}443 \textit{ hours}]\end{aligned}$$

Chapter 7
Data Management

Reliability data is generated by taking measurements on products either during experiments and tests or through observational studies in the field. Appropriate data management is required to realize the maximum value from reliability data. The lifecycle of data usage begins with determining the source of the data and then taking new measurements or collecting existing data. The data are then moved and housed in storage solutions, at which point they can be analyzed. Analysis of reliability data should generate actionable insights that improve the reliability or understanding of the system of products being analyzed. It is critical to consider the entire lifecycle of data when architecting a system to manage said data. Failure to consider the full data lifecycle can lead to inefficiencies in data collection and usage or outright gaps in realizing value from available data. With the advent of digital technologies, such as the internet of things (IoT), data management systems should leverage these digital technologies. Manual data management and analysis, such as hand-drawn statistical process control (SPC) charts, should only be deployed as necessary in situations where digital technologies are inappropriate, such as in a system exposed to extreme cyber security risk.

SOURCES AND USES OF RELIABILITY DATA

> Analyze the sources of reliability data (e.g., prototype, development, test, field, warranty, published data, big data, and the internet of things [IoT]). Recognize the unique advantages and limitations of each for measuring and enhancing product reliability. Consider the importance of normalizing datasets and transformations. (Analyze)
>
> **Body of Knowledge III.B.1**

Reliability data may come from a variety of sources. Throughout a product's lifecycle, the data are used in different ways, and each data source offers a distinct set of advantages and disadvantages. This is summarized in Table 7.1.

Table 7.1 Sources of reliability data.

Source	Comments	Advantages	Limitations	Potential contribution to product reliability
Product Prototypes (Experimental)	Early in design process	Provides visible and testable item	Only a preliminary assessment.	Feasibility and identify potential design flaws
Product Development (Experimental)	Throughout development process	Can track reliability changes with each design	Time/resources required	Identify important design drivers of reliability performance (good and bad)
Final product (Observational)	Test as soon as available from manufacturing	Early basis for warranty calculations	May not be able to simulate environmental conditions	Validation of reliability performance, investigation of process design, establish processing parameters.
Field warranty (Observational)	Need complete data	Customer viewpoint	Data often sketchy	Identify potential issues and costs

Reliability data are generally available to a manufacturer from several sources, both external and internal to the company. Each data source will provide a different type of data, depending on what can be measured at that stage. Data may comprise failure times (possibly censored), measures of product degradation, and covariates or other predictors of performance (e.g., processing parameters).

Published Data

A valuable external source of component reliability data is the supplier. Other users of the same component can also supply reliability data. A comprehensive external data source available to military contractors is the Government–Industry Data Exchange Program (GIDEP). This program generates and distributes reliability data on commercially available off-the-shelf units that are used by the various contractors. Failure data for units such as motors, pumps, relays, and so on are exchanged through the GIDEP program.[1] Professional organizations such as the Institute of Electrical and Electronics Engineers (IEEE) develop and maintain reliability data on hardware of various types.[2] Information on nonelectrical parts (NPRD) can be obtained from the Rome Air Development Center, National Technical Information Service, Springfield, Virginia.

Many times, internally generated data are given higher credibility; the conditions used to generate the data are known and can be controlled. Capabilities of the manufacturing process, the quality control methods employed, and the specified operation environment of the design, as well as other factors specific to the manufacturer, are reflected in internally generated data.

1. Information can be obtained from GIDEP Operations Center, Corona, California.
2. The IEEE can be contacted at 3 Park Avenue 17th Floor, New York, New York 10016–5997.

Product Testing

One example of internally generated data is from the analysis of prototypes. Early in the design process, it is often useful to produce a model of the item and use it for reliability testing. Prototype testing can provide early warning of design flaws, material requirements, and other changes that can improve reliability. Throughout the development process, the reliability engineer can be subjecting proposed parts to preliminary stress studies and other tests. Testing of final products provides another opportunity to collect data for warranty calculations, preventive maintenance, and possible future designs. More information on testing methods may be found in Chapter 9.

Care must be taken to use the data appropriately. The conditions under which data generated internally are developed will be known. The true test time; the test acceleration numbers, if used; the environment of the test; the actual failure mode; and other conditions will be documented. Data that is generated externally might be suspect, as the actual test or use conditions may not be known.

Field Service Data

Another valuable source of reliability data is the field service facility. Most manufacturers maintain data on the repair and maintenance services provided to their customers. This could be warranty data, data from dealers or distributors, or data from customers or users of the product. Each return or maintenance activity should result in a report. These data should be collected and distributed. Failure analyses performed on failed parts, if they are available, will provide data as to the root cause of the failure. It is the function of reliability engineering to collect and distribute field failure data. The data needs to be available for reliability engineering, quality engineering, product design, testing, marketing, service, and other appropriate engineering and support functions.

Warranty Data

Another data source to consider is data coming directly from the customer. Historically, warranty claim data along with focus groups were some of the only ways to collect such data. In the digital age, customer data may also be collected through online reviews, social media, and direct feedback via email or complaint systems.

The data collection process should include an effective method for identifying the failure conditions, including the following information:

- Traceability of each component so that its supplier, date, and conditions of manufacture and assembly are available
- Conditions of use, such as environment, handling, maintenance, and operation
- Failure parameters, including type of failure, symptoms, condition of other components, and action taken (replacement, repair, refund, and so on)

Obtaining useful warranty information may require an expenditure of considerable resources to determine the validity of the warranty claims and the causes of the failures. These expenditures can be justified if the follow-up action results in product improvements that avoid future costs. In addition, the data may yield cost avoidance if current practices result in paying for warranty claims that are not valid or that should properly be assessed to suppliers.

Warranty information can be used in various ways to improve customer satisfaction, including product design changes, process/manufacturing changes, product recalls, improvement in supplier performance, and refinement of reliability prediction methods.

Along with warranty data, the voice of the customer can be obtained through online data sources. A critical difference between these data sources is clarity and validity. A warranty claim likely requires a level of detail and data to prove the claim is valid, whereas an online review may just be an anecdote that lacks detail, clarity, or proof. While an individual warranty claim may be much more valuable than an individual review, it is highly likely that there will be a larger volume of online reviews available. The two types of data do offer different perspectives on reliability, and it is important to leverage both sources in a reliability program.

Manual Versus Automated Data Systems

The IoT approach represents comprehensive data collection both during production and product lifecycles. It is important to distinguish the differences between those two phases. Production applications of IoT may sometimes be referred to as the industrial internet of things (IIoT). This approach to data collection involves real-time monitoring of key process parameters on a production floor through integrated sensor packages. These data sources would most likely be used to improve the reliability of the production system over the product. Commercial IoT applications involve connecting products through internal self-monitoring sensors. An example of this would be when a modern vehicle identifies that it is due for an oil change, or when a smart refrigerator indicates that it is due for a water filter replacement. These data may be connected and shared back to the manufacturer through the internet, providing a broad and comprehensive dataset for product performance.

When designing a data collection plan in a modern digital ecosystem, there are primarily two factors to consider. First is the choice between rational sampling and continuous real-time monitoring. Second is the decision between automated and manual data collection. An automated data collection system may be capable of continuously monitoring system performance in real time, which is a very appealing prospect for reducing downtime and understanding system reliability. Real-time monitoring requires greater capacity for data movement, storage, and analysis. These capacity requirements may increase system costs. If the business case supports these added costs and the choice is made to use continuous real-time monitoring, then data collection must be automated as manual systems cannot sustain real-time data collection. When using rational sampling, automated systems may still be used with specific data collection schedules. Manual data collection executed by a human inspector may also be used. Systems may be designed with both manual and automated elements. The nature and criticality of system parameters should be taken into consideration in developing specific data collection plans.

Figure 7.1 highlights the outcomes of data collection methodologies and when they are appropriate. Manual systems, which may include manual data collection, entry, and analysis, are most useful when real-time data does not present an obvious benefit; however, they may be overwhelmed in data ecosystems in

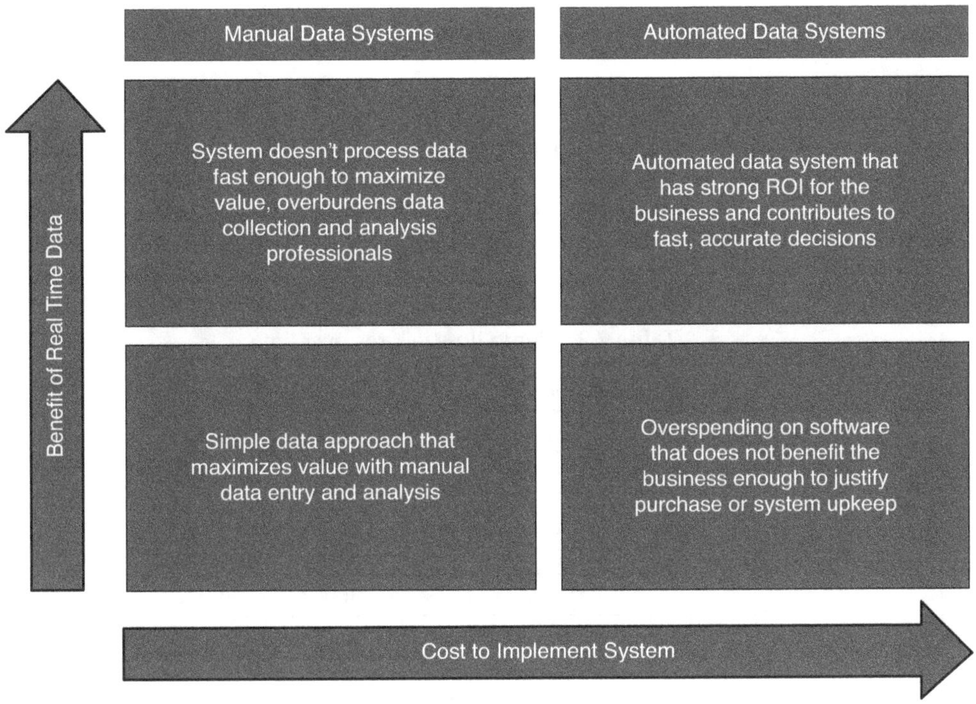

Figure 7.1 Types of data systems and their use cases.

which real-time data would be beneficial. Automated systems, on the other hand, will provide clear return on investment (ROI) in situations where real-time data is useful, but they will likely be unjustifiable costs in situations where real-time data provides no value.

IoT and Big Data

One of the most comprehensive automated approaches to shop-floor data collection is IoT. IoT can be described as an integrated package of data collection sensors that continuously monitor certain data points throughout an entire connected process or target functional area. Typically, this involves continuous monitoring of operational performance on all machines and equipment in a process of functional area. The application of IoT on the shop floor generates more accurate data about the performance of the equipment on that shop floor. This eliminates data gaps that occur from interval checks, as shown in Example 7.1. Generating accurate data for system performance leads to improved resource allocation and greater understanding of failure rates within the system.

IoT sometimes also refers to data collection that is integrated into a finished product. Examples include consumer products that report performance and lifecycle data back to the manufacturer as well as software that reports performance data to the developer. This application of IoT may provide useful data on product lifecycle reliability as opposed to equipment and process reliability within the production facility.

EXAMPLE 7.1

A reliability engineer is tasked with reducing downtime across pumps in a manufacturing facility. Failures are deemed to occur when the flow is below a specified rate. Every hour, a technician checks the flow of ten pumps in the facility, and when flow rates drop, pumps are taken down for unplanned maintenance, with routine maintenance occurring on weekends. The engineer tracks the data on pump inspections over a 12-week period and generates the failure data in Table 7.2.

Table 7.2 Pump flow failures over time.

Week	Failures
1	8
2	12
3	4
4	17
5	10
6	11
7	7
8	8
9	11
10	6
11	13
12	6

Whenever a pump fails, the maintenance response takes an average of 0.25 hours to have the pump back online. Over the 12-week period, the pumps experienced 113 failures, resulting in 28.25 hours of maintenance time. The pumps exhibit a constant failure rate, and the engineer recognizes that pumps can fail any time between inspections, meaning each failure could represent an additional 1–59 minutes of downtime before it is detected.

The engineer assumes an average of 0.5 hours of downtime due to when the pumps break down, based on the constant failure rate, and targets that value for improvement to increase downtime based on the system performance data in Table 7.3.

The reliability engineer calculates the cost of downtime to be $100 per hour and determines that the inspection delay has cost $5650 over the 12-week study and extrapolates an annual cost of $24,483 for inspection delays. She identifies a $500 inline flow meter to install on the outlet of each pump, and the IT department quotes $1000 total to set up the flow meters to notify the maintenance team automatically when the flow drops below acceptable levels.

The engineer executes the project, spending $6000 total to realize a 100% reduction in inspection delays, resulting in a simple payback period of approximately three months and an ROI of approximately 400% in the first year of the system's operation. A maintenance plan is put in place for the flow meters with an acceptable annual cost, and the benefit of inspectors having extra time to take on other manual inspections is also realized.

Table 7.3 Downtime due to pump failures.

Week	Failures	Inspection downtime	Maintenance downtime	Total downtime
1	8	4	2	6
2	12	6	3	9
3	4	2	1	3
4	17	8.5	4.25	12.75
5	10	5	2.5	7.5
6	11	5.5	2.75	8.25
7	7	3.5	1.75	5.25
8	8	4	2	6
9	11	5.5	2.75	8.25
10	6	3	1.5	4.5
11	13	6.5	3.25	9.75
12	6	3	1.5	4.5
Total	**113**	**56.5**	**28.25**	**84.75**

Example 7.1 demonstrates an automated system that is worthwhile, but this is not always the case. If the flowmeters cost $5,000 instead of $500, the system would be very difficult to justify. If the maintenance response couldn't be made faster, the value of automation wouldn't be realized. And, if the downtime cost were $10 per hour instead of $100, the system wouldn't be worth it. How data is to be used should always be considered in a data collection plan.

The data generated out of an IoT system is often described as big data, though the term can apply to data from many sources. Big data is an intuitive term referring to datasets that are exceedingly large. This approach has been pursued more and more as digital technologies are expanding our capabilities to analyze and manage data. Big data often requires preprocessing prior to being useful in an analysis. This may be achieved by compiling smaller datasets into descriptive statistics to reduce the overall volume of data to be handled. An example of preprocessing big data is shown in Example 7.2.

EXAMPLE 7.2

A reliability engineer has analyzed hourly hardness testing data on parts produced on two machines (M1 and M2). That analysis has shown that the two machines are yielding statistically different hardness measurements for parts produced on the previous day. This likely means that machines differ in their processing conditions.

The engineer suspects the samples from M2 with lower hardness are heated to a greater degree during processing. The machines are equipped with continuous temperature sensors, and the engineer extracts 1440 average temperatures for each minute during the day. A simple comparison of the temperature averages (22.02°C for M1 and 22.04°C for M2) does not show a difference in the machines.

Further exploration of the data begins with run charts of two process measures, temperature and vibration, as shown in Figure 7.2.

The graphs immediately show that M1 and M2 are running to the same average temperature, but temperature is not as well controlled for M2, leading to higher temperatures during the heat of the afternoon. The graphs also show greater variation during the afternoon for both temperature and vibration measurements.

A closer examination of the hardness data shows that the lowest hardness values arose in the afternoon, which coincides with M2 running outside of normal operating conditions.

For plant situations involving more sensors, a longer time frame, or more frequent sampling, it becomes challenging to complete this type of graphical analysis. In that case, multivariate statistics and machine learning methods can look for anomalies (e.g., outliers) or differences in variation across the machines.

To illustrate this for the temperature and vibration data, an anomaly detection algorithm on the two machines was run. Just a few anomalies are detected for M1, but for M2 there are many anomalies, and the algorithm provides a list of the afternoon data points for investigation. Since there are two dimensions, the anomalies can be visualized, as shown in Figure 7.3. The anomalous observations are separated from the main body of the data, and plotted in gray. These are the same outlying points seen in the previous plots.

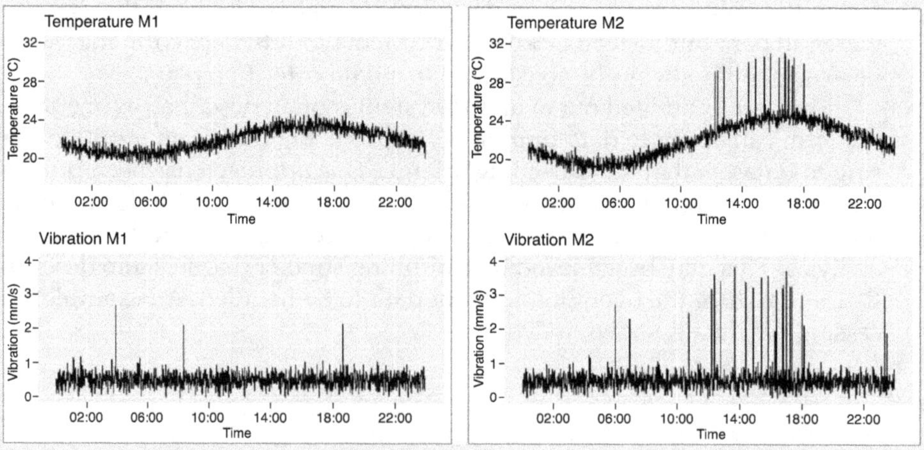

Figure 7.2 Run charts of temperature and vibration.

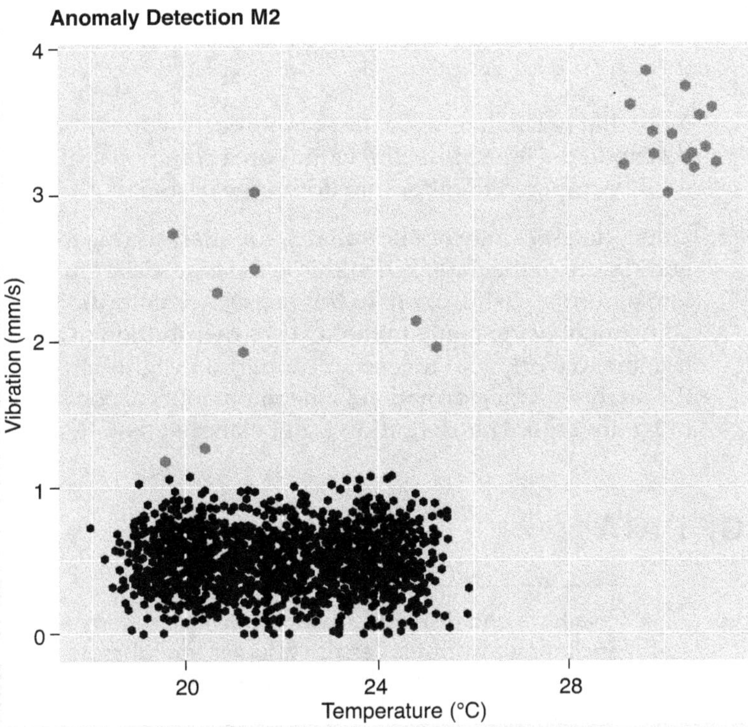

Figure 7.3 Anomaly detection results.

It would seem desirable to model the quality or reliability response as a function of the sensor data to identify process conditions that affect the outcome. However, in practice, the outcome measures are taken much less frequently, and the mismatch in time scales requires some data reduction before proceeding. Readers are referred to a text on machine learning for more in-depth coverage of the topic.

With big data applications like the one detailed previously, often data will be normalized into descriptive statistics as handling large quantities of data can be cumbersome, and in many statistical applications calculating the descriptive statistics is one of the first steps executed mathematically.

Data Normalization and Transformation

The term *normalization* has several meanings. We present two of the most commonly applied definitions here.

1. A useful technique in data processing is *data normalization*, in which we transform data so that it is on a common scale. A normalization technique frequently used is the Z transformation, which results in a unitless, coded dataset that has an average equal to zero and standard deviation equal to 1. We first calculate the sample mean and standard deviation of the data and then apply the transformation to each data point, as shown in Equation 7.1.

Equation 7.1 Normalizing using a Z transformation.

$$\hat{z}_i = \frac{(x_i - \bar{x})}{s}$$

Note that normalizing a dataset in this sense does not make it normally distributed—it just puts the values on a common scale so they can be easily compared to other transformed variables.

2. Often, datasets may not be suitable for analysis due to the distribution and nature of the data. When this is the case, data must be transformed into a normal distribution. In this sense, normalization describes the adjustment being made to the scale or distribution of a dataset to make it feasible to work with to learn meaningful insights and create value from the analysis. Transformation is the mathematical conversion that the data must undergo. This definition is not elaborated on here.

TYPES OF DATA

> Evaluate and distinguish among diverse data types, including attributes versus variables and discrete versus continuous. Determine whether the data is censored or complete and whether it is univariate or multivariate. Describe the implications of deploying parametric and nonparametric statistical methods, identifying the underlying assumptions and applicable conditions of each. Recognize the different scales of measurement, including nominal, ordinal, interval and ratio. Align analysis tools–such as Nevada chart analysis, survival analysis, and the Cox proportional hazard model–with the inherent characteristics and requirements of the respective data types. (Evaluate)
>
> **Body of Knowledge III.B.2**

This section is adapted from *The ASQ Certified Six Sigma Black Belt Handbook, Fourth Edition* (McShane-Vaughn 2022).

When a product is measured as part of a reliability study, we may only record one performance metric. In that case, the dataset has only one variable and is considered *univariate data*. In other cases, we may collect multiple performance measures and/or additional product or process characteristics. The resulting data set has multiple variables and is collectively referred to as *multivariate data*. For both types of data, we then consider the characteristics of each measurement before choosing our analysis method.

Qualitative and Quantitative Data

Data can be categorized into two types: qualitative and quantitative.

Qualitative data are data based on descriptive or verbal information rather than numerical information. This descriptive information can be in terms of the characteristics, attributes, properties, and qualities of an object. For example, an object may be described as "heavy," "blue," "smooth," "defective, "nondefective," and so on. Qualitative data can also be sequences of numbers if they serve as labels instead of measurements. For example, telephone numbers, social security numbers, and bank account numbers are numeric but could be translated into alpha characters with no loss of information. Therefore, they are classified as qualitative data. Verbal information, such as customer reviews and recorded customer service calls, is also classified as qualitative data.

Qualitative data can be collected through observations, interviews, surveys with open-ended questions, and narrative data from social media, for example. Often, the customer feedback gleaned from open-ended survey questions or reviews provides the greatest insights. Qualitative data collected by these methods can be analyzed using advanced techniques, such as computation linguistics and sentiment analysis.

Quantitative data are numerical data that can be measured, verified, and manipulated. For example, measurements such as height, mass, volume, temperature, impedance, and humidity are quantitative. Counts, such as production volumes, inventory levels, and tallies of defect types, are also quantitative. Quantitative data can be further broken into two categories: continuous and discrete.

Discrete and Continuous Data

Discrete data can take on either a finite number of values or a countably infinite number of values. Two examples of finite sets of values follow. Say that the allowable values of a variable X are $\{-3/4, 0, 12\}$, then the variable X is discrete since it can take on only a finite number of distinct values. For the second example, we can consider the outcome of a single roll of a die. Here, the allowable values of X would include $\{1, 2, 3, 4, 5, 6\}$.

On the other hand, if a variable X can take on the values of the set of counting numbers $\{0, 1, 2, 3 \ldots\}$ then it is also considered discrete since the allowable values include distinct numbers that, while infinite, are countable. In statistical process control charts, discrete data is also referred to as attribute or attributes data.

Continuous data are measurements taken from a variable that has a continuous scale, such as length, weight, or temperature. These scales are called continuous because there is an infinite number of values in between any two given values on the scale. For example, there is an infinite number of values between the measurements 1.537 inches and 1.538 inches since there is no limit to the number of decimal places for length. In effect, the number of decimal places we record for a continuous variable is only limited by the discrimination, or resolution, of the instrument we use for our measurement. Continuous data are often referred to as variables data in statistical process control.

Note that continuous data can be transformed into discrete data, but the reverse is not true. We can group continuous data into categories and then count the number in each category. For example, we can measure and record the lengths of parts. If we then classify any parts that exceed the specifications as defective, the tally of the defective parts is a discrete count. Note that the continuous measurements provide more information than the discrete data. Because we measured the parts, we know not only which ones are defective but by how much they exceed the specification. However, if we were only given the count of the defectives, we would lose information about the severity of the nonconformities.

Measurement Scales

To avoid analysis errors, it is important to recognize the type of measurement scale used to collect data. There are four types of measurement scales:

1. Nominal
2. Ordinal
3. Interval
4. Ratio

Nominal scales classify data into categories with no order implied—for example, a list of equipment such as presses, drills, and lathes is considered as nominal. The nominal scale is qualitative and is simply a label. Mathematical operations performed on nominally scaled data will not have any meaning. For example, calculating an average of the telephone numbers in a business office will provide no useful information. When data are classified into two categories, such as male/female, defective/nondefective, and so on, they are called dichotomous.

Ordinal scales refer to positions in a series, where order is important, but precise differences between values aren't defined. An example might be the qualitative, subjective ranking of three light bulbs as bright (third), brighter (second), and brightest (first). Unless the light emitted from each bulb were measured in lumens, the difference between the first and second bulbs and the second and the third bulbs would be unknown. Only the ordering of the bulbs would matter. Ordinal data can also be represented by quantitative ranks, such as first, second, and third place in a 5K race. While these ranks have numerical values, the difference in finishing times between the ranked runners cannot be determined.

Interval scales can be ordered and have meaningful differences but have no inherent zero point. Examples of this measurement scale include temperatures in degrees Fahrenheit and Celsius, since the zero points are arbitrarily chosen values. In addition, calendar years follow an interval scale, since the starting points vary between calendars used throughout the world, such as the Chinese, Gregorian, Islamic, and Jewish calendars. Because there is no inherent zero point, ratios between values on the interval scale are not meaningful. For example, a temperature measured as 20°F cannot be said to be twice as warm as 10°F.

Ratio scales can be ordered, they have meaningful differences, an absolute zero exists, and therefore, ratios are meaningful. For example, a length measured in inches is a ratio scale because zero length is defined as no length, and as a result, 20 inches can be said to be 4 times longer than 5 inches. Other examples of variables measured on a ratio scale include weight, dollars, age, or any variable in which zero means that there is none.

A summary of the four measurement scales, along with their corresponding allowable mathematical operations and appropriate descriptive statistics, is given in Table 7.4.

Table 7.4 Summary table of measurement scales.

Mathematical operations	Measurement scale: Nominal	Ordinal	Interval	Ratio
$=, \neq$	√	√	√	√
$<, >$		√	√	√
$+, -$			√	√
$\times, \div$				√
Types of data	**Qualitative**	**Qualitative or quantitative**	**Quantitative**	**Quantitative**
		Discrete	**Discrete or continuous**	**Discrete or continuous**
Measures of central tendency	Mode	Mode	Mode	Mode
		Median	Median	Median
			Mean	Mean
Measures of dispersion		Range	Range	Range
		Interquartile range	Interquartile range	Interquartile range
			Standard deviation	Standard deviation

Source: The ASQ Certified Six Sigma Black Belt Handbook, fourth edition.

EXAMPLE 7.3

An organization wants to survey its customers using a five-point Likert scale, which, by definition, is ordinal. The general construct of the scale looked like Table 7.5.

The organization could analyze the survey results treating the Likert categories as qualitative and report the percentage of customers who selected "very satisfied," "satisfied," and so on, or the organization could transform the categorical data into discrete data using a 1 to 5 mapping. These data now follow an ordinal scale that is quantitative, and the sample median and range can be reported. Although technically incorrect, in practice, Likert data are often analyzed as interval data, in which the sample mean and standard deviation are calculated.

Table 7.5 Five-point Likert scale.

Very unsatisfied	Unsatisfied	Neutral	Satisfied	Very satisfied
1	2	3	4	5

Discrete (attributes) data are obtained when the characteristic being studied can have either a finite number or countably infinite number of possible values. For example, the results of a leak test might be designated as 0 or 1 to indicate failed or passed. Another example would be the count of the number of scratches on an object. Here, the possible values are 0, 1, 2, . . . , which is a so-called countably infinite set. Attribute control charts (for example, *p*, *np*, *u*, and *c*) are used to monitor discrete data.

Continuous (variables) data are obtained when the characteristic being studied can have any value in a range of numbers. For example, the length of a part can be any value above zero. Between each two values on a continuous scale, there are infinitely many other values. For example, between 2.350 and 2.351 inches, the values 2.3502, 2.350786, and so on, occur. Variables control charts (for example, X and R, mR) are used to monitor continuous data. These charts are used to plot a single variable and are referred to as univariate charts. Output variables such as a dimension, hardness, pH, and so on, and input variables such as pressure, temperature, and voltage are often plotted on these charts. A thorough explanation of when and how to use various control charts and statistics visuals is explored in Chapter 6.

Censored and Complete Data

When collecting failure data, there are several possible ways to measure failure times. Consider the testing of several pumps from Example 7.1. Timers and flowmeters might be placed on each pump so that *exact failure times* can be recorded. This approach is becoming increasingly common as data collection and IT technology become more affordable. If all our measurements are exact, we refer to this as "complete" or "uncensored" data. However, the test might be terminated while some pumps are still functioning. Here, it is known only that these pumps' failure times are longer than the test time. Such data are called *right censored.*

Another testing protocol would require that the pumps be checked every 100 hours. If a pump were operating at hour 400 but not at hour 500, it would only be known that the failure occurred between these two times. These data are called *interval censored.*

If a pump is found to have failed at the first check at hour 100, it is known only that the failure occurred before hour 100. Such data are called *left censored* rather than interval censored because the failure may have occurred before the interval started, assuming a check of the pump was not conducted at time zero.

Parametric and Nonparametric Statistics

When selecting analytical tools to use from those presented in Chapter 6, it is critical to understand the dataset being used. Most common statistical tools are parametric, as parametric statistics stand on the assumption that the data follow a specific probability distribution. In that case, it is important to test this assumption to ensure the validity of the insights gained from the data. Using the wrong analysis for the dataset available poses the risk of making invalid insights and taking the wrong actions.

Nonparametric statistics are less commonly used in practice. Nonparametric statistics are tools that do not require any distributional assumptions. Because of the lack of underlying assumptions about the data, nonparametric statistical analysis is often less powerful than parametric methods. When dealing with non-normal data, a practitioner may choose to use nonparametric statistics, or they may choose to transform and normalize the data. Discernment of when to take which approach is a valuable skill to bring to any data analysis project. Chapter 6 provides a deep dive into the statistical tools available to practitioners.

Data Analysis Tools

A variety of data analysis tools may be used depending on the situation. One of the most important analytical skill sets that any practitioner needs to develop is the discernment of when to use which type of analysis and data visualization. Analysis should result in actionable insights that help improve performance. Common reliability analysis tools include Nevada chart analysis, survival analysis, and the Cox proportional hazard model.

Nevada chart analysis is used to track the performance of products over time while providing time ordered data of when a product was produced against when it failed. This analysis can be used with continuous lifetime measures, as well as discrete measures (e.g., Product A failed in May, or the count of failures in July). The Nevada chart may be used to roll up monthly data, such as warranty claims, as demonstrated in Table 7.6. As the Nevada chart is populated with data, it takes on a similar shape to the state of Nevada.

This chart provides useful insights into the lifecycle performance of products and may be further contextualized by more data sorting categories as appropriate for the situation.

Survival analysis describes the time that is expected to pass before an event occurs, such as maintenance or part failure. There are many types of survival analyses in reliability engineering, such as the hazard function or metrics such as mean time to failure. Being able to adequately describe the expected lifecycle performance of a product or system component is very important for a reliability engineer.

Table 7.6 Nevada chart of number of warranty claims.

Warranty claim count						
Production month	**Claim month**					
	February	**March**	**April**	**May**	**June**	**July**
January	3	5	1	4	2	9
February		1	2	3	6	1
March			2	4	5	9
April				2	3	5
May					5	4
June						1

The *Cox proportional hazard model* reflects a deep dive into survival analysis in which factors contributing to the survival are brought into the analysis to make it more situationally accurate. For example, the hazard function may describe the likelihood of a part surviving at a certain time, and the Cox model incorporates additional variables that may influence the outcome (such as temperature for a system which operates differently under different atmospheric conditions).

DATA COLLECTION METHODS

> Identify and select appropriate data collection methods (e.g. surveys, automated tests, automated monitoring, and reporting tools) to meet various data analysis objectives and data quality needs. (Evaluate)
>
> **Body of Knowledge III.B.3**

As discussed earlier, reliability data is generated by taking measurements on products, either during experiments and tests or through observational studies in the field. Once the data source has been chosen and the measurements have been identified, data collection begins.

Data collection describes the methods by which individual data points will be brought into a system for analysis. A primary consideration in a data collection plan is identifying which data are useful for the overall goals of the system. Methods for data collection should be developed and deployed on a data-point-by-data-point basis with considerations for sensors and automation to increase data availability and simultaneously reduce potential human error in data entry. It is critical to explore available technologies for data collection, as technical capabilities in this area have grown rapidly since the onset of the fourth industrial revolution.

One approach to data collection is to use experimentation and to minimize the influence of outside factors. This can be accomplished by maintaining strict laboratory conditions, perhaps using an environmental chamber. The downside of this course of action is that very few products or processes operate in an environment in which all outside factors are controlled. This technique is most appropriate for early-stage investigation in which fundamental design cause-and-effect relationships are being determined. Some data, such as temperature, time, pressure, and so on, can be collected automatically.

At the other end of the spectrum are field service data. This refers to information gathered from products in use by customers. This information is often impacted by differences in installation, environment, operator procedures, and similar factors that make analysis difficult. However, field service data represent realistic applications of the product and therefore must be considered as part of reliability analysis. In most cases, this approach is used in the feedback process to aid in determining the accuracy of other reliability analysis methods and to provide input for future revision and design initiatives. Various other methods of data collection have been devised. Perhaps the most powerful technique is to study the failure modes and mechanisms that can occur.

In this type of study, special attention must be paid to the conditions under which the product will be transported, stored, and used. These conditions include differences in such factors as geographic location; operator habits; lubricants; possible chemical, radiological, or biologic exposure; vibration; electromagnetic environment; and so on.

Reliability data are often stated in binary terms, such as failed or nonfailed. Some contemplation should also be given to any degradation of satisfaction from the user's viewpoint—short of failure—that may occur. This contemplation should occur in the light of rising user expectations.

It is important to ensure that data being collected is accurate, reflective of reality, and useful to the outcomes desired. Data collection plans may be assessed through the Four Vs of data, which are:

1. Volume—the amount of data being collected
2. Velocity—the speed with which the data makes it through analysis
3. Variety—the amount of different unique data aspects being collected
4. Veracity—the extent to which the data is true and accurate to reality

The Four Vs framework is used to ensure that data collection plans are thorough and that the results of the analysis drawing upon the data can be validated and trusted to drive value for the business. Other considerations include the following:

- Who will collect the information? Data collectors must be familiar with the measuring equipment and the product itself. If failure data are to be collected, a well-defined and understood definition of failure is essential. If possible, more than one person should be involved to provide some cross-checking and coverage when one must be away.
- When will the data be collected? The plan needs to specify at what stage of design or use the data will be collected. Dates and times help the other team members in their planning.
- In what format will the data be collected? The best step here is to design a data collection sheet and specify the format for summarizing the data.
- What measurement equipment will be employed? The plan should specify equipment and recording devices if appropriate.
- What measures will be used to verify data accuracy and integrity? The plan may specify units of measurement, calibration of equipment, or recording of ancillary information that might impact the data. If data are transmitted or stored digitally, the use of error correction systems may be specified.

DATA SUMMARY AND REPORTING

> Analyze the collected data and assess its accuracy, precision, usefulness, and integrity. Employ analytical techniques, such as bad actor analysis. Use artificial intelligence (AI) to delve deeper into the analysis. Select appropriate graphical representations, such as Pareto charts, scatter plots, and box and whisker plots. Summarize, interpret, and present the analyzed data, choosing methods aligned with the varying data types, originating sources, and the specified output requirements. (Create)
>
> **Body of Knowledge III.B.4**

Data Usability and Data Analysis

Data analysis is a key step in the process of communicating results to decision makers. The methods of data analysis must be chosen to clearly communicate the information that is relevant to the questions or decision at hand. However, the best data analysis method will only work if the data is suitable, or useful, for that analysis.

The assessment of usefulness will depend on the analysis, but one general characteristic to look for is the integrity of the data. Was it collected correctly? Are the measurements valid? Were any errors introduced during processing? Once we are sure of the integrity, we can assess the suitability for a given analysis. For example, precision and accuracy are properties of the estimates we will make during analysis. So, we should ask if the sample size is large enough to provide the right level of precision. We should also ensure that the data is representative of the population so that we can make accurate estimates.

Common causes of poor data accuracy are listed in Table 7.7.

Chapter 6 described statistical approaches to reliability data analysis, and any of the tools or methods detailed may be appropriate depending on the use case of the data and particular problem to be solved. Again, understanding the questions to be answered, or decisions to be made, will drive the data analysis plan and allow you to tailor the analysis to the situation. For example, an engineer may want to identify plant equipment that may be problematic. In that case, they may pursue a "bad actor analysis" and design a sampling and analysis plan that compares the performance of different pieces of equipment.

In this era of big data, where large multivariate datasets are increasingly common, the analysis may require computer-intensive statistical methods, such as machine learning or artificial intelligence (AI). These mostly nonparametric methods can help generate hypotheses on reliability issues though pattern identification. They can also be used to fit models, primarily for prediction. It should be

Table 7.7 Common causes of poor data accuracy.

Issue	Comment
Multiple points of entry	When data entry is centralized to a single individual, entry errors are usually minimized over time due to the experience of the individual and, often, the ability of the individual to contact the originator of the data for clarifications. When data originators become responsible for their own data entry, data accuracy usually diminishes, particularly when frequency of data entry decreases or there is high turnover among the data originators.
Limited use of data validation	Such validation would include the presence or absence of data, data ranges, inconsistent data, and other logical checks. Surprisingly, computer technology has not completely eliminated this cause.
Batching input versus real-time input/ synchronization of system updates	This scenario usually results in instantaneous differences, but the differences are usually reconciled with time. Consider the example of an individual inquiring about a checking account balance through an ATM, a bank teller, or online. Because of different cut-off and system update times, the individual will see up to three different account balances.
Multiple means of data correction access to upstream systems	When multiple data correction access points are present, the opportunity to have synchronization issues increases. For example, suppose there are three upstream systems prior to the current system. Label them 4, 3, 2, and 1 for the current system. If corrections are made to 4, then 3, 2, and 1 are updated. However, if corrections are made to 2, only 2 and 1 are updated.
Unclear directions	This issue becomes increasingly important when multiple points of data entry exist.
Lack of training	This issue becomes increasingly important when multiple points of data entry exist.
Ambiguous terminology	Ask an operations manager for monthly data and you get monthly data according to the calendar month. Ask for the same data from an accountant and you get monthly data by fiscal month. Different months' lengths yield different data.
Manual versus automated means of entry	Automated data entry is usually preferred over manual data entry, particularly if you accept the premise that people are fallible.
Rounding	Improper rounding, for example, when rounding up is appropriate instead of rounding down. Also, the cumulative effects of rounding can generate errors. For example, computers store and calculate using numbers to 16, 32, or even 64 digits. An individual using numbers rounded to four digits in the same calculations can arrive at a result that is substantially different than the results arrived at by a computer.

Table 7.7 Common causes of poor data accuracy. (*continued*)

Issue	Comment
Order of calculations	Consider $\frac{36}{(1/9)}$. If you rewrite this equation as $36\left(\frac{9}{1}\right)$ and multiply through, you obtain 324. If you divide out 1/9 and round to four decimals, you obtain 0.1111. Dividing 36/0.1111, you obtain 324.0324. Different order, different answers.
Calculation errors	This is simply an arithmetic error or an error in the specification of the formula to be used. For example, I worked at two organizations that computed "associate turnover" incorrectly.
Inadequate measurement systems	This is an issue commonly known among Six Sigma professionals who have been taught to examine the adequacy of the measurement system first. Examples include Kappa studies and gage repeatability and reproducibility studies.
Units of measure not defined	This type of error occurs when the wrong use of measure is assumed or used, such as "foot" instead of "inch," "yard" instead "foot," or "box" instead of "each." The last example can significantly impact inventory levels.
Failing to use the proper system of measurement	On September 23, 1999, after a 286-day journey from Earth to Mars, NASA lost the $125 million Mars Climate Orbiter after the spacecraft entered orbit 100 kilometers closer than planned. NASA used the metric system, while its partner Lockheed Martin, the organization that designed the navigation system, used English units.
Field inconsistencies/ truncation among systems or data fields	Moving an n character field to an $n-1$ character field or smaller can result in field truncation, which would impact all calculations that use it. For example, 100,789 is moved into a three-character numeric field. The result would be 789, not 100,789.
Similarities of characters	Several letters and numbers can easily be confused, particularly when handwriting is poor or when reading other than the top sheet of a no carbon required form. Common examples include: 0 O (oh) 4 H 8 B 1 l (el) 5 S 9 g (gee) or q 2 Z and occasionally Q 6 G 3 B 7 Z Examples of possible confusion between letters: c-e, C-O, E-F, m-n, M-N, O-Q, P-R, r-v, u-v, U-V, v-W, V-W, v-y
Copying errors	Duplicates of duplicates can result in image degradation and impact the readability of the copy. Also, copy truncation occurs when the image being copied is close to the edge of the paper. Try this with your credit card statement sometime. Ensure that the purchase amounts are adjacent to the edge of the copier glass.

Source: The ASQ Certified Six Sigma Black Belt Handbook, fourth edition.

noted that AI is not appropriate for all situations. These methods require very large training datasets and are useful when repeated predictions are needed (e.g., during automated process monitoring). Smaller datasets and high-risk design decisions will still benefit from the methods outlined in Chapter 6.

Data reporting begins with simple and straightforward analysis techniques; however, when the basics have been explored, deeper analytical methods may be deployed. AI is one of the most powerful types of analysis available today, and just like the statistics presented in Chapter 6, there are numerous types of analyses that may be performed with AI, and they are appropriate for different situations.

In general, AI models tend to function very well at recognizing patterns. They require big data to feed the models, and the implementation process for AI involves training a model with practice data. When all these steps are done correctly, the analytical results can be profound, and any reliability engineer who desires to have strong and applicable data skills for the future may benefit from developing AI skills. Regardless of the chosen statistical analysis, graphing and visualizing data is an important part of any analysis, as it can be a very effective communication tool.

Data Visualization Techniques

It is important to know the most common and standard tools for data visualization. These data visualizations are tried and tested, and they have the benefit of being widely recognized by most audiences. Some common data visualizations used in reliability engineering include Pareto charts, scatter plots, and box-and-whisker plots.

An excellent tool for seeking out and prioritizing potential root causes is the Pareto chart. The purpose of the Pareto chart is to separate the "vital few" causes from the "trivial many." This is often reflected in what is called the 80/20 rule, in which only a few causes are driving most of the problem. This helps the engineer focus on the causes that will produce the largest impact and improvement. This concept is illutrated in Example 7.4.

EXAMPLE 7.4

While awaiting delivery of the flow meters from Example 7.1, the reliability engineer chose to further explore the 113 pump failures incurred over the 12 weeks to look for other improvement opportunities. The engineer uncovered the following six causes for pump failure:

1. Pump overheats
2. Pump clogged from upstream material
3. Pump valve leak
4. Upstream leak prevents pump feed
5. Downstream blockage prevents flow
6. Wrong valve left open/closed

Continued

The engineer summarized the occurrences of each failure in Table 7.8.

Table 7.8 Tally sheet for causes of pump failure.

Issue	Count
Overheat	17
Upstream clog	34
Valve leak	7
Upstream leak	8
Blockage	45
Incorrect valve	2
Total	113

She then elected to use a Pareto chart to organize and visualize the failure data, proceeding to sort the data based on occurrence rate as shown in Table 7.9. This data is visualized in the Pareto chart shown in Figure 7.4.

Table 7.9 Input table for Pareto chart of pump failures.

Issue	Count	% of Failures	Cumulative %
Blockage	45	40%	40%
Upstream clog	34	30%	70%
Overheat	17	15%	85%
Upstream leak	8	7%	92%
Valve leak	7	6%	98%
Incorrect valve	2	2%	100%
Total	113	100%	100%

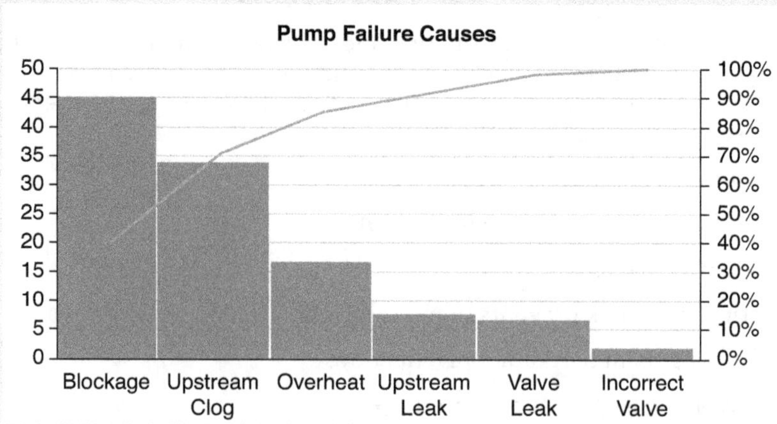

Figure 7.4 Pareto chart for pump failure causes.

Using this chart, the engineer was able to focus her investigation on understanding why blockage and clogs were occurring. The investigation showed that upstream material was introducing solids into the flow of the piping system, which prompted the engineer to install a series of filters. This action led to significant reduction in the overall failure rate of the pumps. After addressing the clogging and blockages, the engineer is now freed up to investigate why the pumps are overheating.

Example 7.4 demonstrates how well-chosen data visualization can lead to action and improvement. Strong data visualization techniques, such as Pareto charts, may often be paired with secondary analytical methods and investigation, such as bad actor analysis. Pareto charts and bad actor analysis pair very well, as bad actor analysis focuses on prioritizing which failures to address based on impact. The initial Pareto analysis highlights the need to focus on blockage and clogging, and it so happened that those issues were related. This may not always be the case, but root cause analysis will clarify the nature of issues when they are identified.

The box-and-whisker plot (also called a box plot), graphically displays the "five number summary" of the data:

1. Minimum value
2. First quartile (Q1)
3. Second quartile (as known as the median)
4. Third quartile (Q3)
5. Maximum value

The interquartile range (Q3-Q1) can also be visually determined from the plot. Commonly, the original box-and-whisker plot is modified to identify outliers.

Depending on the software being used, other descriptive statistics may be available for display as well. Box plots can help us see if the data are compact or scattered, symmetric or skewed, and if any outliers are present. Box-and-whisker plots are quite appealing and provide a lot of information about a data set in just a single diagram. They are also extremely useful as a means of comparing different samples or groups of data, as shown in Example 7.5.

EXAMPLE 7.5

The reliability engineer working on the pump failures wants to understand the operational failures better. They collect temperature and pressure data to look for trends in the failure patterns. The engineer compiles data from an upstream pressure gage and a downstream temperature gage to look for patterns. The pressure data was then presented using the following box-and-whisker plot shown in Figure 7.5.

Continued

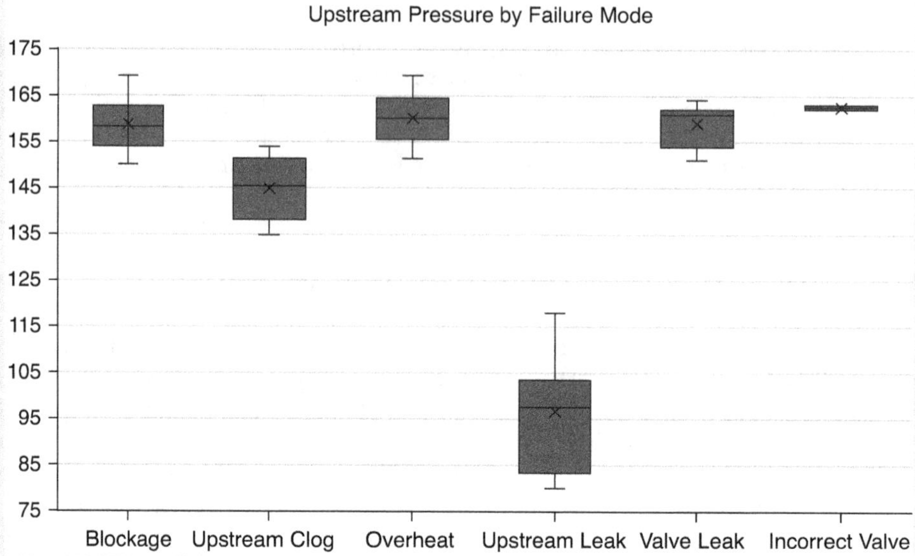

Figure 7.5 Box-and-whisker plot.

The box-and-whisker plot shows that when an upstream leak occurs, the system loses upstream pressure. While this does not provide a clear path to preventing upstream leaks, it does help with diagnostics moving forward. The engineer noted that upstream clogs also had the potential to reduce pressure, but they were able to confidently determine that if the system was reading upstream pressures of less than 125 psi, the maintenance team should shut down that line and look for a leak upstream of the pump.

A *scatter plot* is a plot of two variables, one on the y-axis and the other on the x-axis. The resulting graph allows visual examination for patterns to determine whether the variables show any relationship or if there is just random "scatter." This pattern, or lack thereof, aids in choosing the appropriate type of model for estimation and prediction. It is important to note that evidence of a pattern does not imply that a causal relationship exists between the variables.

Patterns demonstrated by the scatter plot may be linear or nonlinear. If the plotted points suggest a straight line, then we can measure the strength of the linear relationship by using the sample correlation coefficient, denoted by the letter r. A positive correlation means that the line moves from the bottom left to the upper right. A negative correlation means that the line moves from the top left to the bottom right. If all the points fall exactly on a straight line that tips up on the right end, then $r=1$. If all the points fall on a straight line that tips down on the right end, then $r=-1$. The value of r is always between –1 and +1, inclusive. This may be stated as $-1 \leq r \leq 1$. Therefore, the first step in analyzing a set of data for correlation is to construct a scatter diagram. If the points on this diagram tend to form a straight line, it makes sense to calculate the value of r.

The fact that a correlation exists does not imply that a cause-and-effect relationship exists. To conclude that there is a cause-and-effect relationship, we turn to logic, scientific theory, or a designed experiment to determine the cause-and-effect relationships. When we obtain a correlation coefficient high in magnitude, it means that we can predict one variable using the other. When we obtain a correlation coefficient near zero, the predictive ability of the variables is low, and we can generally, but not always, conclude that a cause-and-effect relationship does not exist.

Keep in mind that the correlation coefficient is a statistic. As such, we can place a confidence interval around this statistic and make a probabilistic statement regarding its significance.

EXAMPLE 7.6

Suppose an injection molding machine is producing parts with pitted surfaces. The following potential causes have been suggested:

- Mold pressure
- Coolant temperature
- Mold cooldown time
- Mold squeeze time

Values of each of these variables as well as the quality of the surface finish were collected on ten batches of parts. These data are shown in Table 7.10.

Table 7.10 Data for Example 7.6.

Batch number	Mold pressure, PSI	Coolant temperature, °F	Cooldown time, minutes	Squeeze time, minutes	Surface finish, microinches
1	220	102.5	14.5	0.72	37
2	200	100.8	16.0	0.91	30
3	410	102.6	15.0	0.90	40
4	350	101.5	16.2	0.68	32
5	490	100.8	16.8	0.85	27
6	360	101.4	14.8	0.76	35
7	370	102.5	14.3	0.94	43
8	330	99.8	16.5	0.71	23
9	280	100.8	15.0	0.65	32
10	400	101.2	16.6	0.96	30

Four graphs have been plotted in Figure 7.6. In each graph, "Surface finish" is along the *y*-axis and one variable (that is, mold pressure, coolant temperature, mold cooldown time, or mold squeeze time) is along the *x*-axis. Each data point on a graph represents one batch.

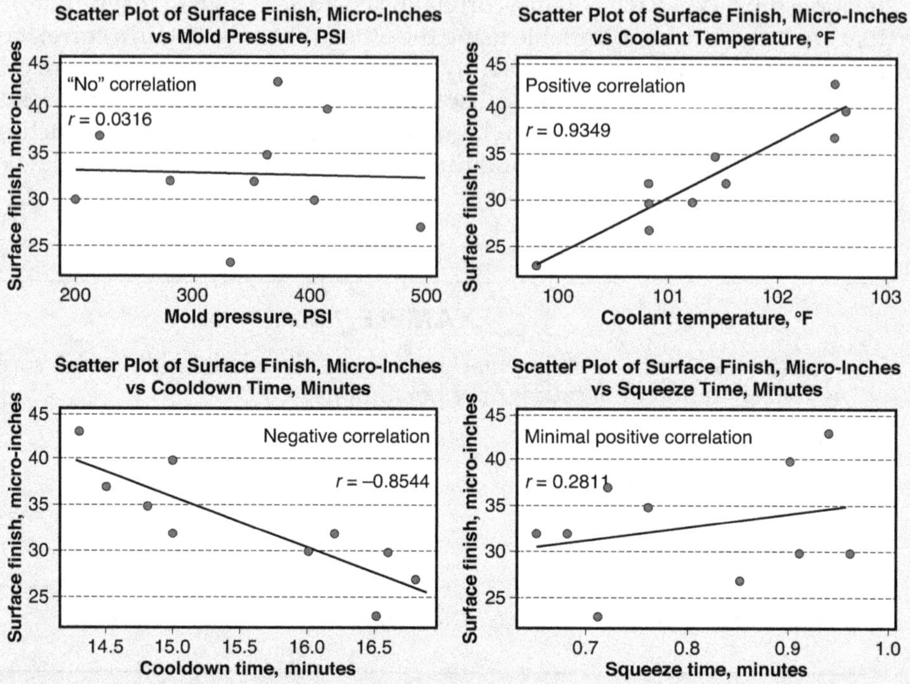

Figure 7.6 Scatter diagrams for Example 7.6.

Let's briefly examine each of the graphs:

- *Scatter plot of surface finish versus mold pressure.* Notice the overall dispersion of the dot points. They are well dispersed and do not appear to present any noticeable pattern. The regression line for the variables is the solid black line in the middle of the graph. Notice that it is almost flat. This indicates that the y-value variation is almost completely unrelated to any change in the x-value. Consequently, there is little to no correlation present.
- *Scatter plot of surface finish versus coolant temperature.* The dot points are more tightly clustered together and clearly present a noticeable linear pattern. The addition of the regression line confirms a visually strong correlation coefficient.
- *Scatter plot of surface finish versus cooldown time.* The dot points in this graph are somewhat polarized in the lower and upper ends of the cooldown times, appearing to contribute to the correlation, which in this case is negative in nature.
- *Scatter plot of surface finish versus squeeze time.* Like the mold pressure, there is a lot of dispersion in the dot points, but a minor appearance of a linear pattern is present. The dot points are polarized and loosely clustered at the ends. An experienced eye might be able to determine the correlation to be minimally positive, but the aid of the regression line is certainly helpful.

Realizing value from data usage often begins with effective data presentation, as knowing the right path forward and securing buy-in do not always go hand in hand. Data visualization is a powerful tool to drive understanding of analytical results and provide information to get good use out of the data. Graphing key reliability metrics is often a good starting point:

- The probability density function (PDF) can be plotted as a histogram showing the number of failures in each time block.
- The hazard function shows the failure rate as a function of time. It is convenient to use Equation 7.2 for the hazard function.

Equation 7.2: Hazard function

$$\lambda(t) = \frac{\textit{Fraction of failures during the time period}}{\textit{Amount of time during the period}}$$

- The reliability function plots reliability as a function of time, using the formula in Equation 7.3.

Equation 7.3: Reliability function

$$R(t) = \frac{\textit{Number surviving at the end of the period}}{\textit{Number of units tested}}$$

FAILURE ANALYSIS METHODS

> Diagnose failures by using a diverse array of failure analysis tools and techniques. Implement physics of failure-based analyses, including methods such as scanning electron microscopy (SEM), scanning acoustic microscopy (SAM), infrared inspection, and radiography analysis. Conduct mechanical and materials analysis, applying specific material testing and measurement techniques like tensile testing, shear testing, viscosity measurement, and vibration testing. Incorporate physical analysis methods including visual, ultrasonic, and electrical analysis. Engage in verification like cycle testing and nondestructive testing (NDT) and employ advanced predictive technologies to anticipate and mitigate potential failures. (Understand)
>
> **Body of Knowledge III.B.5**

It is important to have as much information as possible about each failure. This section discusses some approaches that have proved useful.

The first step in failure analysis is to determine why the item has been classified as failed. This investigation should begin with the warranty report, but it

Table 7.11 Failure analysis methods.

Method	Type	Use case description
Machine learning	Advanced predictive technologies	Teach algorithms to recognize patterns in data
Infrared inspection	Mechanical and materials analysis	Measure image and thermal properties using infrared radiation
Radiography analysis	Mechanical and materials analysis	Evaluate internal structures using special imaging techniques
Ultrasonic inspection	Mechanical and materials analysis	Identify abnormalities in a material by reflecting ultrasonic sound waves off of surfaces
Electrical analysis	Mechanical and materials analysis	Calculate and measure the electrical potential of components and systems
Tensile testing	Physical analysis	Measure material response to pull stress
Shear testing	Physical analysis	Measure material response to shear force
Viscosity measurement	Physical analysis	Measure a material's resistance to flow
Vibration testing	Physics of failure	Observe the vibration patterns in a machine during operation to see operating condition
Lifecycle testing	Physics of failure	Functionally testing a product to failure, often under simulated conditions to accelerate failure
Nondestructive testing	Physics of failure	Inspect parts in ways that do not damage them
Scanning electron microscopy	Verification	Observe materials at high resolution through visual microscopy
Scanning acoustic microscopy	Verification	Create high resolution microscopic images using sound waves
Visual inspection	Verification	Compare a part or system to known good conditions visually

should not end there. The item should be examined for failure in its construction, but further study should be made of the application and installation—things that should be included in a warranty report but often are not. Traceability and other documentation should be examined for clues as to the root cause of the failure. If the item did not physically fail, a thorough study of the environment in which it functioned is required.

If the part has become deformed or broken, two questions arise:

1. Has the part failed to meet the designed strength?
2. Has the part been subjected to stress beyond its designed strength?

To understand how and why parts may have failed, it is critical to bring a knowledge of available measurement methods to the table and when they should be deployed. Many common measurement methods are summarized in Table 7.11.

One approach to resolving the two questions is to subject several identical parts to various stresses. Of course, the identicality, while critical, is impossible to attain. This means a thorough examination of the failed part is necessary to determine whether it has some unique property that makes it more vulnerable to stress. This may involve techniques ranging from a visual study to chemical or metallurgical analysis. Nondestructive tools such as ultrasound, eddy current, X-ray, liquid penetrant, and magnetic particle testing may be helpful. In some cases, grain structure analysis and electron microscope examination are appropriate.

If this analysis shows that the part has been subjected to stress outside its designed strength, new questions are raised, including:

- Is the design strength adequate?
- Was the part applied inappropriately?
- Did the environment (thermal, chemical, electromagnetic, vibration, and so on) contribute to the failure?

The reliability program should support a traceability system that will identify materials, methods, processes, and supplier information for each item. This will help determine whether other items in the field are vulnerable.

After determining the root cause of a failure, a report should be generated that includes any recommendations for changes that will help reduce this type of failure in the future.

FAILURE REPORTING, ANALYSIS, AND CORRECTIVE ACTION SYSTEM (FRACAS)

> Identify elements necessary for FRACAS and demonstrate the importance of a closed-loop process. (Evaluate)
>
> **Body of Knowledge III.B.6**

Despite the efforts that are put into identifying and preventing failure modes, some unanticipated failures occur. These failures are often of an insidious nature, having eluded the team during FMEA activities or else occurring despite corrective/preventive actions that were prescribed. For these reasons, there may be a tendency toward lax reporting of these failures.

Failure reporting and corrective action systems (FRACAS) provide an organized, disciplined approach to this problem. The first step is to establish a failure review board. This group is responsible for implementing these guidelines:

- Require prompt and complete failure reporting.
- Begin the reporting with the earliest tests of products and processes and continue through the life of the product.
- Determine root causes and appropriate corrective or preventive action in the analysis and corrective action phase of each report.

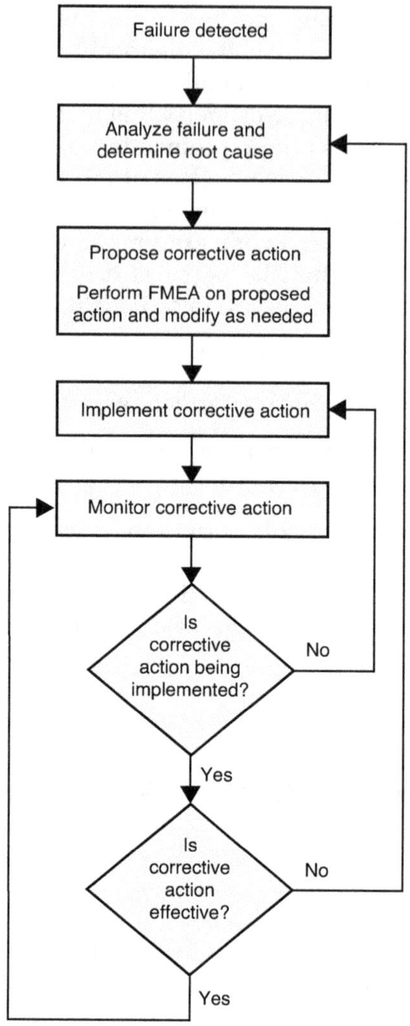

Figure 7.7 FRACAS flowchart.

This phase should be completed within a prescribed period, usually 30 days. It is best to use a closed-loop corrective action process, as shown in Figure 7.7. The RE should also implement audits to determine whether corrective actions have been taken and whether they are effective. When corrective actions do not resolve the problem, the failure review board is responsible for a restudy of the failure mechanism and the implementation of additional corrective actions.

It can be a challenge to describe what every FRACAS system should look like, as they must be equipped to respond to a wide range of different failures. Example 7.7 demonstrates how the reliability engineer dealing with the pump failures would follow through with a FRACAS system.

EXAMPLE 7.7

A reliability engineer desires to implement a structured FRACAS system to continue to improve the performance of the pumps and other systems. They have already identified failures through inspection (and will continue to identify them through automated data collection), identified root causes through Pareto analysis, and proposed and implemented a corrective action.

To follow through with the FRACAS approach, the engineer must perform a failure modes effects analysis (FMEA) on the filters that were implemented to prevent clogs. The following steps are to confirm the implementation and effectiveness of the change. In this scenario, the data from the flowmeters would confirm the effectiveness of the change. The reliability engineer could also perform a hypothesis test over a future 12-week data collection period to confirm that the failure rate has declined.

Part IV

Reliability Planning, Testing, and Modeling

Chapter 8
Reliability Planning

RELIABILITY TEST STRATEGIES

> Evaluate and develop suitable test strategies, such as truncation, test-to-failure, degradation, growth plan, and Test, Analyze, and Fix (TAAF) for various phases of product development. Consider customer profiles, including those reflecting normal and heavy usage. Implement a 'zero failure' test to assess reliability. (Evaluate)
>
> **Body of Knowledge IV.A.1**

The reliability of a product is the probability that it continues to perform as a function of time or another measure of product use, such as cycles. The results of some reliability tests are used to estimate reliability at specified time or cycles, to calculate confidence limits on the estimates, or to demonstrate conformance to a specified reliability requirement. Often, these tests must be performed over an extended period of time. The results of other reliability tests are used to improve the reliability of a product and may not require testing over long periods. If the primary purpose of a testing program is to measure or improve reliability, the test is classified as a reliability test. It is also possible to obtain reliability information from other tests, such as feasibility, functional, or quality assurance tests.

The strategies for various reliability tests are primarily determined by the stage of product development during which the testing is performed. No units are available during the early phases of new product development; therefore, no reliability testing can be performed. Component reliability testing may be planned at this early stage. As the preliminary design becomes available and prototypes or engineering models are built, qualitative testing typically begins. The strategy for this reliability testing program is to stress the units beyond the design limits to cause failures. A reliability improvement program to eliminate the weak components is then initiated based on analyzing these failures. The resulting design will become more robust, and failures due to marginal components will be eliminated, thus improving product reliability.

Once production units become available, reliability tests to estimate the product's failure rate or mean time between failures (MTBF) can begin. These tests, called life tests, accumulate test time on several units while recording failures.

They must run for some time, and quite often the test time is prohibitive, making some acceleration necessary.

If a compliance test is required, it must be performed before the product can be delivered to the customer. The strategy of this type of test is to verify, at a specified level of confidence, that the product conforms to a required minimum reliability value. This may be done by accumulating time on a sample of units, requiring this test to be performed over some time. Compliance testing can only be performed once the final design is set, production tooling is in place, and units have been produced. The strategy of compliance testing is not product improvement.

Planning for an integrated reliability test program must start at the beginning of the project and must be thorough and timely if all the necessary elements are to be in place when they are needed. Planning ensures that the number of units to be tested, the test facilities—including any special equipment—and the necessary time are available when the tests are to be performed. Scheduling the various reliability tests is required so the results can be used promptly based on the test strategy. Making a design change before the product is released to production is much more efficient.

A test plan needs to include the following elements:

- objectives of the testing program
- provision of resources for facilities, test equipment, and equipment calibration information
- the time and necessary personnel to conduct the testing
- test requirements and schedule, including the number of units to be tested and the test environments
- procedures to make changes in the testing program as necessary
- documentation of the test results

Suppose the test strategy is to record failures over time. In that case, a precise definition of product failure must be available to the test engineer, as the output of some products can deteriorate over time.

Many sources list the steps in developing a reliability test plan. The number of steps and the words describing the steps may differ, but most contain essentially the same information. For example, Kececioglu (1993) lists nine steps:

1. Determine the test requirements and objectives.
2. Review existing data to determine if any requirements can be met without testing.
3. Review the list of tests to determine whether combining any tests would be economically feasible.
4. Determine the necessary tests.
5. Allocate the resources required to perform the tests.
6. Develop test specifications and data handling and storage procedures. Review acceptance and qualification criteria. Establish procedures for making future changes in the test specifications.
7. Assign the responsibility for conducting the tests, analyzing the results, and providing the overall integration of the testing program.

8. Develop forms and procedures for reporting the test results.
9. Develop procedures for maintaining the test status information throughout the entire testing program.

Reliability tests can be placed in several different classifications. One way to classify tests is by the development/production cycle phase in which they are conducted. Four major test categories classified in this manner are:

- Product development tests
- Reliability performance tests
- Reliability acceptance tests
- Reliability verification tests

These tests have different objectives and are conducted at various times during product development and production.

Product Development Tests

These tests are performed to take action to improve the design in the event of failures and to evaluate the system design, including the compatibility of subsystems. Results from product development testing may be used to demonstrate the product's functional capabilities but not to determine reliability parameters, such as failure rate or MTBF. The objective of product development testing is to cause failures. Only by analyzing failures and taking appropriate action to change the design can improvement be made to the product. Testing units for long periods and observing no failure is costly and does not result in product improvement.

Reliability Performance Tests

Sometimes referred to as reliability qualification tests, these tests are performed after the design is completed. They will demonstrate that the system can meet the requirements under the specified operation conditions (including environmental). Generally, these tests do not provide the data to determine reliability parameters but only to give assurance that performance under stated conditions can be met.

Reliability Acceptance Tests

Conducted during the production phase, these tests will demonstrate that the design's reliability parameters have been maintained. Successful reliability acceptance tests show that the product continues to meet its reliability requirements and current production may be shipped to customers.

Reliability Verification Tests

These tests are performed to show compliance with stated reliability parameters such as MTBF or failure rate and may be required to demonstrate compliance at a given confidence level. This type of testing is formal and statistical and requires standard procedures. Many units or long periods of test time may be required for these tests, making some form of acceleration necessary. These tests are based on acceptable and rejectable reliability values, such as MTBF, failure rate, and decision risks.

Reliability Test Strategies

Reliability tests can be classified according to how they are conducted and the type of results recorded. The type of data to be taken and how it is reported must be part of the overall reliability test plan. Reliability tests can be continuous or pass/fail. The results of continuous tests are recorded as variable data, and the results of pass/fail tests are recorded as attribute data.

Attribute data are recorded as success or failure for a given test. The units being tested might not have active operating times. These units are referred to as one-shot items. When called on to operate, they are units that perform successfully or fail, such as sensors or fuses. Estimations of the reliability or the probability of success of one-shot items can be made from the attribute data. Confidence limits on the reliability can also be calculated using the data. Attribute data might also be recorded even if the tested units have active operating times. Units being cycled in an environmental chamber may pass or fail the test. Unless each unit is metered separately, once the test is complete, which units were successes and which failed, will be known. The recorded data are the number of units that passed and the number of units that failed the test; the actual time to failure of the units is unknown. Confidence limits and estimates of the reliability of these units for a mission time equal to the equivalent time of the test can be calculated using the data.

The results of continuous tests are recorded as variables. The actual times of failure of each failed unit and the total time accumulated on the nonfailed units are known. Life tests are conducted to determine reliability parameters, such as MTBF or failure rate. Estimates of the parameters and confidence limits can be calculated from the variables data. Continuous testing may be conducted as replacement tests or as non-replacement tests. The advantage of a replacement test is that the number of units on the test is constant, generating data faster. In a non-replacement test, units that fail are not replaced, and the test population becomes smaller.

Compliance tests are performed to show that units have achieved some reliability value. They are also known as reliability demonstration tests. The results of these tests could be recorded as variable or attribute data. Compliance with an MTBF or failure rate can be shown using variable data from fixed-time or sequential testing. Both of these forms of testing are described further in Chapter 9. A fixed-time test is designed assuming a sufficient sample will be observed for a specified time according to the reliability requirements. A sequential test assesses the interim reliability with each new observation. The sequential test may end early if the demonstration is achieved. If required, the requirement could be demonstrated at a given confidence level. The consequence is fewer units required or a decrease in the test time. The disadvantage of sequential testing is that the actual time needed for the test is unknown initially. Attribute data from a pass/fail test can be used to demonstrate compliance to a reliability value or a probability of success value for one-shot items at a given confidence level. The attribute test requiring the fewest number of units is a no-failure test. This is sometimes referred to as success testing.

The time and the number of units required for a life test can become prohibitive. Projects may have limited test units or limited hours available to run the tests. If the product has high inherent reliability, it would not be unusual to need several thousand hours of total unit test time to get meaningful results. Some type of acceleration may be necessary to perform these tests.

Life testing requires the test to be run on several units over time to accumulate a total unit test time. Using the assumption of a constant failure rate or an exponential distribution of time-to-failure, 100 units were tested for 100 hours, 50 units were tested for 200 hours, and 10 units were tested for 1000 hours, resulting in 10,000 hours of total unit test time. The data from these tests can be used to make estimates of product reliability measures, such as failure rate or the MTBF, and to determine the confidence limits of the estimate. The units tested must be typical of the production units. Any changes in the design will affect the testing program, probably requiring more time or test units.

Reliability Test Termination

Before beginning a test, deciding how the test will be terminated is necessary. Several possibilities are available to reliability engineers. Test-to-failure refers to studies that record data about failure type and time. Some such tests continue until all units have failed. Since this can take a very long time, reliability tests can be planned to be truncated. Time truncation means terminating the test after a specified amount of time has elapsed or cycles have been completed. For example, a part or sample of parts may be cycled a million times or cryogenically exposed for 100 hours. The test samples can be inspected periodically during that time or inspected only when the time is completed. Failure truncation means terminating the test after a specified number of failures has occurred. A degradation test terminates when a product reaches a predetermined reduced level of performance. The analysis of data from these different protocols requires special techniques. Statistical software packages are typically used. When a test is truncated before all the units have failed, the resulting data will include right-censored observations.

Reliability Growth Testing

Reliability growth is the improvement in product reliability over time. This improvement in reliability is due to changes in the product design. In the early stage of new product development, problems exist in the design that negatively affect reliability. Early-development reliability activities, such as FMEA, design reviews, reliability prediction, and early testing on prototype and engineering models, will begin to identify these problems. Changes in the product design will begin to eliminate these problems. Tests will then be run on the product of the new design, and other problems will be identified and eliminated. This activity, when repeated, is referred to as *test, analyze, and fix* (TAAF) and will result in reliability growth. *Military Handbook: Reliability Growth Management* (MIL-HDBK-189) provides an understanding of the concepts and principles of reliability growth testing.

The most frequently used models to track reliability growth are the *Duane model,* developed by James Duane in 1962, and the *Army Material Systems Analysis Activity* (AMSAA) *model,* developed by Larry Crowe in 1978. Both use the observed linear relationship between the logarithm of cumulative MTBF and cumulative test time. The Duane model is based on observation and uses a deterministic method to track reliability growth. The AMSAA model allows for statistical procedures to be used to model reliability growth.

The data are cumulated as the TAAF process progresses. A quantity called the cumulative MTBF ($\theta_{C(T)}$) is calculated as:

$$\theta_{C(T)} = T / r(T)$$

where

T = Total unit test time, including all testing

$r(T)$ = Total number of failures, including all testing

As the TAAF process continues and T increases, $\theta_{C(T)}$ will increase, reflecting reliability growth. The Duane model tracks this growth by fitting the following equation to the values of $\theta_{C(T)}$ at different test times:

$$\log \theta_{C(T)} = \mathrm{a} + \mathrm{b} \log T$$

This linear equation in logarithms will be plotted as a straight line on log-log graph paper. In this equation, the value b will be an estimate of the reliability growth rate. In the Duane model, the MTBF as of time T is calculated as $\theta_{C(T)} / (1-\mathrm{b})$.

The growth rate b can be used to compare the growth of a given project with the growth of similar projects. A higher-than-average growth indicates that resources are being used aggressively for that project. A lower-than-average growth shows that the use of reliability improvement resources is limited for the project.

$$b = \frac{\log \frac{\theta_1}{\theta_0}}{\log \frac{T_1}{T_0}}$$

In the Duane model, the true MTBF will grow at the same rate as the cumulative MTBF value. At any time during the testing, the true MTBF (θ) can be found as

$$\theta = \frac{\theta_m}{1-b}$$

The growth rate b can be used to compare the growth of a given project with the growth of similar projects. A higher-than-average growth indicates that resources are being used aggressively for that project. A lower-than-average growth shows that the use of reliability improvement resources is limited for the project.

The average growth rate for most projects is about 0.25 to 0.40. A higher growth rate shows that eliminating design weaknesses has been given top priority. A lower growth rate indicates that reliability improvement actions are taken to eliminate only the most prominent design flaws.

Example 8.1 demonstrates the use of the Duane growth model.

EXAMPLE 8.1

A new speed sensor and control module for an antilock braking system (ABS) are being developed. Units are subjected to three cycles of growth development testing. After each set of tests, the results are analyzed, and corrective action is implemented. After each design change, new units are built for the subsequent test. Each test is conducted as a replacement test using 20 units. The equivalent

Continued

test time for each test is 1000 hours, resulting in a total test time of 20,000 hours. The following results were obtained.

Test Cycle 1

Equivalent test time $T_{(1)}=20{,}000$ hours. The cumulative test time $T=20{,}000$ hours. Number of failures $r_{(1)}=20$. The cumulative number of failures $r=20$.

$$\theta_{C(1)} = \frac{20000}{20} = 1000$$

Test Cycle 2

Equivalent test time $T_{(2)}=20{,}000$ hours. The cumulative test time $T=40{,}000$ hours. Number of failures $r_{(2)}=12$. The cumulative number of failures $r=32$.

$$\theta_{C(2)} = \frac{40000}{32} = 1250 \text{ hours}$$

The $\theta_{C(2)}$ calculation uses the data from test cycles 1 and 2.

Test Cycle 3

Equivalent test time $T_{(3)}=20{,}000$ hours. The cumulative test time $T=60{,}000$ hours. Number of failures $r_{(3)}=8$. The cumulative number of failures $r=40$.

$$\theta_{C(3)} = \frac{60000}{40} = 1500 \text{ hours}$$

The calculation of $\theta_{C(3)}$ uses the data from test cycles 1, 2, and 3.

A plot of the data is shown in Figure 8.1. $\log_{10}$ cumulative MTBF (θ_C) is plotted on the *y*-axis versus $\log_{10}$ cumulative time (T) on the *x*-axis. A linear measure of the slope of the line shows a growth rate of about 0.364, shown in Figure 8.1.

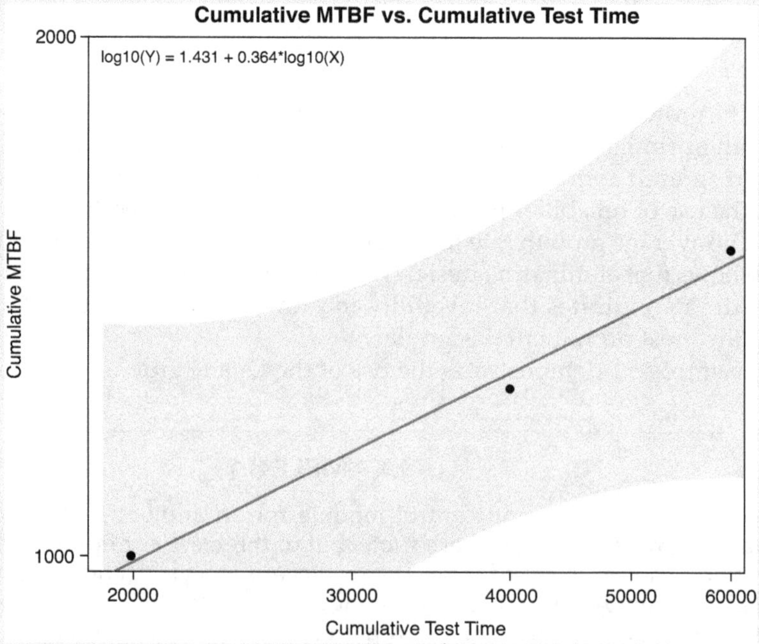

Figure 8.1 Duane growth plot.

Customer Profiles

Reliability test strategies will also depend on the product requirements that depend on customer use profiles. For example, the reliability engineer may determine the distribution from light to nominal to heavy usage of the product. Reliability requirements may be set for different customer profiles, for example, the 10th percentile user, the 50th percentile user, and the 90th percentile user. Then test strategies can be developed to focus on these use profiles.

Planning Zero-Failure Tests to Estimate Reliability

It is possible to plan reliability tests that will demonstrate a specified level of reliability if the test is completed with zero failures. These tests can be planned on an attribute data basis without the assumption of a specific time-to-failure distribution, or they can be planned by assuming that the time-to-failure distribution is Weibull with a known shape (beta) parameter. In both cases, the zero-failure test plan requires the smallest sample size, compared with plans that allow specified numbers of failing units. A challenging aspect of the attribute version of this test is that it can require very large sample sizes. A challenging aspect of both the attribute and Weibull distribution versions of these plans is that to pass and demonstrate a specific level of reliability (say R), the actual level of reliability of the product must be significantly higher than R.

The sample size for the attribute-data zero failure test plan is given by:

$$n = \frac{\log(\alpha)}{\log(R)}$$

where the reliability R at a particular time T is to be demonstrated with $100(1-\alpha)\%$ confidence. If all n samples successfully run to time T, the reliability requirement has been demonstrated.

Sample sizes can be reduced if (1) we can assume a Weibull distribution with a specified shape parameter β for the time-to-failure distribution, and if (2) we can run the test samples beyond the required lifetime T. A zero-failure test plan run for k multiples of the required lifetime T requires a sample size of:

$$n = \frac{1}{k^{\beta}} \times \frac{\log(\alpha)}{\log(R)}$$

Example 8.2 illustrates the procedure.

EXAMPLE 8.2

A new electrical catheter is being developed, and a particular adhesive joint requires a reliability of 0.995 after $t = 10$ clinical cycles. To demonstrate this reliability with 90% confidence, a sample of 460 units would need to be run to $t = 10$ clinical cycles without failure:

$$n = \frac{\log(0.10)}{\log(0.995)} = \frac{-522.735}{-0.002177} = 459.4$$

Now assume that data from a predicate catheter model supports a Weibull distribution model with $\beta = 2.5$ for this lifetime distribution. If it is possible to run

Continued

the catheters to $t=20$ clinical cycles, then a zero-failure test plan with a sample size of 82 can be planned:

$$n=\frac{1}{2^{2.5}}\times\frac{\log(0.1)}{\log(0.995)}=(0.177)x(459.4)=81.2$$

If all 82 samples complete 20 clinical cycles without failure, then the reliability of 0.995 will be demonstrated with 90% confidence.

See Doganaksoy, et al. (2021), Taylor (2017), and Tobias and Trindade (2012) for more information on reliability test planning strategies.

ENVIRONMENTAL FACTORS AND USE CONDITIONS

Analyze various environmental factors and use conditions such as temperature, humidity, and vibration, along with stresses like severity of service, electrostatic discharge (ESD), throughput, and duty cycle to which a product may be exposed. Incorporate the application of multiple stress factors (e.g., highly accelerated life testing [HALT]) to assess their simultaneous impact on a product. (Analyze)

Body of Knowledge IV.A.2

Understanding a product's use-environment and use-condition is a prerequisite for designing effective reliability tests. Note that, by the definition of reliability, a reliable product is a product that functions under its intended use conditions during the required mission time. Therefore, carefully defining and characterizing the product's environmental and use factors will help designers make informed decisions and assist the company in developing a better warranty policy. Other purposes for characterizing environmental and use factors are to:

- Design a product that meets customer expectations
- Determine product safety and durability
- Plan an effective reliability test that can provide useful information for reliability prediction
- Evaluate the product transportation, storage, installation, and use condition
- Design product monitoring schemes
- Effectively mitigate or reduce stress effects

If the product intends to be used outdoors, the environmental stress variables are often climate variables, such as temperature, humidity, UV, wind, hail, salt, precipitation, snow, and so on. Characterizing the use condition of such products requires collecting climate data at their use locations, which are often available from public resources at the National Oceanic and Atmospheric Administration (NOAA) and the Department of Energy (DOE). However, modeling these climate variables may be a complex task. For example, to characterize the use environment of a photovoltaic module, the Jet Propulsion Laboratory (JPL) suggests an environmental cell model that discretizes temperature, humidity, and UV and counts their occurrence at a location over a year. Then, the distributions of these environmental variables can be found. Knowing the stress variable distribution, instead of merely an average value, is important to reliability test planning, as a good test plan should mimic the real use environment and may even elevate the stress level based on the environmental data. Often, reliability engineers would like to know how the product performs under an environmental condition that is more severe than normal, or, in a worst-case scenario, to give a conservative lifetime estimation or to expose any design weakness for future product improvement.

Besides climate-related variables, other environmental stress variables include:

- Pressure
- Electrical current and voltage
- Magnetic
- Vibration
- Friction
- Drop and shock
- Radiation
- Electrostatic discharge (ESD)
- Throughput
- Duty cycle

In general, any external factor that may induce product failure should be investigated. Therefore, identifying environmental factors should be product specific, and the corresponding test plan should be built on specific failure mechanisms.

The use factor of a product also needs to be fully understood before planning a reliability test. Each product is designed for a specific market segment, and the product use intensity of different market segments could be drastically different. For example, a household washing machine may be used once or twice a week and 1 hour per use, but an industrial-grade washing machine that is to be installed in a laundry facility should expect everyday usage and at least 8 hours per day. Therefore, the choice of motor in these machines should be quite different. Without a clear understanding of the use factor, engineers may either under-design or over-design a product.

Traditionally, a customer survey and market analysis are the typical ways to gather data on product use severity. However, with more and more sensors being embedded into products to make them "smart," manufacturers and service providers today have more sophisticated real-time tools for monitoring a product's use condition. For example, a computer can automatically record data such as its power on and off time, CPU unitization and temperature, and memory writing frequency, and these data can be transmitted to a third party for fault diagnosis. The internet-connected automobile, such as the Tesla car, is another example. Aggregating these use data will give a realistic picture of how these products are used in the real world and whether there is a gap between design intention and actual severity of service in use. Similarly, reliability tests can be better designed to simulate the real use conditions.

Last, consider another example—borrowed from Ullman Sails—of environmental and use factors that are primary causes for sail deterioration. Environmental factors that can weaken the sail include UV and saltwater. UV exposure, which affects all sail materials to varying degrees, weakens fibers. Saltwater exposure can result in electrolytic corrosion of sail fastenings. Corrosion, the result of contact of dissimilar metals, causes slides, clew and tack rings, and even headboards to fail over time. Use factors encompass the inevitable stress that sails endure while in use. They include the impacts of luffing, sails hitting the rig, and the typical loads required to harness the power of the wind under sail. Flogging, fluttering, and the sails hitting the rig will damage sailcloth and sail membranes. Understanding these damaging effects of environmental and use factors will help designers take actions to mitigate them. For example, UV inhibitors and top coatings are applied on the sailcloth to improve UV resistance; dissimilar materials are isolated to reduce corrosion potentials. All sailcloth can be subject to various assessments, such as "flutter and impact" tests and "fold and flex" tests, which simulate real-world scenarios and use conditions.

Highly Accelerated Life Test (HALT)

Once a design is set and prototype units are produced, a *highly accelerated life testing* (HALT) program can begin. The strategy for this reliability testing program is to stress the units beyond the design limits to cause failures. A reliability improvement program to eliminate the weak components is then initiated based on analyzing these failures. The resulting design will become more robust, and failures due to marginal components will be eliminated, thus improving product reliability.

Reliability tests can be classified according to the strategy and the types and levels of accelerations used. The strategy for a HALT program is to improve product reliability. The design is typically in the early development stage and has yet to be finalized. Units are tested at one or more stresses exceeding the design limits. The multiple stresses are usually environmental, such as temperature or vibration, but could be other loading stresses, such as voltage. Each stress is increased until some component in the product fails, resulting in product failure. A design change is made to eliminate the weak component. The process is then repeated on the product, incorporating the new design. Reliability improvement is achieved as the marginal design is improved by eliminating the weak components. Sometimes, these tests are also used to extend the beginning of wear-out, known as the end of useful life. The HALT program and other early product development reliability functions will ensure a robust and reliable design is released to production. The test failures are not typical, however, of the normal use of the product. Therefore, the results from the HALT program cannot be used to estimate the product's reliability.

FAILURE CONSEQUENCE

> Describe the importance of recognizing the consequences linked to different failure modes, including the severity and occurrence of such failure modes, when establishing reliability acceptance criteria. (Understand)
>
> **Body of Knowledge IV.A.3**

Reliability analysis provides estimates of the probability of failure. The reliability engineer must go beyond these calculations and examine the consequences of failure. These consequences typically represent the loss of system function and costs to the customer. The customer finds ways of sharing these costs with the producer through the warranty system, loss of business, decrease in reputation, or civil litigation. Therefore, an important reliability function is the anticipation of possible failures and the establishment of reliability acceptance goals that will limit their occurrence and consequent costs.

Identifying failure consequences requires considerable product domain knowledge and the system function view. Most of the time, not only does the direct impact of a failure mode on product functionality need to be assessed, but its second or third order of impacts on the related parts and the surrounding system (including human beings) must also be fully understood. Any potential harm to users due to the failure of a product may lead to severe civil lawsuits. These litigations can damage a company's bottom line and, more importantly, its reputation.

For example, in April and May 2013, multiple car manufacturers recalled the airbags manufactured by the Takata Monclova Plant in Coahuila, Mexico. By that time, at least 16 deaths were linked to defective airbags and a total of 3.6 million cars were recalled (Tabuchi and Jensen 2014). As a result, BMW announced plans to move any orders from the Mexican plant to a Takata plant in Germany (Klayman 2014). On January 13, 2017, the United States charged three Takata executives for the exploding airbag. The company agreed to plead guilty and to pay $1 billion to resolve the investigation, which includes a $25 million fine, $125 million for victim compensation, and $850 million to compensate automobile manufacturers (Reuters 2017). As shown in this example, the consequence of a product failure could be very severe in terms of system failure, human safety endangerment, loss of customers, costs to the company, and so forth. Consumer product failures often invite wide media coverage and regulatory actions that are extremely damaging to the company's reputation, an example of an external cost of poor quality.

Analytical techniques can be used to understand and define the failure consequence. Expert judgment should be considered from groups or individuals with specialized training or knowledge on the subject area, such as product designers and product managers, system engineers, reliability engineers, industry groups and consultants, consumer relation specialists, and customers themselves. Meetings can be held with all stakeholders to brainstorm the possible consequences of a product failure for the whole system and the company's and customers' loss. Each risk needs to be ranked accordingly, based on the probability of its occurrence and its impact. Therefore, identifying the consequences of failure will help prioritize resource allocation to prevent failure.

Both qualitative and quantitative tools have been used for assessing failure consequences, including:

- Document review
- Delphi method
- Root cause analysis
- Probability and impact (PI) matrix
- FMEA

The ASQ Certified Black Belt Handbook (2022) states (p. 536):

> *Using the Delphi method, a questionnaire is sent to experts in the field to collect feedback on [failure consequences], one at a time. After each survey response, the results are anonymized and summarized, and sent out to the next expert until a consensus is achieved. This technique can be time-consuming, but it allows the experts time to think through the problem. In addition, this method avoids team pitfalls such as groupthink, or having a few dominant personalities steer the conversation.*

Root cause analysis can be performed using the 5 Why method or by having the team construct a fishbone diagram. Both techniques are discussed in Chapter 2.

A probability and impact (PI) matrix captures the likelihood of occurrence of potential risks and the magnitude of their impact if they were to occur. The matrix is presented as a grid with probability on the vertical axis and impact on the x-axis with the grid color-coded in green, yellow, and red. The PI matrix is similar to the risk matrix presented in Chapter 4.

In particular, the FMEA technique explicitly requires the numerical ranking of the consequence, or severity, of each failure mode (1 is low, 10 is high), occurrence (1 is low probability, 10 is high probability), and detection (1 is highly effective, 10 is ineffective). These rankings may come from expert judgment or from historical data from the same or similar products or from industrial standards. The severity, occurrence, and detection rankings are multiplied together to yield the risk priority number (RPN). Then, we can use the ranking to prioritize the strategies for preventing or mitigating failures and failure impacts. Product redesign is often mandated for improving reliability, but other strategies, such as specifying use conditions and warranty conditions, are also common in mitigating failure impacts in practice. Another set of rankings may be determined in anticipation of the benefits of the changes. FMEA is presented more fully in Chapter 4.

FAILURE CRITERIA

> Establish failure criteria by considering customer requirements, user needs, system functions, system requirements, and warranty terms and conditions. (Understand)
>
> **Body of Knowledge IV.A.4**

Understanding product failures and failure consequences will lead to a proper definition of product failure criteria; that is, what is defined as a success or failure? First, reliability requirements and goals must be established according to customer requirements, user needs, system functions, industrial standards, and the company's social responsibility. These goals are usually set for the whole system or for the company's overall performance. Therefore, they need to be broken down to the acceptance criteria for individual products or components. The acceptance criteria associated with these goals generally fall into three categories:

- Functional requirements (such as 20 gallons per minute for a pump)
- Environmental requirements (such as temperature, radiation, pH)
- Time requirements (such as failure rate during useful life, time elapsed before wear-out phase begins)

Functional requirements concern the designed functions of a product. Whether a product can be accepted depends upon whether it can deliver these intended functions. The required performance must be quantifiable and measurable. Reliability engineers often need to translate these requirements into specific testing tools and testing goals. Environmental requirements specify the normal or accelerated stress conditions based on the environmental and use factors of the product (as discussed previously). In contrast, time requirements specify the product's expected lifetime under these conditions. Both environmental and time requirements should be connected to the company's warranty program, reflecting what has been promised to customers in the warranty contract. For example, if a product is under full warranty for two years, its reliability requirement should have an expected two-year or longer lifetime. We typically specify the lower percentile time, such as the 5th percentile or 10th percentile of product lifetime, to meet the reliability requirement such that most of the products will be guaranteed to be acceptable. If there is a significant variation in lifetime distribution, this could lead to a very long mean lifetime requirement.

In addition to defining failure criteria, a good warranty program reaches into many organizational functions:

- Product pricing must include allowances for warranty costs.
- Advertising can use warranties as inducements to purchase.
- Supply chain agreements should spell out warranty liabilities and procedures.
- Accounting procedures should include accruals or sinking funds to cover future warranty costs.

- Engineering and reliability functions have responsibilities for determining causes and determining product/process corrective actions as required.

The breadth of warranty considerations implies that improvements in warranty management have the potential for substantial financial impact. Finding and executing these improvements begins with an understanding of terms. Warranties should include:

- Effective initial date and duration of the warranty
- Definitions of parts and defects to be covered
- Warranty procedures, including return routes, replacement/reimbursement policies, and policies that will be followed when no defect is found
- Conditions of use, such as environmental specifications (ranges for temperature, vibration, salinity, electronic parameters) and operation procedures
- Conditions of storage, transportation, and installation

Therefore, the design of the warranty would require the input from engineers on the product's designed functions and its intended use conditions. The influence of warranty policy and failure criteria definition flows in both directions. If suppliers are involved, effective warranty management requires agreement as to all these aspects with each supplier, regardless of tier.

Once component, product, and system reliability goals have been set, a testing protocol should be implemented to provide validation that these goals will impact the failure rates and the associated consequences as planned. These reliability goals typically impose specifications on the product. In anticipation of the start of production, reliability engineers provide further testing procedures to provide verification that these specifications are being met.

TEST ENVIRONMENT

> Evaluate the test environment by considering the system location and operational conditions and incorporate these into the test plan to ensure the implementation of an appropriate test strategy. Validate test capability and conduct the tests following the verified methods. Evaluate and formulate decisions based on the test results, providing rationale and support for each decision made. (Evaluate)
>
> **Body of Knowledge IV.A.5**

The strategy of the reliability test and the product itself will determine the test's environment. As mentioned in "Environmental Factors and Use Conditions," the test environment should consider the usage or system location and operational conditions. If the desired result of the test is to produce failures typical of product use, the test must be conducted in an environment that reflects the environment of use. Sometimes, increased environmental stress can be used to accelerate these tests. Not all products react the same to a given environmental stress.

Solid-state electronics are susceptible to increased temperature. Mechanical units might be affected by vibration or increased contaminants such as salt spray or dust. Magnetic units might be affected by radiation. Electronic circuit boards could be exposed to vibration to detect poor soldering, inadequate strength, or other mechanical defects. Radiation will also affect solid-state electronics and could be used to verify that shielding is adequate. Humidity, rapid temperature ramps, and shock are other environmental stresses used during reliability testing.

If the strategy of the reliability testing is to eliminate design weaknesses from the product, environmental stresses over normal use may be used. The test strategy is to produce failures so that design changes can be introduced to improve reliability. The unit will fail in ways not typical of normal use. The elimination of the weak components will increase product reliability. The level of these stresses can exceed the design limits of the product. No reliability information can be obtained by testing units at stress levels lower than the normal use environment. Test chambers capable of the appropriate test environments are required.

Test equipment is now available to allow the simultaneous application of multiple environments. Combined environmental reliability testing (CERT) simultaneously applies multiple environmental stresses. The simultaneous application of temperature cycling and vibration is an example of CERT. McLean (2002) gives details as to the maximum level of environmental stresses that are appropriate. The maximum levels are specific to the product, but typical operation ranges are:

- Temperature: –70°C to + 100°C
- Temperature rate: 60°C per minute
- Vibration: Up to 30 Grms, or root mean square acceleration

The most advanced vibration systems use six degrees of freedom at frequencies from 2 to 5000 Hz. Low-frequency energy excites components with high mass; higher frequencies excite low-mass components. Solid-state components can be burned in at 150°C. Integrated circuits should withstand temperature ramp rates of 90°C per minute. More typical test environments perform vibration using three degrees of freedom and a temperature ramp rate of 20°C per minute. If the environmental testing is sequential, the order needs to be determined, which might be important. Running combined thermal and vibration tests eliminates this requirement. The results of the testing should be both accurate and precise. Accuracy implies that the data are on target, and precision relates to the tolerance limits of the equipment. This requires that the test chambers, measuring equipment, and data recording devices be calibrated to a standard. Periodic calibration is necessary and should be stated as part of the overall reliability test plan.

As reliability tests are designed for each product requirement, these tests should be validated with evidence that the methods are adequate for their intended purpose.

Chapter 9
Reliability Testing

> Describe the purpose, advantages, and limitations of various hardware and software/firmware tests, emphasizing the different approaches and methodologies required for each. Use common models to develop test plans, evaluate risks, and interpret test results, ensuring comprehensive reliability and functionality assessment.
>
> **Body of Knowledge IV.B**

There are several tests used to assess the reliability of parts, components, and systems, including:

- Accelerated life tests, such as single-stress, multiple-stress, sequential stress, step-stress, highly accelerated life tests (HALT), and margin tests
- Stress screening like environmental-stress screening (ESS), highly accelerated stress screening (HASS), and burn-in tests
- Qualification/demonstration testing such as sequential tests and fixed-length tests.
- Degradation testing, such as wear-to-failure tests

And, for software:

- Software testing such as white box, black box, operational profile, and fault injection

ACCELERATED LIFE TESTS

> Accelerated life tests (e.g., single-stress, multiple-stress, sequential stress, step-stress, HALT, margin tests) (Evaluate)
>
> **Body of Knowledge IV.B.1**

The time required for a reliability life test can be prohibitive. Accelerated life testing can reduce the actual time of the test. The strategy is to increase the rate at which failures occur, not to cause new types of failures. The goal is to compress time to understand the reliability of the tested item. Accelerated life testing assumes that the failure modes are unchanged by the increased stress. Failure analysis can determine whether new modes of failure result from the increased stress. This would be a reason to reduce the level or change the stress that affects the acceleration.

Single-Stress Tests

Some devices can be accelerated in time by simply using them at a rate that exceeds the normal use rate. Automatic test facilities can be built that will continuously operate the units. The time of use and the times of failure can be recorded. For example, a fixture operating at 120 strokes per minute could exercise keys on a computer keyboard one million times in less than six days. A home appliance such as a dishwasher that is cycled every two hours will accumulate 1200 cycles (average for eight years) in 100 days (less than three and one-half months).

Increased stress is often used to accelerate time or cycles. Suppose it is known that under normal use, 5% of the population will fail in a time $t_{(n)}$. If the units are operated under increased environmental stress, the same 5% will fail in a time $t_{(s)}$ where ($t_{(s)} < t_{(n)}$). The ratio $t_{(n)} / t_{(s)}$ is defined as the acceleration factor (AF). The actual time of the test (t) is referred to as the test time or clock time (which can also be expressed as test cycles) of the test. The equivalent time of the test is $t_{(Eq)}$ and is equal to the product of the acceleration factor and the test time:

$$t_{(Eq)} = A_F t$$

The acceleration factor is assumed to remain constant over the time of the test. The models used to accelerate time will depend on several factors. Electrical and mechanical failure modes will need different models. Increasing the failure rate or reducing the time to the onset of wear will also involve different models. A few commonly used models in electronics are the Arrhenius model when temperature is used to accelerate time; the power law model (sometimes referred to as the inverse power law), when stresses other than temperature are used; and multistress life models, for example the *Peck model*, in which the combination of temperature and humidity is used (Peck 1986).

The stress agents are usually derived from the models using physics-of-failure (PoF) analysis for the various failure modes, testing to determine the value of the acceleration factor, and then verifying the acceleration factor. The verification may often be delayed until results from the field can be obtained and analyzed. The determination of acceleration factors can be a lengthy and costly process.

The Arrhenius Model

The Arrhenius acceleration model is based on the Arrhenius equation (Savanti Arrhenius, 1858–1927). The equation states that the rate of a chemical reaction increases as the temperature increases:

$$R = A \times exp\left(\frac{-E_A}{KT}\right)$$

where:

- R is the reaction rate of the chemical reaction
- A is a scaling factor and will divide out of the final result
- K is Boltzmann's constant (8.617×10^{-5} electron volts/degree Kelvin)
- T is the temperature in degrees Kelvin (°C+273°)
- EA is the activation energy in electron volts

Reaction rate can be thought of as being synonymous with life statistics like median or mean life. Increasing the test temperature will increase the failure rate of the units, which has the effect of accelerating time.

The acceleration factor is the ratio of the reaction rate R_S at the increased (stress) temperature (T_S) and the reaction rate R_U at the use temperature (T_U):

$$A_F = \frac{R_S}{R_U}$$

This can be expressed using Equation 9.1.

Equation 9.1 Arrhenius model acceleration factor.

$$A_F = exp\left[\left(\frac{E_A}{K}\right)\left(\frac{1}{T_U} - \frac{1}{T_S}\right)\right]$$

The key to using the Arrhenius model is determining the appropriate activation energy. Without knowing the activation energy, an approximation that can be used is that every 10°C increase in temperature cuts the life to half. Activation energies usually fall from 0.5 electron volts (eV) to 2.0 eV for electronics. The activation energy is assumed to be constant for a given failure mode.

EXAMPLE 9.1

An electronic device that usually operates at 50°C (323°K) is subjected to a stress temperature of 100°C (373°K). The activation energy for the failure mode is 0.8 eV. What is the acceleration factor using the Arrhenius equation?

$$A_F = exp\left[\left(\frac{E_A}{K}\right)\left(\frac{1}{T_U} - \frac{1}{T_S}\right)\right]$$

$$A_F = exp\left[\left(\frac{0.8}{8.617 \times 10^{-5}}\right)\left(\frac{1}{323} - \frac{1}{373}\right)\right] = 47$$

Assuming the increased temperature did not cause new failure modes, two days of testing at 100°C equals about 2 (47 days) = 94 days, which is approximately 3 months of use at 50°C.

To verify the acceleration factor (AF) value, it may be necessary to test units at various stress (i.e., temperature) levels. For example, in the Weibull plot shown in Figure 9.1, units may be tested at a high temperature (C), an intermediate temperature (B), and a low temperature (A).

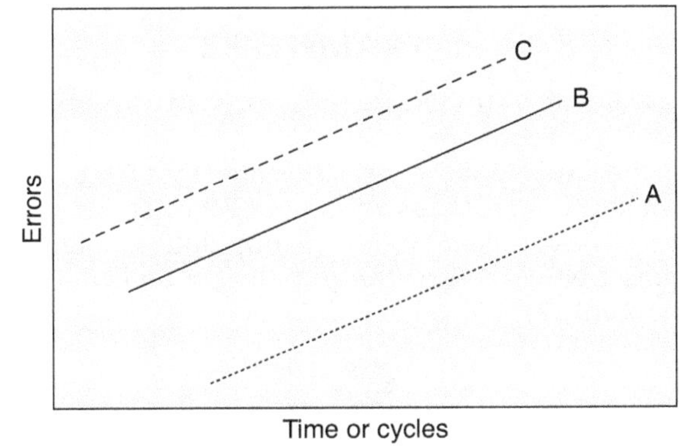

Figure 9.1 Weibull plot for three temperature levels.

The low temperature (A) in Figure 9.1 should represent the normal use value but be high enough to result in a few failures during the test. The high temperature (C) should be near the upper limit over which the failure modes remain unchanged. The intermediate temperature (B) should be significantly different from the two other values. More units need to be assigned to lower temperatures to ensure that failures occur during the test. For example, if 100 units were available for testing, a 4-2-1 allocation would result in 57 units tested at the low temperature, 29 units tested at the intermediate temperature, and 14 units tested at the high temperature. More than one test may be required to find these values.

Failures should occur at a faster rate at higher temperatures. If an analysis of the results shows that the same failure modes occurred at each temperature, then an acceleration factor could be determined by comparing the time for a given percentage of failures to occur. If 5% of failures occur at t_S for the high temperature and 5% at t_L for the low temperature, the acceleration factor is t_L/t_S. This analysis could be done graphically, assuming the Weibull distribution (see Meeker and Hahn 1985).

Note: It is important to perform a thorough root cause analysis after the tests to ensure the root cause of failure is the same. Acceleration can be performed if the root causes of failure are the same. When the root causes of failure are not the same, the analysis needs to be split up into different failure distributions (one distribution per failure mechanism).

The Power Law Model

Example 9.2 illustrates the use of the power law model to determine how to reduce test time by testing at a higher non-thermal stress level. The power law model applies to units subjected to accelerating stresses that are not thermal. The law states that the life of the product is inversely proportional to the increased stress as shown in Equation 9.2.

Equation 9.2 Power law model.

$$\frac{\text{life at rated stress}}{\text{life at accelerated stress}} = \left(\frac{\text{accelerated stress}}{\text{rated stress}}\right)^b$$

Testing at two stress levels will give data to solve for b. It is then assumed that b will remain constant over the applied stress range.

EXAMPLE 9.2

A device that usually operates with 5 volts is tested at two increased stress levels: $V_1 = 15$ volts and $V_2 = 30$ volts.

Analysis of the data from the tests indicates that 5% of the population failed at 150 hours when the high-stress voltage of 30 volts was used. When the stress voltage of 15 volts was applied, 5% of the population failed at 750 hours.

At what time t would 5% of the population be expected to fail under the normal stress of 5 volts?

$$\frac{750}{150} = \left(\frac{30}{15}\right)^b$$

$$\log\left(\frac{750}{150}\right) = b\log\left(\frac{30}{15}\right)$$

$$b = 2.322$$

$$\frac{t}{150} = \left(\frac{30}{5}\right)^{2.322} = 64$$

$$t = 9615$$

This result means that one hour of testing at 30 volts is equivalent to 62 hours of use at 5 volts. This results in an acceleration factor of 62.

Multiple-Stress Tests

Other models can be used with combined stresses. Temperature and humidity, or temperature and vibration, are examples of combined stresses that can be applied to accelerate failures and reduce the actual test time. The *Eyring model* can also be used when temperature is the acceleration stress. The Eyring model, the Arrhenius model, and the power law model can be combined to be used with multiple stresses. For an excellent discussion of combined stress models, see Kececioglu (2001). For example, the Arrhenius-Peck model (Peck, 1986) can be used for accelerated temperature–humidity testing. The relationship is shown in Equation 9.3.

Equation 9.3 Arrhenius-Peck model.

$$A_F = \left(\frac{Humidity_{\text{low}}}{Humidity_{\text{high}}}\right)^{-2.66} \times \exp\left[\left(\frac{Ea}{K}\right)\left(\frac{1}{T_1} - \frac{1}{T_2}\right)\right]$$

where

- $Humidity_{\text{low}}$ = average normal field use humidity level
- $Humidity_{\text{high}}$ = test humidity level
- Ea = activation energy in electron-volts (eV)
- K = Boltzmann's constant (8.617333×10^{-5} eV/°K)
- T = temperature in degrees Kelvin

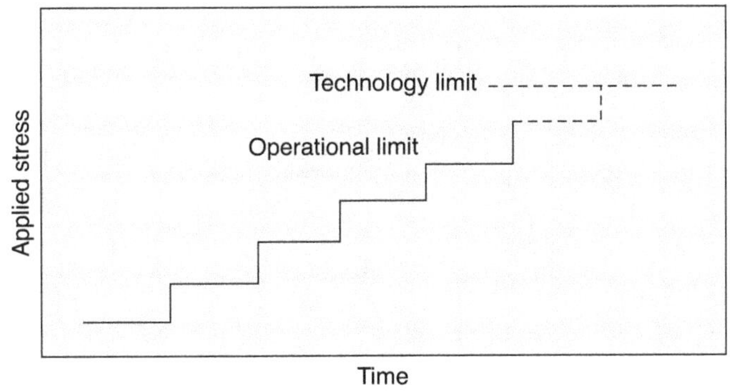

Figure 9.2 Step-stress plot.

- T_1 = field maximum temperature (°K)
- T_2 = test maximum temperature (°K)

EXAMPLE 9.3

A product has normal upper operating limits of (55 + 273.15°K) and 60% relative humidity (RH). What acceleration factor will be achieved if the product is tested at (85 + 273.15°K) and 85% RH, assuming a typical activation energy of 0.70 eV?

$$A_F = \left(\frac{Humidity_{low}}{Humidity_{high}}\right)^{-2.66} \times \exp\left[\left(\frac{Ea}{K}\right)\left(\frac{1}{T_1} - \frac{1}{T_2}\right)\right]$$
$$= \left(\frac{60}{85}\right)^{-2.66} \times \exp\left[\left(\frac{0.7}{8.617333 \times 10^{-5}}\right)\left(\frac{1}{(55+273.15)} - \frac{1}{(85+273.15)}\right)\right]$$
$$= 2.525643 \times \exp[8123.163(0.000255)]$$
$$\approx 20$$

The acceleration factor is 20, meaning one hour of test is equivalent to 20 hours in the field. Testing at these conditions for 500 hours is equivalent to 10,000 hours or

(500 hours) (20) / (24 hrs/day) (365 days/yr) = 1.14 yrs

of life at normal use conditions.

Step-Stress Tests

Step-stress testing is sometimes used to accelerate a reliability test further. This technique involves increasing the stress level at fixed points and recording the failures or degradation levels (see Figure 9.2). For example, a test might apply 5 amps of current for the first 100 hours, then 10 amps for the next 100 hours, and so on, recording product performance at each step. The statistical analysis of the data is rather involved and is covered in Tseng and Wen (2000). Other stressors, such as current, can be used with this technique.

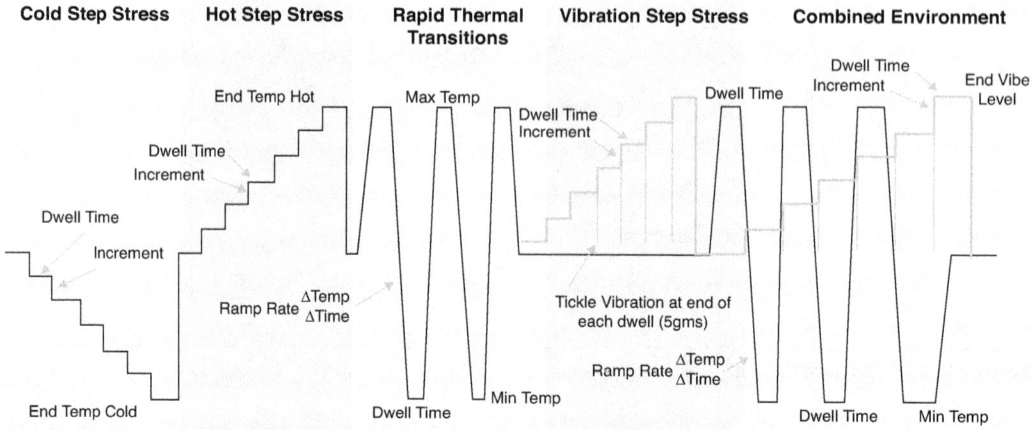

Figure 9.3 Step stresses applied in HALT.
Source: South (2023).

HALT Tests

Highly accelerated life testing (HALT) is a reliability test program used in the early stages of product development. HALT testing grew out of the step-stress or overstress testing programs. The procedure is to continue increasing unit stress until failure occurs. The intent is to identify the weak link in the design and make a change to improve it. The unit is not expected to fail when exposed to stresses within the design limits. The applied stresses exceed the design's specification limits, forcing the unit to fail. Failure analysis and design improvement follow each failure. The process ensures that a robust design is created during the design and development phase and eliminates the need for design changes during the production phase. HALT tests use combined environments using temperature cycling, humidity, and vibration. An example of a HALT test profile is given by South (2023), depicted in Figure 9.3.

The HALT test profile begins with a baseline condition, followed by a sequence of low-temperature steps that gradually introduce cooler conditions to the sample. After this initial phase, the temperature is elevated back to baseline before escalating to high temperatures, which subjects the sample to more intense thermal stress. The profile includes rapid thermal transitions to assess the sample's response to sudden temperature changes. Vibration steps are then introduced, adding mechanical stress intermittently. The profile concludes with a combination of high temperatures and vibration steps, testing the sample under simultaneous thermal and mechanical conditions to evaluate its overall durability and performance.

HALT testing must be employed early in product development to improve the product's reliability. It is much more efficient and cost-effective to make changes before the design is completed, parts are ordered, and the processing tooling is set.

Product failures are analyzed for product improvement purposes. These failures may differ from the product failure modes when operated within its design specifications. The intent is to cause failures to occur quickly, using a small number of units for the test.

Times to failure are not recorded—only stress levels are recorded—and all the products tested might fail. Therefore, reliability measures such as mean time between failures (MTBF), mean time to failure (MTTF), or failure rate cannot be calculated using the test results. The amount of reliability improvement due to the design change cannot be quantified using the results of HALT. The product design will be robust, with a strength that not only exceeds the normal stresses it is expected to experience during use but also the stresses in the tails of the stress distributions. It is these stresses that cause failure in otherwise well-designed products.

Highly accelerated life testing should be a standard component of the design and development phase if there is an expectation that the delivered product will perform with zero or very few failures. The details of the HALT program will differ with the product. The stresses to be used, the limits on the stresses, whether the testing is cycled or static, and other details are product dependent. The HALT stresses used during the design of a new type of stepper motor and radar antenna servo system will differ from those used during the design of an autopilot for a new military aircraft. Testing as an entire unit might be appropriate for some products, while the testing of subsystems might result in quicker reliability improvement for other products. The test plan and procedures must be developed with input from reliability engineering, test engineering, product engineering, and design to ensure the most efficient use of test time for the HALT process.

McLean (2002) describes three distinct phases of the HALT process, as follows.

Phase 1: Pre-HALT Phase

During this phase, testing procedures are documented, test equipment is procured and made ready, data recording procedures are in place, and necessary resources are available. Figure 9.4 provides a view of the item's robustness knowledge before HALT testing.

Phase 2: HALT Phase

Testing is performed during this phase. The test procedures developed in the pre-HALT phase are followed. The operating units are subjected to the test profile as described previously. The tests are monitored, the results analyzed, and the data recorded as previously determined.

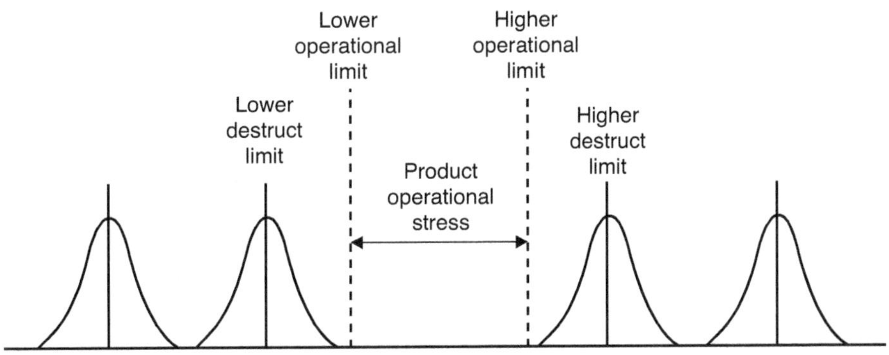

Figure 9.4 Robustness knowledge before HALT testing.

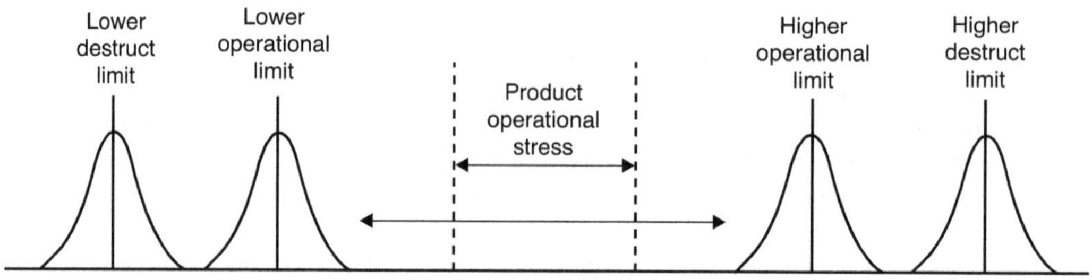

Figure 9.5 Robustness knowledge after HALT testing.

Phase 3: Post-HALT Phase

Each issue uncovered during the HALT phase is subjected to root cause analysis and corrective action. A person is assigned responsibility for each action, which is open-ended until reliability engineering closes it. Phase 2 and Phase 3 can then be repeated as necessary. Figure 9.5 provides a view of robustness knowledge after HALT testing.

Margin Tests

Margin tests apply increasing stress levels to develop a margin above specification for a failure mode. When a failure occurs, the cause is remedied, and the test continues until a design achieves a predetermined margin above the design specification. The larger the margin, the lower the probability that failure mode will occur under design stress.

STRESS SCREENING

> Stress screening (e.g., ESS, HASS, and burn-in tests) (Evaluate)
>
> **Body of Knowledge IV.B.2**

Stress screening, also called *burn-in testing,* describes a type of final test performed on production units before release to the customer. The purpose of burn-in testing is to allow the product to accumulate use time before it is delivered. A burn-in or screening test program will reduce the occurrence of early-life failures after the product is delivered. During the burn-in, manufacturing-related nonconformities that cause early failure will be detected.

Environmental Stress Screening

To accelerate the rate at which the weak units fail (known as early-life failures), some stress (usually environmental) is applied during the test. This practice is referred to as environmental stress screening (ESS). Common screens include thermal cycling, random vibration, immersion, overpressure, or voltage variation (Breneman et al. 2022). This stress should not damage an acceptable unit. Units that succeed will be delivered to customers.

Highly Accelerated Stress Screening

Highly accelerated stress screening (HASS) is a program that ensures that the final product will exceed its environmental requirements. Figure 9.6 provides the location of stress screening in the "bathtub" curve. The design has been proven to be robust using a HALT program. Some of these stress levels may exceed the limits set by the customer requirements and stress the product beyond the levels expected in normal use. It is assumed that good units are unaffected by the test. All the units that pass the test are released for shipment. The product must be robust to the stresses; therefore, HASS testing can only be performed if a HALT program has been part of the product development. HASS testing will remove early-life failure units from the population, but the primary strategy of the test program is to detect any shift in the production process. The environmental stresses of a HASS program may exceed the product design specifications. A HASS program can only be used if a HALT program was used during product development. The HASS program will detect problems in the process that negatively affect product reliability. The HASS program is not intended to detect ongoing process problems. Its purpose is detecting a shift in a process previously shown to be capable. HASS testing is not intended to compensate for a production process that is less than capable. Before HASS can be used successfully, accepted process quality procedures must be applied.

Processes must have capability measures, such as Cpk, at acceptable levels. Suitable statistical process control (SPC) procedures must be in place. These concepts are presented in Chapter 6. The test equipment, including any special test chambers, must be available and in sufficient quantity, and test time must be allotted as part of the process so as not to slow production.

Once the process has demonstrated statistical control, it may be possible to replace the HASS test program performed on 100% of the production units with

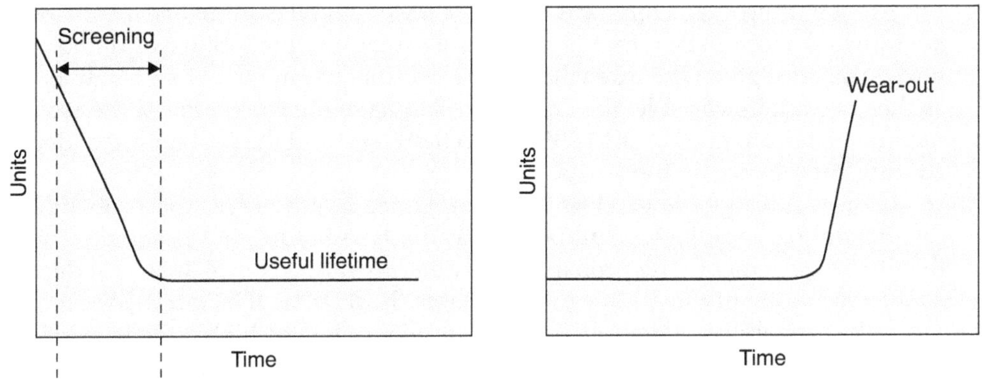

Figure 9.6 Location of stress screening in the bathtub curve.

a stress audit program. This would reduce costs by requiring less test equipment and fewer personnel. It could also reduce time to market for high-volume production. A statistical approach to the audit involving stress screening a sample of the production units is called a highly accelerated stress audit (HASA). A HASA test program will detect a process shift but has the risk of allowing some early-life failures to reach the customer.

QUALIFICATION/DEMONSTRATION TESTING

> Qualification/demonstration testing (e.g., sequential tests and fixed-length tests) (Evaluate)
>
> **Body of Knowledge IV.B.3**

Qualification/demonstration reliability testing is commonly referred to as compliance testing. Compliance tests demonstrate that a product parameter conforms to a given requirement. In reliability compliance testing, the parameter could be MTBF or MTTF, a failure rate (failure intensity), or a reliability value. IEC standard 61124–2006 Reliability Testing—Compliance Tests for Constant Failure Rate and Constant Failure Intensity provides procedures for testing values of failure rate or MTBF/MTTF, and IEC standard 61123–1997 Reliability Testing—Compliance Plans for Success Ratio contains procedures for testing reliability values for success/fail items. IEC standard 61124 contains fixed-time test plans, and IEC 61123 contains fixed-trial test plans. Both standards contain truncated sequential test plans. MIL-HDBK-781 Reliability Test Methods, Plans, and Environments for Engineering Development, Qualification, and Production is the best-known military standard for reliability demonstration or compliance testing. The assumption for all compliance test plans that use time as the continuous variable is that the failure model distribution is exponential. This implies that the failure rate is constant and that MTBF/MTTF is equal to the reciprocal of the failure rate. The assumed failure model for the fixed-trial failure test plans is the binomial.

An operating characteristic (OC) curve describes all compliance test plans. An operating characteristic curve is a graph showing the probability of demonstrating compliance given the true value of the product parameter. For a test with time as the continuous variable, the OC curve will have the true value of the required parameter, failure rate [λ] or MTBF [θ], on the x-axis and the probability of demonstrating compliance on the y-axis.

If the failure rate is used as the compliance requirement, the probability of demonstrating compliance will be high if the true failure rate is equal to or less than the requirement. The probability of demonstrating compliance will decrease as the true value of the failure rate increases. If MTBF is the compliance requirement, the probability of demonstrating compliance will be high if the true MTBF is equal to or greater than the requirement. The probability of demonstrating compliance will decrease as the true MTBF decreases. IEC standard 61124 uses MTBF as the compliance value.

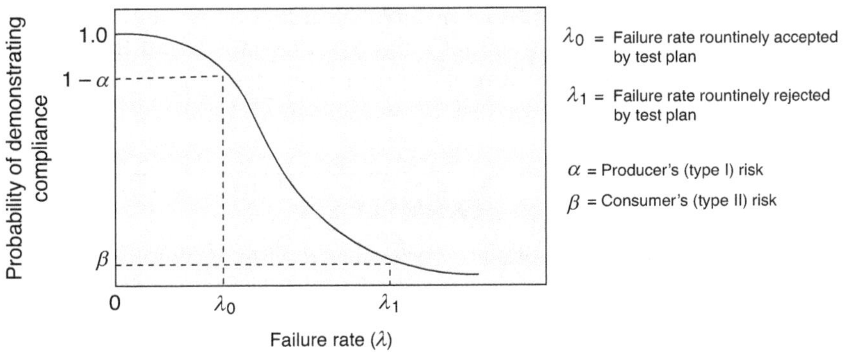

Figure 9.7 Operating characteristic curve for sampling plan.

A typical OC curve for failure rate is shown in Figure 9.7. The OC curve shown uses failure rate (λ) as the compliance value. Two points on the OC curve summarize the test: an accept point defined by λ_0 and $(1 - \alpha)$ and a reject point defined by λ_1 and β. The product complies if the true failure rate is equal to or better (less) than λ_0 and should pass the test. There is a risk of failing the test even though the product complies with the requirement. This is the α risk, also known as the producer's risk, or the probability of a Type I error.

The product is not in compliance if the true failure rate is worse (greater) than λ_1 and should fail the test. There is also a risk of passing the test, even though the product does not comply with the requirement. This is the β risk, also known as the consumer's risk, or probability of a Type II error.

The ratio of the two λ values (λ_1/λ_0) is called the discrimination ratio. As the discrimination ratio approaches 1, the required amount of testing time and units will increase.

MTBF/MTTF could also be used as the compliance value. The accept point for this OC curve is defined by m_0 and $(1 - \alpha)$, and the reject point is determined by m_1 and β. The ratio m_0/m_1 is the discrimination ratio.

Fixed-Time Test Plans

A fixed-time compliance test involves placing units on the test to accumulate test time (T) while recording failures. The criterion for acceptance (compliance) is based on collecting a given amount of test time and having an acceptable number of failures. The maximum number of allowable failures for compliance is usually specified as *c*. The test continues until either the required amount of test time has been accumulated (accept) or the allowable number of failures has been exceeded (reject).

The accept and reject points on the OC curve determine the required test time and the allowable number of failures. Once the two points are defined, the necessary cumulative test time, *T*, and the allowable number of failures, *c*, are set. The cumulative test time *T* is the total of the operating time of all the units during the test, including the units that fail and those that do not fail.

A replacement or nonreplacement test can be performed to acquire the required cumulative test time. In a replacement test, the units that fail are replaced with good units, keeping the same number of units on test at all times. If a nonreplacement test is used, the units that fail are not replaced, and the test population becomes smaller with each failure. The cumulative test time is the total operating times of all the units during the test, no matter which type of test is used.

EXAMPLE 9.4

Ten units are on test. The units are not replaced when they fail. One unit fails at $t_1 = 685$ hours. A second unit fails at $t_2 = 1690$ hours. The test ends at $t = 2500$ hours with no additional failures. What is the total accumulated time?

$$T = 685 + 1690 + 8(2500) = 22{,}375 \text{ hrs}$$

Example 9.5 uses IEC standard 61124.

EXAMPLE 9.5

It is desired to show compliance with a product MTBF. The acceptable value $m_0 = 3000$ hours, and the rejectable value $m_1 = 1000$ hours. The discrimination ratio $m_0/m_1 = (3000)/(1000) = 3$. The α and β risks are both set to 0.10. Test plan B.7 can be used from Table 3 of IEC standard 61124 (see Table 9.1). The OC curve for this plan is shown in Figure B.6 of the standard (see Figure 9.8). The test requires a cumulative test time of 9300 hours, and the allowable number of failures is five.

$$T = 3.1m_0 = 3.1 \times 3000 = 9300$$

$$c = 5$$

The accept and reject points on the OC curve can be verified. Using the constant failure rate concept, the expected number of failures is λT. If the true MTBF is 3000 hours,

$$\lambda T = \frac{1}{3000} \times 9300 = 3.1$$

This is the acceptance point. Using the cumulative Poisson tables, which can be justified because of the constant failure rate, the probability of five or fewer failures is 0.90. At the reject point,

$$\lambda T = \frac{1}{1000} \times 9300 = 9.3$$

Again, from the cumulative Poisson tables, the probability of five or fewer failures is 0.10.

Table 9.1 Table of fixed-time test plans.

Test plan no.	Characteristics of the plan			Test time for termination	Acceptable number of failures	True risks for	
	Normal risks		Discrimination ratio			$m = m_0$	$m = m_1$
	α%	β%	D	T_1/m_0	c	α%	β%
B.1	5	5	1.5	54.10	66	4.96	4.84
B.2	5	5	2	15.71	22	4.97	4.99
B.3	5	5	3	4.76	8	5.35	5.40
B.4	5	5	5	1.88	4	4.25	4.29
B.5	10	10	1.5	32.14	39	10.00	10.20
B.6	10	10	2	9.47	13	10.00	10.07
B.7	10	10	3	3.10	5	9.40	9.90
B.8	10	10	5	1.08	2	9.96	9.48

Source: IEC standard 61124 *Reliability Testing—Compliance Tests for Constant Failure Rate and Constant Failure Intensity.*

B.2 test plans B.5 to B.8

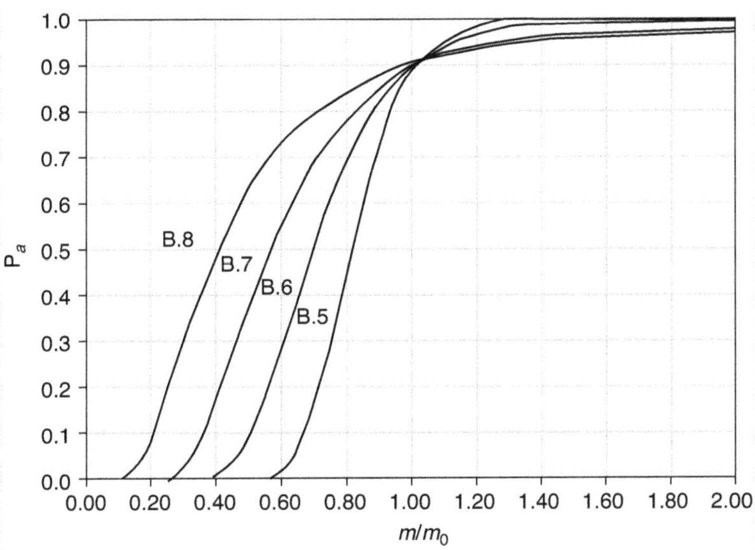

Figure 9.8 OC curves for fixed-time test plans.

Source: IEC standard 61124 *Reliability Testing–Compliance Tests for Constant Failure Rate and Constant Failure Intensity.*

EXAMPLE 9.6

If the rejectable MTBF is changed to 1500 hours ($m_1 = 1500$ hours), the discrimination ratio is 2, and test plan B.6 can be used. The required cumulative test time is 28,410 hours (9.47×3000), and the allowable number of failures is 13. It is important to note the increase in required cumulative test time as the discrimination ratio approaches 1. This result occurs even though the MTBF to be demonstrated remains the same.

The time required to perform a demonstration test can become excessive. If accelerated testing can be used, the actual time required can be reduced. For Example 9.6, the required cumulative test time is 28,410 hours. This test time could be accomplished by placing 20 units on test and running the test for about 1420 hours (approximately two months). If a nonreplacement test is performed, the test time will be extended with each failure. If an acceleration factor of 60 can be applied to the test, the actual test time (t) could be reduced to 23.7 hours to get the same cumulative test time (approximately one day using 20 units on the test):

$$T = A_F tn = 60 \times 23.7 \times 20 \approx 28{,}410$$

Sequential Test Plans

Sequential test or probability ratio sequential test (PRST) plans differ from fixed-time test plans. Accepting a fixed-time test can only be made once the test is completed and the required cumulative test time is reached.

The results of sequential tests are continuously assessed to arrive at one of three alternative decisions. Sequential testing results in a decision to accept compliance, reject compliance, or continue testing at every value of cumulative test time. The intention is to decide whether to accept or reject compliance in the minimum test time. A graph of a sequential test plan is shown in Figure 9.9.

The alternatives described by the three regions on the graph are accept compliance, reject compliance, or continue testing. The three regions are defined by

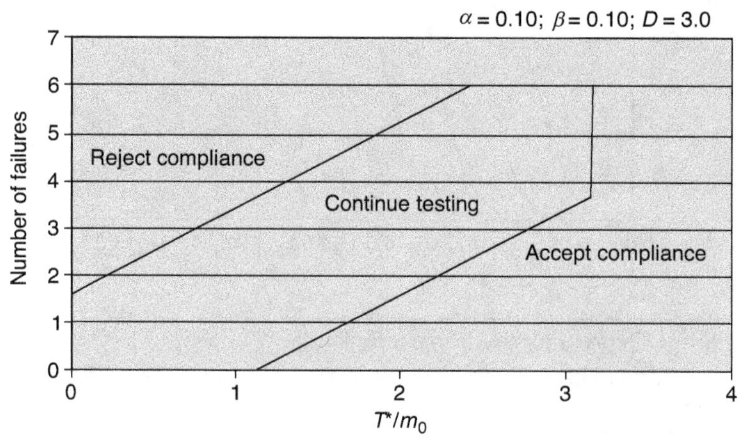

Figure 9.9 Sequential test plan.

parallel straight lines constructed using the values m_0 (acceptable MTBF) and m_1 (rejectable MTBF) along with α and β from a predetermined OC curve if MTBF is to be demonstrated. Acceptable failure rate (λ_0) and rejectable failure rate (λ_1) are used if the compliance value is a failure rate.

There are rules for ending the test to ensure that the test does not continue indefinitely. The test could be terminated at a given time or on a given number of failures if a decision to accept or reject compliance has not been reached. This is referred to as a truncated sequential test. The sequential test plan from IEC standard 61124, shown in Figure 9.9, has α and β risks equal to 0.10 and a discrimination ratio of 3. The OC curve for this test is shown in Figure 9.8 (test B.7). The equations for constructing the straight lines that identify the accept, reject, and continue testing regions, as well as the rules for termination, are included in the standard. On average, an accept or reject decision will be reached with less cumulative test time than with fixed-time test plans.

Reliability Demonstration Test

The reliability demonstration test (RDT), also known as a reliability compliance test, is a critical assessment designed to verify that a product meets its specified reliability target throughout its mission life, with a predefined confidence level and failure criteria. Typically, the goal is to ensure zero failures during the mission. To achieve this, the test design must address two main parameters: the duration of the test and the sample size.

If the sample size is predetermined, the test duration can be calculated to meet the reliability target. Conversely, if the duration of the test is fixed, the required sample size can be determined to ensure that the reliability target is achieved. The Weibull distribution is often employed to model the product's life characteristics due to its flexibility in representing various types of life data. The Weibull distribution is defined by its shape factor, which influences the failure rate over time.

For the reliability demonstration, a shape parameter is chosen for the Weibull distribution based on historical data or industry standards. This shape parameter affects the reliability calculations by altering the distribution's skewness and failure rate behavior. By incorporating these assumptions, the test design can effectively determine the appropriate test parameters—extending the duration or adjusting the sample size—to demonstrate the product's reliability aligns with the target mission life and confidence level, showing its robustness under the defined conditions.

DEGRADATION

Degradation (wear-to-failure) testing (Evaluate)
Body of Knowledge IV.B.4

It may be necessary to test units to determine the end of useful life. This is the beginning of wear-out. Some products may be expected to have extremely long lives. It is common to expect 150,000 miles of operation for some automotive systems or 10 years for some electronic systems. To test for these times, some form of acceleration is necessary to increase the wear rate. Rapid cycling increases the rate of fatigue, salt spray increases the corrosion rate, and temperature increases the degradation rate of solid-state electronics, which are a few acceleration stresses.

The failure rate of solid-state electronics will increase as the temperature increases. Good design practice for solid-state electronics dictates that heat be kept to a minimum and removed from the system. If the objective is to have units fail faster during testing, adding heat causes faster degradation, which can be equated to increasing the test time. The Arrhenius model is often used to determine the equivalent time the unit is on.

It is sometimes possible to extrapolate to a failure time by observing the amount of degradation a unit has experienced in a given time. This would reduce the test time necessary to determine the end of life. The information gained from measuring the amount of wear on a bearing or the intensity of crack propagation due to fatigue at a given time could be used to predict the time to failure if the model for the degradation rate is known or assumed. The linear, exponential, and power law models are often used. Alternatively, the model could be derived using the physics of failure. Each product could be unique.

For example, suppose the specification for the fabric covering of an office chair seat is 100,000 cycles of a 250-pound weight moving across, then off, the fabric. The failure mode is the visible color change when the backing surface appears. Previous testing has shown that the relationship of number of cycles versus fabric thickness lost is approximately linear. The problem is to find the fabric thickness that will meet the specification. To obtain the first estimate for the answer, a fabric is tested for 10,000 cycles, and the thickness loss is measured. This would indicate that the first estimate for the design fabric thickness is 10 times the measured wear loss.

Other techniques for degradation testing are dust testing, where a controlled amount of dust is sprayed in a test room where the product is present, and solar testing, where sun radiation is applied to the product.

SOFTWARE/FIRMWARE RELIABILITY

> Software/firmware reliability (e.g., software design reliability, software reliability prediction, software design and development models, and software reliability allocation) (Understand)
>
> **Body of Knowledge IV.B.5**

Software Design Reliability

According to ANSI/IEEE (1991), *software reliability* is the probability of failure-free software operation for a specified period in a specified environment. Note that if software is installed into a hardware component, it is often referred to as *firmware* and is stored in read-only memory (ROM) in a flash memory chip or other device.

Typical software reliability metrics include the number of defects in a specific time interval (e.g., the number of failures per month), defect density (number of defects per unit of code), mean time between failures, mean time to failure, and availability. Specialized software reliability models are then used to predict the software's future reliability or latent defect count.

The following is adapted from Westfall, *The ASQ Certified Software Quality Engineer Handbook* (2024, pp. 448–449). *Availability* is a quality attribute that describes the extent to which the software or a service is available for use when needed. Availability is closely related to reliability because unreliable software that fails frequently is unavailable for use because of those failures. Availability is also dependent on the ability to *restore* the software product to a working state. Therefore, availability is also closely related to maintainability in cases where a software defect caused the failure, and that defect must be corrected before the operations can continue.

An example of an availability metric is illustrated in Figure 9.10. Here, post-release availability is calculated each month using:

$$\textit{Availability} = \left(1 - \frac{\text{Total minutes of outage}}{\text{Total minutes of operation}}\right) \times 100\%$$

Tracking availability helps relate software failures and reliability issues to their impact on the user community.

When defining reliability and availability metrics for software, care should be taken not to use formulas intended for calculating hardware reliability and availability to evaluate software. This is because, unlike hardware, software does not wear out or fail over time due to aging. And, unlike hardware, where manufacturing can interject the same error multiple times (bent pins, broken wires, solder splashed), once a defect is found and removed from the software, that defect is gone, assuming good configuration management practices do not allow the defective component to be built back into the software by mistake.

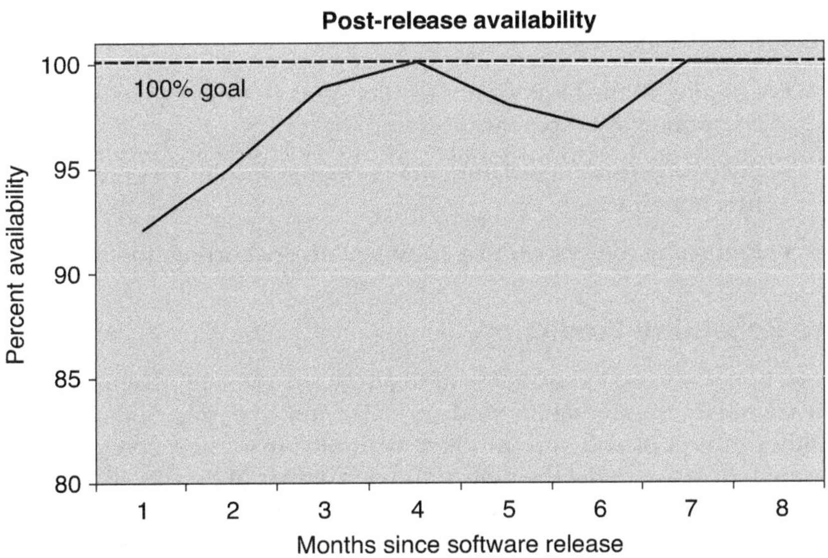

Figure 9.10 Percent availability versus months since software release.
Source: Westfall (2024, p. 448).

The following is adapted from Westfall (2024, pp. 302–303). Software reliability and availability problems occur when defects in the software are encountered during the use of that software in operations, resulting in software failure. Therefore, good requirements, design, and coding practices should be used to prevent as many defects as possible from being introduced into software products. High software complexity is one of the major contributors to software reliability and availability issues. Designing software to increase cohesion and make individual software components as decoupled as possible can help increase the reliability and availability of that software. Good verification and validation practices should also be employed to identify and remove as many defects as possible from software products under development before their release into operations.

Software can also be designed to be more robust and fault tolerant, and therefore more reliable and available. Software components that process inputs and outputs should include appropriate error checking and error handling. This includes inputs and outputs to hardware and databases, as well as those to users.

Fault-detection mechanisms can include periodic ping/echo or heartbeat tactics. In a *ping/echo tactic*, "one component issues a ping and expects to receive back an echo within a predefined time, from the component under scrutiny" (Bass et al. 2013). In a *heartbeat tactic*, one component periodically sends out a heartbeat signal, and another component detects it. If either the ping/echo or heartbeat tactic fails, the nonresponding component is assumed to have a problem.

To increase availability, mechanisms need to be designed and built into the software to recover from faults that occur. One mechanism for accomplishing this is to create distributed or redundant architectures with no single point of failure. A less expensive alternative might be to have spares that can be activated if failures occur. If an external component is suspected of having a problem, for example, that component can be placed out of service.

If an event or transaction does not occur in the expected time frame, it can be timed out or retried. Retries may also be used to attempt to recover from certain types of failures (e.g., failure to allocate memory or buffers). Other recovery mechanisms include:

- Logging of the fault, or notification of the fault (e.g., to users or other components) as soon as it occurs
- Switching to a degraded mode of operations with less capacity or functionality
- Returning the system to a known state or shutting down the system

Software Reliability Prediction

This section was adapted from Westfall (2024, pp.447–448). There are two types of software reliability prediction models: static and dynamic. *Static reliability models* use other project or software product attributes (e.g., size, complexity, programmer capability) to predict the defect rate or number of defects. Typically, information from previous products and projects is used in these models, and the current project or product is viewed as an additional data point in the same population.

Dynamic reliability models are based on collecting multiple data points from the current product or project. Some dynamic models look at defects gathered over the entire lifecycle, while others concentrate on defects found during the formal testing phases at the end of the life cycle. Examples of dynamic reliability models include:

- Jelinski-Moranda (J-M) model
- Basic execution time model
- Littlewood-Verrall (LW) model
- Goel-Okumoto (G-O) model
- Shooman model
- Musa-Okumoto model

Appropriately implementing a software reliability model requires an understanding of the assumptions underlying that model. For example, the J-M model's five assumptions are (Kan 2003):

1. There are *N* unknown software faults at the start of testing.
2. Failures occur randomly, and times between failures are independent.
3. All faults contribute equally to causing a failure.
4. Fix time is negligible.
5. A fix is perfect for each failure; that is, there are no new faults introduced during the correction.

The J-M model is a Markov process model and one of the best-known black-box models. Many variations of the model have been developed to handle variations on the original J-M model assumptions. When selecting a model, consideration must be given to the likelihood that the model's assumptions will be met by the project's software environment.

Software Design and Development Models

Several software development models have been developed. These include the waterfall model, the rapid application development (RAD) model, the verification and validation (V) model, and the breadth of frameworks classified as agile development. Each of these models is briefly introduced here.

Waterfall Model

The waterfall model has five distinct phases that must be completed in order.

1. Requirements analysis and specification
2. Design
3. Implementation and unit testing
4. Integration and system testing
5. Operation and maintenance phase

A schematic of the model is shown in Figure 9.11.

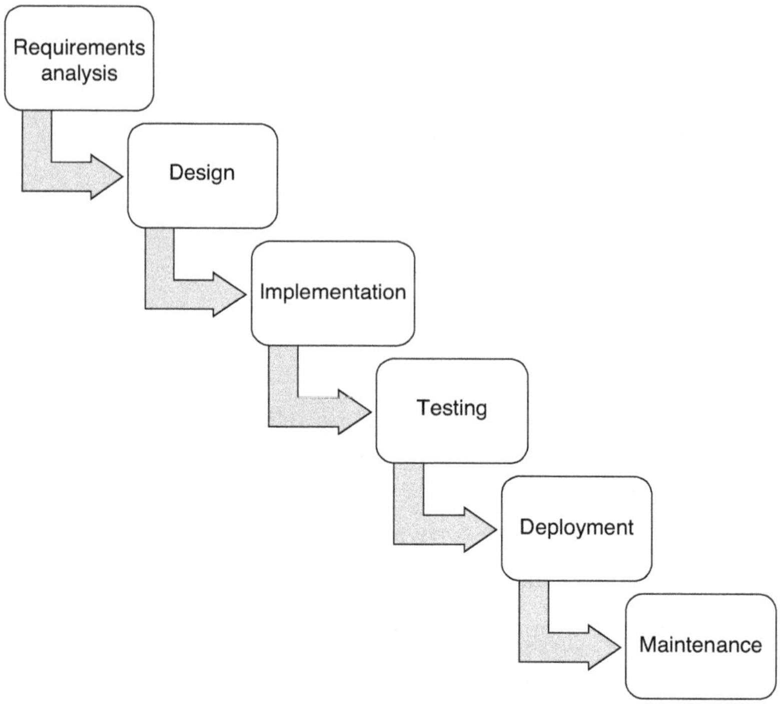

Figure 9.11 Waterfall model.

The waterfall model is a feasible development option when requirements are not in a state of flux, the project is short, and the technology is consistent. The five-phase structure allows the project manager to easily track progress. However, this waterfall model is not flexible since requirements cannot be changed during development.

Rapid Application Development (RAD) Model

The rapid application development (RAD) model is useful when requirements are well prescribed. Features include gathering requirements using focus groups or workshops, prototyping and early user testing, reuse of software components, and a rigidly paced schedule. The phases of the RAD model include business modeling, data modeling, process modeling, application generation, and testing and turnover. These phases are illustrated in Figure 9.12.

Verification and Validation (V) Model

The verification and validation, or V model, is based on phases that must be completed in order, similar to a waterfall model. Testing is planned in parallel with a corresponding state of development. Phases for the verification branch include business requirements analysis, system design, architecture design, module design, and coding phases. For the validation branch, phases include unit testing, integration testing, system testing, and acceptance testing. The phases are shown in Figure 9.13.

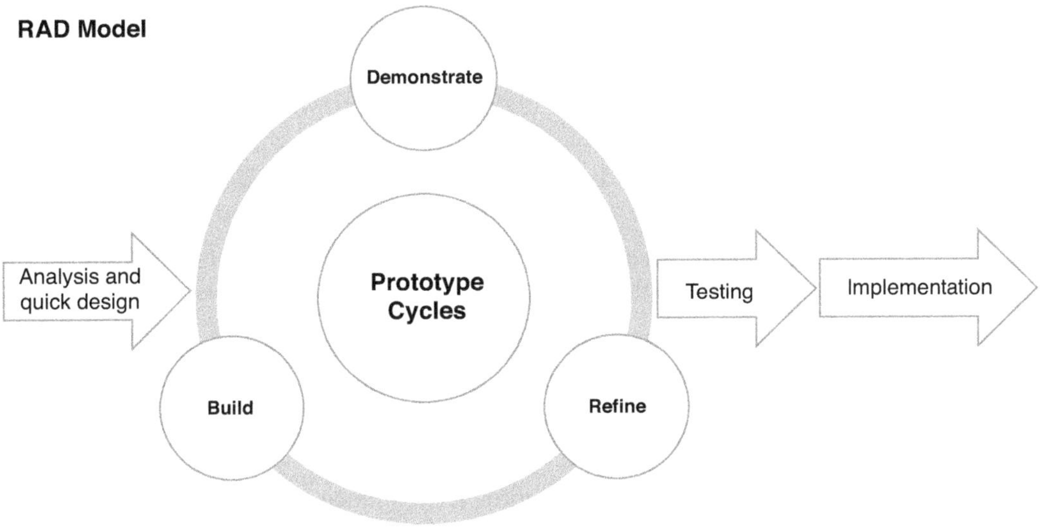

Figure 9.12 Schematic of RAD model phases.

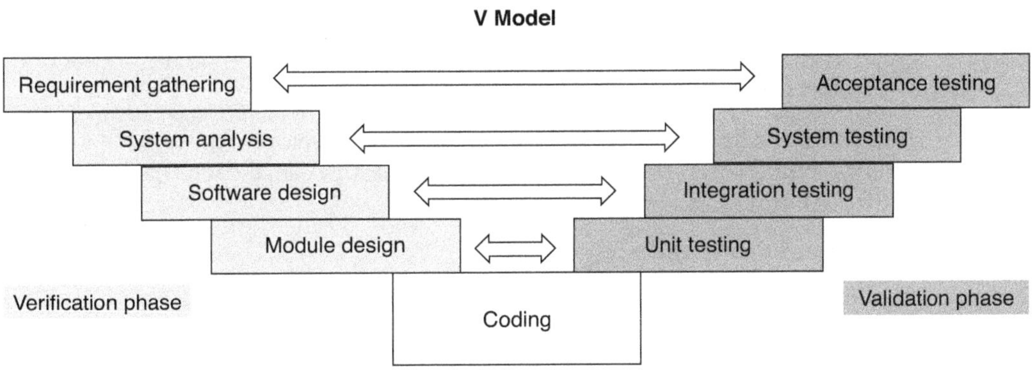

Figure 9.13 Phases of the V model.

Agile Model

This section has been adapted from Westfall (2024, pp. 194–196). There are many agile methods and frameworks, including scrum, extreme programming (XP), test-driven development (TDD), feature driven development (FDD), crystal, lean and Kanban, among others. Each of these frameworks follows the 12 principles that the Agile Alliance (www.agilealliance.org) has defined:

1. Our highest priority is to satisfy the customer through early and continuous delivery of valuable software.
2. Welcome changing requirements, even late in development. Agile processes harness change for the customer's competitive advantage.

3. Deliver working software frequently, from a couple of weeks to a couple of months, with a preference to the shorter timescale.
4. Business people and developers must work together daily throughout the project.
5. Build projects around motivated individuals. Give them the environment and support they need and trust them to get the job done.
6. The most efficient and effective method of conveying information to and within a development team is face-to-face conversation.
7. Working software is the primary measure of progress.
8. Agile processes promote sustainable development. The sponsors, developers, and customers or users should be able to maintain a constant pace indefinitely.
9. Continuous attention to technical excellence and good design enhances agility.
10. Simplicity, the art of maximizing the amount of work not done, is essential.
11. The best architectures, requirements, and designs emerge from self-organizing teams.
12. At regular intervals, the team reflects on how to become more effective, then tunes and adjusts its behavior accordingly.

In general, the agile model is based on iterative development, where tasks are broken into smaller iterations, which are one to four weeks each. Each iteration works through the same cycle:

1. Requirements gathering
2. Requirements design
3. Construction iteration
4. Testing and quality assurance
5. Deployment
6. Feedback

These cycles are illustrated in Figure 9.14.

Software Reliability Allocation

According to Musa et al. (1987), "Reliability allocation deals with the setting of reliability goals for individual components such that a specified system reliability goal is met, and the component goals are well balanced among themselves."

The IEEE Std 1633–2016 presents a summary of software reliability allocation methods that have been developed, shown in Table 9.2. These include allocation by dollars, allocation by equal apportionment, past failure contribution, allocation

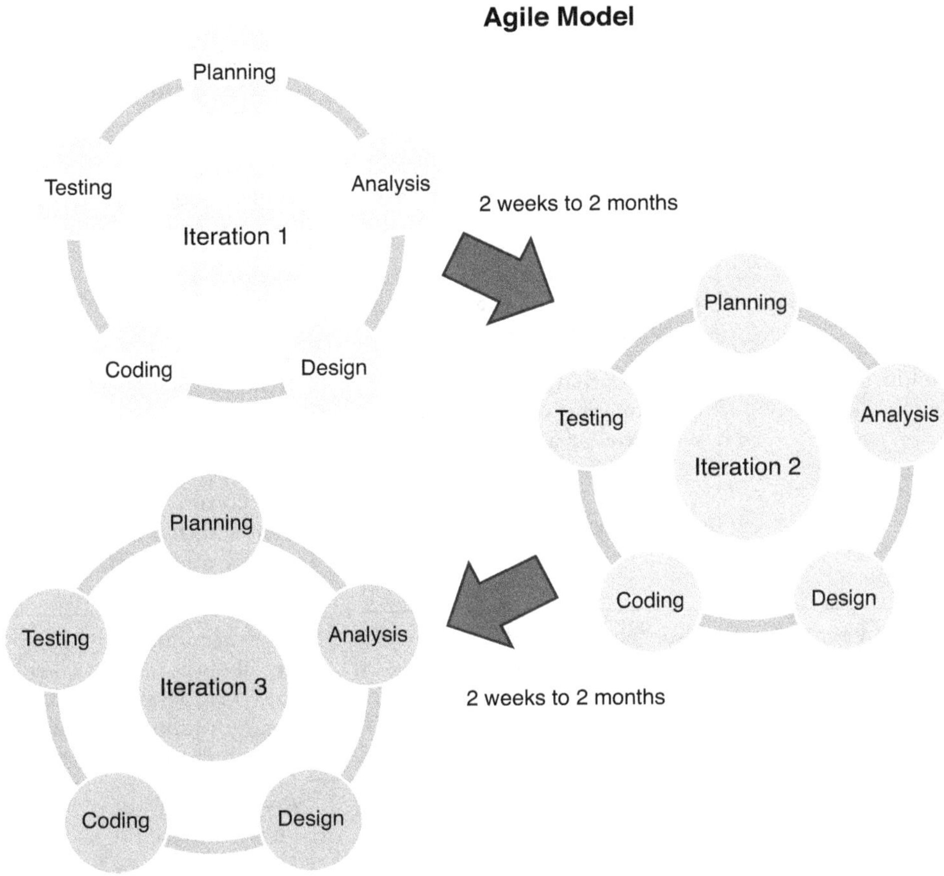

Figure 9.14 Agile model cycles.

by achievable failure rate, the Aeronautical Radio, Incorporated (ARINC) apportionment techniques, feasibility of objectives technique, and the minimization of effort algorithm.

SOFTWARE TESTING

> Software testing (e.g., white-box, black-box, operational profile, fault-injection, software validation and quality assurance, software verification, qualitative versus quantitative reliability testing, built-in testing [BIT], and regression testing to verify code integrity) (Understand)
>
> **Body of Knowledge IV.B.6**

Table 9.2 Software reliability allocation methods.

Method	Description	Benefits/disadvantages
Allocation by $ (Lakey, Neufelder)	Each component's allocation is based on how much R&D funding is required to develop that component.	Easy to compute. Reduces the possibility that software will be allocated zero percent of the system objective.
Allocation by equal apportionment (Lakey, Neufelder)	Assumes for a series of *n* subsystems each are allocated the same reliability requirement goal.	
Past failure contribution (Lakey, Neufelder)	Assumes that the failure rate of a component is relatively equal to what was on a past similar system.	If data is available from the field for similar systems, this is easy to compute. However, it does not take into consideration that systems can change with respect to how much software is in them.
Allocation by achievable failure rate (Lakey, Neufelder)	Uses the predicted failure rate of each component to determine its allocation	More accurate than preceding items because it takes into consideration each configuration item's relative contribution to the system.
ARINC apportionment technique (Von Alven)	Assumes series subsystems with constant failure rates, such that any subsystem failure causes system failure and that subsystem mission time is equal to system mission time.	Requires analysis of each subsystem's OP.
Feasibility of objectives technique (Eng. Des.)	Subsystem allocation factors are computed as a function of Delphi numerical ratings of system intricacy, state of the art, performance time, and environmental conditions.	Method of allocating reliability without repair for mechanical-electrical systems.
Minimization of effort algorithm (MIL-HBBK 338B)	Assumes a system comprised of *n* subsystems in series. The reliability of each subsystem is measured at the present stage of development and apportions reliability such that greater reliability improvement is demanded of the lower reliability system.	Considers minimization of total effort expended to meet system reliability requirements.

Source: IEEE Std 1633–2016 *IEEE Recommended Practice on Software Reliability*, pp. 93–94.

Various software testing strategies are used to verify code integrity such as white-box, black-box, operational profile, fault-injection, software validity and quality assurance, software verification, qualitative versus quantitative reliability testing, built-in testing, and regression testing. Testing should be a function of each development phase to be the most effective at removing faults or "bugs" from the software.

White-Box Testing

White-box or structural testing implies that the tests are designed and implemented with complete knowledge of the system's inner structure. It tests the internal functions that must collectively result in the specified external output. The intent is to cover all paths and test all system branches. White-box testing focuses on how the software works.

Many faults in large software programs are due to poorly stated or misunderstood specifications. The focus of software reliability should be on defect prevention instead of defect detection and removal. Specification and design reviews should be part of every software development project. A testing program is required to ensure that software is released with a minimum number of embedded faults. Testing is the most efficient method of finding and eliminating faults in software. O'Connor (2002) states that more than 50% of programming errors are due to a lack of understanding of the specifications.

Good design controls and well-stated specifications reduce the probability of a programming error. Not introducing an error into the program is much preferred over trying to remove the error from the program.

As faults are found and eliminated, the probability of finding additional faults decreases. This is the early-failure part of the software bathtub curve. This testing activity is a part of the development phase and differs from the early-failure part of the hardware bathtub curve. The early-failure part of the hardware bathtub curve occurs after the production phase. Testing the software package for every possible input under every condition is desirable. Error-free software will then be delivered to the customer if every uncovered fault is eliminated without adding additional faults. For complex software packages, not all possible conditions may be anticipated, and eliminating faults could create new ones. It is generally accepted that the most extensive testing programs will uncover and eliminate about 95% of the faults in a software program.

Software testing should begin as the programs are written. Testing smaller programs is efficient and less complicated. Testing should continue as the programs are integrated into a system. White-box testing is concerned with how the system works. The individual programmers may be best suited for this type of testing as they understand their codes. The program can be immediately corrected as errors are found.

Black-Box Testing

Testing software programs concerning their external specifications is called black-box testing or functional testing. Representative inputs are inserted, and the resulting outputs are compared to the requirements. The intent is to determine whether the software responds as intended and to test the completed system to determine whether the original requirements are met. Black-box testing is not concerned with internal functions. Black-box testing should be performed using the actual interface units in the operating environment.

Errors that the customer finds after the software is released may take time to correct. The customer may use the software in ways not anticipated by the test engineer. If the customer could test the software before release, these faults might be found and eliminated. For some applications, it is possible to involve the customer in testing the software before it is released. This is referred to as beta testing. This testing might be done on a flight simulator instead of an actual aircraft for a flight control software package.

Gray-Box Testing

Gray-box testing is a blending of white-box and black-box testing strategies. Gray-box testing primarily focuses on interfaces and interactions at various levels as units are integrated into programs, programs into subsystems, and subsystems into the software system (Westfall, 2024).

Operational Profile Testing

Operational profile testing exercises the software package under all the users' anticipated conditions. Faults that remain in the software are obscure and difficult to detect. These faults are experienced randomly and will take time to be corrected. This justifies considering the fault rate constant and using the exponential distribution to model the reliability of software after it is released.

Fault-Injection Testing

This section is adapted from Westfall (2024, pp. 543–544). *Fault injection,* also called *fault seeding, fault insertion,* or *defect salting,* estimates the number of undetected defects that escaped from a testing process. Fault injection intentionally inserts known defects into the software. This is typically done by people independent of the developers or testers. The software is then tested, and the defects that are found are analyzed to determine if they are part of the intentionally inserted, known defects or if they are new, unknown defects. The number of unfound, unknown defects is then estimated through the use of the simple ratio shown in Equation 9.4.

Equation 9.4 Ratio for number of unfound, unknown defects.

$$\frac{\textit{Found unknown (unseeded) defects}}{\textit{Found known (seeded) defects}} = \frac{\textit{Unfound unknown (unseeded) defects}}{\textit{Unfound known (seeded) defects}}$$

This technique is illustrated in Example 9.7.

EXAMPLE 9.7

Assume 100 known defects were inserted into the software prior to system test. The software is tested, and 72 of the inserted defects were found and 165 other defects were found. That means that (100 − 72) = 28 of the inserted (known) defects were not found during system test. Find the number of unfound, unknown defects (UUD). Using the following ratio (Equation 9.4), we calculate:

$$\frac{\textit{Found unknown (unseeded) defects}}{\textit{Found known (seeded) defects}} = \frac{\textit{Unfound unknown (unseeded) defects}}{\textit{Unfound known (seeded) defects}}$$

$$\frac{165}{72} = \frac{UUD}{28}$$

$$UUD = 28 \times \frac{165}{72} \cong 64\,defects$$

This technique assumes that the known defects are being found at the same rate as the unknown defects. Care must be taken to "seed" the software with a representative set of defects for this assumption to hold true. For example, if only logic errors are inserted into the code, then only logic-type errors can be reasonably analyzed and predicted. Prior to fault injection, good configuration management is required to make certain that the software can be retrieved for use moving forward. If not, care must be taken to remove all known defects and to ensure that new defects are not introduced by mistake. Note that regression testing must be performed after the completion of the removal activities to verify this removal.

Software Validation and Quality Assurance

Validation is the "confirmation, through the provision of objective evidence, that the requirements for a specific intended use or application have been fulfilled" (IEEE/ISO/IEC/ 2017). Validation is concerned with the process of evaluating the software to make certain that it meets its intended use and matches the stakeholders' needs (Westfall 2024).

Software Verification

The following section is adapted from Westfall (2024, pp. 506–507). According to ISO/IEC/IEEE *Systems and Software Engineering—Vocabulary* (IEEE/ISO/IEC 2017), *verification* is "the process of evaluating a system or component to determine whether the products of a given development phase satisfy the conditions imposed at the start of that phase." Verification is concerned with the process of evaluating the software to make certain that it meets its specified product requirements and adheres to the appropriate standards, practices, and conventions. Verification is implemented by looking one phase back. For example, code is verified against the detailed design, the detailed design against the architectural design, and the architecture against the specified requirements.

The relationship between validation and verification is illustrated in Figure 9.15.

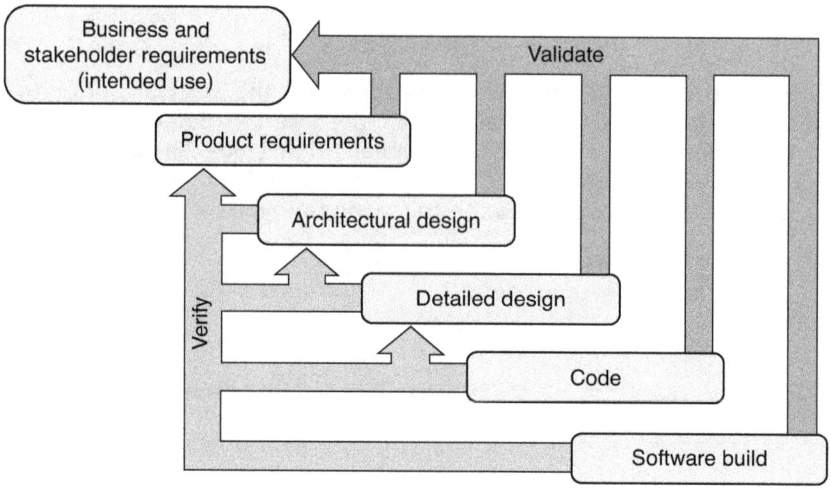

Figure 9.15. Relationship between validation and verification.
Source: Westfall (2024, p. 507).

Verification and validation (V&V) methods include static analysis, structural analysis, mathematical proof, simulation, and automation. The reader is referred to Westfall (2024), *The ASQ Certified Software Quality Engineer Handbook*, for more information.

Performance, Environmental Load, Volume, and Stress Testing

The following section is adapted from Westfall (2024, pp. 554–555). The objective of *performance testing* is to determine if the system has any problems meeting its performance requirements, including throughput (number of transactions per time unit), response time, or capacities (e.g., the number of simultaneous users, terminals, or transactions). Performance testing is usually performed at the system level under full environmental load. Typically, software can perform to specification when not much is going on. However, the performance of some software applications may degrade at full volume, with multiple users interacting with the software and other applications running in the background taking up system resources. Performance tests should be carried out after other testing is complete and the software is relatively problem-free.

Environmental load testing evaluates the software's performance capabilities (throughputs, response times, and capacities) under normal load conditions.

Volume testing, also called *endurance testing* or *soak testing,* is a special type of environmental load testing that subjects the software to heavy loads over long periods of time and evaluates the software's capability. For example, does the software have problems handling the maximum required volumes of data, transactions, users, and peripherals over several days or weeks?

Stress testing is another special type of environmental load testing that subjects the software to surges or spikes in load over short periods of time and evaluates the software's performance.

For example, is the software able to appropriately deal with jumping from no load to a spike of maximum data, transactions, or users? Stress testing can also involve testing for excess load. For example, if the requirement states that the maximum capacity that a telecommunication switch can handle is 10,000, does the software appropriately handle a spike to 11,000 calls by reporting the appropriate messages ("all lines are currently busy, please call back later"), or does it result in an inappropriate failure condition?

Qualitative Versus Quantitative Reliability Testing

Qualitative reliability states a risk probability in terms of the degree of uncertainty, such as almost certain, very likely, likely, unlikely, highly unlikely, or on an ordinal scale, such as a 1 to 5 Likert scale, in which 5 is almost certain. A qualitative loss statement may use terms such as catastrophic, critical, marginal, negligible, or it may be states using a loss index using a 1 to 5 scale. In contrast, a quantitative risk statement will use the probability that the risk will turn into a problem, and losses may be stated in costs, such as dollars or effort hours.

Built-In Testing

Built-in testing (BIT) is especially useful when high availability and reduced maintenance times are crucial to a unit. The built-in test is contained in the module it tests and is often a specialized software routine that exercises the module. Drees and Young (2002) state, "BIT can generally be described as a set of evaluation and diagnostic tests that uses resources that are an integral part of the system under test." BIT tests do not monitor the system continuously. Rather, they take snapshots of the system operation and often give pass/fail feedback. Specifications for BIT are in terms of the amount of coverage to detect faults. For example, a specification might be that the BIT must "detect 95% of all faults, isolate 95% of any fault to one of over 50 replaceable assemblies, and allow false alarms for less than 2% of all detected faults" (Steinmetz 2002).

Regression Testing

This section is adapted from Westfall (2024, pp. 558–599). *Regression testing* is concerned with how changes (fixes or updates) to software affect the unchanged portions of the software or hardware. Regression testing is "selective retesting of a software or software component to verify that modifications have not caused unintended effects and that the system or component still complies with its specified requirements" (IEEE/ISO/IEC 2017).

In the context of software testing, *regression analysis* examines a proposed software change to determine the depth and breadth of the effects the proposed change could have on the software. It determines how extensively a change needs to be regression tested and how many already-executed test cases need to be re-executed.

Regression analysis must balance the risk of releasing the software with undiscovered defects and the software's quality and integrity requirements against the cost of a more extensive testing effort. Of course, the first testing step when software changes is to retest what changed. White-box, gray-box, and black-box testing strategies can be used to test the individual source code modules, components, interfaces, and features that changed.

Regression analysis is then used to determine how extensively other parts of the software should be retested. When deciding what other white-box and gray-box regression tests to execute, consider source code modules or components that share the same local or global variables or that directly call, or are called by, those source code modules or components that changed. More rigorous regression testing might also consider going to a second level. This more rigorous testing includes testing the source code modules or components that call, or are called by, the source code modules or components that were changed. Regression analysis also includes deciding what other black-box regression tests to execute when changes are made to the software. Black-box regression analysis considers functions that:

- Perform similar functions or are closely associated with the changed functions
- Interface with the same external device(s) or user(s)
- Access the same external data or databases

The final step in regression testing is to execute the *regression test suite.* The regression test suite includes white-box, gray-box, and black-box tests that are always repeated no matter what changes in the software. Candidate tests for the regression test suite are tests for the source code modules or components and functions that:

- *Are mission-critical to the success of the software.* For example, if everything else fails, the "save" command in a word processor or the self-destruct in a missile system must work.
- *Are defect-prone.* Software that has been defect-prone in the past is more likely to be defect-prone in the future.
- *Are the most used (operational profile).* Defects in areas of the software that are used most extensively will impact more users.
- *Contain patches.* Since patches are temporary fixes, it is very easy to make a mistake when adding existing patches into a new software build.

Tests included in the regression test suite are prime candidates for automation because those tests will be rerun many times (every time the software is changed). Other well-known methods in software testing (possibly complementary) include:

- *Field testing.* Testing in the customer's final environment
- *Random-input testing (or monkey testing).* Random entering of possible inputs to the software
- *Boundary value testing.* Testing at the boundaries of the software input values
- *Automated software testing:* writing another software routine, replacing the users, or recording user routines (key logger) and playing them back (at a different pace if desired) to stress the software routines automatically and the interconnection of its subsystems. The system reliability can then be modeled using the reliability of the various subsystems.

Chapter 10
Reliability Modeling

This chapter discusses the concepts and techniques for reliability system modeling, in which we seek to understand the reliability aspect of a system. Throughout the chapter, we are specific about the systems and components—they are not arbitrary systems and components. Systems and components are defined in terms of probabilistic behaviors over time. Behaviors include, but are not limited to, failure, being operational, and being available. In the remaining sections of the chapter, we specifically mean a reliability system or reliability component, even if the word "reliability" is not present. In addition, we will not differentiate "subsystem" from "component," unless there is a need to differentiate the two in a relevant context. A subsystem is just a component of a system.

The definition of a reliability system includes at least three major layers or concepts shown in Figure 10.1. The first layer of definition is the dependence structure among individual components. This dependence structure captures the system-level state-change when one or more components change their states. The dependence structure is usually described by a reliability block diagram. The second layer is the definition of reliability aspects of individual components. Each component must have a well-defined time-to-failure distribution given prespecified conditions. The third concept is the change of the system state due to external actions. The third concept is necessary if the reliability system interacts with maintenance operations. This layer is unnecessary if the system does not interact with maintenance operations. To model a reliability system, one needs to configure and define the reliability system at three, or at least the first two, layers of definition. During configuration, one can work at different layers sequentially or alternatively. After the configuration, one needs to conduct numerical analysis to aggregate information from all layers and produce results at a reliability system level. There are different approaches to conducting the numerical analysis.

RELIABILITY BLOCK DIAGRAMS AND MODELS

> Generate and analyze various types of block diagrams and models, including series, parallel, partial redundancy, time dependent, K out of N, and shared load. (Evaluate)
>
> **Body of Knowledge IV.C.1**

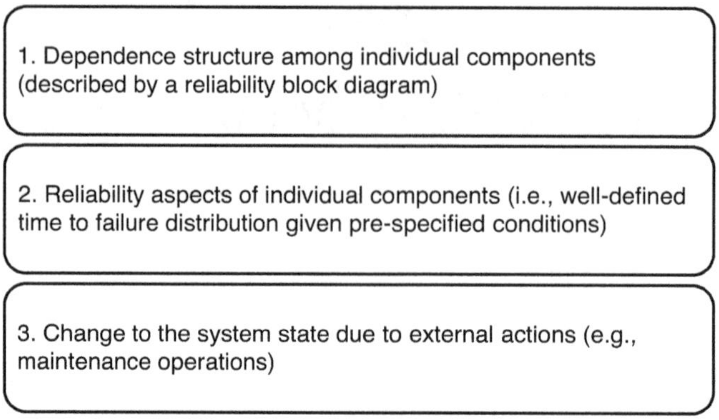

Figure 10.1 Three layers of definition for a reliability system.

A reliability system comprises multiple components that operate together to provide designed functions. Components are in their individual states at any given time. States can be defined in different ways. The simplest and the most important way of defining the states of reliability component is binary. A component is either reliable (being operational) or unreliable (being nonoperational or failed). For simplicity, we call the two states On and Off in the rest of the chapter. Given the binary definition of component states, the state of the system is also binary.

What determines the state of the system includes the states of individual components and the dependence among components. For example, two of the simplest dependence structures are series and parallel structures. In a series structured system, a single component failure will lead to a system failure. In a parallel structured system, a single component failure will not lead to a system failure. The dependence structure can be modeled by a reliability block diagram (RBD). Using the RBD, we can determine the reliability of a system. The system is On if there is a path from one end to the other that only passes through components that are On.

There are also alternative modeling methods. For example, a fault tree is a preferred approach in some use cases. In this chapter, we will focus on RBD and discuss various types of systems and corresponding diagrams that represent some commonly used dependence structures.

Series System

A series system comprises multiple components connected in series. The successful operation of the system depends on the successful operation of all components. The failure of any component will cause the failure of the system. Figure 10.2 shows an RBD representing a three-component series system.

The following formula calculates the reliability of the system in Figure 10.2:

$$R_{system} = R_1 \times R_2 \times R_3,$$

in which R_1, R_2, and R_3 are the corresponding reliabilities of three independent components. Example 10.1 demonstrates the use of the formula.

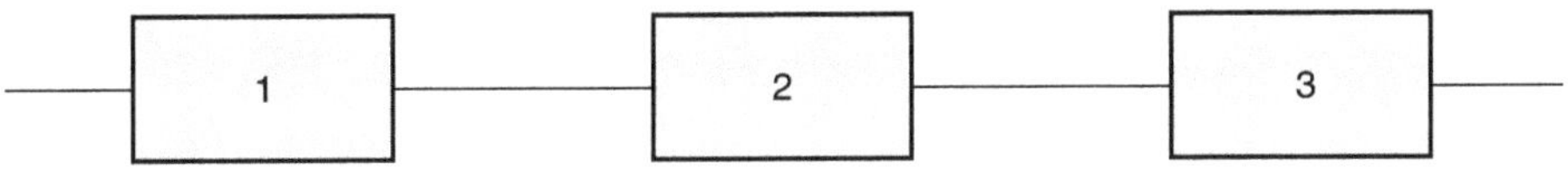

Figure 10.2 Example of a series system.

EXAMPLE 10.1

A system consists of three independent components connected in series. The reliability for each component is $R_1=0.99$, $R_2=0.98$, and $R_3=0.94$. The system reliability is $R_{System}=(0.99)(0.98)(0.94)=0.91$.

The general formula for an arbitrary series system of n independent components is shown in Equation 10.1.

Equation 10.1 Reliability for a series system.

$$R_{system}=\prod_{i=1}^{n} R_i$$

Here, we omit the time t in the formula for calculating reliability to avoid unnecessary complications because the purpose of RBD is to define the dependence structure, and the specific time t is irrelevant.

There are additional assumptions that might be necessary for calculating system reliability using RBD, such as whether components are independent. The independence assumption is necessary to write the series system reliability as the product of component reliability values. Violating such assumptions, however, only invalidates the formula. Using RBD to study dependence structures does not require the independence assumption. We will see components that are not independent later in more complicated nonrepairable systems or repairable systems.

Assuming all components are independent, the series system reliability formula reveals that the system reliability is less than or equal to the reliability of any individual component in a series system. By introducing additional independent components to the system, the system reliability will decrease further. For example, consider a phone charging system consisting of a phone charger and a cable, which form a series system. Traveling with this phone charger to a different country may require the addition of an adapter, which adds a component to the series system. The new system is less reliable than the two-component charger because the adapter is an extra failure point. Unlike adding components, removing components from the series system will increase the system reliability.

The series system reliability is related to terminology, such as competing cause analysis and competing mode analysis, in which competing failure modes or failure causes are equivalent to components in a series system. In such cases, removing a failure cause or failure mode is equivalent to removing a component, which increases the system reliability. For example, a laptop may fail due to at least the battery, monitor, motherboard, or keyboard. Whichever failure occurs, the laptop fails. The reliability system of the laptop can be represented by a series system with those four components.

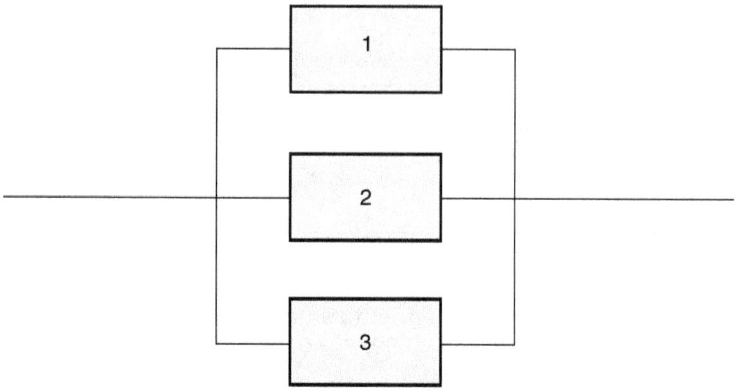

Figure 10.3 Example of a three-component parallel system.

Parallel System

A parallel system comprises multiple components connected in parallel. The successful operation of the system depends on the successful operation of at least one component. The failure of all components will cause the failure of the system. Figure 10.3 shows an RBD representing a three-component parallel system.

We now will calculate the reliability of the three-component parallel system in Figure 10.3.

$$R_{system} = 1 - (1 - R_1) \times (1 - R_2) \times (1 - R_3)$$

in which R_1, R_2, and R_3 are corresponding reliabilities of three components, and $(1 - R_i)$ represents the failure probability F_i of the ith component. The formula is obtained by equating the failure probability of the system to the product of the failure probabilities of three components because the event that the system fails is the intersection of events that individual components fail. The equation of failure probabilities or the 3-component parallel system is as follows:

$$F_{system} = F_1 \times F_2 \times F_3,$$

in which failure probabilities of system or components are complements to their reliability values, respectively.

EXAMPLE 10.2

A parallel system has three independent components. The reliability for each component is $R_1 = 0.99$, $R_2 = 0.98$, and $R_3 = 0.94$. The equation for the reliability of a 3-component parallel system can be used to calculate the reliability of the system:

$$R_{System} = 1 - (1 - 0.99)(1 - 0.98)(1 - 0.94) = 0.99998$$

In general, the formula for an arbitrary parallel system of n independent components is shown in Equation 10.2.

Equation 10.2 General formula for reliability for a parallel system.

$$R_{system} = 1 - \prod_{i=1}^{n}(1 - R_i)$$

It can be shown that, given independent components, the parallel system reliability is greater than or equal to the reliability of any individual component. Parallel system design is usually associated with incorporating redundancy in reliability system design. Redundancy generally increases system reliability if components are independent. For example, a two-engine system on an airplane can be considered as a parallel system. If one engine fails, the plane may still land with one remaining engine. If both engines fail, then the plane is at higher risk. Adding more redundancy to a parallel system should increase reliability, but it may not be practical or economical. For example, engineers don't want to design planes with four or six engines just for reliability concerns. Partial redundancy is used when some, but not all, of the critical components are duplicated.

Series–Parallel System

Series and parallel systems are two basic reliability systems that consist of multiple components. The two types of systems can themselves be components of bigger systems. A big system can sometimes be organized into a composition of series and parallel subsystems. If and only if a reliability system can be represented by a composition of series and parallel subsystems can we call it a series-parallel system. Example 10.3 illustrates this point.

EXAMPLE 10.3

Figure 10.4 shows a series-parallel system. The reliabilities for each independent component are shown in the diagram. Find the reliability of the system.

The reliability of the two-component parallel subsystem is:

$$R_1 = 1 - (1 - 0.99)(1 - 0.99)$$

$$R_1 = 0.9999$$

The reliability of the three-component parallel subsystem is:

$$R_3 = 1 - (1 - 0.89)(1 - 0.92)(1 - 0.90)$$

$$R_3 = 0.99912$$

The system is now an equivalent series system with R_1 and R_3 and $R_2 = 0.999$. The system reliability can be found as:

$$R_{System} = R_1 \times R_2 \times R_3 = (0.9999)(0.999)(0.99912)$$

$$R_{System} = 0.998$$

Continued

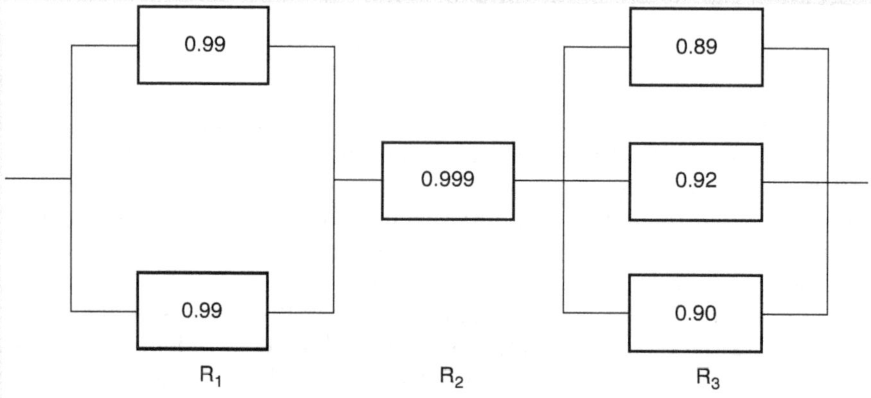

Figure 10.4 A series-parallel system.

It is worth noting that a reliability system does not necessarily mirror the corresponding physical system. For example, consider a small electronic device that requires three AA batteries operated in parallel to supply 6 amps and 1.5 volts. Even if the batteries are physically arranged in parallel, the power system of three batteries forms a series reliability system. This is because a failure of any single battery causes the failure of the entire power supply.

Interestingly, an RBD can be drawn in multiple equivalent ways. Therefore, sometimes, the existence of a series-parallel system is not obvious. For example, Figure 10.5 presents two equivalent series-parallel systems. The diagram on the right reveals that the system on the left is actually a series system of two parallel subsystems.

In practice, unless the system reliability engineer already knows the reliability system is a series-parallel system, this fact may not be obvious in many situations. The recognition of this fact is usually not critical, but doing so may lead to efficient computation with explicit formulas. The system reliability of the series-parallel system in Figure 10.5 can easily be obtained by following the logic in Example 10.3.

Bridge System

There are complex reliability systems that do not have equivalent series-parallel system representations. Figure 10.6 illustrates a so-called bridge system; see "Design Prototyping" later in this chapter, in Meeker et al. (2022). Manual calculation of system reliability for such systems can quickly become overwhelming, and computer software is necessary for large systems. Two methods of calculation are described in the section titled "Computation of Reliability Systems."

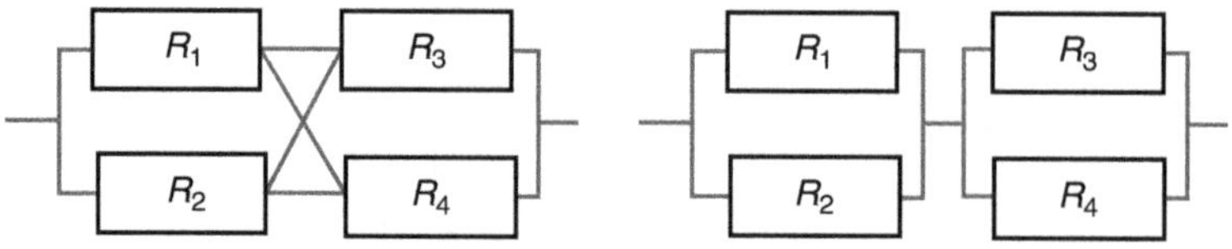

Figure 10.5 Equivalent series-parallel systems.

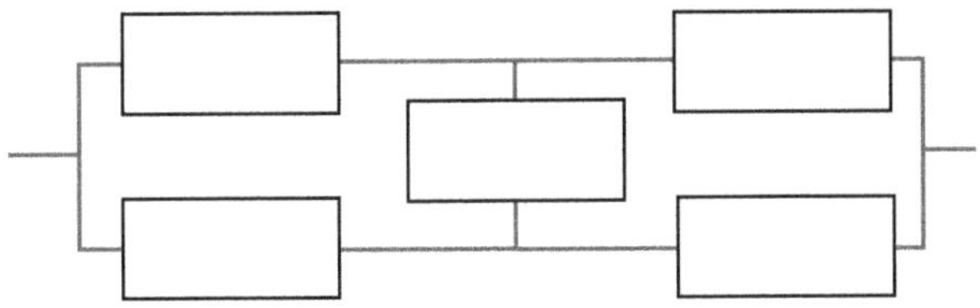

Figure 10.6 Schematic of a bridge system.

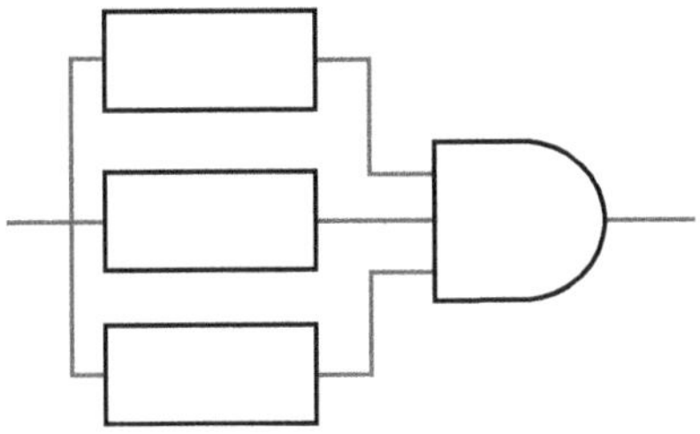

Figure 10.7 Example of a K-out-of-N system.

K-Out-of-N System

In a parallel system of n independent components, the system is reliable if at least one component is reliable. If the requirement for such a system to be reliable is changed to at least k components are reliable, the modified system is a type of "K-out-of-N system," in which individual components are identical. A K-out-of-N system with identical components can be further generalized to allow nonidentical components.

The system can be represented by a special block in an RBD, or one can construct it using fundamental elements of RBD.

Figure 10.7 represents a system of three independent parallel components on the left. The block on the right represents an "And" gate. The gate can be configured such that it is open if at least two out of three components are operational, and this configuration constructs a 2-out-of-3 system.

For a K-out-of-N system with n independent and identical components, the system reliability formula is shown in Equation 10.3:

Equation 10.3 Reliability for a K-out-of-N system.

$$R_{system} = \sum_{i=k}^{n} \binom{n}{i} R^i(1-R)^{n-i},$$

where $\binom{n}{i}$ is the combination formula and R is the reliability for each of the identical components. The use of this formula is shown in Example 10.4. Note that in many cases, it is not practical to compute the reliability of such systems manually: A software program can be necessary to avoid calculation mistakes.

EXAMPLE 10.4

Eight units, each with a reliability of 0.85, are connected in a parallel configuration. At least six units are required for system success. What is the reliability of the system?

The reliability of the system is the probability that six of the units are successful and two of the units fail, or that seven of the units are successful and that one of the units fails, or that all eight of the units are successful. These probabilities are mutually exclusive and can be added to give the probability of system success.

$$\begin{aligned} R_{System}(t) &= C_6^8(0.85)^6(1-0.85)^2 + C_7^8(0.85)^7(1-0.85) + C_8^8(0.85)^8 \\ &= (28)(0.85)^6(0.15)^2 + (8)(0.85)^7(0.15) + (0.85)^8 \\ &= 0.895 \end{aligned}$$

Standby Systems

The parallel and K-out-of-N systems are special cases of the more general standby systems. More specifically, all components in a parallel system are standby redundancies until the last component fails. For K-out-of-N systems, all components are standby redundancies until the *K*th component to the last fails. They are special cases, because all redundancies perform regular functions like any other components until they fail. Because of this, parallel and K-out-of-N systems are sometimes known as hot-standby systems.

Opposite to hot-standby systems, there are cold-standby systems. The RBD diagrams are the same, except the states of individual components need to be configured separately. Furthermore, components are no longer independent. We use the simplest cold-standby system to illustrate. The RBD of the system is identical to a parallel system. At time 0, only one component begins to operate, and the remaining components do not operate. When the operating component fails, another component is chosen to operate until it fails. Then another component is chosen to operate, and so on, until all components fail. Then the system fails.

The system can be represented by a special block in an RBD. Or one can construct it using fundamental elements of RBD. Figure 10.8 is an example.

Figure 10.8 represents a system of three independent parallel components on the left. The block on the right is a switch. The switch will connect to the next available component on the left if the currently connected component fails. This configuration constructs a three-component cold-standby system.

A regular K-out-of-N system with identical components can be similarly generalized to cold-standby K-out-of-N system. If components are not identical, then the scenario will be more complex because the order of operation also matters. The calculation of system reliability of such a system is complex and requires software to avoid calculation errors.

Besides cold-standby systems, in practice, there are also warm-standby systems. Instead of starting each component when they are completely off, warm-standby systems maintain all standby components at a low operational stress. Then one component is made operational under full stress when it is time for the component to perform its full duty. The necessity of warm-standby systems is that cold-standby systems have a risk of failing to start, while the risk is low for a warm-standby system.

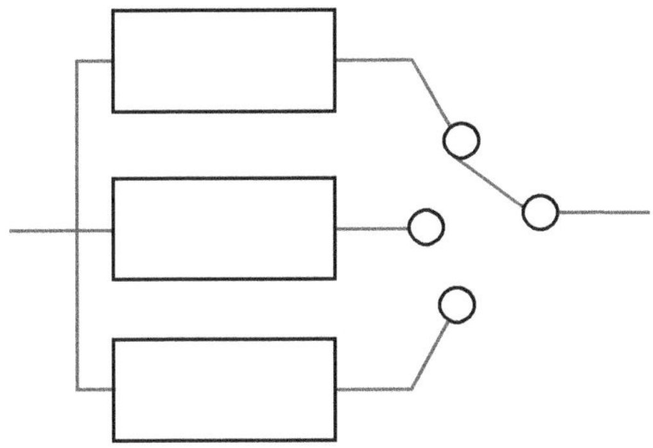

Figure 10.8 Cold-standby system.

Load-Sharing Systems

For parallel systems, we assume that failures of individual components do not impact remaining components that are still operational, such that failures of components are completely independent. The assumption does not hold, or only approximately holds, in some use cases. In those use cases, the components in a parallel system share a fixed amount of load or stress. As the number of operational components decreases, the load or stress in the remaining individual components increases. Thus, increasing load or stress will impact the failure rate of remaining components that are operational.

The system reliability of a load-sharing system involves at least two pieces of additional information. The first is how load or stress is distributed. A relatively simple example is voltage sharing among parallel configured identical protective resistors. The electrical current through an individual resistor is proportional to the reciprocal of the number of resistors. A relatively complex example is a truss system, in which individual substructures share the load in a complex way. Besides how the load or stress is distributed, one still needs to know the mechanism of how increased load or stress would impact failure rate. That is directly related to the accelerated life tests, which have been discussed in Chapter 9. For different acceleration mechanisms, even the load-sharing changes in the same way, and the impact on failure rates will be different.

A simple but practical load-sharing system shares the load proportional to the reciprocal of the number of identical components, and the acceleration mechanism follows a logarithm relationship, or power-law relationship. For example, consider an N-component load-sharing system with equal load-sharing. If the total load is V, an operational component bears V/K, where K is the number of remaining operational components. At the beginning, each component bears V/N load. After one component fails, each component bears $V/(N-1)$. Governed by the logarithm or power law, the time-to-failure of individual component is thus accelerated by a factor of $N/(N-1)$. Overall, load-sharing systems are the most complex systems among what has been discussed so far. Their failure distributions depend on additional information or assumptions.

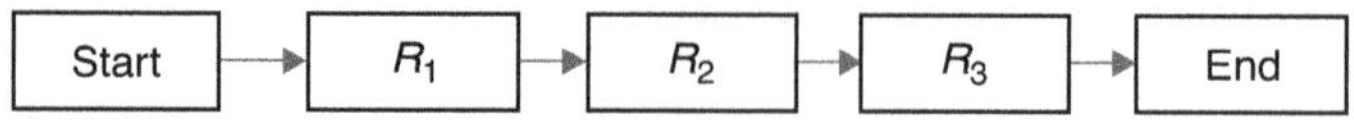

Figure 10.9 Example of a formal definition of the reliability state.

Time Dependent Reliability

In the preceding sections, we have used R to denote the reliability of a component. This is done for simplicity and assumes the reliability values are either constant across time or are being specified at a fixed, common time. When it is necessary to recognize that the reliability may vary with the time of operation, we will write $R(t)$. Note that all the preceding formulas still apply if we replace R with $R(t)$.

Computation of Reliability Systems

We have presented several reliability systems. Some of them have closed-form formulas for calculating their system reliability. The others do not have such formulas because of their complexity. We present two formal methods that can be implemented in software to compute system reliability at a specific time for an arbitrary reliability system with independent components.

Definition of Reliability System's State

We have presented RBDs to illustrate reliability systems. The RBD is not only a graphical representation of a reliability system but is also a formal method to define the system reliability state. The simplest system reliability state is binary, either operational (On) or not (Off). The state is determined by two things, the states of individual components and the dependence structure. To describe the method of determination, we need to introduce additional elements to the RBD. We use a series system of three components to illustrate.

The reliability system in Figure 10.9 is identical to the one in Figure 10.2. The diagram in Figure 10.9, however, is a formal definition of the reliability system. First, two extra nodes are added. One is called "Start," and the other is called "End." In some literature, they may be referred to differently, such as "Source" and "Sink." Second, all the connections between nodes have arrows.

The formal definition of RBD for a reliability system has the following properties:

- There is one and only one Start node. Every edge that connects to the Start node has an arrow that points away from the Start node.
- There is one and only one End node. Every edge that connects to the End node has an arrow that points to the End node.
- All nodes are connected via single arrowed edges. Any trace by following the arrows of edges must be acyclic. In other words, there are no loops in the diagram.

Such a diagram is called a directed acyclic graph (DAG). The name is basic terminology in the field of graph theory in computer science. The diagram has broad applications, and we can find that the term appears in many fields.

Given that a reliability system can be formally defined using a DAG, the binary state of the system can be defined formally as well:

- A reliability system is operational, or in the On state, if there is at least one trace that exists from Start to End.
- A reliability system is not operational, or in the Off state, if there are no traces that exist from Start to End.

The binary state definition is simple but common, especially for nonrepairable systems and components. Repairable systems usually have additional states. For example, availability is a state such that a system or component is not operational but is ready to be turned on at any time. Such a state is different from the Off state, which cannot be turned on in a binary state nonrepairable reliability system. The definitions of multiple states are complex, and calculating the system state requires more complicated analysis. In the remainder of the chapter, we restrict our discussion to the reliability systems with binary states.

System Reliability Computation Methods

Given the definition of a reliability system, the system reliability can be calculated in different ways. In the literature, there are approximate methods for computing the lower and upper bounds of the system reliability. There are also exact methods that compute the exact value of the system reliability. In this chapter, we will describe two exact methods: Boolean algebra and dynamic programming.

System Reliability via Boolean Algebra

In this section, we describe a brute force exact method of computing system reliability. To use this method, one needs to form a table in which every column is associated with a specific component, and the last column is associated with the system. The values in individual cells are either zero or one (or other symbols) denoting Off and On states, respectively. Each row represents a unique combination of component states, such that all rows represent mutually exclusive events that are state combinations. For any row, one can determine the state of the RBD. First, one can remove all the components that are in the Off state. Then, if there exists a trace from State to End, the system is in the On state. Otherwise, the system is in the Off state. The last column of that row then records the probability of being in the On state for that row. The system reliability is then the sum of the probabilities of all rows, because events associated with individual rows are mutually exclusive. Example 10.5 illustrates the calculation through an example.

EXAMPLE 10.5

Find the system reliability of the system in Figure 10.10 using the Boolean algebra method.

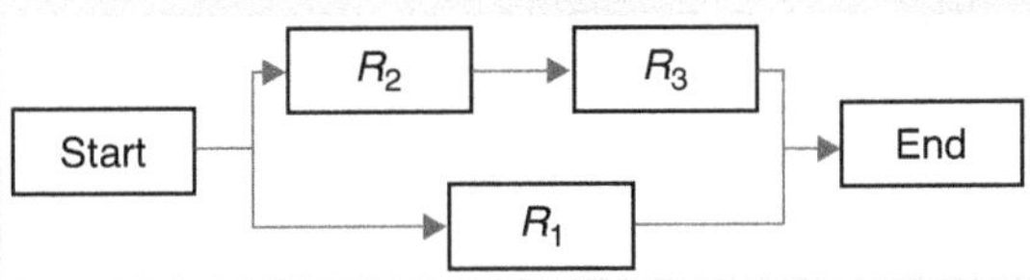

Figure 10.10 Reliability block diagram of the Boolean algebra example.

Continued

Table 10.1 Event probabilities of $S = \text{On}$ states.

Event #	Components R_1	R_2	R_3	System	Event probability
1	S	S	S	S	(0.85)(0.90)(0.92) = 0.7038
2	S	S	F	S	(0.85)(0.90)(0.08) = 0.0612
3	S	F	S	S	(0.85)(0.10)(0.92) = 0.0782
4	S	F	F	S	(0.85)(0.10)(0.08) = 0.0068
5	F	S	S	S	(0.15)(0.90)(0.92) = 0.1242
6	F	S	F	F	
7	F	F	S	F	
8	F	F	F	F	

Component reliability values are $R_1 = 0.85$, $R_2 = 0.90$, and $R_3 = 0.92$. Conduct the Boolean algebra calculation as shown in Table 10.1, using S for the On state and F for the Off state.

The system reliability is the sum of the probabilities of the events that result in system success:

$$R_{\text{System}} = 0.7038 + 0.0612 + 0.0782 + 0.0068 + 0.1242 = 0.9742$$

This can be verified using the series-parallel model:

$$R_{\text{System}} = 1 - [1 - (0.90)(0.92)] \times [1 - 0.85] = 0.9742$$

The drawback of the Boolean algebra approach is that complexity grows exponentially as the number of components increases. There are many combinations that do not result in the On state for the system, and they waste computation resources. The method in the next section is more efficient.

System Reliability via Dynamic Programming

Assume that we have a reliability system of n components, denoted by $C_1, \ldots, C_n$. The failure probability of the system is denoted by F_S, and the components are either On, represented by $C_1, \ldots, C_n$, or Off, represented by $C_1^C, \ldots, C_n^C$.

Using dynamic programming, we choose one component. Without loss of generality, we choose C_1 first. The failure probability of the system is the sum of the failure probabilities of two mutually exclusive states of the system. In one state, C_1 is in the On state. And in the other state, C_1 is in the Off state. Writing the relationship in an equation gives the following:

$$F_S = F_{S|C_1} \times R_{C_1} + F_{S|C_1^C} \times F_{C_1}$$

where $S \mid C_1$ means the system conditioning on an operational C_1, and $S \mid C_1^C$ means the system conditioning on a nonoperational C_1. In terms of RBD, $S \mid C_1$ is the RBD of S, with C_1 replaced by a direct edge going straight through C_1, because C_1 is guaranteed to be operational due to conditioning. On the other hand, the RBD of $S \mid C_1^C$, is the RBD of S, with C_1^C removed. In the previous equation, R_{C_1} is the reliability of C_1 and F_{C_1} is the failure probability of C_1. Both quantities are known.

At this step, however, either $F_{S \mid C_1}$ or $F_{S \mid C_1^C}$, the conditional failure probabilities of the system, are still unknown, if the system is not simple. But both can be further decomposed via the same logic. Suppose the next component to be considered is C_2. $F_{S \mid C_1}$ and $F_{S \mid C_1^C}$ can be expressed as follows:

$$F_{S \mid C_1} = F_{S \mid C_1, C_2} \times R_{C_2} + F_{S \mid C_1, C_2^C} \times F_{C_2}$$
$$F_{S \mid C_1^C} = F_{S \mid C_1^C, C_2} \times R_{C_2} + F_{S \mid C_1^C, C_2^C} \times F_{C_2}$$

The decomposition continues until there is only one component left in the system. At that step, the two equations reduce to one, regardless of what conditions are:

$$F_{S \mid \ldots} = F_{C_n}$$

Then, the failure distribution of the system can be calculated through backward aggregation.

The computation appears complicated, but the efficiency appears when a branching stage can terminate if removing a component from the RBD at that stage results in a nonoperational system. In another words, this calculation does not need to explore all branches, which results in an efficiency improvement. Example 10.6 illustrates the calculation.

EXAMPLE 10.6

Find the system reliability of the system in Figure 10.11 using the dynamic programming method.

Component reliability values are $R_{C1} = 0.85$, $R_{C2} = 0.90$, and $R_{C3} = 0.92$. The first decomposition calculates:

$$F_S = F_{S \mid C_1} \times R_{C_1} + F_{S \mid C_1^C} \times F_{C_1}$$

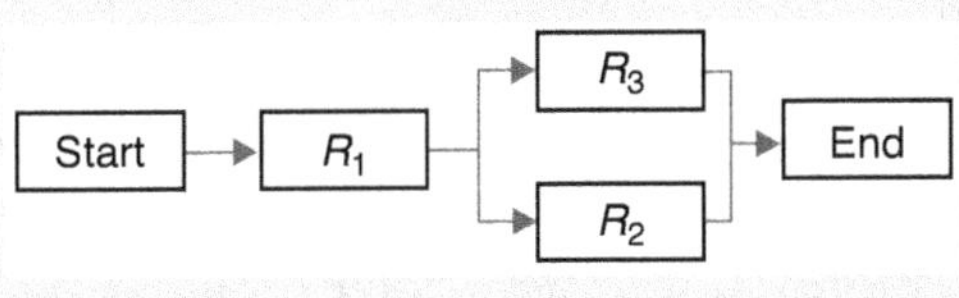

Figure 10.11 Reliability block diagram of the dynamic programming example.

Continued

The RBD of $S|C_1$ is shown in Figure 10.12 after C_1 is replaced by a direct connecting edge:

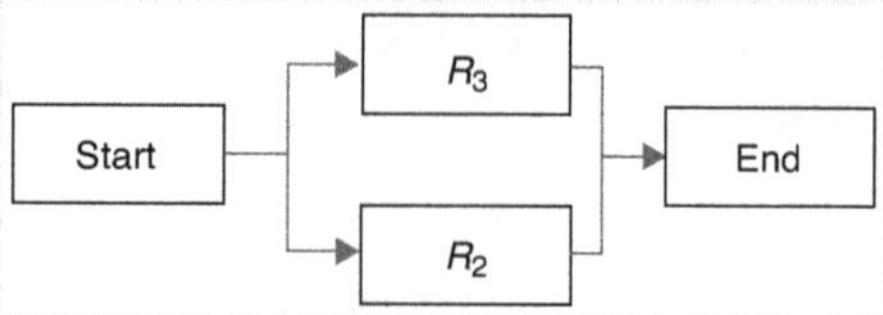

Figure 10.12 Reliability block diagram of $S|C_1$.

Drawing the RBD of $S\,|\,C_1^C$ is unnecessary because the original RBD with C_1 removed results in a nonoperational system. Therefore, $F_{S\,|\,C_1^C} = 1$. This simple case demonstrates that it is unnecessary to decompose the RBD of $S\,|\,C_1^C$ further.

The next step is to decompose $S|C_1$. Although it is an obvious parallel system, we still use the dynamic program method to illustrate our steps.

$$F_{S\,|\,C_1} = F_{S\,|\,C_1,C_2} \times R_{C_2} + F_{S\,|\,C_1,C_2^C} \times F_{C_2}$$

The RBD of $S|C_1$, C_2 is shown in Figure 10.13:

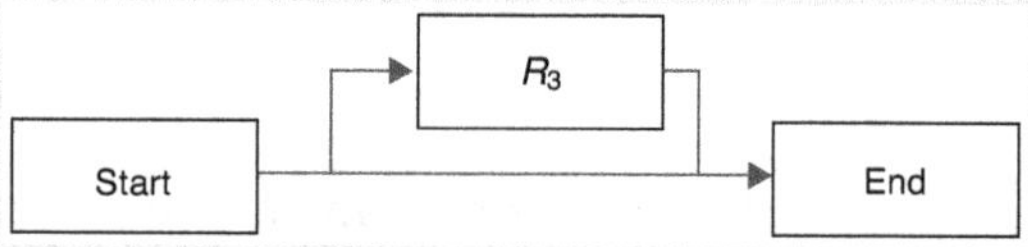

Figure 10.13 Reliability block diagram of $S|C_1$, C_2.

The RBD of $S\,|\,C_1, C_2^C$ is shown in Figure 10.14:

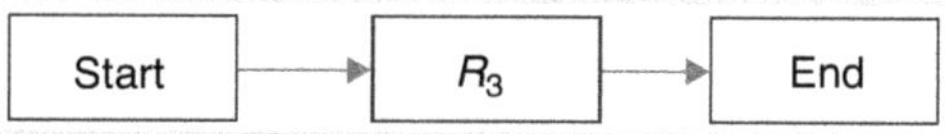

Figure 10.14 Reliability block diagram of $S|C_1$, C_2^C.

At this step, the system is fully decomposed, and the following quantities are known:

$$F_{S\,|\,C_1,C_2} = 0$$

$$F_{S\,|\,C_1,C_2^C} = F_{C_3}$$

$F_{S\,|\,C_1,C_2}$ equals zero, because given C_2 operational, the system is guaranteed to be operational, so that the failure probability is zero. The entire formula for the failure probability of the system can be written as:

$$\begin{aligned} F_S &= \left[F_{S\,|\,C_1,C_2} \times R_{C_2} + F_{S\,|\,C_1,C_2^C} \times F_{C_2}\right] \times R_{C_1} + F_{S\,|\,C_1^C} \times F_{C_1} \\ &= \left[0 \times 0.90 + (1-0.92) \times (1-0.90)\right] \times 0.85 + 1 \times (1-0.85) \\ &= 0.1568 \end{aligned}$$

Therefore, the system reliability is:

$$R_S = 1 - F_S = 0.8432$$

PHYSICS OF FAILURE AND FAILURE MECHANISMS

> Identify various potential failure mechanisms, including fracture, corrosion, memory corruption, excessive deformation, creep, and delamination, and their underlying physical processes. (Apply)
>
> **Body of Knowledge IV.C.2**

In the previous section, we discussed modeling dependence structures, the first layer of reliability system definition, using reliability block diagram (RBD). In this and the next section, we will discuss the second layer of definition, component reliability.

The reliability of a component is defined as the probability that it will perform its intended function under encountered operating conditions for a specified period of time (Meeker et. al. 2022). Obtaining information about this probability distribution is not a trivial task. The information may be collected in different ways. One way is to rely on models that are based solely upon data. The other way is based upon physics models, or physics of failure (PoF), that describe failure mechanisms. According to NASA (2024), PoF is a reliability technique that leverages the knowledge of an in situ design or condition of an item with an understanding of the stresses, processes, and mechanisms that induce its failure to predict its reliability.

One merit of using PoF is that engineers gain insight into not only when a component fails but also why it fails. The study of the PoF of a component can reveal failure mechanisms or failure modes, which are usually intuitive to study from physical perspectives. Also, by studying PoF for a specific problem, it helps to decompose a complex failure phenomenon into different failure modes. Individual failure modes can be studied separately and then combined to configure the reliability system in a rigorous and systematic way. Another important merit of using PoF is that it gives engineers the ability to assess the reliability of a component in unknown environments, because PoF more readily allows extrapolation given reasonable assumptions.

In the rest of this section, we describe a few of these failure mechanisms. The body of PoF is substantial, and many times the relevant materials are not found in the context of reliability but mechanical or chemical system designs. This section does not intend to be comprehensive but only attempts to raise awareness. Reliability engineers need to search for relevant literature for their own use cases.

Fracture

Mechanical fracture is a common failure mechanism in material science. There are different fracture types. Brittle, or fast, fracture is a type of failure that occurs when a material breaks suddenly with little or no crack initiation. Such a failure mode is usually a concern of materials that cannot withstand deformation before breaking. Examples of such materials include ceramics, glasses, cast iron, tile, stone, concrete, certain high-strength steels, and welding joints.

The physical explanation of brittle fracture is easy to understand. When there is a stress exceeding the strength of a piece of brittle material, the material breaks; see Collins et al. (2010). In terms of a simple mathematical form, the condition is Stress ≥ Strength.

We may wish to apply this PoF directly, but this inequality is too simple to be used literally in the real world. Additional conditions and uncertainties arise when one tries to apply this physics of failure explanation in practice. The following are some reasonable concerns.

- How is the strength defined, and is the strength known?
- How is stress applied or loaded—single point or an area?
- Does the geometry of the material matter?

The concerns divide the simple physical rule in many directions: the specificity and uncertainty in strength, stress, and the material, which require further modeling techniques. For example, finite element methods can be applied to understand the deterministic behavior of the material of specific geometry under specific stress. These methods are discussed in Chapter 11. Additional Monte Carlo simulation can be applied on top of deterministic studies by varying the inputs to the deterministic system to determine how materials behave under different conditions in a natural environment.

In contrast to brittle fracture, there is ductile fracture, which has deformation before failure. The deformation process adds additional complexity. For many flexible materials, deformation is expected, but excessive deformation beyond the design limit will eventually cause ductile fracture; see Wulpi (2013). In addition, the likelihood of occurrence for brittle and ductile fracture on the same material changes depending on the temperature. For example, reducing temperature may change the failure from ductile fracture to brittle fracture.

At this point, if one is concerned about fracture, there are at least five factors that require attention. They are types of fracture, stress, strength, material characteristics, and temperature. Even if all relationships are known, the environment is uncertain. But with known relationships, one can at least simulate the environment and make useful calculations, which is usually how PoF is applied. As such, the fracture-induced PoF in a reliability system model can involve a function of other factors for the reliability function.

Corrosion

Corrosion is a process by which metals gradually oxidize. Metal contacting air is a recipe for corrosion. Besides air, additional factors can contribute to the degree and the speed of corrosion, such as moisture, metal type, temperature, electrolyte, and stress.

A simple model that can describe the growth is a zero-order kinetics model in chemistry. Such a model says that the rate of change is constant, meaning that the oxide thickness grows at a constant rate over time. The rate k, however, is a nontrivial function that takes temperature and the material's characteristics (i.e., activation energy) as inputs. For example, k could be calculated using

$$k = A \times e^{\frac{-E_a}{RT}}$$

where A is the pre-exponential factor, E_a is the activation energy, R is universal gas constant, and T is the absolute temperature.

Such an understanding based on PoF allows engineers to extrapolate the corrosion growth in time and across temperatures. For example, assume that the total time can be divided into fixed intervals, with a different temperature for each interval. Then, given temperature and the constant corrosion growth rate, one can calculate the corrosion growth for a fixed time interval. Then the growth may be combined across the intervals to get the growth over the total time.

In addition to the simple model, there are more models that are appropriate to describe corrosion growth, for different situations; see McPherson (2019). These additional factors can quickly introduce complexity but also improve the representation of reality.

Memory Corruption

This type of failure mechanism is specific to computer memory chips. It is usually not within the scope of PoF study. This failure has the same consequence, which is software system crashes or erratic behaviors. There could be multiple causes, either hardware-related or software-related. First, the memory chips can be faulty, such that certain bits in the chips cannot write or read correctly. This could be due to quality issues but could be circumvented through robust hardware or software implementation to avoid the bad bits or to correct the bad bits. Over time, usage can cause hardware to degrade. The most common failure causes of memory corruption are related to software, especially software that is written in low-level programming languages with direct access to memory. Human error in software is a source of such failure. The failure of software is a subject known as software reliability and does not involve PoF. Software reliability is a large area by itself.

Excessive Deformation

The failure mechanism of excessive deformation can be associated with at least three failure mechanisms: ductile fracture, creep under constant stress, and delamination. While creep under constant stress might be a cause that leads to excessive deformation, the other two are the consequences of excessive deformation. Depending on the context, excessive deformation also has specific definitions as described in Dasgupta and Pecht (1991), in which the failure mechanism could be related to small component like electronics, or a large high-precision structure such as space mirrors and antennas, where excessive deformation can trigger unstable vibration modes.

Excessive deformation can cause failures of components or structures themselves or may cause issues for neighboring components in the same structure, eventually leading to unexpected failures. An example is electronics packaging, in which neighboring circuit cards should not contact one another, but bad design could make that possible. A solution is to provide geometric clearance around cards, in which the design needs to consider the range of excessive deformation of the packaging; see Dasgupta and Hu (1992). When studying such a PoF, the objective is no longer to anticipate it as a future reliability issue but to eliminate it at the design phase. After this, such a failure may still have a low probability of occurrence, but the system reliability system will not be greatly affected.

Creep

Creep, in material science, usually means the tendency of a solid material to slowly deform under a constant stress. That is known as creep under constant stress. In some literature, creep could also mean a solid material under a relaxed stress for a fixed deformation. That is known as creep under constant strain. One can find details in McPherson (2019) and other references. Depending on the context, additional factors can be involved. For example, temperature is another important factor for creep failures in materials for pressured vessels.

For metal, significant creep under constant stress usually occurs at a high temperature, and deformation is accelerated. This may lead to excessive deformation causing eventual fracture. For compact granular materials, creep under constant stress can alter the overall structure of the materials. Solid propellant rocket motors exhibit this type of failure; see NASA (1973). When the rocket motors are in storage for a long period, gravity puts a constant pressure on the granular combustible materials and causes structure changes that may lead to firing failures.

Creep under constant strain may sound counterintuitive. Isn't the relaxing stress a good thing? In some applications, it is not. That is specifically important in some clamping applications, for which engineers wish the clamping strength to be kept constant. The clamping strength is considered a stress to the material. But if the material deforms, and no adjustment is made to the clamping strength, creep under constant strain occurs. A simple example of this is a casually fastened bolt, which gradually loosens over time.

Creep is a failure mechanism involving time, and its failure usually involves a threshold. This is a perfect use case to apply degradation to define the failure and use the result from degradation to construct a reliability function as the input to the reliability system.

Delamination

Delamination is a general terminology to describe the fracturing of material into layers. The material could be naturally solid without layers, naturally layered, or layered by manufacturing. Different types would require different PoF theories. The first two fall into the categories of crack and creep that we mentioned previously. The mechanisms are similar, except the type of crack is specific. The third may also further separate into whether there is adhesive material involved. If there is not, the actual failure mechanism may be crack and creep. If there is, the actual failure mechanism usually is more relevant to the adhesive rather than the material itself. In other words, the degradation of the adhesive results in the failure of the material and produces delamination. The degradation of the adhesive usually is a chemical reaction process, which is very similar to the failure mechanism, corrosion, in a later section. Again, degradation analysis can produce a reliability function as the necessary input to the reliability system.

FAILURE MODELS

> Select appropriate theoretical models to assess or predict failure rates, such as Arrhenius, S-N curve, and Coffin-Manson model for temperature cyclic stress. (Analyze)
>
> **Body of Knowledge IV.C.3**

In this section, we will combine PoF with statistics to produce models that can be used to assess reliability. Unless otherwise noted, please refer to the two references for the details of discussed models: Meeker et al. (2022) and Nelson (2004).

We will present four approaches that allow for different amounts of information about the operating conditions or environment. The first approach enables us to specify failure distributions without information about additional factors. These are among the simplest models for failure time, though they can be complicated in situations such as competing failure modes. The second approach uses accelerated test models, which typically include an acceleration factor that is adjusted by the engineers to change the speed of aging of a product or specimen. Such models can be used to derive the failure time distribution under a normal use condition with very long failure times. These models are usually highly related to PoF. The third approach leverages S-N curve models that are appropriate for modeling metal fatigue data. The applications include finding the distribution of number of cycles to fracture or finding the fatigue strength distribution. Some of these models could be related to PoF. The fourth approach uses degradation models, which include both destructive degradation and repeated measures degradation. The objective here is to extrapolate the degradation path and find failure time distributions when it is difficult to observe failures under typical use conditions. These models are usually highly related to PoF.

Failure Distributions Without Additional Factors

Specifying reliability without information about specific operating conditions is useful in many applications. In some applications, the operating condition is stable and unnecessary to be considered in modeling. In other applications, the operating conditions can be too complex. And defining reliability case by case based on conditions can be equally complex. In that case, statistical modeling using the data collected under all conditions can lead to a plausible failure distribution estimate, without extra factors, which is a good approximation to the reality without the granular configurations condition by condition.

Chapter 6, "Probability Distributions," presented several distributions, including the two most important ones used in reliability: lognormal and Weibull. The lognormal is usually used to model failure distributions of metal fatigue failures, electronics and appliance failures, insulation failures, and so forth. The Weibull is usually used to model failure distributions associated with material properties, especially strength, of such components as bearings, ceramics, or electronic appliances. In particular, it is appropriate to model the "weakest-link" type of products, according to extreme value theory, which can be viewed as a PoF if the circumstance is right. Both distributions belong to a family known as the log-location-scale distributions, which all have two parameters: μ for location and σ for scale. The cumulative distribution function is usually denoted by $\Phi(t; \mu, \sigma)$.

In addition to lognormal and Weibull, limited failure population (LFP) distributions are another type of commonly used distributions for semiconductor products. The LFP counterparts of lognormal and Weibull are LFP lognormal and LFP Weibull, respectively. Their cumulative distribution functions (CDFs) have the following form:

$$P(T \leq t) = p \times \Phi(t)$$

where Φ is the corresponding CDF of either lognormal or Weibull, and p is a value between zero and one. The maximum value of an LFP distribution function is p. If a product's failure distribution is an LFP, such a product only has a subpopulation of $100 \times p\%$ that will fail, and the rest will never fail within the duration of interest. The subpopulation is usually small.

Another important type of distribution is used to describe failure mechanisms of a component with competing failure causes or modes. If individual causes or modes are known, and the failure distributions associated with individual causes are known, then the analysis proceeds by using virtual components to represent individual causes and creating a series sub-system of the virtual components to represent the actual component. By assigning failure distributions to corresponding virtual components, the resulting distribution of the series subsystem is the desired failure distribution for the actual component. The details of obtaining failure distributions of competing causes are available in the references. If the information does not allow one to obtain individual competing cause distributions, the task is more challenging and beyond the scope of this chapter.

Accelerated Life Test (ALT) Models

Some of these models are related to PoF and can be considered continuations of the PoF counterparts. If there are no related PoF theories, such models are experimental means for researchers and engineers to understand how to predict failure distributions of high reliability products under the regular use conditions at which one cannot observe failures.

Arrhenius Accelerated Test Model

The Arrhenius model describes the failure distribution as a function of temperature as shown in Equation 10.4:

Equation 10.4 Arrhenius model

$$P(T \leq t) = \Phi\left(t; \mu = \beta_0 + \frac{E_A}{K \times T}, \sigma\right)$$

in which Φ is the CDF function of a log-location-scale distribution, μ is the location parameter, σ is the scale parameter, K is Boltzmann's constant (8.617 x $10^{(-5)}$ eV/degree Kelvin), and T is temperature in Kelvin. The location parameter in this model is a function of two parameters, β_0 and activation energy E_A.

The previous expression is understood well in statistical literature. In engineering literature, the relationship is not usually expressed as a CDF function, but in the form shown in Equation 10.5:

Equation 10.5 Arrhenius model, alternate parameterization

$$t_p(T) = A_0 \exp\left[\frac{E_A}{K \times T}\right],$$

in which t_p is the quantile with probability p of the lifetime distribution, $A_0 = \exp(\beta_0)$, K is Boltzmann's constant, E_A is activation energy, and T is temperature in Kelvin. For most accelerated models, we assume the distribution belongs to the log-location-scale distribution family, and the scale parameter σ is constant across all values of T. Note that t_p in Equation 10.5 refers to a specific quantile, but it does not matter which quantile. For example, one can use the median, or the B10 life, which is the 10% quantile (P_{10}). Furthermore, $t_p(T)$ is a decreasing function of the temperature T. That is, the lower the temperature, the longer the life.

As discussed in Chapter 9, products operating under different temperatures can be compared using the acceleration factor (AF), which can be defined as the ratio of the life distribution quantile $t_p(T)$ between two different operating temperatures. This is expressed using Equation 10.6 using notation for a use condition temperature and accelerated stress temperature.

Equation 10.6 Arrhenius model acceleration factor

$$AF(T) = \frac{t_p(T_U)}{t_p(T_S)} = exp\left[\frac{E_A}{K} \times \left(\frac{1}{T_U} - \frac{1}{T_S}\right)\right]$$

If the temperature values are in Celsius, then a conversion is required. If X is temperature in Celsius, then X + 273.15 is temperature in Kelvin.

EXAMPLE 10.7

An electronic device that normally operates at a temperature of 50°C is subjected to a stress temperature of 80°C. The activation energy for the failure mode is 0.8 electron volts. What is the acceleration factor (*AF*) using the Arrhenius equation? Using Equation 10.6, we obtain:

$$AF(X) = \exp\left[\frac{E_A}{K} \times \left(\frac{1}{X_U + 273.15} - \frac{1}{X_S + 273.15}\right)\right]$$

$$AF(X) = \exp\left[\frac{0.8}{8.617 \times 10^{-5}} \times \left(\frac{1}{50 + 273.15} - \frac{1}{80 + 273.15}\right)\right] \approx 11.5$$

Assuming the model is appropriate, increasing temperature from 50°C to 80°C will increase the aging by a factor of 11.5. In other words, one hour of use under 80°C is equivalent to 11.5 hours of use under 50°C. Or conversely, if the median life under 80°C is 1 year, the median life under 50°C is about 11.5 years.

Inverse Power Accelerated Test Model

While the Arrhenius model is used for temperature-accelerated stresses, the inverse power law model is used for non-thermal stresses, such as voltage. The model can be written as:

$$P(T \leq t) = \Phi(t;\, \mu = \beta_0 + \beta_1 \times \log(V),\, \sigma),$$

where Φ is the CDF function of a log-location-scale distribution, μ is the location parameter, σ is the scale parameter, and V is voltage. In engineering literature, it is often expressed as

$$L = \frac{a}{V^b}, \quad a > 0 \text{ and } b > 0,$$

where $b > 0$ ensures that life L is a decreasing function of voltage V. Therefore, increasing voltage will accelerate the aging process. The acceleration factor for the inverse power law model is discussed in Chapter 9 and shown in Equation 10.7 for voltage.

Equation 10.7 Inverse power acceleration factor, alternate parameterization.

$$AF(V) = \frac{V^b}{V_U^b},$$

where V_U is the use stress, and V is an accelerated stress. An example of the use of this acceleration factor for voltage is given in Example 9.2.

Additional Accelerated Test Models and Some Notes

There are many other accelerated test models in the literature, both statistical and engineering. The (admittedly incomplete) listing that follows deliberately expresses every relationship in two formulations: one that is common in statistical literature, and the other that is common in engineering literature. Anyone who works in the field must be aware of the different expressions. To fit any accelerated test model, an engineer may use statistical software or consult a statistician. It is important for the engineer to recognize the formulation in statistical software, and it is important for a statistician to speak familiar terms with an engineer.

Linear: $\Phi(t;\, \mu = \beta_0 + \beta_1 X,\, \sigma)$, or $L = a \cdot \exp[bX]$.

Log: $\Phi(t;\, \mu = \beta_0 + \beta_1 \log X,\, \sigma)$, or $L = aX^b$.

Square-root: $\Phi\left(t;\, \mu = \beta_0 + \beta_1\sqrt{X}, \sigma\right)$, or $L = a \cdot \exp\left[b\sqrt{X}\right]$.

Eyring: $\Phi(t;\, \mu = \beta_0 + \beta_1 \log(T) + E_a \times \text{Arrhenius}(T),\, \sigma)$, or $L = A_0 T^m \exp\left[\frac{E_a}{k \times T}\right]$.

Coffin-Manson thermal cycling: $\Phi(t;\, \mu = \beta_0 + \beta_1 \log \Delta T,\, \sigma)$, or $L = \frac{a}{\Delta T^b}$.

Eyring Model

The Eyring model has a theoretical basis in chemistry and quantum mechanics and can be used to model acceleration when many stresses are involved. This model has several key features:

- It has a theoretical basis from chemistry and quantum mechanics.
- If a chemical process (chemical reaction, diffusion, corrosion, migration, etc.) is causing degradation leading to failure, the Eyring model describes how the rate of degradation varies with stress or, equivalently, how time to failure varies with stress.
- The model includes temperature and can be expanded to include other relevant stresses.
- The temperature term by itself is very similar to the Arrhenius empirical model, explaining why that model has been so successful in establishing the connection between the ΔH parameter and the quantum theory concept of "activation energy needed to cross an energy barrier and initiate a reaction."

The model for temperature and one additional stress is shown in Equation 10.8.

Equation 10.8 Eyring model for temperature and one additional stress.

$$t_f = AT^{\alpha}\exp\left[\frac{\Delta H}{kT} + \left(B + \frac{C}{T}\right)S_1\right]$$

in which S_1 could be some function of voltage or current or any other relevant stress and the parameters α, ΔH, B, and C determine acceleration between stress combinations. As with the Arrhenius model, k is Boltzmann's constant, and temperature is in degrees Kelvin.

If an additional nonthermal stress term is needed, the model is expanded, as shown in Equation 10.9.

Equation 10.9 Expanded Eyring model.

$$t_f = AT^{\alpha}\exp\left[\frac{\Delta H}{kT} + \left(B + \frac{C}{T}\right)S_1 + \left(D + \frac{E}{T}\right)S_2\right]$$

and as many stresses as are needed may be included by adding similar terms.

Note that the general Eyring model includes terms that have stress and temperature interactions (in other words, the effect of changing temperature varies, depending on the levels of other stresses). Most models do not include interaction terms, so that the relative change in acceleration factors includes only one stress change and does not depend on the level of the other stresses. In models with no interaction, you can compute acceleration factors for each stress and multiply them together. This would not be true if the physical mechanism required interaction terms—but, at least to first approximations, it seems to work for most examples in the literature.

Advantages of the Eyring model are:

- Can handle many stresses
- Can be used to model degradation data as well as failure data
- The ΔH parameter has a physical meaning and has been studied and estimated for many well-known failure mechanisms and materials.

Disadvantages of the Eyring model include:

- Even with just two stresses, there are five parameters to estimate. Each additional stress adds two more unknown parameters.
- Many of the parameters may have only a second-order effect. For example, setting the stress term $\alpha=0$ works quite well since the temperature term then becomes the same as in the Arrhenius model. Also, the constants C and E are only needed if there is a significant temperature interaction effect with respect to the other stresses.
- The form in which the other stresses appear is not specified by the general model and may vary according to the particular failure mechanism. In other words, S_i may be voltage or ln(voltage) or some other function of voltage.

The Eyring relationship temperature acceleration factor is shown in Equation 10.10:

Equation 10.10 Eyring model relationship to the temperature acceleration factor.

$$AF_{Ey}(\text{Temp}, \text{Temp}_U, E_a) = \left(\frac{\text{Temp K}}{\text{Temp}_U K}\right)^m AF_{Ar}(\text{Temp}, \text{Temp}_U, E_a)$$

For use over practical ranges of temperature acceleration and for practical values of m not far from 0, the factor outside the exponential has relatively little effect on the acceleration factor and the additional term is often dropped in favor of the simpler Arrhenius relationship. Example 10.8 provides a comparison using the Arrhenius and the Eyring relationships.

EXAMPLE 10.8

An accelerated life test will be used to study a metallization failure mechanism for a solid-state electronic device. Experience with this type of failure mechanism suggests that the quasi-activation energy should be in the neighborhood of $E_a=1.2$ eV. The usual operating junction temperature for the device is 90°C. The Eyring acceleration factor for testing at 160°C, using $m=1$, is

$$AF_{Ey}(160,90,1.2) = \left(\frac{160+273.15}{90+273.15}\right) AF_{Ar}(160,90,1.2) = 1.1935 \times 494 = 589$$

where

$$AF_{Ar}(160,90,1.2) = \exp\left[\left(\frac{1.2}{8.617\times10^{-5}}\right)\left(\frac{1}{363}-\frac{1}{433}\right)\right] = 494$$

is the Arrhenius acceleration factor. So, for a fixed value of E_a the Eyring relationship predicts an acceleration that is 19% greater than the Arrhenius.

Coffin-Manson Model

Sometimes, even within the same type of literature, the names may not be consistent. For example, if one compares the inverse-power relationship, the log-relationship, and the Coffin-Manson thermal cycling relationship, they are mathematically identical. But one may not immediately recognize other names. Different names are usually associated with different areas of interest. For example, the inverse power is specific to electronics for voltage. Meanwhile, the Coffin-Manson thermal cycling relationship (Nelson 2004, p. 86) models metal fatigue due to thermal cycling. The symbol ΔT in the relationship stands for the temperature range of the thermal cycling. In addition, a single name may represent a class of relationships. For example, the Eyring model is the basic form of many complicated extensions. It beyond the scope of this handbook to go into the details of the extensions. Every extension is particular to some specific conditions.

Example 10.9 demonstrates the use of the Coffin-Manson model.

EXAMPLE 10.9

Consider a small study for a test material in which five samples were cycled to failure under different levels of thermal strain. To develop a relationship between thermal strain and fatigue life over the experimental range, we will fit the Coffin-Manson model to the data. To do that, we fit a linear regression model in the log scale and transform back to the original scale. Figure 10.15 shows the data and the fitted model, on both the log-log scale and the original scale.

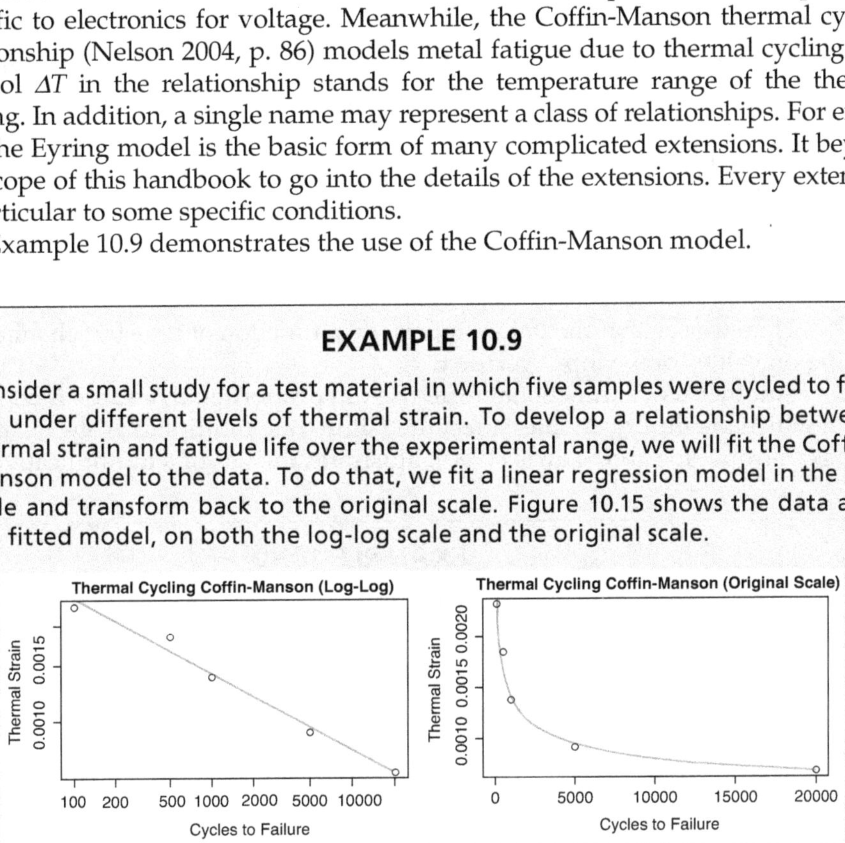

Figure 10.15 Application of Coffin-Manson model.

Proportional Hazards Model

As the name implies, the proportional hazards (PH) model has the property that individual component hazard rate functions are proportional to each other. The proportional hazards model has been used to model the effect of secondary variables on survival. It is more like an acceleration model than a specific life distribution model, and its strength lies in its ability to model and test many inferences about survival without making any specific assumptions about the form of the life distribution model.

Proportional Hazards Model Assumptions

Let $x = \{x_1, x_2, \ldots, x_k\}$ be a vector of $k = 1$ or more explanatory or input variables x believed to affect lifetime. These variables may be continuous (like temperature in engineering studies or dosage level of a particular drug in medical studies), or they may be indicator variables with the value 1 if a given factor or condition is present, and 0 otherwise.

Let the hazard rate for a nominal (or baseline) set $x_0=(x_{10}, x_{20}, \ldots, x_{k0})$ of these variables be given by $h_0(t)$, with $h_0(t)$ denoting a hazard function (failure rate) for some unspecified life distribution model.

The proportional hazards model assumes we can write the changed hazard function for a new instance of x as $h_x(t)=g(x)h_0(t)$. In other words, changing x, the explanatory variable vector, results in a new hazard function that is proportional to the nominal hazard function, and the proportionality constant is a function of x, $g(x)$, independent of the time variable t.

The constant failure rate model for the exponential case is the simplest form, given by:

$$\lambda(\mathbf{x}) = \sum_{i=0}^{k} a_i x_i$$

Here we see that the failure rate is not a function of time but changes depending on values of the input vector x.

Another useful functional form used in proportional hazards modeling is one in which the input variables affect the failure rate multiplicatively. Example 10.10, taken from Ploe and Skewis (1990), illustrates the use of a multiplicative model.

EXAMPLE 10.10

A multiplicative model for predicting the failure rate of ball bearings:

$$\lambda = \lambda_b \left(\frac{L_a}{L_S}\right)^y \left(\frac{A_C}{0.006}\right)^{2.36} \left(\frac{v_0}{v_1}\right)^{0.54} \left(\frac{C_1}{60}\right)^{0.67} \left(\frac{M_b}{M_f}\right) C_w$$

where

λ_b = Base failure rate of a bearing /10^6 hours
L_a = Actual radial load in pounds
L_S = Specification radial load in pounds
y = 3.33 for roller bearings, 3.0 for ball bearings
A_C = Alignment error in radians
v_0 = Specification lubricant viscosity
v_1 = Operating lubricant viscosity
C_1 = Actual contamination level ($\mu g/m^3$)
M_b = "Material factor" (e.g., YS) of base material (psi)
M_f = "Material factor"(e.g., YS) of operating material (psi)
C_W = H_2O contamination factor (leakage of water into the oil lubricant) $= 1+460x$ (for <0.002) $= 2.026+1.029x-0.0647x$ (for $x \geq 0.002$), where x is the percent of H_2O present in the oil (actual or from experience)

Source: Adapted from Ploe and Skewis (1990).

Regardless of the form of the proportional hazards model, the reliability is calculated using Equation 10.11:

Equation 10.11 Reliability, proportional hazards model.

$$R(t) = e^{-\lambda(x)t}$$

S-N Curve Models

S-N curve models are specific to fatigue testing, which models the relationship between number of cycles to failure (N) and stress or strain (S). Graphs of these models look similar to the graphs of accelerated test models, but these models are used when engineers are not necessarily interested in acceleration. In addition, in the literature, some of these models traditionally express stress or strain as a function of number of cycles. That, unfortunately, has caused enormous misunderstanding about the variable of interest in the model.

This section will provide a high-level summary about this class of models. Meeker et al. (2024) provides a detailed explanation. Two such models, the Basquin and Coffin-Manson, are shown, in which N is number of cycles and S is stress or strain, and all other parameters are unknown.

- Basquin: $N = A \cdot S^{-B}$, $B > 0$ and $A > 0$.
- Coffin-Manson: $S = A_{el} \cdot (2N)^{b} + A_{pl} \cdot (2N)^{c}$, $A_{el} > 0$, $A_{pl} > 0$, $b < 0$, *and* $c < 0$.

Notice that it can be shown that the Basquin relationship for the S-N curve has the equivalent form as the inverse-power relationship, or log relationship. Also notice that the Coffin-Manson relationship for the S-N curve here is different from the Coffin-Manson thermal cycling relationship, which is among the accelerated test models. Again, be careful. In the statistics and engineering literature, the same form can have different names, and the same name can mean different things in different contexts.

We first need to emphasize that even though S sometimes appears on the left side of the equations, it is not a response. It is a controlled factor in fatigue test experiments, either stress or strain. The random variable in such an experiment is N, the number of cycles to fracture, and we can observe outcomes of this random variable. The distribution of this random variable is called the fatigue life (FL) distribution in Meeker et al. (2024).

There is another random variable of interest, which is known as the strength of the metal. Each metal specimen may have its own strength, but this random variable is unobservable. The definition of strength is the level of stress that leads to a failure at a given number of cycles N_e. Notice, in the definition of strength, N_e is a predetermined value or the value of a controlled factor. By such a definition, one can never calibrate stress to a level such that the specimen fractures *exactly* at N_e cycles. This is why we say the strength variable is unobservable. But it is in engineers' interest to design their reliability systems using strength distributions, while considering possible stress that the system will endure. Examples of their systems include aircraft and medical devices. All are critical systems. The distribution of strength is called the fatigue strength (FS) distribution.

Historical approaches to fitting S-N curves to fatigue data were archaic, such that run-outs (censored observations) were mishandled, and stress or strength variables were treated as response variables. Meeker et al. (2024) describe a more rigorous approach. In this approach, one can either specify an FL distribution, then fit that distribution to data and derive the FS distribution. Or one can, instead, specify an FS distribution, then derive the FL distribution and fit that distribution to data. The details of the individual S-N curve are beyond the scope of this handbook. But we will use the Basquin relationship to illustrate the main steps. The Basquin relationship is special because the FL and FS distributions can belong to the same type of distributions, such as Weibull or lognormal.

If we specify an FL distribution for the Basquin model, the distribution is:

$$\mathrm{P}(\mathrm{N} \leq \mathrm{n}) = \Phi(n;\, \mu = \beta_0 - B\log(s),\, \sigma)$$

where we use lowercase s to represent a given stress level, Φ is a log-location-scale distribution function with parameters μ and σ, and $\mu = \beta_0 - \mathrm{B}\log(s)$ defines the Basquin relationship. In this expression, we use the same B as the one in the original Basquin relationship and $\beta_0 = \log A$. We can directly fit this distribution to the data by writing down the likelihood function, then use the maximum likelihood method to find our estimates and conduct further inference.

If we specify an FS distribution for the Basquin model, the distribution is:

$$\mathrm{P}(\mathrm{Strength} \leq \mathrm{s}) = \Phi(s;\, \mu = \beta_0 + \beta_1 \log(n),\, \sigma = \sigma_X)$$

in which n is a specified number of cycles to fracture as we mentioned before, lowercase s is the stress level that one needs to calibrate to, Φ is a log-location-scale distribution function with parameters μ and σ, and $\mu = \beta_0 + \beta_1 \log(n)$ defines the Basquin relationship. Unfortunately, we cannot fit this distribution directly to our data, because we don't have observations of strength. But according to Meeker et al. (2024), we can show that the FL distribution is:

$$\mathrm{P}(\mathrm{N} \leq \mathrm{n}) = \Phi\left(n;\, \mu = \frac{\beta_0}{B} - \frac{1}{B}\log(s),\, \sigma = \frac{\sigma_X}{B}\right)$$

One can fit this distribution to the data because we have observations of number of cycles to fracture. In practice, one never specifies an FS distribution for the Basquin model since it is unnecessary. We present the steps here because the formula is relatively simple to illustrate the idea using the Basquin relationship. The formulas of this approach for other S/N curves can be complicated, and in practice, these computations are performed using computer software.

Degradation Models

We have described two types of models that can provide us with estimates of failure time distributions, both of which directly model time-to-failure data, with or without extra factors. For those failures, we may not be able to trace to an underlying degradation process. When degradation processes can be observed, we have new types of data and require different types of statistical models to analyze to get desired estimates of failure time distributions. Here, we need to clarify that there are two types of degradation data, and they should be analyzed differently. One is repeated measures degradation data, and the other is destructive degradation data.

Repeated Measures Degradation

Two important examples of this type of degradation include metal crack growth and degradation of power generation devices. As we have discussed, metal fracture is a major failure mechanism in heavy machinery and aerospace equipment. Undetected crack growth can lead to disasters. A battery degrades over time through charge–discharge cycles. Both manufacturers and consumers have great interest in the reliability of the batteries in electrical vehicles. Figure 10.16 shows plots of two such sets of data. The first shows crack growth data for a metal alloy (Alloy-A), and the second displays battery power output data (Device-B). Both can be found in Meeker et al. (2022).

Both crack growth and power output can be measured repeatedly over time on the same specimen. Every connected curve in the plots is a degradation path of the associated specimen. The Alloy-A data includes specimen ID, time (millions of cycles), and degradation measurement (inches). There is an additional temperature in Device-B, for which the interest is in the failure time distribution under 80°C. This type of Device-B data is known as accelerated degradation data.

There are horizontal lines in both plots. They are thresholds and represent the failure definitions. Failure time is defined as the time that the degradation path crosses the defined threshold. Thus, the crossing-times in the plots are time-to-failure observations. The aim of modeling repeated measures degradation data is to get the distribution of the crossing-time, which is a random variable. For Device-B, the interest is the distribution of the crossing-time under 80°C. And clearly, we don't have data under 80°C, and we must extrapolate the model along the temperature variable to get the desired estimate. The details of such models can be found in Meeker et al. (2022).

Destructive Degradation

For repeated measures degradation, we can measure multiple times on individual specimens. That is not always possible due to the nature of the subject. An example is the degradation of adhesive materials. Due to how the strength of adhesive materials is measured, we can only measure once on an individual specimen. Figure 10.17 shows two plots of such data. They are Adhesive Bond A and Adhesive Bond B data in Meeker et al. (2022).

In both plots, every marker represents a distinct specimen. Individual specimens are measured then discarded. Therefore, there are no tracing curves that connect multiple markers. Like repeated measures degradation, these two plots have their individual threshold lines that define failures. The Adhesive Bond B data has three temperature groups, and interest is in the failure distribution under 25°C, room temperature.

Due to the nature of destructive degradation data, we cannot rely on the concept of crossing time of the degradation path of each specimen. Instead, we model the degradation path of a conditional distribution, which captures the distribution of measures at any given time point. The idea is illustrated in Figure 10.18. Then

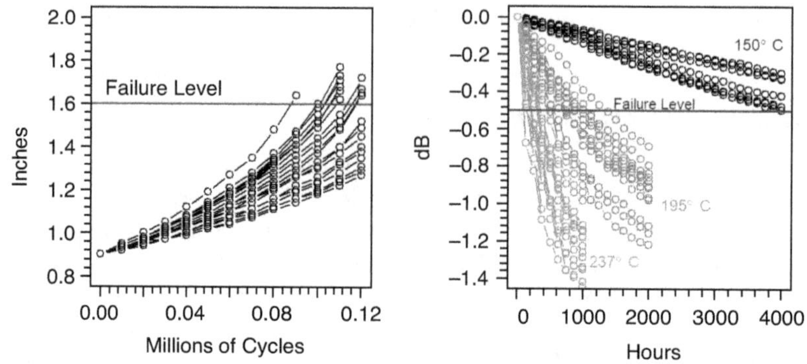

Figure 10.16 Degradation data plots for Alloy-A and Device-B.

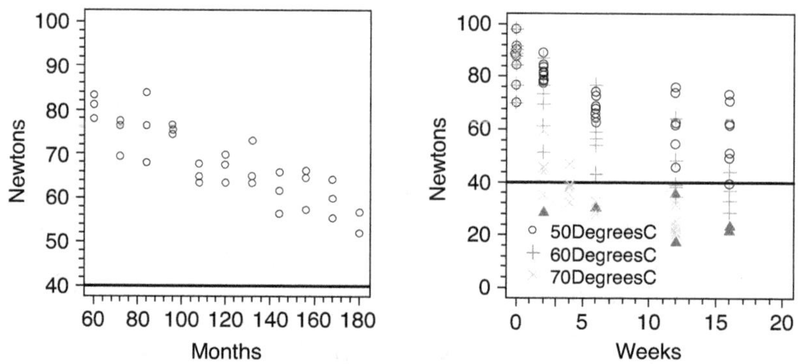

Figure 10.17 Data plots for Adhesive Bond A and Adhesive Bond B.

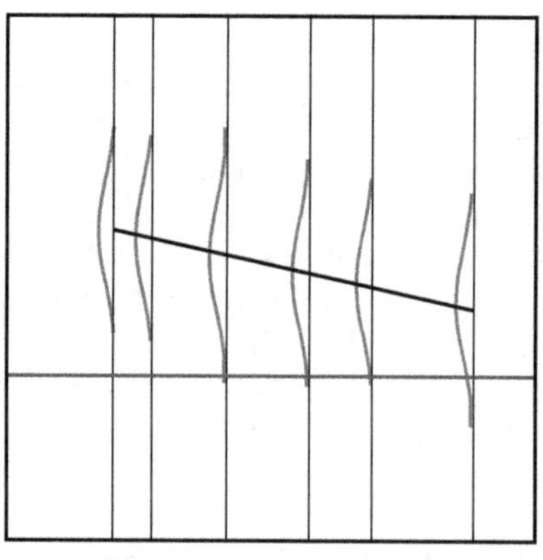

Figure 10.18 Degradation path using the conditional distribution at time *t*.

we can obtain the desired failure time distribution at the threshold, according to Meeker et al. (2022).

RELIABILITY PREDICTION METHODS

> Explain various reliability prediction methods—including Monte Carlo simulation, part stress analysis, parts count prediction, Markov analysis, neural networks, and machine learning—to assess both repairable and non-repairable components and systems and describe the inputs into the model. (Understand)
>
> **Body of Knowledge IV.C.4**

The previous sections built the foundation for constructing reliability system models. The first section provides the framework of describing a system in terms of individual components, describing their dependence structure. The next two sections allow one to configure the structure of the system, component by component. After the configuration, one has a model for the reliability of the system. We can then derive the failure distribution of the system as if it is just another component. With this in hand, we can then make reliability predictions and other inferences about the system.

Not all the methods, however, in the second and third sections, can provide a closed form expression for the needed failure time distribution. For such cases, simulation-based methods are often used.

To this point, we have focused on nonrepairable systems. This section will also describe how to assess a repairable reliability system, which is defined by adding maintenance operations on a nonrepairable reliability system. Methods for analyzing repairable reliability systems, as well as complex nonrepairable systems, generally require simulation-based methods. The rest of this section describes individual simulation methods and how they can be applied to predict the behaviors of either nonrepairable or repairable reliability systems. These simulation methods are sometimes referred to as Monte Carlo simulation methods.

In addition to methods for assessing behaviors of well-configured reliability systems, this section will also discuss methods for assessing behaviors of systems that don't have a reliability system configuration.

Simulating Failure Time with a Known Distribution

One of the tasks among these methods is to simulate random failure times. We can do this easily using the inverse transform theorem. Suppose the failure distribution is known as $\Phi(t)$. We can simulate one random failure time by calculating the inverse of the failure distribution given a random sample from the uniform distribution, ranging from zero to one, that is, $\Phi^{-1}(p)$, where p is drawn from the Uniform (0,1) distribution. The two steps of the technique follow:

1. Draw a random value p from Uniform(0, 1).
2. Calculate $\Phi^{-1}(p)$, which is a random failure time from $\Phi(t)$.

Example 10.11 illustrates this procedure.

EXAMPLE 10.11

Assume we have an engine with several components that are subject to failure. We want to simulate failure times for each of the components to simulate failures for the entire engine. One of those components is the engine belt. As in Example 6.39, we will assume that failure times for the engine belt are distributed as a Weibull random variable with $\beta=0.5$ and $\eta=10{,}000$ hours. To simulate 100 failure times for this engine belt, we first sample 100 random values $P_1, \ldots, P_{100}$ from the Uniform (0, 1) distribution. These are summarized in Figure 10.19(a). We then transform these values into simulated failure times $X_1, \ldots, X_{100}$ from a Weibull distribution by taking each value by applying the formula

$$X_i = -\eta\left[\ln(1-P_i)\right]^{\left(\frac{1}{\beta}\right)}$$

which is the inverse of the Weibull distribution. For the engine belt example, the formula is

$$X_i = -10000\left[\ln(1-P_i)\right]^{\left(\frac{1}{0.5}\right)}$$

The resulting Weibull failure times are summarized in Figure 10.19(b). These component lifetimes can be combined with lifetimes for other engine components to analyze the overall reliability of the engine.

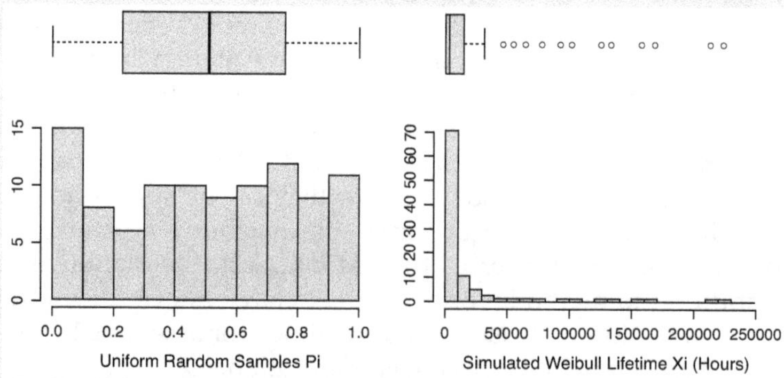

Figure 10.19 Simulation results for engine belt failure. Histogram and boxplot (a) for 100 uniform random samples and (b) for 100 Weibull random samples.

Sometimes the failure distribution does not have closed form, and a different kind of simulation is needed. For example, finite element analysis is a simulation method to get failures from computer programs, and failures can be used to estimate the needed failure time distribution. In the repeated measures degradation modeling, we ultimately need to use the fitted model to simulate sufficient degradation paths for hypothetical specimens to get a sample of cross-times, then get an estimate of the failure time distribution. The general procedure includes the following steps:

1. Simulate a failure time, using whatever method appropriate, such as finite element methods or a degradation model.
2. Repeat step 1 n times and get n failure time samples.
3. Fit a distribution to those n failure time samples and get an estimate of the failure time distribution. The fitted distribution can be used in future applications, such as drawing failure times directly from the fitted distribution, instead of running expensive computer experiments again.

Example 10.12 illustrates the procedure.

EXAMPLE 10.12

We will return to the example of analyzing the reliability of an engine. Here, we will assume that we do not know the distribution for engine belt failures, but we have a physics-based mechanistic computer model that can simulate failures. The model is expensive to run, and so we generated a set of 100 failure times. Experience suggests that this failure mode may lead to failure times that can be modeled by a Weibull distribution. Using Minitab® computer software, we find the Weibull distribution with $\beta=0.46$ and $\eta=5863$ hours provides an adequate fit to the data, as shown in Figure 10.20. This allows us to use the simpler Weibull model for future simulations.

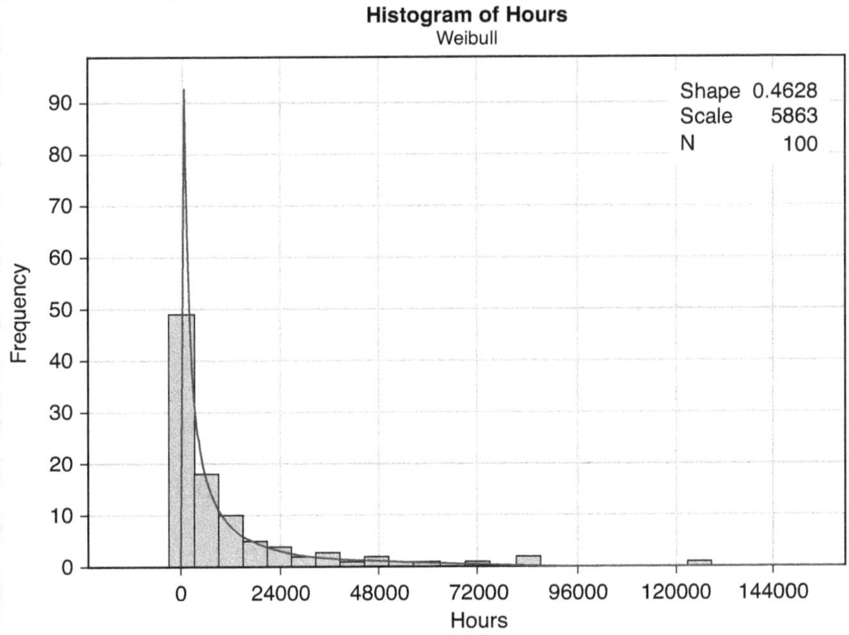

Figure 10.20 Engine belt simulation results.

Simulating Failures Under Modeling Uncertainties

The failure time distributions are usually estimated, and so there are uncertainties in those estimates. We could incorporate modeling uncertainties, such as misspecification of the distribution and errors in the parameter estimates, into our simulation. PoF-based models may also have uncertainties in their parameters, which are usually obtained from data analysis elsewhere. To address such concerns, we need to change how we perform the simulation by taking uncertainties into account. For example, we need to vary parameters accordingly for every simulation.

Simulating Failures of a Nonrepairable Reliability System

Sometimes, the algebraic calculation of the failure distribution of a nonrepairable reliability system is expensive due to complexity. On the other hand, it could be rather straightforward to simulate random failure times of such a system. For example, we can simulate the failure times of all components in the system, and determine which failure determines the failure of the entire system. That can be considered as a simulation of the last resort. A collection of these failures can be used to estimate the failure time distribution of the system. Here are the steps:

1. Given a reliability system, which is defined by a reliability block diagram (RBD) and component reliabilities, one can simulate failure times of all components using the technique described in the previous section, where we simulated failure time with a known distribution.
2. By removing one component at a time in order, starting from the earliest failed component, one can determine which component is the last component that will cause a system failure according to the definition of an RBD. The failure time of that component is the failure time of the system.
3. Repeat above steps n times and get n failure time samples.
4. Fit a distribution to those n failure time samples and get an estimate of the failure time distribution of the entire system.

Simulating Maintenance

There are two fundamental maintenance types: corrective maintenance and preventive maintenance. Corrective maintenance occurs when a component fails, and preventive maintenance occurs at scheduled time points. Maintenance itself can be further categorized. A common maintenance is known as replacement, which replaces the failed component with a brand-new component of the same type, so that the failure distribution of the replacement does not change according to the design. In other words, this type of maintenance makes the component as-good-as-new. Another common maintenance is known as minimal repair, which merely restores the failed component to the state right before it failed. This maintenance makes the component as-bad-as-old.

From the foregoing description, we can see that such a simulation involves the following elements.

- When maintenance starts
- What the maintenance does and its impact on the existing reliability system
- How long the maintenance will take

Maintenance takes time, consumes parts, and requires extra labor. All are constraints that will impact how long it will take to complete the maintenance.

Simulating a Repairable Reliability System

To simulate a repairable reliability system, we iterate the simulation of a nonrepairable reliability system and the simulation of maintenance. In this simulation, failures, prescheduled time, start of a maintenance, and the completion of a maintenance are all called "events." These occur at random time points and can be simulated. Such a characteristic makes repairable reliability system simulation a *discrete event simulation*. There are different types of discrete event simulations. The repairability system simulation is one type and specific to reliability systems. Between events, a reliability system changes its states. For example, between failure events, the system ages. And during maintenance, a system may be deliberately shut down and does not age or may continue to be operational and age. After maintenance, the component that has been maintained may change the system state.

Such a type of simulation can also be a Markovian simulation, but it does not have to be. Markovian simulation is a type of simulation with the property that for any two consecutive states, the later state only depends on the early state, but not earlier history. In other words, given the current state, the future and the history are independent.

A word of recommendation here is that Markov is a very generic methodology. In our description, a simulation of repairable systems is usually Markovian, if we assume that failures are independent, and all other events are assumed independent. But the simulation can be non-Markovian, if the failures are not independent or some events are not independent. For example, in a system with load sharing, a failure of a component may impact how the remaining components degrade or age, which makes the events less likely to be Markovian. Or, if maintenance is grouped or conditioned upon one another, the simulation is unlikely to be Markovian.

Whether a repairable system simulation is Markovian or not, it is not usually a concern. The data collected in a simulation comprise all events, their times, and their consequences or states of the system or components. For any set of assumptions, there are methods to summarize the data and calculate statistics such as system availability probabilities, system downtime probabilities, and system uptime probabilities.

As an illustration of a repairable system analysis, we will consider a complex system that can be analyzed as a discrete Markov process. We will assume that the system can exist in only one state at a time, and except for the immediately preceding state, all future states are independent of past states.

The goal of our Markov analysis is to determine the future probability of a repairable system being in a success state when the failure probability and probability of restoring the system are known. As described previously, our system will either be in a success state (not failed) or in a failed state. If it is in the failed state,

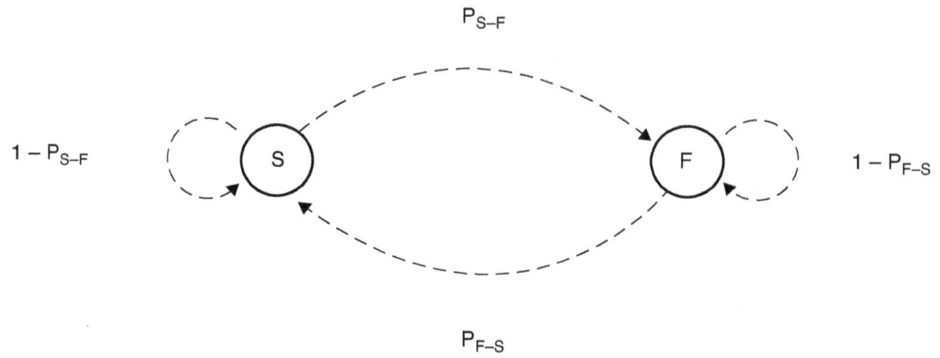

Figure 10.21 A transition diagram for a system with two states.

it is waiting to be returned to the success state. The transition probability from one state to the other is determined by the failure rate and the rate at which the unit can be restored to the success state after a failure (see Figure 10.21). P_{S-F} is the probability of transition from state S (success) to state F (failure) in a given time interval. P_{F-S} is the probability of transition from state F to state S in the same time interval. $1 - P_{S-F}$ is the probability of remaining in state S if the system is in state S. $1 - P_{F-S}$ is the probability of remaining in state F if the system is in state F.

A Markov analysis is not limited to a system with only two states. There can be more than one success state. All the possible states of the system are identified. All the transitions from one state to another are identified, assuming one action at a time. The transition rates from one state to another state are determined.

The probability of the system being in a success state can be determined by considering all possible states and the rates of transition from state to state. A Markov analysis requires the use of differential equations, Laplace transforms, and the solution of a series of linear equations using matrix algebra. Even though the analysis is exact, the assumptions necessary can affect the credibility of the results.

A transition diagram for a system with two states, Success (S) and Failure (F), is shown in Figure 10.21, in which $1 - P_{S-F}$ is the probability of staying in the success state at the next time period. We can also denote this as $P_{S-S.}$ P_{S-F} is the probability of transitioning from S to F, P_{F-S} is the probability of transitioning from F to S at the next time period, and $1 - P_{F-S} = P_{F-F}$ is the probability of remaining in the failure state. Note that each probability path represents a mutually exclusive event.

We now illustrate the probability of a working system being in state S after two time intervals. Let $P_{S-S} = 0.90$, $P_{S-F} = 0.10$, $P_{F-S} = 0.70$, and $P_{F-F} = 0.30$. We can solve for P(S) by enumerating all the possible paths in two time intervals, conditional on beginning in State S at time zero. There are two possible paths to arrive at state S at time period 2 if we begin in state S. We can calculate the probability of these two paths:

$$P(S) = (P_{S-S})\ (P_{S-S}) + (P_{S-F})\ (P_{F-S}) = (0.90)(0.90) + (0.10)(0.70) = 0.88$$

A tree diagram can be helpful in modeling this simple system, as shown in Example 10.13.

EXAMPLE 10.13

A pump operates in one of two states: S (success) or F (Failure). There is a probability of 0.9 that the pump in state S will remain in state S for a given time interval, and there is a probability of 1 − 0.9 = 0.1 that the pump will transition to state F and fail.

If the pump is in state F, there is a probability of 0.7 that it will be restored and transition to state S. There is a probability of 1 − 0.7 = 0.3 that the pump in state F will not be restored in the given time interval and will remain in state F.

Refer to the system illustrated in Figure 10.22.

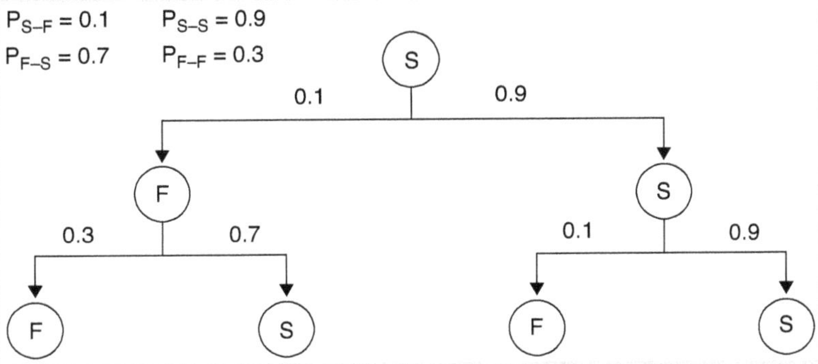

Figure 10.22 Tree diagram for Markov analysis for two time intervals.

The probability of being in the success state (state S) after two time intervals is (0.1)(0.7) + (0.9)(0.9) = 0.88.

As an exercise, expand the tree diagram to cover three time intervals. Verify that the probability of being in the success state (state S) at the end of three time intervals is P(State S) = (0.7)(0.3)(0.1) + (2)(0.9)(0.7)(0.1) + (0.9)(0.9)(0.9) = 0.876.

The tree diagram and manual calculations become complex if the system has several success states. A Markov analysis makes use of a *transition matrix* that can be evaluated to find the state probabilities over many time periods, as illustrated in Example 10.14.

EXAMPLE 10.14

Using the values in Example 10.13, the transition matrix for the system can be written:

$$T = \begin{bmatrix} 1 - P_{S-F} & P_{S-F} \\ P_{F-S} & 1 - P_{F-S} \end{bmatrix} = \begin{bmatrix} 0.90 & 0.10 \\ 0.70 & 0.30 \end{bmatrix}$$

T^k will give the state probabilities after the kth time interval. To find all the state probabilities after two time periods, we calculate:

Continued

$$T^2 = \begin{bmatrix} 0.9 & 0.1 \\ 0.7 & 0.3 \end{bmatrix}^2 = \begin{bmatrix} 0.9 & 0.1 \\ 0.7 & 0.3 \end{bmatrix} \begin{bmatrix} 0.9 & 0.1 \\ 0.7 & 0.3 \end{bmatrix}$$

$$= \begin{bmatrix} 0.9(0.9) + 0.1(0.7) & 0.9(0.1) + 0.1(0.3) \\ 0.7(0.9) + 0.3(0.7) & 0.7(0.1) + 0.3(0.3) \end{bmatrix}$$

$$= \begin{bmatrix} 0.88 & 0.12 \\ 0.84 & 0.16 \end{bmatrix}$$

The first row of the resulting matrix gives the P(S) = 0.88 and P(F) = 0.12 after two time intervals if the system starts in state S. The elements in the second row give P(S) = 0.84 and P(F) = 0.16 after two time intervals if the system begins in state F.

After three time intervals, the transition matrix calculations yield:

$$T^3 = T^2 T = \begin{bmatrix} 0.876 & 0.124 \\ 0.868 & 0.132 \end{bmatrix}$$

As k increases, the state probabilities will converge, becoming the steady-state values of availability for the system. For example, after six time periods, we see that the availability of the system has converged to 0.875, irrespective of the initial state of the system.

$$T^6 = \begin{bmatrix} 0.875 & 0.125 \\ 0.875 & 0.125 \end{bmatrix}$$

Many software programs can perform the transition matrix calculations.

Methods Without Reliability System Modeling

All previous methods to assess or predict behavior of a reliability system assume the complete knowledge of the reliability system model, which includes dependence structure and reliabilities of individual components, with or without maintenance planning.

In practice, there are systems that do not have such a known model, but engineers wish to assess or predict the system reliability in the future. There have been methods that allow engineers to make predictions about reliability system behaviors with limited information. Among them, some are general statistical methods, including traditional time series methods and point process methods. There are also emerging methods from the machine learning communities. There are methods that have been considered outdated. This subsection will go through the three categories to allow the readers to make informed decisions in their own practice.

Part Count Predictions and Part Stress Analysis

A reliability prediction is a design tool to be used early in the design and development stage. To be effective, the prediction is started before the design is completed, and completed before production tooling is set and hardware is procured for production.

Many times, the prediction is done with no actual reliability data on the particular system, although there may be data on some of the components. The reliability prediction can be used to determine the feasibility of the design to meet the system reliability goals, to focus attention on "weak links" in the design, to assess the impact of design changes on the system reliability, to compare the reliability of competing designs early in the development stage, and to assist in establishing maintenance procedures. The reliability prediction is not a reliability estimate, which requires data and should not be used as a measure of the actual achieved system reliability.

The best known and most widely used source of predictive data for electrical/electronic systems is MIL-HDBK-217. The handbook assumes electronics can be modeled using a constant failure rate and contains failure rate data for electrical and electronic components. The handbook contains data for all passive devices: resistors, inductors, capacitors, transformers, and so on. It also includes data for active elements: transistors, diodes, field-effect transistors (FETs), and so on, as well as digital and analog integrated circuits (ICs). There are methods to adjust the base failure rate of the components depending on the number of leads, the number of gates in an IC, the use environment the component will experience, the quality control requirements the customer can impose on the supplier, and other factors that could affect the component failure rate. The multipliers used to adjust the base failure rate are referred to as π (pi) factors. See Example 10.15 and Example 10.16 for illustrations.

EXAMPLE 10.15

The failure rate model for a resistor is $\lambda_P = \lambda_b \times \pi_E \times \pi_R \times \pi_Q$ failures/million hours. Tables in MIL-HDBK-217 contain λ_b, the base failure rate, based on the standard derating used by the designer. Tables also contain the pi factors:

- π_E is the environmental factor reflecting how the resistor is used.
- π_R is a factor based on the value of the resistor.
- π_Q is a quality factor and is dependent on the amount of control the customer has in the production of the component.

These factors and the base failure rate are combined to give the predicted failure rate of the resistor.

EXAMPLE 10.16

A 10,000-ohm carbon resistor is used in a communications receiver located in the crew compartment of a commercial aircraft. Assume a standard design derating of 40% at 60°C.

From MIL-HDBK-217:

- $\lambda_b = 0.0012/10^6$ — Base failure rate
- $\pi_E = 3$ — Environmental stress factor
- $\pi_Q = 1$ — Quality factor
- $\pi_R = 1$ — Resistance factor
- $\lambda_p = (3)(1)(1)(0.0012/10^6) = 0.0036/10^6$ — Predicted failure rate

If the same resistor were used in a radar detector located on the deck of a naval ship:

- $\pi_E = 12$ — Environmental stress factor
- $\lambda_p = (12)(1)(1)(0.0012/10^6) = 0.00144/10^6$ — Predicted failure rate

A circuit board (subsystem) failure rate could be predicted by adding the predicted failure rates of all the components on the board. This assumes a series model for the circuit board and will give a worst-case prediction. This is referred to as the part count method. A system failure rate could be predicted by adding the predicted failure rates of all the subsystems. This also assumes a series model.

Using the assumption of a series model, the system will have a constant failure rate if all the components have a constant failure rate. To overcome some of the deficiencies of MIL-HDBK-217, the PRISM reliability prediction method was developed by the Reliability Analysis Center (RAC).

Instead of using just multipliers to model component failures, PRISM uses combinations of both additive and multiplicative factors. The factors are called π factors, and the model could be of the form

$$\lambda_S = \lambda_1(\pi_1\pi_2 + \pi_3\pi_4 + \pi_5)$$

where λ_i is the initial failure rate and the π values are based on the part. PRISM also allows for consideration of early-life failures as well as failures due to a constant failure rate. The *217(plus) method* is an outgrowth of PRISM. It contains new part-type failure rate models including connectors, switches, and relays.

For predicting the reliability of mechanical systems, the *Handbook for Reliability Prediction for Mechanical Systems* (NSWC-07) is available. Information can be obtained from the Naval Surface Warfare Center, Carderock Division, Bethesda, Maryland. The handbook follows the format of MIL-HDBK-217 and contains base failure rates for mechanical components, which can be adjusted using multipliers or *c* factors depending on the type of material, physical configuration, heat treatment, use environment, and other factors that would affect the probability of failure. Examples of specific mechanical devices included are springs, bearings, brakes and clutches, seals, pumps, and other nonelectrical devices. This handbook assumes the constant failure rate model.

Non-electronic Parts Reliability Data (NPRD-2011) contains failure rates for mechanical and electromechanical parts. This method allows the engineer to consider failures due to wear-out.

Telcordia SR-332 (once known as Bellcore) has prediction data in tables and graphs for communication equipment with commercial applications. These Telcordia models were developed by AT&T Bell Labs, and the method is based on MIL-HDBK-217. Several international standards are also available. The methodology used in these standards is like that in MIL-HDBK-217.

Reliability predictions can also be based on the engineer's experience with components used in previous designs of similar systems. In some cases, the component vendor may supply typical failure rate data. The Government-Industry Data Exchange Program (GIDEP) is a source available to military contractors that contains failure reports on commercially available subsystems such as motors, compressors, pumps, and so on. System reliability values obtained using any of these sources should be treated only as prediction values.

MIL-HDBK-217 and Its Progeny, a Note of Caution

The Military Handbook—Reliability Prediction of Electronic Equipment (MIL-HDBK-217) was first developed in the 1960s. The latest update is MIL-HDBK-217F Notice 2, which was released in 1995. The handbook is now considered unrealistic because of its overly simplified assumptions, and it is also considered outdated due to its information on parts' reliabilities that were only valid up to 1995 or even much earlier. A rigorous critique can be found in Appendix D in the National Research Council (2015). We recommend that engineers seek rigorous approaches to assess reliability systems and use the most up-to-date information.

Continuous Time Process

In this type of processes model, one or more continuous response variables are represented as continuous functions of time, which is also continuous. An example of such a response variable is battery capacity. The continuity in time is emphasized by some models but relaxed by others. If relaxed, sometimes the model of interest is a discretized time version of the continuous time process. A discretized time process means the response is a function of time at a distinct sequence of time points. A typical reason for studying such a process is to determine the distribution of the response at a future time.

Applications of such models include degradation processes and chemical or biological reaction processes. They can all be related to reliability systems as additional metrics to assess the behavior of the systems besides reliability measures. More importantly, additional metrics can provide predictive power to predict system reliability. For example, the continuous time process of modeling battery life, such as the Levy process, is an active subject under prognostics and health monitoring (PHM), which is an extended resource on system reliability.

Discrete Time Processes

Different from continuous time processes, discrete time processes do not emphasize the continuity in time, even if it could be so. A discrete time process may look the same as a discretized continuous time process, but it does not have to be. The well-known models that analyze such processes include autoregressive and moving average (ARMA) models and exponential smoothing models. The interest is usually similar to that of the continuous time processes.

These models are related to Markov analysis, which is a much bigger category. Markov analysis models only assume conditional independence, mentioned earlier. ARMA and exponential smoothing models, however, assume specific structures on how the past can be used to predict the future.

In recent years, as neural networks models have become more accessible to engineers and researchers, such machine learning types of models are natural extension to ARMA models and smoothing models. But machine learning models belong to a separate category that we will describe.

Point Processes

Different from the two types of processes before, the response of a point process is an event together with when the event is observed. There are three such processes that are important to know in the system reliability context: the homogeneous Poisson process (HPP), nonhomogeneous Poisson process (NHPP), and renewal process. These processes model a sequence of events that occur randomly over a period. There are typically two interests. The first is the duration of consecutive events. The second is the count of events in a given period. Using either of the interests, we can distinguish the three types of processes. It is worth noting that these three processes all belong to the category of Markov processes. It is necessary to be specific when one applies a Markov process in the context, since it is such a broad concept.

For an HPP, the durations between consecutive events are identically distributed and follow an exponential distribution. The distribution of the event count of a fixed duration is a Poisson distribution, which is only a function of duration, but not a function of time. In other words, if the failure events of a repairable reliability system can be described by an HPP, this system has a constant failure rate and the system does not degrade over time. HPP is such a strong assumption that one should be careful when it is applied. This is one of the assumptions behind the methods in MIL-HDBK-217.

For an NHPP, the durations between consecutive events can be rather complicated, depending on additional assumptions. The process, however, is usually described through the distribution of the event count of a fixed duration, which is also a Poisson distribution. But this distribution is a function of both duration and time, besides additional assumptions. This type of process is much more flexible to adapt to realistic reliability systems. It can be applied to systems that are improved over time, which means fewer and less frequent failure events over time. Such a situation usually occurs during the development phase of a reliability system. It can also be applied to systems that degrade over time, which means more and more frequent failure events over time. This usually occurs in systems in actual use.

A renewal process relaxes the strong assumption of an HPP but is less complex than an NHPP. The durations between consecutive events are identically distributed and follow a specified distribution, which does not have to be exponential. This process is usually appropriate for repairable reliability systems that get full restoration to the original condition after each failure. For restoration, spare parts should be identical to what was used in the system, and the restored system can be seen as identical to the original system.

Neural Networks and Machine Learning

The subjects of neural networks and machine learning are vast. They can be applied to reliability system modeling as extensions to what we have described in this section. We apply these methods to assess reliability system behaviors with or without complete information about the reliability system that is defined by an RBD, component reliability, and maintenance.

When one models maintenance and tries to optimize the reliability system's performance, such as maximizing the uptime, a class of machine learning models known as reinforcement learning could be useful. Reinforcement learning in system reliability modeling can be seen as a cross-discipline area that involves neural networks, Markov analysis, simulation, and optimization. It is done by simulating the repairable reliability system many times. In each simulation, maintenance is different. And maintenance change is assumed to be a Markovian process. The reinforcement learning system then records all the simulations and evaluates the final performance metric, such as uptime. One then hopes that a reinforcement learning system can learn the optimal maintenance policy by fitting a neural network model to the simulated data. In such applications, machine learning methods incorporate knowledge about the reliability system and try to find an optimal configuration for the system.

When the information of the reliability system is unknown, but one wants to assess the future performance of the system, machine learning and neural networks methods can also be applied, similar to the time process and point process models. But in contrast to simpler model structures, such as those in time processes or point processes, machine learning and neural networks methods have more complex, flexible structures. One should take that with caution because this flexibility has two sides. On one side, it can capture richer dynamics. On the other hand, it has a risk of treating noise as a signal. The general procedure of applying neural networks and machine learning models in place of time processes and point processes is to use more complex prediction formulas for the response. Then fit the model under a rigorous procedure, known as cross-validation, all using the data that are known. The cross-validation procedure is used to try to ensure that the model is applicable to future unknown data. If a model is overfit, that means the model fits the known data very well but will perform poorly on future unknown data. There are different versions and recommendations on how cross-validation should be done. In general, if all observations are independent, a so-called repeated K-fold cross-validation is considered the best practice.

In addition to overfitting, neural networks and machine learning models have additional concerns. Such models adhere to the known data very closely, even if the cross-validation procedure has been correctly applied. There is no guarantee that future data will be like the known data. A rigorous cross-validation procedure can only ensure that the model will be applicable to future unknown data, if and only if future unknown data is similar to the known data. In reality, that is not guaranteed. Therefore, in addition to cross-validation, engineers and researchers who apply neural networks and machine learning models should be constantly evaluating how future data is like the known data, and whether the existing model is applicable to the future data. The matter is especially pertinent to neural networks and machine learning models, compared to more traditional models.

Neural networks and machine learning models have great flexibility to capture patterns in existing data that may not be present in future data.

DESIGN PROTOTYPING

> Describe the advantages and limitations of prototyping and rapid prototyping technologies to enhance product reliability. Employ phase diagrams to model various equipment scenarios across different mission profiles. Determine the correlation of prototype results to simulated results (e.g., digital twins and damage modeling). (Understand)
>
> **Body of Knowledge IV.C.5**

In previous sections, we have described methods of assessing a reliability system through mathematical and statistical modeling. Such a reliability system model approximates and conceptualizes the physical system. The approximation can be used to develop the physical system through prototyping. A prototype can be used to illustrate, test, or verify the design. This section discusses how individual objectives are applied by prototyping. Prototyping takes different forms: physical or digital. The reliability system model is a digital prototype of the physical system. However, sometimes, we need physical prototypes. This section only discusses physical prototyping to avoid confusion.

Reliability System Development Cycle

Figure 10.23 illustrates a typical product development lifecycle, which is applicable to reliability system as well. The main circle starts with the established requirement. Then the process flows through all the steps until the final product is delivered.

The analysis step should include failure mode analysis and other issues that are inputs to design. The design step in the diagram comprises what we have discussed about reliability system design in previous sections. Then engineers produce one or more prototypes. Then the process goes into improvement iterations that consist of redesigning, prototyping, and evaluation. Finally, the design phase ends, and the product goes into production and final delivery. After which, engineers may need to monitor the reliability of shipped systems and may need to go back to the beginning to refine and improve.

Prototyping for Reliability System Modeling

It is common to prototype relatively small systems physically and study their reliability aspects. Examples include mechanical, electrical, electronic, and medical systems. The prototypes may be constructed by using off-the-shelf components, which may or may not be close to the designed components. Depending on how the prototypes are constructed, they can serve different purposes. Or conversely, prototypes can be constructed differently to serve specific purposes. There are at least two purposes for prototyping in reliability system design. First is to understand the system and the design better. Second is to estimate the failure probability of the final product.

Prototyping for Discovering Failure Modes

Such a process is common for developing electronic products. Many products can be prototyped by using off-the-shelf components, which are very close to the designed components. Therefore, prototypes can resemble final products to a high degree. Prototypes can be used to gain useful information about the final product. The prototypes are tested to observe and collect failure information, including failure modes and failure time. To conduct such a study successfully, it is important to pay attention to the following.

If the objective is not to compare different designs, it is important to configure prototypes in an identical way, such as using the same parts, the same construction methods, and so on. It is important to test in meaningful ways, such that testing results can provide insight into meaningful questions. For example, if an accelerated life test is not the objective, tests should be run under expected conditions, under expected load. Tests should not produce failure modes that would never occur.

We must also record all failures and understand individual failures. Every failure mode is a potential failure mode in the final product unless it can be identified because of the prototyping process. The prototyping–redesign iteration then continues until testing results from the latest design and prototype meet the expectations.

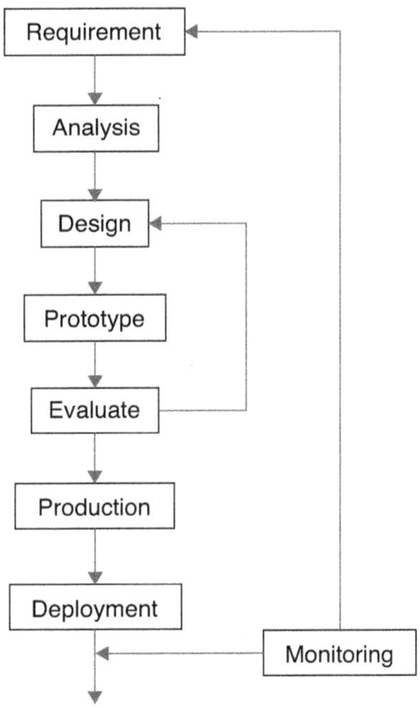

Figure 10.23 Typical product development lifecycle.

Prototyping to Reveal Design Flaws Related to Other Factors

Prototypes are sometimes ideal specimens for testing under extreme conditions, such as extreme temperature, voltage, load. There are testing methods under the name highly accelerated tests (HALT) or other tests for similar purposes. They are not accelerated tests performed for the purpose of model fitting for extrapolation. The purpose of HALT is to break or damage the prototype to reveal vulnerabilities of the design, manufacture, and reasons other than the regular failure mode.

Prototyping for Verifying Design or Model

Prototyping can be used to verify the correctness of a design or a model. Reliability systems can appear in different physical forms. How well does a reliability system represent a real physical system? That is not clear sometimes. And building a prototype is one way to verify the correctness of the designed model.

For example, electronic product prototypes could achieve a high precision representing the reliability of the final products by using off-the-shelf components. It could be practical to require prototype constructions to a high standard and use prototypes to conduct realistic tests that would usually be done on the final products. With prototypes and test results, engineers can obtain useful data to directly predict the performance of the products and compare the results with design.

For another example, chemical or biological reactors can be modeled as reliability systems. They, however, have different complexities from electronic systems. Chemical and biological reactors have become popular applications for digital-twins technology to model, monitor, and control. And reliability system modeling can be built alongside a digital twin. To verify the design, it is wise to build small-scale prototypes to verify both designs, digital-twins and reliability system. The behavior of a smaller reactor will be nonetheless different from that of a full-scale reactor. The relationship between the full-scaled design and small prototype could be rather complicated. In this area, machine learning and neural networks are the methods usually sought after, which can be used to establish or verify associations between the behaviors from the digital design and the behaviors from a small-scale prototype, which are certainly different.

Rapid Prototyping Technologies

3D printing or, more generally, additive manufacturing, became an accessible means to rapid prototyping in the last decade or so. For some products, 3D printing can produce prototypes that are very close to the final product. That is especially true for mechanical products. The actual structure of a 3D-printed material differs from the structure if the material is produced using another method, such as using a traditional extruding method. Such a prototype is not directly relevant to the reliability of the actual product. It, however, provides early information about how the product will be used. This information can reveal design flaws and potential misuse of the product, which would directly cause reliability issues. Rapid prototyping techniques like 3D printing can offer key contributions to reliability system design.

Other rapid prototyping technologies include:

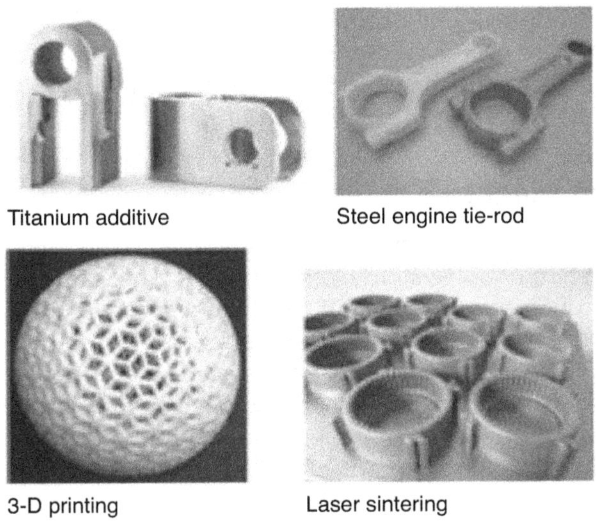

Figure 10.24 Products made with rapid prototyping.

- *Selective laser sintering* (SLS). This technology involves the use of a high-power laser for the fusion of tiny particles of plastic, metal, and so on into a mass that represents a desired 3D object through the help of an SLS machine. This is an additive manufacturing technique. Materials used in this technique are metal powders and thermoplastics.
- *Fused deposition modeling* (FDM). This additive manufacturing technology was invented in the late 1980s by S. Scott Crump and is used for modeling, prototyping, and production applications. Thermoplastics are used extensively in this method.
- *Stereolithography* (SLA). This additive manufacturing technology is used for the production of models and patterns through the stereolithography machine. Photopolymer is the principal material used in this technique.
- *Laminated object manufacturing* (LOM). Paper is the base material used in this technology. In this method, layers of adhesive-coated plastic, paper, or metal laminates are fused together and cut into shape with the aid of a knife or a laser cutter.

Some products made with rapid prototyping are shown in Figure 10.24.

Prototyping Advantages and Limitations

Previous discussions have shown the advantages of prototyping to improve reliability system design and modeling. Prototypes, however, are usually not identical to the final products or systems. Prototypes are usually constructed under limited resources such as financial budget and development time constraint. It is important to understand what kind of test results from prototypes are directly or indirectly related to the final products or systems, and what results are irrelevant. All results might be useful to improve the design, but only from some results can the reliability engineer infer the performance of the final products. We have shown some examples illustrating this in the section.

Reliability Phase Diagrams

Earlier in the chapter, we described reliability block diagrams as a method for representing a system or subsystem of components. This same concept of a directed graph can also be used to represent the changes in operating states of a system over time, or during a specific mission. For example, in a repairable system, the blocks may represent system operation, system maintenance, or system repair. Each block may have a different probability distribution for the time spent in that phase, and the parameters may depend on what occurred in the previous phase. These diagrams are often implemented in reliability software programs to simplify the description of a system and its operation.

By changing the phase diagram, we can model alternative scenarios, such as new system configurations or different maintenance plans. For each phase diagram that is developed, all the methods described in the previous section may be applied for analysis (e.g., simulation and Markov analysis).

Correlation of Physical Prototypes and Digital Models

The goals of the simulation models in "Reliability Prediction Methods" earlier in this chapter and the prototyping methods of this section are generally the same: to predict and improve the reliability of a future production system. Both approaches are extremely useful, as they allow the efficient exploration of the design space before moving forward with a final design and manufacture of a system.

Of course, digital models and physical prototypes are imperfect and developed using different methods. Thus, the results from each approach may differ. As industry continues to move toward greater use of digital models—sometimes called digital twins—there is a need to ensure the results from simulations based on digital models are like the data obtained from testing prototypes.

For example, consider a system in which failure probabilities are driven by material damage due to fracture, deformation, or creep. There are digital PoF models that can be used to simulate these failures. Similarly, we can build prototypes and test them to observe these failures. If we have data from the digital simulation and the physical experiments, we can analyze the relationships between the two to see if the digital simulation can be used to predict the observed data.

A general approach to assessing the correlation between digital and physical follows.

1. Define the reliability measurements that are most important for your application. This would include failures times or measures of degradation. Identify any other measurements of system performance that may be useful. All these characteristics must be measured on the physical prototype and calculated by the digital simulation.
2. Determine the critical design and operating input parameters that likely have the most influence on the reliability of the system. Ensure

that these parameters can be varied on both the physical and digital prototypes.

3. Test the physical prototype by varying design and operating parameters to represent a range of. This is best done using a designed experiment (see Chapter 11). Measure all the responses of interest.
4. Run the digital simulation at each of the design combinations used for physical prototypes testing. Calculate all the responses of interest.
5. Analyze the combined dataset. For each combination of design and operating inputs, you will have data from both the physical and digital system. Now investigate how well the digital results can predict the observed measurements for the physical system, using the methods of Chapter 6.

We end this chapter by showing the implementation of reliability modeling, design improvement, reliability prediction, reliability test, and failure analysis and risk assessment using a high-level case study of a drip coffee maker design, found in Example 10.17.

EXAMPLE 10.17

Consider the schematic of a simple coffee maker shown in Figure 10.25. The design has seven components of interest, including the filter mechanism, carafe, on/off switch, automatic shut-off mechanism, heating mechanism, plug, and pump mechanism.

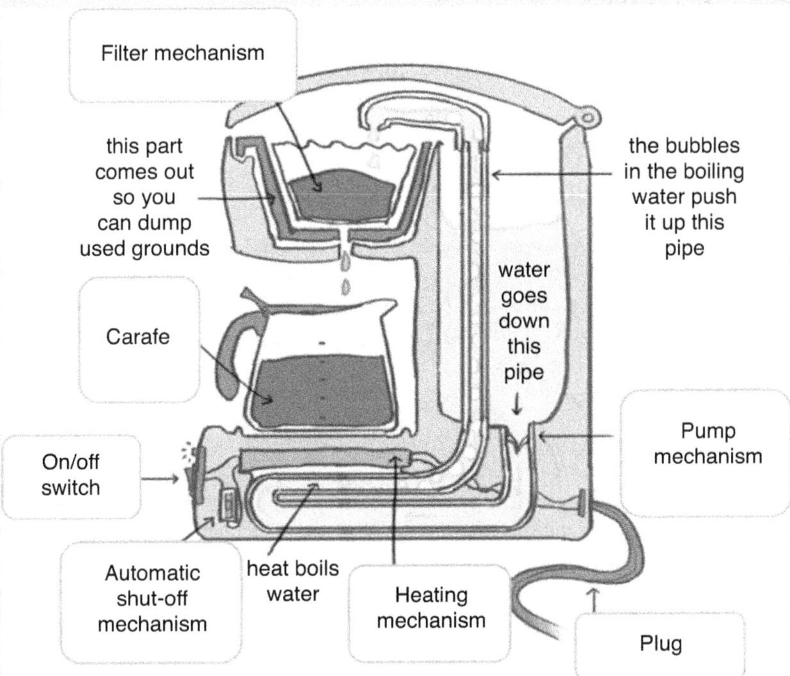

Figure 10.25 Schematic of drip coffeemaker.

Source: https://theartofcoffee.wixsite.com/theartofcoffee/post/how-coffee-makers-work, accessed January 2, 2025.

Reliability Model

The coffee maker can be modeled using a series system model because all subsystems must function for the coffeemaker to brew a pot of coffee. The reliability engineer develops the series reliability block diagram model shown in Figure 10.26.

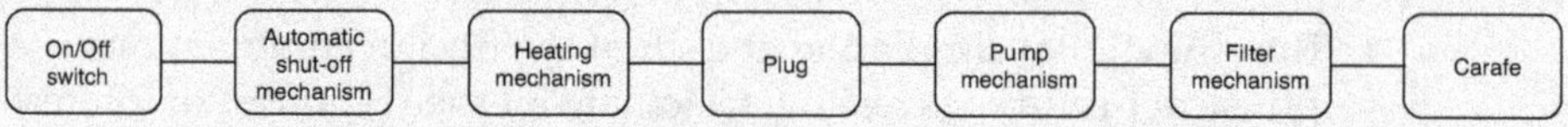

Figure 10.26 Reliability block diagram model of the series system.

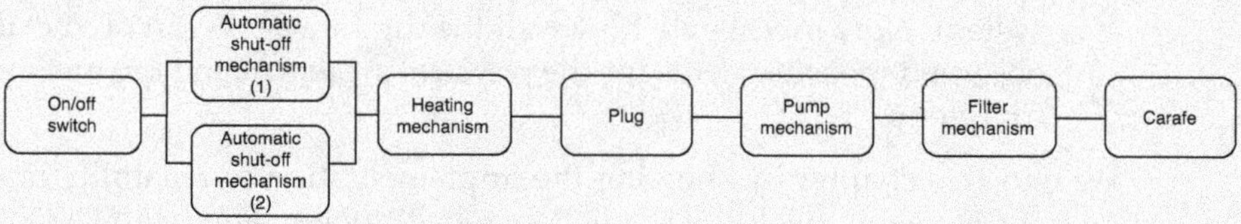

Figure 10.27 Reliability block diagram of improved design with a series/parallel combination.

The equation for the reliability of the series system is thus:

$$R_{\text{System}} = R_{\text{On-Off Switch}} \times R_{\text{Automatic Shut-Off Mechanism}} \times R_{\text{Heating Mechanism}} \times R_{\text{Plug}} \times R_{\text{Pump Mechanism}} \times R_{\text{Filter Mechanism}} \times R_{\text{Carafe}}$$

Design Improvement Model

Now consider that the designer has decided to add another shut-off mechanism as a redundant system. The two shut-off mechanisms will be a parallel system since only one shut-off mechanism needs to function for the coffeemaker to be safe. The model for the new design will follow a combination series/parallel model, as shown in Figure 10.27.

The reliability of the system can be calculated using Equation 10.2.

$$R_{\text{Automatic Shut-off Mechanism 1\&2}} = 1-\left[1-R_{\text{Automatic Shut-off Mechanism 1}}\right] \times [1-R_{\text{Automatic Shut-off Mechanism 2}}]$$

$$R_{\text{System}} = R_{\text{On-Off Switch}} \times R_{\text{Automatic Shut-Off Mechanism 1\&2}} \times R_{\text{Heating Mechanism}} \times R_{\text{Plug}} \times R_{\text{Pump Mechanism}} \times R_{\text{Filter Mechanism}} \times R_{\text{Carafe}}$$

Reliability Prediction

The next step is to predict how well the coffee maker will work. This calculation should be performed before any testing or verification is done. There are several ways to predict the reliability of a system; see Chapter 10, "Reliability Prediction Methods," for the various methods. The starting point is to know the estimated reliability for each subsystem in the model. In this case, the reliability engineer has determined the reliability of each component as shown in Table 10.2.

Table 10.2 Coffeemaker component reliability values.

Component	Reliability
On/off switch	0.95
Automatic shut-off mechanism	0.98
Heating mechanism	0.97
Plug	0.95
Pump mechanism	0.92
Filter mechanism	0.90
Carafe	0.90

The reliability of each component is now added to the reliability model as shown in Figure 10.28.

The reliability of the original system is then predicted using Equation 10.1:

$$R_{System} = 0.95 \times 0.98 \times 0.97 \times 0.95 \times 0.92 \times 0.90 \times 0.90 = 0.64$$

For comparison, consider the improved design with the addition of another, redundant, automatic shut-off mechanism. The new predicted reliability for the automatic shut-off mechanism subsystem is calculated using Equation 10.2:

$$R_{\text{Automatic shut-off mechanism1\&2}} = 1 - [1 - 0.98] \times [1 - 0.98] = 0.99$$

The predicted reliability of the system with a redundant automatic shut-off mechanism is:

$$R_{System} = 0.95 \times 0.99 \times 0.97 \times 0.95 \times 0.92 \times 0.90 \times 0.90 = 0.65$$

Test, Failure Analysis, and Risk Assessment

The prototypes of the improved design are now created, and in-house reliability growth testing is implemented. Reliability growth testing can take on many forms depending on what is needed for the new product development process.

Failures occur during the testing, and the failures are recorded in the failure reporting and corrective action database. Each failure is analyzed by the design team that involves, at a minimum, the design engineer, quality engineer, test engineer, and reliability engineer. During the failure analysis process, the associated hazards, harms, and design failure mode and effects analysis (DFMEA) are reviewed. The respective failures are rated against the associated risks as well as the frequency of the failure modes, as shown in Figure 10.24. The level of risk will determine the team's response and actions.

When the proposed corrective action has been determined, the reliability engineer can predict what the reliability will be after the corrective action is implemented with the cross-functional team, as shown in the predicted column of Figure 10.29. The redesigned devices will then be tested again to verify that the corrective actions were effective. This action will also help to determine if the failure rate is decreasing, increasing, or constant, at which time the reliability can be re-predicted.

Continued

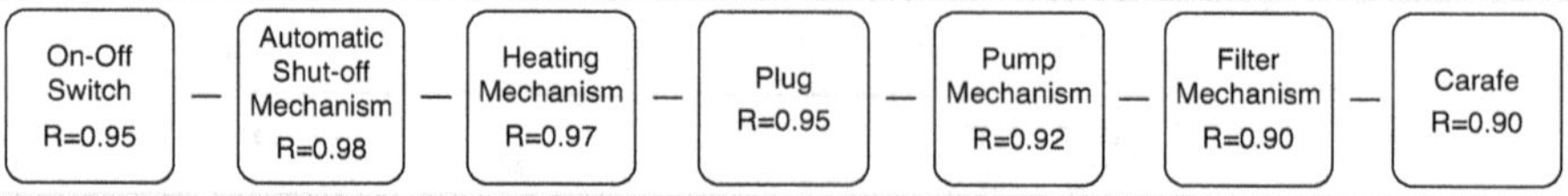

Figure 10.28 Reliability block diagram with component reliability values.

Issue #	Failure mode	# of occurances	Associated Risk Severity 5=High, 1=Minor	Failure rate as tested	Proposed Corrective action	Predicted failure rate	Failure rate with C/A
1001	Coffee is too hot and scalds user	3	4	0.3	Redesign of the the heating mechanism to remove this potential failure	0.005	0.002
1002	Water reservoir is leaking	3	5	0.3	A re-design of the pump mechanism is needed in order to prevent potential hazard of electrical shock	0.005	0.002
1003	Coffee machine damaged during shipping to testing facility	1	1	0.1	Redesign packaging for shipment to prevent damage	0.05	0.04
1004	Coffee machine over heated and caught on fire	1	5	0.1	Redesign the automatic shut-off mechanism to prevent fire	0.05	0.04

Figure 10.29 Testing issue list for the coffeemaker.

This modeling, design, prediction, testing, failure analysis, and corrective action cycle, shown in Figure 10.30, is repeated until either all the failure modes have been eliminated or the reliability of the system is acceptable to the project team.

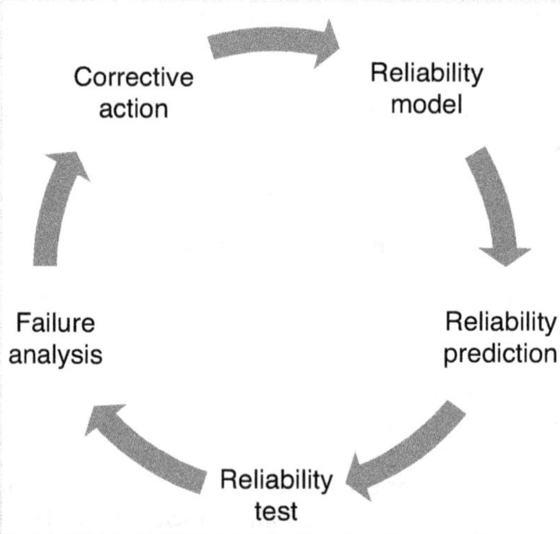

Figure 10.30 Reliability design cycle.

Part V

Lifecycle Reliability

Chapter 11
Reliability Design Techniques

DESIGN EVALUATION TECHNIQUES (VALIDATION AND VERIFICATION)

> Explain how validation, verification, and other review techniques are used to assess the reliability of a product's design at various lifecycle stages. (Apply)
>
> **Body of Knowledge V.A.1**

The reliability engineer performs evaluation functions in each step of the product development process, from concept through design and development, design freeze, design review, pre-production, production, and post-production, as shown in Figure 11.1.

Concept

As the earliest design parameters are established in the concept stage, preliminary estimates of reliability requirements should be made. Factors to consider when creating reliability requirements include what the product is designed to accomplish, who will be using the product, the conditions under which the product will be used, and how long the product will be used.

Design and Development

As the more formal design and development phase is initiated, the reliability engineer should be prepared to provide the design team with guidance and judgment regarding various options. Data on the reliability of proposed components and the implications for product reliability should be documented for the team at each design refinement. The allocation of reliability requirements to various subsystems and components should also be studied. The underlying mindset must be to formally study every potential failure and establish cost-effective ways of preventing them.

The two general approaches to failure prevention are *fault tolerance* and *fault avoidance*. A fault-tolerant approach requires the design of redundant systems so that a fault does not result in a failure. For example, dual master cylinders in an automotive braking system serve as a fault-tolerant system. The fault-avoidance approach

Figure 11.1 Product development process steps.

requires designing the product with components that are sufficiently reliable to guarantee the minimum product reliability. This may be accomplished, for example, by using heavier structural pieces, more reliable components, derating, and other techniques. Reliability growth during the design process should be documented.

The customer's lifecycle cost for the product can be minimized by establishing optimum reliability levels and implementing systems for obtaining them.

Design Verification and Validation

The terms "design validation" and "design verification" are often confused. In the *design verification* step, the reliability engineer confirms that requirements for the product or service are included in the design. During *design validation*, it is confirmed that the final design will meet the customer's requirements. In other words, during design verification, the reliability engineer asks, "Did we design the device right?" During design validation, they ask, "Did we design the right device?"

To undergo verification and validation, the design must be finalized. Samples that are used in verification and validation need to be built to that final design. No further changes are permitted unless issues are found during testing and the design needs to be updated to mitigate or eliminate those issues.

Design Verification

Design verification is a critical step in the product development lifecycle, ensuring that the design outputs meet the specified product requirements. For reliability and risk engineers, the focus is not only on verifying functional performance but also on ensuring that the design minimizes risks and meets reliability objectives.

The product design needs all its design outputs in alignment with its design inputs before testing and, eventually, manufacturing can occur. Design inputs are based on the user needs and intended use of the product and include functional, performance, safety, regulatory, and aesthetic requirements. Design outputs flow from design inputs and provide evidence that the requirements are being met. These outputs can take the form of drawings, specifications, manufacturing instructions, and a list of assemblies, subassemblies, and parts that go into making the device, as well as other documentation specific to the product.

The reliability of the design must be considered. Reliability analysis assesses the product design's ability to perform its intended use over the lifetime of the device. Depending on the nature of the product, this can take multiple forms. For single-use or disposable devices, such as surgical gloves, paper plates, and safety razors, reliability analysis focuses on factors like material integrity and sterility to prevent failures during use. In contrast, for durable goods like refrigerators, automobiles, and industrial robots, the analysis examines long-term performance, identifying potential points of failure and ensuring the product and all its components can withstand repeated use while still maintaining functionality. Testing to substantiate the predicted results may take the form of environmental testing, accelerated life testing, or other methods that stress the product to test its capabilities.

Shelf-life considerations are also crucial when considering product reliability; for single-use devices, this involves ensuring the product remains effective and safe until its expiration date, while for durable goods, it adds the need to assess how operation, handling, and environmental conditions might affect long-term product reliability. By addressing these aspects, reliability analysis helps mitigate risks, enhance product quality, and ensure customer satisfaction.

Risk assessment, as examined in previous chapters, identifies and mitigates risks associated with design failures. These risk mitigations are confirmed to be effective via design verification testing. Depending on the level of risk, these mitigations should be assessed using an appropriate sample size.

Setting sample size requirements for design verification testing involves consideration of both confidence and reliability to ensure robust and meaningful results. The confidence level component of the sample size requirement represents the degree of certainty that the sample accurately reflects the population, while reliability indicates the probability that the product will perform as expected under specified conditions. Greater risks require a higher level of confidence and/or reliability to make the determination that the product is safe and effective.

Confidence and reliability for product risks need to be selected before verification testing is initiated. A high level of risk may require a confidence of 95% and a reliability of 99%, while a relatively low risk may use a confidence of 90% and a reliability of 90%. These values depend on the level of risk, the type of product being tested, and the manner in which it is tested. An attribute (pass/fail) test where zero failures are accepted will be calculated using the binomial distribution as shown in Equation 11.1.

Equation 11.1 Sample size for zero failure acceptance testing.

$$n = \frac{\ln(1 - \textit{confidence level})}{\ln(\textit{reliability})}$$

An example of this for a high-level risk tested at a confidence of 95% and a reliability of 99% would give a sample size of 299, with zero failures allowed.

If the data are continuous and the distribution is known from previous testing (from prototypes or similar products on the market), Equation 11.2 can be used:

Equation 11.2 Sample size for continuous distribution.

$$n = \left(\frac{Z\sigma}{E}\right)^2$$

where Z is the z-score corresponding to the confidence level, sigma is the standard deviation, and E is the margin of error.

Sample sizes should also consider cost, time, and resource availability. If there are constraints on the number of units available for testing, compensation must be put in place to produce data sufficient to show the design meets the requirements. By carefully setting sample size requirements, it ensures that the design verification testing is both efficient and effective, providing high confidence in the product reliability and performance.

Design verification testing should use a documented and validated test method, and the results should be compiled in a formal report.

Design Validation

Design validation is used to check if the intended purpose of the product or service is met by the design. Like verification testing, design validation requires clear criteria to be in place prior to testing. These criteria will vary depending on the product and should be driven by the product's intended use, regulatory requirements, and customer expectations. Once criteria are in place, a test method can be developed to challenge the criteria and determine if the design is appropriate for its use.

Testing typically involves using the product in scenarios that can be expected to occur during regular use. Multiple methods and tools are part of the validation process. For example, validation activities may include usability testing, user acceptance testing, pilot or small-scale trials, simulations, clinical trials, or beta tests in multiple environments to ensure the design is robust to noise.

Design validation activities are most effective when they are used multiple times during the product development process to help identify and correct issues as soon as possible. One of the more common methods is user trials, which provide valuable insights into how the intended users will interact with a product. These trials help ensure that the product meets user needs and performs reliably in real-world conditions. By observing users as they engage with the product, designers can identify usability issues, gather feedback on functionality, and understand user behavior and preferences. When users are consulted on the design early and often, the process development team can enhance product quality, increase customer satisfaction, and reduce the risk of costly post-launch modifications.

The definitive design validation activities take place on a finalized design with finalized user requirements. At this stage of development, the goal of the tests is not to improve the product but to determine if the product meets the user requirements. This can involve surveys and observation of the users as they use the product. Design validation test results are collected, analyzed, and documented prior to recommendations to move the product forward in the development process.

Software Testing

If the product in question contains software, then special considerations must be made for software testing. Software verification and validation are both essential processes in the product development process but serve different purposes and occur at different development stages. Software verification is focused on the correctness, reliability, and quality of the software, while validation is focused on ensuring that the final product meets the intended use and user requirements. Both verification and validation are necessary for delivering high-quality software that is reliable, functional, and user-friendly.

Software verification is the process of evaluating successive software builds to ensure that they meet the specified requirements. Software verification activities typically include code reviews, inspections, walkthroughs, and static analysis. These activities are typically performed during the design and development phase to catch defects early and ensure that the software design and its implementation align with requirements. Verification focuses on the reliability and accuracy of the software, confirming that the internal code and its interactions with system components function as intended and adhere to design specifications.

In contrast, software validation evaluates whether the software meets its users' needs and expectations. Validation involves activities such as system testing, user acceptance testing (UAT), and beta testing. These activities are performed after the software has been fully developed to ensure that it performs correctly in the real-world environment and satisfies user expectations. Validation focuses on the observed performance of the software, ensuring that it delivers the desired outcomes and provides acceptable user experience.

Design Review

Reliability engineers should provide the design team with data from testing the final version of the product, assuring that the design meets the reliability requirements, design requirements, and user needs. Once these results are determined to be satisfactory, the product can be moved to pre-production in preparation for commercialization.

Pre-production

Alternative manufacturing processes and parameters should be studied for their impact on reliability to meet or exceed design requirements. If the process cannot meet the design requirements, then adjustments will need to be made to either the process or the design. If the design is altered because of manufacturing requirements, an assessment must be performed on the design verification and validation testing to determine what testing will need to be repeated to show the new design meets its requirements.

Process design involves defining the commercial manufacturing process based on product and process development studies. At this point, the process should be fully understood and critical process parameters that need to be controlled to ensure product quality should be in place. Detailed documentation and planning are crucial during this stage to establish a robust foundation for the subsequent validation activities.

Production

A testing program should be established to ensure that production output meets reliability requirements. This is to ensure that the process can make products over its lifecycle that will consistently meet requirements. This is often referred to as *process validation.* Process validation activities are essential for ensuring that a manufacturing process consistently produces products that meet product requirements.

Process design takes place in the pre-production stage, as noted previously. The second stage, process qualification, evaluates the process design to determine if it can reproducibly produce products that meet all quality requirements. This stage includes the qualification of equipment and facilities, as well as the performance qualification of the process itself. During this stage, extensive testing is conducted to demonstrate that the process operates within specified limits and consistently produces products that meet quality specifications. Usually, this will involve testing to make sure that the product meets its requirements while the process is operating at its upper and lower limits, as well as testing multiple lots at nominal settings to ensure lot-to-lot consistency. Any deviations or issues identified during this stage must be addressed and resolved to ensure the process is reliable and capable of consistent performance. Once the process is deemed acceptable, the product is allowed to be commercialized.

Even after the product is commercialized, the process will still require monitoring. This continued process verification involves ongoing monitoring and control of the process to ensure it remains in a state of control throughout the product lifecycle. This stage includes regular review of process performance data (via statistical process control or similar methods), routine testing, and periodic requalification to detect any variations or trends that could impact the quality of the products. Using continuous process monitoring, manufacturers can quickly identify and address potential issues, ensuring sustained product quality, compliance with regulatory requirements, and lower material wastage. Frequent and methodical analysis of the process output helps maintain high standards of quality and reliability, leading to consistent customer satisfaction and reduced risk of product recalls.

Post-Production

Once a product is released for production, a system for follow-up should be in place so that any failures, in-house or in the field, can be studied. One approach to doing this is known as *failure reporting, analysis, and corrective action system* (FRACAS).

A system that permits traceability of individual components helps establish sources of failure due to design, production, service, customer misuse, and so on. The thrust should be not to fix the blame but to fix the problem. Thorough, documented design requirements help ensure customer satisfaction. The purpose of design evaluation is to verify that the product meets the requirements at each design stage. The aspects of the requirements relating to product reliability mandate special attention because the issue is not merely "Does it work?" but "How long will it work and under what conditions?" A strong program of testing and documentation can help avoid disappointed customers.

There are four types of evaluation:

1. *Environmental stress screening.* The unit is exposed to the most severe design environmental stresses. Sometimes, accelerated life testing may be used. The purpose is to identify weak components.

2. *Reliability development/growth tests.* A series of tests conducted periodically from design through production phases to demonstrate the impact of corrective actions on reliability.

3. *Reliability qualification tests* (also called *reliability demonstration tests*). Conducted on a sample from production to determine whether production units meet reliability requirements, these tests serve as a basis for production approval.

4. *Product reliability acceptance tests.* A periodic test during production to determine whether the output continues to meet reliability requirements.

A product's lifecycle from the producer's viewpoint may be expanded to include these five stages:

1. *Concept/design.* This is the stage at which reliability testing and analysis can have the most impact, especially on lowering the failure rate during the useful life of the product. Reliability engineers can aid in the

assessment of design validation (that is, fitness-for-use in the intended operating conditions) by conducting accelerated life tests of components and subassemblies.

2. *Introduction.* At this stage, the reliability program gathers and analyzes data on failure rate and proposes measures to reduce it. During this phase, restudy of installation and usage procedures is appropriate. In addition, reliability engineers can aid in the assessment of design validation (that is, compliance with explicit specifications).
3. *Growth.* As the market for the product grows and improvements are implemented, careful traceability becomes critical. It is important to know if these type X failures occurred on products that were produced after improvement Y was in place.
4. *Maturity.* Market growth has plateaued and the enterprise searches for new markets and applications. New environments or functions require a new assessment of reliability.
5. *Decline.* As the market for the product decreases, the costs involved with continued maintenance support must be analyzed. Typically, the support cost per item in use will grow unless carefully monitored.

STRESS-STRENGTH ANALYSIS

> Apply the stress-strength analysis method of calculating probability of failure and interpret the results. (Analyze)
>
> **Body of Knowledge V.A.2**

A part fails when *stress* or *load* on the part exceeds its *strength*. The job of the designer is to design parts such that the material strength is better than the stress for some target design life; otherwise the part will not serve its intended purpose. Accepted design practice incorporates margin in different ways to ensure that part strength is always greater than the stress. Two standard measures used for these structural assessments are *factor of safety* (FS, FoS, Safety Factor, SF) and *margin of safety* (MS, MoS), which are rescaled versions of the same metric. The calculations for factor of safety and margin of safety are shown in Equation 11.3 and Equation 11.4, respectively.

Equation 11.3 Factor of Safety.

$$Factor\ of\ Safety = Safety\ Factor = \frac{Strength}{Stress}$$

Equation 11.4 Margin of Safety.

$$Margin\ of\ Safety = Factor\ of\ Safety - 1 = \frac{Strength}{Stress} - 1$$

In an ideal world, (Strength > Stress) and therefore FoS > 1 and MoS > 0, which yields a "positive margin." Technically, if stress exactly equals strength, the part should not fail, but if FoS < 1 and MoS < 0, we have "negative margin," in which case failure would be expected. However, how much should one rely on this analysis? How much margin should be allowed? What does this mean in terms of risk?

Typically, design standards provide some standard threshold value as a minimum requirement that the FoS (or MoS) must be over, for example, 1.2 (0.2; 20%) or 1.5 (0.5; 50%). These approaches most commonly utilize single-point values to represent both the strength of the part as well as the stress the part will encounter, often under conservative or worst-case assumptions. These conditions are not always well understood but ideally will "bound" anticipated usage, preventing actual part failures from occurring. However, it is important to be aware that FoS or MoS values have no mathematical relationship or connection to probability in any fashion whatsoever and so cannot be translated for use on a hazard risk matrix or a quantitative risk assessment.

For many scenarios, this analysis is adequate for general use, particularly when assumptions concerning part strength and anticipated stresses are well documented. Different organizations may have different philosophies and conventional practices about how to define the design life, for example, that may provide additional conservatism. However, without data, these assumptions may be subject to much discussion. Issues can also arise on both sides. Excessive conservatism can lead to problems of over-design, leading to parts being too heavy, for example, which may have form, fit, or function concerns or cost consequences due to excess material, all of which are undesirable. However, insufficient conservatism leads to under-design issues: poor quality design leads to part failures—if not immediately, then parts will be prone to earlier degradation and issues in the field, which can lead to increases in service or field issues. Multiple objectives may need to be met simultaneously, and/or costs and benefits traded to obtain the most optimal solution.

An alternative approach to this analysis is to view both stress and strength not as deterministic but as random variables that follow a prescribed statistical distribution, known as stress–strength interference analysis. The interference region is the area under the curve where the two distributions overlap, in which stress (X) could randomly exceed strength (Y) (i.e., X > Y) despite having positive margin based on the nominal values alone, as shown in Figure 11.2. Assumptions are still required, but the use of statistical distributions allows results to be calculated as probabilities. If the distributions are far apart, there is a low probability that stress will exceed strength, so greater separation would be reflected in higher reliability. Conversely, greater overlap would be associated with higher risk.

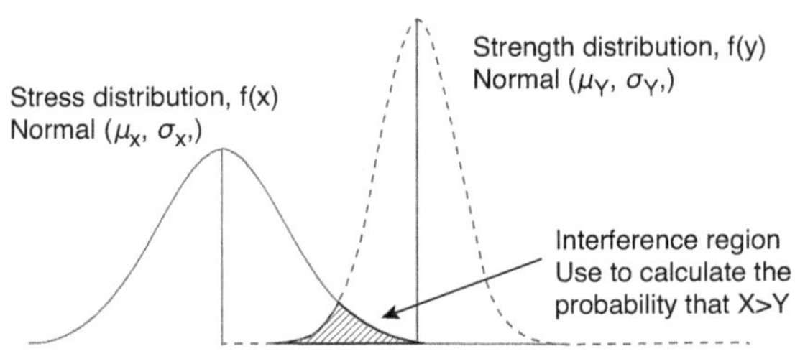

Figure 11.2 Example stress–strength interference region.

The stress–strength distributions highlight the four basic ways the designer can improve reliability. The designer usually has more control over the strength:

1. *Increase the mean strength*. Use different materials or different design
2. *Decrease the strength variation*. Reduce variation in the materials and in the process

The designer may have some control over the stress:

3. *Decrease the mean stress*. Control the loading
4. *Decrease the stress variation*. Limit the use environment

Note that this framework is generalizable beyond stress–strength specifically and can be used for analogous scenarios where some level of exposure is being compared to some measure of part or system capability, such as temperatures or pressures or other operating conditions. Both measures must be available on the same scale.

Stress–Strength Analysis for Normal Distributions

Assuming normal distributions for both stress and strength provides a mathematically friendly formula with a simple closed-form solution, assuming estimated means and standard deviations, as shown previously, but the general approach is extendable to any pair of distributions. Calculations are shown subsequently assuming that stress and strength are both normal distributions. The analysis is performed by taking the difference, $D = (\text{Strength} - \text{Stress})$. If D is negative, it is a failure.

$$\Pr(Fail) = \Pr(D < 0) = \Pr(Strength - Stress < 0) = \Pr(Strength < Stress)$$

Conversely, if $D > 0$ it is a success, and reliability is calculated as:

$$\Pr(Success) = \Pr(D > 0) = \Pr(Strength - Stress > 0) = \Pr(Strength > Stress)$$

The *difference distribution* can be used to calculate the probability of failure. Since the difference between two normally distributed random variables is also normal, we can use rules of mathematical statistics to derive the exact distribution of the difference and use it to calculate the associated probability using a standard normal table (or Student's t as appropriate for small sample sizes). This analysis does assume that stress and strength are independent for any given unit, which seems implicitly reasonable since the inherent strength of a part should not be dependent on the load applied to it at any point.

Thus, since the difference $D = \text{Strength} - \text{Stress} = (Y - X)$, is normally distributed, with mean $\mu_D = (\mu_Y - \mu_X)$ and standard deviation equal to $\sigma_D = \sqrt{\sigma_Y^2 + \sigma_X^2}$, failure occurs when stress is greater than strength, or when $D < 0$. To find the probability of failure, we evaluate

$$\Pr(Failure) = \Pr(D < 0) = \Pr\left(Z < \frac{0 - \mu_D}{\sigma_D}\right)$$

using the standard normal distribution table. Note that the corresponding reliability is equal to $1 - \Pr(\text{Failure})$ or $\Pr(D > 0)$. These calculations are illustrated in Example 11.1.

EXAMPLE 11.1

Let stress be a normally distributed variable with mean $\mu_X = 30,000$ psi and standard deviation $\sigma_X = 5000$ psi. Strength follows a normal distribution with mean $\mu_Y = 40,000$ psi and $\sigma_Y = 3000$ psi. What is the probability of failure?

We first calculate the mean and standard deviation of the difference distribution, or $D = (Y - X)$:

$$\mu_D = (\mu_Y - \mu_X) = 40,000 - 30,000 = 10,000$$

$$\sigma_D = \sqrt{\sigma_Y^2 + \sigma_X^2} = \sqrt{3000^2 + 5000^2} = 5831$$

Failure will occur when D < 0. We find

$$\Pr(Failure) = \Pr(D < 0) = \Pr\left(Z < \frac{0 - \mu_D}{\sigma_D}\right) = \Pr\left(Z < \frac{-10,000}{5831}\right) = \Pr(Z < -1.71)$$

and evaluate using the standard normal distribution table (Appendix D).

The probability of failure is equal to 0.04318, and the reliability is (1 − 0.04318) = 0.95682. This result is corroborated by Figure 11.3.

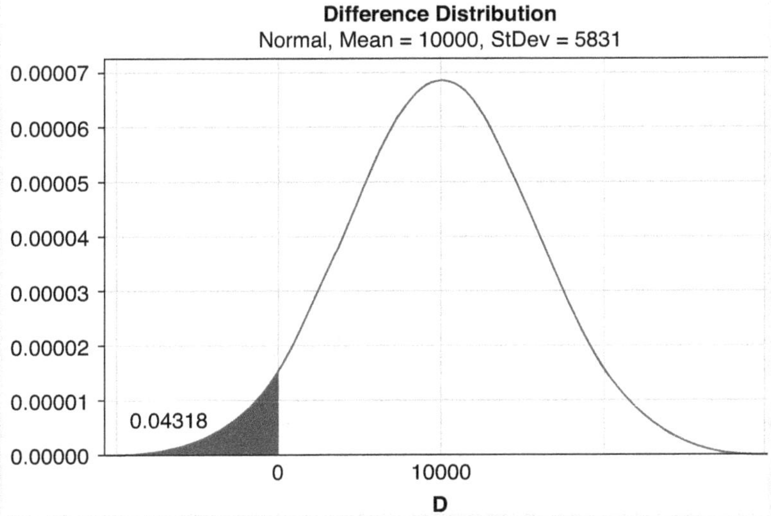

Figure 11.3 Difference distribution showing failure probability.

Stress–Strength Analysis for Exponential Distributions

Let stress (X) be exponentially distributed with parameter β_X and strength (Y) be exponentially distributed with parameter β_Y. The corresponding PDFs of stress and strength are:

$$f(X) = \frac{1}{\beta_X} e^{-X/\beta_X}$$

$$f(Y) = \frac{1}{\beta_Y} e^{-X/\beta_Y}$$

The means of the distributions are β_X and β_Y, respectively. As before, the *difference distribution* can be used to calculate the probability of failure.

$$\Pr(Fail) = \Pr(D < 0) = \Pr(Strength - Stress < 0) = \Pr(Strength < Stress)$$

Assuming stress and strength are independent, we can calculate the probability that D is less than zero using a closed-form solution:

$$\Pr(Failure) = \Pr(D < 0) = \frac{\beta_X}{\beta_X + \beta_Y}$$

$$\text{Reliability} = \Pr(D > 0) = \frac{\beta_Y}{\beta_X + \beta_Y} = \frac{1}{1 + \dfrac{\beta_X}{\beta_Y}}$$

$$\Pr(Failure) = 1 - \Pr\left(Z < \frac{\ln(m_Y/m_X)}{\sqrt{s_X^2 + s_Y^2}}\right)$$

$$\text{Reliability} = \Pr\left(Z < \frac{\ln(m_Y/m_X)}{\sqrt{s_X^2 + s_Y^2}}\right)$$

Conversely, the reliability can be found by calculating $\Pr(D > 0)$

$$\text{Reliability} = \Pr(D > 0) = \frac{\beta_Y}{\beta_X + \beta_Y} = \frac{1}{1 + \dfrac{\beta_X}{\beta_Y}}$$

Using this format of the reliability calculation allows us to relate reliability to the ratio of mean stress to mean strength. We can also relate reliability to the safety factor calculated as β_Y / β_X. Table 11.1 gives reliabilities for various ratios of stress to strength and safety factors.

Table 11.1 Ratio of mean stress to mean strength, safety factors, and corresponding reliability for exponential distributions.

β_X / β_Y	$SF = \beta_Y / \beta_X$	**Reliability**
1	1	0.5
0.8	1.25	0.56
0.75	1.33	0.57
0.6	1.67	0.63
0.5	2	0.67
0.4	2.5	0.71
0.3	3.33	0.77
0.25	4	0.80
0.2	5	0.83
0.1	10	0.91

Stress–Strength Analysis for Lognormal Distributions

The lognormal distribution is often useful in modeling stress and strength because of its positive domain, skewed right tail, and multiplicative properties. In addition, the lognormal distribution is useful when the uncertainty about stress and strength are relatively high.

Let stress (X) and strength (Y) both be lognormally distributed, with parameters μ_X and σ_X, and μ_Y and σ_Y, respectively. It follows that the median (m) of each distribution is:

$$m_X = e^{\mu_X}$$
$$m_Y = e^{\mu_Y}$$

and $\ln(m) = \mu$.

Here, a failure occurs when ($X/Y<1$). Let $W = \ln(X/Y) = \ln(X) - \ln(Y)$. Assuming X and Y are independent, the random variable W is normal with mean $\ln(m_X/m_Y)$ and variance $(\sigma_X^2 + \sigma_Y^2)$.

We can calculate the probability of failure using the standard normal distribution as follows:

$$\Pr(Failure) = 1 - \Pr\left(Z < \frac{\ln(m_Y/m_X)}{\sqrt{s_X^2 + s_Y^2}}\right)$$

And reliability is found using:

$$\text{Reliability} = \Pr\left(Z < \frac{\ln(m_Y/m_X)}{\sqrt{s_X^2 + s_Y^2}}\right)$$

EXAMPLE 11.2

This example is adapted from Ebeling (2010, p. 154).

The capacity of a structure to withstand earthquakes is lognormally distributed, with a median of 8.1 on the Richter scale and $\sigma_Y = 0.07$. Historically, the magnitude of earthquakes in the region has been lognormally distributed with a median of 5.5 and $\sigma_X = 0.15$. Calculate the failure probability of the structure.

$$\begin{aligned}\Pr(Failure) &= 1 - \Pr\left(Z < \frac{\ln(m_Y/m_X)}{\sqrt{s_X^2 + s_Y^2}}\right)\\ &= 1 - \Pr\left(Z < \frac{\ln(8.1/5.5)}{\sqrt{0.15^2 + 0.07^2}}\right)\\ &= 1 - \Pr(Z < 2.33) = 0.01\end{aligned}$$

Stress–Strength Analysis for Weibull Distributions

There is no closed-form solution for calculating failure probabilities when both stress and strength follow Weibull distributions. Instead, practitioners can use Monte Carlo simulation to estimate the failure probability. This technique is illustrated in Example 11.3.

EXAMPLE 11.3

A reliability engineer models stress using a Weibull distribution with shape and scale parameters equal to 3 and 1, respectively. Strength is modeled using a Weibull with shape and scale parameters 2.5 and 3, respectively. Ten thousand observations are generated for each distribution, and the difference between each pair of observations is calculated.

The resulting empirical distributions of stress and strength are shown in Figure 11.4.

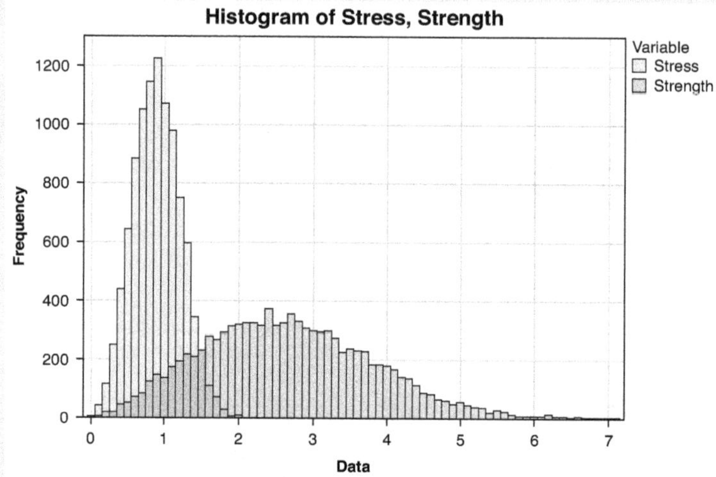

Figure 11.4 Stress and strength simulation results with Stress ~ Weib(3, 1) and Strength ~ Weib(2.5, 3).

A failure will occur when the difference between a strength and stress pair is negative. These occurrences are shown in the black rectangle in the histogram of differences found in Figure 11.5.

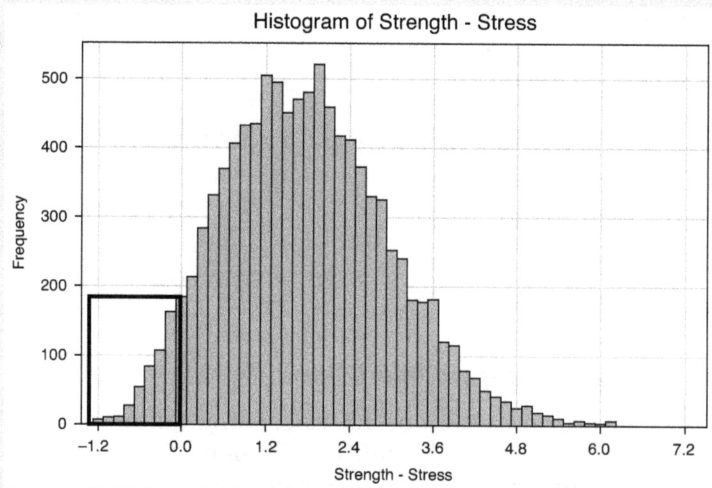

Figure 11.5 Difference between strength and stress for 10,0000 simulated pairs of observations based on Weib(3, 1) and Weib(2.5, 3), respectively.

There are 550 occurrences of (strength < stress) in the 10,000 simulated pairs of observations. The estimated failure probability is thus 550/10,000 = 5.5%, and the reliability is 94.5%.

DESIGN OF EXPERIMENTS (DOE)

Develop and interpret the results of a standard design of experiments (DOE), including both full-factorial and fractional factorial designs, integrating methodologies such as analysis of variance (ANOVA) and blocking for hard to control factors (e.g., planned grouping). Use replication and randomization to optimize the robustness and validity of the experimental findings. (Analyze)

Body of Knowledge V.A.3

Note: This section is adapted from McShane-Vaughn (2022), *The ASQ Six Sigma Black Belt Handbook.*

Design of experiments (DOE) is a statistical approach to gaining empirical knowledge, providing the most time-efficient and economical methods of determining the effect of a set of independent variables, or factors, on a response variable. Knowledge of this relationship permits the experimenter to optimize a process and predict a response variable by setting the factors at specific levels. The aim of a designed experiment is to generate knowledge about a product or process and to establish the mathematical relationship $Y = f(X)$, or empirical model, where X is a list of independent variables and Y is the response variable.

DOE offers many important benefits over ad hoc methods of data collection that might be important to the reliability engineering function. A few of these advantages include:

- Predictable expectations of the experimental outcomes.
- A designed experiment encompasses the reasonable range of input values that might contain the optimal settings.
- The mathematical relationship between the simultaneous changes in multiple inputs and the resulting change in the outcome can be studied and modeled.
- The number of samples necessary to achieve a given level of confidence can be determined while planning the experiment.
- A designed experiment can establish causality by changes in the inputs under controlled conditions.
- The often-important effect of interacting inputs may be accounted for and used to achieve the desired output.

When preparing to conduct an experiment, the first consideration is, "What question are we seeking to answer?" Examples of experimental objectives might be to find:

- The combination of design features that results in the strongest package seal
- The combination of factors that produces the most consistent battery life in the presence of temperature variation
- The combination of valve dimensions that produces the greatest linear output
- The extrusion pressure that maximizes the yield of vascular grafts

Sometimes the experimental objective derives from a question. For example, "What's causing the excess variation in hardness at the rolling mill?" could generate the objective, "Identify the factors responsible for the hardness variation and find the settings that minimize it."

The experimental objective must also be measurable, so the next step is to establish an appropriate measurement system. The measurement system must be reasonably simple and easy to operate.

In simpler terms, the primary objectives of an experimental design are to:

- Identify which independent variables explain the most variation in the response variable
- Determine what levels of the independent variables minimize or maximize the response variable

Once the objective and a measurement system have been determined, the factors and levels are selected. People with the most experience and knowledge about the product or process should be consulted to determine what factors and what levels of each factor are important to achieve the objective. The list of factors and the levels for each factor should flow from their input. But be careful to rule out potential factors based on expert advice in the absence of reproducible and tested data.

DOE for Reliability Improvement

The following sections provide an overview of DOE methodology as it is used broadly for problem solving and process improvement. DOE is also a critical tool for reliability improvement—an efficient and effective way to identify the factors that lead to improvements in reliability performance measures. However, there are some differences in the way DOE is applied in the reliability context, including the choice of performance measurements and managing experimental resources.

Performance Measurements

In most DOE applications, the response variable is a continuous measurement that is well approximated by a normal distribution. Many continuous measures of degradation (e.g., crack size), or measures related to failure modes, such as initial strength, will likely meet these assumptions. Some reliability measures will not meet these assumptions, including discrete pass/fail data, and censored lifetimes.

In these cases, the reliability engineer will need to consult a statistician when planning and analyzing an experiment.

Other reliability measurements may or may not meet these assumptions, such as continuous failure times, or failure counts/rates. These measurements may follow a different distribution (e.g., exponential, Weibull, or Poisson). While care should be exercised, the essential summary statistics and graphical tools of DOE analysis are fairly robust to nonnormal distributions and likely will still be useful for the investigation. However, when formal probability statements are necessary, the reliability engineer should seek the assistance of a statistician to see if it is necessary to incorporate a different distribution into the modeling.

Experimental Resources

Reliability testing is often more time-consuming and expensive when compared to other experimental situations. These costs will create tradeoffs in planning, as they affect the number of experimental runs, the choices for the combinations of input factors, the amount of replication, and the handling of time-dependent noise variables. In some cases, the costs may be reduced by leveraging fractional factorial designs, using accelerated test plans, or collecting degradation measures and other covariates.

While these considerations make reliability DOE more challenging, they do not diminish the importance of the thoughtful choice of performance measures and experimental design in any project. All the following material can be effectively applied to reliability improvement.

Terminology

To better understand the concept of DOE, we will first provide a few fundamental definitions:

- An *experimental design* is a formal plan that details the specifics for conducting an experiment, such as deciding which responses, factors, levels, blocks, treatments, and tools are to be used.
- The *response variable* is the observed result or value of an experimental treatment. It is sometimes known as the *output variable, dependent variable,* or *y variable.* There may be multiple response variables in an experimental design. See the discussion earlier in this section for some considerations when choosing a response variable in the reliability context.
- The *observed value* is a particular value of a response variable determined as a result of a test or measurement.
- A *factor* is an independent variable or assignable cause that might affect the responses, and of which different levels are included in the experiment. There are different kinds of factors. Each kind of factor determines the levels that may be set and the way its effects may be modeled. Factors are also known as *explanatory variables, predictor variables, input variables,* or *x variables.*
- A *continuous factor* may take on any value over a given range. For example, a temperature between 20°C and 35°C is a continuous factor.

A continuous factor effect is modeled as a coefficient of change, or a slope. A discrete numeric factor is also a continuous factor that can be set to a predetermined set of three or more values. For example, a temperature that must be tested at 20°C, 25°C, 30°C, and 35°C is a discrete numeric factor. A discrete numeric effect is modeled as coefficients for all the polynomial terms required.

- A *categorical factor* may take on only one of a given set of values. For example, the choice of suppliers (Supplier A, B, or C) is a categorical factor.
- A *noise factor* is an independent variable that is difficult or too expensive to control as part of standard experimental conditions. In general, it is not desirable to make inferences on noise factors, but they are included in an experiment to develop a set of control factor conditions that are robust against the effects of noise.
- An *effect* is a change in the response when the level of one or more factors changes. A model typically represents the relationship between a factor and a response variable. Note that there can be many factors and response variables. The model includes any terms that are necessary to account for the effect. Specific types of terms include main effect (first-order or additive change); dispersion effect (change in variation); interaction effect (the effect of one factor depends on the level of another factor); and quadratic effect (second-order change). We will deal specifically with main and interaction effects.
- A *main effect* is the impact or influence of a single factor on the mean of the response variable.
- An *interaction effect* occurs when the effect of one factor depends on the level of one or more other factors. The existence of an interaction effect means that the effect of changing factors is more than the sum of the main effects.
- A *level* is the setting or assignment of a factor at a specific value. Factor levels generally change independently of one another. Some situations put constraints on changes or combinations of factor levels. For example, the levels of mixture components do not change independently because they must sum to one.
- A *treatment* is the specific setting or combination of factor levels for an experimental unit.
- An *experimental unit* is the smallest entity receiving a particular treatment that yields a value of the response variable.
- An *experimental run* results from applying a specific treatment combination in the design to an experimental unit.
- *Experimental error* is the variation that occurs in the response variable beyond that which is accounted for by changes in the factors, blocks, or other assignable source when conducting an experiment. This unexplained variation is included in the statistical model as a random term

or effect and estimated as a variance. This variance is used to estimate the standard error of the estimated model coefficients. In practice, this term is truncated to just "error."

- A *factorial design*, according to Montgomery (2019), is an experiment for which "each complete trial or replicate of the experiment all possible combinations of the levels of the factors are investigated."

Design Principles

DOE is an effective tool for determining causal relationships between independent and response variables. However, considerable attention must be given to planning and conducting an experiment. Therefore, it is important to identify and define the critical concepts underlying experimental designs:

- Coded factor levels
- Repeated measures (repetition)
- Replication
- Order
- Randomization
- Blocking
- Hierarchical modeling convention

Each of these concepts will be discussed in the following subsections.

Coded Factor Levels

The estimates of the model parameters depend on the measurement scale of the factor. This dependence makes interpretation difficult and unnecessarily introduces correlation among the parameter estimates. For example, a change in temperature of 1 degree is different if the scale is in degrees Celsius, Fahrenheit, Kelvin, or Rankin. A scale-invariant estimate may be based on *coded factor levels*. The low setting is replaced by a coded value of –1 and the high setting is replaced by +1. The middle of the factor range, or center point, is replaced by 0. The new settings are often represented simply as – and + as seen in the examples that follow.

Using the coded factor levels, the resulting parameter estimates represent the change of one unit on the coded scale, which is now half the range for all factors. The change over the entire range of the factor setting is twice this estimate, or the full effect of the factor. Because the factors are coded, we can compare the relative importance of each effect, and the intercept is now the mean response at the origin of the design space. Coding the factor levels provides additional statistical benefits as well, especially when the model contains higher-order terms for interaction and nonlinear effects.

Statistical software for the analysis of design experiments automatically codes the factor levels during a linear regression analysis, so it is not necessary to perform this coding manually. The parameter estimates may then be doubled to obtain the same effects that are manually calculated in the chapter examples.

Repeated Measure Versus Replication

A *repeated measure*, also known as a *repetition*, is the measurement of a response variable more than once for the same run. Repetition occurs when the responses from each run are measured n times in a row, or measured over time, as with stability studies. Repetition allows one to determine the variation of the measurements.

Replication, on the other hand, occurs when the same treatment is applied to a new experimental unit. Replication of the entire experiment is often performed for a factorial design. Each new run is called a *replicate*. Replication allows an engineer to determine the variation of response, or the experimental error. Replication increases the precision of the estimated effects in an experiment. Note that replication differs from repeated measures because it repeats the entire factorial experiment, not just the measurements for the same run. It is more effective when all elements contributing to the experimental error are included. Replication reflects the sources of variability both within and between runs and adds degrees of freedom to the experiment.

Order

Order refers to the sequence in which the runs of an experiment will be conducted. Generally, we talk about two types of order:

- *Standard order* is also referred to as Yate's order, in which the first factor varies the fastest while the last factor varies the slowest. The first factor changes the most frequently, and the last factor the least. Note that some statistical software packages display a catalog of design choices that are not written in Yate's order.
- *Run order* shows what the order of the runs in an experiment would be if the experiment were run in random order. Random order works to spread the effects of noise variables and to help the data meet some of the important assumptions of regression analysis: independent observations and normally distributed errors.

Randomization

Randomization is used to assign treatments to experimental units so that each unit has an equal chance of being assigned a particular treatment, thus minimizing the effect of variation from uncontrolled noise factors. A completely randomized design is one in which the treatments are assigned at random to the full set of experimental units and all the factors are reset before each run. Randomization might be impractical or impossible, in which case the experiment requires a split-plot design, but such a design and its analysis is beyond the scope of this handbook. No blocks are involved in a completely randomized design.

Blocking

A *block* is a planned grouping of experimental units. Blocks are usually selected to allow for special causes that are part of the experiment but are not from a factor of interest. These special causes may be avoided within blocks, thus providing a more homogeneous response. *Blocking* refers to the method of grouping runs in an experiment to minimize the impact of selected assignable causes. Randomization of the experiment is now restricted because it only occurs within blocks. The *block effect* is the result of changing the block in an experimental design. Existence of a block effect generally means that the method of blocking was appropriate. The regression model may treat the block effect as either a fixed effect or as a random effect. The regression model also does not allow the block effect to interact with factor effects.

Hierarchical Modeling Convention

The *hierarchical modeling convention* requires that the individual effects used in higher-order terms be included in the model, regardless of their significance, to prevent ill-formulated polynomials. For example, a model with a significant interaction term involving factors A and B must also include the main effects of A and B. This convention allows for a standardized interpretation of interaction effects.

Planning Experiments

The activities outlined in the earlier discussion should not be left to happenstance. Instead, they should result from the development of an experimental plan. An *experimental plan* is the assignment of treatments to each experimental unit and the time order in which the treatments are to be applied. Planning an experiment requires several design considerations:

- Factors that are not varied in the design, that is, factors that are held constant
- Noise factors
- Confounding and resolution
- Blocking and randomization
- The cost of running the experiment
- Time considerations—limited time might be available to conduct the experiment

Montgomery (2019) offers guidelines for conducting experiments:

1. *Recognition and statement of the problem.* As discussed earlier, this step might start with a question, but it must eventually be translated into a statistical problem.
2. *Selection of the response variable.* People most familiar with the process should be consulted.
3. *Choice of factors, levels, and range.* As with guideline 2, the subject matter experts (SMEs) who are familiar with the process should be consulted. Experiments might fail to find significant effects or estimate the effects of continuous factors well because the factor range is too narrow to produce a large enough effect. Do not shrink the range because you think you know where the best setting occurs. Keep the range wide so that the model parameters are estimated well and then use the model to find the best setting.

4. *Choice of experimental design.* Choosing an experimental design requires consideration of an experimental plan (discussed previously).
5. *Performing the experiment.* It is critical that the experimenter understand the conditions under which the experiment was conducted. Ideally, these conditions should match those conditions under which the process is routinely performed.
6. *Statistical analysis of the data.* The data must be carefully analyzed to ensure statistical validity.
7. *Conclusions and recommendations.* Statistical conclusions and recommendations must be translated into terms that management and operations people can understand, particularly if they are expected to support and/or fund process changes.

The next few sections discuss various common designs.

Screening Designs

The selection of an experimental design may be constrained by such things as affordability and time available. At this stage, some experimenters establish a budget of time, cost, and other resources that may be used to reach the objective. If production equipment and personnel must be used, how much time is available? How much product and other consumables are available? The choice might also depend on the available prior knowledge and the number of factors involved (complexity).

Choosing the Appropriate Design: Sequential Experimentation

A sequence of experiments is planned to answer all the questions with sufficient detail and accuracy. This strategy is called the *sequential design of experiments*. The initial experiment might only provide the most basic information. For example, the initial experiment may help decide which factors of the many candidates indeed affect the response, separating the "vital few" factors from the "trivial many" factors. The next experiment is run to provide additional data that gives new information and to improve the accuracy of the information. For example, it is possible to add new runs to the initial experiment in what is called an *augmented design*, for more information and with some economy. Typically, 20% of the available budget should be allocated to the first experiment because it seldom meets all the objectives and, in fact, often raises as many questions as it answers. Examples of such new questions include:

- What if an additional level had been used for factor A?
- What if a different factor had been included instead of factor B?
- What if the change in the response is not linear?

Therefore, in the absence of much prior knowledge, rather than designing a large initial experiment involving many variables and levels, it is usually best to begin with a more modest screening design. A *screening design* is an experiment intended as a first step to identify a subset of the potential factors for subsequent study. A screening design is not intended to identify all the possible effects to estimate an unbiased model. Examples of screening designs are regular or minimum-run Resolution IV two-level fractional factorial designs, and *definitive screening designs*. All these designs rely on several key screening principles that generally hold when there is little prior knowledge and there are many factors to study. To learn more about screening designs, readers can refer to Montgomery (2020).

Sparsity of Effects

The principle of the *sparsity of effects* states that while there are many factors and potentially many effects of each factor, most factors will be inactive. Screening designs provide only enough data to fit a model with a minority of the potential factors.

Hierarchy of Effects

The principle of the *hierarchy of effects* states that contribution of effects decreases as the order of the term in the model increases. Main effects tend to be the largest effects, two-factor interaction effects and quadratic effects tend to be the next largest effects, and so on. Screening designs focus on the main effects, but some also consider active second-order effects. None addresses third-order effects or higher.

Heredity of Effects

The principle of the *heredity of effects* states that higher order terms tend to include the same factors that exhibit main effects. *Strong heredity* implies that all the factors included in the interaction effect are also main effects. *Weak heredity* implies that some of the factors included in the interaction effect are also main effects. For example, if the $A \times B$ interaction effect is important, then both A and B are important as main effects under strong heredity, but only A or B is important under weak heredity. Heredity often guides the choice of terms in the model.

Projection

The principle of *projection* states that the design matrix for a low-resolution screening experiment will exhibit higher resolution and support the estimation of more terms of higher order when only a minority of the factors are active. This property relies on the previous three properties. For example, a resolution III design for 9 factors in 16 runs *projects* into a replicated full-factorial (full resolution) for any three factors. A definitive screening design may also support the estimation of the full quadratic model when many of the potential factors have been eliminated.

Screening Experiments

Screening experiments are intended to eliminate the inactive factors from further consideration, so the risk of a type II error (ruling out) is more important than a type I error (ruling in). There are two considerations that can help reduce the risk of a false negative outcome. The first approach is to accept a higher type I error rate. This approach leads to more significant factors that are false positives, but fewer active factors will be determined to be insignificant (false negatives).

The second approach is more quantitative: a power analysis. This analysis requires the specification (prior knowledge) of four quantities that control the type II error rate (β): the level of significance (α), the size of the effect (δ), the sample size (n), and the response variance (σ). The relationship between power and these four experimental variables depends on the statistical hypothesis test that is used to decide significance. In general, power increases when the level of significance increases, the size of the effect increases, the sample size increases, or the variance decreases. High power is desirable, and it should be at least 80% for screening experiments. It is best to use statistical software for the power analysis. Consult the software documentation for the proper setup of this analysis. Screening designs may then be augmented after determining which factors and effects might be active to resolve questions about bias in the model from confounding.

One-Factor Experiments

This section addresses one-factor, completely randomized experiments and the randomized complete block design (RCBD).

One-Factor, Completely Randomized Experiment

A one-factor experiment involves one control variable varies at k levels. The response variable is continuous, and the factor may be either categorical or continuous. For example, an analyst may want to know if there is a significant difference in wall thickness of a plastic part across three different injection molding machines. Here, the factor "machine" is categorical. If the categorical factor includes just two levels, it is often referred to as an A-B comparison.

For the continuous factor case, an engineer may investigate the effect of platen pressure on the seal strength of a paper package. They run an experiment using four different pressure settings. To make this a completely randomized experiment, the pressure settings for each replication of the experiment are run in random order. As mentioned previously, this randomization reduces the effects of any known or unknown noise factors on the seal strength.

The experimental results from these types of experiments are analyzed using a one-way analysis of variance, or ANOVA, to determine if at least one machine produces significantly different wall thicknesses compared to the others, or if at least one pressure setting results in a significantly different mean seal strength. For the continuous case, we can one step further and estimate a regression model that predicts mean seal strength based on pressure.

We first introduce a *dot notation* that is commonly used in many statistical formulas. For readers who may be unfamiliar with this notation, see Table 11.2. This table illustrates how the notation is simply a shorthand that uses subscripts to show completed summations. This notation is convenient when multiple subscripts are being used concurrently.

The subscript i denotes rows, and the subscript j denotes columns. For the one-factor experiment, there are $i = 1, \ldots, a$ levels in the factor, and $j = 1, \ldots, n$ replicates. We replace the subscript i with the factor level index and replace the subscript j with the dot when summing across the replicates. For example, we denote the sum of all replicates for factor level 1 as $y_{1.}$. The total of all observations across factor levels and replicates is denoted as $y_{..}$.

One-Way Analysis of Variance

A one-factor experiment can be analyzed using a one-way ANOVA. The statistical model, formula for the sums of squares, and ANOVA table calculations and for the one-factor experiment have been summarized in Table 11.3.

Table 11.2 Using the dot notation for a one-factor experiment.

Replicates $j = 1, \ldots, n$							
	y_{11}	y_{11}	.	.	.	y_{1n}	$\sum_{j=1}^{n} y_{1j} = y_{1\cdot}$
	y_{21}	y_{22}				y_{2n}	$\sum_{j=1}^{n} y_{2j} = y_{2\cdot}$
Factor levels $i = 1, \ldots, a$	.					.	.
	.					.	.
	.					.	.
	y_{a1}	y_{a2}	.	.	.	y_{an}	$\sum_{j=1}^{n} y_{aj} = y_{a\cdot}$
Total							$\sum_{i=1}^{a} y_{i.} = y_{...}$ $\sum_{i=1}^{a}\sum_{j=1}^{n} y_{ij} = y$

Table 11.3 Statistical model, sum of squares, ANOVA table for the completely randomized one-factor experimental design.

Statistical model			Sums of squares (SS)	
One-way ANOVA (completely randomized design with fixed effects)			$SS_T = \sum_{i=1}^{a}\sum_{j=1}^{n} y_{ij}^2 - \frac{y_{..}^2}{N}$ $SS_{\text{Treatments}} = \sum_{i=1}^{a} \frac{y_{i\cdot}^2}{n} - \frac{y_{..}^2}{N}$ $SS_E = SS_T - SS_{\text{Treatments}}$	
Sources of variation	**Sum of squares**	**Degrees of freedom**	**Mean squares**	**F_0 statistic**
Between treatments	$SS_{\text{Treatments}}$	$a-1$	$MS_B = \frac{SS_B}{a-1}$	$F_0 = \frac{MS_B}{MS_E}$
Error (within treatments)	SS_E (By subtraction)	$N-a$	$MS_E = \frac{SS_E}{N-a}$	
Total	SS_T	$N-1$		

Source: Montgomery (2019).

The one-way ANOVA table is five columns wide and three rows long, as shown at the bottom of the figure. The ANOVA table is built by moving from left to right. The "sources of variation" column identifies the terms for the "between treatment" effect, the error (also known as the "within treatment" effect), and the total. The next column lists the sums of squares, which convey the variation in the response that is associated with each term. Degrees of freedom, which account for how much independent information was contributed to each corresponding sum of squares, are in the next column. The formulas for the degrees of freedom are also shown in the ANOVA table. The mean squares column shows the quotient of the sum of squares and the degrees of freedom, or the mean sum of squares, which is an estimate of the variance associated with between and within treatments.

Finally, the rightmost column is the quotient of the mean sum of squares for a term and the mean sum of squares error, which is an *F* statistic. Recall from Chapter 6 that the hypotheses for the one-way ANOVA are:

$$H_0: \textit{All factor level means are equal}$$
$$H_1: \textit{At least one mean is not equal to the rest.}$$

The *F* statistic in the last column is evaluated against an *F* distribution table value based on the degrees of freedom for the between-treatment effect, the within-treatment effect, and the chosen alpha level. If the calculated *F* from the ANOVA is greater than the *F* statistic from the *F* distribution table, we reject the null hypothesis and conclude that at least one factor level mean is significantly different from the rest. If the null hypothesis is rejected, a test after ANOVA, such as Tukey's test, can determine which means are significantly different from the others.

Typically, terms related to error, such as *sums of squares* and *degrees of freedom*, are calculated by subtraction. More specifically, for the one-factor design we have:

$$SS_E = SS_T - SS_{Treatments}$$
$$DF_E = DF_T - DF_{Treatments}$$

EXAMPLE 11.4

This example is adapted from Montgomery (2019). A medical device manufacturer produces vascular grafts and wants to investigate their reliability. These grafts are produced by extruding resin along with a lubricant into tubes. The developer investigates the effects of four different levels of extrusion pressure (8500, 8700, 8900, and 9100) on the percentage of grafts that do not fail a stress test. The experiment is a one-factor, completely randomized design. The order of the pressure settings was randomized for each of the six replicates in the experiment. The resulting reliability data and the totals across replicates are shown in Table 11.4. We can use this table to calculate the sum of squares for each factor level, or treatment.

Table 11.4 Data for the extrusion pressure one-factor completely randomized experiment.

Pressure	Rep 1	Rep 2	Rep 3	Rep 4	Rep5	Rep 6	Total
8500	90.3	89.2	98.2	93.9	87.4	97.9	**556.9**
8700	92.5	89.5	90.6	94.7	87.0	95.8	**550.1**
8900	85.5	90.8	89.6	86.2	88.0	93.4	**533.5**
9100	82.5	89.5	85.6	87.4	78.9	90.7	**514.6**
Total							**2155.1**

Table 11.5 Squared observations used to calculate the total sum of squares.

Pressure	Squared observations						Total
8500	8154.09	7956.64	9643.24	8817.21	7638.76	9584.41	**51794.35**
8700	8556.25	8010.25	8208.36	8968.09	7569.00	9177.64	**50489.59**
8900	7310.25	8244.64	8028.16	7430.44	7744.00	8723.56	**47481.05**
9100	6806.25	8010.25	7327.36	7638.76	6225.21	8226.49	**44234.32**
Total							**193999.3**

Table 11.5 is used for computing the sum of squares for the total. Each cell that held an original data element is now replaced by the square of that original data element. The bottom right corner shown in bold is $\sum\sum y_{ij}^2 = 193999.3$. The computation for the sum of squares for the total and treatments is shown:

$$SS_T = \sum_{i-1}^{a}\sum_{j=1}^{n} y_{ij}^2 - \frac{y_{..}^2}{N}$$
$$= 193999.3 - \frac{(2155.1)^2}{24}$$
$$= 480.3$$

$$SS_{Treatment} = \sum_{i=1}^{a}\frac{y_{i.}^2}{n} - \frac{y_{..}^2}{N}$$
$$= \frac{556.9^2 + 550.1^2 + 533.5^2 + 514.6^2}{6} - \frac{(2155.1)^2}{24}$$
$$= \frac{1162183.0}{6} - \frac{(2155.1)^2}{24}$$
$$= 178.2$$

At this point, we transfer our sums of squares to the ANOVA table shown in Table 11.6. There are four treatments. Therefore, there are three degrees of freedom for the factor. There are (4×6–1) or 23 degrees of freedom for the total. The degrees of freedom for the error term are obtained by subtraction. The remainder of the table is completed using the procedure described previously.

Table 11.6 ANOVA table for yield experiment.

Source	DF	Adj SS	Adj MS	F-Value
Factor	3	178.2	59.39	3.93
Error	20	302.1	15.11	
Total	23	480.3		

We compare the F statistic from the ANOVA table to an $F(3, 20, 0.05) = 3.10$, found in Appendix F.3. Since the calculated $F = 3.93$ is greater than the table value, we reject H_0 and conclude that at least one pressure setting produces a reliability different from the rest.

Continued

Using Minitab® software for the analysis produces the output in Figure 11.6. The *p* value for the ANOVA table is equal to 0.023, so we reject H_0 and conclude at least one mean reliability is different from the rest. The Tukey pairwise comparison results show that setting the pressure equal to 8500 results in the highest percentage of grafts that survive the stress test and is significantly higher than the reliability at the 9100 pressure settings.

WORKSHEET 1

One-way ANOVA:Yield verus Pressure

Method

Null hypothesis All means are equal
Alternative hypothesis Not all means are equal
Significance level a = 0.05

Equal variances were assumed for the analysis.

Factor information

Factor	Leves	Values
Pressure	4	8500, 8700, 8900, 9100

Analysis of Variance

Source	DF	Adj SS	Adj MS	F-Value	P-Value
Pressure	3	178.2	59.39	3.93	0.023
Error	20	302.1	15.11		
Total	23	480.3			

Model summary

S	R-sq	R-sq(adj)	R-sq(pred)
3.88676	37.10%	27.66%	9.42%

Means

Pressure	N	Mean	StDev	95%CI
8500	6	92.82	4.58	(89.51,96.13)
8700	6	91.68	3.30	(88.37, 94.99)
8900	6	88.92	2.97	(85.61, 92.23)
9100	6	85.77	4.45	(82.46, 89.08)

Pooled StDev = 3.88676

Tukey Pairwise Comparisons

Grouping Information using the Tukey Method and 95% Confidence

Pressure	N	Mean	Grouping	
8500	6	92.82	A	
8700	6	91.68	A	B
8900	6	88.92	A	B
9100	6	85.77		B

Means that do not share a letter are significanty different.

Figure 11.6 Computer output for the yield experiment.

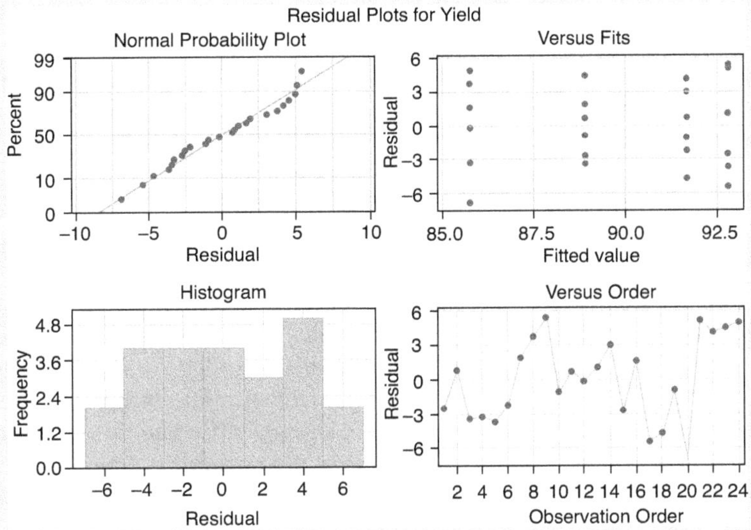

Figure 11.6 (Continued.)

As discussed in Chapter 6, validity of the ANOVA statistics depends on several key assumptions about the *statistical errors*—the difference between the actual and the predicted responses (referred to as *residuals*). The residuals are assumed to be:

- Independent, or uncorrelated
- Identically normally distributed with a mean = 0 and a constant variance, σ^2. That is, the variance of the residuals does not depend on the magnitude of the response.

Various plots of residuals can be used to diagnose appreciable deviations from these assumptions, including:

1. *Normal probability plot* or *normal quantile plot.* This chart plots the residuals versus their predicted probability from a fitted normal distribution. The chart also plots the graph of the cumulative distribution function for the normal distribution for reference. The chart uses a transformation of the *y*-axis such that the residuals and the graph of the normal cumulative distribution function are straight line with a slope equal to σ. Evaluate normality by comparing the pattern residuals to the straight reference line, known as the "fat pencil test." The normal probability plot in Figure 11.6 shows no signs for concern—a fat pencil covers up all the points. More formally, we can use the Anderson-Darling test to determine nonnormality. See Chapter 6 for details.
2. *Residuals versus fitted value.* If the model accounts for all factor influences, the residuals should be random. Thus, each column of points in the plot should be about the same length, and there should be no outliers. The residual plot in Figure 11.6 shows no signs of unequal variance or outliers.

In general, regression analysis can withstand some degree of violation of the underlying assumptions. However, when a large violation is evident, various response transformations, such as a natural log or square root, can be employed to restore normality and equalize variances. When such situations occur, consult a statistician.

Randomized Complete Block Design

An experiment may require the analyst to add one or more blocking variables if they believe that a variable not of experimental interest will influence the responses. The block effects are estimated separately from the factor effect. This separation improves the precision of the estimates of the factor effects because the block effects are additional and undesirable variation in the response. Considering the variability introduced by the blocks helps increase the precision of the analysis. In a *randomized complete block design* (RCBD), each block contains all the treatments, and randomization is restricted within blocks.

Table 11.7 provides the statistical model, the source table, and the definition of the sums of squares for the RCBD. Typically, such designs are analyzed using statistical software. However, the following example will be analyzed manually.

An experimental design consisting of a factor levels and b blocks and assigned via randomization to the experimental units within each block is known as a *randomized block design* and is used to control the variability of experimental units. For completely randomized designs, no stratification of the experimental units is made.

Table 11.7 Model, hypotheses, and ANOVA table for a randomized block design with one factor and one blocking variable.

Randomized complete block design (RCBD)	$y_{ij} = \mu + \tau_i + \beta_j + \varepsilon_{ij} \begin{cases} i = 1,2,\ldots,a \\ j = 1,2,\ldots,b \end{cases}$	$H_0: \mu_1 = \mu_2 = \ldots = \mu_s$	$\varepsilon_y: NID(0, \sigma^2)$
	$y_{ij} = j$ th observation of the i th treatment j th block	$H_A: \mu_i \neq \mu_j$ for at least one pair (i, j) Or	
	μ = overall mean	$H_0: \tau_1 = \tau_2 = \ldots = \tau_a = 0$	
	τ_i = effect of the i th treatment	$H_A: \tau_i \neq 0$ for at least one i	
	β_j = effect of the j th block		
	ε_{ij} = error term		

Sources of variation	**Sum of squares**	**Degrees of freedom**	**Mean squares**	***F* statistic**
Treatments	$SS_{Treatments}$	$a-1$	$MS_{Treatments} = \frac{SS_{Treatments}}{a-1}$	$F_0 = \frac{MS_{Treatments}}{MS_E}$
Blocks	SS_{Blocks}	$b-1$	$MS_{Blocks} = \frac{SS_{Blocks}}{b-1}$	
Error	SS_E	$(a-1)(b-1)$	$MS_E = \frac{SS_E}{(a-1)(b-1)}$	
Total	SST	$N-1$		

In the randomized block design, the treatments are randomly allotted within each block; that is, the randomization is restricted. Like the randomized block design, a *randomized block factorial design* is a design where each block includes a complete set of factorial combinations.

EXAMPLE 11.5

This example is adapted from Montgomery (2019). As in Example 11.4, a medical device manufacturer produces vascular grafts. These grafts are produced by extruding resin along with a lubricant into tubes. Consider now, though, that one batch of resin can only produce enough product to do one full replicate of the experiment (runs at each of the four pressure levels). To run the experiment six times, six batches of resin must be used. The engineer suspects that there is batch-to-batch variation present that will affect the experimental outcome. Consequently, the developer investigates the effects of four different levels of extrusion pressure on reliability using an RCBD design with the batches considered as blocks.

As before, the percentage of grafts surviving a stress test is shown in Table 11.8. Note that these data are different from that in Example 11.4. The column and row totals have been added for treatment and block totals, respectively. In addition, the computations for the sum of squares for both blocks and treatments are shown.

Table 11.8 Data for Example 11.5, randomized block design.

Block (Batch)							
Pressure	1	2	3	4	5	6	Total
8500	90.3	89.2	98.2	93.9	87.4	97.9	**556.9**
8700	92.5	89.5	90.6	94.7	87.0	95.8	**550.1**
8900	85.5	95.5	89.6	86.2	88.0	93.4	**538.2**
9100	82.5	90.2	85.6	90.6	78.9	90.7	**518.5**
Total	**350.8**	**364.4**	**364**	**365.4**	**341.3**	**377.8**	**2163.7**

Computation of $SS_{\text{Treatment}}$ and SS_{Block} are now shown.

$$SS_{Treatment} = \frac{(556.9)^2 + (550.1)^2 + (538.2)^2 + (518.5)^2}{4} - \frac{(2163.7)^2}{24} = 141.6$$

$$SS_{Block} = \frac{(350.8)^2 + (364.4)^2 + (364.0)^2 + (365.4)^2 + (341.3)^2 + (377.8)^2}{6} - \frac{(2163.7)^2}{24} = 203.4$$

Table 11.9 is used for computing the sum of squares for the total. Each cell that held an original data element is now replaced by the square of that original data element. The bottom right corner shown in bold is $\sum\sum y^2 = 195570.3$. The computation for the sum of squares for the total is shown.

Continued

Table 11.9 Squared observations used to compute SS_T.

Block (Batch)

Pressure	1	2	3	4	5	6	**Total**
8500	8154.1	7956.6	9643.2	8817.2	7638.8	9584.4	**51794.35**
8700	8556.3	8010.3	8208.4	8968.1	7569.0	9177.6	**50489.59**
8900	7310.3	9120.3	8028.2	7430.4	7744.0	8723.6	**48356.66**
9100	6806.3	8136.0	7327.4	8208.4	6225.2	8226.5	**44929.71**
Total	**30826.84**	**33223.18**	**33207.12**	**33424.1**	**29176.97**	**35712.1**	**195570.3**

The sum of squares total and error sum of squares are calculated using:

$$SS_T = 195570.3 - \frac{(2163.7)^2}{24} = 195570.3 - 195066.6 = 503.7$$

$$SS_{Error} = SS_T - SS_{Treatment} - SS_{Block} = 503.7 - 141.6 - 203.4 = 158.8$$

At this point, we transfer our sums of squares to the ANOVA table shown in Table 11.10 and begin to complete the table. There are four treatments. Therefore, there are three degrees of freedom for the factor. Similarly, there are six blocks. Therefore, there are five degrees of freedom for the block. There are ($4 \times 6 - 1$) or 23 degrees of freedom for the total. The degrees of freedom for the error term are obtained by subtraction. The remainder of the table is completed using the procedure described previously.

Assuming $\alpha = 0.05$, we must compare our F_0 value for the treatments to $F_{0.05,3,15}$. From Appendix F.3 the F-value is 3.29. Since $4.46 > 3.29$, we reject the null hypothesis and conclude there is a difference in the mean extrusion pressure.

Table 11.10 ANOVA table for randomized block experiment.

Source	Sum of squares	Degrees of freedom	Mean squares	*F* ratio
Treatment	141.6	3	47.20	4.46
Block	203.4	5	40.67	
Error	158.8	15	10.58	
Total	503.7	23		

Note that the F statistic for the block effect is not presented in the ANOVA table. The F statistic for blocks is often calculated by statistical software, but it should not be reported or assessed. The restriction on the randomization (only within blocks) produces an F statistic that does not follow the F distribution, so critical F values or associated p-values cannot be determined. A Tukey test shows that setting pressure at 8500 results in a significantly higher reliability.

Recall that the block itself is not a factor of experimental interest, and it may not be actionable. We leave the blocking variable in the model since it acts as a sort of sponge to absorb any additional variation introduced by the block.

The sum of squares contributed by the block (batches) is 203.4. If we had not accounted for the block effect, the variability contributed by the block would have rolled into the sum of squares error, increasing SS_E from 158.8 to 362.1 and the DF_E to 20. The resulting $MS_E = 18.11$ is 1.7 times higher than the estimate using the correct analysis. The F ratio for pressure for the "incorrect" analysis would then have been 2.61, compared to a table value $F(0.05, 3, 20) = 3.10$. We would fail to reject the null hypothesis in this case.

Tukey Pairwise Comparisons: Pressure

Grouping information using the Tukey method and 95% confidence

Means that do not share a letter are significantly different.

Pressure	N	Mean	Grouping	
8500	6	92.8167	A	
8700	6	91.6833	A	B
8900	6	89.7000	A	B
9100	6	86.4167		B

Figure 11.7 Computer output of Tukey pairwise comparison results.

General Factorial Experiments

A *general factorial design* is an experimental design consisting of all possible combinations of the factor levels of two or more factors, with each factor being studied at two or more levels. For example, an engineer may run an experiment that varies temperature at two levels and pressure at three levels and uses material from four different suppliers. One replicate of the experiment would comprise $2 \times 3 \times 4 = 24$ runs, and the three main effects, three two-factor interactions, and one three-factor interaction would be estimable.

Note that the minimum number of runs for a factorial experiment can be expressed using:

$$n = L^k$$

where

n = Number of runs

L = Number of levels

k = Number of factors

Table 11.11 provides the model, hypothesis, and assumptions for a completely randomized, full-factorial experiment with two factors A and B tested at i and j levels, respectively, with k replicates.

A *two-way analysis of variance* can be used to analyze the results of an experiment with two factors and can accommodate all interaction terms that are estimable by the factorial design. The calculations for the sums of squares for two factors are shown in Table 11.12.

Table 11.11 Statistical model and hypotheses for the two-way ANOVA.

Name	Model	Hypotheses	Assumptions
Two-way ANOVA (completely randomizes design with fixed effects)	$y_{ijk} = \mu + \tau_i + \beta_j + (\tau\beta)_{ij} + \varepsilon_{ijk} \begin{cases} i = 1,2,\ldots,a \\ j = 1,2,\ldots,b \\ k = 1,2,\ldots,n \end{cases}$ y_{ijk} = kth observation in the ith row, jth column μ = Overall mean τ_i = Effect of the ith level of row factor A β_j = Effect of the jth level of column factor B $(\tau\beta)_{ij}$ = Effect of the interaction of the ith level of row factor A and the jth level of column factor B ε_{ijk} = residual for the ijkth observation	$H_0 : \tau_1 = \tau_2 = \cdots = \tau_a = 0$ $H_1 : \tau_i \neq 0$ for at least one i $H_0 : \beta_1 = \beta_2 = \cdots = \beta_b = 0$ $H_1 : \beta_j \neq 0$ for at least one j $H_0 : (\tau\beta)_{ij} = 0 \, f$ or all i, j $H_1 : (\tau\beta)_{ij} \neq 0 \, f$ or at least one i, j	$\varepsilon_{ijk} \sim NID\ (0, \sigma^2)$

Source: Adapted from Montgomery (2019).

Table 11.12 Mathematical model and sums of squares calculations for two-way ANOVA.

Statistical model	Sums of squares (SS)
Two-way ANOVA (completely randomized design with fixed effects)	$SS_T = \sum_{i=1}^{s} \sum_{j=1}^{b} \sum_{k=1}^{n} y_{ijk}^2 - \frac{y^2 \ldots}{abn}$ $SS_A = \frac{1}{bn} \sum_{i=1}^{s} y_{i,.}^2 - \frac{y^2 \ldots}{abn}$ $SS_B = \frac{1}{an} \sum_{j=1}^{b} y_{jj}^2 - \frac{y_i^2 \ldots}{abn}$ $SS_{AB} = \frac{1}{n} \sum_{i=1}^{a} \sum_{j=1}^{b} y_{ij}^2 - \frac{y_{\ldots}^2}{abn} - SS_A - SS_B$ $SS_E = SS_T - SS_A - SS_B - SS_{AB}$

The typical structure of a two-way ANOVA table is illustrated with two factors A and B in Table 11.13. These tables are five columns wide, and the length depends on the number of factors. The table is built moving from left to right. The first column identifies the terms for the main effects, interactions, blocks, and so on. In the next column, the sums of squares are shown, which convey the variation in the response that is associated with each term. The degrees of freedom are listed next. These account for how much independent information was contributed to each

Table 11.13 ANOVA table calculations for a two-factor full-factorial design.

Source of variation	Sum of squares	Degrees of freedom	Mean squares	F_0 statistic
Factor *A*	SS_A	$a-1$	$MS_A = \frac{SS_A}{a-1}$	$F_0 = \frac{MS_A}{MS_E}$
Factor *B*	SS_B	$b-1$	$MS_B = \frac{SS_B}{b-1}$	$F_0 = \frac{MS_B}{MS_E}$
Interaction *AB*	SS_{AB}	$(a-1)(b-1)$	$MS_{AB} = \frac{SS_{AB}}{(a-1)(b-1)}$	$F_0 = \frac{MS_{AB}}{MS_E}$
Error	SS_E By subtraction	By subtraction	$MS_E = \frac{SS_E}{ab(n-1)}$	
Total	SS_T	$abn-1$		

Source: Montgomery (2019).

corresponding sum of squares. Once again, the error sum of squares and degrees of freedom are found via subtraction:

$$SS_E = SS_T - SS_A - SS_B - SS_{AB}$$
$$DF_E = DF_T - DF_A - DF_B - DF_{AB}$$

The Mean Squares column shows the quotient of the sum of squares and the degrees of freedom, or the mean sum of squares, which is an estimate of the variance associated with each term or the errors. The last column is the quotient of the mean sum of squares for a term and the mean sum of squares error, which is an *F* statistic. The variance in the numerator depends on a true effect (alternative hypothesis), but the variance in the denominator does not. Ultimately, the calculated *F* statistics will be compared to an *F*-table value to determine whether each term is significant.

Interaction Effect

Note that the ANOVA table for the general factorial includes an interaction effect. An *interaction effect* occurs when the effect of one factor depends on the level of one or more other factors. The existence of an interaction effect means that the effect of changing factors is more than the sum of the main effects. An *interaction plot* is a graphical depiction that gives the average responses at the combinations of levels of two distinct factors.

Interaction effects are important for two reasons. The first reason is that interaction effects lead to unexpected results when they are unknown or when they are ignored. We cannot predict the change in the response without knowing the level of both factors. The second reason is that they represent an opportunity to mitigate the impact of a hard-to-control factor. We can use the other factor that is involved in the interaction to reduce the effect of the hard-to-control factor. This approach is known as *robust design*, presented later in this section.

Calculations for the general factorial experiment are illustrated in the following example.

EXAMPLE 11.6

This example is adapted from Montgomery (2019). Suppose an engineer is studying the effects of the electrode gap and cathode RF power of a silicon wafer nitride etching process. Furthermore, she wishes to find the settings that will produce the target etch rate, which has been found to correlate with the reliability of the wafer. The gap and power are called the *factors*, or *independent variables*. The silicon nitride etch rate of (Å/m) is called the *response*, or *dependent variable*, because its value depends on the values of the independent variables through the mathematical relationship that will be determined by the experiment. There might be additional independent variables, such as the batch of the etching agent, which influence the dependent variable, but these variables will be controlled.

The experimenter has decided that two continuous factors will be tested at four and two levels each as follows:

- Gap (*G*): 0.8, 1.2, 1.6, 2.0 (cm)
- Power (*P*): 275 and 325 (W)

The experimenter conducts a replicated general factorial experiment to generate the maximum amount of process knowledge. This experiment will test all possible combinations of levels and factors, using one run for each combination. The number of runs in this experiment will be $n = 4 \times 2 \times 2 = 16$ since there is one factor at four levels, one at two levels, and the design is replicated.

Refer to the data given in Table 11.14. Note that this experiment is replicated but not blocked. The first step in calculating the main effects is to average the results for each level of each factor. For example, the mean response when the Gap = 0.8 cm is calculated by averaging the result of the eight runs, in which the electrode was set to 0.8 cm. Note that this method of calculating the main effect is only valid for a balanced design, such as this full-factorial design. Linear regression is required to estimate the effects when the design is not balanced.

Table 11.14 Data for the etch rate general factorial experiment.

Gap	Power	Etch rate
0.8	275	747
0.8	325	1000
1.2	275	665
1.2	325	890
1.6	275	670
1.6	325	755
2.0	275	560
2.0	325	850
0.8	275	730
0.8	325	1030
1.2	275	710
1.2	325	885
1.6	275	680
1.6	325	760
2.0	275	555
2.0	325	800

Table 11.15 shows the calculations for the mean response at each factor level.

Table 11.15 Calculations for the mean response at each factor level.

Condition	Calculation	Mean response
Gap = 0.8 cm	$\frac{747+1000+730+1030}{4}$	876.75
Gap = 1.2 cm	$\frac{665+890+710+885}{4}$	787.5
Gap = 1.6 cm	$\frac{670+755+680+760}{4}$	716.25
Gap = 2.0 cm	$\frac{560+850+555+800}{4}$	691.25
Power = 275 W	$\frac{747+665+670+560+730+71+680+555}{8}$	664.625
Power = 325 W	$\frac{1000+890+755+850+1030+885+760+800}{8}$	871.25

Figure 11.8 shows the plots the main effects for the two factors.

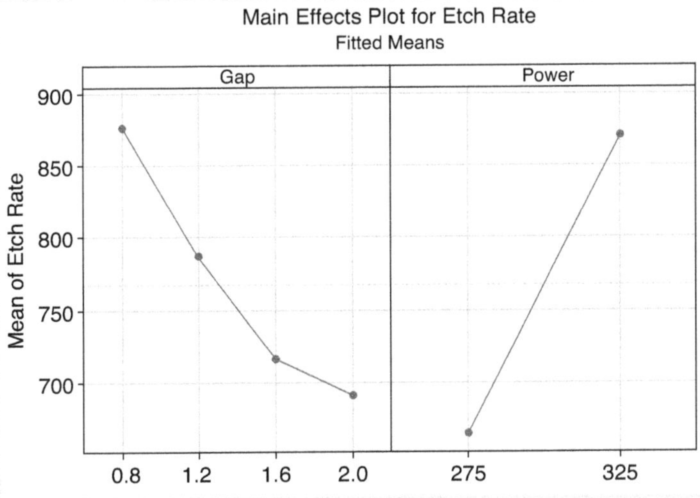

Figure 11.8 Main effects plot for etch rate.

The Minitab® results for the two-way ANOVA follow.

Based on the *p*-values for the *F* test, we conclude that the (Gap × Power) interaction is statistically significant, as are the Gap and Power main effects.

We notice in Figure 11.10 that the Gap and Power curves are not parallel and indicate a large interaction effect, which corroborates the results from the ANOVA. For example, we cannot predict the etch rate at a power setting equal to 325 unless we know the level of the gap.

Continued

WORKSHEET 2

General Factorial Regression: Etch rate versus Gap, Power

Factor Information

Factor	Levels	Values
Gap	4	0.8, 1.2, 1.6, 2.0
Power	2	275, 325

Analysis of Variance

Source	DF	Adj SS	Adj MS	F-Value	P-Value
Model	7	277916	39702	107.87	0.000
Linear	4	253877	63469	172.44	0.000
Gap	3	83102	27701	75.26	0.000
Power	1	170776	170776	463.99	0.000
2-Way Interactions	3	24039	8013	21.77	0.000
Gap*Power	3	24039	8013	21.77	0.000
Error	8	2945	368		
Total	15	280861			

Model Summary

S	R-sq	R-sq(adj)	R-sq(pred)
19.1850	98.95%	98.03%	95.81%

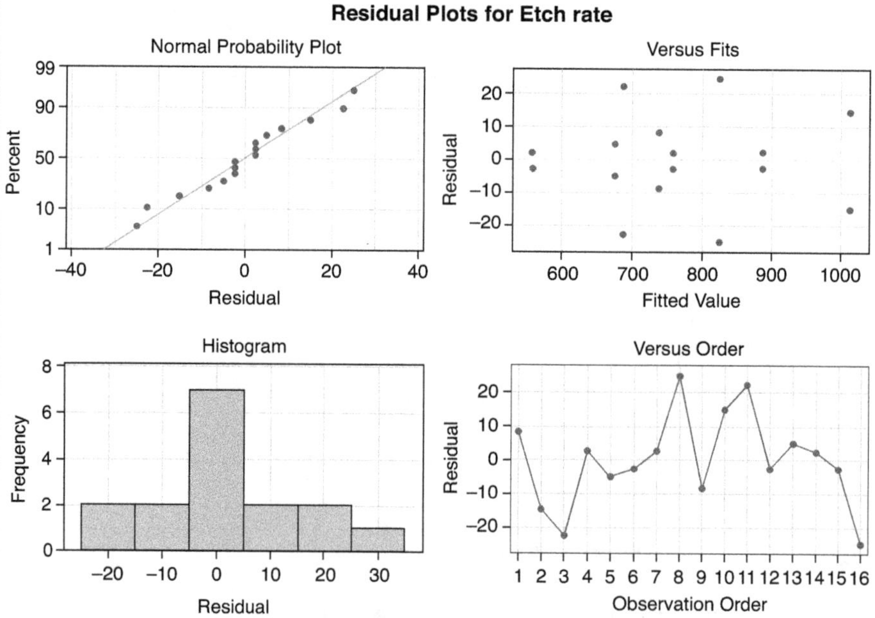

Figure 11.9 Minitab® computer output for etch rate experiment.

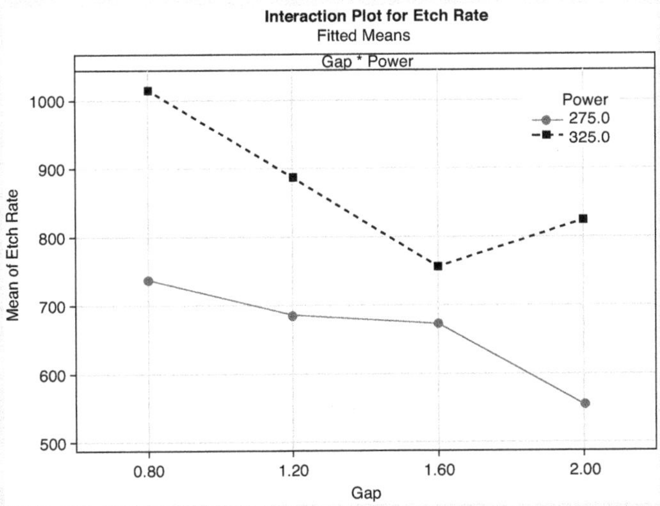

Figure 11.10 Interaction plot for etch rate example.

The engineer would like to achieve a target etch rate equal to 750. Upon examining the interaction plot, she concludes that setting Gap = 1.6 and Power = 325 results in a mean etch rate of 757.50. The confidence and prediction intervals for etch rate at these settings are as shown in Table 11.16.

Table 11.16 Confidence and prediction intervals for etch rate.

Variable	Setting			
Gap	1.6			
Power	325			
Response	**Fit**	**SE Fit**	**95% CI**	**95% PI**
Etch Rate	757.5	13.6	(726.2, 788.8)	(703.3, 811.7)

2^k Full-Factorial Designs

A particularly useful class of full-factorial designs is one in which each of the k factors is varied at two levels. These are referred to as *2^k designs*. The 2^k designs are centered and scaled, meaning each factor is coded using −1 for the low level and +1 for the high level. When factors are scaled, their relative effects on the response variable can be directly compared regardless of the underlying scale of the factor.

Two-level factorial designs are also *orthogonal*, and as a consequence, factor effects are independent. For example, removing an orthogonal factor from the analysis does not change the magnitude or sign of the effects of the other factors that remain in the model.

EXAMPLE 11.7

Here we present the analysis of a 2^2 full-factorial experiment with two replicates. Such an analysis is typically performed by statistical software, but we will hand calculate the effects and the ANOVA table using the formulas in Table 11.17.

A quality engineer is interested in learning how the temperature and pressure settings on a sealing machine affect the burst strength of a paper package. She designs a 2^2 experiment with two replicates and performs the experimental runs in random order. The low and high settings for temperature are 250°F and 300°F, and 35 psi and 45 psi for pressure. Burst strength will be measured in kPa. The experimental results have been tabulated using the standard run order in Table 11.17, with the randomized run order also indicated.

Table 11.17 Burst strength experiment results.

Standard order	Randomized run order	Temperature	Pressure	Burst strength, kPa
1	7	250	35	54
2	4	300	35	60
3	2	250	45	60
4	6	300	45	74
5	5	250	35	58
6	1	300	35	54
7	8	250	45	60
8	3	300	45	73

The average responses for each run are shown in what Minitab® labels as a "cube plot" shown in Figure 11.11.

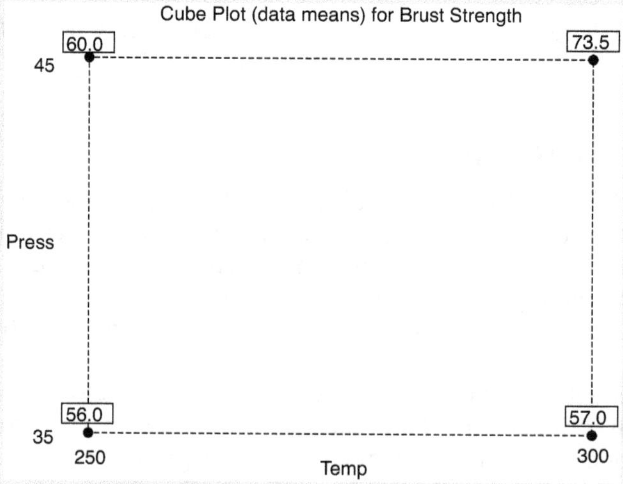

Figure 11.11 "Cube plot" for burst strength experiment results.

The manual ANOVA table calculations were performed using the formulas in Table 11.20 with $a=b=n=2$. The resulting ANOVA table is shown in Figure 11.12. Based on the p-values, we see that there is a significant interaction between temperature and pressure, and that the two main effects are significant as well.

	Pressure		
Temperature	**35**	**45**	**Total**
250	54	60	232
250	58	60	
300	60	74	261
300	54	73	
Total	226	267	493

$$SST = \left(54^2 + 60^2 + \cdots + 73^2\right) - \frac{493^2}{2\times2\times2} = 419.88$$

$$SS_{Temp} = \frac{1}{2\times2}\left(232^2 + 261^2\right) - \frac{493^2}{2\times2\times2} = 105.13$$

$$SS_{Press} = \frac{1}{2\times2}\left(226^2 + 267^2\right) - \frac{493^2}{2\times2\times2} = 210.13$$

$$SS_{Temp\times Press} = \frac{1}{2}\left((54+58)^2 + (60+54)^2 + (60+60)^2 + (74+73)^2\right) - \frac{493^2}{2\times2\times2} - 105.13 - 210.13 = 78.13$$

Analysis of Variance					
Source	**DF**	**Adj SS**	**Adj MS**	***F*-Value**	***P*-Value**
Model	3	393.38	131.125	19.79	0.007
Temp	1	105.13	105.125	15.87	0.016
Press	1	210.13	210.125	31.72	0.005
Temp*Press	1	78.13	78.125	11.79	0.026
Error	4	26.5	6.625		
Total	7	419.88			

Figure 11.12 ANOVA table for burst strength experiment.

The coded coefficients table calculated by Minitab® statistical software as well as the manual calculations are shown in Figure 11.13. We can use the coded effects and coefficients table to determine which factor has the most influence on burst strength. For example, the pressure effect tells us that moving

Continued

the pressure settings from the low to the high level increases the burst strength by an average of 10.25 kPa. The *p*-value column indicates that the interaction between temperature and pressure is significant, as are the temperature and pressure main effects. This result was also shown in the previous ANOVA table. The prediction equation in the original, uncoded units is also given. Note that, based on the hierarchical model convention, we would include the temperature and pressure main effects in the model even if they were not significant because the temperature × pressure interaction is significant.

Coded Coefficients

Term	Effect	Coef	SE Coef	T-Value	P-Value
Constant		61.625	0.91	67.72	0
Temp	7.25	3.625	0.91	3.98	0.016
Press	10.25	5.125	0.91	5.63	0.005
Temp*Press	6.25	3.125	0.91	3.43	0.026

Manual Calculations

$$\text{Effect Temp} = \frac{(60+54+74+73)-(54+58+60+60)}{4} = 7.25$$

$$\text{Coeff Temp} = \frac{7.25}{2} = 3.625$$

$$\text{Effect Press} = \frac{(60+60+74+73)-(54+58+60+54)}{4} = 10.25$$

$$\text{Coeff Temp} = \frac{10.25}{2} = 5.125$$

$$\text{Effect Temp*Press} = \frac{(74+73+54+58)-(60+60+60+54)}{4} = 6.25$$

$$\text{Coeff Temp*Press} = \frac{6.25}{2} = 3.125$$

Regression Equation in Uncoded Units

Burst Strength = 255.7 − 0.855 Temp − 5.85 Press + 0.02500 Temp × Press

Figure 11.13 Calculations of effects and coefficients for burst strength example.

We can check the assumptions behind the ANOVA model: Residuals are independent, identically normally distributed with mean zero and constant variance as shown graphically in Figure 11.14. The normal probability plot of residuals indicates that there is not a significant departure from normality. The standardized residual plot versus the fits indicates that there are no outliers and no major concerns about heterogeneity of the variance. The histogram reflects the normal probability plot, and the residuals-versus-run-order plot does not indicate any autocorrelation in the errors. Hence, the independence assumption is met.

The main effects and interaction plots are shown in Figure 11.15. We see from the interaction plot that when pressure is set at the high level of 45 psi, increasing temperature from 250°F to 300°F has a significant positive impact on the burst strength. On the other hand, when the pressure is set to 35 psi, increasing temperature has minimal effect on the burst strength.

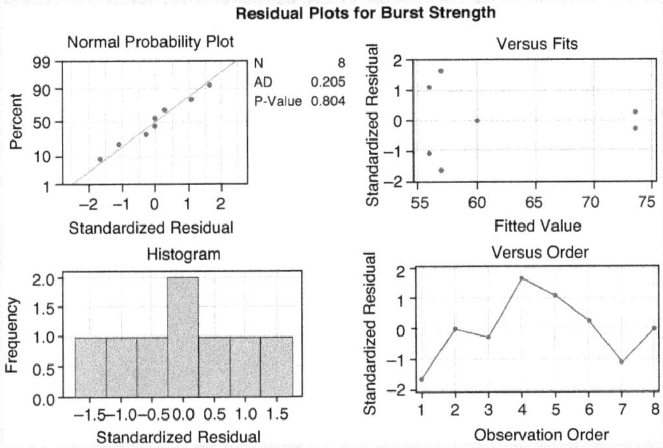

Figure 11.14 Graphical assumption checks for burst strength example.

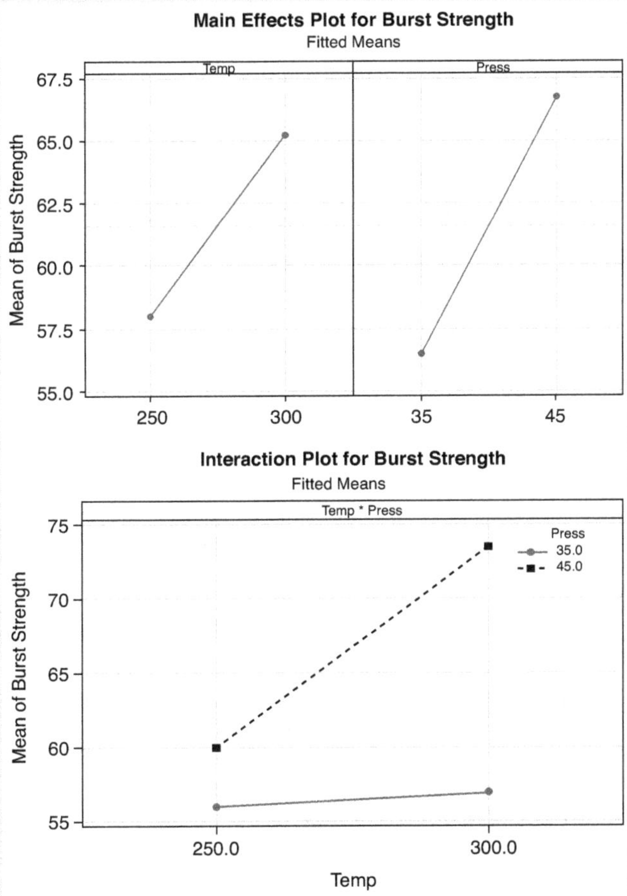

Figure 11.15 Main effects and interaction plots for burst strength experiment.

Continued

The estimated burst strength across the design space is shown in the contour plot in Figure 11.16. Because there is a significant interaction term, the contour plot has a curved gradient. Because we are fitting a model with no quadratic term, the minimum and maximum values for the estimated burst strength will occur at a design, or corner, point. If we wanted to maximize burst strength, for example, then we can easily see that setting temperature to 300°F and pressure at 45 psi would result in the maximum burst strength.

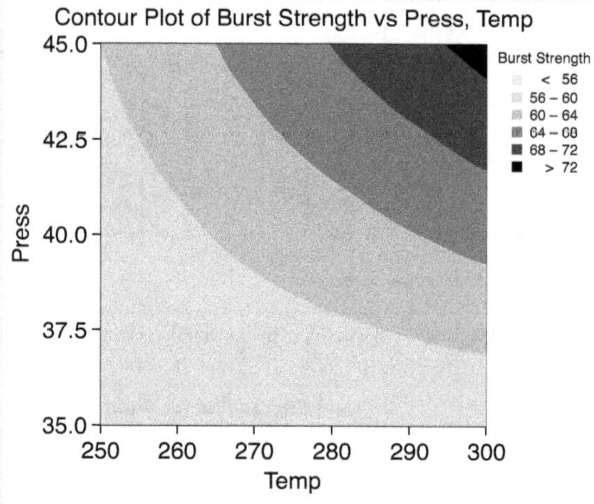

Figure 11.16 Contour plot for burst strength experiment.

However, the experimental results offer much more information beyond "picking the winning design run." The analysis produces an estimated regression equation that allows us to predict burst strengths anywhere inside a circle that circumscribes the design, as shown in Figure 11.17.

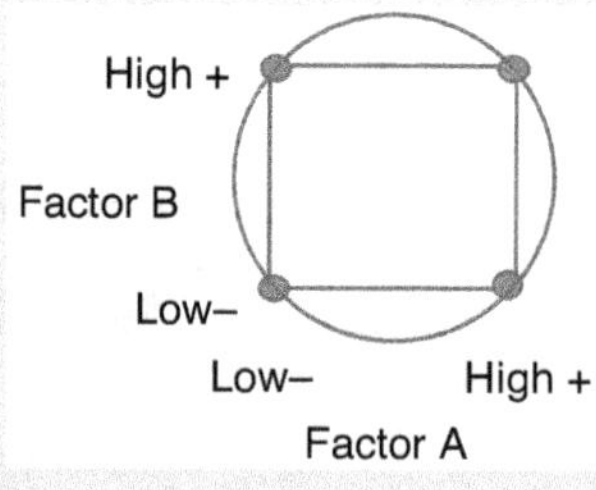

Figure 11.17 Prediction area for the regression equation for burst strength example.

For example, if the required burst strength for individual packages is greater than or equal to 59 kPa, the engineer may not need to run the machine at 300°F to satisfy the specification. Instead, she can use the equation in Figure 11.15 to determine if running the machine at, say, 275°F would meet the standard. The estimated burst strength at a temperature setting of 275°F and 45 psi results in:

$$255.7 - 0.855 \times 275 - 5.85 \times 45 + 0.025 \times 275 \times 45 = 66.75\ \text{kPa}$$

The one-sided prediction interval, shown in Table 11.18 for the individual burst strength at 275°F and 45 psi, has a lower bound equal to 60.6152, which satisfies the requirement.

Table 11.18 One-sided confidence and prediction intervals for burst strength experiment.

Temp	Press	Fit	SE fit	95% lower confidence bound	95% lower prediction bound
275	45	66.75	1.28695	64.0064	60.6152

Two-Level Fractional Factorial Experiments

A *factorial design* is an experimental design consisting of all possible combinations of the factor levels of two or more factors, with each factor being studied at two or more levels. When all combinations are run, the interaction effects as well as the main effects can be estimated. By contrast, a *fractional factorial design* is an experimental design consisting of a subset (fraction) of the factorial design. Typically, the fraction is a simple proportion of the full set of possible treatment combinations. For example, half-fractions, quarter-fractions, and so forth are common. While fractional factorial designs require fewer runs, some degree of confounding occurs.

Full-factorial experiments might require many runs, especially if several factors or several levels are involved. Recall that the minimum number of runs in a full-factorial experiment in which all the factors have the same number of levels is:

$$n = 2^k$$

where

n = Number of runs

L = Number of levels

k = Number of factors

For example, an experiment with eight two-level factors has a minimum of $2^8 = 256$ runs, and an experiment with five three-level factors has a minimum of $3^5 = 243$ runs. Full-factorial experiments can require a significant number of runs, which can be very costly and time-consuming.

Furthermore, if the experiment tests the effect of various factors on product quality in a manufacturing process, the tests typically must be run sequentially rather than simultaneously, so a large full-factorial experiment with several factors or levels might require that a piece of production equipment be taken out of production for a considerable amount of time. Fractional factorial experimental designs were developed because of the extensive resource requirements of full-factorial experiments. Typically, only one replicate of a fractional factorial experiment is run.

For the $2^8 = 256$ run experiment, we can choose to run only one-eighth of the treatment combinations. These treatment combinations are chosen using an algorithm

Table 11.19 Fractional factorial notation for two-level experiments.

Fraction	Fractional-factorial notation
1/2	2^{k-1}
1/4	2^{k-2}
1/8	2^{k-3}
1/16	2^{k-4}

that minimizes the confounding structure of the design. Noting that one-eighth equals $\frac{1}{2^3}$, the resulting fractional factorial experiment would thus have 32 runs:

$$\frac{2^8}{2^3} = 2^{8-3} = 2^5$$

A one-eighth fraction of a two-level experiment is denoted as 2^{k-3}. Table 11.19 lists the notation for other two-level fractional factorial experiments.

Confounding

Confounding means that the effects of factors cannot be distinguished from one another. Confounding occurs when the column of levels in the model matrix that is used to estimate an effect is the same as the column to estimate another effect. In other words, these effects are all *aliases* for the same analysis column. Confounding occurs with fractional factorial designs because such designs do not include every possible combination of factor levels. However, some of the effects of confounding are mitigated because higher-order interactions are generally assumed to not be statistically significant. Domain expertise is valuable in assessing the likelihood of higher-order effects. We can choose which effects are confounded to some extent to permit the estimation of these higher-order terms, but this choice often requires knowledge that is not available when designing the experiment.

Resolution

Resolution in the context of experimental design refers to the level of confounding in a fractional factorial design. Resolution is important when choosing a fractional factorial design. The amount and kind of information available from the experiment must be weighed against the number of runs. Higher resolution yields more information and is generally achieved by using a larger design, but it is less economical. Black Belts must consider this information/cost tradeoff when planning an experiment.

The resolution of a design is generally one more than the smallest order of any term with which a main effect is confounded. It is also the smallest sum of the order of the effects on both sides of the confounding equations. For example, if the first order term for factor A in a three-factor design is confounded with the second order interaction effect $B \times C$, then the confounding equation is $A = B \times C$, and the resolution is $1 + 2 = \text{III}$. Resolution is numbered with Roman numerals. Common resolution designs follow:

- Resolution III
  - —No main effects are confounded with another main effect.
  - —One or more main effects are confounded with two-factor interactions.
  - —Some two-factor interactions may be confounded with other two-factor interactions.
- Resolution IV
  - —No main effects are confounded with another main effect.
  - —No main effects are confounded with two-factor interactions.
  - —Main effects are confounded with three-factor interactions.
  - —Some two-factor interactions are confounded with other two-factor interactions.
- Resolution V
  - —No main effects are confounded with another main effect.
  - —No main effects are confounded with two-factor interactions.
  - —No main effects are confounded with three-factor interactions.
  - —Main effects are confounded with four-factor interactions.
  - —No two-factor interactions are confounded with other two-factor interactions.
  - —Two-factor interactions may be confounded with three-factor interactions.

Higher levels of resolution can also be achieved depending on the specific fractional factorial design. However, the matter of resolution is based on addressing the issue of confounding. Confounded effects cannot be simultaneously estimated. The terms with aliases may not appear together in the same model. If both effects are active, then leaving one of them out of the model will lead to biased estimates for the term that remains in the model.

Some questions an experimenter must ask when choosing a fractional factorial design regarding confounding include:

- Which terms, if any, are confounded?
- Can I live with the confounding? Is the confounding in higher-order interactions that can be assumed to be negligible?
- Does the confounding exist in lower-order interactions that have an impact?
- Can I remove or minimize the confounding through a carefully designed fractional factorial? This change requires prior knowledge that might not be available.

Thoughtful consideration along with subject matter expertise is required before designing fractional factorial designs.

EXAMPLE 11.8

Consider the 2^{4-1} fractional factorial design and data given in Table 11.20. This is an eight-run, resolution IV design in which main effects are confounded with three-factor interactions, and two-factor interactions are confounded with other two-factor interactions. Analyze this design to determine which factors are significant. Assume $\alpha = 0.05$.

Table 11.20 A one-half 2^{4-1} fractional factorial design for Example 11.8.

Std Order	Run #	Temp	Pressure	Time	Matl Type	Y
1	5	−1	−1	−1	−1	10
2	3	+1	−1	−1	+1	83
3	1	−1	+1	−1	+1	45
4	6	+1	+1	−1	−1	205
5	8	−1	−1	+1	+1	12
6	4	+1	−1	+1	−1	57
7	7	−1	+1	+1	−1	65
8	2	+1	+1	+1	+1	187

The following results illustrate the analysis of this design using Design-Expert® software (Stat-Ease, Inc.). Table 11.21 lays out the alias structure from the design evaluation in which A = Temperature, B = Pressure, C = Time, and D = Material Type. All main effects (MEs) are aliased with a three-factor interaction (3fi); for instance, A = BCD. The two-factor interactions (2fi's) are aliased with each other; for instance, AB = CD.

Figure 11.18 shows the half-normal plot with three selected effects labeled—the three largest one (those falling off to the right). Per general DOE practice,

Table 11.21 Alias structure for Example 11.8.

Intercept	= Intercept
A	= A + BCD
B	= B + ACD
C	= C + ABD
D	= D + ABC
AB	= AB + CD
AC	= AC + BD
AD	= AD + BC

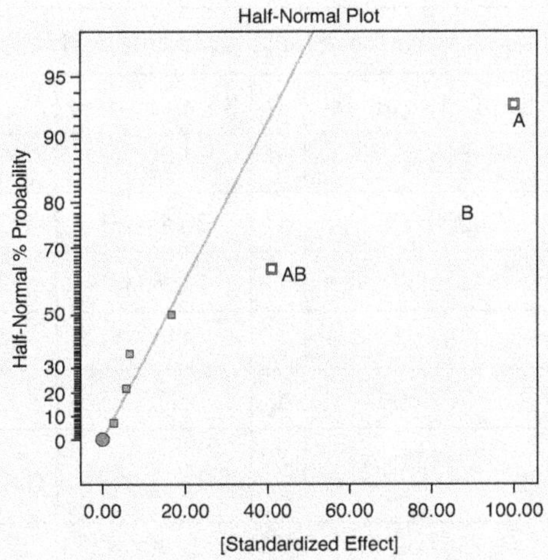

Figure 11.18 Half-normal plot of effects for Example 11.8.

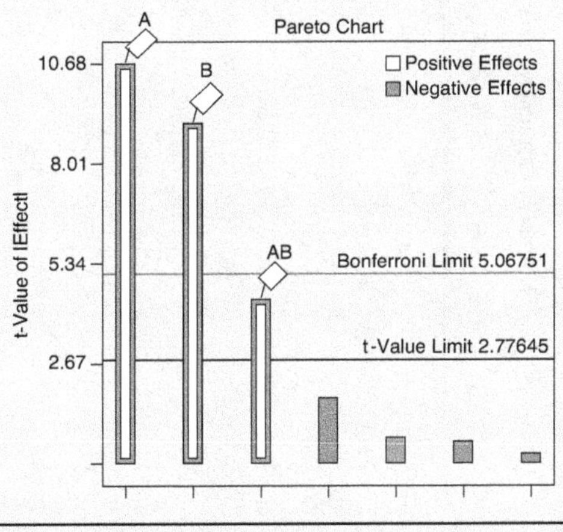

Figure 11.19 Pareto plot of effects for Example 11.8.

the aliasing of the chosen MEs (A and B) is disregarded due to 3FIs (BCD and ABD, respectively) being very unlikely, that is, negligible. Given the emergence of A and B, the AB interaction is chosen over its alias CD per the principle of effect heredity—AB being the "child" of "parents" A and B. The other (unlabeled) effects (C, D, AC, AD, and their aliases) fall in a straight line emanating from the origin at zero effect—presumably due to them exhibiting normal variation, that is, not creating any significant change in the measured response. They will be used to estimate error.

Figure 11.19 displays the resulting Pareto plot of effects. All four of the unchosen (unlabeled) effects fall below the basic *t*-value limit (lower line). The two main effects (A = temperature and B = pressure) exceed the more conservative upper line set with a "Bonferroni" correction for multiple pairwise comparisons. The AB = (temperature × pressure) interaction falls between the two limit-lines but merits being chosen, at least on a provisional basis.

Table 11.22 Analysis of variance table for Example 11.8.

Source	Sum of squares	*df*	Mean square	*F*-value	*p*-value	
Model	37812.00	3	12604.00	71.82	0.0006	significant
Temp	20000.00	1	20000.00	113.96	0.0004	
Press	14450.00	1	14450.00	82.34	0.0008	
Temp × Press	3362.00	1	3362.00	19.16	0.0119	
Residual	702.00	4	175.50			
Corrected Total	38514.00	7				

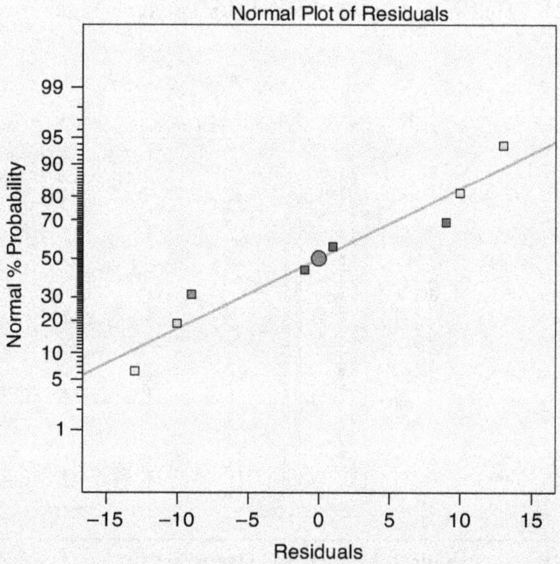

Figure 11.20 Normal plot of residuals for Example 11.8.

Table 11.22 provides the ANOVA for the selected factorial model. It comes out highly significant overall ($p < 0.05$). The individual model terms are also statistically significant, particularly the two main effects (A and B).

Various plots of residuals can be used to diagnose appreciable deviations from these assumptions. A small departure from normality such as that exhibited in Figure 11.20 does not create any cause for alarm—a fat pencil covers up all the points.

The standardized residuals versus run order (Figure 11.21) does not show any outliers.

In general, regression analysis can withstand some degree of violation of the underlying assumptions. However, when a large violation is evident, various response transformations, such as a natural log or square root, can be employed to restore normality and equalize variances. When such situations occur, consult a statistician.

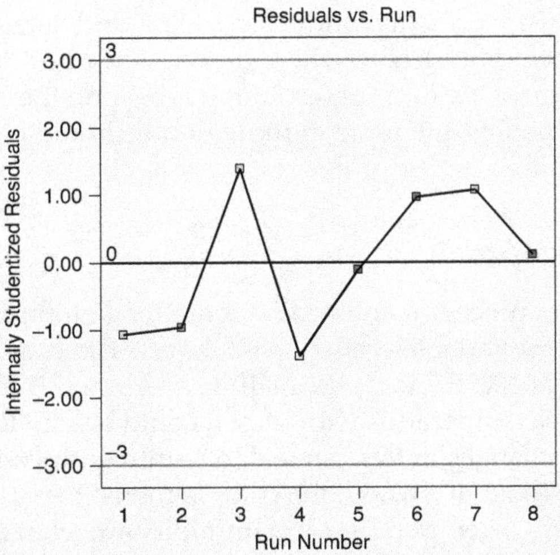

Figure 11.21 Residuals versus run order for Example 11.8.

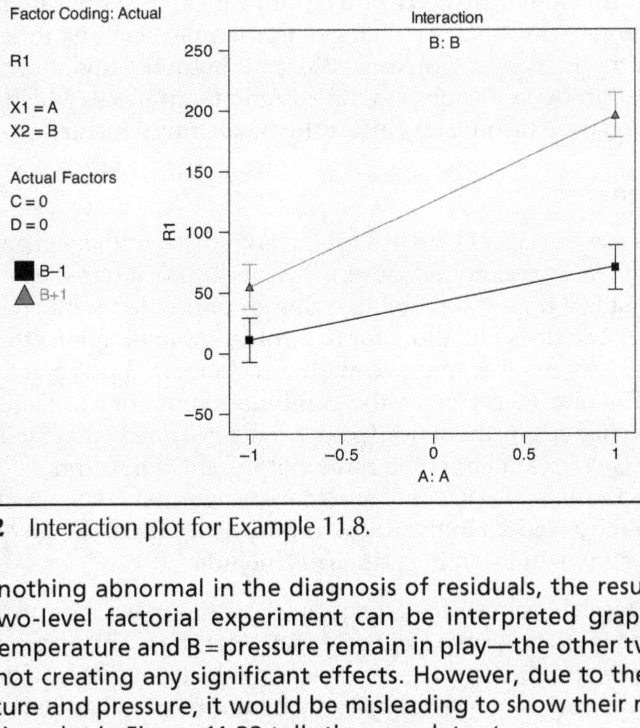

Figure 11.22 Interaction plot for Example 11.8.

Seeing nothing abnormal in the diagnosis of residuals, the results from this fractional two-level factorial experiment can be interpreted graphically. Only factors A = temperature and B = pressure remain in play—the other two factors (C and D) are not creating any significant effects. However, due to the interaction of temperature and pressure, it would be misleading to show their main effects. The interaction plot in Figure 11.22 tells the complete story.

In this case, the effect of A depends on the level of B. When pressure is at its low (minus) level, increasing temperature creates a significantly lesser increase in the response than it does with pressure set high (plus). Only by taking the multi-factor approach of DOE can interactions like this be revealed.

Experimental designs are frequently generated today using computer software. Such software presents candidate designs and detailed information, including resolution, to help you decide which design is the best for your situation. A more detailed coverage of the topic of resolution is beyond the scope of this book. Readers interested in pursuing the topic in more depth should consult Montgomery (2019) or Box et al. (2005).

Robustness Concepts

In robust design, processes and products are made robust, or insensitive to changes in hard-to-control variables called *noise factors*. For example, if brand A chocolate bar is very soft at 100°F and brittle at 40°F, and brand B maintains the same level of hardness at these temperature extremes, it could be said that brand B is more robust to temperature changes in this range. If a painting process produces the same color on moist wood as on dry wood, the color is robust to variation in moisture content. Here, the changes in temperature and humidity are referred to as *noise*. Products and processes that are robust to noise of various kinds are clearly desirable.

In a robust experimental design, these noise factors are varied in the experiment along with the traditionally used control factors. In the analysis, the levels of the control factors are chosen to obtain a mean response that is both close to the desired target and robust to changes in the noise factors, thus reducing the variability in the response. Unlike traditional methods, this robust design technique sets out to produce responses with different variances, which gives the engineer information on which factors affect the mean and which affect the variance.

Taguchi Designs

The Japanese engineer Genichi Taguchi is credited with incorporating noise factors into designed experiments using inner and outer arrays. The inner array of the experiment is a highly fractionated design populated with control factors. Generally, the design does not allow for control factor interaction effects to be estimated. This inner array is then crossed with a full factorial outer array containing noise factors. The aim is to choose the combination of control factors that is the most robust to changes in the noise factors. This is done by calculating signal-to-noise ratios for each treatment combination of the control factors.

Since the late 1980s, the Taguchi experimental design techniques have been scrutinized by leaders in the industrial statistics field; see Box (1988) and Pignatiello et al. (1991). In brief, the criticisms include:

- Taguchi robust design approach uses one large crossed-array experiment to determine optimal factor settings. This practice is in contrast to the widely accepted approach in which several smaller experiments are conducted sequentially as more and more is learned about the factors.
- While all the control-by-noise interaction effects are estimable in a Taguchi design, there are very few degrees of freedom left to estimate control factor interactions.
- There is an emphasis on a pick-the-winner optimization over development of a mathematical model that predicts responses through the design space. If an optimal setting cannot be implemented in production, the mathematical model can be used to choose an alternative.

- Maximizing S/N ratio minimizes the loss function only if the underlying model has multiplicative errors.
- The S/N ratio can obscure the effects of control factors.

Taguchi's *signal-to-noise ratio* (S/N) can be calculated from the experimental results. An analyst then chooses the control factor combination that results in the highest S/N ratio. Note that the S/N ratio is an attempt to find a useful compromise between two competing goals, optimizing y-bar and minimizing the variability in the response due to the noise factors, s. It does not necessarily accomplish either of these goals, so it should be used with a bit of judgment.

There are three basic types of S/N ratios that are dependent upon the goal of the experiment.

1. *Larger is better.* If it is desirable to maximize the response y, the S/N ratio may be calculated for each run, or combination of control factors, i, using the formula:

$$S/N_i = -10\log\left(\frac{1}{n_j}\sum_{j=1}^{n_j}\frac{1}{y_{ij}^2}\right)$$

2. *Smaller is better.* If it is desirable to minimize the response y, the S/N ratio can be calculated using:

$$S/N_i = -10\log\left(\frac{1}{n_j}\sum_{j=1}^{n_j} y_{ij}^2\right)$$

3. *Nominal is best.* If it is desirable to make the response y as close to a target value T as possible, then we calculate the S/N using:

$$S/N_i = 10\log\left(\frac{\bar{y}_i^2}{s_i^2}\right)$$

The Taguchi method uses an *inner and outer array* configuration to study control and noise factors. The inner array is a design for the control factors only. This array is then crossed with an outer array, which is an experiment that varies the noise factors. Generally, the experimental design of the inner array for the control factors is an *orthogonal array*, and the outer array for the noise factors is a full factorial experiment.

The Taguchi orthogonal arrays are denoted by an "L," meaning linear, and a number that indicates the number of runs required. These designs are highly fractionated. For example, an L4 design varies three factors at two levels each, and contains four runs, shown in Table 11.23. This array is comparable to a 2^{3-1} fractional factorial design, which is R_{III}, meaning that each main effect is aliased, or confounded, with an interaction effect. The L8 design can accommodate seven factors and is a 2^{7-4} design (a 16th fraction of a 2^7 design). This fractional design is also R_{III}.

While using a highly fractionated design reduces the number of runs in the inner array, once the array is crossed with the noise factor array, the number of experiments can grow quickly. For example, a design with three control factors and three noise factors, all at two levels, will require $4\times 2^3=32$ runs. Adding another noise factor increases the experiment to 64 runs.

Table 11.23 Taguchi L4 orthogonal array.

Run	a	b	c
1	1	1	1
2	1	2	2
3	2	1	2
4	2	2	1

EXAMPLE 11.9

A process engineer at a fabrication plant wishes to find the machine settings that will produce the best surface finish. Since the machine can be adjusted for feed rate, speed, and coolant temperature, these three factors are considered control factors. Using the Taguchi design method, three factors at two levels each can be run using an L4 array, shown earlier in Table 11.23.

In addition, the engineer has identified two uncontrolled (noise) factors that she suspects may be affecting the variability in surface finish. These two noise factors, hardness and ambient temperature, could conceivably be controlled by enforcing a tighter specification on the hardness of the steel and moving the machine into an environmental enclosure, but these options are cost prohibitive. Instead, two levels of hardness and ambient temperature are added to the experiment to determine settings of the controllable variables that will minimize variation in the output quality characteristic.

The two-pronged goal of the robust design, then, is to find the best combination of feed, speed, and coolant temperature that will simultaneously produce the best surface finish and be robust against changes in two uncontrolled (noise) variables, material hardness, and ambient temperature. These two noise factors will be run using a 2^2 full factorial experiment in the outer array. The experimental factors and their levels are shown in Table 11.24.

Table 11.24 Control and noise factors for the surface finish experiment.

Control factors	**Low**	**High**	**Units**
Feed	0.01	0.04	in/rev
Speed	1300	1800	rev/min
Coolant temp	100	140	C
Noise factors			
Material hardness	54	60	HRC
Ambient temperature	65	80	F

A Taguchi inner and outer array experiment is designed so that each set of control factor levels is run against all the combinations of the noise variables, as shown in Table 11.25. In total, there are 16 runs. When the first run of the design is executed, the feed, speed, and coolant temperature are all set at their low levels, denoted in the inner array section of the table as (1 1 1). One part is made with low-hardness steel and low ambient temperature, and the resulting surface finish of 34 is recorded. The last value in the table, 60, results from running feed and speed at their high levels and coolant temperature at the low level, and both hardness and ambient temperature at the high level. Once the results of the 16 experimental runs have been recorded, the averages and standard deviations for each run of the inner array are calculated, as shown in Table 11.25.

Table 11.25 Inner and outer arrays for the surface finish experiment.

Inner array			Outer array						
Feed	**Speed**	**Coolant**	**Hardness**	—	—	+	+	**y-bar**	*s*
		temp.	**Ambient temp.**	—	+	—	+		
1	1	1		34	37	45	25	35.3	8.3
1	2	2		35	37	33	34	34.8	1.7
2	1	2		25	33	40	21	29.8	8.5
2	2	1		58	54	65	60	59.3	4.6

We will now calculate the S/N ratios for each combination of the control factors using the *larger is better* formula since we want to maximize surface finish.

$$S/N_i = -10\log\left(\frac{1}{n_i}\sum_{j=1}^{n_i}\frac{1}{y_{ij}^2}\right)$$

$$S/N_1 = -10\log\left(\frac{1}{4}\left(\frac{1}{34^2}+\frac{1}{37^2}+\frac{1}{45^2}+\frac{1}{25^2}\right)\right) = 30.35$$

$$S/N_2 = -10\log\left(\frac{1}{4}\left(\frac{1}{35^2}+\frac{1}{37^2}+\frac{1}{33^2}+\frac{1}{34^2}\right)\right) = 30.80$$

$$S/N_3 = -10\log\left(\frac{1}{4}\left(\frac{1}{25^2}+\frac{1}{33^2}+\frac{1}{40^2}+\frac{1}{21^2}\right)\right) = 28.69$$

$$S/N_4 = -10\log\left(\frac{1}{4}\left(\frac{1}{58^2}+\frac{1}{54^2}+\frac{1}{65^2}+\frac{1}{60^2}\right)\right) = 35.40$$

$(S/N)_4$ is the largest signal-to-noise ratio, corresponding to the fourth combination of control factors at high levels of feed and speed, and low level of coolant temperature. The process is thus optimized by setting feed rate to 0.04 in/rev, speed to 1800 rev/min, and coolant temperature at 100°C.

Computer analysis of the experiment yields main effects plots for the mean, standard deviation, and S/N ratios that corroborate the manually calculated results. See Figure 11.23. It appears that speed (B) and coolant temperature (C) have the largest effects on the mean, and that speed (B) produces the largest effect on the variance.

Continued

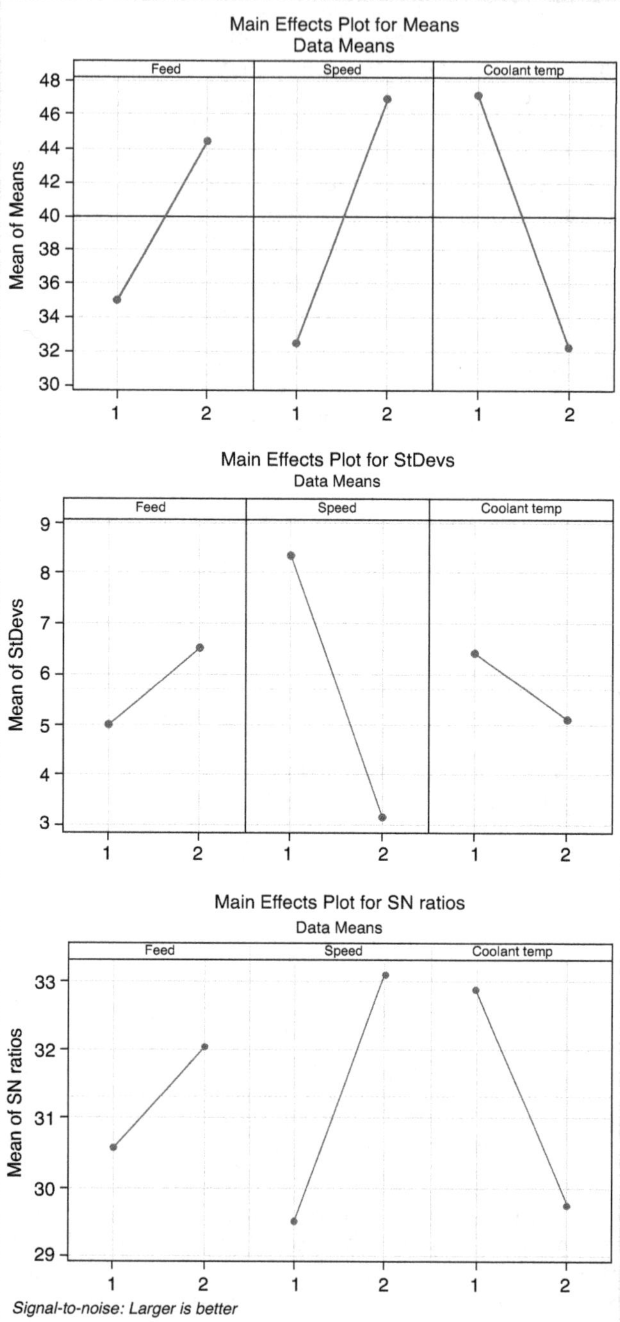

Figure 11.23 Main effects plots for the mean, standard deviation, and S/N ratio.

Combined Arrays and the Response Model Method

Combined arrays were first introduced by Welch et al. in 1990 to overcome the shortfalls of Taguchi designs. Using the combined array, both control and noise factors are varied in standard two-level factorial or fractional factorial design. Shoemaker et al. (1991) noted two advantages of this approach over the Taguchi crossed array. First, the factors can be studied using fewer experimental runs, and the experimenter can estimate control-by-control interactions in addition to control-by-noise interactions.

The combined array is analyzed using either the response model method or the dual response method. We now demonstrate the response model analysis method via an example.

EXAMPLE 11.10

Recall from Example 11.9 that a process engineer at a fabrication plant wishes to find the machine settings that will produce the best surface finish. Since the machine can be adjusted for feed rate, speed, and coolant temperature, these three factors are considered control factors. In addition, the engineer wants to find the best combination of the control factors that will simultaneously produce the best surface finish and be robust against changes in uncontrolled (noise) variables, material hardness, and ambient temperature.

Using the combined array method, the engineer runs a 2^{5-1} experiment that has 16 runs. This experiment is a Resolution V, meaning that all two-factor interactions are estimable. The design layout and results are shown in Table 11.26.

Table 11.26 Combined array design for surface finish experiment.

Run	Feed	Speed	Coolant temp	Matl hardness	Amb temp	Surface finish = y
1	0.01	1300	100	54	80	37
2	0.04	1300	100	54	65	58
3	0.01	1800	100	54	65	55
4	0.04	1800	100	54	80	54
5	0.01	1300	140	54	65	55
6	0.04	1300	140	54	80	33
7	0.01	1800	140	54	80	37
8	0.04	1800	140	54	65	32
9	0.01	1300	100	60	65	45
10	0.04	1300	100	60	80	72
11	0.01	1800	100	60	80	40
12	0.04	1800	100	60	65	65
13	0.01	1300	140	60	80	35
14	0.04	1300	140	60	65	40
15	0.01	1800	140	60	65	33
16	0.04	1800	140	60	80	49

Continued

The significance of the main effect and interactions between both the control factors, between the noise factors, and between the control and noise factors is investigated using a half normal plot (Figure 11.24). Effects falling sufficiently far away from the line are considered significant. We see that control factors A (Feed Rate) and C (Coolant Temp) and the AC interaction are significant. In addition, there is a significant interaction between A and D (material hardness). This result indicates that the feed rate control factor can be used to minimize the variability introduced by the material hardness noise factor.

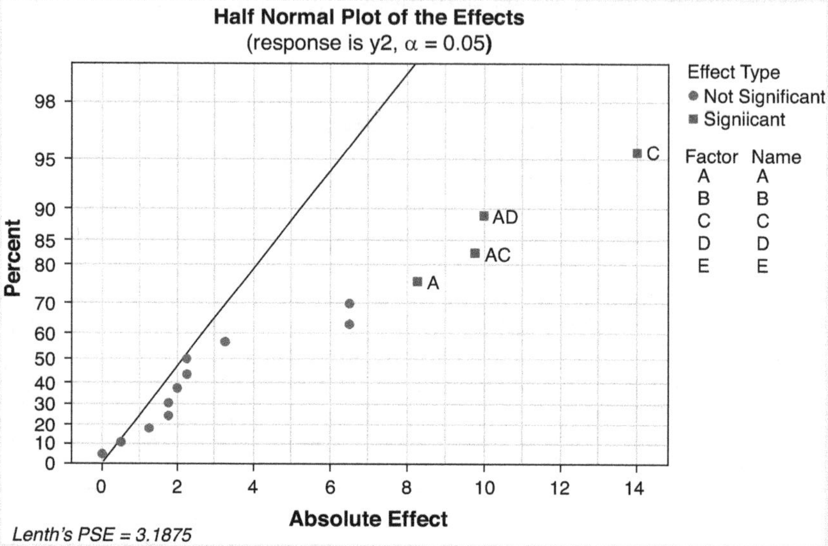

Figure 11.24 Half-normal plot of experimental effects.

We notice that Factor B (Speed) is not significant, which means that it can be set at whichever level is most convenient and economical. However, the Taguchi analysis of this experiment showed that Speed appeared to be the most influential (see Example 11.9). For an L4 inner array, Factor B is confounded with the AC interaction. Hence, the Taguchi analysis led to a spurious conclusion due to the limitations of using an R_{III} inner array.

Using the hierarchical modeling convention, the regression model for the combined array must include the components of any significant two-factor interactions. Therefore, the engineer fits a regression model with A, C, and D main effects and AC and AD interactions, as shown in Figure 11.25.

We see that factor A (feed) affects both the mean of the response as well as the variance, through its interaction with material hardness (D). Factor C (coolant temp) influences only the mean. We note that the response will be maximized if A is set at its high level, and C at its low level. However, setting Factor A to its high level will also increase the variance due to noise, which is not desirable.

We now calculate the expected value and variance of the response using the regression model. The noise factors material hardness (D) and ambient temperature (E) are treated as random variables, and we denote them as z_1 and z_2, respectively. We assume these noise variables are independent and are normally distributed with means equal to zero on the coded scale. We further assume that the low and high levels of each noise factor are mapped to ±1 standard deviation, thus making the variance for each noise factor equal to one on the coded scale.

Regression Equation in Coded Units

$y = 46.25 + 4.13\ A - 7.00\ C + 1.13\ D - 4.88\ A*C + 5.00\ A*D$

Analysis of Variance

Source	*DF*	Adj *SS*	Adj MS	*F*-Value	*P*-Value
Model	5	1856.75	371.35	8.28	0.003
Linear	3	1076.50	358.83	8.01	0.005
A	1	272.25	272.25	6.07	0.033
C	1	784.00	784.00	17.49	0.002
D	1	20.25	20.25	0.45	0.517
2-way interactions	2	780.25	390.13	8.70	0.006
A*C	1	380.25	380.25	8.48	0.015
A*D	1	400.00	400.00	8.92	0.014
Error	10	448.25	44.82		
Total	15	2305.00			

Figure 11.25 Combined array computer output.

We calculate the expected value and variance of the response from the fitted regression model using the *propagation of error technique*, as follows:

$$E(y) = E(46.25 + 4.13A - 7.00C + 1.13D - 4.88AC + 5.00A{\cdot}z_1)$$

Note that the noise factor z_1 (material hardness) does not influence the mean response since we assume that $z_1 \sim N(0,1)$.

$$E(y) = 46.25 + 4.13A - 7.00C - 4.88AC$$

The variance of y is the sum of the variance contributed by the noise variable z_1 and the error variance σ^2_e, also known as the MSE from the ANOVA table.

$$\text{Var}(y) = \text{Var}(46.25 + 4.13A - 7.00C + 1.13\ z_1 - 4.88AC + 5.00A{\cdot}z_1) + \sigma^2_e$$

The control factors are considered fixed and are treated as constants. The noise factor z_1 is a normally distributed random variable. Factoring the regression equation, and noting that $\text{Var}(z_1) = 1$, we calculate:

$$\text{Var}(y) = \text{Var}(z_1\ (1.13 + 5.00A)) + \text{MSE}$$

$$\text{Var}(y) = (1.13 + 5.00A)^2\ (1) + 44.82$$

$$\text{Var}(y) = (1.13^2 + 2\ (1.13)(5.00A) + 5.00^2\ A^2) + 44.82$$

Using the propagation of error formulas, we can calculate the mean and variance of the responses for each experimental run, as shown in Table 11.27.

Continued

Table 11.27 Mean and standard deviation for each run based on propagation of error method.

A = Feed	C = Coolant temp	*y*-bar	*s*
0.01	100	44.24	7.73
0.04	100	62.26	9.08
0.01	100	44.24	7.73
0.04	100	62.26	9.08
0.01	140	40.00	7.73
0.04	140	38.50	9.08
0.01	140	40.00	7.73
0.04	140	38.50	9.08
0.01	100	44.24	7.73
0.04	100	62.26	9.08
0.01	100	44.24	7.73
0.04	100	62.26	9.08
0.01	140	40.00	7.73
0.04	140	38.50	9.08
0.01	140	40.00	7.73
0.04	140	38.50	9.08

We then can use the optimization function in Minitab® to find the settings for feed and coolant temperature that will simultaneously maximize the mean surface finish and minimize the standard deviation. The results of the optimization are shown in Figure 11.26. The desirability function is set up to weigh the mean and standard deviation equally. At the settings A =−0.3131 and C =−1.0, the mean surface finish is 50.43 and the standard deviation is 8.19. These settings translate to A = Feed = 0.02 in/rev and C = Coolant temperature =100° C. The speed can be set to whatever value is most convenient.

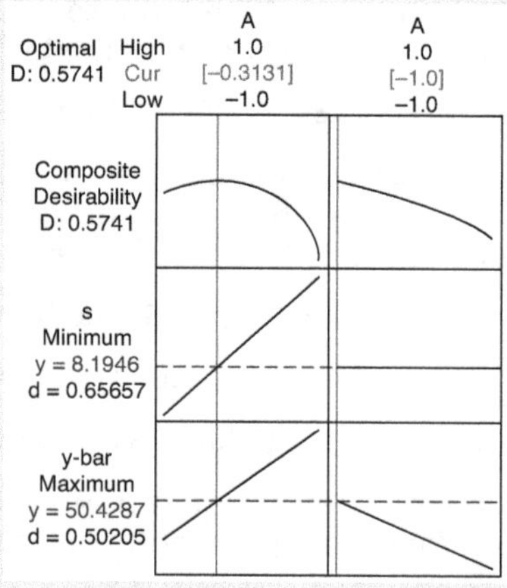

Figure 11.26 Optimization of the mean and standard deviation.

These results are quite different than the ones obtained using the Taguchi S/N method. Using the propagation of error formulas to predict the mean and standard deviation of the Taguchi recommended settings of (A = 1, B = 1, C = −1) yields a mean surface finish equal to 62.26, and a standard deviation equal to 9.08:

$$E(y) = 46.25 + 4\ .13A - 7.00C - 4.88AC$$

$$E(y) = 46.25 + 4.13\ (1) - 7.00\ (-1) - 4.88\ (1)\ (-1) = 62.26$$

$$Var\ (y) = (1.13^2 + 2\ (1.13)(5.00A) + 5.00^2\ A^2) + 44.82$$

$$Var\ (y) = (1.13^2 + 2\ (1.13)(5.00) + 5.00^2) + 44.82 = 82.40$$

$$s(y) = 9.08$$

What have we discovered using the combined array approach?

- While the fractional factorial experiment had the same number of runs as the Taguchi inner and outer array, we were able to glean much more information about the process.
- We were able to fit a regression model to the data to understand the relative effects of the input variables on the response.
- Using the regression model, we can predict the surface finish for any combination of factors within the design space, not just at the corner points.
- We discovered that speed was not a significant control factor, and that there was a significant interaction between feed rate and the noise factor material hardness.
- We determined that the ambient temperature did not interact with the control factors.

RELIABILITY OPTIMIZATION

Employ a diverse range of strategies to enhance reliability, considering user needs while balancing constraints such as cost, schedule, and weight along with other design requirements. (Apply)

Body of Knowledge V.A.4

The reliability engineer involved in the design process will frequently be doing some sort of balancing act. The standard ways to improve reliability, redundancy, derating, and environmental robustness tend to increase cost and weight, and delay the design process. The potential user of the product has an interest in both high reliability and favorable cost, weight and timing factors. The prioritization matrix discussed in Example 11.11 can aid in this balancing act.

EXAMPLE 11.11

A design team is charged with optimization of the motor/drive train for a commercial air handler. The reliability engineer is considering several options, which are listed in the left column of the matrix in Table 11.28. The other columns are self-explanatory, and additional columns may be necessary to cover additional constraints. Once the matrix has been constructed, the decisions become a little more quantitative. Here, the team might decide that the hardened motor can be sold as an alternative but not be included in the standard product. The additional motor might be eliminated from the product line.

Table 11.28 Prioritization matrix.

Option	Advantage	Cost ($)	Weight (kg)	Schedule delay (days)
Derate motor	Increase motor life Handle excess load	900	25	0
Derate drive belt	Extend maintenance schedule	100	2	0
Add second drive belt	Allow operation with failed belt	400	10	5
Harden motor	Allow use in caustic environment	2000	95	10
Add second motor	Allow use after motor failure	2000	300	15

Reliability Optimization Process

The reliability optimization process is a reliability tool that analyzes a system to help determine which components will meet user needs and performance goals at the lowest cost. The reliability optimization process comprises three steps including:

1. Develop a model that represents the entire system by constructing a system reliability block diagram that represents the reliability relationships of the components in the system.
2. Using this model, estimate the reliability and costs that would be incurred in the process of making modifications.
3. Then perform an optimization analysis to find the best combination of component reliability improvements that meet or exceed the performance goals at the lowest cost.

Example 11.12, consisting of Figure 11.27 and Figure 11.28, shows this process for a design represented by a series system.

EXAMPLE 11.12

Suppose we have the reliability block diagram shown in Figure 11.27 representing a three-step process required reliability = 0.90.

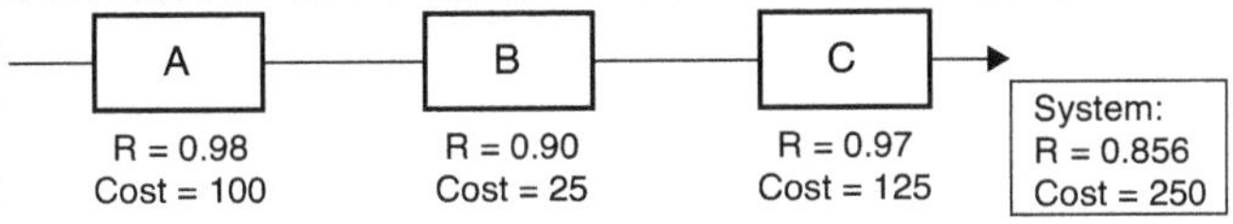

Figure 11. 27 Reliability block diagram for a three-step process.

What should be done to meet the reliability goal? What is the impact on cost? Noting that the required reliability is 0.90, one solution is shown in Figure 11.28.

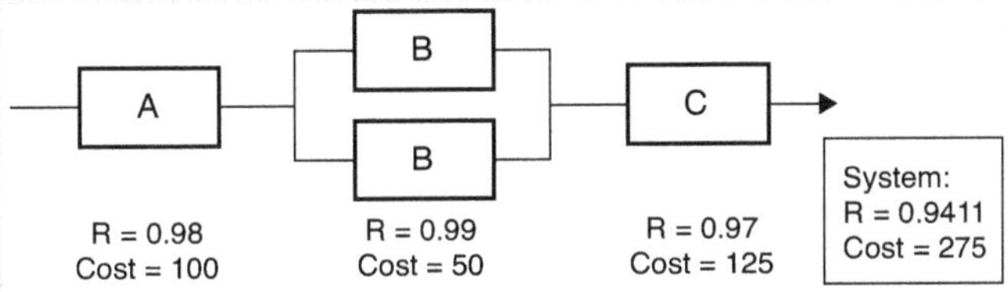

Figure 11.28 One possible solution to meet the reliability goal of 0.90.

HUMAN FACTORS

Describe the relationship between human factors and reliability engineering, considering user safety, understanding user and usage profiles, and analyzing associated failure modes and mechanisms. (Understand)

Body of Knowledge V.A.5

Humans play an important role in the overall reliability of engineering systems. Reliability of the human aspect can be improved by:

- Following human-factors principles in system design
- Careful selection and training of personnel

Stress affects human performance and human-factors related reliability, but not necessarily in a wholly negative way. Stress at a moderate level is useful in increasing human effectiveness to its optimal level. At very low stress levels, a task will be unchallenging and dull, and human performance will not be at its peak, as shown in Figure 11.29.

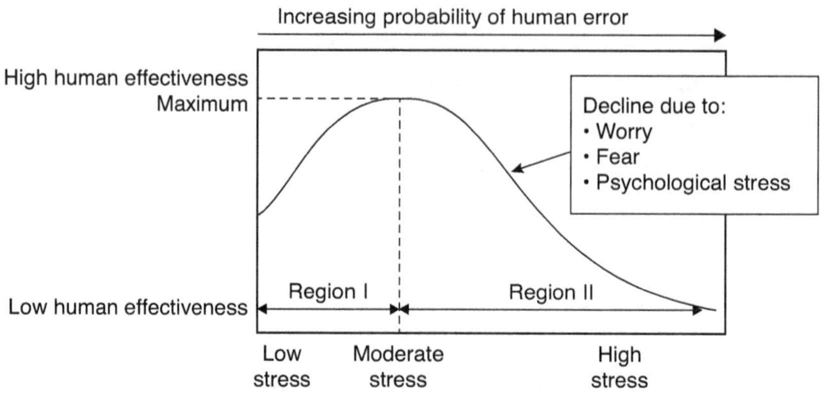

Figure 11.29 The effects of stress on the probability of human failure.

Four types of occupational stressors are:

1. Work overload or work underload
   a. Job overload occurs when requirements exceed individual ability
   b. Job underload occurs when the work performed fails to provide meaningful stimulation
2. Occupational change (e.g., organizational restructuring, promotion, or relocation)
3. Occupational frustration (e.g., job inhibits the meeting of goals because of lack of communication, role ambiguity or bureaucracy difficulties)
4. Other sources (e.g., noise, too little or too much lighting or poor interpersonal relationships)

Human Performance Reliability Modeling

The *human performance reliability* follows analogously from our work on hazard functions:

$$R_e(t) = exp\left[\int_0^t h_e(t)\mathrm{d}t\right]$$

where $R_e(t)$ is the human performance reliability at time t and $h_e(t)$ is the time-dependent human error rate.

This holds for a constant or nonconstant human error rate. In fact, when data has been collected experimentally from time-continuous tasks (observing clock-type displays of lights, where the subjects were asked to press a handheld switch when a failed light occurred), the Weibull distribution fits the data quite well (see Table 11.29 and Figure 11.30).

Table 11.29 Human error probability estimates for selected tasks.

Task description	Human error probability
Reading a chart recorder	0.006
Reading an analog meter	0.003
Reading graphs	0.010
Interpreting incorrectly the indication on the indicator lamps	0.001
Turning a control in the incorrect direction under high stress	0.500
Using the checklist correctly	0.500
Mating a connector	0.010
Choosing an incorrect panel control from a number of similar controls	0.003

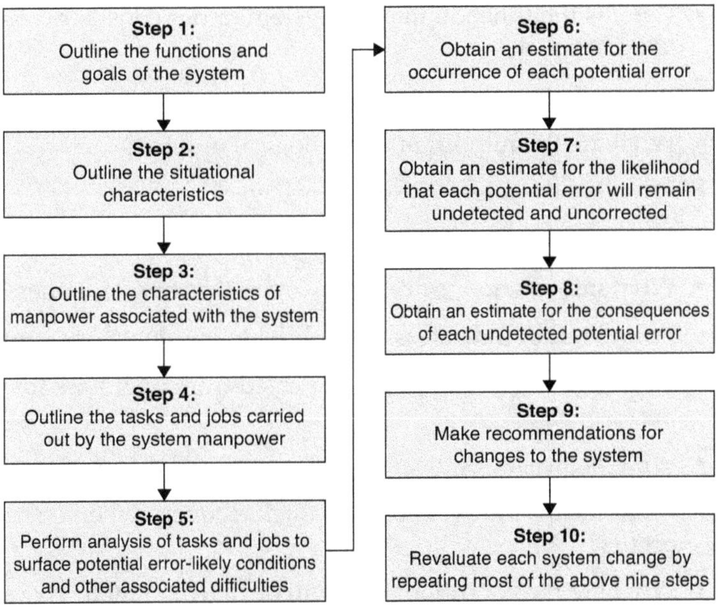

Figure 11.30 Human error prevention: Man–machine systems analysis.

Human Reliability Data Sources

Human reliability analysis (HRA) is the use of systems engineering and behavioral science methods to identify ways to reduce risk.

- Past experience
- Customers' failure-reporting system
- Product development data

- Repair reports
- Field installation testing
- Manufacturing and quality control
- Acceptance tests in the factory
- Field demonstration and environmental qualification tests

Human Factors Considerations in Various Stages of a System Development

There are several human factors to consider when developing a system, including the following:

1. Preliminary design phase
   - Who will be the users of the system under design?
   - What are the environments in which the system is expected to operate?
   - What is the function that the system under consideration is expected to accomplish?
   - Who will operate and maintain the system?
   - What are the important technological options?
   - What are the mission execution stages?
2. Advanced design phase
   - What information is needed by operators or maintainers?
   - What kind of indicators are needed to eliminate operator errors?
   - What type of special skill is expected from operators for effective performance?
   - What is the size of the manpower required during normal operation?
   - What means are available for quick recovery when operator error occurs?
   - What are the conditions that will be detrimental to operator performance?
   - What functions are to be assigned to operating personnel?
   - In what form should the information be presented to the operator?
3. Mock-up or prototype fabrication phase
   - What effect will any proposed changes in configuration have on safety, human performance, and reliability?
   - What quality control methods and procedures will fulfill the task of ensuring the validity of human-factor considerations in the resulting system?

- Are there any options for simplifying any of the instrumentation used in the product?

4. Test and evaluation phase

   - How can the testing and evaluation be made as realistic as possible with respect to operating personnel?
   - What criteria of equipment and operating personnel performance are logical with respect to functions and assigned mission?

System Safety

A typical system safety program has three key elements:

1. *Identification of safety hazards.* The reliability engineer working with the safety organization to achieve a safe product must be innovative and diligent in the discovery of all possible ways that any failure, combination of failures, or other combination of circumstances would present a safety hazard to personnel. Warranty data and other forms of customer feedback should be analyzed. Product testing reports should include safety issues that may have emerged. All available product data should be searched for the occurrence of safety hazards. In addition, if databases for similar products are available, they should also be studied.
2. *Risk analysis.* The standard analysis techniques are failure modes and effects analysis (FMEA); failure mode, effects, and criticality analysis (FMECA); production reliability acceptance test (PRAT); fault tree analysis (FTA); success tree analysis (STA); and failure reporting, analysis, and corrective action system (FRACAS).

   Monte Carlo simulation can be used for those instances where detailed failure times (by mode) are available or where there are competing modes of failure that will affect the component or system. The risk matrix is another approach to quantifying risk. These techniques can be used to estimate the risk associated with various events. It then becomes possible to establish a prioritized list that will provide guidance as the root causes are attacked and resolved.
3. *Correction and prevention.* In the case of safety hazards, there is the additional urgency to avoid harm to personnel. In cases in which flaws in products or processes permit hazardous conditions to occur, an engineering change request (ECR) should be initiated. It is always necessary to consider human error. From one perspective, human error occurs because no system is in place to prevent it. In other words, the onus is on the product/process design community to reduce or eliminate errors. Procedures for preventing/mitigating human error are variously called *mistake-proofing*, *poka-yoke*, and *zero quality control* (ZQC). Human error tends to fall into the following categories: misunderstanding, misidentification, inexperience, inattention, and lack of standards. The ideal way to deal with human mistakes is to incorporate design elements that will prevent them from occurring or prevent their occurrence from causing defects. Principal types of mistake-proofing techniques include:

- *Physical barriers to errors.* The round shaft won't fit through the square hole.
- *Visual reminders.* A photograph of correct and incorrect results is better than a note written on a print or a paragraph of text.
- *Use of automated equipment.* The conveyor will stop if the microswitch detects an error.
- *Standardizing.* The operator takes the same action on a family of parts.

Often, the best way to prevent a reoccurrence of human error is to approach a person who has made the error with the question, "How can we design a system that will make that error impossible to occur?" Activity must be continuous in the three elements of the system safety program. As the higher-priority hazards are corrected or prevented, action can focus on the lower-priority items.

Reliability Testing

Reliability testing must consider variation introduced by differences between people. These differences can be categorized as *use factors, installation factors,* and *process management factors.*

Use Factors

When conducting tests on a home laundry appliance, consideration must be given to the differences in the way people slam doors, choose washing detergent, maintain cleanliness, load the washer, and so on. Automotive components also experience a wide variety of operator habits. If the protocol calls for 100,000 door slams using one particular force, that testing does not account for all the variations that customers may employ.

Installation Factors

Products may be installed in a wide variety of environments. For building materials, automotive spare parts, stationary equipment, and many other products, the useful lifetime is impacted by the quality of the installation and the location parameters. For instance, if a window is installed slightly out of plumb, it may not be as reliable as a properly installed unit. This has implications for design and testing procedures because the window should be robust to minor out-of-plumbness. Knowledge in this area also will impact installation instructions.

Process Management Factors

If reliability testing is conducted on products that are produced by an ideal process, the variation introduced by people in the production process may not be taken into consideration. Operators or process managers can influence product characteristics by the way they hold the paint gun or welding stick, the promptness with which a tank is evacuated, or the delay between molding cycles, and so on.

Recommendations

From the testing viewpoint, it would be easier if all this human variation could be removed by specifying usage, installation, and processing parameters, and in fact, that is typically done. A balance must be struck, of course, between the narrowness of the specifications and the attendant loss in flexibility. If, for instance, an installation specification states that the ambient air must be between 65°F and 75°F, the testing will be simplified, but some customers will be lost to competitors, and others will ignore the specification and be disappointed with product performance. Therefore, the product design team should strive for robustness to the variation introduced by these human factors, keeping the specifications as broad as possible. This means that the reliability testing procedures must include tests throughout the specification spectrum.

DESIGN FOR X (DFX)

> Apply DFX techniques, such as design for manufacturability, testability, and maintainability. (Apply)
>
> **Body of Knowledge V.A.6**

Every design team has constraints within which they must function. In reliability engineering, Design for X is a methodology that defines the priorities or focus of the design team. These priorities are defined by the stakeholders, and a typical approach is to focus on a few areas and approach the others as normal constraints in the product design process. One organization may focus on design for safety, cost, and manufacturing and another may focus on design for test, serviceability, and reuse. The drivers are dependent upon the risk tolerances of the company and their customers.

The priority of each of these characteristics impacts the resulting design:

- *Design for assembly* refers to the conscious thought given to the assembly process by the product design team. Decisions by the team affect the simplicity and ease of the assembly process by giving consideration to visual and mechanical access to fasteners, special tooling or fixturing that may be required, safeguards against incorrect assembly, and so forth.

- *Design for manufacturability* is similar to design for assembly, but it tends to put more emphasis on the fabrication or primary functions that precede assembly. Here, consideration is given to narrowness of tolerances and difficult configurations such as deep dead-end holes, thin walls, and so on.

- *Design for testability* looks at testing procedures and tries to develop a design for which important characteristics are easily and accurately measured.

- *Design for cost* puts emphasis on the final cost of the product and makes this a strong constraint on product or process design.
- *Design for maintainability or serviceability* considers the ease and simplicity of installing replacement parts, as well as standard servicing requirements.
- *Design for reliability* emphasizes the long-term usefulness of the product to the customer.
- *Design for environment* considers the activity beyond the useful life of the product. While considering the various tradeoffs in the design process, it is becoming increasingly important to study issues of recycling, reclamation, and disposal. Issues to consider include contractual obligations, warranties, and regulatory and certification requirements.

Many of these constraints are applied to typical designs, and often the extent of the constraint is a matter of emphasis.

The maintainability of a system is influenced by the ability to detect and repair a system fault. Therefore, the testing requirements for a system should provide a method to detect a system fault and isolate the failed component. For a complex system, the requirement could be met with a built-in test system. This could add additional complexity to the system, and care must be taken to ensure the reliability of the test system.

The testability requirements of systems can also be met by providing easily accessible test points for critical measurements taken using external test equipment. It is important that a maintainability engineering effort is included in the design stage to ensure testability of the system. The testability of a system must be developed as part of the system design. At the system or subsystem level, the testability requirements should reflect the need to detect failure quickly and isolate the failure so that replacement or repair can be accomplished in an acceptable amount of time.

When designing for testability, it is important to consider the various conditions when testing is required, such as during new product development, conducting safety analysis or fatigue analysis, and other times, depending on the product. *Fault detection capability* is a measure of the faults detected by the fault detection system (usually built-in test) compared with the total number of system faults. *Fault isolation capability* is a measure of the percentage of time the failure can be isolated to a given number of replaceable (repairable) components. Fault isolation can be accomplished by a diagnostic analysis, built-in test, or using external test equipment. *False-alarm rate* is a measure of the rate at which the system declares the detection of a failure when no failure has occurred. A *fault recording and analysis system* that identifies specific components will aid in improving reliability. Particular attention should be paid to the possibility of dependence among fault types. The discovery of a cascading pattern among faults permits work on the root cause rather than on symptoms.

When applying multiple Design for X strategies, it is important to keep in mind that optimizing for one characteristic can sub-optimize for another. Consider that the overall optimal design could be one that identifies the minimum requirement in each of the characteristics and simultaneously satisfies all those minimum requirements rather than a design that optimizes one characteristic alone.

DESIGN FOR RELIABILITY (DFR)

> Apply DFR to meet reliability requirements throughout the product or system lifecycle. Integrate tools such as finite element analysis (FEA) to assess and optimize design robustness and to forecast potential areas of failure or stress within the system or product. Make design decisions emphasizing the achievement of built-in reliability and fault tolerance/avoidance as fundamental goals. (Analyze)
>
> **Body of Knowledge V.A.7**

Design for reliability (DFR) is a concurrent engineering approach that begins with the voice of the customer to ensure that a product meets customers' needs over the product's total lifecycle. It comprises linking multiple reliability engineering methods with the design process to ensure robustness and quality of the product or service. Once customer needs are understood, the critical next step is to specify reliability goals consistent with customer expectations. In DFR, the reliability engineer is part of the product development team working to design reliable products with low overall lifecycle costs. A high-level flow of designing for reliability is shown in Figure 11.29.

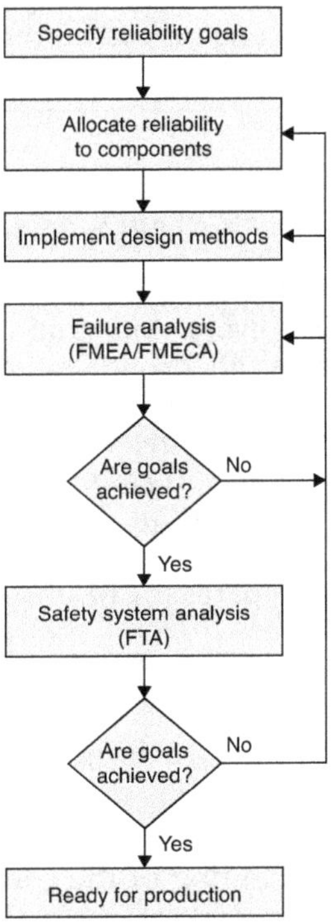

Figure 11.31 High-level flow of DFR.

Considerations in Design for Reliability

Fault tolerance refers to the ability of a system to perform according to specifications even when undesired changes occur in its internal or external environment. The reliability engineer's task here is to use knowledge of system design to prevent a fault from causing a failure. Single-fault tolerance can be affected by a system design that will permit a component (hardware or software) to be removed and replaced without causing system failure. Multiple-fault tolerant systems are designed to continue operating in the presence of more than one concurrent failure. Various techniques have been employed to produce fault-tolerant systems.

Redundancy is a standard technique for mitigation of system failure. The theory is that if two or more parallel channels are available for the system, the likelihood of failure is reduced. Redundancy alone is often insufficient because the common cause variation may overwhelm all branches of the system, for example, improper maintenance of all branches of a parallel system, use of a process material that corrodes all redundant branches, and computer backup systems using identical disk drives in an identical environment (which often results in identical failures). Examples 11.13 and 11.14 describe more specific instances.

EXAMPLE 11.13

A catastrophic failure caused United Airlines Flight 232 to attempt a landing at Sioux City in 1989. The DC-10 was designed with three independent hydraulic systems so that if one or two failed, the surviving system could manage flight control surfaces. Unfortunately, all three lines were at one point in close proximity. Shrapnel from a damaged engine passed through that point and severed all three lines, causing loss of control.

EXAMPLE 11.14

The use of parallel CPUs in computers performing critical tasks is common. Some systems are set up to process every command by two different processors and compare the results. If the results do not match, an error flag is set, and the system is put in a safe mode and maintenance is performed. A problem here is that if the software is defective, the two results may agree, but both could be wrong. The common cause in this case is the software bug. It caused errors in both branches of the parallel design. A variation in this approach is to have simultaneous calculations performed by several processors. When the results are compared, a *voter* module compares the results and goes with the one produced by the majority of the processors.

Diversity and separation are two techniques that have been used to make redundancy more effective. *Diversity* is employed to construct parallel channels that are different in ways that will make them immune to failure by the same cause. This implies an understanding of the common causes—an understanding that will only come from considerable study. In the case of the parallel CPUs mentioned previously, a different piece of software could be employed to make the calculation for one of the CPUs. *Separation* can apply to branches that have separate physical environments, electronic isolation, or other means of preventing a common cause from causing multiple-branch failure. In the case of the DC-10 accident mentioned earlier, the hydraulic lines were not adequately separated.

Reducing variation in design is essential to success and applies not only to physical tolerances but to all portions of the product and process (that is, using robust design principles throughout the design process to ensure all portions operate as closely as possible to nominal). Taguchi and combined array design of experiments methods and Monte Carlo simulation are powerful tools used in parameter design to reduce design variation. Refer to "Design of Experiments (DOE)" for more information.

Finite element analysis (FEA) can be applied to model the performance of a design, helping to make design choices to improve reliability. According to Musa, Kulkarni, and Havanur (2013), FEA is a "numerical procedure that applies to many areas in real-world engineering problems, including structural/stress analysis, fluid flow analysis, heat transfer analysis and electromagnetics analysis." These authors go on to explain that FEA is a method for design and modeling of a physical event. The physical components are replaced in the model with a simplified structure made up of a finite number of "elements" connected at common points called "nodes." Each element has an assumed response to applied loads and the impact of stress, fluid flow, heat transfer, or other phenomena on the nodes can be determined.

Finite element analysis can be used in reliability engineering to model the impact of design choices to assess and optimize the robustness of the product. These models are also used to predict the performance of the design and identify weaknesses or excessive stresses. Using FEA early in the product development process provides opportunities to make choices that increase the ability of the product to meet reliability requirements.

Design for Reliability Processes

DFR methods and processes apply the concepts of requirements, planning, risk identification, risk assessment, risk mitigation, and risk communications. DFR approaches have the following in common:

- Identifying failure modes for the product or service
- Understanding failure modes
- Removing or mitigating failure modes

Illustrations of how DFR tools can be part of the product design process are shown in Figures 11.32 and 11.33.

One description of the DFR process is the reliability nine-step process (Conrad and Stack 2022). This model provides a framework for the delivery of a product or service that meets reliability requirements. The nine steps are:

1. Set customer-focused system reliability targets.
2. Cascade (or allocate) system reliability targets to subsystems and components.
3. Complete FMEA review.
4. Complete reliability planning and risk characterization.
5. Plan system reliability tests.

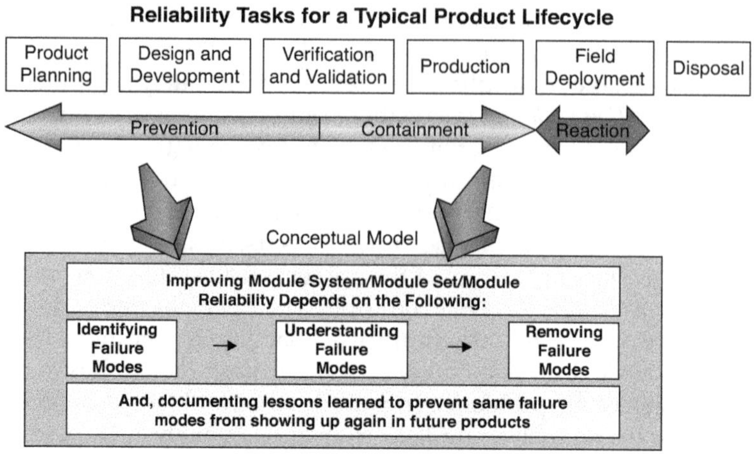

Figure 11.32 Conceptual model of design for reliability.
Source: Conrad and Sack (2022).

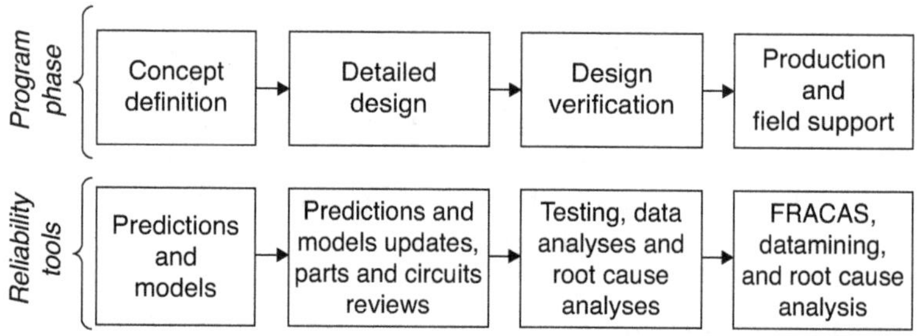

Figure 11.33 Reliability practices in product design.
Source: Raheja and Gullo (2012).

6. Develop accelerated life testing.
7. Complete reliability testing.
8. Apply failure communication and resolution (FRACAS).
9. Verify reliability.

Figure 11.34 illustrates the reliability nine-step process. These steps will have multiple loops when the design has elements that do not meet the reliability requirements. This iterative approach allows the design teams to incorporate learning and develop effective tradeoffs of cost and performance to meet the customer needs with a cost-effective solution.

Step 1: Set Reliability Targets

On the journey to quality and reliability it is essential to define the target reliability of the product or service. This provides a reference for improvement and the ability to identify the areas in the design that require more scrutiny to manage the risk to quality and reliability.

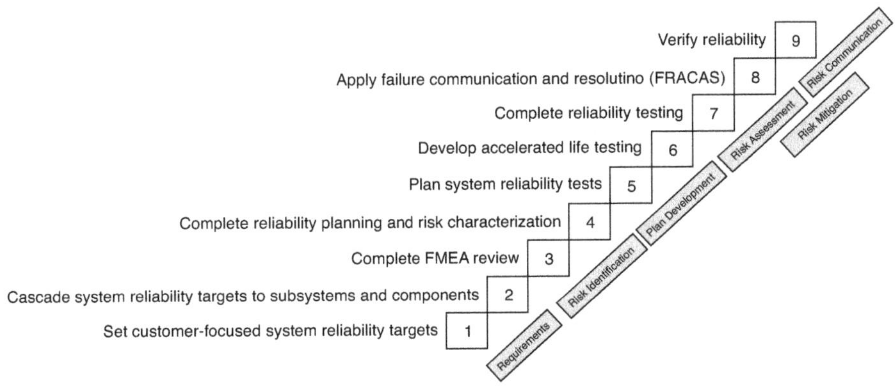

Figure 11.34 Reliability nine-step process.
Source: Conrad and Stack (2022).

Reliability targets start by defining the reliability specification. The reliability specification comprises four elements: *the* ***probability*** *that a product will perform its* ***intended function*** *for a* ***specified period*** *when operating under* ***specified conditions.*** Using these criteria allows the design team, reliability engineer, customers, and other stakeholders to clarify the expectations to be assessed by the reliability program.

Historically, reliability targets may have been set in terms of mean time to failure (MTTF) or mean time between failures (MTBF). Those parameters would only be sufficient in the case of exponential failure distribution, where the failure rate is constant. A better method is to specify required reliability at specific points in time, as described previously. For example, to state that 99% reliability is required after 2 years of operation and 95% reliability is required after 5 years of operation is much more effective than stating an MTTF of 10 years is required.

It can be effective to define two or three reliability design targets. The first target relates to the warranty period. This controls the producer risk, reducing the impact of early life failures on financial performance. The second timeframe is the useful life of the product. For example, the useful life for many consumer products is 10 years of service. Defining the reliability targets at this timeframe allows for a total cost of ownership analysis. The objective is to optimize the minimum cost of ownership, defining the overall consumer risk. A third timeframe recommended is an intermediate timeframe to provide a check on early wear-out failure modes in the design. For example, an intermediate timeframe for most durable products could be 4 years. These failure modes tend to drive product recalls, which are a high cost for the producer and a source of dissatisfaction to the customer. This check point mitigates this risk at the design stage.

EXAMPLE 11.15

As an example of a reliability specification, a spring was determined to be 99.0% reliable at 90% confidence, for 10,000 cycles (design lifetime of the component), under a linear loading of simply supported spring ends. In addition, the displacement was fixed at 130 mm delta for the cycles, and the cycling rate was specified as 10–15 cycles per minute in standard ambient 75°F and 50% RH. By providing

Continued

the level of specification detail, replication of the testing is possible because of the clarity of the requirement.

As business models shift to pay-for-service, one could use uptime or availability as the reliability specification metric. The reliability engineer and design team would cascade this to reliability goals for the design to achieve the availability requirement. They would then apply all the allocation and assessment processes in the design for reliability approaches.

Step 2: Cascade Reliability Targets

Once the system reliability targets are defined, they can be combined with the system architecture to develop and cascade or allocate the reliability requirements to the subsystems and components in the design. This allocation model is typically developed into a reliability block diagram (RBD) for the system. The RBD is used to assess gaps in knowledge on critical to quality (CTQ) components and features. The level of detail is based on both the availability of data and the lowest actionable item. For an example of a purchased component, consider a fan motor. If the motor is not as reliable as it should be, then the reliability engineer's role is to find a motor to meet the reliability requirements. The motor supplier's role is to improve the reliability of their motor to meet our requirements.

Factors to consider for allocation:

- *Likelihood of failure.* Components that have demonstrated a higher reliability in the past can be assigned a higher reliability goal.
- *Complexity.* Higher complexity of the component leads to a lower reliability goal.
- *Criticality.* Critical components will have a higher reliability goal.
- *Cost.* Higher reliability goals can be assigned to less costly components. This is normally aligned with complexity.
- Determine the duty cycles for the reliability target at the lowest actionable item.

The RBD is a model used to analyze the needed subsystem (component) requirements to provide a given system reliability or to determine what system reliability might be expected. An example of an RBD is shown in Figure 11.35. These tools are discussed in more detail in Chapter 10.

There is an impact of complexity on reliability and allocation. If we assume an equal allocation to all components in the system, then the number of components or system complexity (part count) impacts the system level reliability. In Figure 11.36,

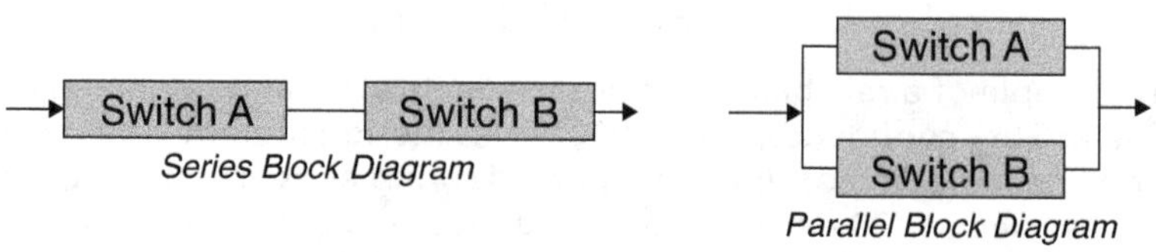

Figure 11.35 Reliability block diagram elements.

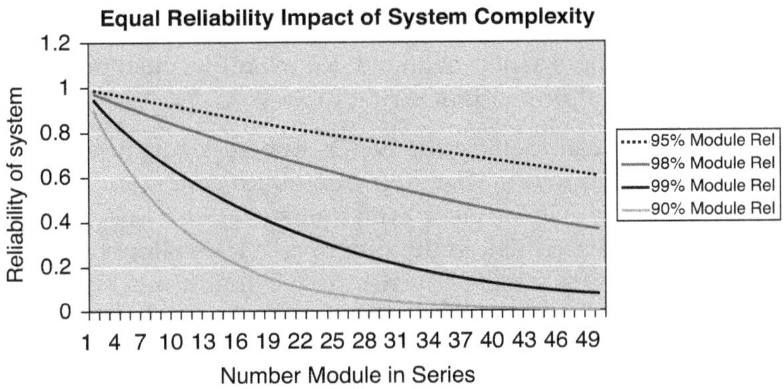

Figure 11.36 Equal reliability impact of system complexity.

we see the impact of the number of components and differing reliability of components on the overall system reliability.

From this graphic, we observe that allocation is a critical aspect of the DFR process. It provides evidence to the design team of where the risks are located and how to optimize tradeoffs of the competing design criteria to meet the reliability objectives.

Step 3: Complete Failure Modes and Effects Analysis (FMEA) Review

A primary risk identification tool used to prevent failures from impacting the design is the FME. The FMEA drives critical thinking in the design process. Both the design FMEA and the process FMEA should be reviewed and updated as needed to ensure the risks are understood and appropriate mitigations are in place. These tools are covered in Chapter 4.

Step 4: Complete Reliability Program Planning and Risk Characterization

As part of scoping for the planning process, the risks need to be characterized. This is used to drive prioritization of work early in the project. Each component or feature is assessed based upon the following criteria:

- *Level I.* Current technology in a known environment
- *Level II.* Current technology in a new environment
- *Level III.* New technology in a known environment
- *Level IV.* New technology in a new environment

To further assess the risk within the DFR project, we would apply the new, unique, difficult/different (NUDs) criteria (Conrad 2023). Identifying a design element as a NUD is an assessment of our knowledge about the design element prior to development work. Does the element have any of these characteristics?:

- New—to us or to the world
- Unique—a specialized element

- Difficult—known to be an issue (internal quality or external quality)
- Different—some aspect changed, for example, customer usage, environment, supplier or manufacturing process

The design team can identify the NUDs to assess and prioritize risk within a product development project. They allow communication and alignment of complexity and risk to the development team and stakeholders. An increased number of NUDs implies increased risk in the program. NUDs allow the team to focus on critical design elements to mitigate the risks to the quality and reliability of a system.

A reliability plan documents the activities the team will implement to achieve the reliability objectives. Planning a reliability program is like any other planning activity: Select or establish the policies, procedures, and activities required to achieve the project's reliability objectives as part of the overall program objectives. The reliability plan is a detailed report that provides a summary of the test and analysis activities that are used to prove product performance of the customer requirements.

Step 5: Plan System Reliability Tests

Reliability testing and analysis are designed to induce failure modes and detect unanticipated failure modes, so corrective action can be implemented. These tests and analyses are used to determine if items or systems meet the reliability requirements that were defined. This is accomplished by comparing estimated rates of failure to actual rates of failure, by monitoring the reliability growth of the design through the design phases, and by overstressing the designs to determine the margin of safety or safety margin inherent in the design.

The reliability test plan should include both virtual (simulated) and physical assessments. Virtual assessments reduce the time and provide confidence in final validation assessments of the design. The plan is used to define the resources the project will require, and it will identify accelerated test beds that need to be developed to support the program. Therefore, the earlier the plan is developed, the better the design team can manage the risk and shorten the development time through effective concurrent engineering approaches. The plan should drive discussion on the relationship of cycles, units on test, time, and number of failures for results. It will also define accelerated test methods required for using smaller sample sizes, but at elevated stress levels to mitigate very large sample sizes.

Analyses that are often utilized to supplement physical testing are:

- Lessons learned
- Warranty analysis of like or similar products
- Parts derating
- Reliability prediction (parts counts)
- Thermal analysis

These methodologies allow the engineer to predict the expected performance of the physical test samples under specific conditions or approaches. These methods are used to help the concurrent design team to select concepts considering the reliability objectives of the project.

Step 6: Develop Accelerated Life Testing

As part of the plan, the team will want to use accelerated test methods such as accelerated life testing (ALT), highly accelerated life testing (HALT), highly accelerated stress testing (HAST), highly accelerated stress audit (HASA), failure mode verification testing (FMVT), and others to reduce the test time as described in Porter (2004) and Shelden (2024). The team will want to use "test to failure" to reduce sample sizes. This is typically done at the component and subsystem level of the design. Once the team has confidence in the components and subsystem, the team will look at system level testing and initiate demonstration testing as well as apply HALT, HAST, and FMVT to assess component interaction robustness.

The FMEA provides inputs to help identify the failure modes and mechanisms that the test fixtures need to utilize to assess the design robustness. FTA is used for reliability and safety analyses. The approach is like the RBD. The aim is to determine possible combinations of causes which can lead to undesirable event(s). The FTA is used to:

- Generate a graphic or logical tree structure for the understanding of how the connections interact
- Identify possible failure causes and their combinations
- Calculate the probability of the undesirable event

Step 7: Complete Reliability Testing

Once the fixtures are developed, reliability assessment and analysis apply to understand and validate the reliability and robustness of the design. Design of experiments is used to evaluate system interactions, component variation, and customer noises (inference space). Assessment of software/hardware interactions and bugs is applied to include software reliability in the overall analysis. The experiment can incorporate customers' noise factors. For example, for a washing machine, noise factors would include load size, additives, grit/sand/dirt, electrical supply, small articles, foreign objects, machine level, temperature, humidity, and interaction time. Incorporating these variations into the reliability assessment helps to ensure the robustness of the design under real customer usage environments. This is part of developing a fault tolerant design.

To visualize and optimize the design concepts, computer aided engineering (CAE) tools such as computational fluid dynamics (CFD) and FEA are used throughout the design process. These virtual analysis tools are powerful allies in the development of robust and reliable products. In Figure 11.37, we see a CFD model of the airflows and temperatures within the merchandiser case (Palaksha and Narasimhamurthy 2022). Simulations like these are used to assess the impact of design features and environmental parameters on the robustness of the design. This can be accomplished using a virtual design of experiment.

FEA is a tool that allows engineers to assess and predict the forces and displacement of structures. In Figure 11.38, we see a displacement plot of a spring bumper design (Narasimhamurthy and Palaksha 2022). In Figure 11.39, we see the stress relationship from a dynamic model (Hack 2022). This allows the durability of the system to be predicted as well as the overall assessment of stress–strength relationships in the design process. Models like this allow the engineers

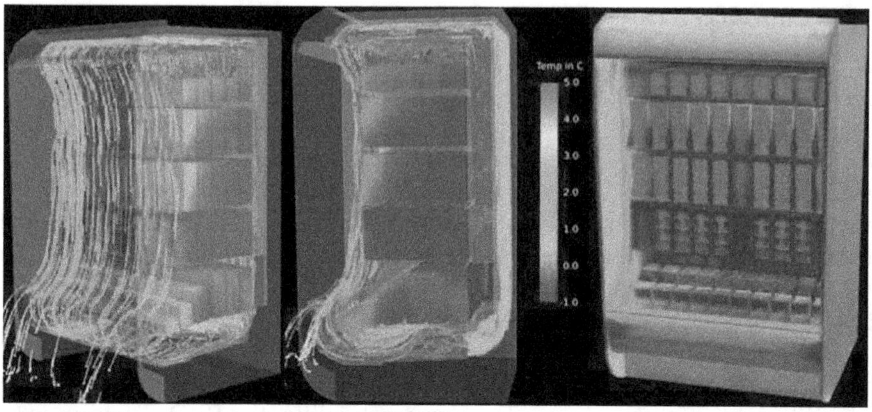

Figure 11.37 Thermal vector and contour plot in °C.

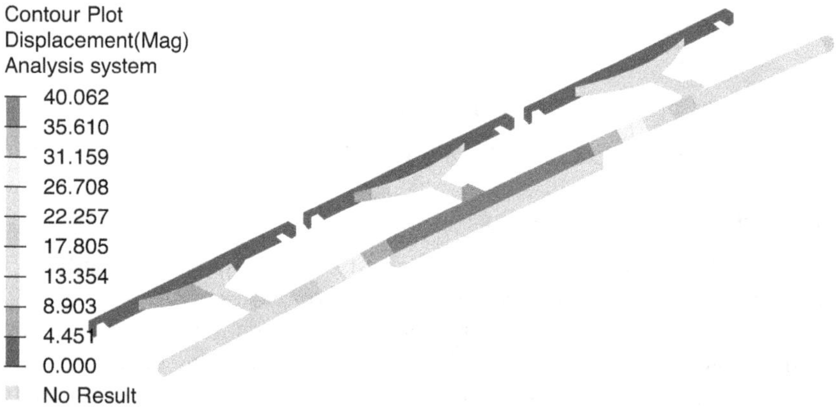

Figure 11.38 Displacement plot of spring bumper design.

Figure 11.39 Dynamic stress simulation, Simcenter.

to evaluate alterative concept designs and to optimize the design to maximize robustness while minimizing cost or weight, as an example. This simulation analysis is done prior to building physical prototypes, which saves time and resources in the DFR project.

Step 8: Apply Failure Communication and Resolution (FRACAS)

The objective of reliability testing is to validate the reliability of a product and its robustness to failure modes. The FMEA identifies potential failure modes. During the assessment process, the weaker elements in the design are identified. These failures are captured in the FRACAS. These failures represent the early failures in the design and can be considered an indication of launch quality and initial warranty assessment. Once these failures are identified, the design team works to mitigate them prior to product launch. The FRACAS process focuses and prioritizes the design team's energy to resolve the most important latent defects in the design. By tracking these issues and their resolution, the risk to the producer and the customer can be communicated effectively.

Step 9: Verify

Finally, once the testing is complete and all issues found have been resolved, a reliability prediction is developed and compared to the allocation targets for the system. This provides leadership with the data needed to define the producer's and consumer's risk. The reliability prediction models will provide an assessment of key subsystems and components at the requisite timeframes. The reliability prediction data need to incorporate interactions between subsystems. These reliability assessments provide confidence in the designed-in reliability at the time of launch.

After launch, the team needs to capture any key learning and share this across the organization for continuous improvement. A lessons-learned document is the beginning of the next DFR project. Table 11.30 contains a list of tools and methodologies used in DFR programs (Yang 2007). The DFR team can use this list to check their plans and timing.

Table 11.30 Design for reliability phases and tools used.

Reliability process phase	Tools used in the phase
Product planning	Organizing a reliability team
	Quality function deployment
	Reliability history analysis
	Reliability planning and specifications.
	Reliability modeling
	Reliability allocation
	Reliability prediction

Continued

Table 11.30 Design for reliability phases and tools used. (Continued)

Reliability process phase	Tools used in the phase
Design and development	Robust design FMEA FTA Design controls Accelerated life testing Accelerated degradation testing Stress derating Failure analysis Reliability estimation Warranty cost prediction Reliability design reviews
Verification and validation	Reliability verification testing Analytical reliability verification Accelerated life testing Accelerated degradation testing Failure analysis Process FMEA Process control plans and charts Process capability analysis
Production	Process control plans and charts Process capability analysis Stress screening Acceptance sampling Failure analysis
Field deployment	Warranty plan development Field failure tracking Warranty data analysis Customer feedback analysis Failure analysis Six-sigma process implementation Lessons learned
Disposal	Reuse and recycle assessment End of useful life assessment

Chapter 12

Parts and Systems Development

MATERIALS, COMPONENTS, EQUIPMENT, AND SOFTWARE SELECTION TECHNIQUES

> Apply techniques (e.g., derating and commercial off-the-shelf [COTS], safe operating area [SOA] criteria) for selecting materials and components to meet reliability goals and requirements. Select software for product and evaluation equipment functionality (e.g., vibration stand and thermal chambers). (Analyze)
>
> **Body of Knowledge V.B.1**

Derating

The *rating* of an electronic component is the maximum level of a stressor that the component can withstand. Examples of these stresses are temperature, voltage, and pressure. *Derating* is a technique employed in the development process of mainly electronic devices, in which the components are operated at less than their rated maximum parameter values. By applying derating on electronic components, the margin of safety between part design limits and applied stresses increases, thereby providing extra protection for the part. By applying derating in an electrical or electronic component, its degradation rate is reduced, and its reliability and life expectancy are improved.

Temperature is generally assumed to be the main contribution of wear-out in semiconductors, as represented by the Arrhenius equation:

$$A_F = \exp[(E_A/K)\,(1/T_U - 1/T_s)]$$

Since the amount of derating depends on several factors, such as the type of application, customer quality expectations, and safety related functions, multiple derating tables or levels can be considered. Often, companies use at least two tables: one that is generally applicable (Level A) and another to be used for more demanding devices (Level B).

Rule 1

Every design derating can be chosen based on customer requirements, the applied level of derating, and the selected derating parameters by part type. Manufacturers will generally provide safe operating area (SOA) information that defines the conditions (voltage and current) in which the component or device is rated to operate without self-damage.

An example derating graph is shown in Figure 12.1.

Figure 12.2 shows an example of a SOA of a MOSFET, or a metal–oxide–semiconductor field-effect transistor, mainly used for amplifying or switching of electronic signals, in terms of drain-to-source voltage (v_{DS}) and drain current (I_{Dmax}).

An example of a safe operating area of a battery is shown in Figure 12.3, in terms of temperature and voltage.

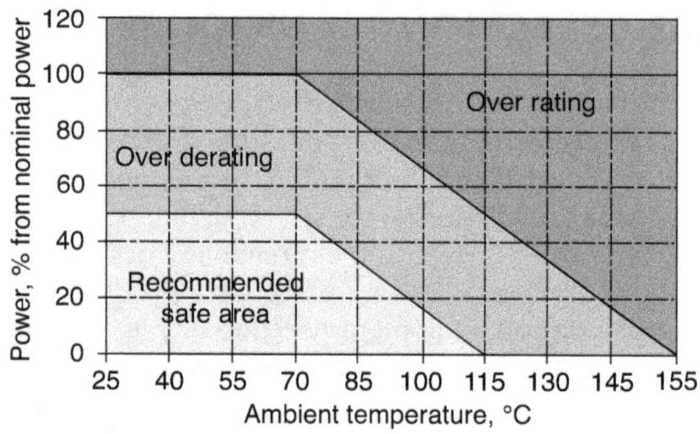

Figure 12.1 Example of derating graph.

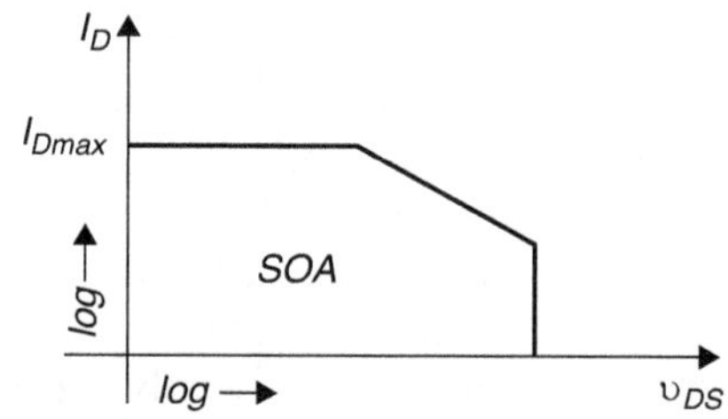

Figure 12.2 Example of a safe operating area of a MOSFET.

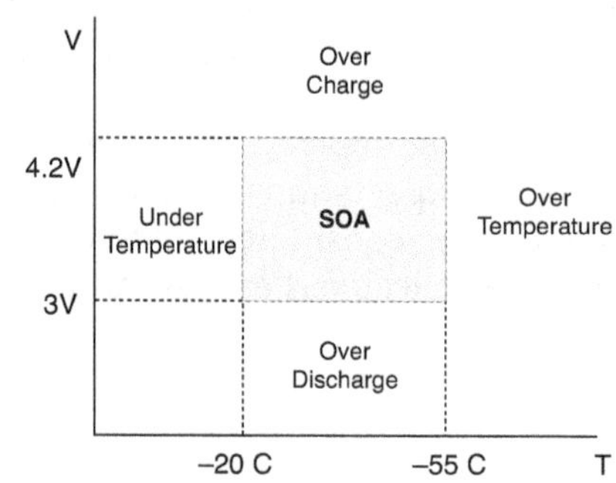

Figure 12.3 Example of a SOA for a battery.

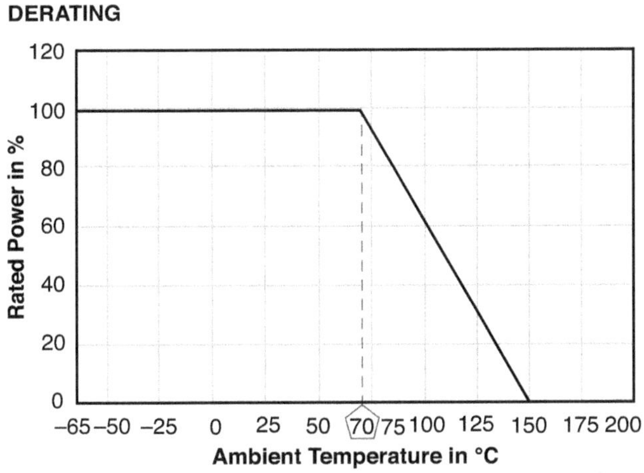

Figure 12.4 Example of a SOA for a metal film resistor.

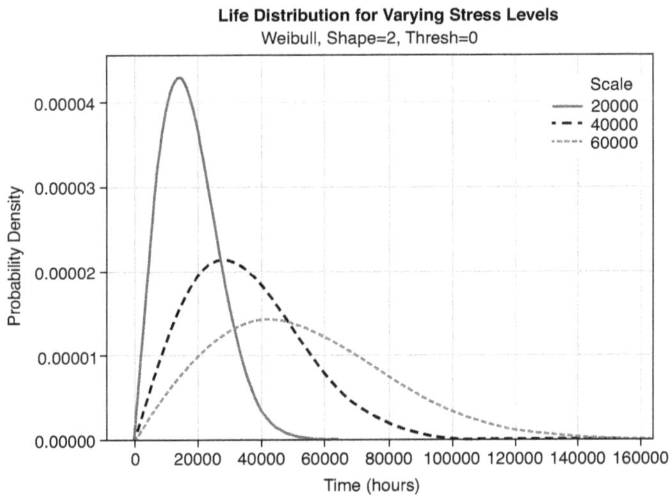

Figure 12.5 Life distribution for components having different stress ratings.

Figure 12.4 is an example of a safe operating area of a metal film resistor in terms of temperature and power.

Stress life distributions for different stress levels are shown in Figure 12.5.

Rule 2 Level A derating figures are used for the design of electronic devices. Level B derating figures are highly advised for appliances:

- Used in industrial, avionics, automotive, transportation or military environments
- Used in safety-related applications
- With lifetimes of more than 7 years
- With low annualized failure rates (0.9% or less) or a long time between failures (one million hours or more)

Stress–Life Relationships

Varying stress ratings will typically affect the life distribution of a component. Figure 12.5 shows the impact of different stress levels on the life distribution of three components.

Examples of derating parameters in electronics include:

- Maximum power
- Maximum temperature
- Maximum voltage
- Maximum reverse voltage
- Maximum current
- Maximum input power
- Fanout (current)
- Timing parameters
- Maximum delta anode index

Available standards and other resources include:

- RADC-TR-84-254 Reliability Derating Procedures
- RL-TR-92-11 Advanced technology component derating
- ECSS-Q-ST-30-11C: Derating—EEE components
- NAVSEA TE000-AB-GTP-010: Parts Derating Requirements and Application Manual for Navy Electronic Equipment
- MIL-HDBK-217: Military Handbook: Reliability Prediction of electronic equipment
- MIL-STD-975M: NASA standard electrical, electronic, and electromechanical (EEE) parts list
- Bellcore/Telcordia (SR-332)
- China 299B

Commercial Off-the-Shelf (COTS) Products

COTS, or commercial-off-the-shelf, is a product that meets the requirements of a design "as is" without customization. COTS hardware is a standard product that already exists and is available from commercial sources. These products are designed to be easily installed and interoperate with existing system components. Some of the benefits of using COTS products are lower costs, reduced development time, faster application of new technology, and lower lifecycle costs resulting from using readily available and up-to-date products. COTS can also be applied to software as a service (SaaS), which in this case are built and delivered by a third party. The benefit of COTS is fast availability and lower cost than in-house development. COTS is generally used for commodity technology/products and not for new, cutting-edge technology products. COTS may not be suitable for all applications, may not meet reliability requirements, and can become obsolete.

An example of a COTS product is the software for the thermal chambers in the lab. This way, the software can be validated by an external specialized design team, and you do not need to invest in the development of a tool. This will also help in an audit, but be aware that the COTS supplier needs to be certified or be able to provide evidence of their capabilities and maturity (see also Krinke and Pai 2024).

Software Selection

An aspect of product and test system development is the selection of the software that is a component of the product or test system. For example, vibration stands and thermal chambers are run using embedded software. This software affects the reliability of the product being tested and needs to be selected by taking reliability into account. Jadhav and Sonar (2008) provide a generic stage-based methodology for evaluating and selecting software packages to meet specified requirements.

- Determine requirements.
- Develop a short list of candidate packages.
- Eliminate most candidate packages that are missing a feature or do not work with existing hardware.
- Use an evaluation technique on the remaining packages.
- Further scrutinize the top contenders with trial copies and pilot testing.
- Negotiate contract terms.
- Purchase and implement.

Systems and tools to accomplish the evaluation step include analytical hierarchical process (AHP), weighted scoring method, and fuzzy multiple criteria decision making (FMCDM).

PARTS STANDARDIZATION AND SYSTEM SIMPLIFICATION

> Describe the importance of standardization, simplification, and parts re-use to meet reliability goals and requirements. Incorporate reliability-centered maintenance (RCM) to maintain the inherent reliability of system components while ensuring optimal performance and longevity. (Apply)
>
> **Body of Knowledge V.B.2**

Standardization

Parts standardization refers to the use of the same components in several products. An example in the automotive industry is the use of the same chassis or engine for several automobile models. In some families of electronic products, the same circuit board may be used, with different functions activated depending on the application. Standardization refers to developing standards for work methods, quantity to use, expected quality, raw material properties, and performance of any product.

Standardization and simplification are related. Most of the standardization should come from simplification while meeting the documented requirements to perform the correct and unambiguous execution. Using standardization to achieve simplification will result in designs that are easier and cheaper to assemble (production) and provide a less complex design at a lower cost.

Simplification

The concept of simplification is closely related to standardization. Simplification is the process of reducing the variety of products manufactured. This provides a reduction of product range, assemblies, parts, materials, and design. For example:

- Using the same resistors throughout the design (simplification) may not reduce the cost per assembly, but it enables the manufacturer to buy in larger quantities and carry a smaller inventory of parts.
- Using the same screws in the full assembly (standardization and simplification) means there is no need for multiple screwdriver sets on the production floor.

Simplification makes a product, assembly, or design less complex and easier to produce. A production line can be simplified when it is stripped of unnecessary complexity and confusion. The availability of suitable standards will assist in simplification. Simplification and standardization also improve maintainability, providing easier access to standard parts and materials and easier access to spare and repair parts.

Maintainability

Maintaining a system should be done in a manner that maximizes profit or minimizes cost. It is assumed that the total cost of a maintenance failure is greater than the cost associated with a scheduled preventive maintenance action. Maintainability is discussed in more detail in Chapter 13.

Reliability Centered Maintenance (RCM)

Reliability centered maintenance (RCM) is a logical way of identifying the equipment in a facility that is required to be maintained on a preventive maintenance basis rather than a let-it-fail-then-fix-it basis, commonly referred to as run-to-failure. RCM can be thought of as a process to determine what must be done to keep assets doing what their operators want them to do in their current operating context.

Some of the common adaptations of RCM are:

- *Total productive maintenance (TPM).* Includes involvement of teams comprised of representatives from operations, maintenance, and engineering. These teams determine what equipment gets worked on and when it gets scheduled.
- *Risk-based maintenance (RBM).* Based on an assessment of the preventive maintenance program. Focused on the critical equipment.
- *Probabilistic safety analysis (PSA) based maintenance.* Methods based on specific safety significance. This method is often applied in nuclear safety.
- *80/20 rule, or the Pareto principle.* Only analyze 20% of the facility—only the objects that are the most important or the critical assets.

A guide for RCM can be SAE JA1011. This standard establishes that, for a process to be acknowledged as RCM, it must follow seven steps in the following order:

1. What are the functions of the asset?
2. What are the functional failures?
3. What are the failure modes?
4. What are the failure effects?
5. What are the failure consequences?
6. What are the (proactive) preventive maintenance tasks?
7. What must be done if a preventive maintenance task cannot be specified?

Possible implementation strategies for RCM can be:

- *Condition based maintenance tasks.* Condition monitoring of the asset so that the maintenance can be performed before the loss of the function.
- *Scheduled tasks.* Time based or scheduled repairs used when there is prior knowledge of the useful lifetime of a component. Refer to the bathtub curve in Chapter 2.

To calculate availability (A) of a system, mean time to failure (MTTF) and mean time to repair (MTTR) can be used as shown in Equation 12.1.

Equation 12. 1 Availability using MTTF and MTTR.

$$A = \frac{MTBF}{MTTR + MTBF}$$

where

$$\lambda = \textit{failure rate} = \mathrm{MTBF} = 1/\lambda;$$

$$v = \textit{repair rate} = \mathrm{MTTR} = 1/v.$$

EXAMPLE 12.1

A repairable system has an MTBF of 2000 hours. The same system has an MTTR of 10 hours. What is the availability of the system? Using Equation 12.1:

$$A = \frac{MTBF}{MTTR + MTBF} = \frac{2000}{10 + 2000} = 0.995$$

The MTTR calculation assumes that spare parts are readily available. This means that spares must be in inventory when they are needed. If there is waiting time to obtain a spare, the unit downtime will increase, and availability will suffer.

System Availability Models

The reliability block diagram (RBD) method shows how to evaluate the reliability (or availability) of the whole system given the reliability of each block. Figure 12.6 shows reliability block diagrams for series and parallel systems. See Chapter 10 for more information on the RBD method.

For optimizing maintenance tasks in a complex RCM system, Markov-chain or Petri net models can be used.

Simplified State Diagram for Markov Chain

Figure 12.7 shows a Markov chain for a nonrepairable component. Degradation and failure are both absorbing states. An absorbing state is a state that, once entered, cannot be left. A Markov chain is an absorbing chain if there is at least one absorbing state, and it is possible to go from any state to at least one absorbing state in a finite number of steps. In an absorbing Markov chain, a state that is not absorbing is called transient.

Figure 12.8 shows the states for a repairable component in which there is a 20% probability that a degraded part can be made operable again. Markov analysis and transition matrices are presented in Chapter 10, "Reliability Prediction Methods."

Petri Net Model

Petri net or place/transition net (PT net) is one of several mathematical modeling languages that are used to model distributed or parallel processes. A Petri net can be drawn either as a diagram (as shown in Figure 12.9) or an activity diagram (shown in Figure 12.10) that shows transitions, connections, and interactions in

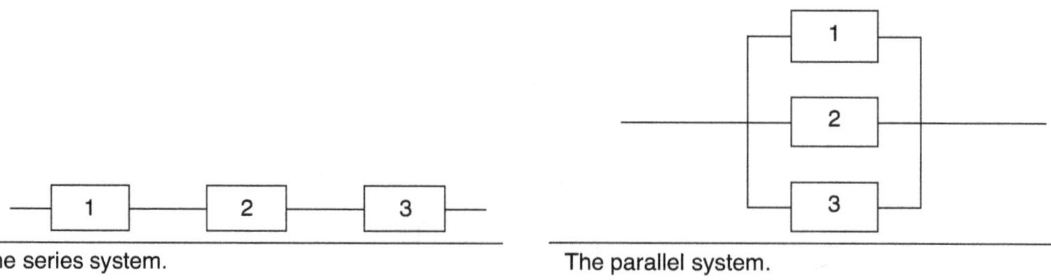

Figure 12.6 Reliability block diagrams for series and parallel systems.

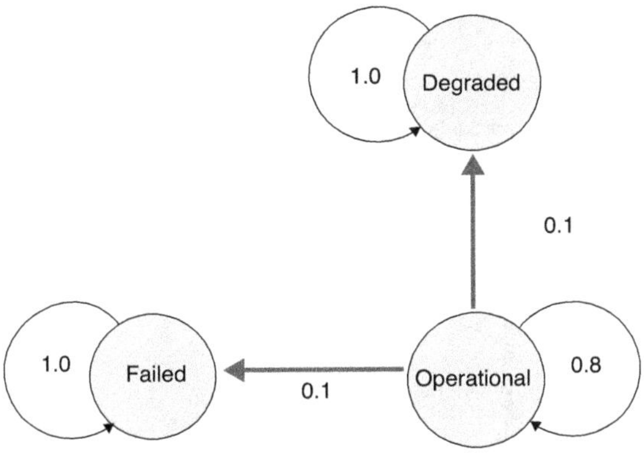

Figure 12.7 Markov chain for a nonrepairable component.

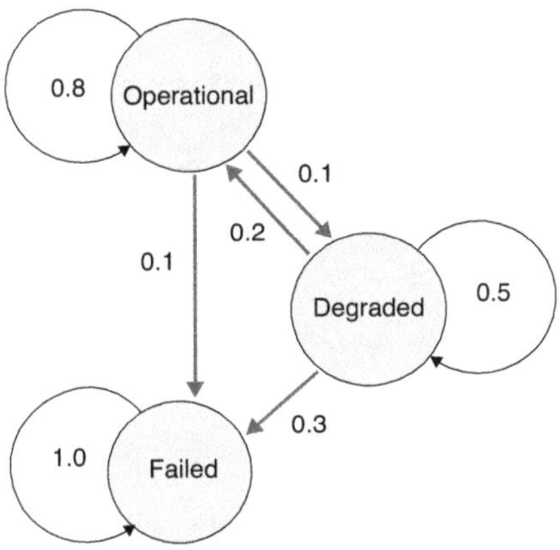

Figure 12.8 Markov chain for a repairable component.

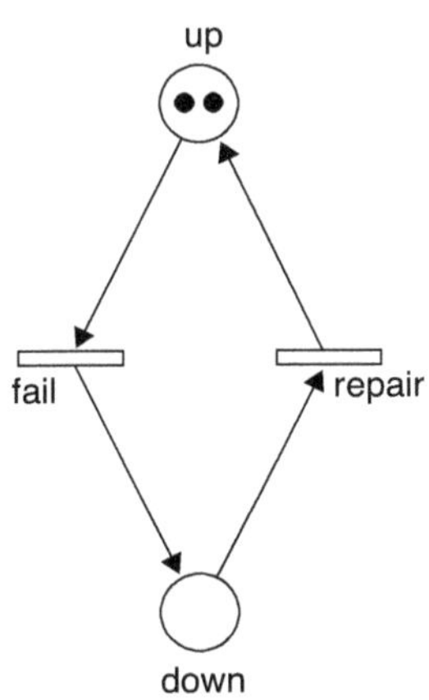

Figure 12.9 Petri net diagram of two repairable components.

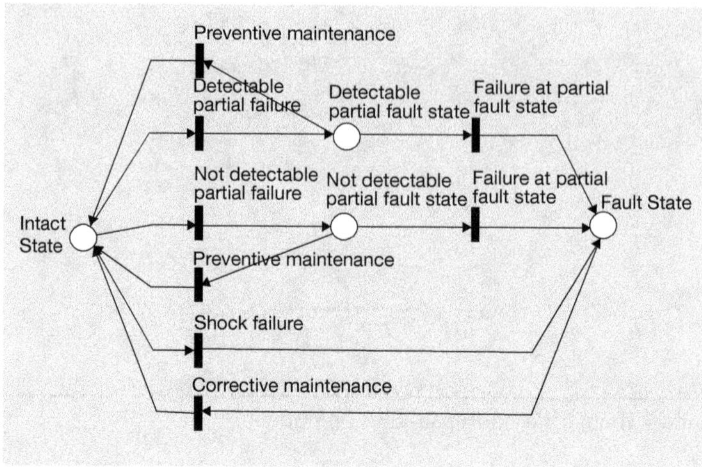

Figure 12.10 Example Petri net activity diagram.

a trajectory. These models are used to perform an analysis of complex systems or networks of systems.

See Reisig (1985) and Trivedi and Bobbio (2024) for further information on Petri Net models for reliability.

Chapter 13
Maintainability

MAINTENANCE STRATEGIES

> Develop a maintenance plan incorporating various strategies (e.g., predictive maintenance, repair or replace decision making, spare parts analysis/ forecasting, and equipment warranties). (Apply)
>
> **Body of Knowledge V.C.1**

Planning

Repairable systems are systems that can be restored to service following a failure. The actions to restore a system to service following a failure are known as *corrective maintenance* actions. Most repairable systems can also have suggested maintenance actions to keep the system in operating condition as well as reduce the probability of failure due to wear-out. These actions are *preventive maintenance* (PM) actions.

Preventive maintenance includes all the actions performed to keep the system in an operating state by performing actions that can prevent a potential failure. PM does not reduce the constant failure rate that is inherent to the system but tends to maintain the system at its inherent level of failure probability. PM actions can be planned and, if possible, carried out when there is no demand for the use of the system.

Corrective maintenance (CM) includes all the actions required to return the system to an operating state once failure has occurred. CM actions cannot be planned and are performed when the system fails. CM actions may be deferred in some cases; however, the system is not operable until the CM is completed.

Availability is a measure of the likelihood that a system will be ready to operate (system is up) when it is called on to operate. Reasons the system would not be ready to operate (system is down) include the possibility that a failure has occurred and the corrective action has not been completed. Also, a system may not be operable because PM actions are necessary to bring to a functional state. There could also be other logistical reasons that the system is not operable.

In general, availability is a function of the number of CM actions necessary and the time required to complete these actions, as well as the amount of maintenance required and the time necessary to restore the unit to service. Thus, availability is a function of *reliability* and *maintainability*. *Reliability* deals with reducing the frequency of unscheduled downtime. *Maintainability* is the ability of an item to be retained in or restored to specified conditions when maintenance actions are necessary. The focus of maintainability is reducing the duration of downtime. This downtime includes both scheduled and unscheduled actions. If PM actions can be planned and performed when there is no demand to use the system, the time to perform these actions will not affect the availability.

There are three types of availability: inherent, achieved, and operational. Inherent availability refers to the availability of a system or component when only considering the inherent reliability and maintainability, excluding any external factors such as logistics, administrative delays, or operational environment. Achieved availability accounts for the inherent availability and includes the effects of PM and CM actions. It still excludes external factors like logistics and administrative delays. Operational availability is the most comprehensive measure of availability, considering all factors that affect the system's performance, including inherent reliability, maintainability, logistics, administrative delays, and operational environment.

In general, availability can be calculated as uptime (the time the system is operable) divided by total time (the time during which there could be a demand to use the system). To increase the inherent availability of a system, the system reliability could be improved by increasing mean time between failures (MTBF) or by decreasing the mean time to repair (MTTR) as shown in Equation 13.1:

Equation 13.1 Availability as a function of MTBF and MTTR.

$$A_i = \frac{MTBF}{MTBF + MTTR}$$

This calculation varies slightly based on the type of availability but keeps the same format. Achieved availability is calculated using Equation 13.2.

Equation 13.2 Achieved availability.

$$A_a = \frac{MTBM}{MTBM + MMT}$$

where:

MTBM = Mean Time Between Maintenance

MMT = Mean Maintenance Time

Operational availability is calculated using Equation 13.3:

Equation 13.3 Operational availability.

$$A_o = \frac{Uptime}{Uptime + Downtime}$$

EXAMPLE 13.1

Metro train 817 runs from 4 A.M. to 12 A.M. seven days a week. Most of the planned maintenance is done during nonoperating time, but the train must be taken out of service for one day each quarter for more intensive maintenance. Train 817 also had multiple stoppages during working hours over the past year, one that lasted two days, two that lasted 4 hours, and eight minor incidents that averaged half an hour of downtime. What is the operational availability of train 817 over the last year?

$$\text{Total time} = 20 \text{ hours/day} \times 365 \text{ days/year} = 7300 \text{ hour/year}$$

$$\text{Uptime} = 7300 \text{ hours/year} - [4 \times (20 \text{ hours}) + 40 \text{ hours} + 2 \times (4 \text{ hours}) + 8 \times (0.5 \text{ hours})] = 7168 \text{ hours/year}$$

$$\text{Availability} = \text{Uptime/Total time} = (7168 \text{ hours/year}) / (7300 \text{ hours/year}) = 0.98$$

The operational availability of train 817 is 98% over the past year.

Maintenance Strategies

Maintenance strategies should be chosen to ensure a high level of availability while controlling costs. Preventive maintenance does not improve the inherent reliability of the system because it does nothing to improve the system or fix intrinsic flaws. Preventive maintenance will maintain the reliability level of useful life, keeping the failure rate low and increasing the availability of the equipment. It will also delay the onset of wear, thus increasing the length of useful life. However, preventive maintenance actions can also introduce a fault into the system, causing the system to fail and need corrective maintenance. A study should be performed to determine the most cost-effective scheduling of maintenance actions necessary to keep the system operational but minimize the opportunities to cause new issues. Maintenance actions that are unnecessary should be eliminated.

System reliability is at the highest level if all units are in their useful life phase. A single unit, such as a pump, could be considered a system. Or the pump could be considered a component of a larger system. Preventive maintenance strategies such as *reliability centered maintenance* and *predictive maintenance* are proactive and require the replacement or refurbishment of the components of the system that are nearing the end of their useful life or entering wear-out. Replacement is made before a failure occurs in the component. *Condition based maintenance* is proactive and is a preventive maintenance strategy. Replacement is made before component failure occurs, based on assessment of physical condition.

The appropriate strategy for replacement needs to be chosen for each unit. The information needed to choose the most appropriate replacement strategy includes:

- The failure distribution of the unit
- The cost associated with the failure of the unit
- Any safety issues associated with the failure
- The cost of the replacement unit
- The cost associated with scheduled replacements
- The cost of inspection or test

Predictive and Reliability Centered Maintenance

Predictive maintenance assumes that the operator can detect the imminent failure of a unit. This detection can be by observation, analysis, or using test equipment. The analysis of an oil sample or the measurement of increased vibration might indicate wear and an increasing failure probability. If the increasing failure probability cannot be detected by the operator, a reliability centered maintenance strategy could be adopted. This strategy uses the predicted failure distributions to determine the optimum replacement time for units about to enter the wear-out phase. Replacement of units that are entering the wear-out portion of the bathtub curve will maintain the system reliability at the useful-life level.

It is possible that any maintenance action can negatively impact reliability by inducing failure modes due to the maintenance action. These maintenance-induced failures are similar to the early-life failures due to the manufacturing or installation of a system. An effort should be made during design to ensure that standard preventive maintenance actions can be performed quickly and with a low risk of introducing problems due to the maintenance actions.

Mean Time Between Failures

The MTBF is a description of the expected time between two failures. MTBF is typically measured in hours and shows the expected operational time of a system before failure occurs. It is commonly used for repairable systems, where the failed component is replaced or repaired, and the system continues to operate. MTBF can also be referred to as the "up time" of the system and is often used as a key reliability metric. The higher the MTBF is for a system, the higher the likelihood that the system will work for a specified period of time before failing. Knowing how often a system will fail allows planning for maintenance and guides decisions such as how many spare parts to keep available.

In general, the MTBF of a system or component is the sum of the durations of the operational periods divided by the number of observed failures in the same time period as shown in Equation 13.4.

Equation 13.4 MTBF of a system or component.

$$MTBF = \frac{\#\ operational\ hours}{\#\ failures}$$

Assuming a constant failure rate (λ), the mean time between failures is the inverse of the failure rate expressed in Equation 13.5.

Equation 13.5 MTBF in terms of failure rate.

$$MTBF = \frac{1}{\lambda}$$

The failure rate, λ, is often assumed to be constant over time in many reliability models, especially under the assumption of an exponential failure distribution. This assumption is common for systems that exhibit random failures, such as electrical or mechanical components, and is used in these cases to simplify the model. This model is inherent and memoryless and is not improved or worsened by preventative maintenance.

In the constant failure rate model, we assume that the system's failure rate does not change over time, meaning the likelihood of failure is independent of how long the system has been operating. Using this assumption, the time between failures follows an *exponential distribution*. The probability of a failure occurring before a certain time t is given by:

$$\Pr(T<t)=1-e^{-\lambda t}$$

where:

- $\Pr(T<t)$ is the probability of failure occurring within time t
- λ is the constant failure rate (failures per unit time),
- t is time (e.g., hours, days).

This model is simple and widely used, especially when the failure process can be described by random or single-fault events.

EXAMPLE 13.2

Consider an industrial mixer that has a constant failure rate of $\lambda=0.002$ failures per hour. If the mixer operates continuously, what is the MTBF of the machine?

$$MTBF=\frac{1}{\lambda}$$

$$MTBF=\frac{1}{0.002}=500\ hrs$$

How long would it take for the mixer to experience five failures on average, assuming all failures are independent?

$$\text{Time for 5 failures}=5\times \text{MTBF}=5\times 500=2500\text{ hours}$$

What is the probability of a failure occurring in the mixer during a week of operation (168 hours)?

Using the cumulative formula for the exponential distribution, we obtain:

$$\begin{aligned}\Pr(T<t)&=1-e^{-\lambda t}\\&=1-e^{-(0.002)(168)}\\&=1-0.715=0.285\end{aligned}$$

The probability of failure over a week of continuous operation is 28.5%.

Maintainability Apportionment and Allocation

Reliability deals with reducing failures of the system and thus reducing the frequency of unscheduled maintenance actions. Maintainability deals with reducing the duration of the downtime that is a result of both scheduled and unscheduled maintenance actions. Maintainability is often thought of as a technique for making repairs easy. It is, however, the engineering activities involved in minimizing the total downtime.

MIL-HDBK-472 gives the following definition of maintainability:

Maintainability is the ability of an item to be retained in or restored to specified conditions when maintenance action is performed by personnel having specified skill levels and using prescribed procedures and resources at each prescribed level of maintenance and repair.

The maintenance actions are to be performed by skilled personnel, using proper procedures, with the proper tools, and having access to standard replacement parts. The objective is best accomplished if an effort is made to eliminate unscheduled downtime and reduce the duration of scheduled downtime.

A system-level maintainability requirement may need to be allocated to lower levels of the system. Maintainability allocation is a continuing process of apportioning requirements at the system level to subsystem levels. This will give values the designer can work to achieve. This is similar to a system reliability requirement being allocated to the various subsystems. To allocate the maintainability requirements, the failure rate and the MTTR are used.

MTTR Allocation

Suppose that a series system has n subsystems. Then the failure rate for subsystem i is λ_i, and the MTTR for subsystem i is t_i for $i=1, \ldots n$.

The system failure rate is calculated using

$$\lambda = \sum_{i=1}^{n} \lambda_i$$

The MTTR of the system is then calculated as shown in Equation 13.6.

Equation 13.6 MTTR of the system.

$$MTTR_{System} = t = \frac{\sum_{i=1}^{n}(\lambda_i t_i)}{\sum_{i=1}^{n}(\lambda_i)} = \frac{\sum_{i=1}^{n}(\lambda_i t_i)}{\lambda}$$

The system level MTTR requirement is denoted as t^*.

In order to meet the maintainability goal, an MTTR allocation to each subsystem must be made if the $MTTR_{(System)} > t^*$. The MTTR allocation to subsystem i is:

$$t_i^* = \left(\frac{t_i}{t}\right) \times t^*$$

The designer can work to subsystem MTTR requirements better than to system-level MTTR requirements.

EXAMPLE 13.3

A series system has $n=3$ subsystems. The failure rates of the subsystems are:

$$\lambda_1 = 10 \times 10^{-6} \text{ failures/hr}$$

$$\lambda_2 = 30 \times 10^{-6} \text{ failures/hr}$$

$$\lambda_3 = 60 \times 10^{-6} \text{ failures/hr}$$

The subsystem MTTR values are:

$$t_1 = 150 \text{ min}$$

$$t_2 = 100 \text{ min}$$

$$t_3 = 50 \text{ min}$$

The system failure rate is calculated using:

$$\lambda = \sum_{i=1}^{n} \lambda_i = (10 + 30 + 60) \times 10^{-6} = 100 \times 10^{-6} \text{ failures/hr}$$

And the system MTTR in minutes is equal to:

$$MTTR_{System} = t = \frac{\sum_{i=1}^{n} (\lambda_i t_i)}{\sum_{i=1}^{n} (\lambda_i)} = \frac{10(150)}{100} + \frac{30(100)}{100} + \frac{60(50)}{100} = 75$$

If system MTTR requirement is $t^* = 60$ min, the attained value is greater than the requirement.

To meet the maintainability goal, an MTTR allocation to each subsystem i is:

$$t_1^* = \left(\frac{t_1}{t}\right) \times t^* = \left(\frac{150}{75}\right) \times 60 = 120 \; min$$

$$t_2^* = \left(\frac{t_2}{t}\right) \times t^* = \left(\frac{100}{75}\right) \times 60 = 80 \; min$$

$$t_3^* = \left(\frac{t_3}{t}\right) \times t^* = \left(\frac{50}{75}\right) \times 60 = 40 \; min$$

Spare Parts Analysis

The required number of spares necessary for a given period is a function of the expected number of failures and the number of spares required for preventive maintenance. The number of failures is a function of the unit operating time and the failure rate. If any spares are required for preventive maintenance, the number is a function of operating time and the preventive maintenance cycle.

The MTTR calculation assumes that spare parts are readily available. This means that spares must be in inventory when they are needed. If there is waiting time to obtain a spare, the unit downtime will increase, and availability will suffer. If the unit has a constant failure rate, the probability of requiring no more than r replacement units can be found using the cumulative Poisson distribution.

The probability of having r or fewer failures in operating time t if the unit is operating with a constant failure rate of λ can be calculated using a Poisson distribution with mean equal to λt.

$$\Pr(r) = \sum_{n=0}^{r} \frac{(\lambda t)^n e^{-\lambda t}}{n!}$$

where

r = the number of failures

λ = the failure rate

t = the operating time

n = the number of spares required for preventative maintenance

Alternatively, readers can find the probability or r or fewer failures using the cumulative Poisson table found in Appendix C.2.

EXAMPLE 13.4

A unit that operates an average of 2000 hours per year has a constant failure rate:

$$\lambda = 100 \times 10^{-6} \text{ failures/hour}$$

The maintenance strategy is to replace the unit every 5000 operating hours. In a 5-year period:

a. How many spares will be necessary for preventive maintenance?

b. With a 0.90 probability, how many spares will be necessary for corrective maintenance?

Solution:

a. The operating time in a 5-year period is $t = 5 \times 2000 = 10{,}000$ hours.

The preventive maintenance cycle is 1/5000 hours. The number of spares required for preventive maintenance is

$$n = (1/5000) \times (10{,}000) = 2$$

b. We want $P(r) \geq 0.90$.

$$\lambda t = (100 \times 10^{-6})\ (10{,}000) = 1.0$$

From the cumulative Poisson tables in Appendix C.2, use the row with λ value of 1 in the left column and find the first element in this row that is greater than 0.90.

$$P(r \leq 2) = 0.92$$
$$\lambda t = 1.0$$

With a probability of 0.92, the number of spares needed for corrective maintenance is no more than two.

If preventive maintenance requires spare parts, the number of spares necessary is a function of the operating time and the maintenance cycle. The number of spares required for preventive maintenance is the product of the maintenance cycle and the total operating time. The total number of spares that must be kept in inventory is a function of the delivery time, the cost of maintaining the inventory, and the availability requirement.

Several methods are in use for forecasting spare parts requirements. Four methods are presented here. Experience will help determine which is best for a given part.

Moving Average

Using the moving average, the forecasted value of the spare parts requirements is the average of the previous n requirements. The value of n is determined by experience. Using a larger n involves requirements farther in the past; using a smaller n involves only more recent requirements. The moving average formula is shown in Equation 13.7.

Equation 13.7 Moving average for spare parts forecasting.

$$R_t = \frac{X_{t-1} + X_{t-2} + \cdots + X_{t-n}}{n}$$

where:

R_t = Forecasted spare parts requirement at time t

X_t = Actual spare parts requirement at time t

n = Number of requirements used in calculation

EXAMPLE 13.5

If the best value of n has been found to be 5, and the last eight weeks' requirements were given as 23, 25, 22, 20, 21, 24, 23, 26, then the predicted spare parts requirement for this part for the ninth week is:

$$R_9 = \frac{20 + 21 + 24 + 23 + 26}{5} \approx 23$$

Weighted Moving Average

This method uses the last n spare parts requirements and weighs the most recent values higher than the previous values. The n-period weighted moving average uses the formula in Equation 13.8.

Equation 13.8 Weighted moving average for spare parts forecasting.

$$R_t = \frac{nX_{t-1} + (n-1)X_{t-2} + \cdots + (1)X_{t-n}}{n + (n-1) + \cdots + 1}$$

where:

R_t = Forecasted spare parts requirement at time t

X_t = Actual spare parts requirement at time t

n = Number of requirements used in calculation

EXAMPLE 13.6

If the best value of n has been found to be 5, and the last eight weeks have seen spare part requirements for a particular part of 23, 25, 22, 20, 21, 24, 23, 26, then the predicted spare parts requirement for this part for the ninth week is:

$$R_9 = \frac{(5)26 + 4(23) + 3(24) + 2(21) + 20}{5 + 4 + 3 + 2 + 1} \approx 24$$

Single Exponential Smoothing

We can also forecast spare parts requirements using single exponential smoothing in which observations are exponentially weighted using a smoothing factor α. This method assumes no trends or seasonality. The formula for predicting the spare parts requirements is shown in Equation 13.9.

Equation 13.9 Single exponential smoothing model for spare parts forecasting.

$$R_t = \alpha X_{t-1} + (1 - \alpha) R_{t-1}$$

where:

R_t = Forecasted spare parts requirement at time t

X_t = Actual spare parts requirement at time t

α = Smoothing factor, usually between 0.1 and 0.4.

Note that unstable demand situations use higher α values. The higher the value of α, the more sensitive the forecast will be to recent values. Thus, an unstable demand situation benefits from the use or higher α values. A lower α value results in a forecast closer to the arithmetic mean. Experience dictates the best value for α.

Note that, to get the time series started, it is common practice to use the forecast for time $t = 2$ to be the actual requirement for time $t = 1$.

EXAMPLE 13.7

Assume time periods of one week are used, and the spare parts requirement for the first week was 23 items. Assume further that an α value of 0.2 is used, and that the actual requirement for the second week was 16 parts. The forecasted spare parts requirement for the third week would be:

$$R_3 = 0.2(16) + 0.8(23) = 21.6 \text{ or } 22 \text{ items}$$

If the actual requirement for week three were 18 items, the forecasted requirement for week four would be

$$R_4 = 0.2(18) + 0.8(21.6) = 20.9 \text{ or } 21 \text{ items}$$

Croston's Method

Croston's method is useful in forecasting intermittent demand. The technique applies exponential smoothing to both the magnitude of the spare parts requirement Z as well as P, the time period between demands for the spare part. If no spare part is needed by the end of the review period at time t, then the values of Z and P for the next period are unchanged.:

$$\begin{aligned}&For\ X_t = 0,\\&Z_{t+1} = Z_t\\&P_{t+1} = P_t\end{aligned}$$

And the demand forecast R_{t+1} is unchanged from the previous period:

$$R_{t+1} = R_t$$

If, however, spare parts are needed at time t (i.e., $X_t > 0$), the revised estimates for Z and P are given by the smoothing formulas:

$$\begin{aligned}&For\ X_t > 0,\\&Z_{t+1} = \alpha X_t + (1-\alpha) Z_t\\&P_{t+1} = \beta G_t + (1-\beta) P_t\end{aligned}$$

And the one-period-ahead demand is forecasted by:

$$R_{t+1} = \frac{Z_t}{P_t}$$

where

X_t = Actual value of demand at time t

Z_t = Forecasted value of demand at time t

G_t = Actual time interval between the last nonzero demand and this demand

P_t = Forecasted value of time between previous nonzero demand and this demand

α, β = Smoothing factors between 0 and 1

To initialize the forecasts, we let Z_2 = magnitude of the first occurrence of nonzero demand, P_2 = (1 + time of first occurrence of non-zero demand), and $R_2 = (X_1/P_1)$. Example 13.8 illustrates the use of Croston's method.

EXAMPLE 13.8

In a maintenance shop, time periods of one week are used. The actual spare parts requirements for weeks 1–3 are shown in Table 13.1.

Table 13.1 Actual weekly spare parts requirements in a maintenance shop.

Week	Actual Spare Parts Required
1	23
2	0
3	16

Continued

Assuming $\alpha=0.2$ and $\beta=0.3$, use Croston's method to forecast the spare parts requirement for the fourth week. To predict demand for week 4, we will calculate the forecasts for weeks 2–3. The Z_2 and P_2 forecasts are set at X_1 and G_1, respectively, to get the forecasts started. Then,

Table 13.2 Croston's method calculations for spare parts forecast.

Week	Actual spare part demand = *X*	*G*	*Z*	*P*	*R* = *Z*/*P*
1	23	1			
2	0		23.00	1.00	23.00
3	16	2	23.00	1.00	23.00
4			21.60	1.30	16.62

The forecast for week 3 remains unchanged from the week 2 forecast since there was no demand at the end of week 2. For week 4, we note that $G_3=2$ (two weeks' elapsed time between demand in week 3 demand and demand in week 1). Then,

$$Z_{t+1}=\alpha X_t+(1-\alpha)Z_t$$
$$Z_4=0.2(X_3)+(1-0.2)Z_3$$
$$Z_4=0.2(16)+(0.8)(23)=21.6$$

$$P_{t+1}=\beta G_t+(1-\beta)P_t$$
$$P_4=(0.3)G_3+(1-0.3)P_3$$
$$P_4=(0.3)(2)+(0.7)(1)=1.3$$

$$R_{t+1}=\frac{Z_t}{P_t}$$
$$R_4=\frac{21.6}{1.3}=16.6$$

PREVENTIVE MAINTENANCE (PM) ANALYSIS

> Define and use PM tasks, optimum PM intervals, and other elements of this analysis. Identify situations when PM is not effective. (Apply)
>
> **Body of Knowledge V.C.2**

The goal of preventive maintenance is to optimize system reliability. Procedures must be established for preventive maintenance, taking into consideration the failure rate pattern of the system or its various subsystems. Cost savings can be realized if preventive maintenance (replacement) is performed on systems with increasing failure rates. A preventive maintenance action can also be advantageous for a unit with if its failure rate will begin to increase if the maintenance is not performed. For example, if a unit in its useful life phase is not lubricated periodically, the failure rate could increase due to excessive wear. If the preventive maintenance action is to replace a working unit, this would only be advantageous if the unit is nearing the end of its useful life or entering wear-out.

It is possible that the maintenance action could induce a failure, causing a negative impact on the reliability of the system. An optimum maintenance interval can be determined by considering the cost of performing the maintenance, the cost of a failure if the maintenance is not performed, and the cost associated with system downtime. If the interval is too short, the maintenance costs increase. If the interval is too long, the costs due to failure increase.

The cost due to system downtime can be minimized if preventive maintenance can be scheduled when there will be no demand to use the system. Downtime can also be reduced if scheduling will allow multiple maintenance actions to be performed once the system is taken off-line.

It is possible that the preventive maintenance action will consist of replacing a unit before failure can occur. O'Connor (2002) lists the following considerations as necessary to determining the optimum replacement time:

1. The time-to-failure distribution and the parameters of that distribution.
2. The effect on the system of the failure.
3. The cost of the failure. This includes the cost of downtime because of the failure and any cost incurred due to safety considerations because of the failure.
4. The cost of the scheduled maintenance, including the cost of the replaced unit.
5. The effect that scheduled maintenance has on reliability. In other words, will the maintenance activity introduce failures into the system?
6. The ability of the operator to detect a potential failure. Can the operator take corrective action before the failure propagates throughout the system, causing other failures?
7. The cost of inspection and testing.

The lognormal distribution is often used to model times to repair. Most repairs can be made in a time less than the mean.

A mean for preventive maintenance actions can be found by replacing the failure rate for each failure mode with the frequency of occurrence of each preventive maintenance action and the time to perform that maintenance action. The steady state or inherent availability of a system with a constant failure rate is

$$A = \frac{MTBF}{MTBF + MTTR}$$

The MTTR is the mean time to repair the system to service after a failure. In other forms of the availability relationship, MTBF is replaced with a term that expresses the mean time between system downtimes for any reason, and MTTR is replaced with a term that includes corrective maintenance, preventive maintenance, and logistic and administrative downtimes. To increase the availability of a system, the MTBF can be increased and/or the MTTR can be decreased.

CORRECTIVE MAINTENANCE ANALYSIS

> Describe and apply the elements of corrective maintenance analysis (e.g., fault-isolation time, repair/replace time, skill level, and crew hours). (Apply)
>
> **Body of Knowledge V.C.3**

Corrective maintenance is performed in the event of a failure or malfunction of the system. Corrective maintenance actions cannot be planned, as the time the system will fail is not known. The goal is to return the system to an operable state while incurring a minimum amount of downtime. The total downtime includes both active maintenance time and inactive or delay time. This active corrective maintenance time can be quantified as time to repair. MTTR can be calculated using the probability that the repair will be necessary and the average time required to perform that repair. The active corrective maintenance time can be determined by summing the times needed to perform seven steps (MIL-HDBK-472). The seven steps are:

1. *Localization.* Determining the system fault without using test equipment
2. *Isolation.* Verification of the system fault using test equipment
3. *Disassembly.* Accessing the fault
4. *Interchange.* Replacing or repairing the fault
5. *Reassembly.* Rebuilding the item
6. *Alignment.* Adjusting the item per specifications
7. *Checkout.* Performing standard quality assurance (QA) checks to verify correct operation

To minimize the repair time, it is necessary to minimize the time required for each step. The time to work on reducing maintenance time is during the design phase of product development. The MTTR for each failure could be found.

Fault Isolation Time

Steps 1 and 2 in the previous list, *localization* and *isolation,* cover the time spent to locate, confirm, and troubleshoot the exact cause of a system or equipment failure. The tools and knowledge used in this process can be very simple or incredibly complex and specialized, depending on the system.

If no specialized tools are available, conduct a thorough review of historical failure data to determine common failure points that may be impacting this situation. If no historical failure information is available or it is not relevant to the current issue, then use root-cause analysis tools to determine the most appropriate place to begin troubleshooting. If the system is complex, additional diagnostic protocols are helpful to streamline determining the point of failure.

The task of isolating faults in complex systems, especially systems that are computer controlled, can be difficult and time-consuming. Computer-controlled automotive engines, flight control systems, electronic-controlled weapons systems, and many other complex systems can be designed to include built-in testing (BIT). The design could include test ports to connect diagnostic testing equipment to the system or have indicators to display the probable fault. BIT can increase the availability of the system by reducing the time to locate and isolate the fault so that the process of restoring the system to service can begin earlier.

Failure of the BIT system can negatively impact reliability. The BIT system in many cases consists of additional hardware. The probability of failure can increase as additional complexity is designed and built into the system.

Improvement in reliability can be achieved if it is possible to use software to monitor the system and report faults instead of using mechanical or electrical sensors. To minimize the total downtime, it is also necessary to minimize the inactive maintenance time. Inactive maintenance time includes waiting time to obtain spare parts, tools, or test equipment; any delay in delivering the failed system to the repair facility; and administrative delay time.

Repair and Replace Time

The remaining steps determine the time it takes to repair and replace the faulty component. Time included in this category includes time for disassembly, actual repair/replacement, wait time for any components, reassembly, and checkout of the repair to determine the system is now functional.

A designer would not intentionally design an automotive system in such a way that it was necessary to remove the engine to replace a broken fan belt. However, unless there is active maintainability engineering input at the design stage, such absurdities can be sent to the customer. Good design practice would dictate that repairs with long MTTR values will rarely be necessary. Equipment that is designed to be maintained must be easy to disassemble, repair, and reassemble.

For systems that are already in place, ways to minimize downtime during this stage of maintenance are standardizing procedures and stocking spare parts for critical components. Time can also be reduced by providing detailed repair manuals and training repair personnel to become proficient in specific repair tasks that either occur frequently or have a history of time overruns.

Crew Hours and Skill Level

Although crew and skill levels are not an official step in the corrective maintenance process, the people that are completing the repair steps will have an impact on the time and cost needed to repair.

The number of people on the repair should be selected to minimize costs. Adding more people to perform the repair reduces downtime to a point. However, at some point, additional people add costs with little value added. When selecting repair crews, optimize crew size to balance efficiency and cost-effectiveness. If the repair takes place across multiple shifts or across different locations, make sure there is clear communication between teams, so all maintenance tasks are completed without duplication. For recurring repairs, the tasks and the time to repair should be monitored to identify patterns and improve resource allocation for future repairs.

In addition to having the right number of people on the job, you need to have the right people on the job. Skill level is an important consideration during unplanned maintenance as a person with the appropriate skill set will be able to complete the fault identification or repair in a shorter time than someone who has a more generalized skill set. Those setting up the repair should match the skill level of technicians with the complexity of maintenance tasks to optimize the repair. If there are no people available with the correct skill set, the repair can be outsourced, but that runs the risk of increasing downtime due to scheduling. For unplanned maintenance, it is more effective to train and certify internal personnel as well as continuing to maintain their training.

EXAMPLE 13.9

A critical pump needs to be repaired. Typical repairs to this pump require a two-person crew to identify and fix the problem. One skilled person who knows how to fix the pump is available. The skilled person costs $60/hr. Other crew members cost $40/hr.

The skilled person, along with another crew member, can identify the problem in an hour and fix the pump in 3 hours. It would take general crew members 2 hours to find the problem and 4 hours to fix it. Which team is more effective at the repair?

- Skilled crew: (1 hr)($60/hr + $40/hr) + (3 hr)($60/hr + $40/hr) = $400
- General crew: (2 hr)($40/hr + $40/hr) + (4 hr)($40/hr + $40/hr) = $480

Not only is the crew with the skilled worker more cost-effective, but this crew also brings the pump back into production 2 hours sooner than the general crew.

EQUIPMENT WARRANTIES

In manufacturing, equipment warranties play a crucial role in ensuring operational efficiency, managing cost, and minimizing the risk of damage to the equipment. A warranty is a formal assurance provided by the equipment manufacturer or supplier guaranteeing the performance, reliability, and quality of the covered equipment over a specified period of time, which will vary based on the manufacturer and the type of equipment. Understanding and leveraging warranties effectively can save manufacturers significant costs associated with repairs, replacements, and downtime.

Key components of equipment warranties include the following:

- *Duration.* The period during which the warranty is valid.
- *Coverage.* Specific parts, systems, maintenance, or failures included in the warranty.

- *Exclusions.* Scenarios or components not covered, such as wear and tear, misuse, or damage.
- *Claim process.* Steps for reporting issues, including documentation and timelines.
- *Service response time.* The expected time frame for warranty services to be rendered.

Warranty Types

Depending on the type of equipment and where it is being used, the organization may opt for different types of warranties.

- *Standard manufacturer warranty.* This covers defects in materials and workmanship for a set duration and also typically includes repair or replacement of faulty components. The duration of the standard warranty may vary for different components on the same piece of equipment. For example, a conveyor system may have a 1-year warranty on the drive system but a 2-year warranty for the belt itself.

  Normal wear and tear on the equipment, accidental damage, misuse or neglect, and unauthorized repairs are typically not covered under a standard warranty. The warranty will specify what the customer must do or not do with the equipment for the supplier to honor the warranty through its duration.
- *Extended warranty.* This is purchased separately to extend the coverage period beyond the standard warranty. An extended warranty may offer additional services like preventive maintenance or priority support. The supplier may offer extended warranties themselves, or they can be purchased through a third-party company. For companies that do not have the in-house ability to repair the equipment, an extended warranty can offer assurance that the equipment will be maintained properly.
- *Performance warranty.* This ensures the equipment will meet specified operational benchmarks, such as production output or energy efficiency over the duration of the warranty. If the equipment does not meet these benchmarks, the backer of the warranty is obliged to repair or replace the equipment and potentially compensate the buyer for the shortfall in performance.
- *Service warranty.* This covers routine maintenance and servicing during the warranty period. This is helpful for organizations that do not have trained personnel or tools required to maintain the equipment in-house. This is a premium warranty that is added to the standard warranty. Repairs or replacement would typically be covered under the standard warranty, which may have a different duration than the service warranty.
- *Implied warranty.* This assumes there are unspoken and unwritten guarantees that are automatically conferred on a sale of goods or services. They ensure that the product sold is fit for its intended purpose and meets certain standards of quality and performance. Implied warranties

provide consumers with essential protections, even if they are not explicitly stated in a contract.

Best Practices for Warranty Use

A warranty, like any other contract, can be rendered void if the organization does not abide by the provisions stated in the warranty. To avoid issues with filing a claim, the organization should consider the following matters.

- *Understand the terms and conditions.* Carefully review warranty documents to ensure clarity on coverage and exclusions. Ensure compliance with warranty stipulations, such as following the prescribed maintenance schedule or not attempting any repairs.
- *Maintain proper documentation.* Keep records of purchase, warranty certificates, and all communication with the supplier or other guarantor of the warranty. Document maintenance activities, training of anyone working with the equipment, and any issues encountered with the equipment.
- *Train staff on proper usage.* Educate operators and maintenance personnel on best practices to prevent voiding warranties. Ensure they know what they can and cannot do to repair the system if an issue arises.
- *Act promptly on issues.* Report malfunctions or failures as soon as they occur to avoid worsening the problem and ensure eligibility for warranty services. Swift reporting of issues will also minimize downtime since the sooner a problem is reported, the faster someone can be scheduled to fix the problem.
- *Leverage warranty services strategically.* Schedule repairs and replacements during planned downtime to minimize disruption to production. With a warranty, the work will typically need to be scheduled with an outside party. Having any outside maintenance scheduled in advance to cover downtime will allow for more availability of the equipment.
- *Understand warranty challenges.* Due to ambiguities in coverage, there can be misunderstandings about what is covered under the warranty. This can lead to disputes with the supplier over what the supplier and the equipment owner are responsible for. Service delays can happen with unplanned maintenance since outside assistance needs to be scheduled and may not be immediately available. Failure to adhere to maintenance schedules or improper equipment usage may void warranties, leading to unplanned delays and expenses.

Part VI

Appendices

Appendix A
Certified Reliability Engineer (CRE) Body of Knowledge

BODY OF KNOWLEDGE (BOK)

The topics in this Body of Knowledge include additional detail in the form of subtext explanations and the cognitive level at which the questions will be written. This information will provide useful guidance for both the Examination Development Committee and the candidates preparing to take the exam. The subtext is not intended to limit the subject matter or be all-inclusive of what might be covered in an exam. It is intended to clarify the type of content to be included in the exam. The descriptor in parentheses at the end of each entry refers to the highest cognitive level at which the topic will be tested. A more comprehensive description of cognitive levels is provided at the end of this document.

I. Reliability Fundamentals (25 questions)

A. Leadership Foundations

1. Benefits of reliability engineering

   Describe the value that reliability has on achieving company goals and objectives, and how reliability engineering techniques and methods improve programs, processes, products, systems, and services. (Understand)

2. Interrelationship of safety, quality, and reliability

   Describe the relationship of and distinguish between reliability and quality. Describe the importance of safety in reliability engineering and how reliability impacts safety. (Understand)

3. Reliability engineer leadership responsibilities

   Describe how to be a reliability champion by influencing program decisions and facilitating cross-functional communication. Understand the fundamentals of reliability strategy, including its mission, vision, objectives, and requirements. (Understand)

4. Reliability engineer role and responsibilities in the product lifecycle

   Describe how the reliability engineer influences the product lifecycle, including their role in anticipating the impact of reliability on risk and costs in the design review process and ensuring performance over time. (Understand)

5. Project management in reliability engineering

   Apply key practices in requirements management, use Gantt charts, and understand the critical path method (CPM). (Apply)

6. Function of reliability in engineering

   Describe how reliability techniques can be used to apply best practices in engineering (e.g., measuring reliability early), how industry standards can impact reliability, and how reliability can inform the decision analysis process. (Analyze)

7. Ethics in reliability engineering

   Identify appropriate ethical behaviors for a reliability engineer in various situations. (Evaluate)

8. Supplier reliability assessments

   Explain how supplier reliability impacts the overall reliability program. Describe key reliability concepts that should be included in supplier reliability assessments. Continuously assess supplier reliability to ensure they meet established reliability standards. (Analyze)

9. Performance monitoring

   Describe the importance of performance monitoring in maintaining compliance with product reliability and safety standards. Identify key points in the product lifecycle where both process and product reliability data are collected and evaluated. Integrate essential reliability elements like defect tracking and reliability growth monitoring (e.g., key performance indicators [KPIs], overall equipment effectiveness [OEE], Duane growth model, and Crow-AMSAA Model) to ensure continuous adherence to these requirements. (Understand)

B. Reliability Foundations

1. Basic reliability terminology

   Clarify fundamental reliability terms and the corresponding metrics, including concepts related to system reliability (e.g., durability, failure rate, mean time to failure [MTTF], mean time between failures [MTBF], mean time between critical failure [MTBCF], and the bathtub curve), maintainability (e.g., service interval, mean time to repair [MTTR], mean time between maintenance [MTBM], and mean downtime [MDT]), and availability (e.g., operational availability, achieved availability, and inherent availability). (Apply)

2. Drivers of reliability requirements and targets

   Describe how customer expectations, industry standards, safety considerations, liability issues, and regulatory concerns drive reliability requirements. Assess expected use conditions or use cases, align reliability requirements with environmental, social, and governance (ESG) policies, and draw valuable insights from lessons learned. (Understand)

3. Corrective and preventive action (CAPA)

   Evaluate specific situations that call for corrective and preventive actions. Explore the implementation of these actions and assess the effectiveness of the measures taken. (Evaluate)

4. Root cause analysis

   Assess root cause analysis by using specialized tools, such as fishbone diagrams, 5 whys, and 8D to investigate the causes of degradation and failure. (Evaluate)

5. Product lifecycle engineering stages

   Examine how different lifecycle stages—concept / design, development / test, introduction, growth, maturity, and decline—affect reliability. Inspect the cost issues related to each of those stages, including aspects like product maintenance (both scheduled and unscheduled), life expectancy, and software defect phase containment. (Understand)

6. Economics of product maintainability and availability

   Describe the cost tradeoffs associated with product maintainability strategies and availability. (Understand)

7. Cost of poor reliability

   Describe the financial and non-financial expenses (e.g., availability, credibility, business operations, and reputation) incurred because of poor reliability. (Understand)

8. Quality triangle

   Describe the relationship between cost, time, and quality with respect to reliability. (Understand)

9. Six sigma methodologies

   Describe how six sigma principles serve as a supportive framework for enhancing reliability engineering practices. Explain how reliability engineering aligns with the principles of the DMAIC (Define, Measure, Analyze, Improve, Control) process, continuous improvement, and lean methodologies. (Understand)

10. Systems engineering and integration

    Comprehend the relationship between reliability engineering and systems engineering, with a focus on integrating components and their interactions within a system. (Understand)

II. Risk Management (25 questions)

A. Identification

1. Risk management techniques

   Use risk management tools and processes to identify, document, and monitor environmental, liability, and security concerns. Incorporate methodologies such as p- diagrams to discern potential failure modes and use cases to conceptualize different real-world scenarios and interactions. Evaluate and prioritize risks related to safety, economics, performance, and customer satisfaction. Address these risks within an appropriate risk management framework. (Evaluate)

2. Risk assessment

   Employ different techniques for assessing and prioritizing risks including qualitative, quantitative, and semi-quantitative methods. Apply risk ranking methods to ascertain the potential impacts, considering likelihood and prioritization factors. Use probabilistic risk assessment (PRA) to categorize risk events and to understand the potential impacts. Perform vulnerability and threat-based assessments to analyze and gauge the susceptibility of the system to various threats. (Analyze)

3. Types of risk

   Examine the various types of operational risks (e.g., technical, scheduling, safety, and environmental), strategic risks (e.g., brand, reputation, stakeholder, and regulatory compliance), financial risks, cybersecurity risks, and analytical risks (e.g., informed and inherent) and their relationship to reliability. (Analyze)

B. Analysis

1. Fault tree analysis (FTA)

   Use fault tree analysis (FTA) to evaluate the potential failures within a product or process. Employ action priority to categorize and prioritize the necessary corrective actions. (Analyze)

2. Failure mode and effects analysis (FMEA)

   Define and distinguish between failure mode and effects analysis (FMEA), failure mode, effects, and criticality analysis (FMECA), functional failure mode and effects analysis (FFMEA), and use failure mode and effects analysis (UFMEA). Evaluate these techniques for use on systems, products, processes, and designs. (Evaluate)

3. Common mode failure analysis

   Describe common mode failure (also known as common cause failure) and how it affects risk. (Understand)

4. Hazard analysis

   Explain how hazard analysis informs the development process and how the information obtained from it is used by reliability engineers. (Understand)

5. Risk matrix

   Explain the application of risk matrices in evaluating risk, focusing on how they aid in assessing the likelihood and the severity of potential adverse events or outcomes. (Understand)

6. System safety

   Identify safety-related issues by analyzing multiple sources including customer feedback, design data, field data, and other pertinent information. Prioritize safety concerns, focusing on those that possess the highest risk of occurrence and impact. Determine steps to minimize the improper or unintended use of equipment, products, or processes, aligning with safety system. Employ tradeoff analysis to determine the optimal risk control measures. (Evaluate)

C. Mitigation

Identify appropriate risk mitigation plans by integrating the guidelines and principles outlined in standards such as ISO 31000 and ISO 55000. Assess risk control plans that incorporate effective controls, aiming to minimize both inherent risks and their subsequent impacts, focusing on safety, liability, and regulatory compliance. Distinguish between the reliability goals of As Low As Reasonably Practical (ALARP), As Low As Reasonably Achievable (ALARA), and As Low As Possible (ALAP). (Evaluate)

III. Probability and Statistics for Reliability (35 questions)

A. Basic Concepts

1. Basic statistics

   Define basic statistical terms such as population, parameter, sample, and statistic. Understand statistical concepts such as sampling and the central limit theorem.

   Differentiate between parametric and non-parametric (e.g., Kaplan-Meier Analysis) statistical methods. Use appropriate methods to estimate and interpret statistical values. (Analyze)

2. Basic probability concepts

   Apply basic probability concepts such as independence, mutually exclusive events, and conditional probability. Calculate probabilities using tools and techniques (e.g., probability trees and expected frequency trees) and interpret their meaning. (Analyze)

3. Probability distributions

   Analyze reliability data using appropriate statistical distributions such as binomial, Poisson, exponential, Weibull, normal, log-normal, chi-square, and Student's t distribution. Interpret the associated probability plots for each continuous distribution. Assess goodness of fit to a distribution. (Analyze)

4. Probability functions

   Assess various probability functions (e.g., cumulative distribution functions [CDFs], probability density functions [PDFs], and hazard functions) and recognize their application in various situations. (Evaluate)

5. Sampling plans for statistics and reliability testing

   Use various theories, tables, and formulas to determine appropriate sample sizes or testing time needed for statistical and reliability testing. Incorporate representative and randomized sampling techniques. (Apply)

6. Statistical process control (SPC) and capability studies (Cp, Cpk, Pp, and Ppk).

   Apply statistical process control (SPC), capability studies, and their associated indices (e.g., Cp, Cpk, Pp, and Ppk) and describe how they relate to reliability. Considering differing sample sizes, select control charts to represent variability over time.

   (Apply)

7. Confidence and tolerance intervals

   Compute confidence intervals and tolerance intervals to analyze reliability with chosen confidence. Assess the use of confidence and tolerance intervals for reliability analysis using Weibull, normal, and lognormal distributions. Describe how point estimates are used to determine the interval. (Evaluate)

B. Data Management

1. Sources and uses of reliability data

   Analyze the sources of reliability data (e.g., prototype, development, test, field, warranty, published data, big data, and the internet of things [IoT]). Recognize the unique advantages and limitations of each for measuring and enhancing product reliability. Consider the importance of normalizing datasets and transformations. (Analyze)

2. Types of data

   Evaluate and distinguish among diverse data types, including attributes versus variables and discrete versus continuous. Determine whether the data is censored or complete and whether it is univariate or multivariate. Describe the implications of deploying parametric and nonparametric statistical methods, identifying the underlying assumptions and applicable conditions of each. Recognize the different scales of measurement, including nominal, ordinal, interval, and ratio. Align analysis tools—such as Nevada chart analysis, survival analysis, and the Cox proportional hazard model—with the inherent characteristics and requirements of the respective data types. (Evaluate)

3. Data collection methods

   Identify and select appropriate data collection methods (e.g., surveys, automated tests, automated monitoring, and reporting tools) to meet various data analysis objectives and data quality needs. (Evaluate)

4. Data summary and reporting

   Analyze the collected data and assess its accuracy, precision, usefulness, and integrity. Employ analytical techniques, such as bad actor analysis. Use artificial intelligence (AI) to delve deeper into the analysis. Select appropriate graphical representations, such as Pareto charts, scatter plots, and box and whisker plots. Summarize, interpret, and present the analyzed data, choosing methods aligned with the varying data types, originating sources, and the specified output requirements. (Create)

5. Failure analysis methods

   Diagnose failures by using a diverse array of failure analysis tools and techniques. Implement physics of failure-based analyses, including methods such as scanning electron microscopy (SEM), scanning acoustic microscopy (SAM), infrared inspection, and radiography analysis. Conduct mechanical and materials analysis, applying specific material testing and measurement techniques like tensile testing, shear testing, viscosity measurement, and vibration testing. Incorporate physical analysis methods including visual, ultrasonic, and electrical analysis. Engage in verification like cycle testing and nondestructive testing (NDT) and employ advanced predictive technologies to anticipate and mitigate potential failures. (Understand)

6. Failure reporting, analysis, and corrective action system (FRACAS)

   Identify elements necessary for FRACAS and demonstrate the importance of a closed-loop process. (Evaluate)

IV. Reliability Planning, Testing, and Modeling (35 questions)

A. Planning

1. Reliability test strategies

   Evaluate and develop suitable test strategies, such as truncation, test-to-failure, degradation, growth plan, and Test, Analyze, and Fix (TAAF), for various phases of product development. Consider customer profiles, including those reflecting normal and heavy usage. Implement a 'zero failure' test to assess reliability. (Evaluate)

2. Environmental factors and use conditions

   Analyze various environmental factors and use conditions such as temperature, humidity, and vibration, along with stresses like severity of service, electrostatic discharge (ESD), throughput, and duty cycle, to which a product may be exposed. Incorporate the application of multiple stress factors (e.g., highly accelerated life testing [HALT]) to assess their simultaneous impact on a product. (Analyze)

3. Failure consequence

   Describe the importance of recognizing the consequences linked to different failure modes, including the severity and occurrence of such failure modes, when establishing reliability acceptance criteria. (Understand)

4. Failure criteria

   Establish failure criteria by considering customer requirements, user needs, system functions, system requirements, and warranty terms and conditions. (Understand)

5. Test environment

   Evaluate the test environment by considering the system location and operational conditions and incorporate these into the test plan to ensure the implementation of an appropriate test strategy.

Validate test capability and conduct the tests following the verified methods. Evaluate and formulate decisions based on the test results, providing rationale and support for each decision made. (Evaluate)

B. Testing

Describe the purpose, advantages, and limitations of various hardware and software / firmware tests, emphasizing the different approaches and methodologies required for each. Use common models to develop test plans, evaluate risks, and interpret test results, ensuring comprehensive reliability and functionality assessment.

1. Accelerated life tests (e.g., single-stress, multiple-stress, sequential stress, step- stress, HALT, and margin tests) (Evaluate)
2. Stress screening (e.g., ESS, HASS, and burn-in tests) (Evaluate)
3. Qualification / demonstration testing (e.g., sequential tests and fixed-length tests) (Evaluate)
4. Degradation (wear-to-failure) testing (Evaluate)
5. Software / firmware reliability (e.g., software design reliability, software reliability prediction, software design and development models, and software reliability allocation) (Understand)
6. Software testing (e.g., white-box, black-box, operational profile, fault-injection, software validation and quality assurance, software verification, qualitative versus quantitative reliability testing, built-in testing [BIT], and regression testing to verify code integrity) (Understand)

C. Modeling

1. Reliability block diagrams and models

   Generate and analyze various types of block diagrams and models, including series, parallel, partial redundancy, time dependent, K out of N, and shared load. (Evaluate)

2. Physics of failure and failure mechanisms

   Identify various potential failure mechanisms including fracture, corrosion, memory corruption, excessive deformation, creep, and delamination, and their underlying physical processes. (Apply)

3. Failure models

   Select appropriate theoretical models to assess or predict failure rates, such as Arrhenius, S-N curve, and Coffin-Manson model for temperature cyclic stress. (Analyze)

4. Reliability prediction methods

   Explain various reliability prediction methods—including Monte Carlo simulation, part stress analysis, parts count prediction, Markov analysis, neural networks, and machine learning—to assess both repairable and non-repairable components and systems and describe the inputs into the model. (Understand)

5. Design prototyping

   Describe the advantages and limitations of prototyping and rapid prototyping technologies to enhance product reliability. Employ phase diagrams to model various equipment scenarios across different mission profiles. Determine the correlation of prototype results to simulated results (e.g., digital twins and damage modeling). (Understand)

V. Lifecycle Reliability (30 questions)

A. Reliability Design Techniques

1. Design evaluation techniques (validation and verification)

   Explain how validation, verification, and other review techniques are used to assess the reliability of a product's design at various lifecycle stages. (Apply)

2. Stress-strength analysis

   Apply the stress-strength analysis method of calculating probability of failure and interpret the results. (Analyze)

3. Design of experiments (DOE)

   Develop and interpret the results of a standard design of experiments (DOE), including both full-factorial and fractional factorial designs, integrating methodologies such as analysis of variance (ANOVA) and blocking for hard to control factors (e.g., planned grouping). Use replication and randomization to optimize the robustness and validity of the experimental findings. (Analyze)

4. Reliability optimization

   Employ a diverse range of strategies to enhance reliability, considering user needs while balancing constraints such as cost, schedule, and weight along with other design requirements. (Apply)

5. Human factors

   Describe the relationship between human factors and reliability engineering, considering user safety, understanding user and usage profiles, and analyzing associated failure modes and mechanisms. (Understand)

6. Design for X (DFX)

   Apply DFX techniques such as design for manufacturability, testability, and maintainability. (Apply)

7. Design for Reliability (DfR)

   Apply DfR to meet reliability requirements throughout the product or system lifecycle. Integrate tools such as finite element analysis (FEA) to assess and optimize design robustness and to forecast potential areas of failure or stress within the system or product. Make design decisions emphasizing the achievement of built-in reliability and fault tolerance / avoidance as fundamental goals. (Analyze)

B. Parts and Systems Development

1. Materials, components, equipment, and software selection techniques

   Apply techniques (e.g., derating and commercial off-the-shelf [COTS], safe operating area [SOA] criteria) for selecting materials and components to meet reliability goals and requirements. Select software for product and evaluation equipment functionality (e.g., vibration stand and thermal chambers). (Analyze)

2. Parts standardization and system simplification

   Describe the importance of standardization, simplification, and parts re-use to meet reliability goals and requirements. Incorporate reliability-centered maintenance (RCM) to maintain the inherent reliability of system components while ensuring optimal performance and longevity. (Apply)

3. Maintenance strategies

   Develop a maintenance plan incorporating various strategies (e.g., predictive maintenance, repair or replace decision-making, spare parts analysis / forecasting, and equipment warranties). (Apply)

4. Preventive maintenance (PM) analysis

   Define and use PM tasks, optimum PM intervals, and other elements of this analysis. Identify situations when PM is not effective. (Apply)

5. Corrective maintenance analysis

   Describe and apply the elements of corrective maintenance analysis (e.g., fault- isolation time, repair / replace time, skill level, and crew hours). (Apply)

Levels of Cognition

Based on Bloom's Taxonomy—Revised (2001)

In addition to **content** specifics, the subtext for each topic in this BOK also indicates the intended **complexity level** of the test questions for that topic. These levels are based on "Levels of Cognition" (from Bloom's Taxonomy—Revised, 2001) and are presented below in rank order, from least complex to most complex.

REMEMBER

Recall or recognize terms, definitions, facts, ideas, materials, patterns, sequences, methods, principles, etc.

UNDERSTAND

Read and understand descriptions, communications, reports, tables, diagrams, directions, regulations, etc.

APPLY

Know when and how to use ideas, procedures, methods, formulas, principles, theories, etc.

ANALYZE

Break down information into its constituent parts and recognize their relationship to one another and how they are organized; identify sublevel factors or salient data from a complex scenario.

EVALUATE

Make judgments about the value of proposed ideas, solutions, etc., by comparing the proposal to specific criteria or standards.

CREATE

Put parts or elements together in such a way as to reveal a pattern or structure not clearly there before; identify which data or information from a complex set is appropriate to examine further or from which supported conclusions can be drawn.

Appendix B.1
Binomial Distribution Table

$$\Pr(X = x) = f(n; x, p) = \binom{n}{k} p^k (1-p)^{n-k}$$

Probability of x occurrences in a sample of size n

n	x	0.05	0.10	0.15	0.20	0.25	0.30	0.35	0.40	0.45	0.50
1	0	0.9500	0.9000	0.8500	0.8000	0.7500	0.7000	0.6500	0.6000	0.5500	0.5000
	1	0.0500	0.1000	0.1500	0.2000	0.2500	0.3000	0.3500	0.4000	0.4500	0.5000
2	0	0.9025	0.8100	0.7225	0.6400	0.5625	0.4900	0.4225	0.3600	0.3025	0.2500
	1	0.0950	0.1800	0.2550	0.3200	0.3750	0.4200	0.4550	0.4800	0.4950	0.5000
	2	0.0025	0.0100	0.0225	0.0400	0.0625	0.0900	0.1225	0.1600	0.2025	0.2500
3	0	0.8574	0.7290	0.6141	0.5120	0.4219	0.3430	0.2746	0.2160	0.1664	0.1250
	1	0.1354	0.2430	0.3251	0.3840	0.4219	0.4410	0.4436	0.4320	0.4084	0.3750
	2	0.0071	0.0270	0.0574	0.0960	0.1406	0.1890	0.2389	0.2880	0.3341	0.3750
	3	0.0001	0.0010	0.0034	0.0080	0.0156	0.0270	0.0429	0.0640	0.0911	0.1250
4	0	0.8145	0.6561	0.5220	0.4096	0.3164	0.2401	0.1785	0.1296	0.0915	0.0625
	1	0.1715	0.2916	0.3685	0.4096	0.4219	0.4116	0.3845	0.3456	0.2995	0.2500
	2	0.0135	0.0486	0.0975	0.1536	0.2109	0.2646	0.3105	0.3456	0.3675	0.3750
	3	0.0005	0.0036	0.0115	0.0256	0.0469	0.0756	0.1115	0.1536	0.2005	0.2500
	4	0.0000	0.0001	0.0005	0.0016	0.0039	0.0081	0.0150	0.0256	0.0410	0.0625
5	0	0.7738	0.5905	0.4437	0.3277	0.2373	0.1681	0.1160	0.0778	0.0503	0.0313
	1	0.2036	0.3281	0.3915	0.4096	0.3955	0.3602	0.3124	0.2592	0.2059	0.1563
	2	0.0214	0.0729	0.1382	0.2048	0.2637	0.3087	0.3364	0.3456	0.3369	0.3125
	3	0.0011	0.0081	0.0244	0.0512	0.0879	0.1323	0.1811	0.2304	0.2757	0.3125
	4	0.0000	0.0005	0.0022	0.0064	0.0146	0.0284	0.0488	0.0768	0.1128	0.1563
	5	0.0000	0.0000	0.0001	0.0003	0.0010	0.0024	0.0053	0.0102	0.0185	0.0313
6	0	0.7351	0.5314	0.3771	0.2621	0.1780	0.1176	0.0754	0.0467	0.0277	0.0156
	1	0.2321	0.3543	0.3993	0.3932	0.3560	0.3025	0.2437	0.1866	0.1359	0.0938
	2	0.0305	0.0984	0.1762	0.2458	0.2966	0.3241	0.3280	0.3110	0.2780	0.2344
	3	0.0021	0.0146	0.0415	0.0819	0.1318	0.1852	0.2355	0.2765	0.3032	0.3125
	4	0.0001	0.0012	0.0055	0.0154	0.0330	0.0595	0.0951	0.1382	0.1861	0.2344
	5	0.0000	0.0001	0.0004	0.0015	0.0044	0.0102	0.0205	0.0369	0.0609	0.0938
	6	0.0000	0.0000	0.0000	0.0001	0.0002	0.0007	0.0018	0.0041	0.0083	0.0156
7	0	0.6983	0.4783	0.3206	0.2097	0.1335	0.0824	0.0490	0.0280	0.0152	0.0078
	1	0.2573	0.3720	0.3960	0.3670	0.3115	0.2471	0.1848	0.1306	0.0872	0.0547
	2	0.0406	0.1240	0.2097	0.2753	0.3115	0.3177	0.2985	0.2613	0.2140	0.1641
	3	0.0036	0.0230	0.0617	0.1147	0.1730	0.2269	0.2679	0.2903	0.2918	0.2734
	4	0.0002	0.0026	0.0109	0.0287	0.0577	0.0972	0.1442	0.1935	0.2388	0.2734
	5	0.0000	0.0002	0.0012	0.0043	0.0115	0.0250	0.0466	0.0774	0.1172	0.1641
	6	0.0000	0.0000	0.0001	0.0004	0.0013	0.0036	0.0084	0.0172	0.0320	0.0547
	7	0.0000	0.0000	0.0000	0.0000	0.0001	0.0002	0.0006	0.0016	0.0037	0.0078

0.55	0.60	0.65	0.70	0.75	0.80	0.85	0.90	0.95	n	x
0.4500	0.4000	0.3500	0.3000	0.2500	0.2000	0.1500	0.1000	0.0500	1	0
0.5500	0.6000	0.6500	0.7000	0.7500	0.8000	0.8500	0.9000	0.9500		1
0.2025	0.1600	0.1225	0.0900	0.0625	0.0400	0.0225	0.0100	0.0025	2	0
0.4950	0.4800	0.4550	0.4200	0.3750	0.3200	0.2550	0.1800	0.0950		1
0.3025	0.3600	0.4225	0.4900	0.5625	0.6400	0.7225	0.8100	0.9025		2
0.0911	0.0640	0.0429	0.0270	0.0156	0.0080	0.0034	0.0010	0.0001	3	0
0.3341	0.2880	0.2389	0.1890	0.1406	0.0960	0.0574	0.0270	0.0071		1
0.4084	0.4320	0.4436	0.4410	0.4219	0.3840	0.3251	0.2430	0.1354		2
0.1664	0.2160	0.2746	0.3430	0.4219	0.5120	0.6141	0.7290	0.8574		3
0.0410	0.0256	0.0150	0.0081	0.0039	0.0016	0.0005	0.0001	0.0000	4	0
0.2005	0.1536	0.1115	0.0756	0.0469	0.0256	0.0115	0.0036	0.0005		1
0.3675	0.3456	0.3105	0.2646	0.2109	0.1536	0.0975	0.0486	0.0135		2
0.2995	0.3456	0.3845	0.4116	0.4219	0.4096	0.3685	0.2916	0.1715		3
0.0915	0.1296	0.1785	0.2401	0.3164	0.4096	0.5220	0.6561	0.8145		4
0.0185	0.0102	0.0053	0.0024	0.0010	0.0003	0.0001	0.0000	0.0000	5	0
0.1128	0.0768	0.0488	0.0284	0.0146	0.0064	0.0022	0.0005	0.0000		1
0.2757	0.2304	0.1811	0.1323	0.0879	0.0512	0.0244	0.0081	0.0011		2
0.3369	0.3456	0.3364	0.3087	0.2637	0.2048	0.1382	0.0729	0.0214		3
0.2059	0.2592	0.3124	0.3602	0.3955	0.4096	0.3915	0.3281	0.2036		4
0.0503	0.0778	0.1160	0.1681	0.2373	0.3277	0.4437	0.5905	0.7738		5
0.0083	0.0041	0.0018	0.0007	0.0002	0.0001	0.0000	0.0000	0.0000	6	0
0.0609	0.0369	0.0205	0.0102	0.0044	0.0015	0.0004	0.0001	0.0000		1
0.1861	0.1382	0.0951	0.0595	0.0330	0.0154	0.0055	0.0012	0.0001		2
0.3032	0.2765	0.2355	0.1852	0.1318	0.0819	0.0415	0.0146	0.0021		3
0.2780	0.3110	0.3280	0.3241	0.2966	0.2458	0.1762	0.0984	0.0305		4
0.1359	0.1866	0.2437	0.3025	0.3560	0.3932	0.3993	0.3543	0.2321		5
0.0277	0.0467	0.0754	0.1176	0.1780	0.2621	0.3771	0.5314	0.7351		6
0.0037	0.0016	0.0006	0.0002	0.0001	0.0000	0.0000	0.0000	0.0000	7	0
0.0320	0.0172	0.0084	0.0036	0.0013	0.0004	0.0001	0.0000	0.0000		1
0.1172	0.0774	0.0466	0.0250	0.0115	0.0043	0.0012	0.0002	0.0000		2
0.2388	0.1935	0.1442	0.0972	0.0577	0.0287	0.0109	0.0026	0.0002		3
0.2918	0.2903	0.2679	0.2269	0.1730	0.1147	0.0617	0.0230	0.0036		4
0.2140	0.2613	0.2985	0.3177	0.3115	0.2753	0.2097	0.1240	0.0406		5
0.0872	0.1306	0.1848	0.2471	0.3115	0.3670	0.3960	0.3720	0.2573		6
0.0152	0.0280	0.0490	0.0824	0.1335	0.2097	0.3206	0.4783	0.6983		7

Probability of x occurrences in a sample of size n *(continued)*											
n	x	**0.05**	**0.10**	**0.15**	**0.20**	**0.25**	**0.30**	**0.35**	**0.40**	**0.45**	**0.50**
8	**0**	0.6634	0.4305	0.2725	0.1678	0.1001	0.0576	0.0319	0.0168	0.0084	0.0039
	1	0.2793	0.3826	0.3847	0.3355	0.2670	0.1977	0.1373	0.0896	0.0548	0.0313
	2	0.0515	0.1488	0.2376	0.2936	0.3115	0.2965	0.2587	0.2090	0.1569	0.1094
	3	0.0054	0.0331	0.0839	0.1468	0.2076	0.2541	0.2786	0.2787	0.2568	0.2188
	4	0.0004	0.0046	0.0185	0.0459	0.0865	0.1361	0.1875	0.2322	0.2627	0.2734
	5	0.0000	0.0004	0.0026	0.0092	0.0231	0.0467	0.0808	0.1239	0.1719	0.2188
	6	0.0000	0.0000	0.0002	0.0011	0.0038	0.0100	0.0217	0.0413	0.0703	0.1094
	7	0.0000	0.0000	0.0000	0.0001	0.0004	0.0012	0.0033	0.0079	0.0164	0.0313
	8	0.0000	0.0000	0.0000	0.0000	0.0000	0.0001	0.0002	0.0007	0.0017	0.0039
9	**0**	0.6302	0.3874	0.2316	0.1342	0.0751	0.0404	0.0207	0.0101	0.0046	0.0020
	1	0.2985	0.3874	0.3679	0.3020	0.2253	0.1556	0.1004	0.0605	0.0339	0.0176
	2	0.0629	0.1722	0.2597	0.3020	0.3003	0.2668	0.2162	0.1612	0.1110	0.0703
	3	0.0077	0.0446	0.1069	0.1762	0.2336	0.2668	0.2716	0.2508	0.2119	0.1641
	4	0.0006	0.0074	0.0283	0.0661	0.1168	0.1715	0.2194	0.2508	0.2600	0.2461
	5	0.0000	0.0008	0.0050	0.0165	0.0389	0.0735	0.1181	0.1672	0.2128	0.2461
	6	0.0000	0.0001	0.0006	0.0028	0.0087	0.0210	0.0424	0.0743	0.1160	0.1641
	7	0.0000	0.0000	0.0000	0.0003	0.0012	0.0039	0.0098	0.0212	0.0407	0.0703
	8	0.0000	0.0000	0.0000	0.0000	0.0001	0.0004	0.0013	0.0035	0.0083	0.0176
	9	0.0000	0.0000	0.0000	0.0000	0.0000	0.0000	0.0001	0.0003	0.0008	0.0020
10	**0**	0.5987	0.3487	0.1969	0.1074	0.0563	0.0282	0.0135	0.0060	0.0025	0.0010
	1	0.3151	0.3874	0.3474	0.2684	0.1877	0.1211	0.0725	0.0403	0.0207	0.0098
	2	0.0746	0.1937	0.2759	0.3020	0.2816	0.2335	0.1757	0.1209	0.0763	0.0439
	3	0.0105	0.0574	0.1298	0.2013	0.2503	0.2668	0.2522	0.2150	0.1665	0.1172
	4	0.0010	0.0112	0.0401	0.0881	0.1460	0.2001	0.2377	0.2508	0.2384	0.2051
	5	0.0001	0.0015	0.0085	0.0264	0.0584	0.1029	0.1536	0.2007	0.2340	0.2461
	6	0.0000	0.0001	0.0012	0.0055	0.0162	0.0368	0.0689	0.1115	0.1596	0.2051
	7	0.0000	0.0000	0.0001	0.0008	0.0031	0.0090	0.0212	0.0425	0.0746	0.1172
	8	0.0000	0.0000	0.0000	0.0001	0.0004	0.0014	0.0043	0.0106	0.0229	0.0439
	9	0.0000	0.0000	0.0000	0.0000	0.0000	0.0001	0.0005	0.0016	0.0042	0.0098
	10	0.0000	0.0000	0.0000	0.0000	0.0000	0.0000	0.0000	0.0001	0.0003	0.0010

0.55	0.60	0.65	0.70	0.75	0.80	0.85	0.90	0.95	*n*	*x*
0.0017	0.0007	0.0002	0.0001	0.0000	0.0000	0.0000	0.0000	0.0000	8	0
0.0164	0.0079	0.0033	0.0012	0.0004	0.0001	0.0000	0.0000	0.0000		1
0.0703	0.0413	0.0217	0.0100	0.0038	0.0011	0.0002	0.0000	0.0000		2
0.1719	0.1239	0.0808	0.0467	0.0231	0.0092	0.0026	0.0004	0.0000		3
0.2627	0.2322	0.1875	0.1361	0.0865	0.0459	0.0185	0.0046	0.0004		4
0.2568	0.2787	0.2786	0.2541	0.2076	0.1468	0.0839	0.0331	0.0054		5
0.1569	0.2090	0.2587	0.2965	0.3115	0.2936	0.2376	0.1488	0.0515		6
0.0548	0.0896	0.1373	0.1977	0.2670	0.3355	0.3847	0.3826	0.2793		7
0.0084	0.0168	0.0319	0.0576	0.1001	0.1678	0.2725	0.4305	0.6634		8
0.0008	0.0003	0.0001	0.0000	0.0000	0.0000	0.0000	0.0000	0.0000	9	0
0.0083	0.0035	0.0013	0.0004	0.0001	0.0000	0.0000	0.0000	0.0000		1
0.0407	0.0212	0.0098	0.0039	0.0012	0.0003	0.0000	0.0000	0.0000		2
0.1160	0.0743	0.0424	0.0210	0.0087	0.0028	0.0006	0.0001	0.0000		3
0.2128	0.1672	0.1181	0.0735	0.0389	0.0165	0.0050	0.0008	0.0000		4
0.2600	0.2508	0.2194	0.1715	0.1168	0.0661	0.0283	0.0074	0.0006		5
0.2119	0.2508	0.2716	0.2668	0.2336	0.1762	0.1069	0.0446	0.0077		6
0.1110	0.1612	0.2162	0.2668	0.3003	0.3020	0.2597	0.1722	0.0629		7
0.0339	0.0605	0.1004	0.1556	0.2253	0.3020	0.3679	0.3874	0.2985		8
0.0046	0.0101	0.0207	0.0404	0.0751	0.1342	0.2316	0.3874	0.6302		9
0.0003	0.0001	0.0000	0.0000	0.0000	0.0000	0.0000	0.0000	0.0000	10	0
0.0042	0.0016	0.0005	0.0001	0.0000	0.0000	0.0000	0.0000	0.0000		1
0.0229	0.0106	0.0043	0.0014	0.0004	0.0001	0.0000	0.0000	0.0000		2
0.0746	0.0425	0.0212	0.0090	0.0031	0.0008	0.0001	0.0000	0.0000		3
0.1596	0.1115	0.0689	0.0368	0.0162	0.0055	0.0012	0.0001	0.0000		4
0.2340	0.2007	0.1536	0.1029	0.0584	0.0264	0.0085	0.0015	0.0001		5
0.2384	0.2508	0.2377	0.2001	0.1460	0.0881	0.0401	0.0112	0.0010		6
0.1665	0.2150	0.2522	0.2668	0.2503	0.2013	0.1298	0.0574	0.0105		7
0.0763	0.1209	0.1757	0.2335	0.2816	0.3020	0.2759	0.1937	0.0746		8
0.0207	0.0403	0.0725	0.1211	0.1877	0.2684	0.3474	0.3874	0.3151		9
0.0025	0.0060	0.0135	0.0282	0.0563	0.1074	0.1969	0.3487	0.5987		10

Appendix B.2

Cumulative Binomial Distribution Table

$$\Pr(X \le x) = F(n; x, p) = \sum_{k=0}^{x} \binom{n}{k} p^k (1-p)^{n-k}$$

		Probability of $\le x$ occurrences in a sample of size n									
n	x	0.05	0.10	0.15	0.20	0.25	0.30	0.35	0.40	0.45	0.50
1	0	0.9500	0.9000	0.8500	0.8000	0.7500	0.7000	0.6500	0.6000	0.5500	0.5000
	1	1.0000	1.0000	1.0000	1.0000	1.0000	1.0000	1.0000	1.0000	1.0000	1.0000
2	0	0.9025	0.8100	0.7225	0.6400	0.5625	0.4900	0.4225	0.3600	0.3025	0.2500
	1	0.9975	0.9900	0.9775	0.9600	0.9375	0.9100	0.8775	0.8400	0.7975	0.7500
	2	1.0000	1.0000	1.0000	1.0000	1.0000	1.0000	1.0000	1.0000	1.0000	1.0000
3	0	0.8574	0.7290	0.6141	0.5120	0.4219	0.3430	0.2746	0.2160	0.1664	0.1250
	1	0.9928	0.9720	0.9393	0.8960	0.8438	0.7840	0.7183	0.6480	0.5748	0.5000
	2	0.9999	0.9990	0.9966	0.9920	0.9844	0.9730	0.9571	0.9360	0.9089	0.8750
	3	1.0000	1.0000	1.0000	1.0000	1.0000	1.0000	1.0000	1.0000	1.0000	1.0000
4	0	0.8145	0.6561	0.5220	0.4096	0.3164	0.2401	0.1785	0.1296	0.0915	0.0625
	1	0.9860	0.9477	0.8905	0.8192	0.7383	0.6517	0.5630	0.4752	0.3910	0.3125
	2	0.9995	0.9963	0.9880	0.9728	0.9492	0.9163	0.8735	0.8208	0.7585	0.6875
	3	1.0000	0.9999	0.9995	0.9984	0.9961	0.9919	0.9850	0.9744	0.9590	0.9375
	4	1.0000	1.0000	1.0000	1.0000	1.0000	1.0000	1.0000	1.0000	1.0000	1.0000
5	0	0.7738	0.5905	0.4437	0.3277	0.2373	0.1681	0.1160	0.0778	0.0503	0.0313
	1	0.9774	0.9185	0.8352	0.7373	0.6328	0.5282	0.4284	0.3370	0.2562	0.1875
	2	0.9988	0.9914	0.9734	0.9421	0.8965	0.8369	0.7648	0.6826	0.5931	0.5000
	3	1.0000	0.9995	0.9978	0.9933	0.9844	0.9692	0.9460	0.9130	0.8688	0.8125
	4	1.0000	1.0000	0.9999	0.9997	0.9990	0.9976	0.9947	0.9898	0.9815	0.9688
	5	1.0000	1.0000	1.0000	1.0000	1.0000	1.0000	1.0000	1.0000	1.0000	1.0000
6	0	0.7351	0.5314	0.3771	0.2621	0.1780	0.1176	0.0754	0.0467	0.0277	0.0156
	1	0.9672	0.8857	0.7765	0.6554	0.5339	0.4202	0.3191	0.2333	0.1636	0.1094
	2	0.9978	0.9842	0.9527	0.9011	0.8306	0.7443	0.6471	0.5443	0.4415	0.3438
	3	0.9999	0.9987	0.9941	0.9830	0.9624	0.9295	0.8826	0.8208	0.7447	0.6563
	4	1.0000	0.9999	0.9996	0.9984	0.9954	0.9891	0.9777	0.9590	0.9308	0.8906
	5	1.0000	1.0000	1.0000	0.9999	0.9998	0.9993	0.9982	0.9959	0.9917	0.9844
	6	1.0000	1.0000	1.0000	1.0000	1.0000	1.0000	1.0000	1.0000	1.0000	1.0000
7	0	0.6983	0.4783	0.3206	0.2097	0.1335	0.0824	0.0490	0.0280	0.0152	0.0078
	1	0.9556	0.8503	0.7166	0.5767	0.4449	0.3294	0.2338	0.1586	0.1024	0.0625
	2	0.9962	0.9743	0.9262	0.8520	0.7564	0.6471	0.5323	0.4199	0.3164	0.2266
	3	0.9998	0.9973	0.9879	0.9667	0.9294	0.8740	0.8002	0.7102	0.6083	0.5000
	4	1.0000	0.9998	0.9988	0.9953	0.9871	0.9712	0.9444	0.9037	0.8471	0.7734
	5	1.0000	1.0000	0.9999	0.9996	0.9987	0.9962	0.9910	0.9812	0.9643	0.9375
	6	1.0000	1.0000	1.0000	1.0000	0.9999	0.9998	0.9994	0.9984	0.9963	0.9922
	7	1.0000	1.0000	1.0000	1.0000	1.0000	1.0000	1.0000	1.0000	1.0000	1.0000

0.55	0.60	0.65	0.70	0.75	0.80	0.85	0.90	0.95	n	x
0.4500	0.4000	0.3500	0.3000	0.2500	0.2000	0.1500	0.1000	0.0500	1	0
1.0000	1.0000	1.0000	1.0000	1.0000	1.0000	1.0000	1.0000	1.0000		1
0.2025	0.1600	0.1225	0.0900	0.0625	0.0400	0.0225	0.0100	0.0025	2	0
0.6975	0.6400	0.5775	0.5100	0.4375	0.3600	0.2775	0.1900	0.0975		1
1.0000	1.0000	1.0000	1.0000	1.0000	1.0000	1.0000	1.0000	1.0000		2
0.0911	0.0640	0.0429	0.0270	0.0156	0.0080	0.0034	0.0010	0.0001	3	0
0.4253	0.3520	0.2818	0.2160	0.1563	0.1040	0.0608	0.0280	0.0073		1
0.8336	0.7840	0.7254	0.6570	0.5781	0.4880	0.3859	0.2710	0.1426		2
1.0000	1.0000	1.0000	1.0000	1.0000	1.0000	1.0000	1.0000	1.0000		3
0.0410	0.0256	0.0150	0.0081	0.0039	0.0016	0.0005	0.0001	0.0000	4	0
0.2415	0.1792	0.1265	0.0837	0.0508	0.0272	0.0120	0.0037	0.0005		1
0.6090	0.5248	0.4370	0.3483	0.2617	0.1808	0.1095	0.0523	0.0140		2
0.9085	0.8704	0.8215	0.7599	0.6836	0.5904	0.4780	0.3439	0.1855		3
1.0000	1.0000	1.0000	1.0000	1.0000	1.0000	1.0000	1.0000	1.0000		4
0.0185	0.0102	0.0053	0.0024	0.0010	0.0003	0.0001	0.0000	0.0000	5	0
0.1312	0.0870	0.0540	0.0308	0.0156	0.0067	0.0022	0.0005	0.0000		1
0.4069	0.3174	0.2352	0.1631	0.1035	0.0579	0.0266	0.0086	0.0012		2
0.7438	0.6630	0.5716	0.4718	0.3672	0.2627	0.1648	0.0815	0.0226		3
0.9497	0.9222	0.8840	0.8319	0.7627	0.6723	0.5563	0.4095	0.2262		4
1.0000	1.0000	1.0000	1.0000	1.0000	1.0000	1.0000	1.0000	1.0000		5
0.0083	0.0041	0.0018	0.0007	0.0002	0.0001	0.0000	0.0000	0.0000	6	0
0.0692	0.0410	0.0223	0.0109	0.0046	0.0016	0.0004	0.0001	0.0000		1
0.2553	0.1792	0.1174	0.0705	0.0376	0.0170	0.0059	0.0013	0.0001		2
0.5585	0.4557	0.3529	0.2557	0.1694	0.0989	0.0473	0.0159	0.0022		3
0.8364	0.7667	0.6809	0.5798	0.4661	0.3446	0.2235	0.1143	0.0328		4
0.9723	0.9533	0.9246	0.8824	0.8220	0.7379	0.6229	0.4686	0.2649		5
1.0000	1.0000	1.0000	1.0000	1.0000	1.0000	1.0000	1.0000	1.0000		6
0.0037	0.0016	0.0006	0.0002	0.0001	0.0000	0.0000	0.0000	0.0000		0
0.0357	0.0188	0.0090	0.0038	0.0013	0.0004	0.0001	0.0000	0.0000	7	1
0.1529	0.0963	0.0556	0.0288	0.0129	0.0047	0.0012	0.0002	0.0000		2
0.3917	0.2898	0.1998	0.1260	0.0706	0.0333	0.0121	0.0027	0.0002		3
0.6836	0.5801	0.4677	0.3529	0.2436	0.1480	0.0738	0.0257	0.0038		4
0.8976	0.8414	0.7662	0.6706	0.5551	0.4233	0.2834	0.1497	0.0444		5
0.9848	0.9720	0.9510	0.9176	0.8665	0.7903	0.6794	0.5217	0.3017		6
1.0000	1.0000	1.0000	1.0000	1.0000	1.0000	1.0000	1.0000	1.0000		7

Probability of ≤ x occurrences in a sample of size n (continued)											
n	x	0.05	0.10	0.15	0.20	0.25	0.30	0.35	0.40	0.45	0.50
8	0	0.6634	0.4305	0.2725	0.1678	0.1001	0.0576	0.0319	0.0168	0.0084	0.0039
	1	0.9428	0.8131	0.6572	0.5033	0.3671	0.2553	0.1691	0.1064	0.0632	0.0352
	2	0.9942	0.9619	0.8948	0.7969	0.6785	0.5518	0.4278	0.3154	0.2201	0.1445
	3	0.9996	0.9950	0.9786	0.9437	0.8862	0.8059	0.7064	0.5941	0.4770	0.3633
	4	1.0000	0.9996	0.9971	0.9896	0.9727	0.9420	0.8939	0.8263	0.7396	0.6367
	5	1.0000	1.0000	0.9998	0.9988	0.9958	0.9887	0.9747	0.9502	0.9115	0.8555
	6	1.0000	1.0000	1.0000	0.9999	0.9996	0.9987	0.9964	0.9915	0.9819	0.9648
	7	1.0000	1.0000	1.0000	1.0000	1.0000	0.9999	0.9998	0.9993	0.9983	0.9961
	8	1.0000	1.0000	1.0000	1.0000	1.0000	1.0000	1.0000	1.0000	1.0000	1.0000
9	0	0.6302	0.3874	0.2316	0.1342	0.0751	0.0404	0.0207	0.0101	0.0046	0.0020
	1	0.9288	0.7748	0.5995	0.4362	0.3003	0.1960	0.1211	0.0705	0.0385	0.0195
	2	0.9916	0.9470	0.8591	0.7382	0.6007	0.4628	0.3373	0.2318	0.1495	0.0898
	3	0.9994	0.9917	0.9661	0.9144	0.8343	0.7297	0.6089	0.4826	0.3614	0.2539
	4	1.0000	0.9991	0.9944	0.9804	0.9511	0.9012	0.8283	0.7334	0.6214	0.5000
	5	1.0000	0.9999	0.9994	0.9969	0.9900	0.9747	0.9464	0.9006	0.8342	0.7461
	6	1.0000	1.0000	1.0000	0.9997	0.9987	0.9957	0.9888	0.9750	0.9502	0.9102
	7	1.0000	1.0000	1.0000	1.0000	0.9999	0.9996	0.9986	0.9962	0.9909	0.9805
	8	1.0000	1.0000	1.0000	1.0000	1.0000	1.0000	0.9999	0.9997	0.9992	0.9980
	9	1.0000	1.0000	1.0000	1.0000	1.0000	1.0000	1.0000	1.0000	1.0000	1.0000
10	0	0.5987	0.3487	0.1969	0.1074	0.0563	0.0282	0.0135	0.0060	0.0025	0.0010
	1	0.9139	0.7361	0.5443	0.3758	0.2440	0.1493	0.0860	0.0464	0.0233	0.0107
	2	0.9885	0.9298	0.8202	0.6778	0.5256	0.3828	0.2616	0.1673	0.0996	0.0547
	3	0.9990	0.9872	0.9500	0.8791	0.7759	0.6496	0.5138	0.3823	0.2660	0.1719
	4	0.9999	0.9984	0.9901	0.9672	0.9219	0.8497	0.7515	0.6331	0.5044	0.3770
	5	1.0000	0.9999	0.9986	0.9936	0.9803	0.9527	0.9051	0.8338	0.7384	0.6230
	6	1.0000	1.0000	0.9999	0.9991	0.9965	0.9894	0.9740	0.9452	0.8980	0.8281
	7	1.0000	1.0000	1.0000	0.9999	0.9996	0.9984	0.9952	0.9877	0.9726	0.9453
	8	1.0000	1.0000	1.0000	1.0000	1.0000	0.9999	0.9995	0.9983	0.9955	0.9893
	9	1.0000	1.0000	1.0000	1.0000	1.0000	1.0000	1.0000	0.9999	0.9997	0.9990
	10	1.0000	1.0000	1.0000	1.0000	1.0000	1.0000	1.0000	1.0000	1.0000	1.0000

0.55	0.60	0.65	0.70	0.75	0.80	0.85	0.90	0.95	n	x
0.0017	0.0007	0.0002	0.0001	0.0000	0.0000	0.0000	0.0000	0.0000	8	0
0.0181	0.0085	0.0036	0.0013	0.0004	0.0001	0.0000	0.0000	0.0000		1
0.0885	0.0498	0.0253	0.0113	0.0042	0.0012	0.0002	0.0000	0.0000		2
0.2604	0.1737	0.1061	0.0580	0.0273	0.0104	0.0029	0.0004	0.0000		3
0.5230	0.4059	0.2936	0.1941	0.1138	0.0563	0.0214	0.0050	0.0004		4
0.7799	0.6846	0.5722	0.4482	0.3215	0.2031	0.1052	0.0381	0.0058		5
0.9368	0.8936	0.8309	0.7447	0.6329	0.4967	0.3428	0.1869	0.0572		6
0.9916	0.9832	0.9681	0.9424	0.8999	0.8322	0.7275	0.5695	0.3366		7
1.0000	1.0000	1.0000	1.0000	1.0000	1.0000	1.0000	1.0000	1.0000		8
0.0008	0.0003	0.0001	0.0000	0.0000	0.0000	0.0000	0.0000	0.0000	9	0
0.0091	0.0038	0.0014	0.0004	0.0001	0.0000	0.0000	0.0000	0.0000		1
0.0498	0.0250	0.0112	0.0043	0.0013	0.0003	0.0000	0.0000	0.0000		2
0.1658	0.0994	0.0536	0.0253	0.0100	0.0031	0.0006	0.0001	0.0000		3
0.3786	0.2666	0.1717	0.0988	0.0489	0.0196	0.0056	0.0009	0.0000		4
0.6386	0.5174	0.3911	0.2703	0.1657	0.0856	0.0339	0.0083	0.0006		5
0.8505	0.7682	0.6627	0.5372	0.3993	0.2618	0.1409	0.0530	0.0084		6
0.9615	0.9295	0.8789	0.8040	0.6997	0.5638	0.4005	0.2252	0.0712		7
0.9954	0.9899	0.9793	0.9596	0.9249	0.8658	0.7684	0.6126	0.3698		8
1.0000	1.0000	1.0000	1.0000	1.0000	1.0000	1.0000	1.0000	1.0000		9
0.0003	0.0001	0.0000	0.0000	0.0000	0.0000	0.0000	0.0000	0.0000	10	0
0.0045	0.0017	0.0005	0.0001	0.0000	0.0000	0.0000	0.0000	0.0000		1
0.0274	0.0123	0.0048	0.0016	0.0004	0.0001	0.0000	0.0000	0.0000		2
0.1020	0.0548	0.0260	0.0106	0.0035	0.0009	0.0001	0.0000	0.0000		3
0.2616	0.1662	0.0949	0.0473	0.0197	0.0064	0.0014	0.0001	0.0000		4
0.4956	0.3669	0.2485	0.1503	0.0781	0.0328	0.0099	0.0016	0.0001		5
0.7340	0.6177	0.4862	0.3504	0.2241	0.1209	0.0500	0.0128	0.0010		6
0.9004	0.8327	0.7384	0.6172	0.4744	0.3222	0.1798	0.0702	0.0115		7
0.9767	0.9536	0.9140	0.8507	0.7560	0.6242	0.4557	0.2639	0.0861		8
0.9975	0.9940	0.9865	0.9718	0.9437	0.8926	0.8031	0.6513	0.4013		9
1.0000	1.0000	1.0000	1.0000	1.0000	1.0000	1.0000	1.0000	1.0000		10

Appendix C.1
Poisson Distribution Table

$$\Pr(X = x) = f(x; \lambda) = \frac{\lambda^x}{x!}e^{-\lambda}$$

						Probability of x occurrences								
λ	0	1	2	3	4	5	6	7	8	9	10	11	12	λ
0.05	0.9512	0.0476	0.0012	0.0000	0.0000	0.0000	0.0000	0.0000	0.0000	0.0000	0.0000	0.0000	0.0000	**0.05**
0.10	0.9048	0.0905	0.0045	0.0002	0.0000	0.0000	0.0000	0.0000	0.0000	0.0000	0.0000	0.0000	0.0000	**0.10**
0.15	0.8607	0.1291	0.0097	0.0005	0.0000	0.0000	0.0000	0.0000	0.0000	0.0000	0.0000	0.0000	0.0000	**0.15**
0.20	0.8187	0.1637	0.0164	0.0011	0.0001	0.0000	0.0000	0.0000	0.0000	0.0000	0.0000	0.0000	0.0000	**0.20**
0.25	0.7788	0.1947	0.0243	0.0020	0.0001	0.0000	0.0000	0.0000	0.0000	0.0000	0.0000	0.0000	0.0000	**0.25**
0.30	0.7408	0.2222	0.0333	0.0033	0.0003	0.0000	0.0000	0.0000	0.0000	0.0000	0.0000	0.0000	0.0000	**0.30**
0.35	0.7047	0.2466	0.0432	0.0050	0.0004	0.0000	0.0000	0.0000	0.0000	0.0000	0.0000	0.0000	0.0000	**0.35**
0.40	0.6703	0.2681	0.0536	0.0072	0.0007	0.0001	0.0000	0.0000	0.0000	0.0000	0.0000	0.0000	0.0000	**0.40**
0.45	0.6376	0.2869	0.0646	0.0097	0.0011	0.0001	0.0000	0.0000	0.0000	0.0000	0.0000	0.0000	0.0000	**0.45**
0.50	0.6065	0.3033	0.0758	0.0126	0.0016	0.0002	0.0000	0.0000	0.0000	0.0000	0.0000	0.0000	0.0000	**0.50**
0.55	0.5769	0.3173	0.0873	0.0160	0.0022	0.0002	0.0000	0.0000	0.0000	0.0000	0.0000	0.0000	0.0000	**0.55**
0.60	0.5488	0.3293	0.0988	0.0198	0.0030	0.0004	0.0000	0.0000	0.0000	0.0000	0.0000	0.0000	0.0000	**0.60**
0.65	0.5220	0.3393	0.1103	0.0239	0.0039	0.0005	0.0001	0.0000	0.0000	0.0000	0.0000	0.0000	0.0000	**0.65**
0.70	0.4966	0.3476	0.1217	0.0284	0.0050	0.0007	0.0001	0.0000	0.0000	0.0000	0.0000	0.0000	0.0000	**0.70**
0.75	0.4724	0.3543	0.1329	0.0332	0.0062	0.0009	0.0001	0.0000	0.0000	0.0000	0.0000	0.0000	0.0000	**0.75**
0.80	0.4493	0.3595	0.1438	0.0383	0.0077	0.0012	0.0002	0.0000	0.0000	0.0000	0.0000	0.0000	0.0000	**0.80**
0.85	0.4274	0.3633	0.1544	0.0437	0.0093	0.0016	0.0002	0.0000	0.0000	0.0000	0.0000	0.0000	0.0000	**0.85**
0.90	0.4066	0.3659	0.1647	0.0494	0.0111	0.0020	0.0003	0.0000	0.0000	0.0000	0.0000	0.0000	0.0000	**0.90**
0.95	0.3867	0.3674	0.1745	0.0553	0.0131	0.0025	0.0004	0.0001	0.0000	0.0000	0.0000	0.0000	0.0000	**0.95**
1.00	0.3679	0.3679	0.1839	0.0613	0.0153	0.0031	0.0005	0.0001	0.0000	0.0000	0.0000	0.0000	0.0000	**1.00**
1.05	0.3499	0.3674	0.1929	0.0675	0.0177	0.0037	0.0007	0.0001	0.0000	0.0000	0.0000	0.0000	0.0000	**1.05**
1.10	0.3329	0.3662	0.2014	0.0738	0.0203	0.0045	0.0008	0.0001	0.0000	0.0000	0.0000	0.0000	0.0000	**1.10**
1.15	0.3166	0.3641	0.2094	0.0803	0.0231	0.0053	0.0010	0.0002	0.0000	0.0000	0.0000	0.0000	0.0000	**1.15**
1.20	0.3012	0.3614	0.2169	0.0867	0.0260	0.0062	0.0012	0.0002	0.0000	0.0000	0.0000	0.0000	0.0000	**1.20**
1.25	0.2865	0.3581	0.2238	0.0933	0.0291	0.0073	0.0015	0.0003	0.0000	0.0000	0.0000	0.0000	0.0000	**1.25**
1.30	0.2725	0.3543	0.2303	0.0998	0.0324	0.0084	0.0018	0.0003	0.0001	0.0000	0.0000	0.0000	0.0000	**1.30**
1.35	0.2592	0.3500	0.2362	0.1063	0.0359	0.0097	0.0022	0.0004	0.0001	0.0000	0.0000	0.0000	0.0000	**1.35**
1.40	0.2466	0.3452	0.2417	0.1128	0.0395	0.0111	0.0026	0.0005	0.0001	0.0000	0.0000	0.0000	0.0000	**1.40**
1.45	0.2346	0.3401	0.2466	0.1192	0.0432	0.0125	0.0030	0.0006	0.0001	0.0000	0.0000	0.0000	0.0000	**1.45**
1.50	0.2231	0.3347	0.2510	0.1255	0.0471	0.0141	0.0035	0.0008	0.0001	0.0000	0.0000	0.0000	0.0000	**1.50**
1.55	0.2122	0.3290	0.2550	0.1317	0.0510	0.0158	0.0041	0.0009	0.0002	0.0000	0.0000	0.0000	0.0000	**1.55**
1.60	0.2019	0.3230	0.2584	0.1378	0.0551	0.0176	0.0047	0.0011	0.0002	0.0000	0.0000	0.0000	0.0000	**1.60**
1.65	0.1920	0.3169	0.2614	0.1438	0.0593	0.0196	0.0054	0.0013	0.0003	0.0000	0.0000	0.0000	0.0000	**1.65**
1.70	0.1827	0.3106	0.2640	0.1496	0.0636	0.0216	0.0061	0.0015	0.0003	0.0001	0.0000	0.0000	0.0000	**1.70**
1.75	0.1738	0.3041	0.2661	0.1552	0.0679	0.0238	0.0069	0.0017	0.0004	0.0001	0.0000	0.0000	0.0000	**1.75**

Probability of x occurrences														
λ	13	14	15	16	17	18	19	20	21	22	23	24	25	λ
0.05	0.0000	0.0000	0.0000	0.0000	0.0000	0.0000	0.0000	0.0000	0.0000	0.0000	0.0000	0.0000	0.0000	**0.05**
0.10	0.0000	0.0000	0.0000	0.0000	0.0000	0.0000	0.0000	0.0000	0.0000	0.0000	0.0000	0.0000	0.0000	**0.10**
0.15	0.0000	0.0000	0.0000	0.0000	0.0000	0.0000	0.0000	0.0000	0.0000	0.0000	0.0000	0.0000	0.0000	**0.15**
0.20	0.0000	0.0000	0.0000	0.0000	0.0000	0.0000	0.0000	0.0000	0.0000	0.0000	0.0000	0.0000	0.0000	**0.20**
0.25	0.0000	0.0000	0.0000	0.0000	0.0000	0.0000	0.0000	0.0000	0.0000	0.0000	0.0000	0.0000	0.0000	**0.25**
0.30	0.0000	0.0000	0.0000	0.0000	0.0000	0.0000	0.0000	0.0000	0.0000	0.0000	0.0000	0.0000	0.0000	**0.30**
0.35	0.0000	0.0000	0.0000	0.0000	0.0000	0.0000	0.0000	0.0000	0.0000	0.0000	0.0000	0.0000	0.0000	**0.35**
0.40	0.0000	0.0000	0.0000	0.0000	0.0000	0.0000	0.0000	0.0000	0.0000	0.0000	0.0000	0.0000	0.0000	**0.40**
0.45	0.0000	0.0000	0.0000	0.0000	0.0000	0.0000	0.0000	0.0000	0.0000	0.0000	0.0000	0.0000	0.0000	**0.45**
0.50	0.0000	0.0000	0.0000	0.0000	0.0000	0.0000	0.0000	0.0000	0.0000	0.0000	0.0000	0.0000	0.0000	**0.50**
0.55	0.0000	0.0000	0.0000	0.0000	0.0000	0.0000	0.0000	0.0000	0.0000	0.0000	0.0000	0.0000	0.0000	**0.55**
0.60	0.0000	0.0000	0.0000	0.0000	0.0000	0.0000	0.0000	0.0000	0.0000	0.0000	0.0000	0.0000	0.0000	**0.60**
0.65	0.0000	0.0000	0.0000	0.0000	0.0000	0.0000	0.0000	0.0000	0.0000	0.0000	0.0000	0.0000	0.0000	**0.65**
0.70	0.0000	0.0000	0.0000	0.0000	0.0000	0.0000	0.0000	0.0000	0.0000	0.0000	0.0000	0.0000	0.0000	**0.70**
0.75	0.0000	0.0000	0.0000	0.0000	0.0000	0.0000	0.0000	0.0000	0.0000	0.0000	0.0000	0.0000	0.0000	**0.75**
0.80	0.0000	0.0000	0.0000	0.0000	0.0000	0.0000	0.0000	0.0000	0.0000	0.0000	0.0000	0.0000	0.0000	**0.80**
0.85	0.0000	0.0000	0.0000	0.0000	0.0000	0.0000	0.0000	0.0000	0.0000	0.0000	0.0000	0.0000	0.0000	**0.85**
0.90	0.0000	0.0000	0.0000	0.0000	0.0000	0.0000	0.0000	0.0000	0.0000	0.0000	0.0000	0.0000	0.0000	**0.90**
0.95	0.0000	0.0000	0.0000	0.0000	0.0000	0.0000	0.0000	0.0000	0.0000	0.0000	0.0000	0.0000	0.0000	**0.95**
1.00	0.0000	0.0000	0.0000	0.0000	0.0000	0.0000	0.0000	0.0000	0.0000	0.0000	0.0000	0.0000	0.0000	**1.00**
1.05	0.0000	0.0000	0.0000	0.0000	0.0000	0.0000	0.0000	0.0000	0.0000	0.0000	0.0000	0.0000	0.0000	**1.05**
1.10	0.0000	0.0000	0.0000	0.0000	0.0000	0.0000	0.0000	0.0000	0.0000	0.0000	0.0000	0.0000	0.0000	**1.10**
1.15	0.0000	0.0000	0.0000	0.0000	0.0000	0.0000	0.0000	0.0000	0.0000	0.0000	0.0000	0.0000	0.0000	**1.15**
1.20	0.0000	0.0000	0.0000	0.0000	0.0000	0.0000	0.0000	0.0000	0.0000	0.0000	0.0000	0.0000	0.0000	**1.20**
1.25	0.0000	0.0000	0.0000	0.0000	0.0000	0.0000	0.0000	0.0000	0.0000	0.0000	0.0000	0.0000	0.0000	**1.25**
1.30	0.000	0.000	0.000	0.000	0.000	0.000	0.000	0.000	0.000	0.000	0.000	0.000	0.000	**1.30**
1.35	0.000	0.000	0.000	0.000	0.000	0.000	0.000	0.000	0.000	0.000	0.000	0.000	0.000	**1.35**
1.40	0.000	0.000	0.000	0.000	0.000	0.000	0.000	0.000	0.000	0.000	0.000	0.000	0.000	**1.40**
1.45	0.000	0.000	0.000	0.000	0.000	0.000	0.000	0.000	0.000	0.000	0.000	0.000	0.000	**1.45**
1.50	0.000	0.000	0.000	0.000	0.000	0.000	0.000	0.000	0.000	0.000	0.000	0.000	0.000	**1.50**
1.55	0.000	0.000	0.000	0.000	0.000	0.000	0.000	0.000	0.000	0.000	0.000	0.000	0.000	**1.55**
1.60	0.000	0.000	0.000	0.000	0.000	0.000	0.000	0.000	0.000	0.000	0.000	0.000	0.000	**1.60**
1.65	0.000	0.000	0.000	0.000	0.000	0.000	0.000	0.000	0.000	0.000	0.000	0.000	0.000	**1.65**
1.70	0.000	0.000	0.000	0.000	0.000	0.000	0.000	0.000	0.000	0.000	0.000	0.000	0.000	**1.70**
1.75	0.000	0.000	0.000	0.000	0.000	0.000	0.000	0.000	0.000	0.000	0.000	0.000	0.000	**1.75**

Probability of x occurrences *(continued)*														
λ	0	1	2	3	4	5	6	7	8	9	10	11	12	λ
1.80	0.1653	0.2975	0.2678	0.1607	0.0723	0.0260	0.0078	0.0020	0.0005	0.0001	0.0000	0.0000	0.0000	**1.80**
1.85	0.1572	0.2909	0.2691	0.1659	0.0767	0.0284	0.0088	0.0023	0.0005	0.0001	0.0000	0.0000	0.0000	**1.85**
1.90	0.1496	0.2842	0.2700	0.1710	0.0812	0.0309	0.0098	0.0027	0.0006	0.0001	0.0000	0.0000	0.0000	**1.90**
1.95	0.1423	0.2774	0.2705	0.1758	0.0857	0.0334	0.0109	0.0030	0.0007	0.0002	0.0000	0.0000	0.0000	**1.95**
2.00	0.1353	0.2707	0.2707	0.1804	0.0902	0.0361	0.0120	0.0034	0.0009	0.0002	0.0000	0.0000	0.0000	**2.00**
2.05	0.1287	0.2639	0.2705	0.1848	0.0947	0.0388	0.0133	0.0039	0.0010	0.0002	0.0000	0.0000	0.0000	**2.05**
2.10	0.1225	0.2572	0.2700	0.1890	0.0992	0.0417	0.0146	0.0044	0.0011	0.0003	0.0001	0.0000	0.0000	**2.10**
2.15	0.1165	0.2504	0.2692	0.1929	0.1037	0.0446	0.0160	0.0049	0.0013	0.0003	0.0001	0.0000	0.0000	**2.15**
2.20	0.1108	0.2438	0.2681	0.1966	0.1082	0.0476	0.0174	0.0055	0.0015	0.0004	0.0001	0.0000	0.0000	**2.20**
2.25	0.1054	0.2371	0.2668	0.2001	0.1126	0.0506	0.0190	0.0061	0.0017	0.0004	0.0001	0.0000	0.0000	**2.25**
2.30	0.1003	0.2306	0.2652	0.2033	0.1169	0.0538	0.0206	0.0068	0.0019	0.0005	0.0001	0.0000	0.0000	**2.30**
2.35	0.0954	0.2241	0.2633	0.2063	0.1212	0.0570	0.0223	0.0075	0.0022	0.0006	0.0001	0.0000	0.0000	**2.35**
2.40	0.0907	0.2177	0.2613	0.2090	0.1254	0.0602	0.0241	0.0083	0.0025	0.0007	0.0002	0.0000	0.0000	**2.40**
2.45	0.0863	0.2114	0.2590	0.2115	0.1295	0.0635	0.0259	0.0091	0.0028	0.0008	0.0002	0.0000	0.0000	**2.45**
2.50	0.0821	0.2052	0.2565	0.2138	0.1336	0.0668	0.0278	0.0099	0.0031	0.0009	0.0002	0.0000	0.0000	**2.50**
2.60	0.0743	0.1931	0.2510	0.2176	0.1414	0.0735	0.0319	0.0118	0.0038	0.0011	0.0003	0.0001	0.0000	**2.60**
2.70	0.0672	0.1815	0.2450	0.2205	0.1488	0.0804	0.0362	0.0139	0.0047	0.0014	0.0004	0.0001	0.0000	**2.70**
2.80	0.0608	0.1703	0.2384	0.2225	0.1557	0.0872	0.0407	0.0163	0.0057	0.0018	0.0005	0.0001	0.0000	**2.80**
2.90	0.0550	0.1596	0.2314	0.2237	0.1622	0.0940	0.0455	0.0188	0.0068	0.0022	0.0006	0.0002	0.0000	**2.90**
3.00	0.0498	0.1494	0.2240	0.2240	0.1680	0.1008	0.0504	0.0216	0.0081	0.0027	0.0008	0.0002	0.0001	**3.00**
3.10	0.0450	0.1397	0.2165	0.2237	0.1733	0.1075	0.0555	0.0246	0.0095	0.0033	0.0010	0.0003	0.0001	**3.10**
3.20	0.0408	0.1304	0.2087	0.2226	0.1781	0.1140	0.0608	0.0278	0.0111	0.0040	0.0013	0.0004	0.0001	**3.20**
3.30	0.0369	0.1217	0.2008	0.2209	0.1823	0.1203	0.0662	0.0312	0.0129	0.0047	0.0016	0.0005	0.0001	**3.30**
3.40	0.0334	0.1135	0.1929	0.2186	0.1858	0.1264	0.0716	0.0348	0.0148	0.0056	0.0019	0.0006	0.0002	**3.40**
3.50	0.0302	0.1057	0.1850	0.2158	0.1888	0.1322	0.0771	0.0385	0.0169	0.0066	0.0023	0.0007	0.0002	**3.50**
3.60	0.0273	0.0984	0.1771	0.2125	0.1912	0.1377	0.0826	0.0425	0.0191	0.0076	0.0028	0.0009	0.0003	**3.60**
3.70	0.0247	0.0915	0.1692	0.2087	0.1931	0.1429	0.0881	0.0466	0.0215	0.0089	0.0033	0.0011	0.0003	**3.70**
3.80	0.0224	0.0850	0.1615	0.2046	0.1944	0.1477	0.0936	0.0508	0.0241	0.0102	0.0039	0.0013	0.0004	**3.80**
3.90	0.0202	0.0789	0.1539	0.2001	0.1951	0.1522	0.0989	0.0551	0.0269	0.0116	0.0045	0.0016	0.0005	**3.90**
4.00	0.0183	0.0733	0.1465	0.1954	0.1954	0.1563	0.1042	0.0595	0.0298	0.0132	0.0053	0.0019	0.0006	**4.00**
4.10	0.0166	0.0679	0.1393	0.1904	0.1951	0.1600	0.1093	0.0640	0.0328	0.0150	0.0061	0.0023	0.0008	**4.10**
4.20	0.0150	0.0630	0.1323	0.1852	0.1944	0.1633	0.1143	0.0686	0.0360	0.0168	0.0071	0.0027	0.0009	**4.20**
4.30	0.0136	0.0583	0.1254	0.1798	0.1933	0.1662	0.1191	0.0732	0.0393	0.0188	0.0081	0.0032	0.0011	**4.30**
4.40	0.0123	0.0540	0.1188	0.1743	0.1917	0.1687	0.1237	0.0778	0.0428	0.0209	0.0092	0.0037	0.0013	**4.40**
4.50	0.0111	0.0500	0.1125	0.1687	0.1898	0.1708	0.1281	0.0824	0.0463	0.0232	0.0104	0.0043	0.0016	**4.50**
4.60	0.0101	0.0462	0.1063	0.1631	0.1875	0.1725	0.1323	0.0869	0.0500	0.0255	0.0118	0.0049	0.0019	**4.60**
4.70	0.0091	0.0427	0.1005	0.1574	0.1849	0.1738	0.1362	0.0914	0.0537	0.0281	0.0132	0.0056	0.0022	**4.70**
4.80	0.0082	0.0395	0.0948	0.1517	0.1820	0.1747	0.1398	0.0959	0.0575	0.0307	0.0147	0.0064	0.0026	**4.80**
4.90	0.0074	0.0365	0.0894	0.1460	0.1789	0.1753	0.1432	0.1002	0.0614	0.0334	0.0164	0.0073	0.0030	**4.90**
5.00	0.0067	0.0337	0.0842	0.1404	0.1755	0.1755	0.1462	0.1044	0.0653	0.0363	0.0181	0.0082	0.0034	**5.00**
5.10	0.0061	0.0311	0.0793	0.1348	0.1719	0.1753	0.1490	0.1086	0.0692	0.0392	0.0200	0.0093	0.0039	**5.10**
5.20	0.0055	0.0287	0.0746	0.1293	0.1681	0.1748	0.1515	0.1125	0.0731	0.0423	0.0220	0.0104	0.0045	**5.20**
5.30	0.0050	0.0265	0.0701	0.1239	0.1641	0.1740	0.1537	0.1163	0.0771	0.0454	0.0241	0.0116	0.0051	**5.30**
5.40	0.0045	0.0244	0.0659	0.1185	0.1600	0.1728	0.1555	0.1200	0.0810	0.0486	0.0262	0.0129	0.0058	**5.40**
5.50	0.0041	0.0225	0.0618	0.1133	0.1558	0.1714	0.1571	0.1234	0.0849	0.0519	0.0285	0.0143	0.0065	**5.50**
5.60	0.0037	0.0207	0.0580	0.1082	0.1515	0.1697	0.1584	0.1267	0.0887	0.0552	0.0309	0.0157	0.0073	**5.60**
5.70	0.0033	0.0191	0.0544	0.1033	0.1472	0.1678	0.1594	0.1298	0.0925	0.0586	0.0334	0.0173	0.0082	**5.70**
5.80	0.0030	0.0176	0.0509	0.0985	0.1428	0.1656	0.1601	0.1326	0.0962	0.0620	0.0359	0.0190	0.0092	**5.80**
5.90	0.0027	0.0162	0.0477	0.0938	0.1383	0.1632	0.1605	0.1353	0.0998	0.0654	0.0386	0.0207	0.0102	**5.90**
6.00	0.0025	0.0149	0.0446	0.0892	0.1339	0.1606	0.1606	0.1377	0.1033	0.0688	0.0413	0.0225	0.0113	**6.00**
6.10	0.0022	0.0137	0.0417	0.0848	0.1294	0.1579	0.1605	0.1399	0.1066	0.0723	0.0441	0.0244	0.0124	**6.10**

Probability of x occurrences *(continued)*														
λ	13	14	15	16	17	18	19	20	21	22	23	24	25	λ
1.80	0.000	0.000	0.000	0.000	0.000	0.000	0.000	0.000	0.000	0.000	0.000	0.000	0.000	**1.80**
1.85	0.000	0.000	0.000	0.000	0.000	0.000	0.000	0.000	0.000	0.000	0.000	0.000	0.000	**1.85**
1.90	0.000	0.000	0.000	0.000	0.000	0.000	0.000	0.000	0.000	0.000	0.000	0.000	0.000	**1.90**
1.95	0.000	0.000	0.000	0.000	0.000	0.000	0.000	0.000	0.000	0.000	0.000	0.000	0.000	**1.95**
2.00	0.000	0.000	0.000	0.000	0.000	0.000	0.000	0.000	0.000	0.000	0.000	0.000	0.000	**2.00**
2.05	0.000	0.000	0.000	0.000	0.000	0.000	0.000	0.000	0.000	0.000	0.000	0.000	0.000	**2.05**
2.10	0.000	0.000	0.000	0.000	0.000	0.000	0.000	0.000	0.000	0.000	0.000	0.000	0.000	**2.10**
2.15	0.000	0.000	0.000	0.000	0.000	0.000	0.000	0.000	0.000	0.000	0.000	0.000	0.000	**2.15**
2.20	0.000	0.000	0.000	0.000	0.000	0.000	0.000	0.000	0.000	0.000	0.000	0.000	0.000	**2.20**
2.25	0.000	0.000	0.000	0.000	0.000	0.000	0.000	0.000	0.000	0.000	0.000	0.000	0.000	**2.25**
2.30	0.000	0.000	0.000	0.000	0.000	0.000	0.000	0.000	0.000	0.000	0.000	0.000	0.000	**2.30**
2.35	0.000	0.000	0.000	0.000	0.000	0.000	0.000	0.000	0.000	0.000	0.000	0.000	0.000	**2.35**
2.40	0.000	0.000	0.000	0.000	0.000	0.000	0.000	0.000	0.000	0.000	0.000	0.000	0.000	**2.40**
2.45	0.000	0.000	0.000	0.000	0.000	0.000	0.000	0.000	0.000	0.000	0.000	0.000	0.000	**2.45**
2.50	0.000	0.000	0.000	0.000	0.000	0.000	0.000	0.000	0.000	0.000	0.000	0.000	0.000	**2.50**
2.60	0.0000	0.0000	0.0000	0.0000	0.0000	0.0000	0.0000	0.0000	0.0000	0.0000	0.0000	0.0000	0.0000	**2.60**
2.70	0.0000	0.0000	0.0000	0.0000	0.0000	0.0000	0.0000	0.0000	0.0000	0.0000	0.0000	0.0000	0.0000	**2.70**
2.80	0.0000	0.0000	0.0000	0.0000	0.0000	0.0000	0.0000	0.0000	0.0000	0.0000	0.0000	0.0000	0.0000	**2.80**
2.90	0.0000	0.0000	0.0000	0.0000	0.0000	0.0000	0.0000	0.0000	0.0000	0.0000	0.0000	0.0000	0.0000	**2.90**
3.00	0.0000	0.0000	0.0000	0.0000	0.0000	0.0000	0.0000	0.0000	0.0000	0.0000	0.0000	0.0000	0.0000	**3.00**
3.10	0.0000	0.0000	0.0000	0.0000	0.0000	0.0000	0.0000	0.0000	0.0000	0.0000	0.0000	0.0000	0.0000	**3.10**
3.20	0.0000	0.0000	0.0000	0.0000	0.0000	0.0000	0.0000	0.0000	0.0000	0.0000	0.0000	0.0000	0.0000	**3.20**
3.30	0.0000	0.0000	0.0000	0.0000	0.0000	0.0000	0.0000	0.0000	0.0000	0.0000	0.0000	0.0000	0.0000	**3.30**
3.40	0.0000	0.0000	0.0000	0.0000	0.0000	0.0000	0.0000	0.0000	0.0000	0.0000	0.0000	0.0000	0.0000	**3.40**
3.50	0.0001	0.0000	0.0000	0.0000	0.0000	0.0000	0.0000	0.0000	0.0000	0.0000	0.0000	0.0000	0.0000	**3.50**
3.60	0.0001	0.0000	0.0000	0.0000	0.0000	0.0000	0.0000	0.0000	0.0000	0.0000	0.0000	0.0000	0.0000	**3.60**
3.70	0.0001	0.0000	0.0000	0.0000	0.0000	0.0000	0.0000	0.0000	0.0000	0.0000	0.0000	0.0000	0.0000	**3.70**
3.80	0.0001	0.0000	0.0000	0.0000	0.0000	0.0000	0.0000	0.0000	0.0000	0.0000	0.0000	0.0000	0.0000	**3.80**
3.90	0.0002	0.0000	0.0000	0.0000	0.0000	0.0000	0.0000	0.0000	0.0000	0.0000	0.0000	0.0000	0.0000	**3.90**
4.00	0.0002	0.0001	0.0000	0.0000	0.0000	0.0000	0.0000	0.0000	0.0000	0.0000	0.0000	0.0000	0.0000	**4.00**
4.10	0.0002	0.0001	0.0000	0.0000	0.0000	0.0000	0.0000	0.0000	0.0000	0.0000	0.0000	0.0000	0.0000	**4.10**
4.20	0.0003	0.0001	0.0000	0.0000	0.0000	0.0000	0.0000	0.0000	0.0000	0.0000	0.0000	0.0000	0.0000	**4.20**
4.30	0.0004	0.0001	0.0000	0.0000	0.0000	0.0000	0.0000	0.0000	0.0000	0.0000	0.0000	0.0000	0.0000	**4.30**
4.40	0.0005	0.0001	0.0000	0.0000	0.0000	0.0000	0.0000	0.0000	0.0000	0.0000	0.0000	0.0000	0.0000	**4.40**
4.50	0.0006	0.0002	0.0001	0.0000	0.0000	0.0000	0.0000	0.0000	0.0000	0.0000	0.0000	0.0000	0.0000	**4.50**
4.60	0.0007	0.0002	0.0001	0.0000	0.0000	0.0000	0.0000	0.0000	0.0000	0.0000	0.0000	0.0000	0.0000	**4.60**
4.70	0.0008	0.0003	0.0001	0.0000	0.0000	0.0000	0.0000	0.0000	0.0000	0.0000	0.0000	0.0000	0.0000	**4.70**
4.80	0.0009	0.0003	0.0001	0.0000	0.0000	0.0000	0.0000	0.0000	0.0000	0.0000	0.0000	0.0000	0.0000	**4.80**
4.90	0.0011	0.0004	0.0001	0.0000	0.0000	0.0000	0.0000	0.0000	0.0000	0.0000	0.0000	0.0000	0.0000	**4.90**
5.00	0.0013	0.0005	0.0002	0.0000	0.0000	0.0000	0.0000	0.0000	0.0000	0.0000	0.0000	0.0000	0.0000	**5.00**
5.10	0.0015	0.0006	0.0002	0.0001	0.0000	0.0000	0.0000	0.0000	0.0000	0.0000	0.0000	0.0000	0.0000	**5.10**
5.20	0.0018	0.0007	0.0002	0.0001	0.0000	0.0000	0.0000	0.0000	0.0000	0.0000	0.0000	0.0000	0.0000	**5.20**
5.30	0.0021	0.0008	0.0003	0.0001	0.0000	0.0000	0.0000	0.0000	0.0000	0.0000	0.0000	0.0000	0.0000	**5.30**
5.40	0.0024	0.0009	0.0003	0.0001	0.0000	0.0000	0.0000	0.0000	0.0000	0.0000	0.0000	0.0000	0.0000	**5.40**
5.50	0.0028	0.0011	0.0004	0.0001	0.0000	0.0000	0.0000	0.0000	0.0000	0.0000	0.0000	0.0000	0.0000	**5.50**
5.60	0.0032	0.0013	0.0005	0.0002	0.0001	0.0000	0.0000	0.0000	0.0000	0.0000	0.0000	0.0000	0.0000	**5.60**
5.70	0.0036	0.0015	0.0006	0.0002	0.0001	0.0000	0.0000	0.0000	0.0000	0.0000	0.0000	0.0000	0.0000	**5.70**
5.80	0.0041	0.0017	0.0007	0.0002	0.0001	0.0000	0.0000	0.0000	0.0000	0.0000	0.0000	0.0000	0.0000	**5.80**
5.90	0.0046	0.0019	0.0008	0.0003	0.0001	0.0000	0.0000	0.0000	0.0000	0.0000	0.0000	0.0000	0.0000	**5.90**
6.00	0.0052	0.0022	0.0009	0.0003	0.0001	0.0000	0.0000	0.0000	0.0000	0.0000	0.0000	0.0000	0.0000	**6.00**
6.10	0.0058	0.0025	0.0010	0.0004	0.0001	0.0000	0.0000	0.0000	0.0000	0.0000	0.0000	0.0000	0.0000	**6.10**

Probability of x occurrences *(continued)*														
λ	0	1	2	3	4	5	6	7	8	9	10	11	12	λ
6.20	0.0020	0.0126	0.0390	0.0806	0.1249	0.1549	0.1601	0.1418	0.1099	0.0757	0.0469	0.0265	0.0137	**6.20**
6.30	0.0018	0.0116	0.0364	0.0765	0.1205	0.1519	0.1595	0.1435	0.1130	0.0791	0.0498	0.0285	0.0150	**6.30**
6.40	0.0017	0.0106	0.0340	0.0726	0.1162	0.1487	0.1586	0.1450	0.1160	0.0825	0.0528	0.0307	0.0164	**6.40**
6.50	0.0015	0.0098	0.0318	0.0688	0.1118	0.1454	0.1575	0.1462	0.1188	0.0858	0.0558	0.0330	0.0179	**6.50**
6.60	0.0014	0.0090	0.0296	0.0652	0.1076	0.1420	0.1562	0.1472	0.1215	0.0891	0.0588	0.0353	0.0194	**6.60**
6.70	0.0012	0.0082	0.0276	0.0617	0.1034	0.1385	0.1546	0.1480	0.1240	0.0923	0.0618	0.0377	0.0210	**6.70**
6.80	0.0011	0.0076	0.0258	0.0584	0.0992	0.1349	0.1529	0.1486	0.1263	0.0954	0.0649	0.0401	0.0227	**6.80**
6.90	0.0010	0.0070	0.0240	0.0552	0.0952	0.1314	0.1511	0.1489	0.1284	0.0985	0.0679	0.0426	0.0245	**6.90**
7.00	0.0009	0.0064	0.0223	0.0521	0.0912	0.1277	0.1490	0.1490	0.1304	0.1014	0.0710	0.0452	0.0263	**7.00**
7.10	0.0008	0.0059	0.0208	0.0492	0.0874	0.1241	0.1468	0.1489	0.1321	0.1042	0.0740	0.0478	0.0283	**7.10**
7.20	0.0007	0.0054	0.0194	0.0464	0.0836	0.1204	0.1445	0.1486	0.1337	0.1070	0.0770	0.0504	0.0303	**7.20**
7.30	0.0007	0.0049	0.0180	0.0438	0.0799	0.1167	0.1420	0.1481	0.1351	0.1096	0.0800	0.0531	0.0323	**7.30**
7.40	0.0006	0.0045	0.0167	0.0413	0.0764	0.1130	0.1394	0.1474	0.1363	0.1121	0.0829	0.0558	0.0344	**7.40**
7.50	0.0006	0.0041	0.0156	0.0389	0.0729	0.1094	0.1367	0.1465	0.1373	0.1144	0.0858	0.0585	0.0366	**7.50**
7.60	0.0005	0.0038	0.0145	0.0366	0.0696	0.1057	0.1339	0.1454	0.1381	0.1167	0.0887	0.0613	0.0388	**7.60**
7.70	0.0005	0.0035	0.0134	0.0345	0.0663	0.1021	0.1311	0.1442	0.1388	0.1187	0.0914	0.0640	0.0411	**7.70**
7.80	0.0004	0.0032	0.0125	0.0324	0.0632	0.0986	0.1282	0.1428	0.1392	0.1207	0.0941	0.0667	0.0434	**7.80**
7.90	0.0004	0.0029	0.0116	0.0305	0.0602	0.0951	0.1252	0.1413	0.1395	0.1224	0.0967	0.0695	0.0457	**7.90**
8.00	0.0003	0.0027	0.0107	0.0286	0.0573	0.0916	0.1221	0.1396	0.1396	0.1241	0.0993	0.0722	0.0481	**8.00**
8.10	0.0003	0.0025	0.0100	0.0269	0.0544	0.0882	0.1191	0.1378	0.1395	0.1256	0.1017	0.0749	0.0505	**8.10**
8.20	0.0003	0.0023	0.0092	0.0252	0.0517	0.0849	0.1160	0.1358	0.1392	0.1269	0.1040	0.0776	0.0530	**8.20**
8.30	0.0002	0.0021	0.0086	0.0237	0.0491	0.0816	0.1128	0.1338	0.1388	0.1280	0.1063	0.0802	0.0555	**8.30**
8.40	0.0002	0.0019	0.0079	0.0222	0.0466	0.0784	0.1097	0.1317	0.1382	0.1290	0.1084	0.0828	0.0579	**8.40**
8.50	0.0002	0.0017	0.0074	0.0208	0.0443	0.0752	0.1066	0.1294	0.1375	0.1299	0.1104	0.0853	0.0604	**8.50**
8.60	0.0002	0.0016	0.0068	0.0195	0.0420	0.0722	0.1034	0.1271	0.1366	0.1306	0.1123	0.0878	0.0629	**8.60**
8.70	0.0002	0.0014	0.0063	0.0183	0.0398	0.0692	0.1003	0.1247	0.1356	0.1311	0.1140	0.0902	0.0654	**8.70**
8.80	0.0002	0.0013	0.0058	0.0171	0.0377	0.0663	0.0972	0.1222	0.1344	0.1315	0.1157	0.0925	0.0679	**8.80**
8.90	0.0001	0.0012	0.0054	0.0160	0.0357	0.0635	0.0941	0.1197	0.1332	0.1317	0.1172	0.0948	0.0703	**8.90**
9.00	0.0001	0.0011	0.0050	0.0150	0.0337	0.0607	0.0911	0.1171	0.1318	0.1318	0.1186	0.0970	0.0728	**9.00**
9.10	0.0001	0.0010	0.0046	0.0140	0.0319	0.0581	0.0881	0.1145	0.1302	0.1317	0.1198	0.0991	0.0752	**9.10**
9.20	0.0001	0.0009	0.0043	0.0131	0.0302	0.0555	0.0851	0.1118	0.1286	0.1315	0.1210	0.1012	0.0776	**9.20**
9.30	0.0001	0.0009	0.0040	0.0123	0.0285	0.0530	0.0822	0.1091	0.1269	0.1311	0.1219	0.1031	0.0799	**9.30**
9.40	0.0001	0.0008	0.0037	0.0115	0.0269	0.0506	0.0793	0.1064	0.1251	0.1306	0.1228	0.1049	0.0822	**9.40**
9.50	0.0001	0.0007	0.0034	0.0107	0.0254	0.0483	0.0764	0.1037	0.1232	0.1300	0.1235	0.1067	0.0844	**9.50**
9.60	0.0001	0.0007	0.0031	0.0100	0.0240	0.0460	0.0736	0.1010	0.1212	0.1293	0.1241	0.1083	0.0866	**9.60**
9.70	0.0001	0.0006	0.0029	0.0093	0.0226	0.0439	0.0709	0.0982	0.1191	0.1284	0.1245	0.1098	0.0888	**9.70**
9.80	0.0001	0.0005	0.0027	0.0087	0.0213	0.0418	0.0682	0.0955	0.1170	0.1274	0.1249	0.1112	0.0908	**9.80**
9.90	0.0001	0.0005	0.0025	0.0081	0.0201	0.0398	0.0656	0.0928	0.1148	0.1263	0.1250	0.1125	0.0928	**9.90**
10.00	0.0000	0.0005	0.0023	0.0076	0.0189	0.0378	0.0631	0.0901	0.1126	0.1251	0.1251	0.1137	0.0948	**10.00**

Probability of x occurrences *(continued)*														
λ	13	14	15	16	17	18	19	20	21	22	23	24	25	λ
6.20	0.0065	0.0029	0.0012	0.0005	0.0002	0.0001	0.0000	0.0000	0.0000	0.0000	0.0000	0.0000	0.0000	**6.20**
6.30	0.0073	0.0033	0.0014	0.0005	0.0002	0.0001	0.0000	0.0000	0.0000	0.0000	0.0000	0.0000	0.0000	**6.30**
6.40	0.0081	0.0037	0.0016	0.0006	0.0002	0.0001	0.0000	0.0000	0.0000	0.0000	0.0000	0.0000	0.0000	**6.40**
6.50	0.0089	0.0041	0.0018	0.0007	0.0003	0.0001	0.0000	0.0000	0.0000	0.0000	0.0000	0.0000	0.0000	**6.50**
6.60	0.0099	0.0046	0.0020	0.0008	0.0003	0.0001	0.0000	0.0000	0.0000	0.0000	0.0000	0.0000	0.0000	**6.60**
6.70	0.0108	0.0052	0.0023	0.0010	0.0004	0.0001	0.0001	0.0000	0.0000	0.0000	0.0000	0.0000	0.0000	**6.70**
6.80	0.0119	0.0058	0.0026	0.0011	0.0004	0.0002	0.0001	0.0000	0.0000	0.0000	0.0000	0.0000	0.0000	**6.80**
6.90	0.0130	0.0064	0.0029	0.0013	0.0005	0.0002	0.0001	0.0000	0.0000	0.0000	0.0000	0.0000	0.0000	**6.90**
7.00	0.0142	0.0071	0.0033	0.0014	0.0006	0.0002	0.0001	0.0000	0.0000	0.0000	0.0000	0.0000	0.0000	**7.00**
7.10	0.0154	0.0078	0.0037	0.0016	0.0007	0.0003	0.0001	0.0000	0.0000	0.0000	0.0000	0.0000	0.0000	**7.10**
7.20	0.0168	0.0086	0.0041	0.0019	0.0008	0.0003	0.0001	0.0000	0.0000	0.0000	0.0000	0.0000	0.0000	**7.20**
7.30	0.0181	0.0095	0.0046	0.0021	0.0009	0.0004	0.0001	0.0001	0.0000	0.0000	0.0000	0.0000	0.0000	**7.30**
7.40	0.0196	0.0104	0.0051	0.0024	0.0010	0.0004	0.0002	0.0001	0.0000	0.0000	0.0000	0.0000	0.0000	**7.40**
7.50	0.0211	0.0113	0.0057	0.0026	0.0012	0.0005	0.0002	0.0001	0.0000	0.0000	0.0000	0.0000	0.0000	**7.50**
7.60	0.0227	0.0123	0.0062	0.0030	0.0013	0.0006	0.0002	0.0001	0.0000	0.0000	0.0000	0.0000	0.0000	**7.60**
7.70	0.0243	0.0134	0.0069	0.0033	0.0015	0.0006	0.0003	0.0001	0.0000	0.0000	0.0000	0.0000	0.0000	**7.70**
7.80	0.0260	0.0145	0.0075	0.0037	0.0017	0.0007	0.0003	0.0001	0.0000	0.0000	0.0000	0.0000	0.0000	**7.80**
7.90	0.0278	0.0157	0.0083	0.0041	0.0019	0.0008	0.0003	0.0001	0.0001	0.0000	0.0000	0.0000	0.0000	**7.90**
8.00	0.0296	0.0169	0.0090	0.0045	0.0021	0.0009	0.0004	0.0002	0.0001	0.0000	0.0000	0.0000	0.0000	**8.00**
8.10	0.0315	0.0182	0.0098	0.0050	0.0024	0.0011	0.0005	0.0002	0.0001	0.0000	0.0000	0.0000	0.0000	**8.10**
8.20	0.0334	0.0196	0.0107	0.0055	0.0026	0.0012	0.0005	0.0002	0.0001	0.0000	0.0000	0.0000	0.0000	**8.20**
8.30	0.0354	0.0210	0.0116	0.0060	0.0029	0.0014	0.0006	0.0002	0.0001	0.0000	0.0000	0.0000	0.0000	**8.30**
8.40	0.0374	0.0225	0.0126	0.0066	0.0033	0.0015	0.0007	0.0003	0.0001	0.0000	0.0000	0.0000	0.0000	**8.40**
8.50	0.0395	0.0240	0.0136	0.0072	0.0036	0.0017	0.0008	0.0003	0.0001	0.0001	0.0000	0.0000	0.0000	**8.50**
8.60	0.0416	0.0256	0.0147	0.0079	0.0040	0.0019	0.0009	0.0004	0.0002	0.0001	0.0000	0.0000	0.0000	**8.60**
8.70	0.0438	0.0272	0.0158	0.0086	0.0044	0.0021	0.0010	0.0004	0.0002	0.0001	0.0000	0.0000	0.0000	**8.70**
8.80	0.0459	0.0289	0.0169	0.0093	0.0048	0.0024	0.0011	0.0005	0.0002	0.0001	0.0000	0.0000	0.0000	**8.80**
8.90	0.0481	0.0306	0.0182	0.0101	0.0053	0.0026	0.0012	0.0005	0.0002	0.0001	0.0000	0.0000	0.0000	**8.90**
9.00	0.0504	0.0324	0.0194	0.0109	0.0058	0.0029	0.0014	0.0006	0.0003	0.0001	0.0000	0.0000	0.0000	**9.00**
9.10	0.0526	0.0342	0.0208	0.0118	0.0063	0.0032	0.0015	0.0007	0.0003	0.0001	0.0000	0.0000	0.0000	**9.10**
9.20	0.0549	0.0361	0.0221	0.0127	0.0069	0.0035	0.0017	0.0008	0.0003	0.0001	0.0001	0.0000	0.0000	**9.20**
9.30	0.0572	0.0380	0.0235	0.0137	0.0075	0.0039	0.0019	0.0009	0.0004	0.0002	0.0001	0.0000	0.0000	**9.30**
9.40	0.0594	0.0399	0.0250	0.0147	0.0081	0.0042	0.0021	0.0010	0.0004	0.0002	0.0001	0.0000	0.0000	**9.40**
9.50	0.0617	0.0419	0.0265	0.0157	0.0088	0.0046	0.0023	0.0011	0.0005	0.0002	0.0001	0.0000	0.0000	**9.50**
9.60	0.0640	0.0439	0.0281	0.0168	0.0095	0.0051	0.0026	0.0012	0.0006	0.0002	0.0001	0.0000	0.0000	**9.60**
9.70	0.0662	0.0459	0.0297	0.0180	0.0103	0.0055	0.0028	0.0014	0.0006	0.0003	0.0001	0.0000	0.0000	**9.70**
9.80	0.0685	0.0479	0.0313	0.0192	0.0111	0.0060	0.0031	0.0015	0.0007	0.0003	0.0001	0.0001	0.0000	**9.80**
9.90	0.0707	0.0500	0.0330	0.0204	0.0119	0.0065	0.0034	0.0017	0.0008	0.0004	0.0002	0.0001	0.0000	**9.90**
10.00	0.0729	0.0521	0.0347	0.0217	0.0128	0.0071	0.0037	0.0019	0.0009	0.0004	0.0002	0.0001	0.0000	**10.00**

Appendix C.2
Cumulative Poisson Distribution Table

$$\Pr(X \le x) = F(x;\lambda) = \sum_{x=0}^{n} \frac{\lambda^x}{x!} e^{-\lambda}$$

						Probability of ≤ x occurrences								
λ	0	1	2	3	4	5	6	7	8	9	10	11	12	λ
0.05	0.9512	0.9988	1.0000	1.0000	1.0000	1.0000	1.0000	1.0000	1.0000	1.0000	1.0000	1.0000	1.0000	**0.05**
0.10	0.9048	0.9953	0.9998	1.0000	1.0000	1.0000	1.0000	1.0000	1.0000	1.0000	1.0000	1.0000	1.0000	**0.10**
0.15	0.8607	0.9898	0.9995	1.0000	1.0000	1.0000	1.0000	1.0000	1.0000	1.0000	1.0000	1.0000	1.0000	**0.15**
0.20	0.8187	0.9825	0.9989	0.9999	1.0000	1.0000	1.0000	1.0000	1.0000	1.0000	1.0000	1.0000	1.0000	**0.20**
0.25	0.7788	0.9735	0.9978	0.9999	1.0000	1.0000	1.0000	1.0000	1.0000	1.0000	1.0000	1.0000	1.0000	**0.25**
0.30	0.7408	0.9631	0.9964	0.9997	1.0000	1.0000	1.0000	1.0000	1.0000	1.0000	1.0000	1.0000	1.0000	**0.30**
0.35	0.7047	0.9513	0.9945	0.9995	1.0000	1.0000	1.0000	1.0000	1.0000	1.0000	1.0000	1.0000	1.0000	**0.35**
0.40	0.6703	0.9384	0.9921	0.9992	0.9999	1.0000	1.0000	1.0000	1.0000	1.0000	1.0000	1.0000	1.0000	**0.40**
0.45	0.6376	0.9246	0.9891	0.9988	0.9999	1.0000	1.0000	1.0000	1.0000	1.0000	1.0000	1.0000	1.0000	**0.45**
0.50	0.6065	0.9098	0.9856	0.9982	0.9998	1.0000	1.0000	1.0000	1.0000	1.0000	1.0000	1.0000	1.0000	**0.50**
0.55	0.5769	0.8943	0.9815	0.9975	0.9997	1.0000	1.0000	1.0000	1.0000	1.0000	1.0000	1.0000	1.0000	**0.55**
0.60	0.5488	0.8781	0.9769	0.9966	0.9996	1.0000	1.0000	1.0000	1.0000	1.0000	1.0000	1.0000	1.0000	**0.60**
0.65	0.5220	0.8614	0.9717	0.9956	0.9994	0.9999	1.0000	1.0000	1.0000	1.0000	1.0000	1.0000	1.0000	**0.65**
0.70	0.4966	0.8442	0.9659	0.9942	0.9992	0.9999	1.0000	1.0000	1.0000	1.0000	1.0000	1.0000	1.0000	**0.70**
0.75	0.4724	0.8266	0.9595	0.9927	0.9989	0.9999	1.0000	1.0000	1.0000	1.0000	1.0000	1.0000	1.0000	**0.75**
0.80	0.4493	0.8088	0.9526	0.9909	0.9986	0.9998	1.0000	1.0000	1.0000	1.0000	1.0000	1.0000	1.0000	**0.80**
0.85	0.4274	0.7907	0.9451	0.9889	0.9982	0.9997	1.0000	1.0000	1.0000	1.0000	1.0000	1.0000	1.0000	**0.85**
0.90	0.4066	0.7725	0.9371	0.9865	0.9977	0.9997	1.0000	1.0000	1.0000	1.0000	1.0000	1.0000	1.0000	**0.90**
0.95	0.3867	0.7541	0.9287	0.9839	0.9971	0.9995	0.9999	1.0000	1.0000	1.0000	1.0000	1.0000	1.0000	**0.95**
1.00	0.3679	0.7358	0.9197	0.9810	0.9963	0.9994	0.9999	1.0000	1.0000	1.0000	1.0000	1.0000	1.0000	**1.00**
1.05	0.3499	0.7174	0.9103	0.9778	0.9955	0.9992	0.9999	1.0000	1.0000	1.0000	1.0000	1.0000	1.0000	**1.05**
1.10	0.3329	0.6990	0.9004	0.9743	0.9946	0.9990	0.9999	1.0000	1.0000	1.0000	1.0000	1.0000	1.0000	**1.10**
1.15	0.3166	0.6808	0.8901	0.9704	0.9935	0.9988	0.9998	1.0000	1.0000	1.0000	1.0000	1.0000	1.0000	**1.15**
1.20	0.3012	0.6626	0.8795	0.9662	0.9923	0.9985	0.9997	1.0000	1.0000	1.0000	1.0000	1.0000	1.0000	**1.20**
1.25	0.2865	0.6446	0.8685	0.9617	0.9909	0.9982	0.9997	1.0000	1.0000	1.0000	1.0000	1.0000	1.0000	**1.25**
1.30	0.2725	0.6268	0.8571	0.9569	0.9893	0.9978	0.9996	0.9999	1.0000	1.0000	1.0000	1.0000	1.0000	**1.30**
1.35	0.2592	0.6092	0.8454	0.9518	0.9876	0.9973	0.9995	0.9999	1.0000	1.0000	1.0000	1.0000	1.0000	**1.35**
1.40	0.2466	0.5918	0.8335	0.9463	0.9857	0.9968	0.9994	0.9999	1.0000	1.0000	1.0000	1.0000	1.0000	**1.40**
1.45	0.2346	0.5747	0.8213	0.9405	0.9837	0.9962	0.9992	0.9999	1.0000	1.0000	1.0000	1.0000	1.0000	**1.45**
1.50	0.2231	0.5578	0.8088	0.9344	0.9814	0.9955	0.9991	0.9998	1.0000	1.0000	1.0000	1.0000	1.0000	**1.50**
1.55	0.2122	0.5412	0.7962	0.9279	0.9790	0.9948	0.9989	0.9998	1.0000	1.0000	1.0000	1.0000	1.0000	**1.55**
1.60	0.2019	0.5249	0.7834	0.9212	0.9763	0.9940	0.9987	0.9997	1.0000	1.0000	1.0000	1.0000	1.0000	**1.60**
1.65	0.1920	0.5089	0.7704	0.9141	0.9735	0.9930	0.9984	0.9997	0.9999	1.0000	1.0000	1.0000	1.0000	**1.65**
1.70	0.1827	0.4932	0.7572	0.9068	0.9704	0.9920	0.9981	0.9996	0.9999	1.0000	1.0000	1.0000	1.0000	**1.70**
1.75	0.1738	0.4779	0.7440	0.8992	0.9671	0.9909	0.9978	0.9995	0.9999	1.0000	1.0000	1.0000	1.0000	**1.75**

	Probability of $\leq x$ occurrences												
λ	13	14	15	16	17	18	19	20	21	22	23	24	25	λ
0.05	1.0000	1.0000	1.0000	1.0000	1.0000	1.0000	1.0000	1.0000	1.0000	1.0000	1.0000	1.0000	1.0000	**0.05**
0.10	1.0000	1.0000	1.0000	1.0000	1.0000	1.0000	1.0000	1.0000	1.0000	1.0000	1.0000	1.0000	1.0000	**0.10**
0.15	1.0000	1.0000	1.0000	1.0000	1.0000	1.0000	1.0000	1.0000	1.0000	1.0000	1.0000	1.0000	1.0000	**0.15**
0.20	1.0000	1.0000	1.0000	1.0000	1.0000	1.0000	1.0000	1.0000	1.0000	1.0000	1.0000	1.0000	1.0000	**0.20**
0.25	1.0000	1.0000	1.0000	1.0000	1.0000	1.0000	1.0000	1.0000	1.0000	1.0000	1.0000	1.0000	1.0000	**0.25**
0.30	1.0000	1.0000	1.0000	1.0000	1.0000	1.0000	1.0000	1.0000	1.0000	1.0000	1.0000	1.0000	1.0000	**0.30**
0.35	1.0000	1.0000	1.0000	1.0000	1.0000	1.0000	1.0000	1.0000	1.0000	1.0000	1.0000	1.0000	1.0000	**0.35**
0.40	1.0000	1.0000	1.0000	1.0000	1.0000	1.0000	1.0000	1.0000	1.0000	1.0000	1.0000	1.0000	1.0000	**0.40**
0.45	1.0000	1.0000	1.0000	1.0000	1.0000	1.0000	1.0000	1.0000	1.0000	1.0000	1.0000	1.0000	1.0000	**0.45**
0.50	1.0000	1.0000	1.0000	1.0000	1.0000	1.0000	1.0000	1.0000	1.0000	1.0000	1.0000	1.0000	1.0000	**0.50**
0.55	1.0000	1.0000	1.0000	1.0000	1.0000	1.0000	1.0000	1.0000	1.0000	1.0000	1.0000	1.0000	1.0000	**0.55**
0.60	1.0000	1.0000	1.0000	1.0000	1.0000	1.0000	1.0000	1.0000	1.0000	1.0000	1.0000	1.0000	1.0000	**0.60**
0.65	1.0000	1.0000	1.0000	1.0000	1.0000	1.0000	1.0000	1.0000	1.0000	1.0000	1.0000	1.0000	1.0000	**0.65**
0.70	1.0000	1.0000	1.0000	1.0000	1.0000	1.0000	1.0000	1.0000	1.0000	1.0000	1.0000	1.0000	1.0000	**0.70**
0.75	1.0000	1.0000	1.0000	1.0000	1.0000	1.0000	1.0000	1.0000	1.0000	1.0000	1.0000	1.0000	1.0000	**0.75**
0.80	1.0000	1.0000	1.0000	1.0000	1.0000	1.0000	1.0000	1.0000	1.0000	1.0000	1.0000	1.0000	1.0000	**0.80**
0.85	1.0000	1.0000	1.0000	1.0000	1.0000	1.0000	1.0000	1.0000	1.0000	1.0000	1.0000	1.0000	1.0000	**0.85**
0.90	1.0000	1.0000	1.0000	1.0000	1.0000	1.0000	1.0000	1.0000	1.0000	1.0000	1.0000	1.0000	1.0000	**0.90**
0.95	1.0000	1.0000	1.0000	1.0000	1.0000	1.0000	1.0000	1.0000	1.0000	1.0000	1.0000	1.0000	1.0000	**0.95**
1.00	1.0000	1.0000	1.0000	1.0000	1.0000	1.0000	1.0000	1.0000	1.0000	1.0000	1.0000	1.0000	1.0000	**1.00**
1.05	1.0000	1.0000	1.0000	1.0000	1.0000	1.0000	1.0000	1.0000	1.0000	1.0000	1.0000	1.0000	1.0000	**1.05**
1.10	1.0000	1.0000	1.0000	1.0000	1.0000	1.0000	1.0000	1.0000	1.0000	1.0000	1.0000	1.0000	1.0000	**1.10**
1.15	1.0000	1.0000	1.0000	1.0000	1.0000	1.0000	1.0000	1.0000	1.0000	1.0000	1.0000	1.0000	1.0000	**1.15**
1.20	1.0000	1.0000	1.0000	1.0000	1.0000	1.0000	1.0000	1.0000	1.0000	1.0000	1.0000	1.0000	1.0000	**1.20**
1.25	1.0000	1.0000	1.0000	1.0000	1.0000	1.0000	1.0000	1.0000	1.0000	1.0000	1.0000	1.0000	1.0000	**1.25**
1.30	1.0000	1.0000	1.0000	1.0000	1.0000	1.0000	1.0000	1.0000	1.0000	1.0000	1.0000	1.0000	1.0000	**1.30**
1.35	1.0000	1.0000	1.0000	1.0000	1.0000	1.0000	1.0000	1.0000	1.0000	1.0000	1.0000	1.0000	1.0000	**1.35**
1.40	1.0000	1.0000	1.0000	1.0000	1.0000	1.0000	1.0000	1.0000	1.0000	1.0000	1.0000	1.0000	1.0000	**1.40**
1.45	1.0000	1.0000	1.0000	1.0000	1.0000	1.0000	1.0000	1.0000	1.0000	1.0000	1.0000	1.0000	1.0000	**1.45**
1.50	1.0000	1.0000	1.0000	1.0000	1.0000	1.0000	1.0000	1.0000	1.0000	1.0000	1.0000	1.0000	1.0000	**1.50**
1.55	1.0000	1.0000	1.0000	1.0000	1.0000	1.0000	1.0000	1.0000	1.0000	1.0000	1.0000	1.0000	1.0000	**1.55**
1.60	1.0000	1.0000	1.0000	1.0000	1.0000	1.0000	1.0000	1.0000	1.0000	1.0000	1.0000	1.0000	1.0000	**1.60**
1.65	1.0000	1.0000	1.0000	1.0000	1.0000	1.0000	1.0000	1.0000	1.0000	1.0000	1.0000	1.0000	1.0000	**1.65**
1.70	1.0000	1.0000	1.0000	1.0000	1.0000	1.0000	1.0000	1.0000	1.0000	1.0000	1.0000	1.0000	1.0000	**1.70**
1.75	1.0000	1.0000	1.0000	1.0000	1.0000	1.0000	1.0000	1.0000	1.0000	1.0000	1.0000	1.0000	1.0000	**1.75**

Probability of $\leq x$ occurrences *(continued)*														
λ	0	1	2	3	4	5	6	7	8	9	10	11	12	λ
1.80	0.1653	0.4628	0.7306	0.8913	0.9636	0.9896	0.9974	0.9994	0.9999	1.0000	1.0000	1.0000	1.0000	**1.80**
1.85	0.1572	0.4481	0.7172	0.8831	0.9599	0.9883	0.9970	0.9993	0.9999	1.0000	1.0000	1.0000	1.0000	**1.85**
1.90	0.1496	0.4337	0.7037	0.8747	0.9559	0.9868	0.9966	0.9992	0.9998	1.0000	1.0000	1.0000	1.0000	**1.90**
1.95	0.1423	0.4197	0.6902	0.8660	0.9517	0.9852	0.9960	0.9991	0.9998	1.0000	1.0000	1.0000	1.0000	**1.95**
2.00	0.1353	0.4060	0.6767	0.8571	0.9473	0.9834	0.9955	0.9989	0.9998	1.0000	1.0000	1.0000	1.0000	**2.00**
2.05	0.1287	0.3926	0.6631	0.8480	0.9427	0.9816	0.9948	0.9987	0.9997	0.9999	1.0000	1.0000	1.0000	**2.05**
2.10	0.1225	0.3796	0.6496	0.8386	0.9379	0.9796	0.9941	0.9985	0.9997	0.9999	1.0000	1.0000	1.0000	**2.10**
2.15	0.1165	0.3669	0.6361	0.8291	0.9328	0.9774	0.9934	0.9983	0.9996	0.9999	1.0000	1.0000	1.0000	**2.15**
2.20	0.1108	0.3546	0.6227	0.8194	0.9275	0.9751	0.9925	0.9980	0.9995	0.9999	1.0000	1.0000	1.0000	**2.20**
2.25	0.1054	0.3425	0.6093	0.8094	0.9220	0.9726	0.9916	0.9977	0.9994	0.9999	1.0000	1.0000	1.0000	**2.25**
2.30	0.1003	0.3309	0.5960	0.7993	0.9162	0.9700	0.9906	0.9974	0.9994	0.9999	1.0000	1.0000	1.0000	**2.30**
2.35	0.0954	0.3195	0.5828	0.7891	0.9103	0.9673	0.9896	0.9971	0.9993	0.9998	1.0000	1.0000	1.0000	**2.35**
2.40	0.0907	0.3084	0.5697	0.7787	0.9041	0.9643	0.9884	0.9967	0.9991	0.9998	1.0000	1.0000	1.0000	**2.40**
2.45	0.0863	0.2977	0.5567	0.7682	0.8978	0.9612	0.9872	0.9962	0.9990	0.9998	0.9999	1.0000	1.0000	**2.45**
2.50	0.0821	0.2873	0.5438	0.7576	0.8912	0.9580	0.9858	0.9958	0.9989	0.9997	0.9999	1.0000	1.0000	**2.50**
2.60	0.0743	0.2674	0.5184	0.7360	0.8774	0.9510	0.9828	0.9947	0.9985	0.9996	0.9999	1.0000	1.0000	**2.60**
2.70	0.0672	0.2487	0.4936	0.7141	0.8629	0.9433	0.9794	0.9934	0.9981	0.9995	0.9999	1.0000	1.0000	**2.70**
2.80	0.0608	0.2311	0.4695	0.6919	0.8477	0.9349	0.9756	0.9919	0.9976	0.9993	0.9998	1.0000	1.0000	**2.80**
2.90	0.0550	0.2146	0.4460	0.6696	0.8318	0.9258	0.9713	0.9901	0.9969	0.9991	0.9998	0.9999	1.0000	**2.90**
3.00	0.0498	0.1991	0.4232	0.6472	0.8153	0.9161	0.9665	0.9881	0.9962	0.9989	0.9997	0.9999	1.0000	**3.00**
3.10	0.0450	0.1847	0.4012	0.6248	0.7982	0.9057	0.9612	0.9858	0.9953	0.9986	0.9996	0.9999	1.0000	**3.10**
3.20	0.0408	0.1712	0.3799	0.6025	0.7806	0.8946	0.9554	0.9832	0.9943	0.9982	0.9995	0.9999	1.0000	**3.20**
3.30	0.0369	0.1586	0.3594	0.5803	0.7626	0.8829	0.9490	0.9802	0.9931	0.9978	0.9994	0.9998	1.0000	**3.30**
3.40	0.0334	0.1468	0.3397	0.5584	0.7442	0.8705	0.9421	0.9769	0.9917	0.9973	0.9992	0.9998	0.9999	**3.40**
3.50	0.0302	0.1359	0.3208	0.5366	0.7254	0.8576	0.9347	0.9733	0.9901	0.9967	0.9990	0.9997	0.9999	**3.50**
3.60	0.0273	0.1257	0.3027	0.5152	0.7064	0.8441	0.9267	0.9692	0.9883	0.9960	0.9987	0.9996	0.9999	**3.60**
3.70	0.0247	0.1162	0.2854	0.4942	0.6872	0.8301	0.9182	0.9648	0.9863	0.9952	0.9984	0.9995	0.9999	**3.70**
3.80	0.0224	0.1074	0.2689	0.4735	0.6678	0.8156	0.9091	0.9599	0.9840	0.9942	0.9981	0.9994	0.9998	**3.80**
3.90	0.0202	0.0992	0.2531	0.4532	0.6484	0.8006	0.8995	0.9546	0.9815	0.9931	0.9977	0.9993	0.9998	**3.90**
4.00	0.0183	0.0916	0.2381	0.4335	0.6288	0.7851	0.8893	0.9489	0.9786	0.9919	0.9972	0.9991	0.9997	**4.00**
4.10	0.0166	0.0845	0.2238	0.4142	0.6093	0.7693	0.8786	0.9427	0.9755	0.9905	0.9966	0.9989	0.9997	**4.10**
4.20	0.0150	0.0780	0.2102	0.3954	0.5898	0.7531	0.8675	0.9361	0.9721	0.9889	0.9959	0.9986	0.9996	**4.20**
4.30	0.0136	0.0719	0.1974	0.3772	0.5704	0.7367	0.8558	0.9290	0.9683	0.9871	0.9952	0.9983	0.9995	**4.30**
4.40	0.0123	0.0663	0.1851	0.3594	0.5512	0.7199	0.8436	0.9214	0.9642	0.9851	0.9943	0.9980	0.9993	**4.40**
4.50	0.0111	0.0611	0.1736	0.3423	0.5321	0.7029	0.8311	0.9134	0.9597	0.9829	0.9933	0.9976	0.9992	**4.50**
4.60	0.0101	0.0563	0.1626	0.3257	0.5132	0.6858	0.8180	0.9049	0.9549	0.9805	0.9922	0.9971	0.9990	**4.60**
4.70	0.0091	0.0518	0.1523	0.3097	0.4946	0.6684	0.8046	0.8960	0.9497	0.9778	0.9910	0.9966	0.9988	**4.70**
4.80	0.0082	0.0477	0.1425	0.2942	0.4763	0.6510	0.7908	0.8867	0.9442	0.9749	0.9896	0.9960	0.9986	**4.80**
4.90	0.0074	0.0439	0.1333	0.2793	0.4582	0.6335	0.7767	0.8769	0.9382	0.9717	0.9880	0.9953	0.9983	**4.90**
5.00	0.0067	0.0404	0.1247	0.2650	0.4405	0.6160	0.7622	0.8666	0.9319	0.9682	0.9863	0.9945	0.9980	**5.00**
5.10	0.0061	0.0372	0.1165	0.2513	0.4231	0.5984	0.7474	0.8560	0.9252	0.9644	0.9844	0.9937	0.9976	**5.10**
5.20	0.0055	0.0342	0.1088	0.2381	0.4061	0.5809	0.7324	0.8449	0.9181	0.9603	0.9823	0.9927	0.9972	**5.20**
5.30	0.0050	0.0314	0.1016	0.2254	0.3895	0.5635	0.7171	0.8335	0.9106	0.9559	0.9800	0.9916	0.9967	**5.30**
5.40	0.0045	0.0289	0.0948	0.2133	0.3733	0.5461	0.7017	0.8217	0.9027	0.9512	0.9775	0.9904	0.9962	**5.40**
5.50	0.0041	0.0266	0.0884	0.2017	0.3575	0.5289	0.6860	0.8095	0.8944	0.9462	0.9747	0.9890	0.9955	**5.50**
5.60	0.0037	0.0244	0.0824	0.1906	0.3422	0.5119	0.6703	0.7970	0.8857	0.9409	0.9718	0.9875	0.9949	**5.60**
5.70	0.0033	0.0224	0.0768	0.1800	0.3272	0.4950	0.6544	0.7841	0.8766	0.9352	0.9686	0.9859	0.9941	**5.70**
5.80	0.0030	0.0206	0.0715	0.1700	0.3127	0.4783	0.6384	0.7710	0.8672	0.9292	0.9651	0.9841	0.9932	**5.80**
5.90	0.0027	0.0189	0.0666	0.1604	0.2987	0.4619	0.6224	0.7576	0.8574	0.9228	0.9614	0.9821	0.9922	**5.90**
6.00	0.0025	0.0174	0.0620	0.1512	0.2851	0.4457	0.6063	0.7440	0.8472	0.9161	0.9574	0.9799	0.9912	**6.00**

Probability of $\leq x$ occurrences *(continued)*													
λ	13	14	15	16	17	18	19	20	21	22	23	24	25	λ
1.80	1.0000	1.0000	1.0000	1.0000	1.0000	1.0000	1.0000	1.0000	1.0000	1.0000	1.0000	1.0000	1.0000	**1.80**
1.85	1.0000	1.0000	1.0000	1.0000	1.0000	1.0000	1.0000	1.0000	1.0000	1.0000	1.0000	1.0000	1.0000	**1.85**
1.90	1.0000	1.0000	1.0000	1.0000	1.0000	1.0000	1.0000	1.0000	1.0000	1.0000	1.0000	1.0000	1.0000	**1.90**
1.95	1.0000	1.0000	1.0000	1.0000	1.0000	1.0000	1.0000	1.0000	1.0000	1.0000	1.0000	1.0000	1.0000	**1.95**
2.00	1.0000	1.0000	1.0000	1.0000	1.0000	1.0000	1.0000	1.0000	1.0000	1.0000	1.0000	1.0000	1.0000	**2.00**
2.05	1.0000	1.0000	1.0000	1.0000	1.0000	1.0000	1.0000	1.0000	1.0000	1.0000	1.0000	1.0000	1.0000	**2.05**
2.10	1.0000	1.0000	1.0000	1.0000	1.0000	1.0000	1.0000	1.0000	1.0000	1.0000	1.0000	1.0000	1.0000	**2.10**
2.15	1.0000	1.0000	1.0000	1.0000	1.0000	1.0000	1.0000	1.0000	1.0000	1.0000	1.0000	1.0000	1.0000	**2.15**
2.20	1.0000	1.0000	1.0000	1.0000	1.0000	1.0000	1.0000	1.0000	1.0000	1.0000	1.0000	1.0000	1.0000	**2.20**
2.25	1.0000	1.0000	1.0000	1.0000	1.0000	1.0000	1.0000	1.0000	1.0000	1.0000	1.0000	1.0000	1.0000	**2.25**
2.30	1.0000	1.0000	1.0000	1.0000	1.0000	1.0000	1.0000	1.0000	1.0000	1.0000	1.0000	1.0000	1.0000	**2.30**
2.35	1.0000	1.0000	1.0000	1.0000	1.0000	1.0000	1.0000	1.0000	1.0000	1.0000	1.0000	1.0000	1.0000	**2.35**
2.40	1.0000	1.0000	1.0000	1.0000	1.0000	1.0000	1.0000	1.0000	1.0000	1.0000	1.0000	1.0000	1.0000	**2.40**
2.45	1.0000	1.0000	1.0000	1.0000	1.0000	1.0000	1.0000	1.0000	1.0000	1.0000	1.0000	1.0000	1.0000	**2.45**
2.50	1.0000	1.0000	1.0000	1.0000	1.0000	1.0000	1.0000	1.0000	1.0000	1.0000	1.0000	1.0000	1.0000	**2.50**
2.60	1.0000	1.0000	1.0000	1.0000	1.0000	1.0000	1.0000	1.0000	1.0000	1.0000	1.0000	1.0000	1.0000	**2.60**
2.70	1.0000	1.0000	1.0000	1.0000	1.0000	1.0000	1.0000	1.0000	1.0000	1.0000	1.0000	1.0000	1.0000	**2.70**
2.80	1.0000	1.0000	1.0000	1.0000	1.0000	1.0000	1.0000	1.0000	1.0000	1.0000	1.0000	1.0000	1.0000	**2.80**
2.90	1.0000	1.0000	1.0000	1.0000	1.0000	1.0000	1.0000	1.0000	1.0000	1.0000	1.0000	1.0000	1.0000	**2.90**
3.00	1.0000	1.0000	1.0000	1.0000	1.0000	1.0000	1.0000	1.0000	1.0000	1.0000	1.0000	1.0000	1.0000	**3.00**
3.10	1.0000	1.0000	1.0000	1.0000	1.0000	1.0000	1.0000	1.0000	1.0000	1.0000	1.0000	1.0000	1.0000	**3.10**
3.20	1.0000	1.0000	1.0000	1.0000	1.0000	1.0000	1.0000	1.0000	1.0000	1.0000	1.0000	1.0000	1.0000	**3.20**
3.30	1.0000	1.0000	1.0000	1.0000	1.0000	1.0000	1.0000	1.0000	1.0000	1.0000	1.0000	1.0000	1.0000	**3.30**
3.40	1.0000	1.0000	1.0000	1.0000	1.0000	1.0000	1.0000	1.0000	1.0000	1.0000	1.0000	1.0000	1.0000	**3.40**
3.50	1.0000	1.0000	1.0000	1.0000	1.0000	1.0000	1.0000	1.0000	1.0000	1.0000	1.0000	1.0000	1.0000	**3.50**
3.60	1.0000	1.0000	1.0000	1.0000	1.0000	1.0000	1.0000	1.0000	1.0000	1.0000	1.0000	1.0000	1.0000	**3.60**
3.70	1.0000	1.0000	1.0000	1.0000	1.0000	1.0000	1.0000	1.0000	1.0000	1.0000	1.0000	1.0000	1.0000	**3.70**
3.80	1.0000	1.0000	1.0000	1.0000	1.0000	1.0000	1.0000	1.0000	1.0000	1.0000	1.0000	1.0000	1.0000	**3.80**
3.90	0.9999	1.0000	1.0000	1.0000	1.0000	1.0000	1.0000	1.0000	1.0000	1.0000	1.0000	1.0000	1.0000	**3.90**
4.00	0.9999	1.0000	1.0000	1.0000	1.0000	1.0000	1.0000	1.0000	1.0000	1.0000	1.0000	1.0000	1.0000	**4.00**
4.10	0.9999	1.0000	1.0000	1.0000	1.0000	1.0000	1.0000	1.0000	1.0000	1.0000	1.0000	1.0000	1.0000	**4.10**
4.20	0.9999	1.0000	1.0000	1.0000	1.0000	1.0000	1.0000	1.0000	1.0000	1.0000	1.0000	1.0000	1.0000	**4.20**
4.30	0.9998	1.0000	1.0000	1.0000	1.0000	1.0000	1.0000	1.0000	1.0000	1.0000	1.0000	1.0000	1.0000	**4.30**
4.40	0.9998	0.9999	1.0000	1.0000	1.0000	1.0000	1.0000	1.0000	1.0000	1.0000	1.0000	1.0000	1.0000	**4.40**
4.50	0.9997	0.9999	1.0000	1.0000	1.0000	1.0000	1.0000	1.0000	1.0000	1.0000	1.0000	1.0000	1.0000	**4.50**
4.60	0.9997	0.9999	1.0000	1.0000	1.0000	1.0000	1.0000	1.0000	1.0000	1.0000	1.0000	1.0000	1.0000	**4.60**
4.70	0.9996	0.9999	1.0000	1.0000	1.0000	1.0000	1.0000	1.0000	1.0000	1.0000	1.0000	1.0000	1.0000	**4.70**
4.80	0.9995	0.9999	1.0000	1.0000	1.0000	1.0000	1.0000	1.0000	1.0000	1.0000	1.0000	1.0000	1.0000	**4.80**
4.90	0.9994	0.9998	0.9999	1.0000	1.0000	1.0000	1.0000	1.0000	1.0000	1.0000	1.0000	1.0000	1.0000	**4.90**
5.00	0.9993	0.9998	0.9999	1.0000	1.0000	1.0000	1.0000	1.0000	1.0000	1.0000	1.0000	1.0000	1.0000	**5.00**
5.10	0.9992	0.9997	0.9999	1.0000	1.0000	1.0000	1.0000	1.0000	1.0000	1.0000	1.0000	1.0000	1.0000	**5.10**
5.20	0.9990	0.9997	0.9999	1.0000	1.0000	1.0000	1.0000	1.0000	1.0000	1.0000	1.0000	1.0000	1.0000	**5.20**
5.30	0.9988	0.9996	0.9999	1.0000	1.0000	1.0000	1.0000	1.0000	1.0000	1.0000	1.0000	1.0000	1.0000	**5.30**
5.40	0.9986	0.9995	0.9998	0.9999	1.0000	1.0000	1.0000	1.0000	1.0000	1.0000	1.0000	1.0000	1.0000	**5.40**
5.50	0.9983	0.9994	0.9998	0.9999	1.0000	1.0000	1.0000	1.0000	1.0000	1.0000	1.0000	1.0000	1.0000	**5.50**
5.60	0.9980	0.9993	0.9998	0.9999	1.0000	1.0000	1.0000	1.0000	1.0000	1.0000	1.0000	1.0000	1.0000	**5.60**
5.70	0.9977	0.9991	0.9997	0.9999	1.0000	1.0000	1.0000	1.0000	1.0000	1.0000	1.0000	1.0000	1.0000	**5.70**
5.80	0.9973	0.9990	0.9996	0.9999	1.0000	1.0000	1.0000	1.0000	1.0000	1.0000	1.0000	1.0000	1.0000	**5.80**
5.90	0.9969	0.9988	0.9996	0.9999	1.0000	1.0000	1.0000	1.0000	1.0000	1.0000	1.0000	1.0000	1.0000	**5.90**
6.00	0.9964	0.9986	0.9995	0.9998	0.9999	1.0000	1.0000	1.0000	1.0000	1.0000	1.0000	1.0000	1.0000	**6.00**

Probability of ≤x occurrences *(continued)*														
λ	0	1	2	3	4	5	6	7	8	9	10	11	12	λ
6.10	0.0022	0.0159	0.0577	0.1425	0.2719	0.4298	0.5902	0.7301	0.8367	0.9090	0.9531	0.9776	0.9900	**6.10**
6.20	0.0020	0.0146	0.0536	0.1342	0.2592	0.4141	0.5742	0.7160	0.8259	0.9016	0.9486	0.9750	0.9887	**6.20**
6.30	0.0018	0.0134	0.0498	0.1264	0.2469	0.3988	0.5582	0.7017	0.8148	0.8939	0.9437	0.9723	0.9873	**6.30**
6.40	0.0017	0.0123	0.0463	0.1189	0.2351	0.3837	0.5423	0.6873	0.8033	0.8858	0.9386	0.9693	0.9857	**6.40**
6.50	0.0015	0.0113	0.0430	0.1118	0.2237	0.3690	0.5265	0.6728	0.7916	0.8774	0.9332	0.9661	0.9840	**6.50**
6.60	0.0014	0.0103	0.0400	0.1052	0.2127	0.3547	0.5108	0.6581	0.7796	0.8686	0.9274	0.9627	0.9821	**6.60**
6.70	0.0012	0.0095	0.0371	0.0988	0.2022	0.3406	0.4953	0.6433	0.7673	0.8596	0.9214	0.9591	0.9801	**6.70**
6.80	0.0011	0.0087	0.0344	0.0928	0.1920	0.3270	0.4799	0.6285	0.7548	0.8502	0.9151	0.9552	0.9779	**6.80**
6.90	0.0010	0.0080	0.0320	0.0871	0.1823	0.3137	0.4647	0.6136	0.7420	0.8405	0.9084	0.9510	0.9755	**6.90**
7.00	0.0009	0.0073	0.0296	0.0818	0.1730	0.3007	0.4497	0.5987	0.7291	0.8305	0.9015	0.9467	0.9730	**7.00**
7.10	0.0008	0.0067	0.0275	0.0767	0.1641	0.2881	0.4349	0.5838	0.7160	0.8202	0.8942	0.9420	0.9703	**7.10**
7.20	0.0007	0.0061	0.0255	0.0719	0.1555	0.2759	0.4204	0.5689	0.7027	0.8096	0.8867	0.9371	0.9673	**7.20**
7.30	0.0007	0.0056	0.0236	0.0674	0.1473	0.2640	0.4060	0.5541	0.6892	0.7988	0.8788	0.9319	0.9642	**7.30**
7.40	0.0006	0.0051	0.0219	0.0632	0.1395	0.2526	0.3920	0.5393	0.6757	0.7877	0.8707	0.9265	0.9609	**7.40**
7.50	0.0006	0.0047	0.0203	0.0591	0.1321	0.2414	0.3782	0.5246	0.6620	0.7764	0.8622	0.9208	0.9573	**7.50**
7.60	0.0005	0.0043	0.0188	0.0554	0.1249	0.2307	0.3646	0.5100	0.6482	0.7649	0.8535	0.9148	0.9536	**7.60**
7.70	0.0005	0.0039	0.0174	0.0518	0.1181	0.2203	0.3514	0.4956	0.6343	0.7531	0.8445	0.9085	0.9496	**7.70**
7.80	0.0004	0.0036	0.0161	0.0485	0.1117	0.2103	0.3384	0.4812	0.6204	0.7411	0.8352	0.9020	0.9454	**7.80**
7.90	0.0004	0.0033	0.0149	0.0453	0.1055	0.2006	0.3257	0.4670	0.6065	0.7290	0.8257	0.8952	0.9409	**7.90**
8.00	0.0003	0.0030	0.0138	0.0424	0.0996	0.1912	0.3134	0.4530	0.5925	0.7166	0.8159	0.8881	0.9362	**8.00**
8.10	0.0003	0.0028	0.0127	0.0396	0.0940	0.1822	0.3013	0.4391	0.5786	0.7041	0.8058	0.8807	0.9313	**8.10**
8.20	0.0003	0.0025	0.0118	0.0370	0.0887	0.1736	0.2896	0.4254	0.5647	0.6915	0.7955	0.8731	0.9261	**8.20**
8.30	0.0002	0.0023	0.0109	0.0346	0.0837	0.1653	0.2781	0.4119	0.5507	0.6788	0.7850	0.8652	0.9207	**8.30**
8.40	0.0002	0.0021	0.0100	0.0323	0.0789	0.1573	0.2670	0.3987	0.5369	0.6659	0.7743	0.8571	0.9150	**8.40**
8.50	0.0002	0.0019	0.0093	0.0301	0.0744	0.1496	0.2562	0.3856	0.5231	0.6530	0.7634	0.8487	0.9091	**8.50**
8.60	0.0002	0.0018	0.0086	0.0281	0.0701	0.1422	0.2457	0.3728	0.5094	0.6400	0.7522	0.8400	0.9029	**8.60**
8.70	0.0002	0.0016	0.0079	0.0262	0.0660	0.1352	0.2355	0.3602	0.4958	0.6269	0.7409	0.8311	0.8965	**8.70**
8.80	0.0002	0.0015	0.0073	0.0244	0.0621	0.1284	0.2256	0.3478	0.4823	0.6137	0.7294	0.8220	0.8898	**8.80**
8.90	0.0001	0.0014	0.0068	0.0228	0.0584	0.1219	0.2160	0.3357	0.4689	0.6006	0.7178	0.8126	0.8829	**8.90**
9.00	0.0001	0.0012	0.0062	0.0212	0.0550	0.1157	0.2068	0.3239	0.4557	0.5874	0.7060	0.8030	0.8758	**9.00**
9.10	0.0001	0.0011	0.0058	0.0198	0.0517	0.1098	0.1978	0.3123	0.4426	0.5742	0.6941	0.7932	0.8684	**9.10**
9.20	0.0001	0.0010	0.0053	0.0184	0.0486	0.1041	0.1892	0.3010	0.4296	0.5611	0.6820	0.7832	0.8607	**9.20**
9.30	0.0001	0.0009	0.0049	0.0172	0.0456	0.0986	0.1808	0.2900	0.4168	0.5479	0.6699	0.7730	0.8529	**9.30**
9.40	0.0001	0.0009	0.0045	0.0160	0.0429	0.0935	0.1727	0.2792	0.4042	0.5349	0.6576	0.7626	0.8448	**9.40**
9.50	0.0001	0.0008	0.0042	0.0149	0.0403	0.0885	0.1649	0.2687	0.3918	0.5218	0.6453	0.7520	0.8364	**9.50**
9.60	0.0001	0.0007	0.0038	0.0138	0.0378	0.0838	0.1574	0.2584	0.3796	0.5089	0.6329	0.7412	0.8279	**9.60**
9.70	0.0001	0.0007	0.0035	0.0129	0.0355	0.0793	0.1502	0.2485	0.3676	0.4960	0.6205	0.7303	0.8191	**9.70**
9.80	0.0001	0.0006	0.0033	0.0120	0.0333	0.0750	0.1433	0.2388	0.3558	0.4832	0.6080	0.7193	0.8101	**9.80**
9.90	0.0001	0.0005	0.0030	0.0111	0.0312	0.0710	0.1366	0.2294	0.3442	0.4705	0.5955	0.7081	0.8009	**9.90**
10.00	0.0000	0.0005	0.0028	0.0103	0.0293	0.0671	0.1301	0.2202	0.3328	0.4579	0.5830	0.6968	0.7916	**10.00**

Probability of ≤ *x* occurrences *(continued)*														
λ	13	14	15	16	17	18	19	20	21	22	23	24	25	λ
6.10	0.9958	0.9984	0.9994	0.9998	0.9999	1.0000	1.0000	1.0000	1.0000	1.0000	1.0000	1.0000	1.0000	**6.10**
6.20	0.9952	0.9981	0.9993	0.9997	0.9999	1.0000	1.0000	1.0000	1.0000	1.0000	1.0000	1.0000	1.0000	**6.20**
6.30	0.9945	0.9978	0.9992	0.9997	0.9999	1.0000	1.0000	1.0000	1.0000	1.0000	1.0000	1.0000	1.0000	**6.30**
6.40	0.9937	0.9974	0.9990	0.9996	0.9999	1.0000	1.0000	1.0000	1.0000	1.0000	1.0000	1.0000	1.0000	**6.40**
6.50	0.9929	0.9970	0.9988	0.9996	0.9998	0.9999	1.0000	1.0000	1.0000	1.0000	1.0000	1.0000	1.0000	**6.50**
6.60	0.9920	0.9966	0.9986	0.9995	0.9998	0.9999	1.0000	1.0000	1.0000	1.0000	1.0000	1.0000	1.0000	**6.60**
6.70	0.9909	0.9961	0.9984	0.9994	0.9998	0.9999	1.0000	1.0000	1.0000	1.0000	1.0000	1.0000	1.0000	**6.70**
6.80	0.9898	0.9956	0.9982	0.9993	0.9997	0.9999	1.0000	1.0000	1.0000	1.0000	1.0000	1.0000	1.0000	**6.80**
6.90	0.9885	0.9950	0.9979	0.9992	0.9997	0.9999	1.0000	1.0000	1.0000	1.0000	1.0000	1.0000	1.0000	**6.90**
7.00	0.9872	0.9943	0.9976	0.9990	0.9996	0.9999	1.0000	1.0000	1.0000	1.0000	1.0000	1.0000	1.0000	**7.00**
7.10	0.9857	0.9935	0.9972	0.9989	0.9996	0.9998	0.9999	1.0000	1.0000	1.0000	1.0000	1.0000	1.0000	**7.10**
7.20	0.9841	0.9927	0.9969	0.9987	0.9995	0.9998	0.9999	1.0000	1.0000	1.0000	1.0000	1.0000	1.0000	**7.20**
7.30	0.9824	0.9918	0.9964	0.9985	0.9994	0.9998	0.9999	1.0000	1.0000	1.0000	1.0000	1.0000	1.0000	**7.30**
7.40	0.9805	0.9908	0.9959	0.9983	0.9993	0.9997	0.9999	1.0000	1.0000	1.0000	1.0000	1.0000	1.0000	**7.40**
7.50	0.9784	0.9897	0.9954	0.9980	0.9992	0.9997	0.9999	1.0000	1.0000	1.0000	1.0000	1.0000	1.0000	**7.50**
7.60	0.9762	0.9886	0.9948	0.9978	0.9991	0.9996	0.9999	1.0000	1.0000	1.0000	1.0000	1.0000	1.0000	**7.60**
7.70	0.9739	0.9873	0.9941	0.9974	0.9989	0.9996	0.9998	0.9999	1.0000	1.0000	1.0000	1.0000	1.0000	**7.70**
7.80	0.9714	0.9859	0.9934	0.9971	0.9988	0.9995	0.9998	0.9999	1.0000	1.0000	1.0000	1.0000	1.0000	**7.80**
7.90	0.9687	0.9844	0.9926	0.9967	0.9986	0.9994	0.9998	0.9999	1.0000	1.0000	1.0000	1.0000	1.0000	**7.90**
8.00	0.9658	0.9827	0.9918	0.9963	0.9984	0.9993	0.9997	0.9999	1.0000	1.0000	1.0000	1.0000	1.0000	**8.00**
8.10	0.9628	0.9810	0.9908	0.9958	0.9982	0.9992	0.9997	0.9999	1.0000	1.0000	1.0000	1.0000	1.0000	**8.10**
8.20	0.9595	0.9791	0.9898	0.9953	0.9979	0.9991	0.9997	0.9999	1.0000	1.0000	1.0000	1.0000	1.0000	**8.20**
8.30	0.9561	0.9771	0.9887	0.9947	0.9977	0.9990	0.9996	0.9998	0.9999	1.0000	1.0000	1.0000	1.0000	**8.30**
8.40	0.9524	0.9749	0.9875	0.9941	0.9973	0.9989	0.9995	0.9998	0.9999	1.0000	1.0000	1.0000	1.0000	**8.40**
8.50	0.9486	0.9726	0.9862	0.9934	0.9970	0.9987	0.9995	0.9998	0.9999	1.0000	1.0000	1.0000	1.0000	**8.50**
8.60	0.9445	0.9701	0.9848	0.9926	0.9966	0.9985	0.9994	0.9998	0.9999	1.0000	1.0000	1.0000	1.0000	**8.60**
8.70	0.9403	0.9675	0.9832	0.9918	0.9962	0.9983	0.9993	0.9997	0.9999	1.0000	1.0000	1.0000	1.0000	**8.70**
8.80	0.9358	0.9647	0.9816	0.9909	0.9957	0.9981	0.9992	0.9997	0.9999	1.0000	1.0000	1.0000	1.0000	**8.80**
8.90	0.9311	0.9617	0.9798	0.9899	0.9952	0.9978	0.9991	0.9996	0.9998	0.9999	1.0000	1.0000	1.0000	**8.90**
9.00	0.9261	0.9585	0.9780	0.9889	0.9947	0.9976	0.9989	0.9996	0.9998	0.9999	1.0000	1.0000	1.0000	**9.00**
9.10	0.9210	0.9552	0.9760	0.9878	0.9941	0.9973	0.9988	0.9995	0.9998	0.9999	1.0000	1.0000	1.0000	**9.10**
9.20	0.9156	0.9517	0.9738	0.9865	0.9934	0.9969	0.9986	0.9994	0.9998	0.9999	1.0000	1.0000	1.0000	**9.20**
9.30	0.9100	0.9480	0.9715	0.9852	0.9927	0.9966	0.9985	0.9993	0.9997	0.9999	1.0000	1.0000	1.0000	**9.30**
9.40	0.9042	0.9441	0.9691	0.9838	0.9919	0.9962	0.9983	0.9992	0.9997	0.9999	1.0000	1.0000	1.0000	**9.40**
9.50	0.8981	0.9400	0.9665	0.9823	0.9911	0.9957	0.9980	0.9991	0.9996	0.9999	0.9999	1.0000	1.0000	**9.50**
9.60	0.8919	0.9357	0.9638	0.9806	0.9902	0.9952	0.9978	0.9990	0.9996	0.9998	0.9999	1.0000	1.0000	**9.60**
9.70	0.8853	0.9312	0.9609	0.9789	0.9892	0.9947	0.9975	0.9989	0.9995	0.9998	0.9999	1.0000	1.0000	**9.70**
9.80	0.8786	0.9265	0.9579	0.9770	0.9881	0.9941	0.9972	0.9987	0.9995	0.9998	0.9999	1.0000	1.0000	**9.80**
9.90	0.8716	0.9216	0.9546	0.9751	0.9870	0.9935	0.9969	0.9986	0.9994	0.9997	0.9999	1.0000	1.0000	**9.90**
10.00	0.8645	0.9165	0.9513	0.9730	0.9857	0.9928	0.9965	0.9984	0.9993	0.9997	0.9999	1.0000	1.0000	**10.00**

Appendix D

Cumulative Standard Normal Distribution Table

Standard Normal Z Table

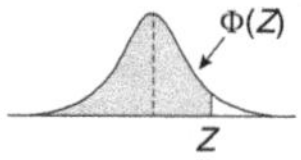

Table values give $\Phi(z) = \Pr(Z < z)$ for negative values of z

z	0.0	0.01	0.02	0.03	0.04	0.05	0.06	0.07	0.08	0.09
−4.0	0.0000	0.0000	0.0000	0.0000	0.0000	0.0000	0.0000	0.0000	0.0000	0.0000
−3.9	0.0000	0.0000	0.0000	0.0000	0.0000	0.0000	0.0000	0.0000	0.0000	0.0000
−3.8	0.0001	0.0001	0.0001	0.0001	0.0001	0.0001	0.0001	0.0001	0.0001	0.0001
−3.7	0.0001	0.0001	0.0001	0.0001	0.0001	0.0001	0.0001	0.0001	0.0001	0.0001
−3.6	0.0002	0.0002	0.0001	0.0001	0.0001	0.0001	0.0001	0.0001	0.0001	0.0001
−3.5	0.0002	0.0002	0.0002	0.0002	0.0002	0.0002	0.0002	0.0002	0.0002	0.0002
−3.4	0.0003	0.0003	0.0003	0.0003	0.0003	0.0003	0.0003	0.0003	0.0003	0.0002
−3.3	0.0005	0.0005	0.0005	0.0004	0.0004	0.0004	0.0004	0.0004	0.0004	0.0003
−3.2	0.0007	0.0007	0.0006	0.0006	0.0006	0.0006	0.0006	0.0005	0.0005	0.0005
−3.1	0.0010	0.0009	0.0009	0.0009	0.0008	0.0008	0.0008	0.0008	0.0007	0.0007
−3.0	0.0013	0.0013	0.0013	0.0012	0.0012	0.0011	0.0011	0.0011	0.0010	0.0010
−2.9	0.0019	0.0018	0.0018	0.0017	0.0016	0.0016	0.0015	0.0015	0.0014	0.0014
−2.8	0.0026	0.0025	0.0024	0.0023	0.0023	0.0022	0.0021	0.0021	0.0020	0.0019
−2.7	0.0035	0.0034	0.0033	0.0032	0.0031	0.0030	0.0029	0.0028	0.0027	0.0026
−2.6	0.0047	0.0045	0.0044	0.0043	0.0041	0.0040	0.0039	0.0038	0.0037	0.0036
−2.5	0.0062	0.0060	0.0059	0.0057	0.0055	0.0054	0.0052	0.0051	0.0049	0.0048
−2.4	0.0082	0.0080	0.0078	0.0075	0.0073	0.0071	0.0069	0.0068	0.0066	0.0064
−2.3	0.0107	0.0104	0.0102	0.0099	0.0096	0.0094	0.0091	0.0089	0.0087	0.0084
−2.2	0.0139	0.0136	0.0132	0.0129	0.0125	0.0122	0.0119	0.0116	0.0113	0.0110
−2.1	0.0179	0.0174	0.0170	0.0166	0.0162	0.0158	0.0154	0.0150	0.0146	0.0143
−2.0	0.0228	0.0222	0.0217	0.0212	0.0207	0.0202	0.0197	0.0192	0.0188	0.0183
−1.9	0.0287	0.0281	0.0274	0.0268	0.0262	0.0256	0.0250	0.0244	0.0239	0.0233
−1.8	0.0359	0.0351	0.0344	0.0336	0.0329	0.0322	0.0314	0.0307	0.0301	0.0294
−1.7	0.0446	0.0436	0.0427	0.0418	0.0409	0.0401	0.0392	0.0384	0.0375	0.0367
−1.6	0.0548	0.0537	0.0526	0.0516	0.0505	0.0495	0.0485	0.0475	0.0465	0.0455
−1.5	0.0668	0.0655	0.0643	0.0630	0.0618	0.0606	0.0594	0.0582	0.0571	0.0559
−1.4	0.0808	0.0793	0.0778	0.0764	0.0749	0.0735	0.0721	0.0708	0.0694	0.0681
−1.3	0.0968	0.0951	0.0934	0.0918	0.0901	0.0885	0.0869	0.0853	0.0838	0.0823
−1.2	0.1151	0.1131	0.1112	0.1093	0.1075	0.1056	0.1038	0.1020	0.1003	0.0985
−1.1	0.1357	0.1335	0.1314	0.1292	0.1271	0.1251	0.1230	0.1210	0.1190	0.1170
−1.0	0.1587	0.1562	0.1539	0.1515	0.1492	0.1469	0.1446	0.1423	0.1401	0.1379
−0.9	0.1841	0.1814	0.1788	0.1762	0.1736	0.1711	0.1685	0.1660	0.1635	0.1611
−0.8	0.2119	0.2090	0.2061	0.2033	0.2005	0.1977	0.1949	0.1922	0.1894	0.1867

Standard Normal Z Table

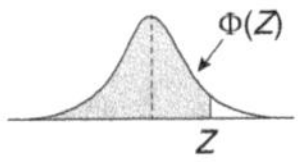

Table values give $\Phi(z) = \Pr(Z < z)$ for positive values of z

z	0.0	0.01	0.02	0.03	0.04	0.05	0.06	0.07	0.08	0.09
−0.7	0.2420	0.2389	0.2358	0.2327	0.2296	0.2266	0.2236	0.2206	0.2177	0.2148
−0.6	0.2743	0.2709	0.2676	0.2643	0.2611	0.2578	0.2546	0.2514	0.2483	0.2451
−0.5	0.3085	0.3050	0.3015	0.2981	0.2946	0.2912	0.2877	0.2843	0.2810	0.2776
−0.4	0.3446	0.3409	0.3372	0.3336	0.3300	0.3264	0.3228	0.3192	0.3156	0.3121
−0.3	0.3821	0.3783	0.3745	0.3707	0.3669	0.3632	0.3594	0.3557	0.3520	0.3483
−0.2	0.4207	0.4168	0.4129	0.4090	0.4052	0.4013	0.3974	0.3936	0.3897	0.3859
−0.1	0.4602	0.4562	0.4522	0.4483	0.4443	0.4404	0.4364	0.4325	0.4286	0.4247
−0	0.5000	0.4960	0.4920	0.4880	0.4840	0.4801	0.4761	0.4721	0.4681	0.4641
0.0	0.5000	0.5040	0.5080	0.5120	0.5160	0.5199	0.5239	0.5279	0.5319	0.5359
0.1	0.5398	0.5438	0.5478	0.5517	0.5557	0.5596	0.5636	0.5675	0.5714	0.5753
0.2	0.5793	0.5832	0.5871	0.5910	0.5948	0.5987	0.6026	0.6064	0.6103	0.6141
0.3	0.6179	0.6217	0.6255	0.6293	0.6331	0.6368	0.6406	0.6443	0.6480	0.6517
0.4	0.6554	0.6591	0.6628	0.6664	0.6700	0.6736	0.6772	0.6808	0.6844	0.6879
0.5	0.6915	0.6950	0.6985	0.7019	0.7054	0.7088	0.7123	0.7157	0.7190	0.7224
0.6	0.7257	0.7291	0.7324	0.7357	0.7389	0.7422	0.7454	0.7486	0.7517	0.7549
0.7	0.7580	0.7611	0.7642	0.7673	0.7704	0.7734	0.7764	0.7794	0.7823	0.7852
0.8	0.7881	0.7910	0.7939	0.7967	0.7995	0.8023	0.8051	0.8078	0.8106	0.8133
0.9	0.8159	0.8186	0.8212	0.8238	0.8264	0.8289	0.8315	0.8340	0.8365	0.8389
1.0	0.8413	0.8438	0.8461	0.8485	0.8508	0.8531	0.8554	0.8577	0.8599	0.8621
1.1	0.8643	0.8665	0.8686	0.8708	0.8729	0.8749	0.8770	0.8790	0.8810	0.8830
1.2	0.8849	0.8869	0.8888	0.8907	0.8925	0.8944	0.8962	0.8980	0.8997	0.9015
1.3	0.9032	0.9049	0.9066	0.9082	0.9099	0.9115	0.9131	0.9147	0.9162	0.9177
1.4	0.9192	0.9207	0.9222	0.9236	0.9251	0.9265	0.9279	0.9292	0.9306	0.9319
1.5	0.9332	0.9345	0.9357	0.9370	0.9382	0.9394	0.9406	0.9418	0.9429	0.9441
1.6	0.9452	0.9463	0.9474	0.9484	0.9495	0.9505	0.9515	0.9525	0.9535	0.9545
1.7	0.9554	0.9564	0.9573	0.9582	0.9591	0.9599	0.9608	0.9616	0.9625	0.9633
1.8	0.9641	0.9649	0.9656	0.9664	0.9671	0.9678	0.9686	0.9693	0.9699	0.9706
1.9	0.9713	0.9719	0.9726	0.9732	0.9738	0.9744	0.9750	0.9756	0.9761	0.9767
2.0	0.9772	0.9778	0.9783	0.9788	0.9793	0.9798	0.9803	0.9808	0.9812	0.9817
2.1	0.9821	0.9826	0.9830	0.9834	0.9838	0.9842	0.9846	0.9850	0.9854	0.9857
2.2	0.9861	0.9864	0.9868	0.9871	0.9875	0.9878	0.9881	0.9884	0.9887	0.9890
2.3	0.9893	0.9896	0.9898	0.9901	0.9904	0.9906	0.9909	0.9911	0.9913	0.9916
2.4	0.9918	0.9920	0.9922	0.9925	0.9927	0.9929	0.9931	0.9932	0.9934	0.9936
2.5	0.9938	0.9940	0.9941	0.9943	0.9945	0.9946	0.9948	0.9949	0.9951	0.9952
2.6	0.9953	0.9955	0.9956	0.9957	0.9959	0.9960	0.9961	0.9962	0.9963	0.9964
2.7	0.9965	0.9966	0.9967	0.9968	0.9969	0.9970	0.9971	0.9972	0.9973	0.9974
2.8	0.9974	0.9975	0.9976	0.9977	0.9977	0.9978	0.9979	0.9979	0.9980	0.9981
2.9	0.9981	0.9982	0.9982	0.9983	0.9984	0.9984	0.9985	0.9985	0.9986	0.9986
3.0	0.9987	0.9987	0.9987	0.9988	0.9988	0.9989	0.9989	0.9989	0.9990	0.9990
3.1	0.9990	0.9991	0.9991	0.9991	0.9992	0.9992	0.9992	0.9992	0.9993	0.9993
3.2	0.9993	0.9993	0.9994	0.9994	0.9994	0.9994	0.9994	0.9995	0.9995	0.9995
3.3	0.9995	0.9995	0.9995	0.9996	0.9996	0.9996	0.9996	0.9996	0.9996	0.9997
3.4	0.9997	0.9997	0.9997	0.9997	0.9997	0.9997	0.9997	0.9997	0.9997	0.9998
3.5	0.9998	0.9998	0.9998	0.9998	0.9998	0.9998	0.9998	0.9998	0.9998	0.9998
3.6	0.9998	0.9998	0.9999	0.9999	0.9999	0.9999	0.9999	0.9999	0.9999	0.9999
3.7	0.9999	0.9999	0.9999	0.9999	0.9999	0.9999	0.9999	0.9999	0.9999	0.9999
3.8	0.9999	0.9999	0.9999	0.9999	0.9999	0.9999	0.9999	0.9999	0.9999	0.9999
3.9	1.0000	1.0000	1.0000	1.0000	1.0000	1.0000	1.0000	1.0000	1.0000	1.0000
4.0	1.0000	1.0000	1.0000	1.0000	1.0000	1.0000	1.0000	1.0000	1.0000	1.0000

Appendix E
Student's *t* Distribution Table

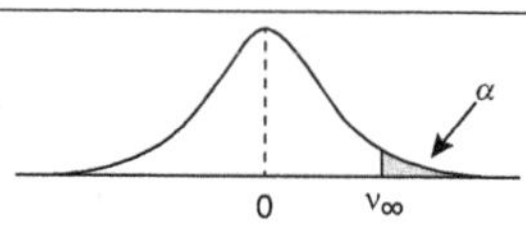

Student's *t* Cumulative Distribution Table

Degrees of freedom	α				
v	**0.10**	**0.05**	**0.025**	**0.01**	**0.005**
1	3.078	6.314	12.706	31.821	63.657
2	1.886	2.920	4.303	6.965	9.925
3	1.638	2.353	3.182	4.541	5.841
4	1.533	2.132	2.776	3.747	4.604
5	1.476	2.015	2.571	3.365	4.032
6	1.440	1.943	2.447	3.143	3.707
7	1.415	1.895	2.365	2.998	3.499
8	1.397	1.860	2.306	2.896	3.355
9	1.383	1.833	2.262	2.821	3.250
10	1.372	1.812	2.228	2.764	3.169
11	1.363	1.796	2.201	2.718	3.106
12	1.356	1.782	2.179	2.681	3.055
13	1.350	1.771	2.160	2.650	3.012
14	1.345	1.761	2.145	2.624	2.977
15	1.341	1.753	2.131	2.602	2.947
16	1.337	1.746	2.120	2.583	2.921
17	1.333	1.740	2.110	2.567	2.898
18	1.330	1.734	2.101	2.552	2.878
19	1.328	1.729	2.093	2.539	2.861
20	1.325	1.725	2.086	2.528	2.845
21	1.323	1.721	2.080	2.518	2.831
22	1.321	1.717	2.074	2.508	2.819
23	1.319	1.714	2.069	2.500	2.807
24	1.318	1.711	2.064	2.492	2.797
25	1.316	1.708	2.060	2.485	2.787
26	1.315	1.706	2.056	2.479	2.779
27	1.314	1.703	2.052	2.473	2.771
28	1.313	1.701	2.048	2.467	2.763
29	1.311	1.699	2.045	2.462	2.756
30	1.310	1.697	2.042	2.457	2.750
31	1.309	1.696	2.040	2.453	2.744
32	1.309	1.694	2.037	2.449	2.738
33	1.308	1.692	2.035	2.445	2.733

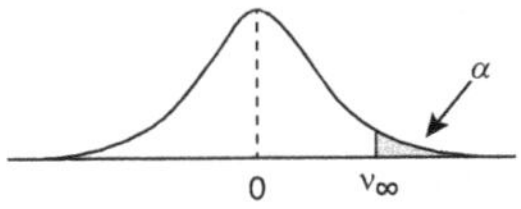

Student's *t* Cumulative Distribution Table

Degrees of freedom	α				
v	**0.10**	**0.05**	**0.025**	**0.01**	**0.005**
34	1.307	1.691	2.032	2.441	2.728
35	1.306	1.690	2.030	2.438	2.724
36	1.306	1.688	2.028	2.434	2.719
37	1.305	1.687	2.026	2.431	2.715
38	1.304	1.686	2.024	2.429	2.712
39	1.304	1.685	2.023	2.426	2.708
40	1.303	1.684	2.021	2.423	2.704
41	1.303	1.683	2.020	2.421	2.701
42	1.302	1.682	2.018	2.418	2.698
43	1.302	1.681	2.017	2.416	2.695
44	1.301	1.680	2.015	2.414	2.692
45	1.301	1.679	2.014	2.412	2.690
46	1.300	1.679	2.013	2.410	2.687
47	1.300	1.678	2.012	2.408	2.685
48	1.299	1.677	2.011	2.407	2.682
49	1.299	1.677	2.010	2.405	2.680
50	1.299	1.676	2.009	2.403	2.678
55	1.297	1.673	2.004	2.396	2.668
60	1.296	1.671	2.000	2.390	2.660
65	1.295	1.669	1.997	2.385	2.654
70	1.294	1.667	1.994	2.381	2.648
75	1.293	1.665	1.992	2.377	2.643
80	1.292	1.664	1.990	2.374	2.639
85	1.292	1.663	1.988	2.371	2.635
90	1.291	1.662	1.987	2.368	2.632
95	1.291	1.661	1.985	2.366	2.629
100	1.290	1.660	1.984	2.364	2.626
120	1.289	1.658	1.980	2.358	2.617
∞	1.282	1.645	1.960	2.326	2.576

Appendix F
Cumulative *F* Distribution

The following eight appendices contain the critical values for the *F* distribution based on numerator and denominator degrees of freedom and alpha value. The tables are arranged so that the upper tail (F_{α}) tables appear first, since these are the most common look-up values. Then the tables for the lower tails ($F_{1-\alpha}$) are presented:

Upper Tails:

Appendix F.1: $F_{0.01}$
Appendix F.2: $F_{0.025}$
Appendix F.3: $F_{0.05}$
Appendix F.4: $F_{0.10}$

Lower Tails:

Appendix F.5: $F_{0.90}$
Appendix F.6: $F_{0.95}$
Appendix F.7: $F_{0.975}$
Appendix F.8: $F_{0.99}$

Appendix F.1

$F(0.01)$ Distribution Table

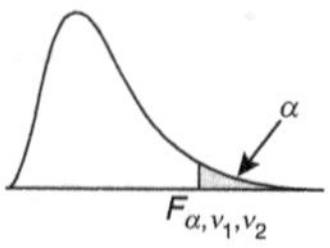

Degrees of freedom for the numerator (v_1)

Degrees of freedom for the denominator (v_2)

(v_2)	1	2	3	4	5	6	7	8	9
1	4052.18	4999.50	5403.35	5624.58	5763.65	5858.99	5928.36	5981.07	6022.47
2	98.50	99.00	99.17	99.25	99.30	99.33	99.36	99.37	99.39
3	34.12	30.82	29.46	28.71	28.24	27.91	27.67	27.49	27.35
4	21.198	18.000	16.694	15.977	15.522	15.207	14.976	14.799	14.659
5	16.258	13.274	12.060	11.392	10.967	10.672	10.456	10.289	10.158
6	13.745	10.925	9.780	9.148	8.746	8.466	8.260	8.102	7.976
7	12.246	9.547	8.451	7.847	7.460	7.191	6.993	6.840	6.719
8	11.259	8.649	7.591	7.006	6.632	6.371	6.178	6.029	5.911
9	10.561	8.022	6.992	6.422	6.057	5.802	5.613	5.467	5.351
10	10.044	7.559	6.552	5.994	5.636	5.386	5.200	5.057	4.942
11	9.646	7.206	6.217	5.668	5.316	5.069	4.886	4.744	4.632
12	9.330	6.927	5.953	5.412	5.064	4.821	4.640	4.499	4.388
13	9.074	6.701	5.739	5.205	4.862	4.620	4.441	4.302	4.191
14	8.862	6.515	5.564	5.035	4.695	4.456	4.278	4.140	4.030
15	8.683	6.359	5.417	4.893	4.556	4.318	4.142	4.004	3.895
16	8.531	6.226	5.292	4.773	4.437	4.202	4.026	3.890	3.780
17	8.400	6.112	5.185	4.669	4.336	4.102	3.927	3.791	3.682
18	8.285	6.013	5.092	4.579	4.248	4.015	3.841	3.705	3.597
19	8.185	5.926	5.010	4.500	4.171	3.939	3.765	3.631	3.523
20	8.096	5.849	4.938	4.431	4.103	3.871	3.699	3.564	3.457
21	8.017	5.780	4.874	4.369	4.042	3.812	3.640	3.506	3.398
22	7.945	5.719	4.817	4.313	3.988	3.758	3.587	3.453	3.346
23	7.881	5.664	4.765	4.264	3.939	3.710	3.539	3.406	3.299
24	7.823	5.614	4.718	4.218	3.895	3.667	3.496	3.363	3.256
25	7.770	5.568	4.675	4.177	3.855	3.627	3.457	3.324	3.217
26	7.721	5.526	4.637	4.140	3.818	3.591	3.421	3.288	3.182
27	7.677	5.488	4.601	4.106	3.785	3.558	3.388	3.256	3.149
28	7.636	5.453	4.568	4.074	3.754	3.528	3.358	3.226	3.120
29	7.598	5.420	4.538	4.045	3.725	3.499	3.330	3.198	3.092
30	7.562	5.390	4.510	4.018	3.699	3.473	3.304	3.173	3.067
40	7.314	5.179	4.313	3.828	3.514	3.291	3.124	2.993	2.888
60	7.077	4.977	4.126	3.649	3.339	3.119	2.953	2.823	2.718
100	6.895	4.824	3.984	3.513	3.206	2.988	2.823	2.694	2.590

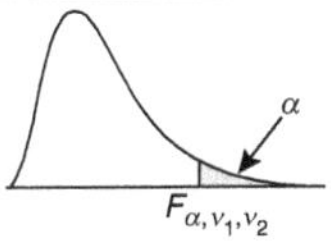

Degrees of freedom for the numerator (v_1)

10	11	12	13	14	15	16	17	(v_2)
6055.85	6083.32	6106.32	6125.86	6142.67	6157.28	6170.10	6181.43	**1**
99.40	99.41	99.42	99.42	99.43	99.43	99.44	99.44	**2**
27.23	27.13	27.05	26.98	26.92	26.87	26.83	26.79	**3**
14.546	14.452	14.374	14.307	14.249	14.198	14.154	14.115	**4**
10.051	9.963	9.888	9.825	9.770	9.722	9.680	9.643	**5**
7.874	7.790	7.718	7.657	7.605	7.559	7.519	7.483	**6**
6.620	6.538	6.469	6.410	6.359	6.314	6.275	6.240	**7**
5.814	5.734	5.667	5.609	5.559	5.515	5.477	5.442	**8**
5.257	5.178	5.111	5.055	5.005	4.962	4.924	4.890	**9**
4.849	4.772	4.706	4.650	4.601	4.558	4.520	4.487	**10**
4.539	4.462	4.397	4.342	4.293	4.251	4.213	4.180	**11**
4.296	4.220	4.155	4.100	4.052	4.010	3.972	3.939	**12**
4.100	4.025	3.960	3.905	3.857	3.815	3.778	3.745	**13**
3.939	3.864	3.800	3.745	3.698	3.656	3.619	3.586	**14**
3.805	3.730	3.666	3.612	3.564	3.522	3.485	3.452	**15**
3.691	3.616	3.553	3.498	3.451	3.409	3.372	3.339	**16**
3.593	3.519	3.455	3.401	3.353	3.312	3.275	3.242	**17**
3.508	3.434	3.371	3.316	3.269	3.227	3.190	3.158	**18**
3.434	3.360	3.297	3.242	3.195	3.153	3.116	3.084	**19**
3.368	3.294	3.231	3.177	3.130	3.088	3.051	3.018	**20**
3.310	3.236	3.173	3.119	3.072	3.030	2.993	2.960	**21**
3.258	3.184	3.121	3.067	3.019	2.978	2.941	2.908	**22**
3.211	3.137	3.074	3.020	2.973	2.931	2.894	2.861	**23**
3.168	3.094	3.032	2.977	2.930	2.889	2.852	2.819	**24**
3.129	3.056	2.993	2.939	2.892	2.850	2.813	2.780	**25**
3.094	3.021	2.958	2.904	2.857	2.815	2.778	2.745	**26**
3.062	2.988	2.926	2.871	2.824	2.783	2.746	2.713	**27**
3.032	2.959	2.896	2.842	2.795	2.753	2.716	2.683	**28**
3.005	2.931	2.868	2.814	2.767	2.726	2.689	2.656	**29**
2.979	2.906	2.843	2.789	2.742	2.700	2.663	2.630	**30**
2.801	2.727	2.665	2.611	2.563	2.522	2.484	2.451	**40**
2.632	2.559	2.496	2.442	2.394	2.352	2.315	2.281	**60**
2.503	2.430	2.368	2.313	2.265	2.223	2.185	2.151	**100**

Degrees of freedom for the denominator

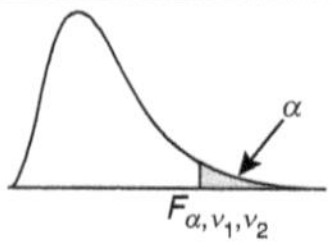

Degrees of freedom for the numerator (v_1)

Degrees of freedom for the denominator

(v_{21})	18	19	20	21	22	23	24	25	26
1	6191.53	6200.58	6208.73	6216.12	6222.84	6228.99	6234.63	6239.83	6244.62
2	99.44	99.45	99.45	99.45	99.45	99.46	99.46	99.46	99.46
3	26.75	26.72	26.69	26.66	26.64	26.62	26.60	26.58	26.56
4	14.080	14.048	14.020	13.994	13.970	13.949	13.929	13.911	13.894
5	9.610	9.580	9.553	9.528	9.506	9.485	9.466	9.449	9.433
6	7.451	7.422	7.396	7.372	7.351	7.331	7.313	7.296	7.280
7	6.209	6.181	6.155	6.132	6.111	6.092	6.074	6.058	6.043
8	5.412	5.384	5.359	5.336	5.316	5.297	5.279	5.263	5.248
9	4.860	4.833	4.808	4.786	4.765	4.746	4.729	4.713	4.698
10	4.457	4.430	4.405	4.383	4.363	4.344	4.327	4.311	4.296
11	4.150	4.123	4.099	4.077	4.057	4.038	4.021	4.005	3.990
12	3.909	3.883	3.858	3.836	3.816	3.798	3.780	3.765	3.750
13	3.716	3.689	3.665	3.643	3.622	3.604	3.587	3.571	3.556
14	3.556	3.529	3.505	3.483	3.463	3.444	3.427	3.412	3.397
15	3.423	3.396	3.372	3.350	3.330	3.311	3.294	3.278	3.264
16	3.310	3.283	3.259	3.237	3.216	3.198	3.181	3.165	3.150
17	3.212	3.186	3.162	3.139	3.119	3.101	3.084	3.068	3.053
18	3.128	3.101	3.077	3.055	3.035	3.016	2.999	2.983	2.968
19	3.054	3.027	3.003	2.981	2.961	2.942	2.925	2.909	2.894
20	2.989	2.962	2.938	2.916	2.895	2.877	2.859	2.843	2.829
21	2.931	2.904	2.880	2.857	2.837	2.818	2.801	2.785	2.770
22	2.879	2.852	2.827	2.805	2.785	2.766	2.749	2.733	2.718
23	2.832	2.805	2.781	2.758	2.738	2.719	2.702	2.686	2.671
24	2.789	2.762	2.738	2.716	2.695	2.676	2.659	2.643	2.628
25	2.751	2.724	2.699	2.677	2.657	2.638	2.620	2.604	2.589
26	2.715	2.688	2.664	2.642	2.621	2.602	2.585	2.569	2.554
27	2.683	2.656	2.632	2.609	2.589	2.570	2.552	2.536	2.521
28	2.653	2.626	2.602	2.579	2.559	2.540	2.522	2.506	2.491
29	2.626	2.599	2.574	2.552	2.531	2.512	2.495	2.478	2.463
30	2.600	2.573	2.549	2.526	2.506	2.487	2.469	2.453	2.437
40	2.421	2.394	2.369	2.346	2.325	2.306	2.288	2.271	2.256
60	2.251	2.223	2.198	2.175	2.153	2.134	2.115	2.098	2.083
100	2.120	2.092	2.067	2.043	2.021	2.001	1.983	1.965	1.949
∞	1.93	1.90	1.88	1.85	1.83	1.81	1.79	1.77	1.76

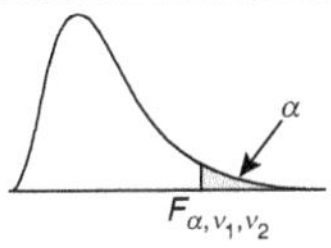

Degrees of freedom for the numerator

(v_1)

27	28	29	30	40	60	100	∞	(v_2)
6249.07	6253.20	6257.05	6260.65	6286.78	6313.03	6334.11	6365.86	1
99.46	99.46	99.46	99.47	99.47	99.48	99.49	99.50	**2**
26.55	26.53	26.52	26.50	26.41	26.32	26.24	26.13	**3**
13.878	13.864	13.850	13.838	13.745	13.652	13.577	13.46	**4**
9.418	9.404	9.391	9.379	9.291	9.202	9.130	9.02	**5**
7.266	7.253	7.240	7.229	7.143	7.057	6.987	6.88	**6**
6.029	6.016	6.003	5.992	5.908	5.824	5.755	5.65	**7**
5.234	5.221	5.209	5.198	5.116	5.032	4.963	4.86	**8**
4.685	4.672	4.660	4.649	4.567	4.483	4.415	4.31	**9**
4.283	4.270	4.258	4.247	4.165	4.082	4.014	3.91	**10**
3.977	3.964	3.952	3.941	3.860	3.776	3.708	3.60	**11**
3.736	3.724	3.712	3.701	3.619	3.535	3.467	3.36	**12**
3.543	3.530	3.518	3.507	3.425	3.341	3.272	3.17	**13**
3.383	3.371	3.359	3.348	3.266	3.181	3.112	3.00	**14**
3.250	3.237	3.225	3.214	3.132	3.047	2.977	2.87	**15**
3.137	3.124	3.112	3.101	3.018	2.933	2.863	2.75	**16**
3.039	3.026	3.014	3.003	2.920	2.835	2.764	2.65	**17**
2.955	2.942	2.930	2.919	2.835	2.749	2.678	2.57	**18**
2.880	2.868	2.855	2.844	2.761	2.674	2.602	2.49	**19**
2.815	2.802	2.790	2.778	2.695	2.608	2.535	2.42	**20**
2.756	2.743	2.731	2.720	2.636	2.548	2.475	2.36	**21**
2.704	2.691	2.679	2.667	2.583	2.495	2.422	2.31	**22**
2.657	2.644	2.632	2.620	2.535	2.447	2.373	2.26	**23**
2.614	2.601	2.589	2.577	2.492	2.403	2.329	2.21	**24**
2.575	2.562	2.550	2.538	2.453	2.364	2.289	2.17	**25**
2.540	2.526	2.514	2.503	2.417	2.327	2.252	2.13	**26**
2.507	2.494	2.481	2.470	2.384	2.294	2.218	2.10	**27**
2.477	2.464	2.451	2.440	2.354	2.263	2.187	2.06	**28**
2.449	2.436	2.423	2.412	2.325	2.234	2.158	2.03	**29**
2.423	2.410	2.398	2.386	2.299	2.208	2.131	2.01	**30**
2.241	2.228	2.215	2.203	2.114	2.019	1.938	1.80	**40**
2.068	2.054	2.041	2.028	1.936	1.836	1.749	1.60	**60**
1.934	1.919	1.906	1.893	1.797	1.692	1.598	1.43	**100**
1.74	1.72	1.71	1.70	1.59	1.47	1.36	1.00	∞

Degrees of freedom for the denominator

Appendix F.2

$F(0.025)$ Distribution Table

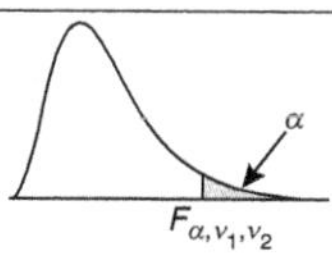

Degrees of freedom for the numerator (v_1); degrees of freedom for the denominator (v_2)

(v_2)	1	2	3	4	5	6	7	8	9	10	11	12	13	14	15	16	17
1	647.79	799.50	864.16	899.58	921.85	937.11	948.22	956.66	963.28	968.63	973.03	976.71	979.84	982.53	984.87	986.92	988.73
2	38.51	39.00	39.17	39.25	39.30	39.33	39.36	39.37	39.39	39.40	39.41	39.41	39.42	39.43	39.43	39.44	39.44
3	17.44	16.04	15.44	15.10	14.88	14.73	14.62	14.54	14.47	14.42	14.37	14.34	14.30	14.28	14.25	14.23	14.21
4	12.22	10.65	9.98	9.60	9.36	9.20	9.07	8.98	8.90	8.84	8.79	8.75	8.71	8.68	8.66	8.63	8.61
5	10.01	8.43	7.76	7.39	7.15	6.98	6.85	6.76	6.68	6.62	6.57	6.52	6.49	6.46	6.43	6.40	6.38
6	8.81	7.26	6.60	6.23	5.99	5.82	5.70	5.60	5.52	5.46	5.41	5.37	5.33	5.30	5.27	5.24	5.22
7	8.07	6.54	5.89	5.52	5.29	5.12	4.99	4.90	4.82	4.76	4.71	4.67	4.63	4.60	4.57	4.54	4.52
8	7.57	6.06	5.42	5.05	4.82	4.65	4.53	4.43	4.36	4.30	4.24	4.20	4.16	4.13	4.10	4.08	4.05
9	7.21	5.71	5.08	4.72	4.48	4.32	4.20	4.10	4.03	3.96	3.91	3.87	3.83	3.80	3.77	3.74	3.72
10	6.94	5.46	4.83	4.47	4.24	4.07	3.95	3.85	3.78	3.72	3.66	3.62	3.58	3.55	3.52	3.50	3.47
11	6.72	5.26	4.63	4.28	4.04	3.88	3.76	3.66	3.59	3.53	3.47	3.43	3.39	3.36	3.33	3.30	3.28
12	6.55	5.10	4.47	4.12	3.89	3.73	3.61	3.51	3.44	3.37	3.32	3.28	3.24	3.21	3.18	3.15	3.13
13	6.41	4.97	4.35	4.00	3.77	3.60	3.48	3.39	3.31	3.25	3.20	3.15	3.12	3.08	3.05	3.03	3.00
14	6.30	4.86	4.24	3.89	3.66	3.50	3.38	3.29	3.21	3.15	3.09	3.05	3.01	2.98	2.95	2.92	2.90
15	6.20	4.77	4.15	3.80	3.58	3.41	3.29	3.20	3.12	3.06	3.01	2.96	2.92	2.89	2.86	2.84	2.81
16	6.12	4.69	4.08	3.73	3.50	3.34	3.22	3.12	3.05	2.99	2.93	2.89	2.85	2.82	2.79	2.76	2.74
17	6.04	4.62	4.01	3.66	3.44	3.28	3.16	3.06	2.98	2.92	2.87	2.82	2.79	2.75	2.72	2.70	2.67
18	5.98	4.56	3.95	3.61	3.38	3.22	3.10	3.01	2.93	2.87	2.81	2.77	2.73	2.70	2.67	2.64	2.62
19	5.92	4.51	3.90	3.56	3.33	3.17	3.05	2.96	2.88	2.82	2.76	2.72	2.68	2.65	2.62	2.59	2.57
20	5.87	4.46	3.86	3.51	3.29	3.13	3.01	2.91	2.84	2.77	2.72	2.68	2.64	2.60	2.57	2.55	2.52
21	5.83	4.42	3.82	3.48	3.25	3.09	2.97	2.87	2.80	2.73	2.68	2.64	2.60	2.56	2.53	2.51	2.48
22	5.79	4.38	3.78	3.44	3.22	3.05	2.93	2.84	2.76	2.70	2.65	2.60	2.56	2.53	2.50	2.47	2.45
23	5.75	4.35	3.75	3.41	3.18	3.02	2.90	2.81	2.73	2.67	2.62	2.57	2.53	2.50	2.47	2.44	2.42
24	5.72	4.32	3.72	3.38	3.15	2.99	2.87	2.78	2.70	2.64	2.59	2.54	2.50	2.47	2.44	2.41	2.39
25	5.69	4.29	3.69	3.35	3.13	2.97	2.85	2.75	2.68	2.61	2.56	2.51	2.48	2.44	2.41	2.38	2.36
26	5.66	4.27	3.67	3.33	3.10	2.94	2.82	2.73	2.65	2.59	2.54	2.49	2.45	2.42	2.39	2.36	2.34
27	5.63	4.24	3.65	3.31	3.08	2.92	2.80	2.71	2.63	2.57	2.51	2.47	2.43	2.39	2.36	2.34	2.31
28	5.61	4.22	3.63	3.29	3.06	2.90	2.78	2.69	2.61	2.55	2.49	2.45	2.41	2.37	2.34	2.32	2.29
29	5.59	4.20	3.61	3.27	3.04	2.88	2.76	2.67	2.59	2.53	2.48	2.43	2.39	2.36	2.32	2.30	2.27
30	5.57	4.18	3.59	3.25	3.03	2.87	2.75	2.65	2.57	2.51	2.46	2.41	2.37	2.34	2.31	2.28	2.26
40	5.42	4.05	3.46	3.13	2.90	2.74	2.62	2.53	2.45	2.39	2.33	2.29	2.25	2.21	2.18	2.15	2.13
60	5.29	3.93	3.34	3.01	2.79	2.63	2.51	2.41	2.33	2.27	2.22	2.17	2.13	2.09	2.06	2.03	2.01
100	5.18	3.83	3.25	2.92	2.70	2.54	2.42	2.32	2.24	2.18	2.12	2.08	2.04	2.00	1.97	1.94	1.91
∞	5.02	3.69	3.12	2.79	2.57	2.41	2.29	2.19	2.11	2.05	1.99	1.94	1.90	1.87	1.83	1.80	1.78

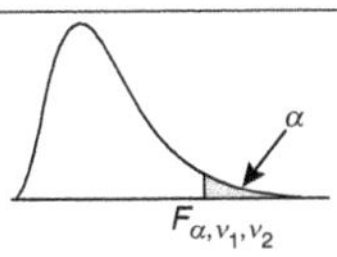

Degrees of freedom for the numerator

(v_1)

18	19	20	21	22	23	24	25	26	27	28	29	30	40	60	100	∞	(v_2)
90.35	991.80	993.10	994.29	995.36	996.35	997.25	998.08	998.85	999.56	1000.22	1000.84	1001.41	1005.60	1009.80	1013.17	1018.26	**1**
39.44	39.45	39.45	39.45	39.45	39.45	39.46	39.46	39.46	39.46	39.46	39.46	39.46	39.47	39.48	39.49	39.50	**2**
14.20	14.18	14.17	14.16	14.14	14.13	14.12	14.12	14.11	14.10	14.09	14.09	14.08	14.04	13.99	13.96	13.90	**3**
8.59	8.58	8.56	8.55	8.53	8.52	8.51	8.50	8.49	8.48	8.48	8.47	8.46	8.41	8.36	8.32	8.26	**4**
6.36	6.34	6.33	6.31	6.30	6.29	6.28	6.27	6.26	6.25	6.24	6.23	6.23	6.18	6.12	6.08	6.02	**5**
5.20	5.18	5.17	5.15	5.14	5.13	5.12	5.11	5.10	5.09	5.08	5.07	5.07	5.01	4.96	4.92	4.85	**6**
4.50	4.48	4.47	4.45	4.44	4.43	4.41	4.40	4.39	4.39	4.38	4.37	4.36	4.31	4.25	4.21	4.14	**7**
4.03	4.02	4.00	3.98	3.97	3.96	3.95	3.94	3.93	3.92	3.91	3.90	3.89	3.84	3.78	3.74	3.67	**8**
3.70	3.68	3.67	3.65	3.64	3.63	3.61	3.60	3.59	3.58	3.58	3.57	3.56	3.51	3.45	3.40	3.33	**9**
3.45	3.44	3.42	3.40	3.39	3.38	3.37	3.35	3.34	3.34	3.33	3.32	3.31	3.26	3.20	3.15	3.08	**10**
3.26	3.24	3.23	3.21	3.20	3.18	3.17	3.16	3.15	3.14	3.13	3.13	3.12	3.06	3.00	2.96	2.88	**11**
3.11	3.09	3.07	3.06	3.04	3.03	3.02	3.01	3.00	2.99	2.98	2.97	2.96	2.91	2.85	2.80	2.72	**12**
2.98	2.96	2.95	2.93	2.92	2.91	2.89	2.88	2.87	2.86	2.85	2.85	2.84	2.78	2.72	2.67	2.60	**13**
2.88	2.86	2.84	2.83	2.81	2.80	2.79	2.78	2.77	2.76	2.75	2.74	2.73	2.67	2.61	2.56	2.49	**14**
2.79	2.77	2.76	2.74	2.73	2.71	2.70	2.69	2.68	2.67	2.66	2.65	2.64	2.59	2.52	2.47	2.40	**15**
2.72	2.70	2.68	2.67	2.65	2.64	2.63	2.61	2.60	2.59	2.58	2.58	2.57	2.51	2.45	2.40	2.32	**16**
2.65	2.63	2.62	2.60	2.59	2.57	2.56	2.55	2.54	2.53	2.52	2.51	2.50	2.44	2.38	2.33	2.25	**17**
2.60	2.58	2.56	2.54	2.53	2.52	2.50	2.49	2.48	2.47	2.46	2.45	2.44	2.38	2.32	2.27	2.19	**18**
2.55	2.53	2.51	2.49	2.48	2.46	2.45	2.44	2.43	2.42	2.41	2.40	2.39	2.33	2.27	2.22	2.13	**19**
2.50	2.48	2.46	2.45	2.43	2.42	2.41	2.40	2.39	2.38	2.37	2.36	2.35	2.29	2.22	2.17	2.09	**20**
2.46	2.44	2.42	2.41	2.39	2.38	2.37	2.36	2.34	2.33	2.33	2.32	2.31	2.25	2.18	2.13	2.04	**21**
2.43	2.41	2.39	2.37	2.36	2.34	2.33	2.32	2.31	2.30	2.29	2.28	2.27	2.21	2.14	2.09	2.00	**22**
2.39	2.37	2.36	2.34	2.33	2.31	2.30	2.29	2.28	2.27	2.26	2.25	2.24	2.18	2.11	2.06	1.97	**23**
2.36	2.35	2.33	2.31	2.30	2.28	2.27	2.26	2.25	2.24	2.23	2.22	2.21	2.15	2.08	2.02	1.94	**24**
2.34	2.32	2.30	2.28	2.27	2.26	2.24	2.23	2.22	2.21	2.20	2.19	2.18	2.12	2.05	2.00	1.91	**25**
2.31	2.29	2.28	2.26	2.24	2.23	2.22	2.21	2.19	2.18	2.17	2.17	2.16	2.09	2.03	1.97	1.88	**26**
2.29	2.27	2.25	2.24	2.22	2.21	2.19	2.18	2.17	2.16	2.15	2.14	2.13	2.07	2.00	1.94	1.85	**27**
2.27	2.25	2.23	2.22	2.20	2.19	2.17	2.16	2.15	2.14	2.13	2.12	2.11	2.05	1.98	1.92	1.83	**28**
2.25	2.23	2.21	2.20	2.18	2.17	2.15	2.14	2.13	2.12	2.11	2.10	2.09	2.03	1.96	1.90	1.81	**29**
2.23	2.21	2.20	2.18	2.16	2.15	2.14	2.12	2.11	2.10	2.09	2.08	2.07	2.01	1.94	1.88	1.79	**30**
2.11	2.09	2.07	2.05	2.03	2.02	2.01	1.99	1.98	1.97	1.96	1.95	1.94	1.88	1.80	1.74	1.64	**40**
1.98	1.96	1.94	1.93	1.91	1.90	1.88	1.87	1.86	1.85	1.83	1.82	1.82	1.74	1.67	1.60	1.48	**60**
1.89	1.87	1.85	1.83	1.81	1.80	1.78	1.77	1.76	1.75	1.74	1.72	1.71	1.64	1.56	1.48	1.35	**100**
1.75	1.73	1.71	1.69	1.67	1.66	1.64	1.63	1.61	1.60	1.59	1.58	1.57	1.48	1.39	1.30	1.00	**∞**

Degrees of freedom for the denominator

Appendix F.3

$F(0.05)$ Distribution Table

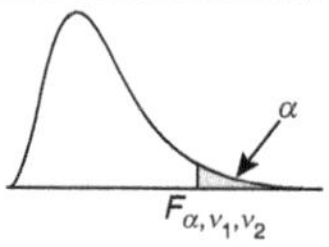

Degrees of freedom for the numerator (v_1); Degrees of freedom for the denominator (v_2)

(v_2)	1	2	3	4	5	6	7	8	9	10	11	12	13	14	15	16	17
1	161.45	199.50	215.71	224.58	230.16	233.99	236.77	238.88	240.54	241.88	242.98	243.91	244.69	245.36	245.95	246.46	246.92
2	18.51	19.00	19.16	19.25	19.30	19.33	19.35	19.37	19.38	19.40	19.40	19.41	19.42	19.42	19.43	19.43	19.44
3	10.13	9.55	9.28	9.12	9.01	8.94	8.89	8.85	8.81	8.79	8.76	8.74	8.73	8.71	8.70	8.69	8.68
4	7.71	6.94	6.59	6.39	6.26	6.16	6.09	6.04	6.00	5.96	5.94	5.91	5.89	5.87	5.86	5.84	5.83
5	6.61	5.79	5.41	5.19	5.05	4.95	4.88	4.82	4.77	4.74	4.70	4.68	4.66	4.64	4.62	4.60	4.59
6	5.99	5.14	4.76	4.53	4.39	4.28	4.21	4.15	4.10	4.06	4.03	4.00	3.98	3.96	3.94	3.92	3.91
7	5.59	4.74	4.35	4.12	3.97	3.87	3.79	3.73	3.68	3.64	3.60	3.57	3.55	3.53	3.51	3.49	3.48
8	5.32	4.46	4.07	3.84	3.69	3.58	3.50	3.44	3.39	3.35	3.31	3.28	3.26	3.24	3.22	3.20	3.19
9	5.12	4.26	3.86	3.63	3.48	3.37	3.29	3.23	3.18	3.14	3.10	3.07	3.05	3.03	3.01	2.99	2.97
10	4.96	4.10	3.71	3.48	3.33	3.22	3.14	3.07	3.02	2.98	2.94	2.91	2.89	2.86	2.85	2.83	2.81
11	4.84	3.98	3.59	3.36	3.20	3.09	3.01	2.95	2.90	2.85	2.82	2.79	2.76	2.74	2.72	2.70	2.69
12	4.75	3.89	3.49	3.26	3.11	3.00	2.91	2.85	2.80	2.75	2.72	2.69	2.66	2.64	2.62	2.60	2.58
13	4.67	3.81	3.41	3.18	3.03	2.92	2.83	2.77	2.71	2.67	2.63	2.60	2.58	2.55	2.53	2.51	2.50
14	4.60	3.74	3.34	3.11	2.96	2.85	2.76	2.70	2.65	2.60	2.57	2.53	2.51	2.48	2.46	2.44	2.43
15	4.54	3.68	3.29	3.06	2.90	2.79	2.71	2.64	2.59	2.54	2.51	2.48	2.45	2.42	2.40	2.38	2.37
16	4.49	3.63	3.24	3.01	2.85	2.74	2.66	2.59	2.54	2.49	2.46	2.42	2.40	2.37	2.35	2.33	2.32
17	4.45	3.59	3.20	2.96	2.81	2.70	2.61	2.55	2.49	2.45	2.41	2.38	2.35	2.33	2.31	2.29	2.27
18	4.41	3.55	3.16	2.93	2.77	2.66	2.58	2.51	2.46	2.41	2.37	2.34	2.31	2.29	2.27	2.25	2.23
19	4.38	3.52	3.13	2.90	2.74	2.63	2.54	2.48	2.42	2.38	2.34	2.31	2.28	2.26	2.23	2.21	2.20
20	4.35	3.49	3.10	2.87	2.71	2.60	2.51	2.45	2.39	2.35	2.31	2.28	2.25	2.22	2.20	2.18	2.17
21	4.32	3.47	3.07	2.84	2.68	2.57	2.49	2.42	2.37	2.32	2.28	2.25	2.22	2.20	2.18	2.16	2.14
22	4.30	3.44	3.05	2.82	2.66	2.55	2.46	2.40	2.34	2.30	2.26	2.23	2.20	2.17	2.15	2.13	2.11
23	4.28	3.42	3.03	2.80	2.64	2.53	2.44	2.37	2.32	2.27	2.24	2.20	2.18	2.15	2.13	2.11	2.09
24	4.26	3.40	3.01	2.78	2.62	2.51	2.42	2.36	2.30	2.25	2.22	2.18	2.15	2.13	2.11	2.09	2.07
25	4.24	3.39	2.99	2.76	2.60	2.49	2.40	2.34	2.28	2.24	2.20	2.16	2.14	2.11	2.09	2.07	2.05
26	4.23	3.37	2.98	2.74	2.59	2.47	2.39	2.32	2.27	2.22	2.18	2.15	2.12	2.09	2.07	2.05	2.03
27	4.21	3.35	2.96	2.73	2.57	2.46	2.37	2.31	2.25	2.20	2.17	2.13	2.10	2.08	2.06	2.04	2.02
28	4.20	3.34	2.95	2.71	2.56	2.45	2.36	2.29	2.24	2.19	2.15	2.12	2.09	2.06	2.04	2.02	2.00
29	4.18	3.33	2.93	2.70	2.55	2.43	2.35	2.28	2.22	2.18	2.14	2.10	2.08	2.05	2.03	2.01	1.99
30	4.17	3.32	2.92	2.69	2.53	2.42	2.33	2.27	2.21	2.16	2.13	2.09	2.06	2.04	2.01	1.99	1.98
40	4.08	3.23	2.84	2.61	2.45	2.34	2.25	2.18	2.12	2.08	2.04	2.00	1.97	1.95	1.92	1.90	1.89
60	4.00	3.15	2.76	2.53	2.37	2.25	2.17	2.10	2.04	1.99	1.95	1.92	1.89	1.86	1.84	1.82	1.80
100	3.94	3.09	2.70	2.46	2.31	2.19	2.10	2.03	1.97	1.93	1.89	1.85	1.82	1.79	1.77	1.75	1.73
∞	3.84	3.00	2.60	2.37	2.21	2.10	2.01	1.94	1.88	1.83	1.79	1.75	1.72	1.69	1.67	1.64	1.62

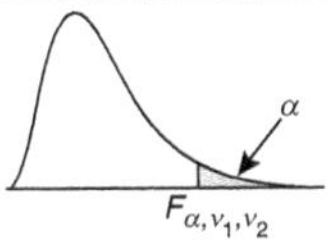

Degrees of freedom for the numerator

(v_1)

18	19	20	21	22	23	24	25	26	27	28	29	30	40	60	100	∞	(v_2)
247.32	247.69	248.01	248.31	248.58	248.83	249.05	249.26	249.45	249.63	249.80	249.95	250.10	251.14	252.20	253.04	254.31	**1**
19.44	19.44	19.45	19.45	19.45	19.45	19.45	19.46	19.46	19.46	19.46	19.46	19.46	19.47	19.48	19.49	9.48	**2**
8.67	8.67	8.66	8.65	8.65	8.64	8.64	8.63	8.63	8.63	8.62	8.62	8.62	8.59	8.57	8.55	5.13	**3**
5.82	5.81	5.80	5.79	5.79	5.78	5.77	5.77	5.76	5.76	5.75	5.75	5.75	5.72	5.69	5.66	3.76	**4**
4.58	4.57	4.56	4.55	4.54	4.53	4.53	4.52	4.52	4.51	4.50	4.50	4.50	4.46	4.43	4.41	3.10	**5**
3.90	3.88	3.87	3.86	3.86	3.85	3.84	3.83	3.83	3.82	3.82	3.81	3.81	3.77	3.74	3.71	2.72	**6**
3.47	3.46	3.44	3.43	3.43	3.42	3.41	3.40	3.40	3.39	3.39	3.38	3.38	3.34	3.30	3.27	2.47	**7**
3.17	3.16	3.15	3.14	3.13	3.12	3.12	3.11	3.10	3.10	3.09	3.08	3.08	3.04	3.01	2.97	2.29	**8**
2.96	2.95	2.94	2.93	2.92	2.91	2.90	2.89	2.89	2.88	2.87	2.87	2.86	2.83	2.79	2.76	2.16	**9**
2.80	2.79	2.77	2.76	2.75	2.75	2.74	2.73	2.72	2.72	2.71	2.70	2.70	2.66	2.62	2.59	2.06	**10**
2.67	2.66	2.65	2.64	2.63	2.62	2.61	2.60	2.59	2.59	2.58	2.58	2.57	2.53	2.49	2.46	1.97	**11**
2.57	2.56	2.54	2.53	2.52	2.51	2.51	2.50	2.49	2.48	2.48	2.47	2.47	2.43	2.38	2.35	1.90	**12**
2.48	2.47	2.46	2.45	2.44	2.43	2.42	2.41	2.41	2.40	2.39	2.39	2.38	2.34	2.30	2.26	1.85	**13**
2.41	2.40	2.39	2.38	2.37	2.36	2.35	2.34	2.33	2.33	2.32	2.31	2.31	2.27	2.22	2.19	1.80	**14**
2.35	2.34	2.33	2.32	2.31	2.30	2.29	2.28	2.27	2.27	2.26	2.25	2.25	2.20	2.16	2.12	1.76	**15**
2.30	2.29	2.28	2.26	2.25	2.24	2.24	2.23	2.22	2.21	2.21	2.20	2.19	2.15	2.11	2.07	1.72	**16**
2.26	2.24	2.23	2.22	2.21	2.20	2.19	2.18	2.17	2.17	2.16	2.15	2.15	2.10	2.06	2.02	1.69	**17**
2.22	2.20	2.19	2.18	2.17	2.16	2.15	2.14	2.13	2.13	2.12	2.11	2.11	2.06	2.02	1.98	1.66	**18**
2.18	2.17	2.16	2.14	2.13	2.12	2.11	2.11	2.10	2.09	2.08	2.08	2.07	2.03	1.98	1.94	1.63	**19**
2.15	2.14	2.12	2.11	2.10	2.09	2.08	2.07	2.07	2.06	2.05	2.05	2.04	1.99	1.95	1.91	1.61	**20**
2.12	2.11	2.10	2.08	2.07	2.06	2.05	2.05	2.04	2.03	2.02	2.02	2.01	1.96	1.92	1.88	1.59	**21**
2.10	2.08	2.07	2.06	2.05	2.04	2.03	2.02	2.01	2.00	2.00	1.99	1.98	1.94	1.89	1.85	1.57	**22**
2.08	2.06	2.05	2.04	2.02	2.01	2.01	2.00	1.99	1.98	1.97	1.97	1.96	1.91	1.86	1.82	1.55	**23**
2.05	2.04	2.03	2.01	2.00	1.99	1.98	1.97	1.97	1.96	1.95	1.95	1.94	1.89	1.84	1.80	1.53	**24**
2.04	2.02	2.01	2.00	1.98	1.97	1.96	1.96	1.95	1.94	1.93	1.93	1.92	1.87	1.82	1.78	1.52	**25**
2.02	2.00	1.99	1.98	1.97	1.96	1.95	1.94	1.93	1.92	1.91	1.91	1.90	1.85	1.80	1.76	1.50	**26**
2.00	1.99	1.97	1.96	1.95	1.94	1.93	1.92	1.91	1.90	1.90	1.89	1.88	1.84	1.79	1.74	1.49	**27**
1.99	1.97	1.96	1.95	1.93	1.92	1.91	1.91	1.90	1.89	1.88	1.88	1.87	1.82	1.77	1.73	1.48	**28**
1.97	1.96	1.94	1.93	1.92	1.91	1.90	1.89	1.88	1.88	1.87	1.86	1.85	1.81	1.75	1.71	1.47	**29**
1.96	1.95	1.93	1.92	1.91	1.90	1.89	1.88	1.87	1.86	1.85	1.85	1.84	1.79	1.74	1.70	1.46	**30**
1.87	1.85	1.84	1.83	1.81	1.80	1.79	1.78	1.77	1.77	1.76	1.75	1.74	1.69	1.64	1.59	1.38	**40**
1.78	1.76	1.75	1.73	1.72	1.71	1.70	1.69	1.68	1.67	1.66	1.66	1.65	1.59	1.53	1.48	1.29	**60**
1.71	1.69	1.68	1.66	1.65	1.64	1.63	1.62	1.61	1.60	1.59	1.58	1.57	1.52	1.45	1.39	1.21	**100**
1.60	1.59	1.57	1.56	1.54	1.53	1.52	1.51	1.50	1.49	1.48	1.47	1.46	1.39	1.32	1.24	1.00	**∞**

Degrees of freedom for the denominator

Appendix F.4

F(0.10) Distribution Table

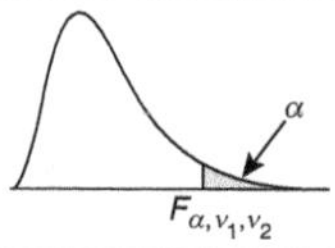

Degrees of freedom for the numerator (v_1); Degrees of freedom for the denominator (v_2)

(v_2)	1	2	3	4	5	6	7	8	9	10	11	12	13	14	15	16	17	(v_2)
1	39.86	49.50	53.59	55.83	57.24	58.20	58.91	59.44	59.86	60.19	60.47	60.71	60.90	61.07	61.22	61.35	61.46	**1**
2	8.53	9.00	9.16	9.24	9.29	9.33	9.35	9.37	9.38	9.39	9.40	9.41	9.41	9.42	9.42	9.43	9.43	**2**
3	5.54	5.46	5.39	5.34	5.31	5.28	5.27	5.25	5.24	5.23	5.22	5.22	5.21	5.20	5.20	5.20	5.19	**3**
4	4.54	4.32	4.19	4.11	4.05	4.01	3.98	3.95	3.94	3.92	3.91	3.90	3.89	3.88	3.87	3.86	3.86	**4**
5	4.06	3.78	3.62	3.52	3.45	3.40	3.37	3.34	3.32	3.30	3.28	3.27	3.26	3.25	3.24	3.23	3.22	**5**
6	3.78	3.46	3.29	3.18	3.11	3.05	3.01	2.98	2.96	2.94	2.92	2.90	2.89	2.88	2.87	2.86	2.85	**6**
7	3.59	3.26	3.07	2.96	2.88	2.83	2.78	2.75	2.72	2.70	2.68	2.67	2.65	2.64	2.63	2.62	2.61	**7**
8	3.46	3.11	2.92	2.81	2.73	2.67	2.62	2.59	2.56	2.54	2.52	2.50	2.49	2.48	2.46	2.45	2.45	**8**
9	3.36	3.01	2.81	2.69	2.61	2.55	2.51	2.47	2.44	2.42	2.40	2.38	2.36	2.35	2.34	2.33	2.32	**9**
10	3.29	2.92	2.73	2.61	2.52	2.46	2.41	2.38	2.35	2.32	2.30	2.28	2.27	2.26	2.24	2.23	2.22	**10**
11	3.23	2.86	2.66	2.54	2.45	2.39	2.34	2.30	2.27	2.25	2.23	2.21	2.19	2.18	2.17	2.16	2.15	**11**
12	3.18	2.81	2.61	2.48	2.39	2.33	2.28	2.24	2.21	2.19	2.17	2.15	2.13	2.12	2.10	2.09	2.08	**12**
13	3.14	2.76	2.56	2.43	2.35	2.28	2.23	2.20	2.16	2.14	2.12	2.10	2.08	2.07	2.05	2.04	2.03	**13**
14	3.10	2.73	2.52	2.39	2.31	2.24	2.19	2.15	2.12	2.10	2.07	2.05	2.04	2.02	2.01	2.00	1.99	**14**
15	3.07	2.70	2.49	2.36	2.27	2.21	2.16	2.12	2.09	2.06	2.04	2.02	2.00	1.99	1.97	1.96	1.95	**15**
16	3.05	2.67	2.46	2.33	2.24	2.18	2.13	2.09	2.06	2.03	2.01	1.99	1.97	1.95	1.94	1.93	1.92	**16**
17	3.03	2.64	2.44	2.31	2.22	2.15	2.10	2.06	2.03	2.00	1.98	1.96	1.94	1.93	1.91	1.90	1.89	**17**
18	3.01	2.62	2.42	2.29	2.20	2.13	2.08	2.04	2.00	1.98	1.95	1.93	1.92	1.90	1.89	1.87	1.86	**18**
19	2.99	2.61	2.40	2.27	2.18	2.11	2.06	2.02	1.98	1.96	1.93	1.91	1.89	1.88	1.86	1.85	1.84	**19**
20	2.97	2.59	2.38	2.25	2.16	2.09	2.04	2.00	1.96	1.94	1.91	1.89	1.87	1.86	1.84	1.83	1.82	**20**
21	2.96	2.57	2.36	2.23	2.14	2.08	2.02	1.98	1.95	1.92	1.90	1.87	1.86	1.84	1.83	1.81	1.80	**21**
22	2.95	2.56	2.35	2.22	2.13	2.06	2.01	1.97	1.93	1.90	1.88	1.86	1.84	1.83	1.81	1.80	1.79	**22**
23	2.94	2.55	2.34	2.21	2.11	2.05	1.99	1.95	1.92	1.89	1.87	1.84	1.83	1.81	1.80	1.78	1.77	**23**
24	2.93	2.54	2.33	2.19	2.10	2.04	1.98	1.94	1.91	1.88	1.85	1.83	1.81	1.80	1.78	1.77	1.76	**24**
25	2.92	2.53	2.32	2.18	2.09	2.02	1.97	1.93	1.89	1.87	1.84	1.82	1.80	1.79	1.77	1.76	1.75	**25**
26	2.91	2.52	2.31	2.17	2.08	2.01	1.96	1.92	1.88	1.86	1.83	1.81	1.79	1.77	1.76	1.75	1.73	**26**
27	2.90	2.51	2.30	2.17	2.07	2.00	1.95	1.91	1.87	1.85	1.82	1.80	1.78	1.76	1.75	1.74	1.72	**27**
28	2.89	2.50	2.29	2.16	2.06	2.00	1.94	1.90	1.87	1.84	1.81	1.79	1.77	1.75	1.74	1.73	1.71	**28**
29	2.89	2.50	2.28	2.15	2.06	1.99	1.93	1.89	1.86	1.83	1.80	1.78	1.76	1.75	1.73	1.72	1.71	**29**
30	2.88	2.49	2.28	2.14	2.05	1.98	1.93	1.88	1.85	1.82	1.79	1.77	1.75	1.74	1.72	1.71	1.70	**30**
40	2.84	2.44	2.23	2.09	2.00	1.93	1.87	1.83	1.79	1.76	1.74	1.71	1.70	1.68	1.66	1.65	1.64	**40**
60	2.79	2.39	2.18	2.04	1.95	1.87	1.82	1.77	1.74	1.71	1.68	1.66	1.64	1.62	1.60	1.59	1.58	**60**
100	2.76	2.36	2.14	2.00	1.91	1.83	1.78	1.73	1.69	1.66	1.64	1.61	1.59	1.57	1.56	1.54	1.53	**100**
∞	2.71	2.30	2.08	1.94	1.85	1.77	1.72	1.67	1.63	1.60	1.57	1.55	1.52	1.50	1.49	1.47	1.46	**∞**

Degrees of freedom for the numerator (v_1)

(v_2)	18	19	20	21	22	23	24	25	26	27	28	29	30	40	60	100	∞	(v_2)
1	61.57	61.66	61.74	61.81	61.88	61.95	62.00	62.05	62.10	62.15	62.19	62.23	62.26	62.53	62.79	63.01	63.33	1
2	9.44	9.44	9.44	9.44	9.45	9.45	9.45	9.45	9.45	9.45	9.46	9.46	9.46	9.47	9.47	9.48	9.49	2
3	5.19	5.19	5.18	5.18	5.18	5.18	5.18	5.17	5.17	5.17	5.17	5.17	5.17	5.16	5.15	5.14	5.13	3
4	3.85	3.85	3.84	3.84	3.84	3.83	3.83	3.83	3.83	3.82	3.82	3.82	3.82	3.80	3.79	3.78	5.76	4
5	3.22	3.21	3.21	3.20	3.20	3.19	3.19	3.19	3.18	3.18	3.18	3.18	3.17	3.16	3.14	3.13	3.10	5
6	2.85	2.84	2.84	2.83	2.83	2.82	2.82	2.81	2.81	2.81	2.81	2.80	2.80	2.78	2.76	2.75	2.72	6
7	2.61	2.60	2.59	2.59	2.58	2.58	2.58	2.57	2.57	2.56	2.56	2.56	2.56	2.54	2.51	2.50	2.47	7
8	2.44	2.43	2.42	2.42	2.41	2.41	2.40	2.40	2.40	2.39	2.39	2.39	2.38	2.36	2.34	2.32	2.29	8
9	2.31	2.30	2.30	2.29	2.29	2.28	2.28	2.27	2.27	2.26	2.26	2.26	2.25	2.23	2.21	2.19	2.16	9
10	2.22	2.21	2.20	2.19	2.19	2.18	2.18	2.17	2.17	2.17	2.16	2.16	2.16	2.13	2.11	2.09	2.06	10
11	2.14	2.13	2.12	2.12	2.11	2.11	2.10	2.10	2.09	2.09	2.08	2.08	2.08	2.05	2.03	2.01	1.97	11
12	2.08	2.07	2.06	2.05	2.05	2.04	2.04	2.03	2.03	2.02	2.02	2.01	2.01	1.99	1.96	1.94	1.90	12
13	2.02	2.01	2.01	2.00	1.99	1.99	1.98	1.98	1.97	1.97	1.96	1.96	1.96	1.93	1.90	1.88	1.85	13
14	1.98	1.97	1.96	1.96	1.95	1.94	1.94	1.93	1.93	1.92	1.92	1.92	1.91	1.89	1.86	1.83	1.80	14
15	1.94	1.93	1.92	1.92	1.91	1.90	1.90	1.89	1.89	1.88	1.88	1.88	1.87	1.85	1.82	1.79	1.76	15
16	1.91	1.90	1.89	1.88	1.88	1.87	1.87	1.86	1.86	1.85	1.85	1.84	1.84	1.81	1.78	1.76	1.72	16
17	1.88	1.87	1.86	1.86	1.85	1.84	1.84	1.83	1.83	1.82	1.82	1.81	1.81	1.78	1.75	1.73	1.69	17
18	1.85	1.84	1.84	1.83	1.82	1.82	1.81	1.80	1.80	1.80	1.79	1.79	1.78	1.75	1.72	1.70	1.66	18
19	1.83	1.82	1.81	1.81	1.80	1.79	1.79	1.78	1.78	1.77	1.77	1.76	1.76	1.73	1.70	1.67	1.63	19
20	1.81	1.80	1.79	1.79	1.78	1.77	1.77	1.76	1.76	1.75	1.75	1.74	1.74	1.71	1.68	1.65	1.61	20
21	1.79	1.78	1.78	1.77	1.76	1.75	1.75	1.74	1.74	1.73	1.73	1.72	1.72	1.69	1.66	1.63	1.59	21
22	1.78	1.77	1.76	1.75	1.74	1.74	1.73	1.73	1.72	1.72	1.71	1.71	1.70	1.67	1.64	1.61	1.57	22
23	1.76	1.75	1.74	1.74	1.73	1.72	1.72	1.71	1.70	1.70	1.69	1.69	1.69	1.66	1.62	1.59	1.55	23
24	1.75	1.74	1.73	1.72	1.71	1.71	1.70	1.70	1.69	1.69	1.68	1.68	1.67	1.64	1.61	1.58	1.53	24
25	1.74	1.73	1.72	1.71	1.70	1.70	1.69	1.68	1.68	1.67	1.67	1.66	1.66	1.63	1.59	1.56	1.52	25
26	1.72	1.71	1.71	1.70	1.69	1.68	1.68	1.67	1.67	1.66	1.66	1.65	1.65	1.61	1.58	1.55	1.50	26
27	1.71	1.70	1.70	1.69	1.68	1.67	1.67	1.66	1.65	1.65	1.64	1.64	1.64	1.60	1.57	1.54	1.49	27
28	1.70	1.69	1.69	1.68	1.67	1.66	1.66	1.65	1.64	1.64	1.63	1.63	1.63	1.59	1.56	1.53	1.48	28
29	1.69	1.68	1.68	1.67	1.66	1.65	1.65	1.64	1.63	1.63	1.62	1.62	1.62	1.58	1.55	1.52	1.47	29
30	1.69	1.68	1.67	1.66	1.65	1.64	1.64	1.63	1.63	1.62	1.62	1.61	1.61	1.57	1.54	1.51	1.46	30
40	1.62	1.61	1.61	1.60	1.59	1.58	1.57	1.57	1.56	1.56	1.55	1.55	1.54	1.51	1.47	1.43	1.38	40
60	1.56	1.55	1.54	1.53	1.53	1.52	1.51	1.50	1.50	1.49	1.49	1.48	1.48	1.44	1.40	1.36	1.29	60
100	1.52	1.50	1.49	1.48	1.48	1.47	1.46	1.45	1.45	1.44	1.43	1.43	1.42	1.38	1.34	1.29	1.21	100
∞	1.44	1.43	1.42	1.41	1.40	1.39	1.38	1.38	1.37	1.36	1.35	1.35	1.34	1.30	1.24	1.18	1.00	∞

Degrees of freedom for the denominator

Appendix F.5

F(0.90) Distribution Table

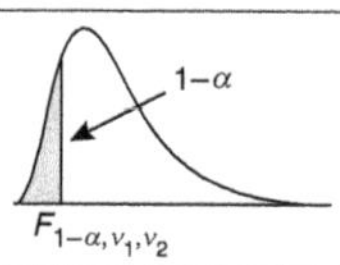

Degrees of freedom for the numerator (v_1)

Rows: Degrees of freedom for the denominator (v_2)

(v_2)	1	2	3	4	5	6	7	8	9	10	11	12	13	14	15	16	17	(v_2)
1	0.03	0.12	0.18	0.22	0.25	0.26	0.28	0.29	0.30	0.30	0.31	0.31	0.32	0.32	0.33	0.33	0.33	**1**
2	0.02	0.11	0.18	0.23	0.26	0.29	0.31	0.32	0.33	0.34	0.35	0.36	0.36	0.37	0.37	0.37	0.38	**2**
3	0.02	0.11	0.19	0.24	0.28	0.30	0.33	0.34	0.36	0.37	0.38	0.38	0.39	0.40	0.40	0.41	0.41	**3**
4	0.02	0.11	0.19	0.24	0.28	0.31	0.34	0.36	0.37	0.38	0.39	0.40	0.41	0.42	0.42	0.43	0.43	**4**
5	0.02	0.11	0.19	0.25	0.29	0.32	0.35	0.37	0.38	0.40	0.41	0.42	0.43	0.43	0.44	0.45	0.45	**5**
6	0.02	0.11	0.19	0.25	0.29	0.33	0.35	0.37	0.39	0.41	0.42	0.43	0.44	0.45	0.45	0.46	0.46	**6**
7	0.02	0.11	0.19	0.25	0.30	0.33	0.36	0.38	0.40	0.41	0.43	0.44	0.45	0.46	0.46	0.47	0.48	**7**
8	0.02	0.11	0.19	0.25	0.30	0.34	0.36	0.39	0.40	0.42	0.43	0.45	0.46	0.46	0.47	0.48	0.49	**8**
9	0.02	0.11	0.19	0.25	0.30	0.34	0.37	0.39	0.41	0.43	0.44	0.45	0.46	0.47	0.48	0.49	0.49	**9**
10	0.02	0.11	0.19	0.26	0.30	0.34	0.37	0.39	0.41	0.43	0.44	0.46	0.47	0.48	0.49	0.49	0.50	**10**
11	0.02	0.11	0.19	0.26	0.30	0.34	0.37	0.40	0.42	0.43	0.45	0.46	0.47	0.48	0.49	0.50	0.51	**11**
12	0.02	0.11	0.19	0.26	0.31	0.34	0.37	0.40	0.42	0.44	0.45	0.47	0.48	0.49	0.50	0.50	0.51	**12**
13	0.02	0.11	0.19	0.26	0.31	0.35	0.38	0.40	0.42	0.44	0.46	0.47	0.48	0.49	0.50	0.51	0.52	**13**
14	0.02	0.11	0.19	0.26	0.31	0.35	0.38	0.40	0.43	0.44	0.46	0.47	0.48	0.49	0.50	0.51	0.52	**14**
15	0.02	0.11	0.19	0.26	0.31	0.35	0.38	0.41	0.43	0.45	0.46	0.48	0.49	0.50	0.51	0.52	0.52	**15**
16	0.02	0.11	0.19	0.26	0.31	0.35	0.38	0.41	0.43	0.45	0.46	0.48	0.49	0.50	0.51	0.52	0.53	**16**
17	0.02	0.11	0.19	0.26	0.31	0.35	0.38	0.41	0.43	0.45	0.47	0.48	0.49	0.50	0.51	0.52	0.53	**17**
18	0.02	0.11	0.19	0.26	0.31	0.35	0.38	0.41	0.43	0.45	0.47	0.48	0.49	0.51	0.52	0.52	0.53	**18**
19	0.02	0.11	0.19	0.26	0.31	0.35	0.38	0.41	0.43	0.45	0.47	0.48	0.50	0.51	0.52	0.53	0.53	**19**
20	0.02	0.11	0.19	0.26	0.31	0.35	0.39	0.41	0.44	0.45	0.47	0.49	0.50	0.51	0.52	0.53	0.54	**20**
21	0.02	0.11	0.19	0.26	0.31	0.35	0.39	0.41	0.44	0.46	0.47	0.49	0.50	0.51	0.52	0.53	0.54	**21**
22	0.02	0.11	0.19	0.26	0.31	0.35	0.39	0.41	0.44	0.46	0.47	0.49	0.50	0.51	0.52	0.53	0.54	**22**
23	0.02	0.11	0.19	0.26	0.31	0.35	0.39	0.42	0.44	0.46	0.48	0.49	0.50	0.51	0.53	0.53	0.54	**23**
24	0.02	0.11	0.19	0.26	0.31	0.35	0.39	0.42	0.44	0.46	0.48	0.49	0.50	0.52	0.53	0.54	0.54	**24**
25	0.02	0.11	0.19	0.26	0.31	0.36	0.39	0.42	0.44	0.46	0.48	0.49	0.51	0.52	0.53	0.54	0.55	**25**
26	0.02	0.11	0.19	0.26	0.31	0.36	0.39	0.42	0.44	0.46	0.48	0.49	0.51	0.52	0.53	0.54	0.55	**26**
27	0.02	0.11	0.19	0.26	0.31	0.36	0.39	0.42	0.44	0.46	0.48	0.49	0.51	0.52	0.53	0.54	0.55	**27**
28	0.02	0.11	0.19	0.26	0.31	0.36	0.39	0.42	0.44	0.46	0.48	0.50	0.51	0.52	0.53	0.54	0.55	**28**
29	0.02	0.11	0.19	0.26	0.31	0.36	0.39	0.42	0.44	0.46	0.48	0.50	0.51	0.52	0.53	0.54	0.55	**29**
30	0.02	0.11	0.19	0.26	0.32	0.36	0.39	0.42	0.44	0.46	0.48	0.50	0.51	0.52	0.53	0.54	0.55	**30**
40	0.02	0.11	0.19	0.26	0.32	0.36	0.39	0.42	0.45	0.47	0.49	0.50	0.52	0.53	0.54	0.55	0.56	**40**
60	0.02	0.11	0.19	0.26	0.32	0.36	0.40	0.43	0.45	0.47	0.49	0.51	0.53	0.54	0.55	0.56	0.57	**60**
100	0.02	0.11	0.19	0.26	0.32	0.36	0.40	0.43	0.46	0.48	0.50	0.52	0.53	0.55	0.56	0.57	0.58	**100**
∞	0.02	0.11	0.19	0.27	0.32	0.37	0.40	0.44	0.46	0.49	0.51	0.53	0.54	0.56	0.57	0.58	0.59	∞

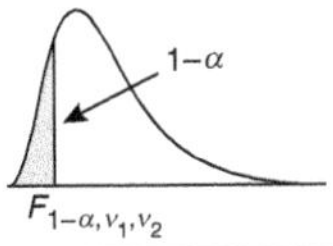

Degrees of freedom for the numerator

(v_1)

(v_2)	18	19	20	21	22	23	24	25	26	27	28	29	30	40	60	100	∞	(v_2)
1	0.33	0.33	0.34	0.34	0.34	0.34	0.34	0.34	0.34	0.34	0.35	0.35	0.35	0.35	0.36	0.36	0.37	1
2	0.38	0.38	0.39	0.39	0.39	0.39	0.39	0.40	0.40	0.40	0.40	0.40	0.40	0.41	0.42	0.42	0.43	**2**
3	0.41	0.42	0.42	0.42	0.43	0.43	0.43	0.43	0.43	0.44	0.44	0.44	0.44	0.45	0.46	0.47	0.48	**3**
4	0.44	0.44	0.44	0.45	0.45	0.45	0.46	0.46	0.46	0.46	0.46	0.47	0.47	0.48	0.49	0.50	0.51	**4**
5	0.46	0.46	0.46	0.47	0.47	0.47	0.48	0.48	0.48	0.48	0.48	0.49	0.49	0.50	0.51	0.52	0.54	**5**
6	0.47	0.47	0.48	0.48	0.49	0.49	0.49	0.49	0.50	0.50	0.50	0.50	0.50	0.52	0.53	0.55	0.56	**6**
7	0.48	0.49	0.49	0.49	0.50	0.50	0.50	0.51	0.51	0.51	0.51	0.52	0.52	0.53	0.55	0.56	0.58	**7**
8	0.49	0.50	0.50	0.50	0.51	0.51	0.52	0.52	0.52	0.52	0.53	0.53	0.53	0.55	0.56	0.58	0.60	**8**
9	0.50	0.50	0.51	0.51	0.52	0.52	0.52	0.53	0.53	0.53	0.54	0.54	0.54	0.56	0.58	0.59	0.61	**9**
10	0.51	0.51	0.52	0.52	0.53	0.53	0.53	0.54	0.54	0.54	0.54	0.55	0.55	0.57	0.59	0.60	0.63	**10**
11	0.51	0.52	0.52	0.53	0.53	0.54	0.54	0.54	0.55	0.55	0.55	0.55	0.56	0.58	0.60	0.61	0.64	**11**
12	0.52	0.52	0.53	0.53	0.54	0.54	0.55	0.55	0.55	0.56	0.56	0.56	0.56	0.58	0.60	0.62	0.65	**12**
13	0.52	0.53	0.53	0.54	0.54	0.55	0.55	0.56	0.56	0.56	0.56	0.57	0.57	0.59	0.61	0.63	0.66	**13**
14	0.53	0.53	0.54	0.54	0.55	0.55	0.56	0.56	0.56	0.57	0.57	0.57	0.58	0.60	0.62	0.64	0.66	**14**
15	0.53	0.54	0.54	0.55	0.55	0.56	0.56	0.56	0.57	0.57	0.57	0.58	0.58	0.60	0.62	0.64	0.67	**15**
16	0.53	0.54	0.55	0.55	0.56	0.56	0.56	0.57	0.57	0.58	0.58	0.58	0.59	0.61	0.63	0.65	0.68	**16**
17	0.54	0.54	0.55	0.55	0.56	0.56	0.57	0.57	0.58	0.58	0.58	0.59	0.59	0.61	0.63	0.65	0.69	**17**
18	0.54	0.55	0.55	0.56	0.56	0.57	0.57	0.58	0.58	0.58	0.59	0.59	0.59	0.62	0.64	0.66	0.69	**18**
19	0.54	0.55	0.55	0.56	0.57	0.57	0.58	0.58	0.58	0.59	0.59	0.59	0.60	0.62	0.64	0.66	0.70	**19**
20	0.54	0.55	0.56	0.56	0.57	0.57	0.58	0.58	0.59	0.59	0.59	0.60	0.60	0.62	0.65	0.67	0.70	**20**
21	0.55	0.55	0.56	0.57	0.57	0.58	0.58	0.58	0.59	0.59	0.60	0.60	0.60	0.63	0.65	0.67	0.71	**21**
22	0.55	0.56	0.56	0.57	0.57	0.58	0.58	0.59	0.59	0.60	0.60	0.60	0.61	0.63	0.66	0.68	0.71	**22**
23	0.55	0.56	0.56	0.57	0.58	0.58	0.59	0.59	0.59	0.60	0.60	0.60	0.61	0.63	0.66	0.68	0.72	**23**
24	0.55	0.56	0.57	0.57	0.58	0.58	0.59	0.59	0.60	0.60	0.60	0.61	0.61	0.64	0.66	0.68	0.72	**24**
25	0.55	0.56	0.57	0.57	0.58	0.58	0.59	0.59	0.60	0.60	0.61	0.61	0.61	0.64	0.66	0.69	0.73	**25**
26	0.56	0.56	0.57	0.58	0.58	0.59	0.59	0.60	0.60	0.60	0.61	0.61	0.62	0.64	0.67	0.69	0.73	**26**
27	0.56	0.56	0.57	0.58	0.58	0.59	0.59	0.60	0.60	0.61	0.61	0.61	0.62	0.64	0.67	0.69	0.73	**27**
28	0.56	0.57	0.57	0.58	0.58	0.59	0.59	0.60	0.60	0.61	0.61	0.62	0.62	0.64	0.67	0.70	0.74	**28**
29	0.56	0.57	0.57	0.58	0.59	0.59	0.60	0.60	0.61	0.61	0.61	0.62	0.62	0.65	0.68	0.70	0.74	**29**
30	0.56	0.57	0.58	0.58	0.59	0.59	0.60	0.60	0.61	0.61	0.62	0.62	0.62	0.65	0.68	0.70	0.75	**30**
40	0.57	0.58	0.59	0.59	0.60	0.60	0.61	0.61	0.62	0.62	0.63	0.63	0.64	0.66	0.70	0.72	0.77	**40**
60	0.58	0.59	0.60	0.60	0.61	0.62	0.62	0.63	0.63	0.64	0.64	0.65	0.65	0.68	0.72	0.75	0.84	**60**
100	0.59	0.60	0.61	0.61	0.62	0.63	0.63	0.64	0.64	0.65	0.65	0.66	0.66	0.70	0.74	0.77	0.84	**100**
∞	0.60	0.61	0.62	0.63	0.64	0.65	0.65	0.66	0.67	0.67	0.68	0.68	0.69	0.73	0.77	0.82	1.00	∞

Degrees of freedom for the denominator

Appendix F.6

$F(0.95)$ Distribution Table

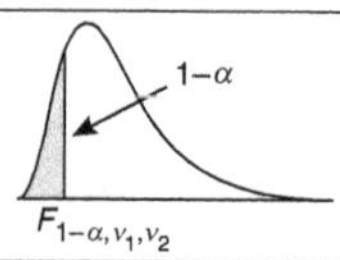

Degrees of freedom for the numerator (v_1)

Degrees of freedom for the denominator (v_2)

(v_2)	1	2	3	4	5	6	7	8	9	10	11	12	13	14	15	16	17	(v_2)
1	0.01	0.05	0.10	0.13	0.15	0.17	0.18	0.19	0.20	0.20	0.21	0.21	0.21	0.22	0.22	0.22	0.22	**1**
2	0.01	0.05	0.10	0.14	0.17	0.19	0.21	0.22	0.23	0.24	0.25	0.26	0.26	0.27	0.27	0.28	0.28	**2**
3	0.00	0.05	0.11	0.15	0.18	0.21	0.23	0.25	0.26	0.27	0.28	0.29	0.29	0.30	0.30	0.31	0.31	**3**
4	0.00	0.05	0.11	0.16	0.19	0.22	0.24	0.26	0.28	0.29	0.30	0.31	0.31	0.32	0.33	0.33	0.34	**4**
5	0.00	0.05	0.11	0.16	0.20	0.23	0.25	0.27	0.29	0.30	0.31	0.32	0.33	0.34	0.34	0.35	0.36	**5**
6	0.00	0.05	0.11	0.16	0.20	0.23	0.26	0.28	0.30	0.31	0.32	0.33	0.34	0.35	0.36	0.36	0.37	**6**
7	0.00	0.05	0.11	0.16	0.21	0.24	0.26	0.29	0.30	0.32	0.33	0.34	0.35	0.36	0.37	0.38	0.38	**7**
8	0.00	0.05	0.11	0.17	0.21	0.24	0.27	0.29	0.31	0.33	0.34	0.35	0.36	0.37	0.38	0.39	0.39	**8**
9	0.00	0.05	0.11	0.17	0.21	0.24	0.27	0.30	0.31	0.33	0.35	0.36	0.37	0.38	0.39	0.39	0.40	**9**
10	0.00	0.05	0.11	0.17	0.21	0.25	0.27	0.30	0.32	0.34	0.35	0.36	0.37	0.38	0.39	0.40	0.41	**10**
11	0.00	0.05	0.11	0.17	0.21	0.25	0.28	0.30	0.32	0.34	0.35	0.37	0.38	0.39	0.40	0.41	0.41	**11**
12	0.00	0.05	0.11	0.17	0.21	0.25	0.28	0.30	0.33	0.34	0.36	0.37	0.38	0.39	0.40	0.41	0.42	**12**
13	0.00	0.05	0.11	0.17	0.21	0.25	0.28	0.31	0.33	0.35	0.36	0.38	0.39	0.40	0.41	0.42	0.42	**13**
14	0.00	0.05	0.11	0.17	0.22	0.25	0.28	0.31	0.33	0.35	0.37	0.38	0.39	0.40	0.41	0.42	0.43	**14**
15	0.00	0.05	0.11	0.17	0.22	0.25	0.28	0.31	0.33	0.35	0.37	0.38	0.39	0.41	0.42	0.43	0.43	**15**
16	0.00	0.05	0.12	0.17	0.22	0.25	0.29	0.31	0.33	0.35	0.37	0.38	0.40	0.41	0.42	0.43	0.44	**16**
17	0.00	0.05	0.12	0.17	0.22	0.26	0.29	0.31	0.34	0.36	0.37	0.39	0.40	0.41	0.42	0.43	0.44	**17**
18	0.00	0.05	0.12	0.17	0.22	0.26	0.29	0.32	0.34	0.36	0.37	0.39	0.40	0.41	0.42	0.43	0.44	**18**
19	0.00	0.05	0.12	0.17	0.22	0.26	0.29	0.32	0.34	0.36	0.38	0.39	0.40	0.42	0.43	0.44	0.45	**19**
20	0.00	0.05	0.12	0.17	0.22	0.26	0.29	0.32	0.34	0.36	0.38	0.39	0.41	0.42	0.43	0.44	0.45	**20**
21	0.00	0.05	0.12	0.17	0.22	0.26	0.29	0.32	0.34	0.36	0.38	0.39	0.41	0.42	0.43	0.44	0.45	**21**
22	0.00	0.05	0.12	0.17	0.22	0.26	0.29	0.32	0.34	0.36	0.38	0.40	0.41	0.42	0.43	0.44	0.45	**22**
23	0.00	0.05	0.12	0.17	0.22	0.26	0.29	0.32	0.34	0.36	0.38	0.40	0.41	0.42	0.44	0.45	0.45	**23**
24	0.00	0.05	0.12	0.17	0.22	0.26	0.29	0.32	0.34	0.37	0.38	0.40	0.41	0.43	0.44	0.45	0.46	**24**
25	0.00	0.05	0.12	0.17	0.22	0.26	0.29	0.32	0.35	0.37	0.38	0.40	0.41	0.43	0.44	0.45	0.46	**25**
26	0.00	0.05	0.12	0.17	0.22	0.26	0.29	0.32	0.35	0.37	0.39	0.40	0.42	0.43	0.44	0.45	0.46	**26**
27	0.00	0.05	0.12	0.17	0.22	0.26	0.29	0.32	0.35	0.37	0.39	0.40	0.42	0.43	0.44	0.45	0.46	**27**
28	0.00	0.05	0.12	0.17	0.22	0.26	0.30	0.32	0.35	0.37	0.39	0.40	0.42	0.43	0.44	0.45	0.46	**28**
29	0.00	0.05	0.12	0.17	0.22	0.26	0.30	0.32	0.35	0.37	0.39	0.40	0.42	0.43	0.44	0.45	0.46	**29**
30	0.00	0.05	0.12	0.17	0.22	0.26	0.30	0.32	0.35	0.37	0.39	0.41	0.42	0.43	0.45	0.46	0.47	**30**
40	0.00	0.05	0.12	0.17	0.22	0.26	0.30	0.33	0.35	0.38	0.40	0.41	0.43	0.44	0.45	0.46	0.48	**40**
60	0.00	0.05	0.12	0.18	0.23	0.27	0.30	0.33	0.36	0.38	0.40	0.42	0.44	0.45	0.46	0.47	0.49	**60**
100	0.00	0.05	0.12	0.18	0.23	0.27	0.31	0.34	0.36	0.39	0.41	0.43	0.44	0.46	0.47	0.48	0.49	**100**
∞	0.00	0.05	0.12	0.18	0.23	0.27	0.31	0.34	0.37	0.39	0.42	0.44	0.45	0.47	0.48	0.50	0.51	∞

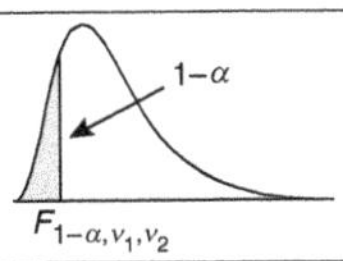

Degrees of freedom for the numerator

(v_1)

(v_2)	18	19	20	21	22	23	24	25	26	27	28	29	30	40	60	100	∞	(v_2)
1	0.23	0.23	0.23	0.23	0.23	0.23	0.23	0.24	0.24	0.24	0.24	0.24	0.24	0.24	0.25	0.25	0.26	1
2	0.28	0.28	0.29	0.29	0.29	0.29	0.29	0.30	0.30	0.30	0.30	0.30	0.30	0.31	0.32	0.32	0.33	2
3	0.32	0.32	0.32	0.33	0.33	0.33	0.33	0.33	0.34	0.34	0.34	0.34	0.34	0.35	0.36	0.37	0.38	3
4	0.34	0.35	0.35	0.35	0.36	0.36	0.36	0.36	0.36	0.37	0.37	0.37	0.37	0.38	0.40	0.41	0.42	4
5	0.36	0.36	0.37	0.37	0.38	0.38	0.38	0.38	0.39	0.39	0.39	0.39	0.39	0.41	0.42	0.43	0.45	5
6	0.38	0.38	0.38	0.39	0.39	0.40	0.40	0.40	0.40	0.41	0.41	0.41	0.41	0.43	0.44	0.46	0.48	6
7	0.39	0.39	0.40	0.40	0.41	0.41	0.41	0.42	0.42	0.42	0.42	0.43	0.43	0.44	0.46	0.48	0.50	7
8	0.40	0.40	0.41	0.41	0.42	0.42	0.42	0.43	0.43	0.43	0.44	0.44	0.44	0.46	0.48	0.49	0.52	8
9	0.41	0.41	0.42	0.42	0.43	0.43	0.43	0.44	0.44	0.44	0.45	0.45	0.45	0.47	0.49	0.51	0.53	9
10	0.41	0.42	0.43	0.43	0.44	0.44	0.44	0.45	0.45	0.45	0.46	0.46	0.46	0.48	0.50	0.52	0.55	10
11	0.42	0.43	0.43	0.44	0.44	0.45	0.45	0.45	0.46	0.46	0.46	0.47	0.47	0.49	0.51	0.53	0.56	11
12	0.43	0.43	0.44	0.44	0.45	0.45	0.46	0.46	0.47	0.47	0.47	0.48	0.48	0.50	0.52	0.54	0.57	12
13	0.43	0.44	0.44	0.45	0.46	0.46	0.46	0.47	0.47	0.48	0.48	0.48	0.48	0.51	0.53	0.55	0.58	13
14	0.44	0.44	0.45	0.46	0.46	0.47	0.47	0.47	0.48	0.48	0.48	0.49	0.49	0.51	0.54	0.56	0.59	14
15	0.44	0.45	0.45	0.46	0.46	0.47	0.47	0.48	0.48	0.49	0.49	0.49	0.50	0.52	0.54	0.57	0.60	15
16	0.44	0.45	0.46	0.46	0.47	0.47	0.48	0.48	0.49	0.49	0.49	0.50	0.50	0.53	0.55	0.57	0.61	16
17	0.45	0.46	0.46	0.47	0.47	0.48	0.48	0.49	0.49	0.50	0.50	0.50	0.51	0.53	0.56	0.58	0.62	17
18	0.45	0.46	0.46	0.47	0.48	0.48	0.49	0.49	0.50	0.50	0.50	0.51	0.51	0.54	0.56	0.59	0.62	18
19	0.45	0.46	0.47	0.47	0.48	0.49	0.49	0.49	0.50	0.50	0.51	0.51	0.51	0.54	0.57	0.59	0.63	19
20	0.46	0.46	0.47	0.48	0.48	0.49	0.49	0.50	0.50	0.51	0.51	0.51	0.52	0.54	0.57	0.60	0.64	20
21	0.46	0.47	0.47	0.48	0.49	0.49	0.50	0.50	0.51	0.51	0.51	0.52	0.52	0.55	0.58	0.60	0.64	21
22	0.46	0.47	0.48	0.48	0.49	0.49	0.50	0.50	0.51	0.51	0.52	0.52	0.52	0.55	0.58	0.61	0.65	22
23	0.46	0.47	0.48	0.48	0.49	0.50	0.50	0.51	0.51	0.52	0.52	0.52	0.53	0.55	0.58	0.61	0.65	23
24	0.47	0.47	0.48	0.49	0.49	0.50	0.50	0.51	0.51	0.52	0.52	0.53	0.53	0.56	0.59	0.61	0.66	24
25	0.47	0.47	0.48	0.49	0.50	0.50	0.51	0.51	0.52	0.52	0.52	0.53	0.53	0.56	0.59	0.62	0.66	25
26	0.47	0.48	0.48	0.49	0.50	0.50	0.51	0.51	0.52	0.52	0.53	0.53	0.53	0.56	0.59	0.62	0.67	26
27	0.47	0.48	0.49	0.49	0.50	0.50	0.51	0.52	0.52	0.52	0.53	0.53	0.54	0.57	0.60	0.63	0.67	27
28	0.47	0.48	0.49	0.49	0.50	0.51	0.51	0.52	0.52	0.53	0.53	0.54	0.54	0.57	0.60	0.63	0.68	28
29	0.47	0.48	0.49	0.50	0.50	0.51	0.51	0.52	0.52	0.53	0.53	0.54	0.54	0.57	0.60	0.63	0.68	29
30	0.47	0.48	0.49	0.50	0.50	0.51	0.52	0.52	0.53	0.53	0.54	0.54	0.54	0.57	0.61	0.64	0.69	30
40	0.48	0.49	0.50	0.51	0.52	0.52	0.53	0.53	0.54	0.54	0.55	0.55	0.56	0.59	0.63	0.66	0.72	40
60	0.50	0.51	0.51	0.52	0.53	0.54	0.54	0.55	0.55	0.56	0.57	0.57	0.57	0.61	0.65	0.69	0.76	60
100	0.51	0.52	0.52	0.53	0.54	0.55	0.56	0.56	0.57	0.57	0.58	0.58	0.59	0.63	0.68	0.72	0.80	100
∞	0.52	0.53	0.54	0.55	0.56	0.57	0.58	0.58	0.59	0.60	0.60	0.61	0.62	0.66	0.72	0.78	1.00	∞

Degrees of freedom for the denominator

Appendix F.7

F(0.975) Distribution Table

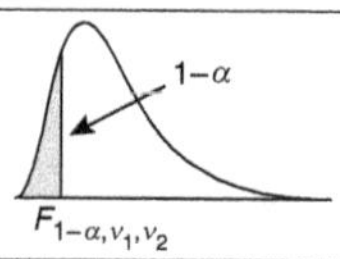

Degrees of freedom for the numerator (v_1)

Degrees of freedom for the denominator

(v_2)	1	2	3	4	5	6	7	8	9	10	11	12	13	14	15	16	17	(v_2)
1	0.00	0.03	0.06	0.08	0.10	0.11	0.12	0.13	0.14	0.14	0.15	0.15	0.16	0.16	0.16	0.16	0.17	1
2	0.00	0.03	0.06	0.09	0.12	0.14	0.15	0.17	0.17	0.18	0.19	0.20	0.20	0.21	0.21	0.21	0.22	2
3	0.00	0.03	0.06	0.10	0.13	0.15	0.17	0.18	0.20	0.21	0.22	0.22	0.23	0.24	0.24	0.25	0.25	3
4	0.00	0.03	0.07	0.10	0.14	0.16	0.18	0.20	0.21	0.22	0.23	0.24	0.25	0.26	0.26	0.27	0.27	4
5	0.00	0.03	0.07	0.11	0.14	0.17	0.19	0.21	0.22	0.24	0.25	0.26	0.27	0.27	0.28	0.29	0.29	5
6	0.00	0.03	0.07	0.11	0.14	0.17	0.20	0.21	0.23	0.25	0.26	0.27	0.28	0.29	0.29	0.30	0.31	6
7	0.00	0.03	0.07	0.11	0.15	0.18	0.20	0.22	0.24	0.25	0.27	0.28	0.29	0.30	0.30	0.31	0.32	7
8	0.00	0.03	0.07	0.11	0.15	0.18	0.20	0.23	0.24	0.26	0.27	0.28	0.30	0.30	0.31	0.32	0.33	8
9	0.00	0.03	0.07	0.11	0.15	0.18	0.21	0.23	0.25	0.26	0.28	0.29	0.30	0.31	0.32	0.33	0.34	9
10	0.00	0.03	0.07	0.11	0.15	0.18	0.21	0.23	0.25	0.27	0.28	0.30	0.31	0.32	0.33	0.33	0.34	10
11	0.00	0.03	0.07	0.11	0.15	0.18	0.21	0.24	0.26	0.27	0.29	0.30	0.31	0.32	0.33	0.34	0.35	11
12	0.00	0.03	0.07	0.11	0.15	0.19	0.21	0.24	0.26	0.28	0.29	0.31	0.32	0.33	0.34	0.35	0.35	12
13	0.00	0.03	0.07	0.11	0.15	0.19	0.22	0.24	0.26	0.28	0.29	0.31	0.32	0.33	0.34	0.35	0.36	13
14	0.00	0.03	0.07	0.12	0.15	0.19	0.22	0.24	0.26	0.28	0.30	0.31	0.32	0.34	0.35	0.35	0.36	14
15	0.00	0.03	0.07	0.12	0.16	0.19	0.22	0.24	0.27	0.28	0.30	0.31	0.33	0.34	0.35	0.36	0.37	15
16	0.00	0.03	0.07	0.12	0.16	0.19	0.22	0.25	0.27	0.29	0.30	0.32	0.33	0.34	0.35	0.36	0.37	16
17	0.00	0.03	0.07	0.12	0.16	0.19	0.22	0.25	0.27	0.29	0.30	0.32	0.33	0.34	0.36	0.37	0.37	17
18	0.00	0.03	0.07	0.12	0.16	0.19	0.22	0.25	0.27	0.29	0.31	0.32	0.34	0.35	0.36	0.37	0.38	18
19	0.00	0.03	0.07	0.12	0.16	0.19	0.22	0.25	0.27	0.29	0.31	0.32	0.34	0.35	0.36	0.37	0.38	19
20	0.00	0.03	0.07	0.12	0.16	0.19	0.22	0.25	0.27	0.29	0.31	0.33	0.34	0.35	0.36	0.37	0.38	20
21	0.00	0.03	0.07	0.12	0.16	0.19	0.22	0.25	0.27	0.29	0.31	0.33	0.34	0.35	0.36	0.38	0.38	21
22	0.00	0.03	0.07	0.12	0.16	0.19	0.23	0.25	0.27	0.30	0.31	0.33	0.34	0.36	0.37	0.38	0.39	22
23	0.00	0.03	0.07	0.12	0.16	0.19	0.23	0.25	0.28	0.30	0.31	0.33	0.34	0.36	0.37	0.38	0.39	23
24	0.00	0.03	0.07	0.12	0.16	0.20	0.23	0.25	0.28	0.30	0.32	0.33	0.35	0.36	0.37	0.38	0.39	24
25	0.00	0.03	0.07	0.12	0.16	0.20	0.23	0.25	0.28	0.30	0.32	0.33	0.35	0.36	0.37	0.38	0.39	25
26	0.00	0.03	0.07	0.12	0.16	0.20	0.23	0.25	0.28	0.30	0.32	0.33	0.35	0.36	0.37	0.38	0.39	26
27	0.00	0.03	0.07	0.12	0.16	0.20	0.23	0.26	0.28	0.30	0.32	0.33	0.35	0.36	0.37	0.39	0.40	27
28	0.00	0.03	0.07	0.12	0.16	0.20	0.23	0.26	0.28	0.30	0.32	0.34	0.35	0.36	0.38	0.39	0.40	28
29	0.00	0.03	0.07	0.12	0.16	0.20	0.23	0.26	0.28	0.30	0.32	0.34	0.35	0.36	0.38	0.39	0.40	29
30	0.00	0.03	0.07	0.12	0.16	0.20	0.23	0.26	0.28	0.30	0.32	0.34	0.35	0.37	0.38	0.39	0.40	30
40	0.00	0.03	0.07	0.12	0.16	0.20	0.23	0.26	0.29	0.31	0.33	0.34	0.36	0.37	0.39	0.40	0.41	40
60	0.00	0.03	0.07	0.12	0.16	0.20	0.24	0.26	0.29	0.31	0.33	0.35	0.37	0.38	0.40	0.41	0.42	60
100	0.00	0.03	0.07	0.12	0.16	0.20	0.24	0.27	0.29	0.32	0.34	0.36	0.37	0.39	0.40	0.42	0.43	100
∞	0.00	0.03	0.07	0.12	0.17	0.21	0.24	0.27	0.30	0.32	0.35	0.37	0.39	0.40	0.42	0.43	0.44	∞

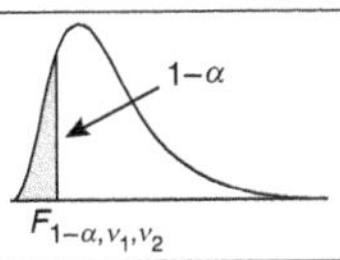

Degrees of freedom for the numerator

(v_1)

(v_2)	18	19	20	21	22	23	24	25	26	27	28	29	30	40	60	100	∞	(v_2)
1	0.17	0.17	0.17	0.17	0.17	0.17	0.17	0.18	0.18	0.18	0.18	0.18	0.18	0.18	0.19	0.19	0.20	1
2	0.22	0.22	0.22	0.23	0.23	0.23	0.23	0.23	0.23	0.24	0.24	0.24	0.24	0.25	0.25	0.26	0.27	2
3	0.25	0.26	0.26	0.26	0.26	0.27	0.27	0.27	0.27	0.27	0.28	0.28	0.28	0.29	0.30	0.31	0.32	3
4	0.28	0.28	0.28	0.29	0.29	0.29	0.30	0.30	0.30	0.30	0.30	0.31	0.31	0.32	0.33	0.34	0.36	4
5	0.30	0.30	0.30	0.31	0.31	0.31	0.32	0.32	0.32	0.32	0.33	0.33	0.33	0.34	0.36	0.37	0.39	5
6	0.31	0.32	0.32	0.32	0.33	0.33	0.33	0.34	0.34	0.34	0.34	0.35	0.35	0.36	0.38	0.39	0.42	6
7	0.32	0.33	0.33	0.34	0.34	0.34	0.35	0.35	0.35	0.36	0.36	0.36	0.36	0.38	0.40	0.41	0.44	7
8	0.33	0.34	0.34	0.35	0.35	0.36	0.36	0.36	0.37	0.37	0.37	0.37	0.38	0.40	0.41	0.43	0.46	8
9	0.34	0.35	0.35	0.36	0.36	0.37	0.37	0.37	0.38	0.38	0.38	0.39	0.39	0.41	0.43	0.45	0.47	9
10	0.35	0.35	0.36	0.37	0.37	0.37	0.38	0.38	0.39	0.39	0.39	0.40	0.40	0.42	0.44	0.46	0.49	10
11	0.36	0.36	0.37	0.37	0.38	0.38	0.39	0.39	0.39	0.40	0.40	0.40	0.41	0.43	0.45	0.47	0.50	11
12	0.36	0.37	0.37	0.38	0.38	0.39	0.39	0.40	0.40	0.41	0.41	0.41	0.41	0.44	0.46	0.48	0.51	12
13	0.37	0.37	0.38	0.38	0.39	0.40	0.40	0.40	0.41	0.41	0.42	0.42	0.42	0.44	0.47	0.49	0.53	13
14	0.37	0.38	0.38	0.39	0.40	0.40	0.41	0.41	0.41	0.42	0.42	0.42	0.43	0.45	0.48	0.50	0.54	14
15	0.37	0.38	0.39	0.39	0.40	0.41	0.41	0.41	0.42	0.42	0.43	0.43	0.43	0.46	0.49	0.51	0.55	15
16	0.38	0.39	0.39	0.40	0.40	0.41	0.41	0.42	0.42	0.43	0.43	0.44	0.44	0.46	0.49	0.52	0.55	16
17	0.38	0.39	0.40	0.40	0.41	0.41	0.42	0.42	0.43	0.43	0.44	0.44	0.44	0.47	0.50	0.52	0.56	17
18	0.39	0.39	0.40	0.41	0.41	0.42	0.42	0.43	0.43	0.44	0.44	0.44	0.45	0.47	0.50	0.53	0.57	18
19	0.39	0.40	0.40	0.41	0.42	0.42	0.43	0.43	0.44	0.44	0.44	0.45	0.45	0.48	0.51	0.54	0.58	19
20	0.39	0.40	0.41	0.41	0.42	0.42	0.43	0.43	0.44	0.44	0.45	0.45	0.46	0.48	0.51	0.54	0.59	20
21	0.39	0.40	0.41	0.42	0.42	0.43	0.43	0.44	0.44	0.45	0.45	0.46	0.46	0.49	0.52	0.55	0.59	21
22	0.40	0.40	0.41	0.42	0.42	0.43	0.44	0.44	0.45	0.45	0.45	0.46	0.46	0.49	0.52	0.55	0.60	22
23	0.40	0.41	0.41	0.42	0.43	0.43	0.44	0.44	0.45	0.45	0.46	0.46	0.47	0.49	0.53	0.56	0.60	23
24	0.40	0.41	0.42	0.42	0.43	0.43	0.44	0.45	0.45	0.46	0.46	0.46	0.47	0.50	0.53	0.56	0.61	24
25	0.40	0.41	0.42	0.42	0.43	0.44	0.44	0.45	0.45	0.46	0.46	0.47	0.47	0.50	0.54	0.56	0.62	25
26	0.40	0.41	0.42	0.43	0.43	0.44	0.45	0.45	0.46	0.46	0.47	0.47	0.47	0.50	0.54	0.57	0.62	26
27	0.40	0.41	0.42	0.43	0.44	0.44	0.45	0.45	0.46	0.46	0.47	0.47	0.48	0.51	0.54	0.57	0.63	27
28	0.41	0.41	0.42	0.43	0.44	0.44	0.45	0.45	0.46	0.46	0.47	0.47	0.48	0.51	0.55	0.58	0.63	28
29	0.41	0.42	0.42	0.43	0.44	0.44	0.45	0.46	0.46	0.47	0.47	0.48	0.48	0.51	0.55	0.58	0.63	29
30	0.41	0.42	0.43	0.43	0.44	0.45	0.45	0.46	0.46	0.47	0.47	0.48	0.48	0.51	0.55	0.58	0.64	30
40	0.42	0.43	0.44	0.45	0.45	0.46	0.47	0.47	0.48	0.48	0.49	0.49	0.50	0.53	0.57	0.61	0.67	40
60	0.43	0.44	0.45	0.46	0.47	0.47	0.48	0.49	0.49	0.50	0.51	0.51	0.52	0.55	0.60	0.64	0.72	60
100	0.44	0.45	0.46	0.47	0.48	0.49	0.49	0.50	0.51	0.51	0.52	0.53	0.53	0.57	0.63	0.67	0.77	100
∞	0.46	0.47	0.48	0.49	0.50	0.51	0.52	0.52	0.53	0.54	0.55	0.55	0.56	0.61	0.67	0.74	1.00	∞

Degrees of freedom for the denominator

Appendix F.8

$F(0.99)$ Distribution Table

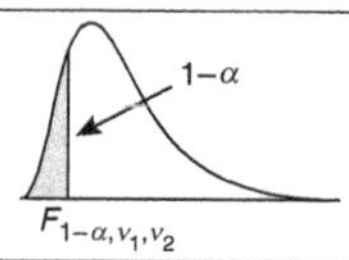

Degrees of freedom for the numerator (v_1)

Degrees of freedom for the denominator

(v_2)	1	2	3	4	5	6	7	8	9	10	11	12	13	14	15	16	17	(v_2)
1	0.00	0.01	0.03	0.05	0.06	0.07	0.08	0.09	0.09	0.10	0.10	0.11	0.11	0.11	0.12	0.12	0.12	**1**
2	0.00	0.01	0.03	0.06	0.08	0.09	0.10	0.12	0.12	0.13	0.14	0.14	0.15	0.15	0.16	0.16	0.16	**2**
3	0.00	0.01	0.03	0.06	0.08	0.10	0.12	0.13	0.14	0.15	0.16	0.17	0.17	0.18	0.18	0.19	0.19	**3**
4	0.00	0.01	0.03	0.06	0.09	0.11	0.13	0.14	0.16	0.17	0.18	0.18	0.19	0.20	0.20	0.21	0.21	**4**
5	0.00	0.01	0.04	0.06	0.09	0.11	0.13	0.15	0.17	0.18	0.19	0.20	0.21	0.21	0.22	0.23	0.23	**5**
6	0.00	0.01	0.04	0.07	0.09	0.12	0.14	0.16	0.17	0.19	0.20	0.21	0.22	0.22	0.23	0.24	0.24	**6**
7	0.00	0.01	0.04	0.07	0.10	0.12	0.14	0.16	0.18	0.19	0.20	0.22	0.23	0.23	0.24	0.25	0.25	**7**
8	0.00	0.01	0.04	0.07	0.10	0.12	0.15	0.17	0.18	0.20	0.21	0.22	0.23	0.24	0.25	0.26	0.26	**8**
9	0.00	0.01	0.04	0.07	0.10	0.13	0.15	0.17	0.19	0.20	0.22	0.23	0.24	0.25	0.26	0.26	0.27	**9**
10	0.00	0.01	0.04	0.07	0.10	0.13	0.15	0.17	0.19	0.21	0.22	0.23	0.24	0.25	0.26	0.27	0.28	**10**
11	0.00	0.01	0.04	0.07	0.10	0.13	0.15	0.17	0.19	0.21	0.22	0.24	0.25	0.26	0.27	0.28	0.28	**11**
12	0.00	0.01	0.04	0.07	0.10	0.13	0.15	0.18	0.20	0.21	0.23	0.24	0.25	0.26	0.27	0.28	0.29	**12**
13	0.00	0.01	0.04	0.07	0.10	0.13	0.16	0.18	0.20	0.22	0.23	0.24	0.26	0.27	0.28	0.29	0.29	**13**
14	0.00	0.01	0.04	0.07	0.10	0.13	0.16	0.18	0.20	0.22	0.23	0.25	0.26	0.27	0.28	0.29	0.30	**14**
15	0.00	0.01	0.04	0.07	0.10	0.13	0.16	0.18	0.20	0.22	0.24	0.25	0.26	0.27	0.28	0.29	0.30	**15**
16	0.00	0.01	0.04	0.07	0.10	0.13	0.16	0.18	0.20	0.22	0.24	0.25	0.26	0.28	0.29	0.30	0.31	**16**
17	0.00	0.01	0.04	0.07	0.10	0.13	0.16	0.18	0.20	0.22	0.24	0.25	0.27	0.28	0.29	0.30	0.31	**17**
18	0.00	0.01	0.04	0.07	0.10	0.13	0.16	0.18	0.21	0.22	0.24	0.26	0.27	0.28	0.29	0.30	0.31	**18**
19	0.00	0.01	0.04	0.07	0.10	0.13	0.16	0.19	0.21	0.23	0.24	0.26	0.27	0.28	0.29	0.30	0.31	**19**
20	0.00	0.01	0.04	0.07	0.10	0.14	0.16	0.19	0.21	0.23	0.24	0.26	0.27	0.29	0.30	0.31	0.32	**20**
21	0.00	0.01	0.04	0.07	0.10	0.14	0.16	0.19	0.21	0.23	0.25	0.26	0.27	0.29	0.30	0.31	0.32	**21**
22	0.00	0.01	0.04	0.07	0.11	0.14	0.16	0.19	0.21	0.23	0.25	0.26	0.28	0.29	0.30	0.31	0.32	**22**
23	0.00	0.01	0.04	0.07	0.11	0.14	0.16	0.19	0.21	0.23	0.25	0.26	0.28	0.29	0.30	0.31	0.32	**23**
24	0.00	0.01	0.04	0.07	0.11	0.14	0.16	0.19	0.21	0.23	0.25	0.26	0.28	0.29	0.30	0.31	0.32	**24**
25	0.00	0.01	0.04	0.07	0.11	0.14	0.17	0.19	0.21	0.23	0.25	0.27	0.28	0.29	0.31	0.32	0.33	**25**
26	0.00	0.01	0.04	0.07	0.11	0.14	0.17	0.19	0.21	0.23	0.25	0.27	0.28	0.29	0.31	0.32	0.33	**26**
27	0.00	0.01	0.04	0.07	0.11	0.14	0.17	0.19	0.21	0.23	0.25	0.27	0.28	0.30	0.31	0.32	0.33	**27**
28	0.00	0.01	0.04	0.07	0.11	0.14	0.17	0.19	0.21	0.23	0.25	0.27	0.28	0.30	0.31	0.32	0.33	**28**
29	0.00	0.01	0.04	0.07	0.11	0.14	0.17	0.19	0.21	0.23	0.25	0.27	0.28	0.30	0.31	0.32	0.33	**29**
30	0.00	0.01	0.04	0.07	0.11	0.14	0.17	0.19	0.22	0.24	0.25	0.27	0.29	0.30	0.31	0.32	0.33	**30**
40	0.00	0.01	0.04	0.07	0.11	0.14	0.17	0.20	0.22	0.24	0.26	0.28	0.29	0.31	0.32	0.33	0.34	**40**
60	0.00	0.01	0.04	0.07	0.11	0.14	0.17	0.20	0.22	0.24	0.26	0.28	0.30	0.31	0.33	0.34	0.35	**60**
100	0.00	0.01	0.04	0.07	0.11	0.14	0.17	0.20	0.23	0.25	0.27	0.29	0.31	0.32	0.34	0.35	0.36	**100**
∞	0.00	0.01	0.04	0.07	0.11	0.15	0.18	0.21	0.23	0.26	0.28	0.30	0.32	0.33	0.35	0.36	0.38	∞

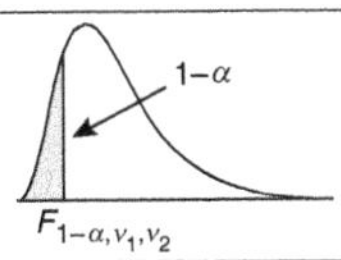

Degrees of freedom for the numerator

(v_1)

(v_2)	18	19	20	21	22	23	24	25	26	27	28	29	30	40	60	100	∞	(v_2)
1	0.12	0.12	0.12	0.12	0.13	0.13	0.13	0.13	0.13	0.13	0.13	0.13	0.13	0.14	0.14	0.15	0.15	1
2	0.17	0.17	0.17	0.17	0.17	0.18	0.18	0.18	0.18	0.18	0.18	0.18	0.19	0.19	0.20	0.21	0.22	2
3	0.20	0.20	0.20	0.21	0.21	0.21	0.21	0.21	0.22	0.22	0.22	0.22	0.22	0.23	0.24	0.25	0.26	3
4	0.22	0.22	0.23	0.23	0.23	0.23	0.24	0.24	0.24	0.24	0.25	0.25	0.25	0.26	0.27	0.28	0.30	4
5	0.24	0.24	0.24	0.25	0.25	0.25	0.26	0.26	0.26	0.26	0.27	0.27	0.27	0.28	0.30	0.31	0.33	5
6	0.25	0.25	0.26	0.26	0.27	0.27	0.27	0.28	0.28	0.28	0.28	0.29	0.29	0.30	0.32	0.33	0.36	6
7	0.26	0.27	0.27	0.27	0.28	0.28	0.29	0.29	0.29	0.30	0.30	0.30	0.30	0.32	0.34	0.35	0.38	7
8	0.27	0.28	0.28	0.29	0.29	0.29	0.30	0.30	0.30	0.31	0.31	0.31	0.32	0.33	0.35	0.37	0.40	8
9	0.28	0.28	0.29	0.29	0.30	0.30	0.31	0.31	0.31	0.32	0.32	0.32	0.33	0.35	0.37	0.39	0.42	9
10	0.29	0.29	0.30	0.30	0.31	0.31	0.32	0.32	0.32	0.33	0.33	0.33	0.34	0.36	0.38	0.40	0.43	10
11	0.29	0.30	0.30	0.31	0.31	0.32	0.32	0.33	0.33	0.33	0.34	0.34	0.34	0.37	0.39	0.41	0.44	11
12	0.30	0.30	0.31	0.32	0.32	0.33	0.33	0.33	0.34	0.34	0.35	0.35	0.35	0.38	0.40	0.42	0.46	12
13	0.30	0.31	0.31	0.32	0.33	0.33	0.34	0.34	0.34	0.35	0.35	0.36	0.36	0.38	0.41	0.43	0.47	13
14	0.31	0.31	0.32	0.33	0.33	0.34	0.34	0.35	0.35	0.35	0.36	0.36	0.36	0.39	0.42	0.44	0.48	14
15	0.31	0.32	0.32	0.33	0.34	0.34	0.35	0.35	0.36	0.36	0.36	0.37	0.37	0.40	0.43	0.45	0.49	15
16	0.31	0.32	0.33	0.33	0.34	0.35	0.35	0.36	0.36	0.36	0.37	0.37	0.38	0.40	0.43	0.46	0.50	16
17	0.32	0.32	0.33	0.34	0.34	0.35	0.35	0.36	0.36	0.37	0.37	0.38	0.38	0.41	0.44	0.46	0.51	17
18	0.32	0.33	0.33	0.34	0.35	0.35	0.36	0.36	0.37	0.37	0.38	0.38	0.38	0.41	0.44	0.47	0.52	18
19	0.32	0.33	0.34	0.34	0.35	0.36	0.36	0.37	0.37	0.38	0.38	0.38	0.39	0.42	0.45	0.48	0.52	19
20	0.32	0.33	0.34	0.35	0.35	0.36	0.37	0.37	0.38	0.38	0.38	0.39	0.39	0.42	0.45	0.48	0.53	20
21	0.33	0.34	0.34	0.35	0.36	0.36	0.37	0.37	0.38	0.38	0.39	0.39	0.40	0.43	0.46	0.49	0.54	21
22	0.33	0.34	0.35	0.35	0.36	0.37	0.37	0.38	0.38	0.39	0.39	0.40	0.40	0.43	0.46	0.49	0.55	22
23	0.33	0.34	0.35	0.35	0.36	0.37	0.37	0.38	0.38	0.39	0.39	0.40	0.40	0.43	0.47	0.50	0.55	23
24	0.33	0.34	0.35	0.36	0.36	0.37	0.38	0.38	0.39	0.39	0.40	0.40	0.41	0.44	0.47	0.50	0.56	24
25	0.34	0.34	0.35	0.36	0.37	0.37	0.38	0.38	0.39	0.39	0.40	0.40	0.41	0.44	0.48	0.51	0.56	25
26	0.34	0.35	0.35	0.36	0.37	0.37	0.38	0.39	0.39	0.40	0.40	0.41	0.41	0.44	0.48	0.51	0.57	26
27	0.34	0.35	0.36	0.36	0.37	0.38	0.38	0.39	0.39	0.40	0.40	0.41	0.41	0.45	0.48	0.52	0.57	27
28	0.34	0.35	0.36	0.36	0.37	0.38	0.38	0.39	0.40	0.40	0.41	0.41	0.41	0.45	0.49	0.52	0.58	28
29	0.34	0.35	0.36	0.37	0.37	0.38	0.39	0.39	0.40	0.40	0.41	0.41	0.42	0.45	0.49	0.52	0.58	29
30	0.34	0.35	0.36	0.37	0.37	0.38	0.39	0.39	0.40	0.40	0.41	0.41	0.42	0.45	0.49	0.53	0.59	30
40	0.35	0.36	0.37	0.38	0.39	0.39	0.40	0.41	0.41	0.42	0.42	0.43	0.43	0.47	0.52	0.56	0.63	40
60	0.36	0.37	0.38	0.39	0.40	0.41	0.42	0.42	0.43	0.44	0.44	0.45	0.45	0.50	0.54	0.59	0.68	60
100	0.37	0.38	0.39	0.40	0.41	0.42	0.43	0.44	0.44	0.45	0.46	0.46	0.47	0.52	0.57	0.63	0.74	100
∞	0.39	0.40	0.41	0.42	0.43	0.44	0.45	0.46	0.47	0.48	0.48	0.49	0.50	0.55	0.62	0.70	1.00	∞

Degrees of freedom for the denominator

Appendix G
Chi-Square Distribution Table

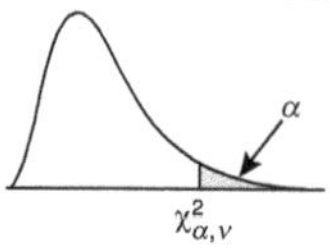

	α										
υ	**0.995**	**0.990**	**0.975**	**0.950**	**0.900**	**0.100**	**0.050**	**0.025**	**0.010**	**0.005**	**υ**
1	0.00004	0.00016	0.00098	0.00393	0.01579	2.70554	3.84146	5.02389	6.63490	7.87944	**1**
2	0.010	0.020	0.051	0.103	0.211	4.605	5.991	7.378	9.210	10.597	**2**
3	0.072	0.115	0.216	0.352	0.584	6.251	7.815	9.348	11.345	12.838	**3**
4	0.207	0.297	0.484	0.711	1.064	7.779	9.488	11.143	13.277	14.860	**4**
5	0.412	0.554	0.831	1.145	1.610	9.236	11.070	12.833	15.086	16.750	**5**
6	0.676	0.872	1.237	1.635	2.204	10.645	12.592	14.449	16.812	18.548	**6**
7	0.989	1.239	1.690	2.167	2.833	12.017	14.067	16.013	18.475	20.278	**7**
8	1.344	1.646	2.180	2.733	3.490	13.362	15.507	17.535	20.090	21.955	**8**
9	1.735	2.088	2.700	3.325	4.168	14.684	16.919	19.023	21.666	23.589	**9**
10	2.156	2.558	3.247	3.940	4.865	15.987	18.307	20.483	23.209	25.188	**10**
11	2.603	3.053	3.816	4.575	5.578	17.275	19.675	21.920	24.725	26.757	**11**
12	3.074	3.571	4.404	5.226	6.304	18.549	21.026	23.337	26.217	28.300	**12**
13	3.565	4.107	5.009	5.892	7.042	19.812	22.362	24.736	27.688	29.819	**13**
14	4.075	4.660	5.629	6.571	7.790	21.064	23.685	26.119	29.141	31.319	**14**
15	4.601	5.229	6.262	7.261	8.547	22.307	24.996	27.488	30.578	32.801	**15**
16	5.142	5.812	6.908	7.962	9.312	23.542	26.296	28.845	32.000	34.267	**16**
17	5.697	6.408	7.564	8.672	10.085	24.769	27.587	30.191	33.409	35.718	**17**
18	6.265	7.015	8.231	9.390	10.865	25.989	28.869	31.526	34.805	37.156	**18**
19	6.844	7.633	8.907	10.117	11.651	27.204	30.144	32.852	36.191	38.582	**19**
20	7.434	8.260	9.591	10.851	12.443	28.412	31.410	34.170	37.566	39.997	**20**
21	8.034	8.897	10.283	11.591	13.240	29.615	32.671	35.479	38.932	41.401	**21**
22	8.643	9.542	10.982	12.338	14.041	30.813	33.924	36.781	40.289	42.796	**22**
23	9.260	10.196	11.689	13.091	14.848	32.007	35.172	38.076	41.638	44.181	**23**
24	9.886	10.856	12.401	13.848	15.659	33.196	36.415	39.364	42.980	45.559	**24**
25	10.520	11.524	13.120	14.611	16.473	34.382	37.652	40.646	44.314	46.928	**25**
26	11.160	12.198	13.844	15.379	17.292	35.563	38.885	41.923	45.642	48.290	**26**
27	11.808	12.879	14.573	16.151	18.114	36.741	40.113	43.195	46.963	49.645	**27**
28	12.461	13.565	15.308	16.928	18.939	37.916	41.337	44.461	48.278	50.993	**28**
29	13.121	14.256	16.047	17.708	19.768	39.087	42.557	45.722	49.588	52.336	**29**
30	13.787	14.953	16.791	18.493	20.599	40.256	43.773	46.979	50.892	53.672	**30**
40	20.707	22.164	24.433	26.509	29.051	51.805	55.758	59.342	63.691	66.766	**40**
50	27.991	29.707	32.357	34.764	37.689	63.167	67.505	71.420	76.154	79.490	**50**
60	35.534	37.485	40.482	43.188	46.459	74.397	79.082	83.298	88.379	91.952	**60**
70	43.275	45.442	48.758	51.739	55.329	85.527	90.531	95.023	100.425	104.215	**70**
80	51.172	53.540	57.153	60.391	64.278	96.578	101.879	106.629	112.329	116.321	**80**
90	59.196	61.754	65.647	69.126	73.291	107.565	113.145	118.136	124.116	128.299	**90**
100	67.328	70.065	74.222	77.929	82.358	118.498	124.342	129.561	135.807	140.169	**100**

Appendix H.1

Normal Distribution Tolerance Interval Factors, One-Sided

	Confidence Level								
	0.90			0.95			0.99		
	Coverage Proportion			Coverage Proportion			Coverage Proportion		
n	0.90	0.95	0.99	0.90	0.95	0.99	0.90	0.95	0.99
4	3.188	3.957	5.438	4.162	5.144	7.042	7.380	9.084	12.387
5	2.742	3.400	4.666	3.407	4.203	5.741	5.362	6.578	8.939
6	2.494	3.092	4.243	3.006	3.708	5.062	4.411	5.406	7.335
7	2.333	2.894	3.972	2.755	3.400	4.642	3.859	4.728	6.412
8	2.219	2.754	3.783	2.582	3.187	4.354	3.497	4.285	5.812
9	2.133	2.650	3.641	2.454	3.031	4.143	3.240	3.972	5.389
10	2.066	2.568	3.532	2.355	2.911	3.981	3.048	3.738	5.074
12	1.966	2.448	3.371	2.210	2.736	3.747	2.777	3.410	4.633
15	1.867	2.329	3.212	2.068	2.566	3.520	2.522	3.102	4.222
20	1.765	2.208	3.052	1.926	2.396	3.295	2.276	2.808	3.832
25	1.702	2.132	2.952	1.838	2.292	3.158	2.129	2.633	3.601
30	1.657	2.080	2.884	1.777	2.220	3.064	2.030	2.516	3.447
40	1.598	2.010	2.793	1.697	2.126	2.941	1.902	2.364	3.249
50	1.560	1.965	2.735	1.646	2.065	2.862	1.821	2.269	3.125
60	1.532	1.933	2.694	1.609	2.022	2.807	1.764	2.202	3.038
80	1.495	1.890	2.638	1.559	1.964	2.733	1.688	2.114	2.924
100	1.470	1.861	2.601	1.527	1.927	2.684	1.639	2.056	2.850

Computed with the R package "tolerance" (Statistical Tolerance Intervals and Regions), version 3.0.0, published 2024-04-17.

Appendix H.2

Normal Distribution Tolerance Interval Factors, Two-Sided

	Confidence Level								
	0.90			0.95			0.99		
	Coverage Proportion			Coverage Proportion			Coverage Proportion		
n	0.90	0.95	0.99	0.90	0.95	0.99	0.90	0.95	0.99
4	4.157	4.913	6.372	5.368	6.341	8.221	9.416	11.118	14.405
5	3.499	4.143	5.387	4.291	5.077	6.598	6.655	7.870	10.220
6	3.141	3.723	4.850	3.733	4.422	5.758	5.383	6.374	8.292
7	2.913	3.456	4.509	3.390	4.020	5.241	4.658	5.520	7.191
8	2.754	3.270	4.271	3.156	3.746	4.889	4.189	4.968	6.479
9	2.637	3.132	4.094	2.986	3.546	4.633	3.860	4.581	5.980
10	2.546	3.026	3.958	2.856	3.393	4.437	3.617	4.294	5.610
12	2.414	2.871	3.759	2.670	3.175	4.156	3.279	3.896	5.096
15	2.286	2.720	3.565	2.492	2.965	3.885	2.967	3.529	4.621
20	2.158	2.570	3.372	2.319	2.760	3.621	2.675	3.184	4.175
25	2.081	2.479	3.254	2.215	2.638	3.462	2.506	2.984	3.915
30	2.029	2.417	3.173	2.145	2.555	3.355	2.394	2.851	3.743
40	1.961	2.336	3.069	2.055	2.448	3.216	2.253	2.684	3.524
50	1.918	2.286	3.003	1.999	2.382	3.129	2.166	2.580	3.390
60	1.889	2.250	2.956	1.960	2.335	3.068	2.106	2.509	3.297
80	1.849	2.203	2.895	1.908	2.274	2.988	2.028	2.417	3.175
100	1.823	2.172	2.855	1.875	2.234	2.936	1.978	2.357	3.098

Computed with the R package "tolerance" (Statistical Tolerance Intervals and Regions), version 3.0.0, published 2024-04-17; using the EXACT method for the two-sided case.

Glossary

1 – α—*See* confidence level.

1 – β—*See* power.

Φ—The power of testing a hypothesis is 1—β. It is the probability of correctly rejecting the null hypothesis, H_0.

2^k factorial design—Factorial design in which k factors are studied, each of them at exactly two levels.

5 whys—A persistent questioning technique to probe deeper to reach the root cause of a problem. *See also* 5Ws & 1H.

5S (Five S)—A term that derives its name from the five terms beginning with "s" that are used to create a workplace suited for visual control and lean production. Collectively, 5S is about how to create a workplace that is visibly organized, free of clutter, neatly arranged, and sparkling clean. The 5S's are defined as follows: *seiri* (sort; also sifting) means to separate needed tools, parts, and instructions from unneeded materials and to remove the latter; *seiton* (set in order) means to neatly arrange and identify parts and tools for ease of use; *seiso* (shine; also sanitize or scrub) means to conduct a cleanup campaign; *seiketsu* (standardize) means to conduct seiri, seiton, and seiso at frequent, indeed daily, intervals to maintain a workplace in perfect condition; and *shitsuke* (sustain; also self-discipline) means to form the habit of always following the first four S's. *See also* individual listing for each of the 5S's.

5Ws & 1H—Addressing the *who, what, where, when, why,* and *how* questions is a useful technique to help develop an objective and a concise statement of a problem. *See also* 5 whys.

6Ms—Typically the primary categories of the cause-and-effect diagram: machines, manpower, materials, measurements, methods, and Mother Nature. Using these categories as a structured approach provides some assurance that few causes will be overlooked. *See also* 7Ms.

7Ms—The 6Ms with the addition of management. *See also* 6Ms.

α (alpha)—(1) The maximum probability, or risk, of making a type I error when dealing with the significance level of a test. (2) The probability or risk of incorrectly deciding that a shift in the process mean has occurred when the process is unchanged (when referring to α in general or as the p-value obtained in the test). (3) α is usually designated as producer's risk.

β (beta)—The maximum probability, or risk, of making a type II error; see also α *(alpha)*. The probability or risk of incorrectly deciding that a shift in the process mean has not occurred when the process has changed. β is usually designated as consumer's risk. See *power curve.*

λ (lambda)—Failure rate.

ρ (rho)—The population correlation coefficient. *See also* correlation coefficient.

σ (sigma)—See *standard deviation.*

$\hat{\sigma}$ (sigma-hat)—In general, any estimate of the population standard deviation. There are various ways to get this estimate depending on the particular application.

σ^2 (sigma square)—See *variance.*

$\sigma_{\bar{x}}$ (sigma x-bar)—The standard deviation (or standard error) of $\bar{x}$.

A

ABC—*See* activity-based costing.

accelerated life testing—A technique in which units are tested at stress levels higher than they were designed for in an effort to cause failures sooner.

acceptance control chart—A control chart intended primarily to evaluate whether the plotted measure can be expected to satisfy specified tolerances.

acceptance quality limit (AQL)—The quality level that is the worst tolerable product average when a continuing series of lots is submitted for acceptance sampling. Note the following: (1) This concept applies only when an acceptance sampling scheme with rules for switching and discontinuation is used. (2) Although individual lots with quality as bad as the AQL can be accepted with fairly high probability, the designation of an AQL does not suggest that this is a desirable quality level. (3) Acceptance sampling schemes found in standards with their rules for switching and discontinuation of sampling inspection are designed to encourage suppliers to have process averages consistently better than the AQL. If suppliers fail to do so, there is a high probability of being switched from normal inspection to tightened inspection, where lot acceptance becomes more difficult. Once on tightened inspection, unless corrective action is taken to improve product quality, it is very likely that the rule requiring discontinuation of sampling inspection pending such improvement will be invoked unless action is taken to improve the process. (4) The use of the abbreviation "AQL" to mean *acceptable quality level* is no longer recommended since modern thinking is that no fraction defective is really acceptable. Using "acceptance quality limit" rather than "acceptable quality level" indicates a technical value where acceptance occurs.

accuracy—The closeness of agreement between a test result or measurement result and the true or reference value.

activity network diagram (AND)—A tool used to illustrate a sequence of events or activities (nodes) and the interconnectivity of such nodes. It is used for scheduling and especially for determining the critical path through nodes. It is also known as an *arrow diagram.* The activity network diagram is one of the seven management and planning tools.

activity—An action of some type that requires a time duration for accomplishment.

adjusted coefficient of determination (R^2_{adj})—A statistic that takes into consideration the number of predictor variables in the model and is helpful when comparing models with a different number of predictor variables.

affinity diagram—A tool used to organize information and help achieve order out of the chaos that can develop in a brainstorming session. Large amounts of data, concepts, and ideas are grouped based on their natural relationship to one another. It is more a creative process than a logical process. Also known as the *KJ method*, the affinity diagram is one of the seven management and planning tools.

agile approach—*See* lean.

AIAG—Automotive Industry Action Group.

alias—An effect that is completely confounded with another effect due to the nature of the designed experiment. Aliases are the result of confounding, which may or may not be deliberate.

alpha (α) risk—*See* type I error.

alpha (α)—*See* α (alpha)

alternative hypothesis, H1—A hypothesis that is accepted if the null hypothesis (H_0) is rejected. Example 1: Consider the null hypothesis that the statistical model for a population is a normal distribution. The alternative hypothesis to this null hypothesis is that the statistical model of the population is *not* a normal distribution. Note 1: The alternative hypothesis is a statement that contradicts the null hypothesis. The corresponding test statistic is used to decide between the null and alternative hypotheses.

analysis of covariance (ANCOVA)—A technique for estimating and testing the effects of treatments when one or more concomitant variables influence the response variable. Note: Analysis of covariance can be viewed as a combination of regression analysis and analysis of variance.

analysis of means (ANOM)—A statistical procedure for troubleshooting industrial processes and analyzing the results of experimental designs with factors at fixed levels. It provides a graphical display of data. Ellis R. Ott developed the procedure in 1967 because he observed that nonstatisticians had difficulty understanding analysis of variance. Analysis of means is easier for quality practitioners to use because it is an extension of the control chart. In 1973, Edward G. Schilling further extended the concept, enabling analysis of means to be used with nonnormal distributions and attributes data where the normal approximation to the binomial distribution does not apply. This is referred to as *analysis of means for treatment effects*.

analysis of variance (ANOVA)—A basic statistical technique for analyzing experimental data. It subdivides the total variation of a data set into meaningful component parts associated with specific sources of variation to test a hypothesis on the parameters of the model or to estimate variance components. There are three model types: fixed, random, and mixed.

AND—*See* activity network diagram.

ANOM—*See* analysis of means.

ANOVA—*See* analysis of variance.

ANSI—American National Standards Institute.

appraisal costs—The costs associated with measuring, evaluating, or auditing products or services to ensure conformance to quality standards and performance requirements. These include costs such as incoming and source inspection/test of purchased material; in-process and final inspection/test; product, process, or service audit; calibration of measuring and test equipment; and the cost of associated supplies and materials.

AQL—*See* acceptance quality limit.

Arrhenius model—A technique in accelerated life testing using relationships between temperature and failure rates.

arrow diagram—*See* activity network diagram.

AS-9100—A standard for the aeronautics industry embracing the ISO 9001 standard.

ASQ—American Society for Quality.

assignable cause—A specifically identified factor that contributes to variation and is detectable. Eliminating assignable causes so that the points plotted on the control chart remain within the control limits helps achieve a state of statistical control. Note: Although assignable cause is sometimes considered synonymous with special cause, a special cause is assignable only when it is specifically identified. *See also* special cause.

assumption—A condition that must be true for a statistical procedure to be valid.

attribute data—Characteristic or property that is measured in terms of pass/fail or go/no-go.

attribute—A countable or categorized quality characteristic that is qualitative rather than quantitative in nature.

availability—The probability that a system or equipment is operating satisfactorily at any point in time when used under stated conditions. The total time considered includes operating time, active repair time, administrative time, and logistic time.

avoidance, risk—Refers to the practice of eliminating the risk factor.

axiomatic design—A theory by Dr. Nam Pyo Suh that stresses each functional requirement be designed to achieve robustness without affecting other functional requirements.

B

balanced design—A design where all treatment combinations have the same number of observations. If replication in a design exists, it would be balanced only if the replication was consistent across all the treatment combinations. In other words, the number of replicates of each treatment combination is the same.

balanced incomplete block (BIB) design—Incomplete block design in which each block contains the same number (k) of different levels from the (l) levels of the principal factor arranged so that every pair of levels occurs in the same number (l) of blocks from the b blocks. Note: This design implies that every level of the principal factor appears the same number of times in the experiment.

batch—A definite quantity of some product accumulated under conditions considered uniform or accumulated from a common source. This term is sometimes synonymous with *lot*.

bathtub curve—A graphic representation of the relationship of the life of a product versus the probable failure rate. The curve contains three phases: early or infant failure (break-in), a stable rate during normal use, and wear-out. The bathtub curve is also known as the life-history curve.

benchmark—An organization, a part of an organization, or a measurement that serves as a reference point or point of comparison.

benchmarking—An improvement process in which an organization measures its performance against that of best-in-class organizations (or others that are good performers), determines how those organizations achieved their performance levels, and uses the information to improve its own performance. Areas that can be benchmarked include strategies, operations, processes, and procedures.

best subsets—A very useful tool to help select a regression model among many other viable models. It is based on the concept that for a model comprising m predictor variables, there are $2^m - 1$ possible subsets of predictor variables possible. Each model can be compared using such measures as VIF, Mallows' C_p, PRESS, R^2_{adj}, and squared error to aid in the model selection.

beta (β) risk—*See* type II error.

beta (β)—*See* β (beta).

beta testing—The release of a preliminary version of a product to a restricted set of users.

bias—A systematic difference between the mean of the test result or measurement result and a true or reference value. For example, if one measures the lengths of 10 pieces of rope that range from 1 foot to 10 feet and always concludes that the length of each piece is 2 inches shorter than the true length, then the individual is exhibiting a bias of 2 inches. Bias is a component of accuracy.

BIB—*See* balanced incomplete block design.

bimodal—Having two distinct statistical modes.

binomial distribution—A two-parameter discrete distribution involving the mean μ and the variance $\sigma 2$, of the variable x with probability p, where p is a constant $0 \le p \le 1$, and sample size n. Mean $= np$ and variance $= np(1-p)$.

Black Belt (BB)—A Six Sigma role associated with an individual who is typically assigned full time to train and mentor Green Belts as well as lead improvement projects using specified methodologies such as define, measure, analyze, improve, and control (DMAIC) and design for Six Sigma (DFSS).

blemish—An imperfection that causes awareness but does not impair function or usage.

block diagram—A diagram that describes the operation, interrelationships, and interdependencies of components in a system. Boxes, or blocks (hence the name), represent the components; connecting lines between the blocks represent interfaces. There are two types of block diagrams: a functional block diagram, which shows a system's subsystems and lower-level products, their interrelationships, and interfaces with other systems; and a reliability block diagram, which is similar to the functional block diagram except that it is modified to emphasize those aspects influencing reliability.

block effect—An effect resulting from a block in an experimental design. Existence of a block effect generally means that the method of blocking was appropriate and that an assignable cause has been found.

block—A collection of experimental units more homogeneous than the full set of experimental units. Blocks are usually selected to allow for special causes, in addition to those introduced as factors to be studied. These special causes may be avoidable within blocks, thus providing a more homogeneous experimental subspace.

blocking—Method of including blocks in an experiment in order to broaden the applicability of the conclusions or to minimize the impact of selected assignable causes. The randomization of the experiment is restricted and occurs within blocks.

brainstorming—A problem-solving tool that teams use to generate as many ideas as possible that are related to a particular subject. Team members begin by offering all their ideas; the ideas are not discussed or reviewed until after the brainstorming session.

B*X* life—The time at which *X*% of the units in a population will have failed. For example, if an item has a B10 life of 100 hours, that means that 10% of the population will have failed by 100 hours of operation.

C

c **(count)**—The number of events (often nonconformities) of a given classification occurring in a sample of fixed size.

calibration—The comparison of a measurement instrument or system of unverified accuracy to a measurement instrument or system of known accuracy to detect any variation from the true value.

capability index—See *process capability index.*

capability—Performance of a process demonstrated to be in a state of statistical control.

causal factor—A variable that when changed or manipulated in some manner serves to influence a given effect or result.

cause—An identified reason for the presence of a symptom, defect, or problem. See *effect.*

cause-and-effect analysis—The process of identifying the likely causes of any outcome. Some cause-and-effect analyses produce an output like a fishbone diagram. Others produce a cause-and-effect tree with multiple sub-branches.

cause-and-effect diagram—A diagram resembling a fish skeleton that is used to illustrate the main causes and sub-causes leading to an effect. One of the seven basic tools of quality, the cause-and-effect diagram is also known as the *Ishikawa diagram* or the *fishbone diagram.*

central limit theorem—Statistical principle that the distribution of sample averages is approximately normal even if the population from which the sample is drawn is not normally distributed. The approximation improves as the sample size increases.

central tendency—The propensity of data collected on a process to concentrate around a value situated approximately midway between the lowest and highest values. Three common measures of central tendency include (arithmetic) mean, median, and mode. *See also* mean, median, mode.

Certified Reliability Engineer (CRE)—An ASQ certification.

chance cause—*See* random cause.

chance variation—Variation due to chance causes. *See also* random variation or noise.

changeover time—The time interval between the last good piece off the current run and the first good piece off the next run.

check sheet—A simple data recording device typically used to collect the frequency of occurrences of nominal data by category. Whenever the user observes the occurrence of one of the categories, they place a mark next to the category. When the observation process is complete, final counts by category can then be obtained. The user custom designs the check sheet so that they can readily interpret the results. The check sheet is occasionally considered to be one of the seven basic tools of quality. Note: Check sheets are often confused with checklists. *See also* checklist.

checklist—A quality tool that is used for processes where there is a large human element. Examples are audit checklists and an airplane pilot's preflight checklist. Checklists serve as reminders and, depending on design, evidence that important items have been observed or appropriate actions have been taken. A well-designed checklist helps gather information to support findings and observations and serves as a guide and a place to record information. Note: Checklists are often confused with check sheets. *See also* check sheet.

chi-square distribution (χ^2 distribution)—A positively skewed distribution that varies with the degrees of freedom with a minimum value of zero.

chi-square statistic (χ^2 statistic)—A value obtained from the χ^2 distribution at a given percentage point and specified degrees of freedom.

chi-square test (χ^2 test)—A statistic used in testing a hypothesis concerning the discrepancy between observed and expected results.

circle diagram—A tool used to show linkages between various items. The circle diagram is constructed by evenly placing item descriptors around a circle. Such item descriptors might include products, services, organizations, individuals, and so forth. Arrows are drawn from each item to other items where a flow exists. Circle diagrams are highly useful in that they readily depict predecessor and successor relationships as well as potential bottlenecks. Too many inputs or outputs from any given descriptor around the circumference of the circle may indicate a limiting function. A circle diagram is also known as a *hand-off map*.

closed-loop feedback system—When the manipulated variable is adjusted by the final control element, and that variable, in turn, affects the next measurement, the system is said to be a closed-loop feedback system.

CLT—*See* central limit theorem.

coefficient of determination (*R*2)—A statistic that indicates how much variation in the response variable is accounted for by a simple regression model. $0 \leq R^2 \leq 1$. The coefficient of determination is the fraction of the variation in y that is explained by the fitted equation.

coefficient of multiple determination—The coefficient of determination for a multiple linear regression model. *See* coefficient of determination.

coefficient of variation (CV)—Measures relative dispersion. It is the standard deviation divided by the mean and is commonly reported as a percentage.

complete block—Block that accommodates a complete set of treatment combinations.

completely randomized design—A design in which the treatments are assigned at random to the full set of experimental units. No blocks are involved in a completely randomized design.

completely randomized factorial design—A factorial design in which all the treatments are randomly assigned to the full set of experimental units. *See also* completely randomized design.

concomitant variable—A variable or factor that cannot be accounted for in the data analysis or design of the experiment but whose effect on the results should be accounted for.

concurrent engineering—A way to reduce cost, improve quality, and shrink cycle time by simplifying a product's system of lifecycle tasks during the early concept stages. Concurrent engineering is a process to get all departments including engineering, purchasing, marketing, manufacturing, and finance to work on a new design at once to speed development. The emphasis is on upstream prevention versus downstream correction. Concurrent engineering is also known as *simultaneous engineering*.

confidence coefficient (1 – α)—See *confidence level.*

confidence interval—An estimate of the interval between two statistics that includes the true value of the parameter with some probability. This probability is called the *confidence level* of the estimate. Confidence levels typically used are 90%, 95%, and 99%. Either the interval contains the parameter or it does not.

confidence level (1 – α)—The probability that (1) the confidence interval described by a set of confidence limits actually includes the population parameter, and (2) an interval about a sample statistic actually includes the population parameter. The confidence level is also known as the *confidence coefficient*.

confidence limits—The endpoints of the interval about the sample statistic that is believed, with a specified confidence level, to include the population parameter. *See also* confidence interval.

confounding—Indistinguishably combining an effect with other effects or blocks. When done deliberately, higher-order effects are systematically aliased so as to allow estimation of lower-order effects. Sometimes, confounding results from inadvertent changes to a design during the running of an experiment or from poor planning of the design. This can diminish or even invalidate the effectiveness of the experiment.

consumer's risk (β)—Probability of acceptance when the quality level has a value stated by the acceptance sampling plan as unsatisfactory. Note 1: Such acceptance is a type II error. Note 2: Consumer's risk is usually designated as β (beta).

continuous distribution—Distribution where data is from a continuous scale. Examples of continuous scales are the normal, *t*, and *F* distributions.

continuous scale—A scale with a continuum of possible values. Note: A continuous scale can be transformed into a discrete scale by grouping values, but this leads to some loss of information.

continuous variable—A variable whose possible values form an interval set of numbers such that between each two values in the set another member of the set occurs.

control chart—A chart that plots a statistical measure of a series of samples in a particular order to steer the process regarding that measure and to control and reduce variation. The control chart comprises the plotted points, a set of upper and lower control limits, and a centerline. Specific rules are used to determine when the control chart goes out of control. Note: (1) The order is either time or sample number order based, and (2) the control chart operates most effectively when the measure is a process characteristic correlated with an ultimate product or service characteristic. The control chart is one of the seven basic tools of quality.

control limit—A line on a control chart used for judging the stability of a process. Note: (1) Control limits provide statistically determined boundaries for the deviations from the centerline of the statistic plotted on a Shewhart control chart due to random causes alone; (2) control limits (with the exception of the acceptance control chart) are based on actual process data, not on specification limits; (3) other than points outside the control limits, out-of-control criteria can include runs, trends, cycles, periodicity, and unusual patterns within the control limits; (4) the calculation of control limits depends on the type of control chart.

control plan—A document describing the system elements to be applied to control variation of processes, products, and services in order to minimize deviation from their preferred values.

controlled variable—In the context of automated process control, the controlled variable is a variable that must be maintained at a set or specified level. The control variable is also known as the *process variable* or simply the *measurement*. In lean Six Sigma terminology, it is the *Y* variable.

COPQ—*See* cost of poor quality.

COQ—*See* cost of quality.

corrective action—Action taken to eliminate the root cause(s) and symptom(s) of an existing deviation or nonconformity to prevent recurrence.

corrective maintenance (CM)—The actions required to return the system to an operating state once failure has occurred.

correlation coefficient (*r*)—A statistic that measures the degree of linear relationship between two sets of numbers

correlation—Measures the linear association between two variables. It is commonly measured by the sample correlation coefficient, *r*.

cost of poor quality (COPQ)—The costs associated with the production of nonconforming material. These costs include both external and internal failure costs. *See also* external failure costs; internal failure costs.

cost of quality (COQ)—The costs specifically associated with the achievement or nonachievement of product or service quality, including all product or service requirements established by the organization and its contracts with customers and society. More specifically, quality costs are the total costs incurred by (1) investing in the prevention of nonconformances to requirements, (2) appraising a product or service for conformance to requirements, and (3) failing to meet requirements. These can then be categorized as prevention, appraisal, and failure. *See also* appraisal costs; failure costs; prevention costs.

covariance—Measures the relationship between pairs of observations from two variables.

C_p (process capability index)—An index describing process capability in relation to specified tolerance of a characteristic divided by the natural process variation for a process in a state of statistical control:

C_{pk} (minimum process capability index)—The smaller of the upper process capability index C_{pkU} and the lower process capability index C_{pkL}. The C_{pk} considers the mean of the process with respect to the upper and lower specifications. The C_{pk} can be converted to a sigma level using sigma level = $3C_{pk}$.

C_{pkL} (lower process capability index; CPL)—An index describing process capability in relation to the lower specification limit.

C_{pkU} (upper process capability index; CPU)—An index describing process capability in relation to the upper specification limit.

CPL—*See* C_{pkL}.

C_{pm} (process capability index of the mean)—An index that takes into account the location of the process average relative to a target value.

CPM—*See* critical path method.

CPU—*See* C_{pkU}.

critical path method (CPM)—An activity-oriented project management technique that uses arrow-diagramming methods to demonstrate both the time and the cost required to complete a project. It provides one time estimate: normal time.

critical path—The sequence of tasks that takes the longest time and determines a project's completion date.

critical success factor (CSF)—*See* key business drivers.

critical to quality (CTQ)—Characteristic of a product or service that is essential to ensure customer satisfaction.

critical value—The numerical values of the test statistic that determine the rejection region.

criticality—An indication of the consequences expected to result from a failure.

critical-to-quality (CTQ)—A characteristic of a product or service that is essential to ensure customer satisfaction.

cross-functional team—A group consisting of members from more than one department that is organized to accomplish a project.

CTQ—See *critical to quality.*

cube point—In a design, experimental runs that are in the corner points of the design space. In general, a factorial design consists of cube points only. They also exist in fractional factorial designs and some central composite designs.

cumulative frequency distribution—The sum of the frequencies accumulated up to the upper boundary of a class in the distribution.

cumulative sum (CUSUM) control chart—A control chart on which the plotted value is the cumulative sum of deviations of successive samples from a target value. The ordinate of each plotted point represents the algebraic sum of the previous ordinate and the most recent deviations from the target.

CUSUM—*See* cumulative sum (CUSUM) control chart.

CV—See *coefficient of variation.*

cycle time—The time required to complete one cycle of an operation. If cycle time for every operation in a complete process can be reduced to equal takt time, products can be made in single-piece flow. *See also* takt time.

cycle-time reduction—A method for reducing the amount of time it takes to execute a process or build a specific product. Elimination of duplicate or unnecessary tasks also reduces the time that is necessary to execute a process. Improving hand-offs, which tend to be points of substantial discontinuity, also eliminates rework or duplicate efforts.

D

decision matrix—*See* Pugh matrix.

defective—A unit of product that contains one or more defects with respect to the quality characteristic(s) under consideration.

defect—Nonfulfillment of a requirement related to an intended or specified use.

defects per million opportunities (DPMO)—Measure of capability for discrete (attribute) data found by dividing the number of defects by the opportunities for defects times one million. It allows for comparison of different types of product.

defects per unit (DPU)—Measure of capability for discrete (attribute) data found by dividing the number of defects by the number of units.

defect—The nonfulfillment of a requirement related to an intended or specified use for a product or service. In simpler terms, a defect is anything not done correctly the first time. Note: The distinction between the concepts *defect* and *nonconformity* is important because it has legal connotations, particularly those associated with product liability issues. Consequently, the term "defect" should be used with extreme caution.

degrees of freedom (*v*, *df*)—In general, the number of independent comparisons available to estimate a specific parameter, which serves as a means of entering certain statistical tables.

demerit—A weighting assigned to a classification of an event or events to provide a means of obtaining a weighted quality score.

Deming cycle—*See* plan–do–study–act (PDSA) cycle.

dependability—The degree to which a product is operable and capable of performing its required function at any randomly chosen time during its specified operating time, provided that the product is available at the start of that period. (Non-operation-related influences are not included.)

dependent events—Two events, *A* and *B*, are dependent if the probability of one event occurring is higher given the occurrence of the other event.

derating—The practice of using components at lower stress levels than those stated by the component specifications.

descriptive statistics—The collection of tools and techniques for displaying and summarizing data.

design failure mode and effects analysis (DFMEA)—An analysis process used to identify and evaluate the relative risk associated with a particular design.

design for reliability (DFR)—An engineering program in which reliability engineering is considered during the development cycle to ensure reliability and cost goals are achieved.

design for Six Sigma (DFSS)—A structured methodology that focuses on designing new products or services with the intent of achieving Six Sigma quality levels. *See also* DMADV, DMADOV, DMEDI, and IDOV.

design for X (DFX)—An umbrella term whereby *X* is a variable that can assume multiple descriptions such as maintainability, manufacturability, producibility, quality, testability, or whatever component of design, logistics, production, quality, and so forth, of the product needs to be emphasized. Design for X involves the principle of designing products so that they are cost-effective and achieve the objectives set forth.

design of experiments (DOE, DOX)—The arrangement in which an experimental program is to be conducted, including the selection of factor combinations and their levels. Note: The purpose of designing an experiment is to provide the most efficient and economical methods of reaching valid and relevant conclusions from the experiment. The selection of the design is a function of many considerations, such as the type of questions to be answered, the applicability of the conclusions, the homogeneity of experimental units, the randomization scheme, and the cost of running the experiment. A properly designed experiment will permit simple interpretation of valid results.

design review—Documented, comprehensive, and systematic examination of a design to evaluate its capability to fulfill the requirements for quality.

design space—The multidimensional region of possible treatment combinations formed by the selected factors and their levels.

detection—The likelihood of finding a failure once it has occurred. Detection may be evaluated based on a ten-point scale. On the lowest end of the scale (1), it is assumed that a design control will detect a failure with certainty. On the highest end of the scale (10), it is assumed that a design control will not detect a failure when a failure occurs.

DFMaint—An acronym for design for maintainability. *See* design for X.

DFM—An acronym for design for manufacturability. *See* design for X.

DFMEA—*See* design failure mode and effects analysis.

DFP—An acronym for design for producibility. *See* design for X.

DFQ—An acronym for design for quality. *See* design for X.

DFR—An acronym for design for reliability. *See* design for X.

DFSS—*See* design for Six Sigma.

DFT—An acronym for design for testability. *See* design for X.

DFX—*See* design for X.

discrete scale—A scale with only a set or sequence of distinct values. Examples include defects per unit, events in a given time period, types of defects, and number of orders on a truck.

discrete variable—A variable whose possible values form a finite or, at most, countably infinite set.

discrimination—The capability of the measurement system to detect and indicate small changes in the characteristic measured. Discrimination is also known as *resolution*. For example, a tape measure with gradations of one inch would be unable to distinguish between lengths of objects that fall in between the inch marks. Hence, we would say the measurement system could not properly discriminate between the objects. If an object to be measured is 2.5 inches, the measurement system (that is, tape measure) would produce a value of 2 or 3 inches depending on how the individual decided to round. Therefore, to measure an object that is 2.5 inches, a tape measure with finer gradations would be required.

dispersion effect—The influence of a single factor on the variance of the response variable.

dispersion—A quantifiable variation of measurements within a population.

DMADV—A structured DFSS methodology. DMADV is an acronym for define, measure, analyze, design, and verify. Variations of DMADV exist. *See also* design for Six Sigma, DMADOV, DMEDI, and IDOV.

DMAIC—A data driven quality strategy for improving processes and an integral part of a Six Sigma quality initiative. DMAIC is an acronym for define, measure, analyze, improve, and control.

DOE—*See* design of experiments.

DPMO—*See* defects per million opportunities.

DPU—*See* defects per unit.

E

effect (design of experiments usage)—A relationship between factor(s) and a response variable(s). Specific types include main effect, dispersion effect, or interaction effect.

eight types of waste—The seven types of waste plus an additional waste: people. The eighth category was added by Taiichi Ohno, as he saw the underutilization of employees (for example, brainpower, skills, experience, and talents) as a waste. *See also* seven types of waste.

eighty-twenty (80–20) rule—A term referring to the Pareto principle, which was first defined by J. M. Juran in 1950. The principle suggests that most effects come from relatively few causes; that is, 80% of the effects come from 20% of the possible causes. *See also* Pareto principle.

element—See *unit.*

error detection—A hybrid form of error-proofing whereby a bad part can still be made but will be caught immediately, and corrective action will be taken to prevent another bad part from being produced. An error-detection device is used to spot the error and stop the process when a bad part is made. This is used when error-proofing is too expensive or not easily implemented.

error—(1) Error in measurement is the difference between the indicated value and the true value of a measured quantity. (2) A fault resulting from defective judgment, deficient knowledge, or carelessness. It is not to be confused with measurement error, which is the difference between a computed or measured value and the true or reference value.

error-proofing—The use of process or design features to prevent the acceptance or further processing of nonconforming products. *See also* fool-proofing, mistake-proofing, and poka-yoke.

event—An occurrence of some attribute or outcome. In the quality field, events are often nonconformities.

evolutionary operation (EVOP)—A sequential form of experimentation conducted in production facilities during regular production. The range of variation of the factors is usually quite small in order to avoid extreme changes in settings, so it often requires considerable replication and time.

EVOP—See *evolutionary operation.*

expected value—The mean of a random variable.

experimental design—A formal plan that details the specifics for conducting an experiment, such as which responses, factors, levels, blocks, treatments, and tools are to be used. *See also* design of experiments.

experimental error—Variation in the response variable beyond that accounted for by the factors, blocks, or other assignable sources in the conduct of the experiment.

experimental run—A single performance of the experiment for a specific set of treatment combinations.

experimental unit—The smallest entity receiving a particular treatment, subsequently yielding a value of the response variable.

explanatory variable—*See* predictor variable.

exploratory data analysis (EDA)—Technique that isolates patterns and features of the data and reveals these forcefully to the analyst.

external failure costs—The failure costs occurring after delivery or shipment of the product, and during or after furnishing of a service, to the customer. Examples are the costs of processing customer complaints, customer returns, warranty claims, and product recalls.

Eyring model—A model used in accelerated life testing using temperature as the accelerant.

F

***F* distribution**—A continuous distribution that is a useful reference for assessing the ratio of independent variances. See Appendices F, G, and H for the actual values of the distribution.

***F* test**—A statistical test that uses the *F* distribution. It is most often used when dealing with a hypothesis related to the ratio of independent variances.

facilitator—An individual responsible for creating favorable conditions that will enable a team to reach its purpose or achieve its goals by bringing together the necessary tools, information, and resources to get the job done.

factor level—See *level.*

factor—An independent variable or assignable cause that may affect the responses and of which different levels are included in the experiment. Factors are also known as *explanatory, independent, regressor,* or *predictor* variables.

factorial design—Experimental design consisting of all possible treatments formed from two or more factors, each being studied at two or more levels. When all combinations are run, the interaction effects as well as main effects can be estimated.

failure analysis—The process of breaking down a failure to determine its cause and to put measures in place to prevent future problems.

failure costs—The costs resulting from product or services not conforming to requirements or customer/user needs. Failure costs are further divided into internal and external categories. *See also* external failure costs, internal failure costs.

failure mechanism—The physical, chemical, or mechanical process that caused the defect or failure.

failure mode effects analysis (FMEA)—A systematized group of activities to recognize and evaluate the potential failure of a product or process and its effects, identify actions that could eliminate or reduce the occurrence of the potential failure, and document the process.

failure mode effects and criticality analysis (FMECA)—A procedure performed after a failure mode effects analysis to classify each potential failure effect according to its severity and probability of occurrence.

failure mode—The type of defect contributing to a failure.

failure rate—The number of failures per unit time (for equal time intervals).

failure reporting, analysis, and corrective action system (FRACAS)—A methodology that provides a systematic process for reporting, classifying, and analyzing failures, as well as planning corrective actions.

failure—The inability, because of defect(s), of an item, product, or service to perform its required functions as needed.

fault tree analysis (FTA)—A "top down" approach that is used to identify the cause, causes, or combinations of causes of failures.

first quartile (Q_1 or lower quartile)—One quarter of the data lies below. See *quartiles*.

first-pass yield (FPY)—The percentage of units that complete a process and meet quality guidelines without being scrapped, rerun, retested, returned, or diverted into an offline repair area. FPY is calculated by dividing the units entering the process minus the defective units by the total number of units entering the process. First-pass yield is also known as the *quality rate* or *first-time yield* (FTY).

first-time yield (FTY)—*See* first-pass yield.

fishbone diagram—*See* cause-and-effect diagram.

fitness for use—A term used to indicate that a product or service fits the customer's or user's defined purpose for that product or service.

five S—*See* 5S.

five whys—*See* 5 whys.

fixed factor—A factor that only has a limited number of levels that are of interest.

fixed model—A model that contains only fixed factors.

flowchart—A basic quality tool that uses graphical representation for the steps in a process. Effective flowcharts include decisions, inputs, and outputs as well as the sequence of process steps. The flowchart is occasionally considered to be one of the seven basic tools of quality.

FMEA—*See* failure mode and effects analysis.

FPY—*See* first-pass yield.

fractional factorial design—Experimental design consisting of a subset (fraction) of the factorial design.

frequency distribution—A set of all the various values that individual observations may have and the frequency of their occurrence in the sample or population.

frequency—Number of occurrences or observed values in a specified class, sample, or population.

FTY—*See* first-pass yield.

G

gage R&R study—A type of measurement system analysis done to evaluate the performance of a test method or measurement system. Such a study quantifies the capabilities and limitations of a measurement instrument, often estimating its repeatability and reproducibility.

Gantt chart—A type of bar chart used in process/project planning and control to display planned work and finished work in relation to time. Also called a *milestone chart.*

gap analysis—A technique that compares an organization's existing state with its desired state (typically expressed by its long-term plans) to help determine what needs to be done to remove or minimize the gap.

Gaussian distribution—See *normal distribution.*

gemba walk—Gemba is a Japanese term that means "place of work" or "the place where the truth can be found." Still others may call it "the value proposition." A gemba walk is a method of obtaining voice of customer information that requires the design team to walk the process and observe how the customer uses the product in their environment.

generator—In experimental design, a generator is used to determine the level of confounding and the pattern of aliases in a fractional factorial design.

geometric distribution—Case of the negative binomial distribution where $c=1$ (c is integer parameter). The geometric distribution is a discrete distribution.

Gopertz model—Use in calculating reliability growth.

GR&R study—*See* Gage repeatability and reproducibility (GR&R) study.

H

Hawthorne effect—The effect in an experiment that occurs when humans perform better than they normally would because they are being measured or observed.

hazard rate—Instantaneous failure rate at some specified time t.

highly accelerated life test (HALT)—A process for uncovering design defects and weaknesses in electronic and mechanical assemblies using a vibration system combined with rapid high and low temperature changes. The purpose of HALT is to optimize product reliability by identifying the functional and destructive limits of a product at an early stage in product development.

highly accelerated stress audit (HASA)—A technique in which a sample of parts (as opposed to 100% of the production as in HASS) is subjected to stresses similar to the levels and duration for HALT. In monitoring the production process, the intent of HASA is to detect slight shifts in the attributes of the product so corrective actions can be taken and implemented before the performance of outgoing product approaches the specifications.

highly accelerated stress screening (HASS)—A technique for production screening that rapidly exposes process or production flaws in products. Its purpose is to expose a product to optimized production screens without affecting product reliability. Unlike HALT, HASS uses nondestructive stresses of extreme temperatures and temperature change rates with vibration.

histogram—A plot of a frequency distribution in the form of rectangles (cells) whose bases are equal to the class interval and whose areas are proportional to the frequencies. A histogram is a graphic summary of variation in a set of data. The graphical nature of the histogram permits easier visualization of patterns of variation that are difficult to detect in a simple table of numbers. The histogram is one of the seven basic tools of quality.

house of quality—A diagram named for its house-shaped appearance that clarifies the relationship between customer needs and product features. It helps correlate market or customer requirements and analysis of competitive products with higher-level technical and product characteristics and makes it possible to bring several factors into a single figure. The house of quality is also known as *quality function deployment* (QFD).

hypothesis testing—A statistical hypothesis that is a conjecture about a population parameter. There are two statistical hypotheses for each situation—the null hypothesis (H_0) and the alternative hypothesis (H_a). The null hypothesis proposes that there is no difference between the population of the sample and the specified population; the alternative hypothesis proposes that there is a difference between the sample and the specified population.

hypothesis—Statement about a population to be tested. *See* null hypothesis, alternative hypothesis, and hypothesis testing.

I

ideation—According to Jonson (2005) it is "the creative process of generating, developing, and communicating new ideas, where an idea is understood as a basic element of thought that can be visual, concrete, or abstract."

IDOV—A structured methodology as an alternative to design for Six Sigma. IDOV is an acronym for identify, design, optimize, and validate. *See also* design for Six Sigma.

iid—Independent and identically distributed.

imprecision—A measure of precision computed as a standard deviation of the test or measurement results. Less precision is reflected by a larger standard deviation.

incomplete block design—A design in which the design space is subdivided into blocks in which there are insufficient experimental units available to run a complete replicate of the experiment.

incomplete block—A block that accommodates only a subset of treatment combinations.

in-control process—A condition where the existence of special causes is no longer indicated by a Shewhart control chart. This does not indicate that only random causes remain, nor does it imply that the distribution of the remaining values is normal (Gaussian). It indicates (within limits) a predictable and stable process.

independent events—Two events, *A* and *B*, are called independent if the probability that they both occur is the product of the probabilities of their individual occurrences.

independent variable—*See* predictor variable.

inferential statistics—Techniques for reaching conclusions about a population on the basis of analysis of data from a sample.

inherent process variation—The variation in a process when the process is operating in a state of statistical control.

inspection—The process of measuring, examining, testing, gauging, or otherwise comparing the unit with the applicable requirements.

interaction effect—The effect for which the apparent influence of one factor on the response variable depends on one or more other factors. Existence of an interaction effect means that the factors cannot be changed independently of each other.

interaction plot—A plot providing the average responses at the combinations of levels of two distinct factors.

internal failure costs—The failure costs occurring prior to delivery or shipment of the product, or the furnishing of a service, to the customer. Examples are the costs of scrap, rework, reinspection, retesting, material review, and downgrading.

interval scale—A quantitative scale in which meaningful differences exist, but where multiplication and division are not allowed since there is not absolute zero. An example is temperature measured in °F, because 20°F isn't twice as warm as 10°F.

Ishikawa diagram—*See* cause-and-effect diagram.

ISO 14000 series—A set of individual, but related, international standards and guidelines relevant to developing and sustaining an environmental management system.

ISO 9000 series—A set of individual, but related, international standards and guidelines on quality management and quality assurance developed to help organizations effectively document the quality system elements to be implemented to maintain an efficient quality system. The standards were developed by the International Organization for Standardization, a specialized international agency for standardization composed of the national standards bodies of nearly 100 countries.

ISO—"Equal" (Greek). A prefix for a series of standards published by the International Organization for Standardization.

J

JIT—*See* just-in-time manufacturing.

just-in-time manufacturing (JIT manufacturing)—A material requirement planning system for a manufacturing process in which there is little or no manufacturing material inventory on hand at the manufacturing site and little or no incoming inspection.

K

kaizen blitz/event—An intense team approach to employ the concepts and techniques of continuous improvement in a short time frame (for example, to reduce cycle time, increase throughput).

kaizen—A Japanese term that means gradual unending improvement by doing little things better and setting and achieving increasingly higher standards. The kaizen approach is usually implemented as a small, intensive event or project over a relatively short duration, such as a week.

kanban—A Japanese term for a system that signals the need to replenish stock or materials or to produce more of an item. Kanban is also known as a "pull" approach. Kanban systems need not be elaborate, sophisticated, or even computerized to be effective. Taiichi Ohno of Toyota developed the concept of a kanban system after a visit to a U.S. supermarket.

key performance indicator (KPI)—Quantifiable financial and nonfinancial measurements that reflect an organization's key business drivers (KBDs) and are usually long term.

KPI—*See* key performance indicator.

L

lag—The number of time units between differenced values.

latent defects—Hidden defects in either material and/or workmanship that may cause a component to malfunction or fail, but is not discoverable through inspection.

Latin square design—A design involving three factors in which the combination of the levels of any one of them with the levels of the other two appears once and only once.

lean approach—*See* lean.

lean manufacturing—Applying the lean approach to improving manufacturing operations.

Lean Six Sigma—A fact-based, data-driven philosophy of improvement that values defect prevention over defect detection. It drives customer satisfaction and bottom-line results by reducing variation, waste, and cycle time, while promoting the use of work standardization and flow, thereby creating a competitive advantage. It applies anywhere variation and waste exist, and every employee should be involved.

lean—A systematic approach to identifying and eliminating waste (non-value-added activities) through continuous improvement by flowing the product at the pull of the customer in pursuit of perfection.

least squares, method of—A technique of estimating a parameter that minimizes the sum of the difference squared, where the difference is between the observed value and the predicted value (residual) derived from the model.

level of significance—See *significance level.*

level—The setting or assignment of a factor at a specific value.

lifecycle cost—The cost to purchase, operate, and maintain the product during its useful lifetime.

linear programming—A constrained optimization technique comprising both a linear objective function and linear constraints.

linear regression coefficients—Values associated with each predictor variable in a linear regression equation that tell how the response variable changes with each unit increase in the predictor variable. The term "linear" refers to the fact that the estimated coefficients are linear combinations of the responses. It does not refer to the form of the regression model: for example, a linear regression model can be a quadratic function.

linear regression equation—A mathematical function or model that indicates the predictive relationship between a set of predictor variables and a response variable.

linearity (general sense)—The degree to which a pair of variables follows a straight-line relationship. Linearity can be measured by the correlation coefficient.

linearity (measurement system sense)—The difference in bias through the range of measurement. A measurement system that has good linearity will have a constant bias no matter the magnitude of measurement. If one views the relation between the observed measurement result on the y-axis and the true value on the x-axis, an ideal measurement system would have a line of slope = 1. Linearity is a component of accuracy.

Lloyd-Lipow model—Used in reliability growth when fatigue is the major cause of failure.

logistic regression—A type of regression that is used when the response variable is discrete. Although there are many types of logistic regression, three common types include binary, nominal, and ordinal.

lognormal distribution—If log x is normally distributed, it is a lognormal distribution.

μ (mu)—See *population mean.*

M

MAD—*See* mean absolute deviation.

main effect—The influence of a single factor on the mean of the response variable.

main effects plot—The plot giving the average responses at the various levels of individual factors.

maintainability—The measure of the ability of an item to be retained or restored to a specified condition when maintenance is performed by personnel having specified skill levels, using prescribed procedures and resources, at each prescribed level of maintenance and repair.

mean absolute deviation (MAD)—A measure of accuracy of the fitted time series values.

mean absolute percentage error (MAPE)—A measure of accuracy of the fitted time series values.

mean life—The arithmetic average of the lifetimes of all items considered. A lifetime may consist of time between malfunctions, time between repairs, time to remove or replace parts, or any other desired interval of observation.

mean squared deviation (MSD)—A measure of accuracy of the fitted time series values.

mean time between failures (MTBF)—A basic measure of reliability for repairable items. The mean number of life units during which all parts of an item perform within their specified limits during a particular measurement interval under stated conditions.

mean time to failure (MTTF)—Measure of system reliability for nonrepairable items. The total number of life units of an item divided by the total number of failures within that population during a particular measurement interval under stated conditions. Expectation of the time to failure.

mean time to repair (MTTR)—Measure of maintainability. The sum of corrective maintenance times at any specific level of repair, divided by the total number of failures within an item repaired at that level, during a particular interval under stated conditions.

mean—A measure of central tendency and the arithmetic average of all measurements in a data set.

measurement error—The difference between the actual value and the measured value of a quality characteristic.

median—The value for which half the data is larger and half is smaller. The median provides an estimator that is insensitive to very extreme values in a dataset, whereas the average is affected by extreme values. Note: For an odd number of units, the median is the middle measurement; for an even number of units, the median is the average of the two middle units.

meta-analysis—Use of statistical methods to combine the results of multiple studies into a single conclusion.

method of least squares—A technique for estimating a parameter that minimizes the sum of the squared differences between the observed and the predicted values derived from the model.

metrology—The science and practice of measurements.

midrange—(Highest value + Lowest value)/2.

mistake-proofing—The use of process or design features to prevent manufacture of nonconforming product. It is typically engineering techniques that make the process or product sufficiently robust so as to avoid failing. *See also* poka-yoke.

mitigation—Minimizing the impact of the risk.

mixture design—A design constructed to handle the situation in which the predictor variables are constrained to sum to a fixed quantity, such as proportions of ingredients that make up a formulation or blend.

model—Description relating the response variable to predictor variable(s) and including attendant assumptions.

mode—The value that occurs most frequently in a data set.

monitoring—Continuing to observe low risks where the cost of mitigation or avoidance is too high.

moving average—Let $x_1, x_2, \ldots$ denote individual observations. The moving average of span w at time i.

MSD—*See* mean squared deviation.

MTBF—*See* mean time between failures.

MTTF—*See* mean time to failure.

MTTR—*See* mean time to repair.

muda—A Japanese term for an activity that consumes resources but creates no value; the seven categories are correction, overprocessing, inventory, waiting, overproduction, internal transport, and motion. *See also* seven types of waste.

multicollinearity—Multicollinearity is said to occur when two or more predictor variables in a multiple regression model are correlated. Multicollinearity may cause the coefficient estimates and significance tests for each predictor variable to be underestimated and difficult to interpret.

multimodal—More than one mode.

multiphase planning—The process of dissecting the planning time frame into smaller time elements or phases to recognize or even celebrate significant events or accomplishments and to demonstrate progress.

multiple analysis of variance (MANOVA)—A statistical technique that can be used to analyze both balanced and unbalanced experimental designs. MANOVA is used to perform multivariate analysis of variance for balanced designs when there is more than one dependent variable. The advantage of MANOVA over multiple one-way ANOVAs is that MANOVA controls the family error rate. By contrast, multiple one-way ANOVAs work to increase the alpha error rate.

multiple linear regression—A linear regression equation in which multiple predictor variables are used.

multi-vari analysis—A graphical technique for viewing multiple sources of process variation. Different sources of variation are categorized into families of related causes and quantified to reveal the largest causes.

multivariate control chart—A variables control chart that allows plotting of more than one variable. These charts make use of the T^2 statistic to combine information from the dispersion and mean of several variables.

N

natural tolerance—For a normally distributed process output, the range from the mean plus three standard deviations to the mean minus three standard deviations.

n—*See* sample size.

negative binomial distribution—A two-parameter, discrete distribution.

noise factor—In robust parameter design, a noise factor is a predictor variable that is hard to control or is not desirable to control as part of the standard experimental conditions. In general, it is not desirable to make inference on noise factors, but they are included in an experiment to broaden the conclusions regarding control factors.

nominal scale—A scale with unordered, labeled categories, or a scale ordered by convention. Examples include type of defect, breed of dog, and complaint category. Note: It is possible to count by category but not by order or measure.

nonconformance—*See* defective.

nonconformity—*See* defect.

nonlinear regression—With linear regression, the coefficients are linear combinations of the responses, and have a closed-form solution. In nonlinear regression, the estimated coefficients are nonlinear functions of the responses and cannot be written in a closed form. Instead, the coefficient estimates must be calculated using iterative optimization techniques that may or may not converge.

non-value-added—A term that describes a process step or function that is not required for the direct achievement of process output. This step or function is identified and examined for potential elimination.

normal distribution (Gaussian distribution)—A continuous, symmetrical, bell-shaped frequency distribution for variables that is the basis of the control charts for variables.

null hypothesis, H_0—The hypothesis that there is no difference (null) between the population of the sample and the specified population (or between the populations associated with each sample). The null hypothesis can never be proved true, but it can be shown (with specified risks of error) to be untrue; that is, that a difference exists between the populations. Example: In a random sample of independent random variables with the same normal distribution with unknown mean and unknown standard deviation, a typical null hypothesis for the mean μ is that the mean is less than or equal to a given value μ_0. The hypothesis is written as $H_0 = \mu \leq \mu_0$.

null hypothesis—A hypothesis that is formulated without considering any new information or formulated based on the belief that what already exists is true.

NVA—*See* non-value-added.

O

objective—A quantitative statement of future expectations and an indication of when the expectations should be achieved; it flows from goal(s) and clarifies what people must accomplish.

observational study—Analysis of data collected from a process without imposing changes on the process.

observation—The process of determining the presence or absence of attributes or measuring a variable. Also, the result of the process of determining the presence or absence of attributes or measuring a variable.

observed value—The particular value of a response variable determined as a result of a test or measurement.

OC curve—See *operating characteristic curve.*

occurrence—The likelihood of a failure occurring. Occurrence is evaluated using a ten-point scale. On the lowest end of the scale (1), it is assumed the probability of a failure is unlikely. On the highest end of the scale (10), it is assumed the probability of a failure is nearly inevitable.

ogive—A type of graph that represents the cumulative frequencies for the classes in a frequency distribution.

one-tailed test—A hypothesis test that involves only one of the tails of a distribution. Example: We wish to reject the null hypothesis H_0 only if the true mean is *larger* than μ_0.

OOC—*See* out of control.

operating characteristic curve (OC curve)—A curve showing the relationship between the probability of acceptance of product and the incoming quality level for a given acceptance sampling plan.

order—There are two types of order that are considered when creating experimental designs: (1) standard order shows what the order of the runs in the experiment would be if the experiment were done in standard order (also called Yates's order), and (2) run order shows what the order of the runs in the experiment would be if the experiment were run in random order.

ordinal scale—A scale with ordered, labeled categories. Note: (1) The borderline between ordinal and discrete scales is sometimes blurred. When subjective opinion ratings such as excellent, very good, neutral, poor, and very poor are coded (as numbers 1–5), the apparent effect is conversion from an ordinal to a discrete scale. However, such numbers should not be treated as ordinary numbers, because the distance between 1 and 2 may not be the same as that between 2 and 3, 3 and 4, and so forth. On the other hand, some categories that are ordered objectively according to magnitude—such as the Richter scale, which ranges from 0 to 8 according to the amount of energy release—could equally well be related to a discrete scale. (2) Sometimes, nominal scales are ordered by convention. An example is the blood groups *A*, *B*, and *O*, which are always stated in this order. It's the same case if different categories are denoted by single letters; they are then ordered by convention, according to the alphabet.

orthogonal design—A design in which all pairs of factors at particular levels appear together an equal number of times.

out of control—A process is described as operating out of control when special causes are present.

outlier—An extremely high or an extremely low data value compared to the rest of the data values. Great caution must be used when trying to identify an outlier.

output variable—Variable representing the outcome of the process.

P

parameter—A constant or coefficient describing some characteristic of a population (for example, standard deviation, mean).

Pareto chart—A graphical tool based on the Pareto principle for ranking causes from most significant to least significant. It utilizes a vertical bar graph in which the bar height reflects the frequency or impact of causes. The graph is distinguished by the inclusion of a cumulative percentage line that identifies the vital few opportunities for improvement. It is also known as the *Pareto diagram*. The Pareto chart is one of the seven basic tools of quality.

Pareto principle—An empirical rule named after the nineteenth-century economist Vilfredo Pareto that suggests that most effects come from relatively few causes; that is, about 80% of the effects come from about 20% of the possible causes. The Pareto principle is also known as the *80–20 rule*.

parts per billion (ppb)—The number of times an occurrence happens in one billion chances. In a typical quality setting it usually indicates the number of times a defective part will happen in a billion parts produced; the calculation is projected into the future on the basis of past performance. Parts per billion allows for comparison of different types of product. Two ppb corresponds to a Six Sigma level of quality assuming there is no shift in the mean.

parts per million (PPM or ppm)—One part per million (or one part per 106).

Pearson's correlation coefficient—See *correlation coefficient*.

percentile—Division of the dataset into 100 equal groups.

parts per million (ppm)—The number of times an occurrence happens in one million chances. In a typical quality setting it usually indicates the number of times a defective part will happen in a million parts produced; the calculation is often projected into the future on the basis of past performance. Parts per million allows for comparison of different types of product. A ppm of 3.4 corresponds to a Six Sigma level of quality assuming a 1.5σ shift of the mean.

PDCA cycle—*See* plan–do–check–act (PDCA) cycle.

PDSA cycle—*See* plan–do–study–act (PDSA) cycle.

PERT—*See* program evaluation and review technique.

PFMEA—*See* process failure mode and effects analysis.

physics of failure—A technique used for design and problem solving (root cause analysis) that uses the knowledge, science, and understanding of mechanisms that induce failure.

plan–do–check–act (PDCA) cycle—A four-step process for quality improvement. In the first step (plan), a plan to effect improvement is developed. In the second step (do), the plan is carried out, preferably on a small scale. In the third step (check), the effects of the plan are observed. In the last step (act), the results are studied to determine what was learned and what can be predicted. The plan–do–check–act cycle is sometimes referred to as the *Shewhart cycle*, after Walter A. Shewhart.

plan–do–study–act (PDSA) cycle—A variation on the plan–do–check–act (PDCA) cycle, with the variation indicating that additional study is required after a change is made. The PDSA cycle is attributed to W. Edwards Deming.

point estimate—The single value used to estimate a population parameter. Point estimates are commonly referred to as the points at which the interval estimates are centered; these estimates give information about how much uncertainty is associated with the estimate.

Poisson distribution—Describes occurrences of isolated events in a continuum of time or space. It is a one-parameter, discrete distribution depending only on the mean.

poka-yoke—A Japanese term that means to mistake-proof a process by building safeguards into the system that avoid or immediately find errors. A poka-yoke device prevents incorrect parts from being made or assembled and easily identifies a flaw or error. The term comes from the Japanese *poka*, which means "error," and *yokeru*, which means "to avoid." *See also* error-proofing, mistake-proofing.

pooled standard deviation—A standard deviation value resulting from some combination of individual standard deviation values.

population mean (μ)—The true mean of the population, represented by μ (mu).

population standard deviation—See *standard deviation*.

population variance—See *variance*.

population—Entire set (totality) of units, quantity of material, or observations under consideration. A population may be real and finite, real and infinite, or completely hypothetical. See *sample*.

potential process capability—*See* P_{pk}.

power (1 − β)—The probability of correctly rejecting the null hypothesis when the alternative is true. We write this as $P(\text{rejecting } H_0 \mid H_0 \text{ is not true}) = 1 - \beta$. The stronger the power of a hypothesis test, the more sensitive the test is to detecting small differences.

power curve—The curve showing the relationship between the probability $(1 - \beta)$ of rejecting the hypothesis that a sample belongs to a given population with a given characteristic(s) and the actual population value of that characteristic(s).

power—Equivalent to one minus the probability of a type II error $(1 - \beta)$. A higher power is associated with a higher probability of finding a statistically significant difference. Lack of power usually occurs with smaller sample sizes.

Pp (process performance index)—Index describing process performance in relation to specified tolerance. The standard deviation is calculated across all the data, which captures between, within, and special cause variation. Note: A state of statistical control is not required.

ppb—*See* parts per billion.

P_{pk} (minimum process performance index)—The smaller of the upper process performance index P_{pkU} and the lower process performance index P_{pkL}. The P_{pk}, similar to the C_{pk}, considers the mean of the process with respect to the upper and lower specifications. The P_{pk} is known as the potential process capability and is used in the validation stage of a new product launch. A state of statistical control is not required.

P_{pkL} (lower process performance index or P_p)—Index describing process performance in relation to the lower specification limit. For a symmetrical normal distribution.

P_{pkU} (upper process performance index or P_p)—Index describing process performance in relation to the upper specification limit. For a symmetrical normal distribution.

PPM (or ppm)—See *parts per million.*

PPU—*See* P_{pkU}.

precision—The closeness of agreement among randomly selected individual measurements or test results. It is that aspect of measure that addresses repeatability or consistency when an identical item is measured several times.

predicted value—The prediction of future observations based on the formulated model.

prediction interval—Similar to a confidence interval. It is an interval based on the predicted value that is likely to contain the values of future observations. It will be wider than the confidence interval because it contains bounds on individual observations rather than a bound on the mean of a group of observations.

predictor variable—Variable that can contribute to the explanation of the outcome of an experiment.

prevention costs—The cost of all activities specifically designed to prevent poor quality in products or services. Examples are the cost of new product review, quality planning, supplier capability surveys, process capability evaluations, quality improvement team meetings, quality improvement projects, and quality education and training.

preventive maintenance (PM)—Actions performed to keep the system in an operating state by preventing wear-out failures.

prioritization matrix—A tool used to choose between several options that have many useful benefits, but where not all of them are of equal value. The choices are prioritized according to known weighted criteria and then narrowed down to the most desirable or effective one(s) to accomplish the task or problem at hand. The prioritization matrix is one of the seven management and planning tools.

probability distribution—A function that completely describes the probabilities with which specific values occur. The values may be from a discrete scale or a continuous scale.

probability plot—Plot of ranked data versus the sample cumulative frequency on a special vertical scale. The special scale is chosen (that is, normal, lognormal, and so on) so that the cumulative distribution is a straight line.

process—A series of interrelated steps consisting of resources and activities that transform inputs into outputs and work together to a common end. A process can be graphically represented using a flowchart. A process may or may not add value.

process capability index—A single-number assessment of ability to meet specification limits on the quality characteristic(s) of interest. The indices compare the variability of the characteristic to the specification limits. Three basic process capability indices are C_p, C_{pk}, and C_{pm}.

process capability—Calculated inherent variability of a characteristic of a product. It represents the best performance of the process over a period of stable operations.

process control—Process management focused on fulfilling process requirements. Process control is also the methodology for keeping a process within boundaries and minimizing the variation of a process.

process failure mode and effects analysis (PFMEA)—An analysis process used to identify and evaluate the relative risks associated with a particular process.

process mapping—The flowcharting of a process in detail that includes the identification of inputs and outputs.

process performance index—A single-number assessment of ability to meet specification limits on the quality characteristic(s) of interest. The indices compare the variability of the characteristic to the specification limits. Three basic process capability indices are P_p, P_{pk}, and P_{pm}.

process performance—A statistical measure of the outcome of a characteristic from a process that may not have demonstrated to be in a state of statistical control. Note: Use this measure cautiously since it may contain a component of variability from special causes of unpredictable value. It differs from process capability because a state of statistical control is not required.

process quality—A statistical measure of the quality of product from a given process. The measure may be an attribute (qualitative) or a variable (quantitative). A common measure of process quality is the fraction or proportion of nonconforming units in the process.

producer's risk (α)—The probability of nonacceptance when the quality level has a value stated by the acceptance sampling plan as acceptable. Note: (1) Such nonacceptance is a type I error; (2) producer's risk is usually designated as alpha (α); (3) quality level could relate to fraction nonconforming and acceptable to AQL; (4) interpretation of producer's risk requires knowledge of the stated quality level.

program evaluation and review technique (PERT)—An event-oriented project management planning and measurement technique that utilizes an arrow diagram or road map to identify all major project events and demonstrates the amount of time (critical path) needed to complete a project. It provides three time estimates: optimistic, most likely, and pessimistic.

project—A project exists for producing results. Three components must be present for an activity to be defined as a project: scope, schedule, and resources.

project management—The discipline of planning, organizing, and managing resources to bring about the successful completion of specific project goals and objectives, categorized into five component processes (sometimes called life cycle phases): initiating, planning, executing, controlling, and closing. The Project Management Institute notes that project management is the "application of knowledge, skills, tools, and techniques" to activities to meet the requirements of a specific project.

project plan—Documents that contain the details of why the project is to be initiated, what the project is to accomplish, when and where it is to be implemented, who will be responsible, how implementation will be carried out, how much it will cost, what resources are required, and how the project's progress and results will be measured.

project prioritization—A selected project will be independently assessed against specific criteria in order to establish a rank score. Rank scores of other selected projects will be compared and a final ranking of the projects will be established. This final ranking will set the precedence and order for project assignment.

propagation of error—Also known as *transmission of error*; has application in two major areas of lean Six Sigma: (1) response model analysis where the analysis will address how random noise in the x_i values affects the predicted value of y and (2) tolerance stack-up analysis where the analysis will address tolerance stack-up problems.

proportions, tests for—Tests for proportions include the binomial distribution.

Pugh matrix—Also known as a *decision matrix*; appropriate to use when a single option must be selected from several and when multiple criteria are to be used.

***p*-value**—The probability of obtaining the statistical result or something more extreme assuming the null hypothesis is true. When the *p*-value is less than the chosen α level, we reject the null hypothesis.

Q

Q_1—See *first quartile.*

QFD—*See* quality function deployment.

qualitative data—See *attribute data.*

quality—Degree to which a set of inherent characteristics fulfills requirements.

quality assurance—The planned or systematic actions necessary to provide adequate confidence that a product or service will satisfy given needs.

quality control—The operational techniques and the activities that sustain a quality of product or service that will satisfy given needs; also the use of such techniques and activities.

quality function deployment (QFD)—A method used to translate voice of customer information into product requirements/CTQs and to continue deployment (for example, cascading) of requirements to parts and process requirements. Quality function deployment is also known as the *house of quality*.

quality improvement—The actions taken throughout an organization to increase the effectiveness and efficiency of activities and processes in order to provide added benefits to both the organization and its customers.

quality loss function—A parabolic approximation of the quality loss that occurs when a quality characteristic deviates from its target value. The quality loss function is expressed in monetary units. The cost of deviating from the target increases as a quadratic function the farther the quality characteristic moves from the target. The formula used to compute the quality loss function depends on the type of quality characteristic used. Genichi Taguchi first introduced the quality loss function in this form.

quality management—Coordinated activities to direct and control an organization with regard to quality. Such activities generally include establishment of the quality policy, quality objectives, quality planning, quality control, quality assurance, and quality improvement.

quartiles—Division of the distribution into four groups, denoted by Q_1 (first quartile), Q_2 (second quartile), and Q_3 (third quartile). Note that Q_1 is the same as the 25th percentile, Q_2 is the same as the 50th percentile and the median, and Q_3 corresponds to the 75th percentile.

R

***R*-bar**—The average range calculated from the set of subgroup ranges under consideration. See *range*.

***R* chart**—See *range chart*.

R^2_{adj}—*See* adjusted coefficient of determination.

r^2—*See* coefficient of determination.

R^2—*See* coefficient of multiple determination.

random cause—Source of process variation that is inherent in a process over time. Also called *common cause* or *chance cause*.

random sampling—Sampling where a sample of n sampling units is taken from a population in such a way that each of the possible combinations of n sampling units has a particular probability of being taken.

random variable—A variable whose value depends on chance.

random variation—Variation from random causes.

randomization—Process used to assign treatments to experimental units so that each experimental unit has an equal chance of being assigned a particular treatment.

randomized block design—An experimental design consisting of b blocks with t treatments assigned via randomization to the experimental units within each block. This is a method for controlling the variability of experimental units. For the completely randomized design, no stratification of the experimental units is made. In the randomized block design, the treatments are randomly allotted within each block; that is, the randomization is restricted.

randomized block factorial design—A factorial design run in a randomized block design where each block includes a complete set of factorial combinations.

range (*R*)—A measure of dispersion that is the absolute difference between the highest and lowest value in a given subgroup: R = highest observed value—lowest observed value.

range chart (*R* chart)—A variables control chart that plots the range of a subgroup to detect shifts in the subgroup range. See *range (R)*.

rapid prototyping—Refers to a sample that can be produced in a much shorter time than the standard production process.

ratio scale—A scale where meaningful differences are shown, where absolute zero exists, and where multiplication and division are permitted. One example of a ratio scale is length in inches, because zero length is defined as having no length, and 20 inches is twice as long as 10 inches.

rational subgroup—A subgroup that is expected to be as free as possible from assignable causes (usually consecutive items). In a rational subgroup, variation is presumed to be only from random cause.

redundancy—The existence of more than one means for accomplishing a given function. Each means of accomplishing the function need not necessarily be identical.

reengineering—The process of completely redesigning or restructuring a whole organization, an organizational component, or a complete process. It's a "start all over again from the beginning" approach, sometimes called a *breakthrough*. In terms of improvement approaches, reengineering is contrasted with incremental improvement (kaizen).

regression analysis—A technique that typically uses continuous predictor variable(s) (that is, regressor variables) to predict the variation in a continuous response variable. Regression analysis uses the method of least squares to determine the values of the linear regression coefficients and the corresponding model.

rejection region—The numerical values of the test statistic for which the null hypothesis will be rejected.

relative frequency—Number of occurrences or observed values in a specified class divided by the total number of occurrences or observed values.

reliability growth—The improvement in product reliability over a period of time.

reliability—The probability that an item can perform its intended function for a specified interval under stated conditions.

repairability—The probability that a failed system will be restored to operable condition in a specified active repair time.

repeatability—The precision under conditions where independent measurement results are obtained with the same method on identical measurement items by the same appraiser (that is, operator) using the same equipment within a short period of time. Although misleading, repeatability is often referred to as equipment variation (*EV*).

It is also referred to as within-system variation when the conditions of measurement are fixed and defined (that is, equipment, appraiser, method, and environment). Repeatability is a component of precision.

repeated measures—The measurement of a response variable more than once under similar conditions. Repeated measures allow one to determine the inherent variability in the measurement system. Repeated measures are also known as *duplication* or *repetition*.

replicate—A single repetition of the experiment. See also *replication*.

replication—The performance of an experiment more than once for a given set of predictor variables. Each repetition of the experiment is called a *replicate*. Replication differs from repeated measures in that it is a repeat of the entire experiment for a given set of predictor variables, not just a repeat of measurements on the same experiment. Note: Replication increases the precision of the estimates of the effects in an experiment. It is more effective when all elements contributing to the experimental error are included.

representative sample—Sample that by itself or as part of a sampling system or protocol exhibits characteristics and properties of the population sampled.

reproducibility—Precision under conditions where independent measurement results are obtained with the same method on identical measurement items with different operators using different equipment.

residual analysis—Method of using residuals to determine appropriateness of assumptions made by a statistical method.

residual plot—A plot used in residual analysis to determine appropriateness of assumptions made by a statistical method.

residuals—The difference between the observed result and the predicted value (estimated treatment response) based on an empirically determined model.

resolution—(1) The smallest measurement increment that can be detected by the measurement system. (2) In the context of experimental design, resolution refers to the level of confounding in a fractional factorial design. For example, in a resolution III design, the main effects are confounded with the two-way interaction effects.

response surface design—A design intended to investigate the functional relationship between the response variable and a set of predictor variables. It is generally most useful when the predictor variables are continuous.

response surface methodology (RSM)—A methodology that uses design of experiments, regression analysis, and optimization techniques to determine the best relationship between the response variable and a set of predictor variables.

response variable—A variable that shows the observed results of an experimental treatment. It is sometimes known as the *dependent* variable or y variable. There may be multiple response variables in an experimental design.

resubmitted lot—A lot that previously has been designated as not acceptable and that is submitted again for acceptance inspection after having been further tested, sorted, reprocessed, and so on.

risk assessment/management—The process of determining what risks are present in a situation and what actions might be taken to eliminate or mediate them.

risk, consumer's (β)—See *consumer's risk (β).*

risk, producer's (α)—See *producer's risk (α).*

risk—The probability of not achieving a project goal or schedule, budget, or resource target because something did or didn't occur. Risk may be negative in the form of threats or positive in the form of opportunities.

robust—A characteristic of a statistic or statistical method. A robust statistical method still gives reasonable results even though the standard assumptions are not met. A robust statistic is unchanged by the presence of unusual data points or outliers.

robust designs—Products or processes that continue to perform as intended in spite of manufacturing variation and extreme environmental conditions during use.

robustness—The condition of a product or process design that remains relatively stable with a minimum of variation even though factors that influence operations or usage, such as environment and wear, are constantly changing.

robust parameter design—A design that aims at reducing the performance variation of a product or process by choosing the setting of its control factors to make it less sensitive to the variability from noise factors.

rolled throughput yield (RTY)—The probability of a unit of product passing through an entire process defect-free. The rolled throughput yield is determined by multiplying first-pass yield from each subprocess of the total process.

root cause—A factor (that is, original cause) that, through a chain of cause and effect, causes a defect or nonconformance to occur. Root causes should be permanently eliminated through process improvement. Note: Several root causes may be present and may work either together or independently to cause a defect.

root cause analysis—The process of identifying causes. Many systems are available for analyzing data to ultimately determine the root cause.

r—*See* correlation coefficient.

R—*See* range.

RSM—*See* response surface methodology.

RTY—*See* rolled throughput yield.

run chart—A chart showing a line connecting consecutive data points collected from a process running over a period of time. The run chart is occasionally considered to be one of the seven basic tools of quality. Note: A trend is indicated when the series of collected data points head up or down.

S

s is used for standard deviation since both random and special causes may be present. Note: A state of statistical control is not required. Also note the similarity to the formula for C_p.

s^2—See *sample variance.*

safe operating area (SOA)—The conditions (voltage and current) in which the component or device is rated to operate without self-damage.

sample—A group of units, portions of material, or observations taken from a larger collection of units, quantity of material, or observations that serves to provide information that may be used as a basis for making a decision concerning the larger quantity.

sample mean—The sample mean (or average) is the sum of random variables in a random sample divided by the number in the sum.

sample size (*n*)—The number of units in a sample. Note: In a multistage sample, the sample size is the total number of units at the conclusion of the final stage of sampling.

sample standard deviation—*See* standard deviation.

sample variance—*See* variance.

sampling interval—In systematic sampling, the fixed interval of time, output, running hours, and so on, between samples.

sampling plan (acceptance sampling usage)—A specific plan that states the sample size(s) to be used and the associated criteria for accepting the lot. Note: The sampling plan does not contain the rules on how to take the sample.

scatter diagram—A plot of two variables, one on the y-axis and the other on the x-axis. The resulting graph allows visual examination for patterns to determine if the variables show any relationship or if there is just random "scatter." This pattern, or lack thereof, aids in choosing the appropriate type of model for estimation. Note: Evidence of a pattern does not imply that a causal relationship exists between the variables. The scatter diagram is one of the seven tools of quality.

scatterplot—*See* scatter diagram.

serviceability—The ease or difficulty with which equipment can be repaired.

seven basic tools of quality—A set of both qualitative and quantitative tools that help organizations understand and improve. Unfortunately, there is no longer a unified set of seven tools that are universally agreed on, but most authors agree that five of the seven basic tools include the cause-and-effect diagram, control charts, histogram, Pareto chart, and scatter diagram. Additional tools suggested as part of the seven basic tools of quality include the check sheet, flowchart, graphs, run chart, trend chart, and stratification. Note: (1) graphs are sometimes included with control charts and are distinct at other times; (2) run charts are sometimes included with control charts and are distinct at other times; (3) run charts are sometimes synonymous with trend charts and are distinct at other times; (4) run charts are sometimes included with graphs and are distinct at other times. Some authors are known to have eight basic tools of quality. *See also* individual entries.

seven management and planning tools—The seven commonly recognized management and planning tools are the affinity diagram, tree diagram, process decision program chart (PDPC), matrix diagram, interrelationship diagram, prioritization matrices, and the activity network diagram. *See also* individual entries.

seven tools of quality—*See* seven basic tools of quality.

severity—An indicator of the degree of a failure should a failure occur. Severity can be evaluated based on a ten-point scale. On the lowest end of the scale (1), it is assumed a failure will have no noticeable effect. On the highest end of the scale (10), it is assumed a failure will impact safe operation or violate compliance with regulatory mandate.

signal—An indication on a control chart that a process is not stable or that a shift has occurred. Typical indicators are points outside control limits, runs, trends, cycles, patterns, and so on.

significance level—Maximum probability of rejecting the null hypothesis when in fact it is true. Note: The significance level is usually designated by α and should be set before beginning the test.

simple linear regression—A linear regression equation in which one predictor variable is used.

simulation—The act of modeling the characteristics or behaviors of a physical or abstract system.

simultaneous engineering—*See* concurrent engineering.

single-minute exchange of die (SMED)—A series of techniques pioneered by Shigeo Shingo for changeovers of production machinery in less than ten minutes. The long-term objective is always zero setup, in which changeovers are instantaneous and do not interfere in any way with continuous flow. Setup in a single minute is not required but used as a reference. SMED is also known as *rapid exchange of tooling and dies* (RETAD).

SIPOC—A macro-level analysis of the suppliers, inputs, process, outputs, and customers.

Six Sigma—Can take on various definitions across a broad spectrum depending on the level of focus and implementation. (1) Philosophy—the philosophical perspective views all work as processes that can be defined, measured, analyzed, improved, and controlled (DMAIC). Processes require inputs and produce outputs. If you control the inputs, you will control the outputs. This is generally expressed as the $Y=f(X)$ concept. (2) Set of tools—Six Sigma as a set of tools includes all the qualitative and quantitative techniques used by the Six Sigma practitioner to drive process improvement through defect reduction and the minimization of variation. A few such tools include statistical process control (SPC), control charts, failure mode and effects analysis, and process mapping. There is probably little agreement among Six Sigma professionals as to what constitutes the tool set. (3) Methodology—this view of Six Sigma recognizes the underlying and rigorous approach known as DMAIC. DMAIC defines the steps a Six Sigma practitioner is expected to follow, starting with identifying the problem and ending with implementing long-lasting solutions. While DMAIC is not the only Six Sigma methodology in use, it is certainly the most widely adopted and recognized. (4) Metrics—in simple terms, Six Sigma quality performance means 3.4 defects per million opportunities (accounting for a 1.5-sigma shift in the mean).

skewness—A measure of symmetry about the mean. For the normal distribution, skewness is zero since it is symmetric.

slack time—The time an activity can be delayed without delaying the entire project; it is determined by calculating the difference between the latest allowable date and the earliest expected date.

slope—See *regression analysis*.

SMART—SMART stands for *specific, measurable, achievable, relevant*, and *timely* and is used generally in regard to the creation of goal statements.

SMED—*See* single-minute exchange of die.

spaghetti chart—A before-improvement chart of existing steps in a process and the many back-and-forth interrelationships (can resemble a bowl of spaghetti); used to see the redundancies and other wasted movements of people and material.

SPC—*See* statistical process control.

special cause—A source of process variation other than inherent process variation. Note: (1) sometimes special cause is considered synonymous with assignable cause, but a special cause is assignable only when it is specifically identified; (2) a special cause arises because of specific circumstances that are not always present. Therefore, in a process subject to special causes, the magnitude of the variation over time is unpredictable. *See also* assignable cause.

specification—An engineering requirement used for judging the acceptability of a particular product/service based on product characteristics such as appearance, performance, and size. In statistical analysis, specifications refer to the document that prescribes the requirements within which the product or service has to perform.

specification limit(s)—Limiting value(s) stated for a characteristic. See *tolerance.*

spread—A term sometimes synonymous with variation or dispersion.

SQC—*See* statistical quality control.

s—*See* standard deviation.

stability (of a measurement system)—The change in bias of a measurement system over time and usage when that system is used to measure a master part or standard. Thus, a stable measurement system is one in which the variation is in statistical control, which is typically demonstrated through the use of control charts. Stability is a component of accuracy.

stable process—A process that is predictable within limits; a process that is subject only to random causes. (This is also known as a state of statistical control.) Note: (1) A stable process will generally behave as though the results are simple random samples from the same population; (2) this state does not imply that the random variation is large or small, within or outside specification limits, but rather that the variation is predictable using statistical techniques; (3) the process capability of a stable process is usually improved by fundamental changes that reduce or remove some of the random causes present and/or adjusting the mean toward the target value.

stakeholders—The people, departments, and groups that have an investment or interest in the success or actions taken by the organization.

standard—A statement, specification, or quantity of material against which measured outputs from a process may be judged as acceptable or unacceptable.

standard deviation—A measure of the spread of the process output or the spread of a sampling statistic from the process. When working with the population, the standard deviation is usually denoted by σ (sigma). When working with a sample, the standard deviation is usually denoted by s.

standard error of predicted values—A measure of the variation of individual predicted values of the dependent variable about the population value for a given value of the predictor variable. This includes the variability of individuals about the sample line

about the population line. It measures the variability of individual observations and can be used to calculate a prediction interval.

standard error—The standard deviation of a sample statistic or estimator. When dealing with sample statistics, we either refer to the standard deviation of the sample statistic or to its standard error.

statistic—A quantity calculated from a sample of observations, most often to form an estimate of some population parameter.

statistic—A value calculated from or based on sample data (for example, a subgroup average or range), used to make inferences about the process that produced the output from which the sample came. A quantity calculated from a sample of observations, most often to form an estimate of some population parameter.

statistical control—A process is considered to be in a state of statistical control if variation among the observed sampling results can be attributed to a constant system of chance causes.

statistical measure—A statistic or mathematical function of a statistic.

statistical process control (SPC)—The use of statistical techniques such as control charts to reduce variation, increase knowledge about the process, and steer the process in the desired way. Note: (1) SPC operates most efficiently by controlling variation of process or in-process characteristics that correlate with a final product characteristic, and/or by increasing the robustness of the process against this variation; (2) a supplier's final product characteristic can be a process characteristic to the next downstream supplier's process.

statistical tolerance interval—An interval estimator determined from a random sample so as to provide a specified level of confidence that the interval covers at least a specified proportion of the sampled population.

stratification—The layering of objects or data; also, the process of classifying data into subgroups based on characteristics or categories. Stratification is occasionally considered to be one of the seven tools of quality.

subsystem—A combination of sets, groups, and so on, that performs an operational function within a system and its major subdivision of the system.

SWOT analysis—An assessment of an organization's key strengths, weaknesses, opportunities, and threats. It considers factors such as the organization's industry, competitive position, functional areas, and management.

SWOT—Strengths, weaknesses, opportunities, threats.

system—A composite of equipment, skills, and techniques capable of performing or supporting an operational role, or both. A complete system includes all equipment, related facilities, material, software, services, and personnel required for its operation and support to the degree that it can be considered self-sufficient in its intended operating environment.

systems thinking—A problem-solving approach that expands the view of something under analysis to determine how it interacts with other elements of the system of which it is a part.

T

***t* distribution**—A theoretical distribution widely used in practice to evaluate the sample mean when the population standard deviation is estimated from the data. Also known as *Student's t distribution.*

Taguchi design—See *robust parameter design.*

Taguchi methods—The American Supplier Institute's trademarked term for the quality engineering methodology developed by Genichi Taguchi. In this engineering approach to quality control, Taguchi calls for off-line quality control, online quality control, and a system of experimental designs to improve quality and reduce costs.

takt time—A term derived from the German word *taktzeit,* meaning "clock cycle." Takt time is the available production time divided by the rate of customer demand. Operating under takt time sets the production pace to customer demand.

target value—Preferred reference value of a characteristic stated in a specification.

temperature–humidity model—Used in accelerated life testing when temperature and humidity are the major accelerants.

temperature–nonthermal models—Used in accelerated life testing when temperature and another factor are the major accelerants.

test statistic—A statistic calculated using data from a sample. It is used to determine whether the null hypothesis will be rejected.

testing—A means of determining the ability of an item to meet specified requirements by subjecting the item to a set of physical, chemical, environmental, or operating actions and conditions.

theory of constraints (TOC)—Goldratt's theory deals with techniques and tools for identifying and eliminating the constraints (bottlenecks) in a process.

throughput time—The total time required (processing + queue time) from concept to launch, from order received to delivery, or from raw materials received to delivery to customer.

tolerance—Difference between upper and lower specification limits.

tolerance limits—See specification limit(s).

TOC—*See* theory of constraints.

tolerance design (Taguchi)—A rational grade limit for components of a system; determines which parts and processes need to be modified and to what degree it is necessary to increase their control capacity; a method for rationally determining tolerances.

tolerance—The difference between upper and lower specification limits.

total productive maintenance (TPM)—A methodology pioneered by Nippondenso (a member of the Toyota group) that works to ensure that every machine in a production process is always able to perform its required tasks such that production is never interrupted. TPM maximizes equipment effectiveness by using a preventive maintenance program throughout the life of the equipment.

TPM—*See* total productive maintenance.

transfer, risk—Refers to moving the risk to another party or individual and allows them to assume the risk responsibility.

transformation—A re-expression of the data aimed toward achieving normality.

transmission of error—*See* propagation of error.

treatment—The specific setting of factor levels for an experimental unit.

tree diagram—A tool that depicts the hierarchy of tasks and subtasks needed to complete an objective. The finished diagram resembles a tree. The tree diagram is one of the seven management and planning tools.

trend chart—A control chart in which the deviation of the subgroup average from an expected trend in the process level is used to evaluate the stability of a process. The trend chart is also known as the *trend control chart*. The trend chart is occasionally considered to be one of the seven basic tools of quality.

TRIZ—A Russian acronym for "theory of inventive problem solving"; a systematic means of inventing and solving design conflicts. TRIZ involves three items to solve technical problems: (1) various tricks, (2) methods based on utilizing physical effects and phenomena, and (3) complex methods.

true value—A value for a quantitative characteristic that does not contain any sampling or measurement variability. (The true value is never exactly known; it is a hypothetical concept.)

***t*-test**—A test of significance that uses the t distribution to compare a sample statistic to a hypothesized population mean or to compare two means.

two-tailed test—A hypothesis test that involves two tails of a distribution. Example: We wish to reject the null hypothesis H_0 if the true mean is within minimum and maximum (two tails) limits.

Type I error—An error that occurs when we reject the null hypothesis when it is true. We refer to the Pr(Type I error) = Pr(rejecting H_0 | H_0 is true) = α. A type I error is also known as an *alpha* (α) *error* and *error of the first kind*. The Pr(Type I error) is also known as *α—value, producer's risk, level of significance*, and *significance level*.

Type II error—An error that occurs when we fail to reject the null hypothesis when it is false. We refer to the Pr(Type II error) = P(not rejecting H_0 | H_0 is false) = β. A Type II error is also known as a *beta* (β) *error* and *error of the second kind*. The Pr(Type II error) is also known as *β—value* and *consumer's risk*.

U

uncertainty—A parameter that characterizes the dispersion of the values that could reasonably be attributed to the particular quantity subject to measurement or characteristic. Uncertainty indicates the variability of the measured value or characteristic that considers two major components of error: (1) bias and (2) the random error from the imprecision of the measurement process.

unique lot—Lot formed under conditions peculiar to that lot and not part of a routine sequence.

unit—A quantity of product, material, or service forming a cohesive entity on which a measurement or observation can be made.

universe—A group of populations, often reflecting different characteristics of the items or material under consideration.

V

validation—Refers to the effectiveness of the design process itself and is intended to ensure the design process is capable of meeting the requirements of the final product or process.

validity—In the context of training, a valid measurement process is one that measures what it is designed to measure. The most commonly used method for training programs is the content-validity approach. This is a highly subjective approach by which subject matter experts provide opinions regarding whether the process measures what is needed on the job.

value engineering (VE)—An engineering discipline responsible for analyzing the components and processes that create a product, with an emphasis on minimizing costs while maintaining standards required by the customer.

value stream mapping—A technique for following the production path for a product or service from beginning to end while drawing a visual representation of every process in the material and information flows. Subsequently, a future state map is drawn of how value should flow.

value stream—All activities, both value-added and non-value-added, required to bring a product from a raw material state into the hands of the customer, a customer requirement from order to delivery, or a design from concept to launch.

value-added (VA)—The tasks or activities that convert resources into products or services consistent with customer requirements. The customer can be internal or external to the organization. Value-added activities add worth to the product or service from the customer's perspective and typically change form, fit, or function.

value—The net difference between customer-perceived benefits and burdens; it is sometimes expressed as a ratio of benefits to burdens or a ratio of worth to cost.

variables data—Data resulting from the measurement of a parameter or a variable. The resulting measurements may be recorded on a continuous scale.

variance—A measure of the variation in the data. When working with the entire population, the population variance is used; when working with a sample, the sample variance is used.

variation—Difference between values of a characteristic. Variation can be measured and calculated in different ways, such as range, standard deviation, or variance. Also known as *dispersion* or *spread.*

verification—Refers to the design meeting customer requirements and ensures that the design yields the correct product or process.

visual controls—The collection of approaches and techniques that permit one to visually determine the status of a system, factory, or process at a glance and to prevent or minimize process variation. To some degree, it can be viewed as a minor form of mistake-proofing. *Visual controls* are sometimes referred to as the visual factory.

VOC—*See* voice of the customer (VOC).

voice of the customer (VOC)—An organization's efforts to understand the customers' needs and expectations ("voice") and to provide products and services that truly meet them.

voice of the process (VOP)—The 6σ spread between the upper and lower control limits as determined from an in-control process. The VOP is also known as *natural process variation*.

VOP—*See* voice of the process (VOP).

W

warning limits—There is a high probability that the statistic under consideration is in a state of statistical control when it is within the warning limits (generally 2σ) of a control chart.

waste—Any activity that consumes resources but does not add value to the product or service a customer receives. Waste is also known as *muda*.

Weibull distribution—Widely used in reliability engineering. Depending on the shape parameter, it is a distribution that can take on the characteristics of other types of distributions.

Y

$Y=f(X)$—A foundational concept of Six Sigma, this equation represents the idea that process outputs are a function of process inputs.

References

Ahmed, Sajid, Wajid Ahmed Channa, and Ubaidullah Alias Kashif. 2021, May–June. "Comparative Analysis of Software Process Models in Software Development," *International Journal of Advanced Trends in Computer Science and Engineering* 10(3).

AIAG. 2008. *Potential Failure Mode and Effects Analysis*. 4th ed. New York: Automotive Industry Action Group.

Akinsola, J. E. T., A. S. Ogunbanwo, O. J. Okesola, I. J. Odun-Ayo, F. D. Ayegbusi, and A. A. Adebiyi. 2020. "Comparative Analysis of Software Development Life Cycle Models (SDLC)." In R. Silhavy (ed.), *Intelligent Algorithms in Software Engineering*. CSOC 2020. *Advances in Intelligent Systems and Computing*, Vol. 1224. Springer, Cham. https://doi.org/10.1007/978-3-030-51965-0_27

ANSI/ASQ Z1.4-2003 (R2018). *Sampling Procedures and Tables for Inspection by Attributes.*

ANSI/ASQ Z1.9-2003 (R2018). *Sampling Procedures and Tables for Inspection by Variables for Percent Nonconforming.*

ANSI/IEEE, Standard Glossary of Software Engineering Terminology, STD-729–1991, ANSI/IEEE, Washington, DC, 1991.

ASQ Eight Disciplines. 2025. https://asq.org/quality-resources/eight-disciplines-8d.

Bass, Len, Paul Clements, and Risk Kazman. 2013. *Software Architecture in Practice*. 3rd ed. SEI Series in Software Engineering. Upper Saddle River, NJ: Prentice Hall PTR.

Box, G. (1988). Signal-to-noise ratios, performance criteria, and transformations. *Technometrics* 30(1): 1–17.

Box, George E. P, J. Stuart Hunter, and William G. Hunter.2005. *Statistics for Experimenters: Design, Innovation, and Discovery*. 2nd ed. Hoboken, NJ: John Wiley & Sons.

Breneman, J. E., C. Sahay, and E. E. Lewis. 2022. *Introduction to Reliability Engineering*. 3rd edition. New York: John Wiley & Sons.

Cathcart, Timothy, et al. 2010. "Systems Engineering in Military Projects." In *Mission-Critical and Safety-Critical Systems Handbook* (pp.461–570). Newnes, an imprint of Elsevier. https://doi.org/10.1016/B978-0-7506-8567-2.00006-8.

CDC. 2024. "Guidelines for ALARA—As Low As Reasonably Achievable." 2024. https://www.cdc.gov/radiation-health/safety/alara.html .. Accessed September 10, 2024.

Coffin, L. F. (1954). "A Study of the Effects of Cyclic Thermal Stresses on the Ductile Metal." *Transactions of the ASME* 76: 931–950.

Collins J. A., H. Busby, and George Staab. 2010. *Mechanical Design of Machine Elements and Machines*. 2nd ed. New York: Wiley.

Collins, J. A., H. Busby, and G. Staab. 2003. *Mechanical Design of Machine Elements and Machines: A Failure Prevention Perspective*. New York: John Wiley & Sons.

Conrad, Daniel C. 2023. "Is It New, Unique, or Difficult?" Reliability, Maintenance and Managing Risk Conference (RMMR), July 27–28, https://asqrrd.org/rmmr-2023-presentations/.

Conrad, Daniel C. and Jerry Stack. 2022. "Reliability 9-Step Process." Reliability, Maintenance and Managing Risk Conference (RMMR), August 2–3, https://asqrrd.org/rmmr-2022-presentations/.

Crow, L. H. 1975. *Reliability Analysis for Complex, Repairable Systems.* AMSAA Technical Report No. 138. Maryland: US AMSAA, Aberdeen Proving Ground.

Dasgupta, A. and M. Pecht. 1991. "Material Failure Mechanisms and Damage Models." *IEEE Transactions on Reliability* 40(5).

Dasgupta, A., and J. M. Hu. 1992. "Failure-Mechanism Models for Excessive Elastic Deformation." *IEEE Transactions on Reliability* 41(1).

de Bono, Edward. 1967. *The Use of Lateral Thinking.* International Center for Creative Thinking.

Dieter, G., and L. Schmidt. 2009. *Engineering Design.* 4th ed. Boston: McGraw-Hill Higher Education.

Doganaksoy, N., W. Q. Meeker, and G. J. Hahn. 2021. *Achieving Product Reliability: A Key to Business Success.* Boca Raton FL: CRC Press.

Drees, R., and N. Young. 2002. "Built-in-Test in Support System Maintenance." *IEEE Instrumentation and Measurement Magazine* 5(3): 25–29.

Duane, J. T. 1964. "Learning Curve Approach to Reliability Monitoring," *IEEE Transactions on Aerospace* 2: 563–566.

Durivage, M. A. 2014. *Practical Engineering, Process, and Reliability Statistics.* Milwaukee, WI: Quality Press.

Ebeling, Charles E. 2010. *An Introduction to Reliability and Maintainability Engineering.* 2nd ed. Long Grove, IL: Waveland Press.

Engineering Design Handbook: Design for Reliability. 1976. AMCP 706–196, ADA 027370.

FDA 21 CFR Part 120 HACCP.

Govardhan, A. 2010. "A Comparison Between Five Models of Software Engineering." *IJCSI* International Journal of Computer Science 1694–0814(7):94–101.

Hack, Michael. 2022, December 13. "The Three Challenges of Strength and Durability Analysis." *Siemens blog.* https://blogs.sw.siemens.com/en-US/simcenter/the-three-challenges-of-strength-and-durability-analysis/.

Hahn, Gerald J. and Samual S. Shapiro. 1994. *Statistical Models in Engineering.* Hoboken, NJ: Wiley-Interscience.

Handbook for Reliability Prediction for Mechanical Systems (NSWC-07). Bethesda, MD: Naval Surface Warfare Center, Caderock Division. http://everyspec.com/USN/NSWC/NSWC-10_RELIABILITY_HDBK_JAN2010_45818.

Harry, M. J., and R. Schroeder. 2000. *Six Sigma: The Breakthrough Management Strategy Revolutionizing the World's Top Corporations* New York: Doubleday.

Hollander, Myles, Douglas A. Wolfe, and Eric Chicken. 2014. *Nonparametric Statistical Methods.* 3rd ed. Hoboken, NJ: John Wiley & Sons.

HASpod. 2023. "The ALARP Principle with 5 Real World Examples." https://www.haspod.com/blog/management/alarp-principle-and-examples Accessed September 10, 2024.

IEC 60812 *Analysis Techniques for System Reliability—Procedure for Failure Mode and Effects Analysis (FMEA).* 2006. International Electrotechnical Commission.

IEC 61025 *Fault Tree Analysis.* 2012. International Electrotechnical Commission.

IEC 61123 *Reliability Testing—Compliance Plans for Success Ratio.* 2019. International Electrotechnical Commission.

IEC 61124 *Reliability Testing—Compliance Tests for Constant Failure Rate and Constant Failure Intensity.* 2023. International Electrotechnical Commission.

IEC 61508 *Functional Safety of Electrical/Electronic/Programmable Electronic Safety- related Systems.* 2010. International Electrotechnical Commission.

IEC 61511 *Functional Safety—Safety Instrumented Systems for the Process Industry Sector*. 2016. International Electrotechnical Commission.

IEEE Std 1633–2016, *IEEE Recommended Practice on Software Reliability*.

Ilhan, I., R. ten Klooster, and I. Gibson. 2023. "Observing the Effect of Pressure and Temperature on the Seal Integrity of Critical Seal Regions of Various Flexible Bag Designs." *Food Packaging and Shelf Life* 37: 101088. https://doi.org/10.1016/j.fpsl.2023.101088

Ireson, W. G., C. F. Combs, Jr., and R. Y. Moss. 1995. *The Handbook of Reliability Engineering and Management*. 2nd ed. New York: McGraw Hill.

ISO 13849–1:2015 (en) *Safety of Machinery—Safety-related Parts of Control Systems—Part 1: General Principles for Design*. https://www.iso.org/obp/ui/#!iso:std:69883:en.

IEEE/ISO/IEC. 2017. *Systems and Software Engineering—Vocabulary*.

ISO 31000, Risk Management- Principles and Guidelines. 2018. Section 6..5.2, p. 13

ISO 9001 Document Control (4.2.3)

Jadhav, Anil S. and Rajendra M. Sonar. 2009. "Evaluating and Selecting Software Packages: A Review." *Information and Software Technology* 51: 555–563.

Johnson, B. 2005. "Design Ideation: The Conceptual Sketch in the Digital Age." *Design Studies*, 26, 613–624.

Kan, Stephen H. 2003. *Metrics and Models in Software Quality Engineering*. 2nd ed. Boston, MA: Pearson Education.

Kececioglu, D. 2001. *Reliability and Life Testing Handbook*. Volume II. Tucson, AZ: University of Arizona.

Kececioglu, D. 1993. *Reliability and Life Testing Handbook*. Volume I. Englewood Cliffs, NJ: Prentice Hall.

Khan, M. A., A. Parveen and M. Sadiq. 2014. "A Method for the Selection of Software Development Life Cycle Models Using Analytic Hierarchy Process." 2014 International Conference on Issues and Challenges in Intelligent Computing Techniques (ICICT), Ghaziabad, India, pp. 534–540. https://doi.org/10.1109/ICICICT.2014.6781338.

Kim, Steven B. and Jeffrey O. Wand. 2019. "Sequential Testing in Reliability and Validity Studies with Repeated Measurements per Subject," *International Journal of Statistics and Probability* 8(1).

Klayman, B. 2014, November 19. "BMW Says Its Takata Inflator Work Moving from Mexico to Germany." *Automotive News*.

Kneuper, R. 2017. "Sixty Years of Software Development Life Cycle Models." *IEEE Annals of the History of Computing* 39(3): 41–54. https://doi.oprg/10.1109/MAHC.2017.3481346.

Krajcuskova, Z. 2007. "Software Reliability Models." 17th International Conference Radioelektronika, Brno, Czech Republic, pp. 1–4. https://doi.org/10.1109/RADIOELEK.2007.371428.

Krinke, T. A. and D. K. Pai. 2024. "COTS/ROTS for Mission-Critical Systems." General Dynamics Information Systems. Accessed December 30, 2024, https://nepp.nasa.gov/pages/pubs.cfm.

Kubiak, T. M. and Donald W. Benbow. 2010. *The ASQ Six Sigma Black Belt Handbook*. 3rd ed. Milwaukee, WI: Quality Press.

Kudchadker, Rajat. 2024. "Radiation Vault Design and Shielding." https://international.anl.gov/training/materials/AX/Kudchadker/1-Room-Shield-Design.pdf, Accessed September 10, 2024.

Lakey, Peter, and A. M. Neufelder. 1997. *System and Software Reliability Assurance Notebook*. Rome: Laboratory.

Laman, Scott. 2022. The ASQ Certified Quality Engineer Handbook. 5th ed. Milwaukee, WI: ASQ Quality Press.

Linderman, K., R. G. Schroeder, S. Zaheer, and A. S. Chao. 2003. "Six Sigma: A Goal Theoretic Perspective." *Journal of Operations Management* 21: 193–203.

Littlewood, B., and J. T. Verrall. 1973. "A Bayesian Reliability Growth Model for Computer Software," *Journal of the Royal Statistical Society, Series C (Applied Statistics)* 22(3): 332–346.

Magnusson, K., D. Kroslid, and B. Bergman. 2003. *Six Sigma, The Pragmatic Approach*. 2nd ed. Professional Pub Service.

Manuele, J. 2003. *On the Practice of Safety*. Hoboken, NJ: John Wiley & Sons.

McLean, H. W. 2002. *HALT, HASS, and HASA Explained: Accelerated Reliability Techniques* Milwaukee, WI: ASQ Quality Press.

McPherson, J. W. 2019. *Reliability Physics and Engineering*. 3rd ed. Springer.

McShane-Vaughn, Mary. 2016. *The Probability Handbook*. Milwaukee, WI: ASQ Quality Press.

McShane-Vaughn, Mary. 2022. *The ASQ Certified Six Sigma Black Belt Handbook*. 4th ed. Milwaukee, WI: ASQ Quality Press.

Meeker, W. Q., and G. J. Hahn. 1985. *How to Plan an Accelerated Life Test—Some Practical Guidelines*. Volume 10. Milwaukee, WI: ASQC Quality Press.

Meeker W. Q, G. J. Hahn, and L. A. Escobar. 2017, *Statistical Intervals: A Guide to Practitioners and Researchers*, 2nd Edition, Hoboken, New Jersey: John Wiley and Sons, Inc.

Meeker, W. Q., L. A. Escobar, and F.G. Pascual. 2022. *Statistical Methods for Reliability Data*. 2nd ed. New York: Wiley.

Meeker, W. Q., L. A. Escobar, F. G. Pascual, Y. Hong, P. Liu, W. M. Falk, and B. Ananthasayanam. 2024. "Modern Statistical Models and Methods for Estimating Fatigue-Life and Fatigue-Strength Distributions from Experimental Data". *Statistical Science* (in press). https://arxiv.org/abs/2212.04550.

MIL-HDBK-338B. 1998, October 1. Military Handbook: Electronic Reliability Design Handbook.

MIL-HDBK-217. 1991, December 02. Military Handbook: Reliability Prediction of Electronic Equipment.

MIL-HDBK-781. 2015, January 30. *Military Handbook: Reliability Test Methods, Plans and Environments for Engineering Development, Qualification and Production*.

MIL-HNBK-189C. 2011, June 14. *Department of Defense Handbook of Reliability Growth Management*.

MIL-STD-882. 2012, *Standard Practice System Safety*. Department of Defense. http://www.system-safety.org/Documents/MIL-STD-882E.pdf.

Montgomery, Douglas C. 2005. *Introduction to Statistical Quality Control*. 5th ed. Hoboken, NJ: John Wiley & Sons.

Montgomery, Douglas C. 2019. *Design and Analysis of Experiments*.10th ed. Hoboken, NJ: John Wiley & Sons.

Montgomery, Douglas C. 2022. *Design of Experiments for Reliability Achievement*. New York: Wiley.

Montgomery, Douglas C., and George C. Runger. 2010. *Applied Statistics and Probability for Engineers*. 5th ed. Hoboken, NJ: John Wiley.

Moser, Barry K., and Gary R. Stebens, 1992. "Homogeneity of Variance in the Two-Sample Means Test," *The American Statistician*, Vol. 46, No. 1 (Feb.1992), pp. 19–21.

Munassar, Nabil M. A., and A. Govardhan. 2010, September. "A Comparison Between Five Models of Software Engineering," *IJCSI International Journal of Computer Science Issues* 7(5).

Musa, J., D. Iannino, and K. Okumoto. 1987. *Software Reliability: Measurement, Prediction, Application*. New York: McGraw-Hill.

Musa, Sarhan M., A.V. Kulkarni and V.K. Havanur. 2013. *Finite Element Analysis*. Boston: Mercury Learning and Information.

Narasimhamurthy, Shivanna, and Sandeep Palaksha. 2022. "The Impact Damage Study of a Refrigerated Display Case with Spring Bumper." India Altair Technology Conference, September 13–14.

NASA, 1973. *Solid Propellant Grain Structural Integrity Analysis.*

NASA. 2024. *NASA Methodology for Physics of Failure-Based Reliability Assessments Handbook.*

National Research Council. (2015). *Reliability Growth: Enhancing Defense System Reliability.* Washington, DC: The National Academies Press. https://doi.org/10.17226/18987.

Nelson, W. B. 2004. *Accelerated Testing: Statistical Models, Test Plans, and Data Analysis.* New York: Wiley.

NIST, 2019. "Lean, ISO and Six Sigma." https://www.nist.gov/baldrige/lean-iso-and-six-sigma. Accessed January 14, 2025.

NPRD-2011. 2011. *Non-electronic Parts Reliability Data.* Utica, NY: Reliability Information Analysis Center. https://roads.quanterion.com/text/NPRD-2011-1.pdf.

NUREG/CR-5485 Guidelines on Modeling Common-Cause Failures in Probabilistic Risk Assessment. 1998. https://nrcoe.inel.gov/resultsdb/publicdocs/CCF/NUREGCR-5485_Guidelines%20on%20Modeling%20Common-Cause%20Failures%20in%20PRA.pdf.

O'Connor, P. 2002. *Practical Reliability Engineering.* 4th ed. West Sussex, England: John Wiley & Sons.

O'Connor, Patrick D.T. and Andre V. Kleyner. 2025. *Practical Reliability Engineering.* 6th ed. West Sussex, England: John Wiley & Sons.

Palaksha, Sandeep, and Shivanna Narasimhamurthy. 2022. "Improving Energy Efficiency in a Refrigerated Display Case Using Aero-Foil Design." India Altair Technology Conference, September 13–14.

Pecht, M., and J. Gu. 2009. "Physics-of-Failure-Based Prognostics for Electronic Product." *Transactions of the Institute for Measurement and Control* 31(3/4): 309–22.

Peck, D. S. 1986. "Comprehensive Model for Humidity Testing Correlation," 24th International Reliability Physics Symposium, Anaheim, CA, USA, pp. 44–50. https://doi.org/10.1109/IRPS.1986.362110.

Pignatiello, J. J. Jr., and J. S. Ramberg. 1991. "Top Ten Triumphs and Tragedies of Genichi Taguchi." *Quality Engineering* 4(2): 211–225.

Ploe, R. J., and W. H. Skewis. 1990. *Handbook of Reliability Prediction Procedures for Mechanical Equipment.* DTRC-90/010. Bethesda, MD: United States Navy-David Taylor Research Center.

Porter, Alex. 2004. *Accelerated Testing and Validation.* Boston: Newnes.

Raheja, Dev G., and Louis J. Gullo, eds. 2012. *Design for Reliability.* Hoboken, NJ: John Wiley & Sons.

Reisig, Wolfgang. 1985. *Petri Nets: An Introduction.* Berlin: Springer-Verlag.

Reliasoft. 2024. "Reliability Growth and Repairable System Analysis." https://help.reliasoft.com/reference/reliability_growth_and_repairable_system_analysis/rg_rsa/rga_overview.html

Reuters. 2017, January 14. "Takata to Plead Guilty, Pay $1 Billion U.S. Penalty over Air Bag Defect."

Rigdon, S. E., R. Pan, D.C. Montgomery, & L. J. Freeman. 2022. *Design of Experiments for Reliability Achievement.* New York: John Wiley & Sons.

Saidi, M. I. M., M. A. Isa, D. N. A. Jawawi, and L. F. Ong, 2015. "A Survey of Software Reliability Growth Model Selection Methods for Improving Reliability Prediction Accuracy," 9th Malaysian Software Engineering Conference (MySEC), Kuala Lumpur, Malaysia, pp. 200–205. https://doi.org/10.1109/MySEC.2015.7475221.

Shah, R., and P. T. Ward. 2007, June. "Defining and Developing Measures of Lean Production." *Journal of Operations Management* 25(4): 785–805.

Shelden, David, Caleb Cope, and Daniel C. Conrad. 2024. "A Structural HALT/HAST Test for Doors." Reliability, Maintenance and Managing Risk Conference (RMMR), July 25–26, https://asqrrd.org/rmmr-2024-presentations/, accessed January 9, 2025.

Shoemaker, A. C., K. L. Tsui, and C. F. J. Wu. 1991. "Economical Experimentation Methods for Robust Designs," *Technometrics* 33: 415–427.

South, Chris. 2024. "What Is Highly Accelerated Life Testing (HALT)?" *Ansys.com* [blog]. https://www.ansys.com/it-it/blog/planning-a-halt

Steinmetz M. 2002. "Built-in-Test Instrumentation and 21 Rules of Thumb." *IEEE Instrumentation and Measurement Magazine* 5(3): 31.

Tabuchi, H., and C. Jensen. 2014. "Now the Air Bags Are Faulty, Too." *New York Times*, June 23.

Taguchi, G. 1986. *Orthogonal Arrays and Linear Graphs*. Dearborn, MI: American Supplier Institute.

Tague, Nancy R. 2023. *The Quality Toolbox*. 3rd ed. Milwaukee: Quality Press.

Taylor, W. A. 2017. *Statistical Procedures for the Medical Device Industry*. Taylor Enterprises.

Tobias, P. A., and D. C. Trindade. 2012. *Applied Reliability*. 3rd ed. Boca Raton, FL: CRC Press.

Tong, Xuhang, Hoo Zhang, and D. Y. Li. 2015. "Effect of Annealing Treatment on Mechanical Properties of Nonocrystalline α-Iron: An Atomistic Study," *Science Reports* 5: 8459 https://doi.org/10.1038/srep08459.

Tranquillo, Joseph, Jay Goldberg, and Robert Allen (2023). *Biomedical Engineering Design*. London: Academic Press.

Trivedi, K., & Bobbio, A. (2024). "Reliability and Availability Analysis in Practice: Toward Multilevel Models for Complex Systems." *Computer* 57(4): 82–92.

Tseng, S-T., and Z-C. Wen. 2000. "Step-Stress Accelerated Degradation Analysis for Highly Reliable Products." *Journal of Quality Technology* 32(3): 209–216.

Tutoline.com. 2024. "Software Engineering SLDC Models." https://tutoline.com/software-engineering-sdlc-models/

Tutoline.com. 2024. "Software Engineering Software Reliability Models." https://tutoline.com/software-engineering-software-reliability-metrics/

Tutoline.com. 2024. "Software Reliability Metrics." https://tutoline.com/software-relaibility-metrics/

Ullman Sails. n.d. "Environmental and Use Factors." https://ullmansails.com/sail-care/environmental-use-factors/.

Vasumathi, B., and A. Saradha. 2015, October. "Enhancement of Intermittent Demands in Forecasting for Spare Parts Industry," *Indian Journal of Science and Technology* 8(25).

Von Alven, W. H. 1964. *Reliability Engineering*. Englewood Cliffs, NJ: Prentice Hall.

Welch, W. J., T. K. Yu, S. M. Kang, and J. Sacks. 1990. "Computer Experiments for Quality Control by Parameter Design," *Journal of Quality Technology* 22(1): 15–22.

Westfall, Linda. 2024. *The ASQ Certified Software Engineer Handbook*. 3rd ed. Milwaukee, WI: ASQ Quality Press.

Wulpi, D. J. 2013. *Understanding How Components Fail*. 3rd ed. ASM International.

Xu, Qingzheng, Na Wang, and Heping Shi. 2012. " A Review of Croston's Method for Intermittent Demand Forecasting. Proceedings—2012 9th International Conference on Fuzzy Systems and Knowledge Discovery.

Yang, Guangbin. 2007. *Life Cycle Reliability Engineering*. Hoboken, NJ: John Wiley & Sons.

Zahedi, F., and N. Ashrafi, 1991. "Software Reliability Allocation Based on Structure, Utility, Price, and Cost." *IEEE Transactions on Software Engineering* 17(4): 345–356. https://doi.org/10.1109/32.90434.

Index

A

B

D

E

F

G

H

I

J

K

L

M

Q

R

S

T

U

V

W

X

Y

Z

About the Editors

Karen J. Hulting is a statistician in the medical device industry. After beginning her career in the automotive and metals industries, she joined Medtronic and enjoyed a 28-year career there, providing leadership for the use of statistical tools in R&D, quality, reliability, and manufacturing. In 2019, she was inducted into the Bakken Society, the highest technical honor at the company. She continues at Medtronic as a consultant and devotes time to teaching, mentoring, and professional activities. Karen is an adjunct instructor at St. Cloud State University (St. Cloud, MN) in the Medical Technology Quality Graduate program.

Karen is an ASQ certified Quality Engineer, Reliability Engineer, and Six Sigma Black Belt. She is currently a member of the CQE Exam Committee, the Shewhart Medal Committee, and the Fall Technical Conference Program Committee. She is a past chair of the CRE and CSSBB Exam Committees.

Dr. Hulting earned her PhD and MS in statistics from Iowa State University and her BA in Mathematics from St. Olaf College. She is a Fellow of ASQ and a member of the American Statistical Association.

Mary McShane-Vaughn is principal at University Training Partners, a company that designs and delivers Six Sigma training. Previously, Dr. McShane-Vaughn was a tenured professor at Southern Polytechnic State University (now Kennesaw State) where she directed the MS in Quality Assurance program. Before her career in academics, Mary worked for 15 years as a quality engineer and statistician in the automotive, medical device, consumer product testing, and revenue management industries.

Mary is currently the chair of the ASQE Certification Council and a member of the Technical Program and Quality Press Peer Review committees. She is currently the editor of the *Lean & Six Sigma Review*. Mary has published extensively, serving as editor, author, or coauthor for *The ASQ Certified Six Sigma Black Belt Handbook, The Six Sigma Black Belt Study Guide, The Probability Handbook, The Probability Workbook, The Certified Quality Inspector Handbook, The Handbook of Quality Equations, Leadership Tools for Black Belts,* and several Six Sigma course companion texts. She is an ASQ certified Reliability Engineer, Quality Engineer, and Six Sigma Black Belt.

Dr. McShane-Vaughn earned her PhD in industrial engineering and MS in applied statistics from the Georgia Institute of Technology, and a BS in industrial engineering from Kettering University. She is a senior member of ASQ and a member of the Institute of Industrial and System Engineers and the American Statistical Association.

www.ingramcontent.com/pod-product-compliance
Lightning Source LLC
LaVergne TN
LVHW080645210926
848485LV00003B/329
9781636941783